Contents

Population Limitation
in Birds

Ian Newton
Institute of Terrestrial Ecology, Monks Wood
Cambridgeshire, UK

Drawings by Keith Brockie

ACADEMIC PRESS

An imprint of Elsevier

Amsterdam Boston Heidelberg London New York Oxford
Paris San Diego San Francisco Singapore Sydney Tokyo

Copyright © 1998 by ACADEMIC PRESS LIMITED
Illustrations © Keith Brockie

ISBN 0–12–517366–0

First published 1998
Reprinted 2003

ACADEMIC PRESS
525B Street
Suite 1990
San Diego, California 92101-4495, USA
http://www.academicpress.com

ACADEMIC PRESS
84 Theobald's Road
LONDON
WC1X 8RR, UK
http:// www.academicpress.com

A catalogue record of this book is available from the British Library

Typeset by Phoenix Photosetting, Chatham, Kent
Transferred to digital printing, 2007
Printed and bound by CPI Antony Rowe, Eastbourne

Bullfinch *Pyrrula pyrrhula* eating ash seed.

Preface

An understanding of the processes of population limitation, whether in birds or in other organisms, is central in the study of ecological science. It also provides the basis for the practical management of wild populations, whether for conservation or for sustained human exploitation. The aim of this book is to provide a coherent synthesis of findings on bird populations from various parts of the world, which I hope will give a sound basis both for further research and for more effective management. The book is based almost entirely on results from field studies, set within a contemporary theoretical framework. It is written in simple language in the hope that it will be of value not only to the professional ecologist, but to anyone with an interest in birds. Technical jargon and mathematical expression are kept to a minimum.

This is not the first time that such a synthesis has been attempted. Two pioneering books by David Lack, namely 'The Natural Regulation of Animal Numbers' (1954) and 'Population studies of birds' (1966), did much to stimulate research on bird populations years ago. Since these books were written, much new information and some new ideas have emerged, so a reassessment of the issues seems timely. The emphasis throughout this book is on density limitation, and larger scale aspects, such as migration and geographical range, will be discussed in a later volume. Furthermore, only the minimum of detail on field and analytical methodology is given, as these aspects have been treated in depth by other authors elsewhere.

The scientific literature on bird populations is now so vast, in the English-speaking world alone, that no one person can expect to have seen it all, let alone to have read it. Inevitably, then, I have had to refer selectively to this literature. For important points, on which a crucial argument depends, I have tabulated a large number of relevant studies, but for more firmly established points I have given only a few examples, referring wherever possible to review articles that themselves quote more original sources. Despite these attempts at curtailment, the reference list at the end still contains more than 1800 titles, extending up to 1996 (with a few later ones). I am well aware, however, that many relevant studies have necessarily been excluded.

The book begins with a preview, in which the scope and general conceptual framework are laid out. The rest is arranged in three main parts, each comprising several chapters. The first part deals with the role of behaviour in density regulation, the second with the effects of various external limiting factors, such as food, predators and parasites, and the third with the effects of various human impacts, such as hunting, pesticide use and pollution. A final chapter deals with extinction. With this form of treatment, some topics inevitably arise in more than one chapter, but I have tried to keep the repetition to a minimum, and to cross refer wherever possible. Because so many species are mentioned, from diverse parts of the globe, scientific names are given at the first mention in every chapter, and elsewhere if confusion would otherwise be likely.

Many of the diagrams in this book are redrawn from previously published sources, as acknowledged in the captions. For permission to reproduce this subject matter, I am grateful to Academic Press, Blackwell Scientific Publications, British Ecological Society (*Journal of Animal Ecology, Journal of Applied Ecology*), British Trust for Ornithology (*Bird Study*), The Game Conservancy Trust, Oxford University Press, The Royal Society (Proceedings: Biological Sciences) and the Peregrine Fund.

As in any active field of science, individuals differ in their interpretation of some of the same data, and hold different views about the importance of different processes in the dynamics of bird populations. This book naturally reflects my own views, although I have tried throughout to present as objective and balanced a picture as possible. On particular issues, I have benefited greatly over the years from discussion with friends and colleagues, in particular with M.P. Harris, D. Jenkins, M. Marquiss, C.M. Perrins and P. Rothery, and I would like to thank them all, together with the following, for commenting helpfully on particular chapters: J.P. Dempster (5), S. Hinsley (2, 3, 6, 16), R.E. Kenward (10), K. Lakhani (5), M. Marquiss (10, 13), R.E. Newton (10), S. Redpath (8, 14), T. Royama (5) and P. Rothery (5). In addition, I owe a special debt to David Jenkins, who has read and commented critically on the whole book in draft. However, neither he nor any other colleagues are in any way responsible for its shortcomings.

In addition, I am grateful to Doreen Wade for typing the manuscript many times over, as well as to Lois Dale for help in producing the diagrams and to Keith Brockie for his beautiful illustrations. Finally, I thank my wife, Halina, for encouragement throughout, and for checking the whole book in draft.

White Storks *Ciconia ciconia*.

Chapter 1
Preview

This book is about bird numbers, and about why these numbers vary in the way they do, from year to year or from place to place. It is therefore concerned with the various factors that influence bird numbers: with the role of food-supplies and other resources, of predators, parasites and pathogens, and of various human impacts. These are all matters of fundamental interest to ecologists, including those involved with the management of bird populations, whether for conservation, crop protection or sport-hunting. Although birds form a tiny fraction of all animal species – perhaps less than 0.2% of the total – we have no reason to suppose that the numbers of birds are limited in ways different from those of other animals. Yet for some aspects of study birds offer distinct advantages.

In the first place, most bird species are diurnal, and can be readily distinguished from one another in the field. This makes them easy to watch and to count. Indeed, in the breeding season many species obligingly force themselves upon our attention by singing their species-specific songs. Secondly, most birds can be trapped and marked with leg rings or other tags. Such marked birds then remain as identifiable individuals throughout their lives, enabling their movements and life-histories to be followed. Many species, too, are large enough to carry radio-transmitters, facilitating detailed studies of their day-to-day activities. Thirdly, most bird species raise their young in discrete nests. Their individual reproductive rates can therefore be measured in a way that is not possible for most other creatures with no parental care, and in which the young from different parents soon intermingle. Fourthly, the highly developed social structure of birds has enabled the role of territorial and other dominance behaviour in population regulation to be studied in greater detail than for most other animals. All these features combine to make birds popular with their human observers. In some parts of the world, through the combined efforts of birdwatchers, changes in the numbers and distributions of birds have been monitored with precision over long periods and wide geographical areas. For these various reasons, then, the population ecology of birds is perhaps better known than that of any other group of organisms, and the general principles that have emerged have much wider relevance in the science of ecology.

The aim of this book, as stated in the Preface, is to review current ideas on population limitation of birds, illustrating the issues involved with examples from different species and from different regions of the world. For the most part, different aspects of the subject are covered in different chapters. The purpose of this introductory chapter is to provide a preview, outlining the scope and conceptual framework within which subsequent chapters lie.

BIRD NUMBERS

Compared with many other organisms, birds are well-known taxonomically. On recent estimates, there are about 9672 known species of birds in the world (Sibley & Monroe 1990). Each decade, usually through human action, a few species may become extinct, while a similar number of previously undescribed species may come to light.

These bird species vary from extremely widespread and abundant, numbering many millions of individuals, to extremely localised and rare, consisting of a few individuals on the verge of oblivion. Among landbirds, the most numerous species probably include the House Sparrow *Passer domesticus* and European Starling *Sturnus vulgaris*, both of which are widespread in the Old World and, through introductions, in the New World too. They also include the Red-billed Quelea *Quelea quelea*, a seed-eating weaver bird which breeds over much of Africa south of the Sahara. The total populations of all these species have been estimated at more than 500 millions. Nightly roosts of more than a million individuals are regular in Quelea and Starlings, as well as in Sand Martins *Riparia riparia* in Europe and Red-winged Blackbirds *Agelaius phoeniceus* in North America. The record for individual roosts, however, is probably held by the Brambling *Fringilla montifringilla*, whose nightly gatherings in areas of good Beech *Fagus sylvatica* crops in central Europe have sometimes numbered more than 20 million individuals, as noted at least from the 15th century onwards (Jenni 1987).

Many species of seabirds range widely over the oceans, but while nesting gather in vast colonies at traditional locations, where they can be counted. Perhaps the most numerous seabird in the world is the Wilson's Petrel *Oceanites oceanicus* of the Southern Ocean which is thought to number more than 50 million pairs, distributed in a range of colonies. The second most numerous is probably the Arctic-nesting Little Auk *Alle alle*, with an estimated 30 million pairs. However, estimating the numbers of small birds that nest below ground in remote places is fraught with problems, and these figures are little more than guesses. Among more conspicuous seabird species, some individual colonies of Adelie Penguins *Pygoscelis adeliae* in Antarctica have been found to contain more than a million pairs, while colonies of Sooty Terns *Sterna fuscata* on some tropical islands may contain up to a staggering ten million pairs. Moreover, some seabirds in their colonies achieve phenomenal densities (Warham 1990). One colony of the Short-tailed Shearwater *Puffinus tenuirostris* was estimated to number 250 000 pairs on about 21 ha (at about one pair per m^2). In the Sooty Shearwater *P. griseus*, an estimated 2.75 million occupied burrows were found on 328 ha of the Snares Islands (about 1.2 burrows per m^2), and in the Great Shearwater *P. gravis*, two million occupied burrows were found on 200 ha of Nightingale Island (about one per m^2). Even these figures are low, however, compared with those for a Galapagos Storm Petrel *Oceanodroma tethys* colony, in which 200 000 pairs occupied an area of 2500 m^2, giving a mean density of 8 pairs per m^2 (Harris 1969).

At the other extreme, some landbird species, restricted to small oceanic islands, have naturally small populations, consisting in some cases of fewer than 100 breeding pairs. This is especially true of some birds-of-prey, which live only at low densities, and have probably never been much more numerous than they are today. But most rare species owe their present status to human action. More than 1000 bird species (about 11% of the total) now give cause for concern, and in recent decades the populations of several have persisted for some years at fewer than 20 individuals (Chapter 16). Extreme examples include the Mauritius Kestrel *Falco punctatus* (once reduced to four individuals) and the Chatham Island Black Robin *Petroica traversi* (reduced to seven individuals), both of which have since increased as a result of conservation measures (Butler & Merton 1992, Jones *et al.* 1995). For if rare species are protected, they usually recover again, providing that suitable habitat remains.

The overall abundance of any bird species depends on the extent of its geographical range, the amount of suitable habitat within that range, and the mean density achieved within that habitat. In many habitats, small birds tend to be more abundant than large ones. In various natural bird communities in North America, for which complete censuses had been taken, the most numerous species were those weighing around 40 g, and maximum densities dropped markedly in species that were either below or above this weight (Brown & Maurer 1987). The relationship held independently among species in different dietary categories, including herbivores, nectivores, omnivore-insectivores and carnivores. It also held among birds in Britain, but again only in broad terms: among birds in all weight classes some species were rare (Blackburn & Lawton 1994). Body size is thought to influence density partly because it affects resource needs. Individuals of larger species require more food and space, so must live at lower densities than small ones. Larger species also eat larger food-items, on average, and larger prey occur at lower densities than small ones, the link between body size and density occurring among a wide range of organisms and not just among birds (Blackburn *et al.* 1990, 1992). In fact, body size is related to so many aspects of species biology that it is hard to tell which are most relevant to density.

A broad relationship between body size and abundance probably holds among the birds of most habitats, but some types of habitat support generally larger birds than others. Most woodland species tend to be small (mostly <100 g) compared with those in other habitats, whereas most aquatic species tend to be large, with most seabird species weighing more than 500 g. The differences can be explained partly in terms of the living conditions offered by different habitats: birds living among dense twiggy foliage benefit from being small, whereas those ranging over vast areas of open ocean benefit from being large.

COUNTING BIRDS

Except for very localised species, it is seldom possible to study the whole population of any bird. Instead the researcher has usually to work within a defined area, whose avian occupants may form a tiny part of a much wider population. Individuals may move freely in and out of the study area, and those birds that breed there may occupy a wider area, or even a different part of the world, outside the breeding season. Hence, most 'population studies' of birds have been concerned with the numbers found in a defined area at a specific time. The general applicability of the conclusions depends partly on how typical is the study area of the bird's range as a whole, and on whether the area is big enough compared with the scale over which the factors that influence the overall population level operate.

The annual cycles of most birds fall into two distinct phases: a breeding (summer) season, when numbers increase because reproduction exceeds mortality, and a non-breeding (winter) season when numbers decline because of mortality. Numbers therefore reach a seasonal peak immediately after one breeding season, and a seasonal trough early in the next **(Figure 1.1)**. The extent of these seasonal fluctuations depends mainly on the reproductive and mortality rates of the species concerned, being most pronounced in the most prolific species. Most bird species can be counted most easily when they are breeding, because they are conspicuous then, and tied for long periods to specific locations where they nest. Knowledge of population changes in many bird species, from year to year or from place to place, is based entirely on counts of breeding numbers, or on indices of breeding numbers, such as displaying males or nests. Counts of breeding numbers are of particular value, however, because it is at the beginning of the breeding season that numbers reach their annual low, and it is upon breeders that future additions to the population depend. Hence, throughout this book, unless stated otherwise, I shall be concerned primarily with the limitation of bird breeding numbers. This is an important point, because different factors can limit numbers at different times of year. Many birds suffer high predation of eggs and young, so that predation may limit their post-breeding numbers, but in winter many individuals may starve, so that food-shortage becomes the main factor limiting subsequent breeding numbers.

In the breeding season, as well as territorial nesting pairs, some bird populations contain numbers of non-territorial, non-breeding individuals. In some species, such as crows or swans, non-breeders usually occur in flocks, which can be readily seen and counted. In other species, including many songbirds and raptors, the non-breeders

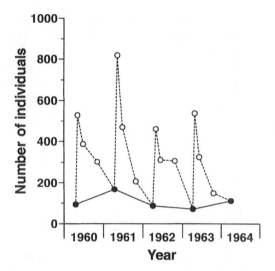

Figure 1.1   Changes in the numbers of Great Tits *Parus major* within five successive years in Marley Wood, Oxford, showing how the rise in numbers during each breeding season was followed by a decline to the start of the next breeding season. The continuous line joins the numbers of breeding adults each year, and the dashed line follows, respectively, the total of adults plus eggs, the total of adults plus fledged young and the estimated autumn numbers. From Perrins 1979.

live secretive solitary lives in and around the territories of breeders, or continually move around from place to place. Such non-breeders are therefore not much in evidence, and in some species almost impossible to count. However, two lines of evidence have repeatedly confirmed their existence. Firstly, if territorial birds are removed from their territories, they are often replaced within a few days by other individuals which then proceed to nest (Chapter 4). Secondly, in species that are limited by nest-sites, such as many cavity-nesters, the provision of artificial sites (such as nest-boxes) often leads to an immediate rise in breeding density (Chapter 8). The implication is that these incoming birds were previously present in the vicinity but unseen, and would not have nested if opportunities had not been provided. In both types of experiment, however, care is needed to be sure that incoming birds have not simply moved from other breeding sites elsewhere.

Despite the advantages of knowing the numbers of breeders, some bird species are most easily counted in winter. This is true, for example, of many waterfowl and shorebirds which nest in remote or inaccessible places, but migrate to winter at lower latitudes. The fact that such species often gather in large numbers at traditional sites makes the job of counting that much easier. In some species, adults can be distinguished in the field from young of the year, so that counts can give some idea of both overall numbers and breeding success. In other species, separation of age-groups is possible only in the hand, so field counts are less revealing.

Yet other bird species can be readily counted on migration. This is true of some raptors and other soaring birds, in which huge numbers from large areas pass specific concentration sites each spring or autumn. Annual variations in weather may affect the proportions of populations that pass such sites from year to year, so counts do not necessarily reflect year-to-year changes in numbers, but they do reflect long-term trends (Newton 1979). Various other constant effort or observational schemes have been used to detect year-to-year changes in bird numbers, and sometimes it has been possible to compare the results from different monitoring methods on the same populations (Furness & Greenwood 1993).

POPULATION TRENDS

If counts of breeding birds in the same area are examined year by year, almost any pattern of fluctuation can be found **(Figure 1.2)**. In some species breeding numbers normally remain fairly stable through time. Extreme examples are provided by some birds-of-prey, such as the Golden Eagle *Aquila chrysaetos*, in which the density of territorial pairs in some large areas may vary by no more than 15% on either side of the mean level over several decades (Newton 1979). Most small-bird populations fluctuate somewhat more, perhaps halving or doubling in size from one year to the next, or declining after hard winters or droughts, and increasing again in subsequent years. Small finches and other birds, which depend on tree-seeds, can fluctuate in particular areas by 20-fold or more from year to year, depending on the size of the seed-crop. But even these fluctuations are small compared with those in some other animals, such as certain insects, in which annual changes of 100-fold or more are not uncommon.

In most bird species the year-to-year fluctuations in numbers are irregular, but some northern gallinaceous birds, such as grouse and ptarmigan, undergo regular cycles of abundance, with peaks in numbers occurring about every 3–10 years, depending on region (Chapter 7). These cycles in game bird numbers are matched by similar fluctuations in their predators, whose densities typically go up and down from year to year in step with their prey (Chapter 9). Yet within suitable habitats, the mean level of abundance of these cyclic species does not necessarily change much over periods of several decades.

Some patterns of fluctuation in bird populations can be detected only by long-term study. The ten-year cycle is an obvious example, as several decades of counts or hunting records were needed to discern its regularity with certainty. Even longer periods would be needed to detect the multi-year changes that occur in some North American warblers that respond to an approximate 35-year cycle in Spruce Budworm *Choristoneura fumiferana* abundance, itself revealed mainly from historical records, coupled with patterns in tree growth rings (Blais 1965). It thus seems reasonable to suppose that other long-term cycles in animal (including bird) abundance remain as yet undetected. In addition, although ecologists have long been aware that population densities of some species might show chaotic fluctuations, the evidence for chaos in natural populations is sparse and unconvincing, but this may be partly because of the technical difficulties in detecting it.[1]

Although general stability in bird populations is probably the norm, it would be

---

[1]*Deterministic non-linear mathematical models can produce fluctuations which are erratic and apparently random. This behaviour is called chaos, a feature of which is that population size is sensitive to initial conditions, and therefore unpredictable in the long term (see Brown & Rothery 1993).*

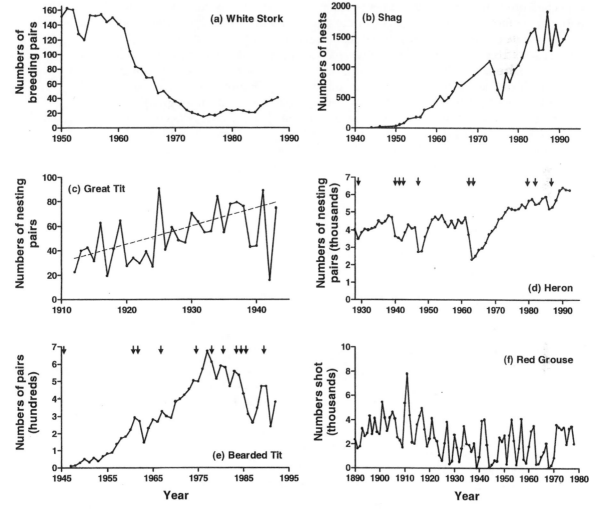

Figure 1.2   Patterns of long-term trends in different bird populations.

(a) Long-term decline in the numbers of White Storks *Ciconia ciconia* in Baden-Württemberg, Germany. Based on annual counts of occupied nests, 1950–88. Redrawn from Bairlein 1996.

(b) Long-term increase in the numbers of Shags *Phalacrocorax aristotelis* on the Isle of May, Scotland, showing how numbers increased and then levelled off, but with large annual fluctuations. Based on annual counts of occupied nests, 1944–92. Redrawn from Harris *et al.* 1994.

(c) Irregular annual fluctuations superimposed on a long-term increase in the numbers of Great Tit *Parus major* pairs in gradually maturing woodland at Oranje Nord, Netherlands. Based on annual counts of occupied nest boxes, 1912–43. From Kluijver 1951.

(d) Numbers of nesting Grey Herons *Ardea cinerea* in Britain, showing declines after hard winters (shown by arrows), followed by recoveries. Based on annual nest counts, 1928–92. Redrawn from Greenwood *et al.* 1994.

(e) Weather-driven fluctuations in the numbers of Bearded Tits *Panurus biarmicus* in southern England. Trends were influenced mainly by the frequency of cold winters (shown by-arrows). Based on annual counts of breeding pairs 1947–92. Redrawn from Campbell *et al.* 1996.

(f) Cyclic fluctuations in the numbers of Red Grouse *Lagopus l. scoticus*, Scotland. Based on numbers shot, 1890–1978. Redrawn from Hudson 1992.

expected only in stable environments. Many bird species occupy successional habitats, preferring particular stages in the development of forest or other vegetation. They reach large densities in particular localities only for the few years that habitat is suitable there, and then decline again as the habitat develops and becomes unsuitable. But while they are declining in one area, they may be increasing in another, where the habitat is reaching a favourable stage **(Figure 1.3)**. These changes, dependent on vegetation succession, can be regarded as natural, although for forest and other habitats, areas at early successional stages have probably become more widespread as a result of human activities. In other species, notably wetland birds, the amount of habitat available in some parts of the world varies greatly from year to year, or over periods of years, according to rainfall (Chapter 11). Typically, such

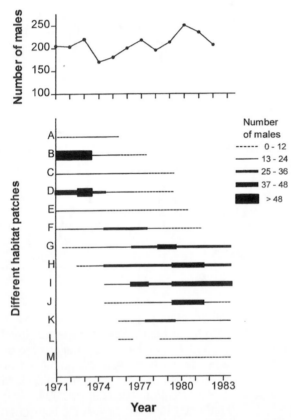

Figure 1.3   Numbers of the rare Kirtland's Warbler *Dendroica kirtlandii*, both overall (upper) and in individual habitat patches (lower). This species is concentrated in a few breeding areas in Michigan, each comprising a patch of dense young Jack Pine *Pinus banksiana* forest, with trees 2-5 m tall. Any one forest patch remains suitable for only 10-14 years, during which warbler numbers first rise then fall as the growing forest becomes increasingly then decreasingly suitable. The total population is the sum of all birds in all occupied habitat patches and thus depends on both the total extent and the suitability of all potential habitat patches. This situation is typical for birds of early successional habitats. Redrawn from Probst 1986.

populations expand greatly during wet periods, only to decline again during droughts, tracking the trends in their habitats. Even more pronounced fluctuations occur in some bird species as a result of occasional catastrophic losses, as when large proportions of a population suddenly die from an epidemic or climatic extreme, and then take several years to recover.

Whatever the type of fluctuation, the degree of variability in any population tends to increase with the span of years over which counts are made. This is partly because the chances of including an unusual year increase with the length of study, and also because the year-to-year fluctuations may be superimposed on a long-term upward or downward trend. In fact, whenever bird species have been studied over periods of several decades, their abundance and distribution patterns are often found to have changed greatly. Some such changes would be expected to occur naturally, perhaps in response to climate changes, but in recent times most such changes can be attributed to human action. One of the best examples is provided by the White Stork *Ciconia ciconia* whose conspicuous nests have been counted in parts of Europe each year for more than a century, revealing in most areas a long-term decline **(Figure 1.2)**. This has been linked with falling food-supplies, resulting from the progressive draining of meadows in Europe and from the increasing frequency of droughts in the African winter quarters (Chapter 11). During the last 100 years in Europe, agricultural developments have caused the widespread decline of first the Corncrake *Crex crex*, then the Partridge *Perdix perdix*, and latterly the Lapwing *Vanellus vanellus* and various seed-eaters (Chapter 7). Over the same period, however, most seabird species have increased greatly, in some species because of reduced human persecution, and in others because human fishing activities have made their foods more available (Chapter 13). Large gulls, especially Herring Gulls *Larus argentatus*, have also benefited from foraging on the growing numbers of rubbish dumps, resulting from the increasing affluence of an expanding human population.

Other birds have greatly extended their ranges in recent decades, becoming common breeders in areas where they were formerly absent. The most striking example is the Collared Dove *Streptopelia decaocto*, which until 1930 extended no further west than the Balkans, but then spread rapidly across Europe, reaching Britain about 1955. For many years in newly colonised regions, its numbers increased exponentially (by a constant percentage each year), until habitats became filled and numbers levelled off (Novak 1971, Hudson 1972, Hengeveld 1988). Most species that have spread in this way clearly benefit from human activities, which may therefore have facilitated the spread. Extreme range changes occurred in other bird species in the past, long before there were ornithologists to record them. Indeed some of the most familiar birds of today in 'developed' regions may have been rare or absent in these regions only 500 years ago. In other species, range expansions appear wholly natural, in that the birds concerned do not depend entirely on man-made habitats. Early in the present century some Cattle Egrets *Bubulcus ibis* crossed the Atlantic from the Old World to the New, and have subsequently spread over much of North and South America (Hengeveld 1988, Maddock & Geering 1994).

Within a human lifetime, such spectacular expansions in range and numbers are exceptional, however, as are precipitous declines to extinction. Yet every species when unchecked has an intrinsic rate of natural increase (r) until it reaches a level (K) determined by the carrying capacity of its habitat. Thereafter, in the absence of marked environmental changes, the species tends to remain fairly constant in numbers over the years, fluctuating within 'extremely narrow limits compared to what is theoretically possible' (Lack 1954).

The patterns that emerge from studies of bird populations thus depend partly on the timescale over which the studies are made. They also depend on the spatial scale. The main question that arises from studying birds in small areas is knowing whether the changes recorded there, either year to year or longer term, are typical of those occurring in the same species over a wider area. Some species, such as the White Stork *Ciconia ciconia*, are so conspicuous and so closely associated with human habitation, that population changes are apparent to everyone over the whole geographical range of the species. Widescale studies of bird populations, involving large numbers of observers, have increasingly enabled the question of spatial synchrony in bird population changes to be addressed, and in an increasing range of species. In general, if the factors that influence bird abundance themselves operate over wide areas, such as aspects of weather, fluctuations on particular study plots can be expected to parallel those occurring on a wider scale. But if the factors are plot-specific, such as local caterpillar outbreaks, then the patterns of fluctuation in bird numbers are likely to vary from place to place across a region (for examples, see Holmes & Sherry 1988). Aggregated results from many different study plots can clearly give a more representative picture of regional dynamics than can studies in a single plot, but the loss of resolution at the larger scale can sometimes hamper attempts to understand the causes of fluctuations.

ENVIRONMENTAL LIMITING FACTORS

To understand what limits bird populations within the habitats they occupy, it is helpful to distinguish between the external (environmental) factors that influence populations and the intrinsic (demographic) features that these factors affect. External limiting factors include resources (notably food-supplies), competing species and natural enemies (predators and parasites). Any of these factors can be considered limiting if it prevents a population from increasing or causes it to decline. Particular populations may be affected by more than one, perhaps all, of these different factors, but often one factor emerges as of overriding importance at any one time. Intrinsic (demographic) features include the rates of births and deaths, immigration and emigration, the net effects of which mediate the influence of external factors to determine local population trends. Both extrinsic and intrinsic factors can be considered as 'causing' population changes, the former as ultimate factors and the latter as proximate ones. Thus, within suitable habitat, a population might be said to decline because of food-shortage (the ultimate cause) or because of the resulting mortality (the proximate cause).

If we are to understand what determines the average level of populations, and why this level varies from year to year and from place to place, we must study the external factors. In our attempts to manage bird populations, it is the external factors that must be altered before any desired change in population level can be achieved. However, long-term trends in bird numbers are not necessarily caused by the same factors that cause the year-to-year fluctuations about the trend, as explained later.

Because studies of bird numbers are normally confined to particular study areas, as well as to particular time periods, the best we can usually hope to show is that some specific factor, or combination of factors, limits bird density at a specified place and time. It is necessary to define the area, for while density may be limited within this area, some individuals may move out and survive elsewhere, so that total numbers are unaffected. The emigrants may or may not return at a later date. It is important to define the time period because the numbers in any population may be limited by different factors at different times, either from season to season or from year to year.

Resources

To be limited by food, a species need not be up against the food limit all the time. Shortages may occur only at certain times of year, under specific weather conditions, or every few years. Some species of high latitudes are cut back so severely by food becoming unavailable in hard winters that they can then experience several years of increase until they stabilise or decline again (for Grey Heron *Ardea cinerea* see **Figure 1.2**). Similarly, many species of arid regions tend to decline through food-shortage during periods of drought, only to increase again as food becomes more available during the intervening 'wet' periods. The season when limitation occurs may also change over time, as feeding conditions alter. For example, the period of greatest food-shortage for Woodpigeons *Columba palumbus* in Britain changed from late winter (Murton *et al.* 1964) to summer (Feare 1991), following changes in the types of crop grown on farmland, where the birds mainly feed.

The evidence that certain bird species can be limited by food is mostly circumstantial: (1) densities are higher in areas where food is abundant than where it is scarce; (2) densities are higher in years when food is abundant than when it is scarce; and (3) long-term or stepwise changes in densities often accompany long-term or stepwise changes in food supply. In a few species, moreover, experimental manipulations of food supplies have been followed by appropriate changes in numbers, as explained in Chapter 7. However, the relationship between birds and their food is not always straightforward. With certain types of food (such as tree flowers), the amount eaten in one year may have little or no influence on the amount available the next. But with other types of food (such as invertebrates), the amount eaten in one year can greatly influence the amount available the next. In this way, prey and predator can interact in various ways, with both short-term and long-term consequences to the population levels of both. Secondly, it is not only the quantity and availability of food that are important, but also its quality, especially for herbivores whose plant-food can vary greatly through the year in digestibility and nutrient content. In addition, some birds can store large amounts of food as reserves within their bodies, so that (at least in the larger species) the food acquired at one time of year can affect performance (breeding or survival) at another (Chapter 11). Other species can store substantial amounts of food externally in caches, and can thus survive periods when food would otherwise be short (Chapter 7). Other resources that in some areas can limit bird numbers include water and other components of habitat, such as cover, roost-sites and nest-sites (Chapters 8 and 12).

Inter-specific competition

Most bird species share part of their food and other resources with other species, both closely related and more distant ones, including other kinds of animals. The seeds eaten by finches, for example, are also consumed by many other creatures, including other birds, mammals and insects. Overlap in food and other resources provides the potential for competition, because some of the resources removed by one species might otherwise have been available for a second. If either of such species is limited by resources, then it follows that the numbers of the one could influence the numbers of the other. Similarly, many species require special nest-sites, such as tree-cavities. If such cavities are in short supply, their use by one species can limit the numbers of other species that can breed. In such competitive situations, changes of either resources or competitor numbers can lead to changes in the numbers of the species of interest. In any pair of competing species, the effects of competition usually fall more heavily on one species than on the other, but the same species may gain in some circumstances and lose in others (Chapter 12).

In any form of competition, it is useful to distinguish between exploitation (or depletion) and interference (Miller 1967). In exploitation competition, individuals have free access to a resource, but its use by one species reduces the amount for another. In interference competition, individuals of one species are denied access to a resource by the aggressive or other disruptive behaviour of another. In an extreme case, birds of one species kept from a food-source by a second, more dominant one, could starve in the midst of plenty. Both forms of interaction can promote spatial or temporal separation between species.

Predators and parasites

With or without competitors, food and other resources can provide a ceiling on bird numbers. In some species, however, breeding numbers are held well below that ceiling by natural enemies (Chapters 9 and 10). In many areas, for example, the numbers of Grey Partridges *Perdix perdix* are held at a low level by predators, and when predators are removed, Partridges can increase to much higher levels (Chapter 9; Potts 1980, Aebischer 1991, Tapper *et al.* 1996). Similarly, the numbers of ducks and game birds are sometimes greatly reduced by parasites or pathogens, so that for years their numbers remain below the level that might otherwise be expected. In assessing the effects of any limiting factor on populations the size of study area can be important, but this is especially true for natural enemies. Because of spatial variations in impact, any predator or parasite can have devastating effects in a small area, but averaged over a larger area the impact might be small.

Relationships between resources and natural enemies

For many birds it is hard to tell whether breeding numbers are limited primarily by resources or by natural enemies, without a field experiment in which one or other is manipulated against an appropriate control. Studying the causes of mortality will not necessarily help to find the limiting factor. Imagine that the density of a territorial species was limited by habitat quality or food-supply, so that surplus individuals were forced into unsuitable habitat, where they were eaten by predators (the 'doomed surplus' model of Errington 1946). From a study of mortality, one would conclude that predators limited numbers because virtually all the deaths occurred through predation; however, the real limiting factor was habitat quality or food supply, which influenced the density of territories in optimal habitat. To change density in the long term would entail a change of habitat, not of predators, and in each case surplus individuals would be removed by predators or any other mortality agents available locally. The key point is that the factors that cause most mortality in a bird population are not necessarily those that ultimately determine the population level. The importance of experiments in clarifying the true limiting factor is obvious.

In any case, assessing the causes of mortality in a bird population is itself not always straightforward. A bird weakened by food-shortage may succumb to disease, but just before death it may fall victim to a predator. For this bird, food-shortage is the underlying cause of death, while predation is the immediate cause. Nor can we assume that a population may be limited by only one factor: food-shortage, predation or disease, as the case may be. In reality, no one factor is likely to account wholly for a given population level. This is because reproduction and survival are seldom influenced by one factor alone, but by several which may act independently or in combination. Thus, whenever a particular factor is described as limiting, it is likely to be one of several factors affecting a population; it may be essential to account for a given population level but only in conjunction with other factors. The primary limiting factor can be considered as the one that, once removed, will permit the biggest rise in numbers. Again this factor can best be revealed by appropriate field experiments.

The various ways in which populations exposed to more than one limiting factor might respond to a change in resource levels over a period of years are shown in **Figure 1.4**. In this Figure, resources are shown, by way of illustration, as continuously increasing, but the same relationships could hold if resources declined or fluctuated. The patterns differ mainly in how closely changes in numbers over time track changes in resource levels. Again, not all the patterns shown can be readily distinguished without the help of experiments.

Another complication is that different limiting factors may interact (Chapter 13). For example, if food were abundant, a bird with many gut parasites might be able to keep itself just as well nourished as one without parasites, simply by eating more. But if food were scarce, a parasitised bird – through having part of its food-intake absorbed by the parasites – might die of starvation, while one without parasites might survive. In this case parasites accentuate the effects of food-shortage. Similarly, if food were plentiful, a bird might be able to obtain its needs in only a small part of each day, keeping a constant watch for predators and foraging only in safe sites, thereby reducing its risk of being caught. But if food were scarce, the same bird might have to expose itself for longer each day, reduce its vigilance, and feed in less safe places, thereby increasing its risk of predation. In cases like these, alterations in either food-supplies or predator or parasite numbers could have marked effects on bird numbers. Hence, single-factor explanations of mortality may not always be realistic.

As a further complication, different limiting factors may not necessarily all act negatively on populations (Chapter 13). Imagine that predators selectively removed diseased individuals from a prey population. Predators might thereby prevent an epidemic, enabling the prey population to remain at a higher level than in their absence. Similarly, many species of birds have a non-renewing stock of food to last the winter, so that the removal of a proportion of individuals by predators or parasites at the start of winter might reduce the rate at which food is depleted, and enable more birds to survive the winter than might otherwise have done so. Although both these possibilities have emerged from field studies, neither is proven, and investigation of the combined effects of different limiting factors has only

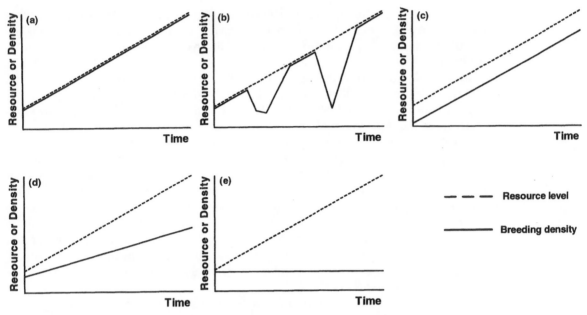

Figure 1.4 Models of how population density might change in relation to resource levels over a period of years.

(a) Breeding density increases in parallel with resource levels, remaining throughout close to the resource ceiling. Implies that resources are the only important factor influencing numbers.

(b) Breeding density varies up to the resource ceiling, so that the upper limit (and the average level) of the population increases with resources. Implies that resources limit numbers at certain times and other factors at other times.

(c) Breeding density increases in parallel with resources, but remains at a lower level throughout. Implies that resources influence numbers, but that other factors consistently hold densities below the resource ceiling.

(d) Breeding density increases with resources but not in parallel. Implies that resources influence density, but that other factors become increasingly (or decreasingly) important as density rises.

(e) Breeding density does not change in relation to resources. Implies that resources at the levels prevailing have no influence on population density, which is limited entirely by other factors.

recently begun. It is a research area in which mathematical modelling, leading to renewed field experimentation, could prove especially useful. The important point is that, just because a particular factor kills individuals and reduces their numbers in the short term, paradoxically – through interactions with other factors – that same factor may actually increase numbers in the longer term (Chapter 13).

Species that show regular cycles of abundance have attracted particular attention because, in principle, their fluctuations could result from interactions with food-supply (quality or quantity), predators or parasites (Watson & Moss 1979). Their fluctuations have also been attributed to regular genetic or behavioural changes in the individuals comprising those populations, leading numbers to rise and fall in regular patterns (Chitty 1967, Mountford *et al.* 1990, Watson *et al.* 1994). Although some evidence has accrued for each of these hypotheses, no one explanation has proved widely applicable, and it could well be that different factors are involved at different places and times, or that two or more factors acting together are involved (Chapters 7, 9, 10).

Weather impacts

Extremes of weather clearly affect the numbers of birds and other animals. At high latitudes, where life is usually temperature-limited, the numbers of many resident bird species are lower than average after unusually hard winters, and higher after mild ones. At lower latitudes, where life in arid regions may be water-limited, the numbers of many birds decline during extreme drought years, and recover again in more normal years. For the most part, weather patterns influence birds indirectly, by affecting their habitats and food-supplies. Some such weather effects can be delayed or long term, as when rain stimulates plant-growth which only later will provide food for birds, while other weather effects can be immediate or short term, as when a snowfall suddenly makes ground-based foods unavailable. Sometimes, however, extreme weather can reduce bird numbers directly, as when individuals die of hypothermia or heat stress or, in some parts of the world, are battered by hailstorms or hurricanes (Chapter 11). Such extremes can sometimes cause sudden catastrophic declines in bird numbers, after which several years are needed for recovery.

Human impacts

In addition to various natural limiting factors, human activities are imposing increasing losses in bird populations, sufficient to cause widespread population declines. The most obvious impact is through habitat destruction, as forests are felled and marshes drained, and converted to farmland or other human use. Because the overall numbers of any species depend on the amount of habitat available to it, destruction of habitat results in reductions in both local and overall population levels (Chapter 3). It is not only the overall loss of habitat that is important, but also its fragmentation, as a result of which species become increasingly patchily distributed. They then become vulnerable to local extinctions and reduced immigration, which can in turn lower their overall populations below the level expected from the amount of habitat remaining (Chapter 6). Other losses imposed by human activity are more direct. Examples include the heavy mortalities inflicted on certain species by overhunting, overhead wires and fishing gear (Chapter 14), and on other species by various pesticides and pollutants (Chapter 15). Some pollutants have also affected birds and other animal populations over wide areas by influencing the physical or chemical structure of their habitats, the dieback of forest and acidification of lakes being obvious examples.

DEMOGRAPHIC PARAMETERS

From the foregoing, it is clear that to understand what determines the level of any population, a knowledge of the external limiting factors is required. Detailed studies of reproductive and mortality rates are not essential to this understanding, but they can provide useful supplementary information. Remember, though, that two populations may be identical in rates of births, deaths and movements, and yet persist indefinitely at quite different densities, if resource levels differ between areas (see below). Conversely, populations at the same density may have different rates of births, deaths and movements. Providing that in any one population, the inputs (from births and immigration) equal the losses (from deaths and emigration), numbers will remain constant through time, regardless of density. For stability to be maintained, a change in one parameter must be compensated by a corresponding change in another. For example, Blue Tits *Parus caeruleus* reared up to three times as many young per pair per year in an area in southern France as they did in another area in Corsica. However, the breeding populations in both areas remained approximately stable, evidently because after leaving the nest the juveniles survived better in Corsica than in France and began breeding at an earlier average age (Blondel *et al.* 1992). Similarly, Blackbirds *Turdus merula* in southern England had better breeding success in suburban than in rural areas (2.0 versus 1.7 young per year), but this difference was compensated by differential survival, so that both populations were balanced in the longer term (Batten 1973).

Where fecundity and survival rates have been studied in the same population during periods of both increase (or stability) and decline, the comparison has provided useful pointers to where the cause of the decline lay. Thus, if reproduction declined while survival stayed the same, the problem would lie in the breeding areas, but if survival had decreased the problem could lie in breeding or wintering areas, depending on when the extra deaths occurred. Decline in a Golden Plover *Pluvialis apricaria* population was associated with a decline in survival rate but no change in reproduction, while a decline in a Lapwing *Vanellus vanellus* population was associated with a decline in reproduction but no change in survival (Parr 1979, Peach *et al.* 1994). In other species, long-term population decline was associated with reduction in both breeding and survival **(Table 1.1)**. The White Stork *Ciconia ciconia*, as already mentioned, has suffered from reduced reproduction on its European breeding grounds, caused by drainage and pesticide-induced food-shortages, and from reduced survival on its African wintering grounds, caused by drought-induced food-shortages (Dallinga & Schoenmakers 1989, Kanyamibwa *et al.* 1993, Bairlein 1996). While changes in the demography of populations mainly result from changes in environment, they can also result from changes in intrinsic features of the population, such as age-structure or genetic composition (Cooke 1987, van Noordwijk 1987, Newton 1989).

In some monogamous bird species, the sex ratio among potential breeding adults is markedly unequal, so that the scarcer sex can limit reproduction by the other. In most species that have been studied, notably waterfowl and game birds, males are in surplus (Chapter 9), so a proportion of males cannot breed each year, however plentiful food and other resources are. As a result of competition for mates, the mean age of first breeding is often higher in males than in females.

Relation between post-breeding and breeding numbers

Each year, as already mentioned, the numbers of any bird species reach their highest level at the end of breeding, and then decline to reach their lowest level near the start of the next breeding season. To what extent, then, do numbers at the end of one breeding season influence those still present at the start of the next?

The answer to this question may depend largely on how close the population is to the limit imposed by the habitat (or resources). Typically the numbers of any species

Table 1.1 Species in which temporal changes in breeding density have been linked with changes in previous overwinter survival (A), with changes in previous breeding rate (B), or with changes in both previous survival and breeding rate (C).

Species	Location	Long-term upward or downward trend	Annual fluctuations	Source
A. Associated with change in overwinter survival				
Wren *Troglodytes troglodytes*	England		+	Peach *et al.* 1995b
Great Tit *Parus major*	England		+	Perrins 1979
Willow Tit *Parus montanus*	Sweden		+	Ekman *et al.* 1981, Ekman 1984
Song Thrush *Turdus philomelos*	Britain		+†	Thomson *et al.* 1997
Sedge Warbler *Acrocephalus schoenobaenus**	England		+	Peach et al. 1991
Pied Flycatcher *Ficedula hypoleuca**	England		+	Stenning *et al.* 1988
Blackcap *Sylvia atricapilla**	Britain		+	Baillie & Peach 1992
Whitethroat *Sylvia communis**	Britain	+	+	Winstanley *et al.* 1974, Baillie & Peach 1992
Willow Warbler *Phylloscopus trochilus**	Britain	+	+	Baillie & Peach 1992, Peach *et al.* 1995a
Barn Swallow *Hirundo rustica**	Denmark		+	Møller 1989a
Sand Martin *Riparia riparia**	Britain		+	Cowley 1979, Bryant & Jones 1995
Avocet *Recurvirostra avosetta**	England		+	Hill 1988
Golden Plover *Pluvialis apricaria**	Scotland	+		Parr 1992
Puffin *Fratercula arctica*	Scotland	+	+	Harris & Wanless 1991
Night Heron *Nycticorax nycticorax**	France		+	den Held 1981
Grey Heron *Ardea cinerea*	Britain		+	Lack 1966
Purple Heron *Ardea purpurea**	Netherlands		+	den Held 1981, Cavé 1983
Mute Swan *Cygnus olor*	England		+	Perrins *et al.* 1994
B. Associated with change in breeding rate				
Black-throated Blue Warbler *Dendroica caerulescens**	New Hampshire	+	+	Holmes *et al.* 1991, 1996
Wilson's Warbler *Wilsonia pusilla**	California		+	Chase *et al.* 1997
Pied Flycatcher *Ficedula hypoleuca**	Finland		+	Virolainen 1984
	England		+	Stenning *et al.* 1988
American Redstart *Setophaga ruticilla**	New Hampshire		+	Sherry & Holmes 1991
Swainson's Thrush *Catharus ustulatus**	California		+	Johnson & Geupel 1996
Wood Thrush *Hylocichla mustelinus**	Delaware		+	Roth & Johnson 1993
Mallard *Anas platyrhynchos**			+	Reynolds 1987
Dunlin *Calidris alpina**	Fennoscandia	+	+	Soikkeli 1970, Jönsson 1991
Lapwing *Vanellus vanellus*	Britain	+		Peach *et al.* 1994
Willow Grouse *Lagopus lagopus*	Norway		+	Myrberget 1972, 1985
Black Grouse *Tetrao tetrix*	Britain	+	+	Baines 1991
Grey Partridge *Perdix perdix*	England	+	+	Potts & Aebischer 1991, Aebischer 1991
Attwater's Prairie Chicken *Tympanuchus cupido attwateri*	Texas	+		Peterson & Silvy 1996
Various duck species*	Iceland		+	Gardarsson & Einarsson 1994
C. Associated with change in both overwinter survival and previous breeding rate				
Barn Owl *Tyto alba*	Scotland		+	Taylor 1994
Tawny Owl *Strix aluco*	England		+	Petty 1992
Great Horned Owl *Bubo virginianus*	Yukon		+	Rohner 1996
White Stork *Ciconia ciconia**	France-Germany	+	+	Dallinga & Schoenmakers 1989, Kamyamibwa *et al.* 1993, Bairlein 1996
Great Skua *Stercorarius skua**	Scotland	+	+	Klomp & Furness 1992

*Summer visitors to the region concerned, wintering elsewhere
†Survival of first-year birds only

introduced to a new but favourable area increase rapidly to begin with, and then gradually level off, as numbers approach the maximum that the area will support (called K – the carrying capacity) (Chapter 5). During the increase phase, a large proportion of any young produced may survive to breed, giving an initial linear relationship between post-breeding and subsequent breeding numbers **(Figure 1.5)**. But as the growing population begins to level off, and competition intensifies, mortality among young may increase, so that a progressively smaller proportion survive to breed. At this stage, the population could withstand a substantial reduction in post-breeding numbers (breeding success) without any reduction in subsequent breeding numbers. On this model, then, the closer a population is to the maximum level that the habitat will support, the less will be the influence of any given reduction in breeding success on next year's breeding numbers.

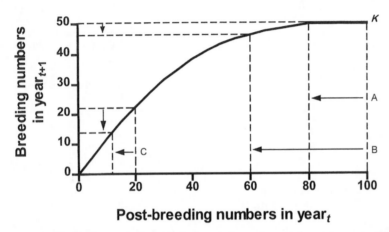

Post-breeding numbers in year$_t$

Figure 1.5   Model showing relationship between breeding and previous post-breeding numbers in an increasing population. In the situation shown, at their maximum level, post-breeding numbers could be reduced by 20% (A) without causing any reduction in subsequent breeding numbers, and by 40% (B) to cause only a 10% reduction in breeding numbers. However, at much lower post-breeding numbers (C), a 40% reduction would cause a similar percentage decline in breeding numbers. The effect of breeding success on subsequent breeding numbers thus depends partly on how close the population is to the carrying capacity (K) of its environment.

In some bird populations that have been studied, breeding numbers increased in years that followed a good breeding season and decreased in years that followed a poor breeding season (**Table 1.1, Figure 1.6**). It seems that summer conditions in breeding areas had most influence on subsequent year-to-year changes in breeding numbers, and that such populations were therefore below K for the winter habitat. In other bird populations, breeding numbers varied from year to year according to previous winter conditions, implying that such populations were close to K for the winter habitat. Over several years, as indicated above, the same population could change from one state to another, as its status with respect to available habitat changed. In yet other bird populations, changes in breeding numbers from year to year were associated with changes in both breeding and survival, which acted together to cause

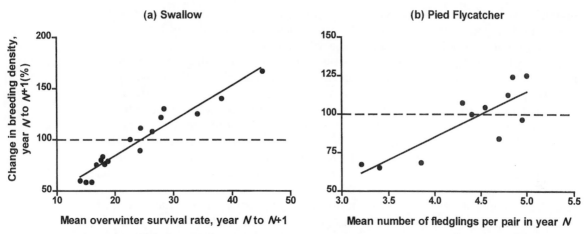

Figure 1.6   Demography and population change. (a) Relationship between annual survival of adult Barn Swallows *Hirundo rustica* and annual change in breeding density. From Møller 1989a. (b) Relationship between annual breeding output of Pied Flycatchers *Ficedula hypoleuca* and annual change in breeding density. From Virolainen 1984. Significance of relationships: Barn Swallow, $b = 3.44$, $r^2 = 0.93$, $P<0.001$; Pied Flycatcher, $b = 0.15$, $r^2 = 0.74$, $P<0.001$.

an increase or decrease. They included species that ate the same type of food year-round in the same area (such as some owls dependent on voles; Chapter 7) and other species in which changes in breeding numbers were affected by conditions in both breeding and wintering areas (such as the White Stork, mentioned above).

Movements

Movements form an integral part of density limitation, because they facilitate rapid changes in local densities in response to changes in conditions. They enable birds to escape areas where survival or reproductive prospects are poor and to find other areas where conditions are better. Indeed, movements are often crucial in the maintenance of local densities. In any widespread species, we can expect that in some areas (called 'source' areas), production of young is more than enough to offset mortality (net exports), whereas in other areas (called 'sink' areas) the reverse is true, so that densities can be maintained only by continued immigration (net imports) (Pulliam 1988).

Failure to appreciate the importance of dispersal can easily lead to an exaggerated view of the role of birth and death rates in influencing local population trends. Most studies of bird populations have been concerned with the dynamics of local numbers. Yet almost all species have geographical ranges larger than an observer's study plot, and individual birds continually move in and out, on local or longer journeys. Such movements can produce much more rapid and pronounced rises in local density than could any increase in fecundity or survival. Of course, the smaller the study area, with respect to the entire geographical range of a species, the more important immigration/emigration are likely to be in influencing the local dynamics. Moreover, as natural habitats become ever more fragmented by human activities, movements are likely to play an increasingly important role in maintaining local populations and in ensuring genetic continuity between them. The term metapopulation has come into use for species which live in fragmented habitats, as subpopulations linked by dispersal (Chapter 6).

The most spectacular of bird movements is long-distance migration, in which birds move each autumn from regular breeding to wintering areas, then back again in spring. At high latitudes, bird numbers increase greatly in spring, as migrants arrive to take advantage of the summer flush of food, returning to lower latitudes for the winter. Tropical areas, already rich in resident birds, may thus receive migrants from the northern hemisphere for half the year (the northern winter) and from the southern hemisphere for the other half (the southern winter). Migration is more prevalent in birds than in other animals, and each year results in huge seasonal shifts in the distribution of birds over the earth's surface.

Conditions in both breeding and wintering areas influence the population levels of migrants, and changes in conditions in one or other area can affect these levels (Chapter 3). On the North American prairies, rainfall varies greatly from year to year, and influences the amount of wetland habitat available to breeding waterfowl (Chapter 11). In wet periods populations increase and in dry periods they decline. So important are these prairie wetlands as breeding habitat that they influence the entire continental wintering population of several species, including American Coot *Fulica americana* **(Figure 1.7)**. This is a clear example of conditions in the breeding areas largely determining year-to-year fluctuations in the total population. In contrast, the numbers of several migrant songbird species counted each spring on their European breeding areas have fluctuated according to rainfall (and food-supplies) in their

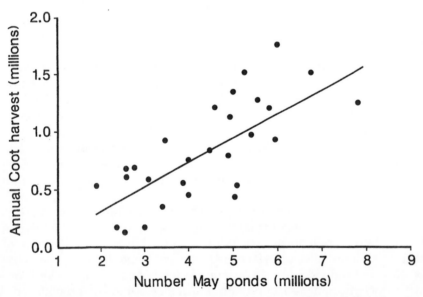

Figure 1.7   Relationship between the numbers of American Coots *Fulica americana* shot each year in the United States (reflecting total population size) and the number of ponds on the prairies in the preceding summer (reflecting the habitat and feeding conditions). Significance of relationship: $b = 0.21$, $r^2 = 0.54$, $P<0.001$. Redrawn from Alisauskas & Arnold 1994.

African wintering areas. Examples include Sedge Warbler *Acrocephalus schoenobaenus*, Whitethroat *Sylvia communis* and Purple Heron *Ardea purpurea* (Chapter 11). They provide examples of conditions in wintering areas largely determining year-to-year fluctuations in the total population. Other examples of summer-influenced and winter-influenced population changes among migrant species may be found in **Table 1.1**.

DENSITY-DEPENDENCE

General year-to-year stability in bird numbers is usually attributed to limiting factors acting in a density-dependent 'regulating' manner (Chapter 5). Because no population increases indefinitely and species only rarely become extinct, regulation can be assumed to occur, tending to keep numbers within restricted limits. Any regulating factor that tends to cause numbers to increase when they are low and decrease when they are high can act in this way. Potential regulating processes include competition for food and other resources, parasitism (or infectious disease) and predation, all of which could affect a greater proportion of individuals as their numbers rise. Emigration might be a secondary factor, mitigating the effects of any of these. Such factors contrast with density-independent factors, such as severe weather and other natural disasters, which can affect a large proportion of a population regardless of its size. Density-independent factors tend to destabilise numbers, causing large and unpredictable fluctuations.

Not all factors that act in a manner correlated with density necessarily promote stability in numbers. They might, for example, be 'inversely density-dependent', affecting a larger proportion of individuals as their numbers fall. This would tend to accentuate fluctuations rather than dampen them. Alternatively, they might act after a time lag, causing a 'delayed density-dependent' response, which in some circumstances can cause populations to fluctuate in regular cycles (Chapter 5). In whatever way they act, however, both density-dependent and density-independent factors can affect the rates of breeding and mortality, immigration and emigration.

Some years ago, arguments were frequent on whether animal numbers were limited primarily by density-dependent or by density-independent factors, some people being impressed by the stability of populations and others by their instability. It now seems more sensible to ask, for any one species, what is the relative importance of the two types of factors in causing the changes observed? Abundance is determined by all the factors that impinge on the population, whether they act in a manner that is dependent or independent of density. In general, the more dominant the influence of regulating density-dependent factors, the more stable the population through time, but even in populations that fluctuate widely, some aspect of breeding or mortality must by logical necessity be density-dependent, otherwise such populations could increase indefinitely or die out. Density-dependence is thus one of the central concepts of population ecology (Chapter 5). It underlies many important ecological problems, including the dynamics of predation and disease, hunting and other human impacts (Chapters 9, 10 and 14).

The fact that many of the bird species that have been monitored over several decades have changed greatly in abundance does not argue against regulation. It highlights the fact that environmental conditions change, so as to alter the population level around which regulation occurs. To use an analogy, populations behave like the temperature of a room in which the controlling thermostat is periodically set at different levels.

Throughout this book, I have used the term 'limiting' for any factor that can restrict density, whether or not it acts in a density-dependent manner. Following the usual convention, the term 'regulating' is confined to factors that operate in a direct density-dependent manner, acting to prevent both indefinite increase and extinction. In some species, densities remain more stable (and hence more regulated) in main habitats, and are more fluctuating (less regulated) in subsidiary habitats (Chapter 3).

Equilibrium models

As explained above, for a population to persist, the gains (from births and immigration) or the losses (from deaths and emigration) must vary in a density-dependent manner. The outcome of the interaction between gains and losses can therefore be expressed in simple diagrams (**Figure 1.8**). Imagine, for illustration, that

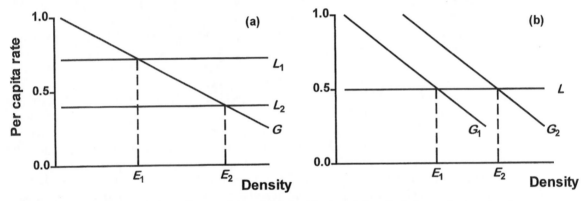

Figure 1.8   Models depicting effects of density-dependent and density-independent processes on equilibrium population sizes. In (a), the loss rate (L) from deaths and emigration is constant with respect to density (i.e. is density-independent), but is set at two levels L_1 and L_2, while the gain rate (G) from births and immigration varies with density (i.e. is density-dependent). Equilibrium population sizes occur where the lines cross and are shown by E_1 and E_2. In (b), the loss rate is again density-independent but the gain rate can be density-dependent over two different density ranges (as, for example, if food-supply varied between two areas). Equilibrium population sizes again occur where the lines cross, and are shown by E_1 and E_2. In this example, a population could have identical gain and loss rates, yet stabilise at markedly different densities. This situation holds because density is expressed in the usual way, as numbers per unit area. If density was expressed as numbers per unit of food-supply, the two E lines would coincide.

In both these models the equilibrium point for the population is set by both density-dependent and density-independent factors. Both types of factor are limiting and affect the equilibrium position. The density-dependent (regulating) factors tend to return the population to its equilibrium position. In these simple models, per capita losses are considered constant, but the conclusions do not alter if losses are density-dependent and gains are density-independent, or if both gains and losses are density-dependent. The strength (slope) of the density-dependent relationship and whether it is linear (as shown) or curvilinear, can greatly affect the equilibrium density.

losses are density-independent and gains are density-dependent. In a stable environment, after any perturbation, the population will tend to return to the point at which gains and losses are equal; in other words the population will tend to stabilise. This is because, if numbers move above or below the equilibrium, the density-dependent gain rate will respond so as to return the population towards that level. However, the equilibrium point would itself alter if the density-independent loss rate was changed by some external factor: for example, if hard weather caused an increase in mortality **(Figure 1.8a)**. The equilibrium point would also alter if the density-dependent gain rate was changed with respect to density by some external factor, for example if an improved food-supply raised the density at which competition set in, or if a growth of cover gave more protection against predation **(Figure 1.8b)**. In fact, any external factor that affects the equilibrium level of a population must be limiting in some degree. The challenge is to find to what extent particular factors change the equilibrium position and in what ways. If we take a large enough area, such as the entire geographical range of a species, immigration and emigration become irrelevant, and the trend of a population depends entirely on the ratio of births to deaths.

PEST CONTROL AND HUNTING

Of all the density-dependent processes in nature, the ones that concern us most are those that occur in species that we kill as pests or food-items (Chapter 14). In pest control we remove all or part of a population in order to reduce the harm to resources that we wish to protect. To be effective, the number of pest individuals killed per unit time must exceed the ability of the population to compensate through improved survival and reproduction of the remaining birds, or through immigration. The killing must also be sustained year after year if numbers are to remain low, for relaxation in control efforts can quickly lead to population recovery. In game management, by contrast, we attempt to 'crop' or 'harvest' a population, using the individuals we remove and leaving others to reproduce and provide future harvests. Sustained harvesting depends on the fact that depleted populations can in time restore their size, through density-dependent changes in births or deaths (Chapter 14). To take the maximum sustainable harvest from any bird population, it is important that breeding success is good, in order to ensure a large post-breeding surplus, otherwise hunters are likely to 'overharvest', removing more individuals per unit time than the population can stand, causing long-term decline.

Among all game species, therefore, an important question in management concerns the extent to which mortality from hunting is additive to natural mortality, rather than compensatory (that is, offset by subsequent reduction in natural loss). If hunting mortality were additive, previously stable populations could decline when hunted, but if it were compensatory they could maintain their numbers. For hunting loss to be compensated by reduced natural loss, three conditions must normally be fulfilled:

(1) the hunting loss must be less than the natural loss;
(2) the natural loss must be density-dependent; and
(3) the hunting loss must occur before or during, but not after, the main period of natural loss.

If hunting is concentrated in autumn, there is then ample time for at least some compensation through reduced natural loss before the next spring. But if hunting is concentrated in late winter, when much natural mortality has already occurred, then it may become mainly 'additive', resulting in population decline. The extent to which hunting loss is additive to natural loss in any species therefore depends not only on the extent of hunting loss, but also on the degree of density-dependence in the total loss and on the relative timing of the two types of loss. The same considerations influence the extent to which any density-dependent and density-independent form of loss can oppose or compensate one another.

SOCIAL BEHAVIOUR

Confusion has long centred on the role of social interactions in population regulation. Whenever individual birds come into contact with one another, they usually reveal some form of dominance hierarchy or 'peck-order' (Chapter 2). This can lead to an uneven sharing of resources, and when supplies are limited, it can provide a basis for the regulation of density, through influencing which individuals survive and reproduce and which do not. Territorial behaviour is an extreme example, as it serves to space birds out through the available habitat, and to exclude surplus individuals once the habitat is saturated (Chapter 4). Such behaviour can thus be regarded as a form of contest, in which each individual attempts to commandeer from a limited supply the resources it needs to survive and breed. Territorialism can also be adjusted to correspond with resources, such as food-supplies, as pairs defend smaller areas in food-rich habitats than in poor ones (Chapter 3). In such cases, food-supply is the environmental factor influencing density, while territorialism is the behavioural mechanism through which this adjustment is made. The experimental removal of territory holders often reveals a surplus of birds able to occupy the vacant sites (Chapter 4). This does not mean that territorial behaviour evolved in order to limit population density. Ownership of a territory confers great advantage on the individuals concerned, including the ability to breed, and natural selection acts on individuals, not on populations. Hence, the limitation of density is best regarded as an incidental consequence of behaviour evolved to benefit the individual.

Even in species that are not markedly territorial, social behaviour can still act as a mediator of the effects of density-dependent competition, concentrating the effects of shortages on particular individuals, and resulting in continual re-adjustment of numbers (or breeding) to resource levels (Chapter 3). Thus stated, territorial and other dominance behaviour serves as the mechanism by which the effects of changes in food and other resource levels are translated into changes in demography and population density. The whole sequence of events between changes in resources and population may be depicted as follows:

Change in resource level → change in proportion of birds excluded or disadvantaged by dominance behaviour → change in rates of birth or death or movement → change in local population density.

Although dominance gives some individuals greater access to resources and in turn improves their survival and reproductive prospects, it is a status that not all individuals can achieve. So for the time being, such individuals must remain subordinate in social interactions and suffer the consequences of their low status. There would be no advantage in subordinates starting a fight if they had no chance of winning. If they survive, their status may improve in later life, as they gain in body size and experience.

LIFE-HISTORY FEATURES

Birds span a wide range of life-history types, from short-lived, mostly small-bodied species, to long-lived, mostly large-bodied ones. Flying birds range in weight from about 2 g (the Bee Hummingbird *Calypte helenae*) to 18 kg (Great Bustard *Otis tarda*), while flightless birds extend up to 150 kg (Ostrich *Struthio camelus*). The heaviest bird species is thus 75 000 times heavier than the smallest.

Among birds as a whole, body size correlates with life-history features. The larger the species: (1) the longer it tends to live; (2) the later the age at which it begins breeding; (3) the longer the breeding cycle; and (4) the fewer the young produced at each attempt. Near one extreme, the small Blue Tit *Parus caeruleus* lives up to about eight years and begins breeding in its first year. It lays 10–12 eggs in a clutch, with one-day intervals between each egg, and has incubation, nestling and post-fledging periods lasting about 14, 16 and 10 days, respectively, bringing the total breeding period from the first egg to about 50 days. Moreover, two broods may be reared by a pair in a single year. The maximum increase in population possible in a year is thus 10–12 times the breeding population, assuming that both parents survive, and even greater if a second brood is reared. Such fast-breeding species are often said to be '*r*-selected', from the technical notation for the intrinsic rate of increase, r.

At the other extreme, the large Wandering Albatross *Diomedia exulans* can probably live for more than 50 years. It does not begin breeding until it is about ten years old, lays only one egg at a time, and has incubation and nestling periods lasting about two months and up to 12 months, respectively, so that annual breeding is impossible. Ignoring the non-breeding immatures, the maximum possible increase in a population of such birds is 50% in two years, or 25% in one year. Such slow-breeding species are often said to be '*K*-selected' from the technical notation for carrying capacity, K.

These two contrasting species lie at opposite ends of a spectrum, between which other bird species show a continuum of variation in life-history types. Although in general, life-history features relate to body size, exceptions occur. Thus swifts are fairly small, but are long-lived with small clutches, while grouse-like birds are medium-sized but short-lived with large clutches. Also, the smallest of all birds (the hummingbirds) lay only two eggs, while the largest (Ostriches) lay up to ten. Overall, however, the main life-history features among birds are correlated, so that long life, delayed maturity, long breeding cycles and small broods usually go together, as do short life, early maturity, short breeding cycles and large broods. Similar patterns occur in mammals, where they also relate to body size.

Within this general trend is another: tropical species tend to be more K in character than their high-latitude equivalents. Thus tropical and sub-tropical birds are generally longer-lived, with smaller clutches and longer breeding cycles, than similar temperate-zone species. Such trends are apparent within both passerines and non-passerines. They hold even within species, if they are widely distributed, and in many such species the average clutch size and annual mortality increase with increase in latitude (Lack 1954, Rowley & Russell 1991).

Although the basis of these trends is far from understood, it is clear that the particular life-history strategies shown by different species greatly affect their population dynamics. In big long-lived species, population turnover is generally slow, with more overlap between generations and a more stable age-structure, all of which tend to dampen short-term fluctuations in numbers. There also tends to be a large non-breeding sector, consisting mainly of immatures. Less than half the total population may breed in any one year, producing only a small number of young. In small short-lived species, by contrast, population turnover is rapid, with less overlap between generations, a less stable age-structure, and a high production of young, all of which facilitate short-term fluctuations in numbers. Most individuals that survive a winter will be able to breed in spring, so that the non-breeding sector usually remains small. And because of their fast breeding rates, such species can recover from a population reduction more quickly than can larger species. They are therefore better able to withstand heavy predation and persecution by humans.

The r and K species present somewhat different problems for conservation. The small, short-lived species often live at high densities (see above) but, because of greater vulnerability to various mortality agents, they are prone to greater fluctuation, and hence to local extirpations. Marked declines in reproduction may soon be followed by declines in adult numbers. In general, however, large, long-lived species present greater problems. The low densities at which most such species live makes them especially vulnerable to human impact, because of the large areas of habitat required to sustain their populations. This is especially true for large raptors (Newton 1979), but also holds for the foraging areas of long-lived seabirds. Also, because long-lived species do not breed until they are several years old, their populations usually contain a large proportion of 'immatures', which in some species occupy different habitat from breeders. These immatures have to be catered for in conservation, in addition to the breeding adults. Perhaps most importantly, however, because of low reproductive rate and long-deferred maturity, long-lived species take much longer to recover their numbers after a catastrophic decline than do short-lived species. Even in the best conditions, the California Condor *Gymnogyps californianus*, now numbering less than 100 individuals, would take more than a century to re-occupy to the full the whole of its historical range.

CONCLUDING REMARKS

In assessing research findings from birds or other animals, a note of caution is needed. Almost all studies have been undertaken in parts of the world where the flora and fauna have been greatly modified by human action. While there is no reason to suspect that such studies mislead us on the general principles and mechanisms behind population limitation in birds, the factors affecting individual species may be quite different in human-modified landscapes from those that operated in the natural habitats in which those species evolved. Studies in natural forest have shown how different are the species compositions, densities, population dynamics, limiting factors, interactions and behaviour from what is seen in the same species in man-made landscapes (Tomialojc & Wesolowski 1990). One of the most urgent requirements in avian population research is for more work in extensive areas of natural habitat before such areas are gone forever. Only then can we tell the full extent of human influence on population phenomena, and view our findings in a more appropriate perspective.

In this introductory chapter, I have tried to give some idea of the general content of the book and of the main ideas that underlie it. In the following chapters, different aspects are developed in more detail, beginning with social behaviour and density-dependence, then the main limiting factors of resources, predation and parasitism, followed by human impact and extinction.

SUMMARY

The most numerous bird-species in the world, such as the House Sparrow *Passer domesticus*, have been estimated to number more than 500 million individuals, while the least numerous contain only a few individuals on the verge of extinction. In their year-to-year changes in breeding density, birds show almost any pattern, from relative stability in some species, through marked irregular or regular fluctuations in others. In addition, over periods of several decades, many species have also shown long-term upward or downward trends in average abundance.

Among the various factors that affect bird numbers, it is useful to distinguish between the extrinsic factors (such as food-supply, predation and disease) and the intrinsic demographic factors (such as birth rates, death rates and movements). Extrinsic factors affect intrinsic factors, causing changes in population level. Some extrinsic factors act in a direct density-dependent manner, affecting a greater proportion of individuals as their numbers rise, while other extrinsic factors act in a density-independent manner, sometimes causing big changes in numbers regardless of density. Sustained harvesting of populations (as in natural predation or human hunting) is contingent upon density-dependent changes in birth or death rates, so that the removal of some individuals is compensated by improved reproduction or survival among the remainder.

In general among birds, with some exceptions, large species start breeding at a later age, and have lower annual reproductive and mortality rates than small ones. Associated with these differences, large species tend to show less annual variation in abundance than small ones, have larger proportions of non-breeding immatures, and slower potential rates of increase.

Part One
Behaviour and Density Regulation

Brent Geese *Branta bernicla nigricans* interacting on feeding areas.

Chapter 2
Social Systems and Status

Social organisation depends on the behavioural relationships between individuals and on their dispersion through the habitat. As a group, birds show a wide range of social systems, in some of which individuals may live alone or in pairs and in others gregariously, spending all their lives in the close company of others. The social system shown by any one species at a particular time depends partly on features of the species itself, such as its mobility and diet, whether breeding or non-breeding, and partly on local conditions, such as food-supply. Social systems are important in a population context because they provide the behavioural framework within which the regulation of density and distribution occurs.

Central to such regulation is the concept of social rank. Anyone who watches birds as they feed together can easily discern that some individuals are dominant over others. The dominants may drive subordinates from food-sources, but are seldom challenged themselves. Social status thus gives priority of access to resources, and can thereby affect the survival and reproductive success of individuals. In competitive situations, it can provide a means by which numbers are continually adjusted to the resources available, whether food, nest-sites, roost-sites, mates or any other requirement that is in short supply or of variable quality.

Aggressive interactions between individuals, whether momentary over food-items or long term over territory defence, provide a form of 'contest competition', in which resources are shared unequally among individuals. In consequence, the effects of resource shortages fall unevenly through the population. As the total numbers of competitors rise, or as resources dwindle, increasing proportions of individuals fail to meet their needs, and leave or die. If the effects of shortage were to fall evenly on all individuals, as in 'scramble competition', this could result in all succumbing. Contest competition through dominance could thus result in more individuals surviving a food-shortage than would occur under scramble competition, and could give much greater stability to populations (Lomnicki 1980). It also has a profound influence on which individuals survive (the most dominant) and which do not (the most subordinate). Among birds, older individuals of a species tend to be dominant over younger ones, and in most species, males (being larger) over females of similar age. Within winter flocks of geese and swans, however, the family is the main competitive unit, with large families dominating small ones, which in turn dominate pairs and single birds (Boyd 1953, Raveling 1970, Black & Owen 1989). Dominance status is one of two aspects of individual behaviour that can influence the feeding efficiency of individuals, the other being foraging skill.

SOCIAL SYSTEMS

The division of bird social systems into solitary and gregarious is somewhat arbitrary, and in practice gradations exist between exclusive highly defended territories at one extreme, through overlapping and sometimes clumped home ranges, to widely spaced but dense nesting colonies, communal roosts or feeding flocks at the other extreme. Our understanding of the ecological basis of these different systems is still poor. From comparisons between species, dispersion systems have been linked with patterns of food distribution: territoriality with evenly spread and dependable food-supplies, and colonial-flocking systems with patchy and unpredictable supplies (Crook 1965, Lack 1968). From comparisons within species, it is clear that other factors are involved too. In particular, individuals can defend territories only when it is practical for them to do so, and when the benefits outweigh the costs – the concept of 'economic defendability' (Brown 1969). If food-supplies are too thinly or sporadically distributed, available at one place on one day and somewhere else the next, then over a period of days individuals must range over areas too large to defend. Similarly, if the numbers of intruders on a territory increase, the costs of defence for the owner become prohibitive, or flock-feeding may become more profitable (Tye 1986). While a bird might for a time defend a fruiting tree, for example (and indeed some birds do), so many other birds may be attracted to the feast that defence becomes impossible.

Cost–benefit analyses of territoriality have been done mainly for species that defend feeding territories for short periods of hours, days or weeks, and make continual adjustments to the area defended, as do some nectivorous birds (Gill & Wolf 1975, Kodric-Brown & Brown 1978, Gass 1979). In such species, calorific benefits and costs are relatively easy to quantify, and to manipulate, as individual birds can be given extra food in the form of sugar solution in special feeders. Territorial activity in Anna Hummingbirds *Calypte anna* was measured while energy availability on the territory was varied. On days when energy supplies were unlimited, residents defended highly exclusive territories, and primarily by expensive defence behaviour, such as chases. As energy supplies decreased, territories became less exclusive, defence changed towards less expensive (and less effective) methods, and owners spent more of their time feeding elsewhere (Ewald & Carpenter 1978). Hence, food-supply influenced both the intensity of defence and the degree of territorial overlap between neighbours. The same was found by experimental food provision in several other species, including Great Tit *Parus major* and Carolina Wren *Thryothorus ludovicianus* (Ydenberg 1984, Strain & Mumme 1988).

In some species, rather precise figures on the relationship between energy budgets and behaviour have been obtained. In the nectar-eating Common Amakihi *Hemignathus virens* on Hawaii, individuals established territories only in areas that could provide at least 2000 µl of nectar per day (van Riper 1984). Experimental provision of extra food enabled birds to breed where they had not bred previously. Up to a point, breeding success was influenced by the levels of nectar available, with better breeding in areas with more nectar. But beyond a level of about 35 000 µl of nectar per day within the territory, successful breeding became impossible because, with higher intrusion rates, the owners spent too much time protecting their nectar sources. In another type of experiment, in which food availability was held constant while intruder pressure was increased (by installing extra feeders), Black-chinned Hummingbirds *Archilochus alexandri* defended smaller areas (Norton *et al.* 1982).

At very high resource levels, birds may cease to defend territories for another reason, namely that they have no need of the extra resources gained by defence. For example, in nectar-feeders, one advantage of territorial defence is that it raises the amount of nectar per available flower (by exclusion of nectar-thieves) and thus increases foraging efficiency (Gill & Wolf 1975). But if nectar levels are already high, the extra increment gained by territorial defence saves so little foraging time that the benefits of defence disappear.

In species that defend longer-term territories, over months or years, cost-benefit analyses are complicated by the fact that, while defence costs are incurred continuously, benefits may accrue with time: for example, if by defence now, the bird protects a food-supply or a nest-site for later use. Moreover, the benefits of territoriality cannot always be measured by net energy gain alone; they are likely in the longer term to involve other advantages, such as the ability to breed. In any interaction, owners may place a higher value on the territory (because of their higher investment and local knowledge), leading to some asymmetry in owner–intruder relations. Despite these considerations, however, the fact remains that territoriality is sustainable only if the defence costs do not plunge the bird into a prolonged energy deficit.

As food-sources become increasingly widely separated, or far from nesting areas, close individual attachment to an exclusive feeding area becomes impracticable. Social organisation may then shift to a more aggregated pattern, based on 'central places', such as nesting colonies or roosting sites, from which individuals range over wide areas for food. At such food-sources, birds may defend feeding territories if it is profitable to do so, or forage in flocks. Many rich food-sources, including fruiting trees, fish shoals or large carcasses, are sporadic, but for a short time provide enough food in one place for many birds to feed together. Other benefits gained from a gregarious lifestyle include information on feeding sites or techniques gleaned from other individuals, or increased protection against predation (Krebs *et al.* 1972, Powell 1974, Brown & Brown 1996). Such protection has several aspects: (1) the dilution effect in which the larger the group the less likely any one individual is to be captured during an attack, with birds in the centre of a flock being more secure than those on the edges; (2) the protection gained by group-mobbing and other collective defence behaviour; (3) the extra vigilance provided by more eyes. By sharing the vigilance costs, individuals in a flock also gain more time for feeding (Caraco 1979). These various advantages hold whether individuals feed in single-species or mixed-species flocks (Metcalfe 1989). Disadvantages of flocking include the facts that, in some circumstances, flocks are more likely to attract predators than are solitary individuals, and are more likely to facilitate the transmission of parasites and pathogens.

Typical flock-sizes vary between species, as does the spacing of individuals within flocks (Goss-Custard 1970a). Large or dense flocks may give better individual protection against predators, but lead to more interference in feeding, especially where food-items are clumped, large or evasive. In short, any social system is probably a compromise between the costs and benefits prevailing, and can change with the needs of the bird (whether breeding or not) and with food sources, predation pressure, and other environmental circumstances at the time (Goss-Custard 1985, Krebs & Davies 1987). Moreover, birds can respond quickly to external changes that affect their feeding rates. Some shorebirds may change from territories to flocks during a single tidal cycle according to where, and on what, the tidal flow allows them to feed (Myers 1982).

While breeding, most bird species that have been studied live in monogamous pairs, consisting of one male and one female, but in other species some individuals live in polygynous groups consisting of one male and two or more females, as polyandrous groups consisting of one female and two or more males, or in poly-gynandrous groups containing up to several males and several females, depending on circumstances. Moreover, whereas some species (or populations) retain the same social system year-round, others change between the breeding and non-breeding seasons. In particular, many species are found as pairs while breeding and as groups or flocks at other times, another response to changing circumstances.

Experiments have helped to confirm the links between social organisation and environmental conditions. By laying out apples in an open field, Tye (1986) caused some Fieldfares *Turdus pilaris* to leave their flocks and establish large feeding territories; and by altering the distribution of food, Zahavi (1971) caused White Wagtails *Motacilla alba* to switch from flock-feeding to defence of a food-supply. Food-provision for various tit species reduced both the size of feeding flocks and the frequency of mixed-species flocks (Berner & Grubb 1985, Székely *et al.* 1989). This finding was held to support the view that flocking developed partly in response to reduced feeding success. Food-provision also shifted the mating system from monogamy to polygyny in Red-winged Blackbirds *Agelaius phoeniceus* (Ewald & Rohwer 1982, Wimberger 1988) and from polyandry to monogamy or polygynandry in Dunnocks *Prunella modularis* (Davies & Lundberg 1984). In each case, food-provision enabled particular males to spend more time on defending their access to females, thus increasing their mating success at the expense of other males. Social organisation has also been changed in response to predator pressure. After experimental exposure to a Goshawk *Accipiter gentilis*, various tits in one area began to feed in bigger, more mixed-species flocks than did those in a control area (Székely *et al.* 1989). This change would be expected if flocking gave greater protection.

In general, experiments have served to confirm the influence of external circumstances on social organisation, and the flexibility of individual birds. If there is an underlying generalisation, it is that individuals will defend and monopolise any resource that they can, but defence becomes increasingly impracticable as other activities intervene, as the numbers of other birds intent on exploiting that resource increase, or as the risks of predation increase. Under high predation risk, gregarious systems may give the best individual protection.

ADVANTAGES OF DOMINANCE

Many studies of particular species have shown that dominant adults often feed in more food-rich areas, or gain more food per unit time, than subordinate juveniles (Ekman & Askenmo 1984, Goss-Custard *et al.* 1984, Bautista *et al.* 1995). In extreme cases, the feeding rates of dominants may be more than twice as high as those of subordinates. This need not matter for subordinates if they can compensate by feeding for longer each day with no other costs, but if food is scarce they can starve. Dominants may also gain safer feeding and roosting sites, and thus benefit from reduced predation risk (Hogstad 1989, Cresswell 1994).

Several studies have revealed how the advantages of dominance are translated into improved survival or breeding success. In a study of Silvereyes *Zosterops lateralis* in eastern Australia, marked individuals were classed as dominant, intermediate or subordinate, depending on how often they won in skirmishes with other individuals (Kikkawa 1980). Some 94% of dominants survived the winter, compared with 90% of intermediates and 82% of subordinates, giving a statistically significant association between dominance and survival ($G=7.82$, $P<0.02$) (Kikkawa 1980). Similar results were found in many other species, including Dark-eyed Junco *Junco hyemalis* (Fretwell 1969), Song Sparrow *Melospiza melodia* (Arcese & Smith 1985), Black-capped Chickadee *Parus atricapillus* (Smith 1976, Desrochers *et al.* 1988) and Willow Tit *Parus montanus* (Ekman 1984, Hogstad 1989). Among Jackdaws *Corvus monedula*, dominant individuals obtained more food, provisioned their young at a faster rate, and raised larger broods (Henderson & Hart 1995). In polygynous species, dominant males achieve most matings, as shown among Black Grouse *Tetrao tetrix* and other birds, as well as in many mammals (Krebs & Davies 1981, Alatalo *et al.* 1996). In a study of lifetime reproduction in Red Junglefowl *Gallus gallus*, the three most dominant hens in a flock of 28 hens produced more offspring of breeding age than all the rest together (Collias *et al.* 1994), while 11 hens raised no young at all. Dominant hens achieved higher lifetime success partly because they lived longer than other hens, on average, and partly because they raised more young per year. Social rank has emerged as a correlate of lifetime breeding success in various other bird species, wherever it was studied (Newton 1989). However, not all studies have revealed such clear-cut differences in survival and breeding success between dominants and subordinates (e.g. Patterson 1977). In general, the benefits of dominance are most obvious when resources are clumped or in short supply, leading to intense aggression or physical combat.

PERFORMANCE OF TERRITORIAL AND NON-TERRITORIAL INDIVIDUALS

In many species of birds, confronted with limited breeding habitat, dominant individuals acquire territories while subordinates do not. This is true whether a territory is a large area providing all the bird requires or a small area containing only a nest-site or a mating-site. Several researchers have examined the difference in survival and reproduction between territorial and non-territorial individuals in the same area. Among Red Grouse *Lagopus l. scoticus,* for example, which defend large territories, almost all tagged territorial birds survived overwinter compared with hardly any tagged non-territorial birds **(Table 2.1)**. This difference held regardless of age: both first-year and older birds were represented among the territorial and non-territorial sectors (Watson 1985). Moreover, when birds changed social class, their survival rate altered to that typical of their new class. This happened when non-territorial birds acquired territories as replacements for territory owners that were killed or experimentally removed, or when birds that had been territorial in one year failed to regain a territory during the following year. The acquisition of a territory evidently conferred survival benefits over and above those accruing from dominance alone. Many of the non-territorial birds were subsequently found dead in the area, having succumbed to starvation, predation, disease or accident, while others may have moved out. Similar, but less marked, differences in survival or emigration between territorial and non-territorial individuals have been observed among many other resident bird species, including Australian Magpie *Gymnorhina tibicen* (Carrick 1963) and Black-capped Chickadee *Parus atricapillus* (Smith 1976), and among wintering Wood Thrush *Hylocichla mustelina* (Winker *et al.* 1990). In the latter, the subordinate flocking birds occupied less favoured habitats, and were subject to greater predation. In other resident species, no such survival differences were noted, and the advantage of year-round territoriality was a subsequent opportunity to breed (Matthysen 1989).

In the breeding season, it is normally only the territorial individuals in a population that can reproduce. This again was shown strikingly in the Red Grouse *Lagopus l. scoticus,* in which all territorial individuals nested while no non-territorial ones did so **(Table 2.2)**. In some other species, however, the difference between the two groups is not always so clear cut, as non-territorial birds can sometimes gain some reproductive success, albeit usually less than that of territorial ones. They

Table 2.1   Overwinter fate and subsequent breeding success of Red Grouse *Lagopus l. scoticus* that were classed as territorial or non-territorial in October. The data were for individually-marked birds in two areas of eastern Scotland (Glen Esk and Kerloch). The territorial birds survived well and many reared young, whereas the non-territorial birds nearly all died and none reared young. From Watson 1985.

| | Glen Esk (1957-61) | | | | Kerloch (1961-67) | | | |
| | Territorial | | Non-Territorial | | Territorial | | Non-Territorial | |
	Cocks	Hens	Cocks	Hens	Cocks	Hens	Cocks	Hens
Numbers present in late October	212	162	153	213	204	130	136	175
Numbers (%) remaining in April	205(97)	159(98)	7(5)	2(1)	198(97)	130(100)	2(1)	14(8)
Numbers (% of April totals) raising young	104(51)	73(46)	0(0)	0(0)	40(22)	21(16)	0(0)	0(0)

The figures refer to the pooled data from several different years, but the marked difference between territorial and non-territorial birds held in every individual year. Many of the non-territorial birds were found dead in the area, having died of various causes including predation, but many others simply disappeared and may have emigrated. Most losses occurred in the period January-April.

might, for example, indulge in extra-pair copulations (sneak matings) or lay eggs in the nests of other individuals (dump laying), or they might continue to challenge territorial birds and acquire a territory late in the season but in time to raise some young (for Song Sparrow *Melospiza melodia*, see Smith & Arcese 1989; for Tree Swallow *Tachycineta bicolor*, see Stutchbury & Robertson 1985, 1987). In the Great Tit *Parus major*, some non-territorial birds formed pairs and nested surreptitiously within the territories of other individuals, refraining throughout from any territorial advertisement (Dhondt & Schillemans 1983). This behaviour is evidently rare in Great Tits, and as yet unknown in any other species. All these activities give some reproductive success to non-territorial birds, but less than that of those with territories. In general, these various observations have revealed, not only that the territorial mode is preferred (at least in the breeding season), but that it results from local dominance, and confers better survival and reproductive prospects. Non-territorialism is second best, although it may be a stage through which many individuals have to pass before getting a territory of their own. In so-called cooperative breeders, in which birds live year-round in group territories, it is normally only the dominant pair that breeds in any one year, although other group members may help to raise the young (Emlen 1978, Stacey & Koenig 1990).

Because in many bird-species the same individuals may adopt different lifestyles at different stages of their lives, switching from non-territorial to territorial, the merits of different lifestyles are best assessed on the basis of lifetime reproductive success. Among Song Sparrows *Melospiza melodia* living on Mandarte Island, British Columbia, males that (through their dominance) acquired a territory and a mate in their first year raised, on average, twice as many breeding offspring in their lifetimes as did other males that acquired neither until their second year (Smith & Arcese 1989). The adoption of an inferior lifestyle, such as non-territoriality, could thus be viewed as a 'conditional strategy', maintained only until a better option presents itself, providing the bird lives long enough. Some individuals die so young that they can never become dominant or acquire a territory. In any comparison of the two lifestyles, care must be taken (as in the examples above) to separate the confounding effects of age on performance from those of social status. Generalising, we can conclude that (1) the existence of two lifestyles in the same population at the same time in the same area is often due to competition, (2) one lifestyle confers greater benefits than the other, and (3) the inferior lifestyle is imposed on subordinates at least partly by the behaviour of dominants. The alternative view, that both lifestyles are equally advantageous and offer similar lifetime success, both existing simultaneously in the population in an evolutionarily stable state, is not tenable on the information currently available, even allowing for sneak matings and dump laying. Further evidence on these matters is given in later chapters.

ADVANTAGES OF SUBMISSION

Aggressive interactions have costs, in that they increase metabolic rates as well as injury risks (Hogstad 1987, 1989; Bryant and Newton, 1994). These costs are especially high for subordinates that have little chance of winning a fight, so might benefit from submission. The optimal behaviour for any individual, therefore, might be to accept whatever status will give it the greatest net benefit at the time. This in turn will depend on its own attributes (including size and weight) in relation to those of other individuals with which it is interacting. Because both these attributes change with age, so does dominance. Although work with captive birds has shown that aggressiveness can be inherited (Guhl *et al.* 1960, Ficken *et al.* 1978), aggressiveness (the tendency to start fights) is not necessarily correlated with dominance (the ability to win). In any wild population, much of the variation in dominance is related to features not under genetic control (such as age), or to features that are under environmental as well as genetic control (such as body size and weight). In some species, parasites may also influence competitive ability. Among Red Grouse *Lagopus l. scoticus*, individuals that are heavily infected with nematodes *Trichostrongylus tenuis* are less likely to acquire a territory (Jenkins *et al.* 1963). Sick birds cannot compete well, whatever their other attributes. However, while dominant individuals are often found to be larger, heavier or less parasitised than subordinates of the same age (Fretwell 1969, Hogstad 1987), it is not always obvious whether these attributes are causes or, at least in part, consequences of dominance. Thus a large male might become dominant because of his size, but then gain more weight because of his dominance, in which case his final weight would be a consequence of both size and dominance.

Another important feature of dominance is that, for any one individual, it is a relative measure which depends on other individuals present. A bird that is subordinate in one social group could achieve higher status (and performance) by moving to another group, consisting of younger or weaker individuals. Such 'status-adjusting' behaviour is another adaptive strategy for individuals, which may well contribute to the spatial variations in the age and sex composition of bird populations discussed below. Increasing experience at both feeding and fighting may alter the balance, so that the optimum strategy changes with age, at different rates in different individuals, but causing the same individual to behave differently at different stages of its life.

SOME CONSEQUENCES OF DOMINANCE RELATIONSHIPS

In addition to differential survival and breeding success, dominance relations within a population can also promote spatial variation in age and sex ratios. Such variation is evident at a range of spatial scales, from single feeding flocks to large geographical areas. At the local level, good competitors tend to accumulate in the richest feeding sites, and gain the highest rates of intake, while weak competitors end up in poor sites, with lower rates of intake. This is not necessarily because poor competitors are driven from the richer sites, but because if they moved to richer sites, they would suffer more interference, and have even lower intake. For Herring Gulls *Larus argentatus* feeding on a rubbish tip, Monaghan (1980) distinguished a main area in which most of the rubbish was dumped and a secondary area nearby. In the main area the density of gulls and their average intake was about five times higher, but there were also three times as many aggressive encounters. As the total numbers of gulls on the tip increased, a higher proportion of the immatures used the secondary area where the social environment was less hostile. Age ratios therefore varied from one part of the tip to another. Similarly, among White-winged Crossbills *Loxia leucoptera* feeding in conifers, the amount of fighting increased with flock-size. This was enough to reduce the feeding rates of females and juveniles, but not of adult males. Correspondingly, females and juveniles tended to avoid flocks consisting mostly of adult males, resulting in some spatial separation of age and sex groups in the same area. This segregation was apparent only in the non-breeding season (Benkman 1997).

Among coastal Oystercatchers *Haematopus ostralegus*, juveniles occupied favoured mussel beds in summer when adults were away in breeding areas, but abandoned these beds in the autumn when the adults returned and interactions increased (Goss-Custard *et al.* 1984). In winter, therefore, adults occurred mainly on favoured mussel beds, and juveniles on less favoured beds nearby. Among Snow Buntings *Plectrophenax nivalis* wintering in northeast Scotland, the proportions of adults were greatest on mountains, while the proportions of juveniles (especially females) were greatest on coasts (R.D. Smith *et al.* 1993). The high altitude sites were thought to be preferred, and the distributional differences were again attributed to competition. Likewise, local studies of various wintering duck species have shown that males predominated in the best feeding areas (Nichols & Haramis 1980, Hepp & Hair 1984). For example, among Canvasbacks *Aythya valisineria* in South Carolina, males formed 74% of birds in areas where food was concentrated (giving higher feeding rates), but only 48% elsewhere (Alexander & Hair 1979). Sex differences in habitat, with males concentrated in the best areas, have also been noted among wintering songbirds and other bird species (Marra *et al.* 1993).

In the face of predation, dominants can also concentrate in the safest sites, leading subordinates to feed where the risks are greater (Hinsley *et al.* 1996). Among Willow Tits *Parus montanus*, dominants usually occupied sites near the centres of trees, while subordinates were forced to the outer parts, where they spent more time on vigilance, but were still more often caught by Pygmy Owls *Glaucidium passerinum* (Ekman *et al.* 1981). The subordinates thus suffered from both reduced feeding rate and from increased predation. In Black-capped Chickadees *Parus atricapillus*, the temporary removal of males from three flocks enabled females to forage in the secure inner parts of trees, only to be forced to the edges again when the males were released (Desrochers 1989). The same result was obtained in a similar experiment with Downy Woodpeckers *Picoides pubescens*, when females occupied the best feeding sites only after males were removed (Peters & Grubb 1983). In ground-feeding species, too, the juveniles often end up on the edges of flocks and are more often caught by predators (Whitfield 1985).

Birds also compete over roost-sites which vary in the degree of protection and shelter they provide. Among species that roost communally, the distribution of individuals within a roost can be markedly non-random with respect to age, sex and body condition. Among Starlings *Sturnus vulgaris*, for example, adult males predominated in the centre of a roost, which was safer and warmer, while first-year females were most frequent around the periphery. Furthermore, birds in the centre were relatively heavier and fatter than those on the edge. That this distribution could have resulted through competition was supported by observations on captive birds, among which known dominants got the best sites (Feare *et al.* 1995). Statistically significant segregation within roosts, according to sex or age, has been noted in many other species, including Brambling *Fringilla montifringilla* (Jenni 1993), Rook *Corvus frugilegus* (Swingland 1977), Red-winged Blackbird *Agelaius phoeniceus* (Weatherhead & Hoysak 1984), Chough *Pyrrhocorax pyrrhocorax* (Still *et al.* 1987) and Dunlin *Calidris alpina* (Ruiz *et al.* 1989). In each species, the adults predominated in the safer and more sheltered sites and the juveniles in the more exposed sites.

In all these species, dominance with respect to particular feeding and roosting sites was associated with some degree of spatial segregation between adults and juveniles, and in some species outside the breeding season, between males and females. Such differences have also emerged over much wider spatial scales. The composition of several Oystercatcher roosts on the Dutch Waddensea differed according to the area of adjacent mud flats (Swennen 1984). Roosts near the smallest mud flats were smaller, but also included a higher proportion of juveniles, immatures and males (which are smaller than females). There was other evidence that these roosts contained birds of poorer quality – birds of each age-class were, on average, significantly below their expected weight – and also contained more birds with deformed bills. During severe weather, a significantly greater proportion of the birds from these roosts died, compared with the larger roosts. This study was done on two islands and the same pattern emerged on each.

Dominance relations also influence local movements, even within a single age-group (De Laet 1985). In the Great Tit *Parus major*, for example, young leaving the nest later in the season are smaller on average, lose a disproportionate number of disputes (Garnett 1981), and disperse further than do older fledglings (Dhondt & Hublé 1968). But later young moved less far when Kluijver (1971) removed 90% of the first-brood young. Moreover, young Great Tits disperse further from their natal areas in years of high population density, as found in both resident and migratory populations (Berndt & Henss 1967, Greenwood *et al.* 1979, O'Connor 1980). One consequence of such differential dispersal is that dominant birds tend to remain in the same areas, whereas subordinate ones move elsewhere.

On a larger scale, some species of birds migrate chiefly in years of high population or low food supply, when competition would be expected to be strong (Chapter 8). In both these and in some regular migrants, females move in greater proportion and further than males, and juveniles in greater proportion and further than adults (Newton 1972, Gauthreaux 1978, 1982). Over the whole wintering range, this gives rise to latitudinal gradients in sex and age ratios, with adult males usually predominating in areas nearest the breeding grounds and adult females and juveniles further away. Linnaeus named the Chaffinch *Fringilla coelebs* (meaning bachelor), because it was chiefly the males that remained to winter in Sweden where he lived, whereas in the extreme south and west of Europe females predominated (Newton 1972). Such winter gradients have been well described in the Dark-eyed Junco *Junco hyemalis* in North America (Ketterson & Nolan 1976) and in the Pochard *Aythya ferina* in western Europe **(Figure 2.1)**, but they are known in many other species. Moreover, the implication that the same individuals may move further in their first than in subsequent winters has been confirmed in some species from ringing results (Newton 1979).

While these patterns could be explained by dominance relations, assuming that the favoured habitat lay nearest the breeding areas, they could also be explained in other ways (Ketterson & Nolan 1983). Because in most species males are the larger sex, males may be able to winter in harsher climates, and for this reason stay nearer their breeding grounds in winter. In addition, because males need to be first to return to breeding areas to obtain a territory, they might gain by migrating less far. With several plausible explanations, special interest attaches to species in which females are larger. In those raptors and owls that have been studied, the males usually fly furthest, and the females stay nearer the breeding areas, a difference that

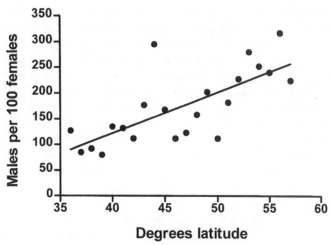

Figure 2.1   Sex ratio among wintering Pochard *Aythya ferina* in relation to latitude in western Europe. Redrawn from Carbone & Owen 1995.

could be explained by dominance or by winter climate (Belopolskij 1971, Clark 1985, Kerlinger & Lein 1986, Kjellén 1994). In shorebirds, in contrast, the larger females winter no further north than the males and in some species further south (Myers 1981, Meltofte 1996). There is probably no single explanation for differential migration, but to judge from small-scale distributions, dominance is probably involved. It could influence movements through direct effects on individuals at the time of migration, or through evolution, by providing a consistent selection pressure for subordinate sex- and age-groups to migrate furthest. The migratory behaviour in captive male and female Dark-eyed Juncos *Junco hyemalis* was compared when they were given the same photoperiod and access to food. In the migration season, both sexes showed the same intensity of migratory behaviour but, on average, females showed migratory restlessness for 21 more nights than males. This implied a strong genetic component, which would have led females to migrate further from their breeding areas (Holberton 1993).

These various observations suggest that the effects of dominance on bird densities and distributions could operate at a range of spatial scales from the most local to the most widespread, including the entire wintering range in some migrants. Whatever the spatial scale, however, it is probably not just a case of subordinates being 'driven' from the best areas, but of their opting to leave because, in prevailing conditions, they can do better elsewhere, under reduced competition. It is only at small spatial scales that the role of dominance in distribution has been studied in detail, and experimentally tested. It is hard to imagine how this could be done at larger scales. In consequence, the importance of dominance, relative to other factors, in influencing geographical gradients in sex and age ratios remains uncertain.

OTHER INEQUALITIES

In addition to dominance, foraging efficiency is influenced by several other features of the individual, including age, sex and morphology. Regardless of social status, adult birds can usually feed at higher rates than juveniles, a difference assumed to result mainly from experience. Age differences in feeding efficiency have now been shown in a wide range of bird species (reviews: Burger 1990, Wunderle 1991), and in general, the more skill required to obtain food, the less successful juveniles are relative to adults. Thus, juvenile Laughing Gulls *Larus atricilla* were found to be much less successful than adults when plunge diving for fish, slightly less successful when aerial dipping, and equally successful when catching offal in the air (Burger & Gochfield 1983). As an example of sexual differences, male Goldfinches *Carduelis carduelis* have slightly longer beaks than females, and are better able to reach the seeds of teasels *Dipsacus fullonum* which lie at the bottom of long spiked tubes. Males can thus exploit this food-plant better than females which, with their shorter beaks, are probably better equipped to deal with other foods (Newton 1967a). Similarly, the sexes of raptor species that show marked size dimorphism also take prey of partly different sizes (Newton 1979). With such sex differences, it is not a question of one sex being generally more efficient than the other, but of the two sexes being most efficient on different foods, leading to some segregation in feeding habits between them.

Within sex- and age-groups, individual morphology can vary and influence feeding habits. An African estrildid finch, the Black-bellied Seedcracker *Pyrenestes ostrinus*, is polymorphic in bill size. Smith (1987) found that the morphs, which occur in both sexes, differed in feeding efficiency on hard and soft sedge seeds, a difference that influenced their diets. Likewise, Norton-Griffiths (1968) found that Oystercatcher *Haematopus ostralegus* individuals differed in the way they opened mussels, some hammering the closed shells to break them, and others stabbing into the open shell to snip the adductor mussel. The two types had differently shaped beaks (as a result of differential wear), and the method of opening shells appeared to be culturally inherited, with young learning from their parents. The hammerers were restricted to mussels with soft shells, but unlike the stabbers, had access to closed mussels as well as to open ones. Some bird populations can thus be viewed as consisting of a series of specialised individuals, presumably affected differently by shortages of particular food-types (Partridge 1976). Together with dominance relations, such differences mean that not all individuals are affected at the same time or to the same extent by dwindling stocks of particular foods.

SUMMARY

Birds show a wide range of social systems from solitary to gregarious. Particular species can change from one system to another depending on breeding commitments, and on external factors, such as food-sources, competitor pressure and predation risks. Birds also show marked individual variation in competitive ability, influenced largely by dominance relations, which ensure that resources are divided unequally – whether food, territories, nest-sites, roost-sites or mates. Dominants have prior access, and often survive and breed better than subordinates. The advantages of dominance are obvious, but in the short term submission pays in individuals that would have little chance of winning a fight. If they survive, individuals normally gain in dominance as they age, and in most species males (being larger) are dominant to females of similar age. Such differences could contribute to patterns in the distribution of different sex- and age-groups, evident outside the breeding season, both locally within feeding and roosting areas, and more widely within large geographical regions, thereby influencing the spatial structure of populations. In addition to dominance relations, other variations in feeding efficiency between individuals result from experience (or age) and from sexual and individual differences in morphology. They too ensure that not all individuals are affected equally by shortage of some particular food.

Shags *Phalacrocorax aristotelis.*

Chapter 3
Habitat and Density Regulation

In conjunction with dominance behaviour, spatial variation in habitat is involved in the regulation of bird numbers. The central idea is based on two generally accepted tenets. Firstly, for any species, habitat varies in quality from place to place; that is in the benefits in terms of survival and reproduction that it confers on its occupants. Secondly, as numbers rise, good habitat is occupied first and in preference to poorer habitat. Such a pattern depends on individuals distinguishing and preferring good habitat, and beyond a certain level of occupation, deterring settlement by further individuals, which then move into poorer habitat. If these conditions hold, it follows that, as total numbers rise, an increasing proportion of individuals is pushed down the habitat gradient, where (on the definition above) their survival or breeding success is reduced. Some individuals may be excluded from nesting habitat altogether, and accumulate as a non-breeding contingent or die. As a result of these processes, the mean per capita performance (survival or reproduction) for the population as a whole declines progressively in a density-dependent manner. In the absence of other constraints, total numbers (of breeders plus non-breeders) will stabilise when the mean per capita mortality rate of the entire population exactly balances the mean per capita reproductive rate. In this way, variation in habitat could function in regulating the overall population level. Resources and other limiting factors are involved in this regulation because they influence the quality and carrying capacity of habitats.

LOCAL SETTLEMENT IN RELATION TO RESOURCE LEVELS

Territorial systems

Imagine that birds returning from migration are occupying an area of habitat and establishing territories. The first arrivals encounter no resistance and can settle with little conflict. They establish large territories which they might later contract to some extent under pressure from later settlers. Then, as more birds arrive, established birds can compress their territories no further, so that newcomers find it increasingly difficult to find a place, and eventually numbers reach a plateau **(Figure 3.1)**. The area is then occupied to capacity, and no further birds can settle unless they displace others already there. Both numbers and subsequent breeding output are thereby limited. This pattern of spring settlement has been observed in a wide range of bird species (reviews: Tinbergen 1957, Klomp 1972, Davies 1978). It provides a clear behavioural mechanism that could not only limit densities in a given year, but also regulate densities over a number of years because it acts in a density-dependent manner. The mechanism ensures greater resistance to further settlement as territorial numbers rise, and at the same time could lead to the exclusion of a greater proportion of potential settlers as total numbers rise. Moreover, as explained later, the level at which density stabilises is sensitive to food-supply and other features of the habitat itself.[1]

Many studies have shown that, within particular bird species, territory sizes are influenced by food-supplies, being larger where food densities are lower **(Figure 3.2)**. This holds when comparing different territories in the same area **(Figure 3.2)**, or

---

[1] *In theory, there could be a further stage beyond that at which territory numbers stabilise. If birds unable to obtain territories remain locally, rather than moving elsewhere, they might interfere with territory owners so much that they reduce breeding success. They might even push up the defence costs to such a level that territoriality ceases to be feasible, causing many birds to resort to a non-territorial existence, and not breed. The interference generated by rising numbers of contenders could thus force down the numbers of territory holders, to nil if the process continued. Probably, this stage is reached only rarely and the evidence is as yet rather skimpy. However, some shorebirds were seen to establish territories when they first arrived on their wintering grounds, only to surrender them later in favour of flocking as further birds arrived. Myers et al. (1979) found that wintering Sanderling Calidris alba gave up attempts to defend territories in the best feeding areas where intruder pressure was greatest. Moreover, in several species, interference by non-territorial birds has been shown to reduce the feeding and breeding success of territorial ones (Jenkins 1961, van Riper 1984, Jenny 1992, Haller 1996, Komdeur 1996) and in certain birds-of-prey to increase the mortality rate of territory owners (Bowman et al. 1995, Haller 1996).*

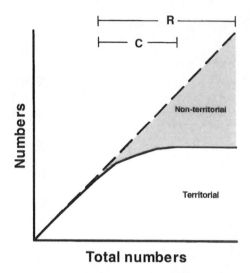

Total numbers

Figure 3.1 Model showing how the numbers of territorial and non-territorial birds might change according to change in the total numbers of potential settlers. C shows the range of total numbers over which territories could be compressed under pressure from further contenders, and R shows the range of total numbers over which replacements of dead (or removed) territorial birds could be expected.

At low numbers, all individuals can establish a territory, and territorial behaviour serves merely to space out birds within the habitat. At higher numbers (zone C) an increasing *number* of birds can establish territories, but an increasing *proportion* is excluded from doing so, providing a mechanism for the regulation of local density. Territorial behaviour can thus limit density from the start of zone C, even though higher densities can be reached (to the end of zone C) under pressure from further rise in the number of contenders. Beyond C, no further increase in territorial numbers occurs, despite further rise in the numbers of contenders. At this high level, for every additional territorial bird that settles, one must die or leave.

Over short periods of years, any one population would normally fluctuate within only part of the density range shown on the horizontal axis.

mean territory sizes from one district to the next (Newton 1979, Watson *et al.* 1992). The variation in territory sizes within species can be great – up to 100-fold in the Rufous Hummingbird *Selasphorus rufus* (Gass *et al.* 1976, Kodric-Brown & Brown 1978). Moreover, in many species, individuals change the size of their territories, in line with changes in food-supplies, day to day in some nectar-feeders outside the breeding season (Gass 1979), and month to month or year to year in other species (Lockie 1955, Pitelka *et al.* 1955). In some long-lived raptors, territories may be maintained for many years, despite some annual fluctuation in food-supply, but they may change slowly in size over several years in response to long-term change in average food abundance (Newton 1979). These relationships do not necessarily imply that territory owners respond directly to the food-supply; owners may be forced to take smaller territories in areas of high food density, because many other birds are attracted to such areas, thereby increasing the defence costs of larger territories (Myers *et al.* 1979).

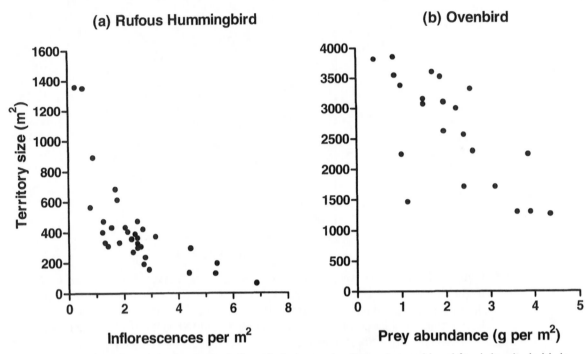

Figure 3.2   Examples of the inverse relationship between territory size and local food density in birds.
(a) Rufous Hummingbird *Selasphorus rufus* which fed on *Castilleja* flowers in a meadow. Redrawn from Gass 1979.
(b) Ovenbird *Seiurus aurocapillus* which fed on invertebrates from among forest ground litter. Redrawn from Smith & Shugart 1987.

Experimental evidence has confirmed a causal link between food-supply and territory sizes. Among Rufous Hummingbirds *Selasphorus rufus*, studied on alpine meadows in California, the experimental removal of 50% of nectar-bearing flowers resulted within one day in an approximate doubling of territory sizes, so that the number of flowers per defended territory was restored to about the original value. Just as rapidly, territories shrank following a recovery in flower density (Hixon *et al.* 1983). Likewise, experimental food-provision was followed by shrinkage of territories (and increase in density) in other nectar-feeding species, as well as in other birds, such as Nuthatch *Sitta europaea* (Enoksson & Nilsson 1983) and Dunnock *Prunella modularis* (Davies & Lundberg 1984), while experimental reduction of food-supply led to expansion of territories (and decrease in density) in Rufous-sided Towhees *Pipilo erythrophthalmus* (Wasserman 1983).

In this way, as individuals defend territories of a size corresponding to food-content, bird densities can become adjusted to food-supplies over wide areas. However, while food-supply clearly has a major influence on territory sizes, it is not the only factor involved, as later examples show, and food manipulation has not always led to a change in territory size (Yom-Tov 1974, Franzblau & Collins 1980). Food within the territory is clearly unimportant in species that obtain their food elsewhere, or that for other reasons hold territories larger than needed for feeding.

Non-territorial systems

It is perhaps easiest to envisage the behavioural limitation of density in territorial species, which aggressively protect individual areas, and thereby exclude other individuals. However, any other competitive interaction can have the same effect, providing it results in an uneven sharing of resources or predation risks among competing individuals, leading some to remain and others to leave or die. In this way, the proportion excluded could increase as total numbers rise. Among flock-feeding birds, two such competitive processes have been recognised (Miller 1968). One results from aggressive and other interactions over food or feeding sites (interference competition). Such interactions include the disturbance of prey items, making them less available, together with avoidance, threats, fighting and food-robbing. These interactions can reduce the feeding rates of all individuals but, because some are dominant over others, the effects of aggression fall most heavily on subordinates (Murton 1968, Goss-Custard *et al.* 1982, 1984). On any given food-supply, interactions tend to increase as bird density rises, and to decrease as bird density falls **(Figure 3.3)**.

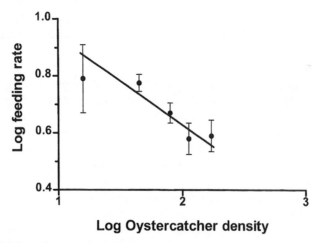

Figure 3.3   The interference experienced by Oystercatchers *Haematopus ostralegus*, expressed as decline in mean feeding rate (± s.e.) with increasing density of competitors. Interference arose largely because birds stole mussels from each other. Redrawn from Zwarts & Drent 1981.

The second competitive process results from the depletion of food supplies through bird feeding (depletion or 'exploitation' competition). This process again gradually lowers the feeding rates of all individuals, but especially of the less efficient and subordinate ones. The food-intake of any individual in a given situation is thus likely to depend both on its social status and on its competence at food collection **(Figure 3.4)**. In both respects, experienced adults are likely to fare better than inexperienced juveniles. Among juveniles alone, however, the two attributes are not necessarily correlated, some dominant individuals being relatively inefficient feeders and vice versa (Goss-Custard *et al.* 1984).

The relative importance of interference and depletion competition varies between

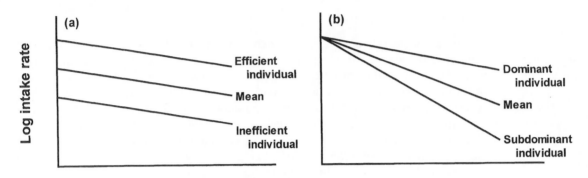

Log density of competitors

Figure 3.4   Two ways in which individuals may differ in their rates of food intake under increasing interference.
(a) Different individuals of similar dominance but different food-gathering efficiencies. All individuals are affected similarly by increased interference.
(b) Different individuals of varying social status, and experiencing different degrees of interference. The most dominant individuals are affected less than the subordinates.
Modified from Sutherland & Parker (1985), based partly on field studies of Oystercatchers *Haematopus ostralegus* by Ens & Goss-Custard (1984).

species and the foods they are eating. Birds that eat items too large to be quickly swallowed are liable to robbery, while birds that eat mobile and evasive prey are liable to the disturbing effects of other individuals. In contrast, birds that eat small or immobile items suffer much less interference, so that depletion becomes the main competitive impact. Moreover, species change their behaviour to suit their food, so experience the effects of interference in some situations and of depletion in others. The effects of interference and depletion have usually been studied separately, but in most flock-feeding situations both would be expected.

Like territorial behaviour, these competitive processes can again act as proximate behavioural mechanisms through which bird numbers are continually adjusted to resource levels **(Figure 3.5)**. Several studies have shown that, under rising bird densities or diminishing food-supplies, competitive interactions increase and individual food intake declines, less in some individuals than in others, thus providing a clear regulating mechanism (Murton 1968, Goss-Custard *et al.* 1982, Sutherland & Koene 1982, Goss-Custard & Durell 1988). Once a bird's feeding rate falls below the minimum necessary for survival, whether through interference or depletion, the bird must leave or die (Goss-Custard 1985).

In one of the first relevant studies, individually-tagged Woodpigeons *Columba palumbus* were watched as they fed in fields on spilled grain or clover leaves (Murton 1968). Within the flocks, dominant birds fed consistently at the maximum rate the food-supply allowed, but the subordinates – which spent much of their time avoiding the dominants – fed only two-thirds as fast, as judged by the number of pecks per minute. This was still faster than when they fed alone (which few did), however, because they then spent more time watching for hawks.

For a wide range of food densities, subordinate Woodpigeons, with their lower feeding rates, could compensate by feeding longer each day. At high food densities,

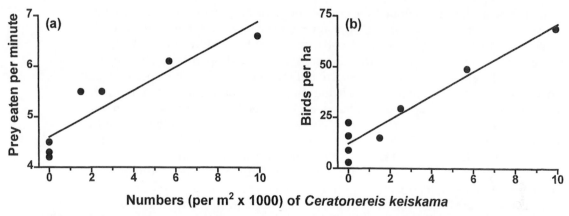

Figure 3.5   Relationships between (a) feeding rates and prey densities, and (b) bird densities and prey densities in Curlew Sandpiper *Calidris ferruginea* feeding on nereid worms *Ceratonereis keiskama*. Redrawn from Kalejta & Hockey 1996.

therefore, flock sizes bore little relationship to local food-supply, and overall numbers and survival rates were not obviously limited by food. At lower food densities, however, when the daily intake of subordinates fell below the threshold for survival, such birds had to move elsewhere or die **(Figure 3.6)**. In this way, local numbers declined in response to dwindling food-stocks. Subordinates were not actively driven from preferred areas; they seemed to leave because they could achieve higher intake rates elsewhere.

When shortage was widespread, some subordinate pigeons moved from flock to flock and eventually died of starvation, predation or disease. Through this mechanism

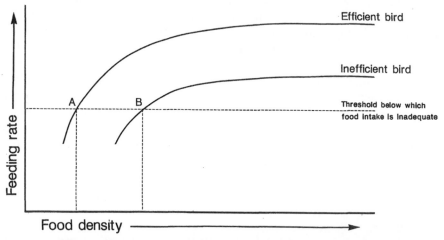

Figure 3.6   Theoretical response curves showing feeding rate in relation to food density for birds with high and low intake rates. The food density at which intake becomes inadequate for survival is given by the vertical lines. Birds whose intake rates are high can survive at lower food densities than birds whose intake is low. Derived from field studies of Woodpigeons *Columba palumbus* feeding on cereal grains and clover leaves. From Murton (1968).

of continual exclusion resulting from competition, the proportion of birds starving at any one time was usually low, and adjustment was continuous. However, a sudden massive reduction in food-stocks, caused for example by snowfall, could greatly increase the number of starving birds and cause a marked drop in population level.

This early study showed how social interaction in a flocking species could function in density regulation, ensuring that local numbers were continually adjusted to local food-supplies. Similar mechanisms have since been shown in other flocking species, notably Oystercatchers *Haematopus ostralegus* and other shorebirds (Goss-Custard & Durell 1988); in each case adults were normally dominant over juveniles at feeding sites, had higher feeding and survival rates, and were less likely to move around in response to changes in food densities. Adjustments were not merely to decreasing food-supplies, but also to increasing ones: when competition was relaxed, more birds could move in. This was found in observational studies and in experiments involving the provision of extra food (Chapter 7).

EFFECT OF HABITAT VARIATION ON SETTLEMENT PATTERN

Theoretical base

The model of local density limitation in **Figure 3.1** can be developed further to allow for consistent variation in habitat quality, and for the fate of excluded birds **(Figure 3.7)**. Assume that birds have a hierarchy of habitat preferences, based on the survival and reproductive success they can achieve there. Thus, habitat A is preferred because it is better than B, and birds will occupy A first, and will move to B only when A is saturated. Similarly, as population size grows, B will fill, and then when B is saturated, birds will settle in C, which permits survival but not breeding. Birds accumulate in C as a non-breeding surplus. If a reduction in overall population level occurs through some external event, such as a hard winter, then birds will leave C in favour of B and then B in favour of A. In this way, habitat C serves as a 'buffer' habitat with respect to B, and B with respect to C, so that over a period of years, numbers remain least stable in C and most stable in A.

On this system, therefore, it is not merely that secondary habitats are populated by surplus birds evicted from primary habitats. These same birds provide a pool of potential immigrants able to fill any gaps that arise in primary habitats (Brown 1969, Fretwell 1972). In this way, densities in primary habitats can be buffered against dramatic fluctuations in overall numbers, and the stability of densities in good habitat depends on the presence of birds in nearby poor habitat.[2] Although the

---

[2]*Further modification of the above model was proposed by Fretwell & Lucas (1970) for situations where there was density-dependent decline in performance within habitats, a situation that need not necessarily hold in all populations. Where it does hold, the performance of new settlers in any given habitat area may depend not only on intrinsic features of the habitat, but also on the density of individuals already present. With increasing density in the favoured habitat, the feeding rates, survival or reproductive prospects of individuals within that habitat may decline, eventually to the point when new settlers would do better by moving to poorer but less crowded habitat, and so on down the habitat gradient. The term 'zero density suitability', can be used to describe the relative ranking of habitats when no competitors are present. Fretwell & Lucas (1970) distinguished two extreme scenarios, called: (1) the 'ideal free distribution' in which individuals had equal competitive abilities, and were free to move between habitats and settle*

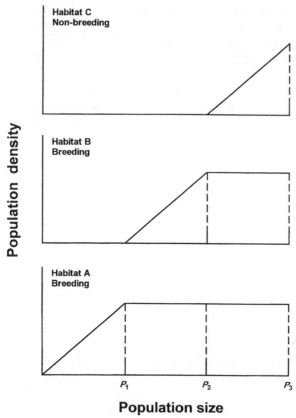

Figure 3.7   Sequential fill of habitats of decreasing quality, A-C. Habitats A and B permit survival and breeding, whereas C permits survival only. Habitat A is preferred and fills first (P_1) , followed by B (P_2) and then C (P_3). Conversely, if population level falls, birds abandon C in favour of B, and B in favour of A. Modified from Brown (1969).

wherever the fitness benefits at the time of settling seemed greatest, so all individuals would have equal reward, regardless of where they settled; and (2) the 'ideal despotic distribution', in which dominant individuals took precedence, had greater fitness, and prevented other individuals from settling where was best for them. Since the distinction was proposed, the evidence for birds is increasingly against the ideal-free model in its strict form, and in favour of the despotic or some intermediate model which allows for unequal competitors (Bernstein et al. 1991). The inadequacy of the ideal-free model arises largely because the variation in dominance and other competitive abilities of birds ensure that the resources are divided unequally. Moreover, the average performance of individuals is often demonstrably different between habitats, whether measured by feeding rates (Parker & Sutherland 1986), or by reproduction and survival rates, as shown in tits (Krebs 1971, Dhondt et al. 1992, Minot 1981), corvids (Andrén 1990, Goodburn 1991), flycatchers (Lundberg & Alatalo 1992), warblers (Holmes et al. 1996), waders (Ens et al. 1992), raptors (Newton 1986), and others. On the other hand, density-dependent decline in breeding success within habitats in territorial species is by no means universal, occurring in some populations but not in others (O'Connor 1985, Hunt 1988, Newton 1986, Brawn et al. 1987, Goss-Custard et al. 1995a, b). When it does occur, it is often linked with habitat variation and due to occupancy of poorer locales at higher population levels (Nilsson 1987, Dhondt et al. 1992). As a further development, because in territorial systems each individual pre-empts occupancy by another, Pulliam & Danielson (1991) suggested the term 'ideal pre-emptive distribution', as a category of the ideal despotic model.

original model of Brown (1969) was developed for territorial birds, in which competition for space is most apparent, it could apply equally to flocking birds, given the competitive interactions described above (Sutherland & Parker 1985, Parker & Sutherland 1986). Habitat variation may of course occur at a range of relevant spatial scales, from one territory to the next in the same area, to large tracts of different quality habitats distributed across a region.

In practice, as mentioned already, habitats are seldom likely to fall into distinct categories, but instead lie on a continuum from the best, where individuals achieve the highest survival or reproductive success, to the worst where breeding or long-term survival is not possible. Also, habitats are unlikely to fill in strict sequence, but in an overlapping manner, so that occupation of the second best begins sometime before the best is filled. Thirdly, in some species there are likely to be obligate non-breeders in the population even at the lowest densities (before breeding habitats are saturated), if only because of unequal sex ratios or immaturity. Fourthly, site-fidelity may interfere with the flow of individuals from secondary to primary habitat, producing a lag in redistribution between habitats whenever overall numbers decline. This means that even poor habitats can normally be expected to hold some birds.

This whole process, involving habitat variation, provides more than density limitation in particular habitat patches. As explained at the start of this chapter, it gives a means of population regulation over a wider area, in that, as total numbers rise, an increasing proportion of individuals is pushed down the habitat gradient so that the mean per capita survival or breeding rate in the overall population declines.

On such a system, the overall status of a population, the range of habitats occupied and the size of the non-territorial (non-breeding) component, all depend on the survival and reproductive rates of breeders. Anything that happens to lower these rates, such as a widespread increase in predation or food-shortage, could depress the whole population, shrink the range of habitats occupied, and reduce the ratio of non-breeders to breeders. At any one time, we can expect to find bird populations at different points along the continuum of conditions shown in Figure 3.7, from tiny remnants which hang on only in primary habitats, through full occupancy of primary habitats and variable occupancy of secondary habitats, to full occupancy of all potential nesting habitats, and a large non-territorial, non-breeding contingent in the breeding season. It is the latter situation that can give the greatest stability in overall breeding numbers, which persists by virtue of a large surplus of potential breeders. A substantial decline in non-breeder numbers could occur before any decline became apparent in breeder numbers. As conditions change from year to year, or from region to region, populations will also change in status. This is essentially what we see in nature, and the sections below consider some of the evidence.

Sequential habitat occupancy

The quality of habitat has been assessed from measurements either of cover, disturbance, internal food-supplies, or proximity to good feeding areas (whichever was most relevant), or more directly from the feeding rates, survival or reproductive rates of previous occupants. On the basis of such assessments, evidence for the sequential fill of habitats, in order of good to poor, is of three kinds. The first involves the process of settlement as birds arrive on their breeding grounds in spring or on their wintering grounds in autumn: in general, areas classed as best on previous evidence are occupied first. For example, among migrant Willow Ptarmigan *Lagopus lagopus* in Alaska, the first birds to arrive settled near a creek, while later settlers were forced by territorial competition onto adjacent hillsides, where food was shown to be poorer (Moss 1972). In a previous year, when numbers were lower, only the creek area was occupied. Similarly, Pied Flycatchers *Ficedula hypoleuca* arriving in central Sweden in spring settled in deciduous areas in preference to coniferous (**Figure 3.8**). In deciduous areas, food-supply increased earlier in spring, the birds settled there at greater density, laid earlier and raised more young. Also, the males (but not the females) that occupied deciduous areas were larger than those in coniferous areas, a difference attributed to the outcome of competitive interactions (Lundberg *et al.* 1981).

Similar patterns of spring settlement, with the best areas occupied first, have been described in many other bird species, including Song Sparrow *Melospiza melodia* (Arcese 1987), Great Tit *Parus major* (Drent 1984), Wheatear *Oenanthe oenanthe* (Brooke 1979), Painted Bunting *Passerina ciris* (Lanyon & Thompson 1986), Great Reed Warbler *Acrocephalus arundinaceus* (Bensch & Hasselquist 1991) and Savi's Warbler *Locustella luscinioides* (Aebischer *et al.* 1996). In several such species, the same sequence of territory settlement held from year to year, even though the occupants changed, and even though some early settlers were displaced by former owners which returned later. In some species, however, the sequential pattern was disrupted by site-fidelity, as some returning birds settled in poor areas where they had previously bred, despite the presence of vacancies in better habitat (Lanyon & Thompson 1986). Site-fidelity sometimes led individuals to occupy for several further years habitats that had deteriorated (Hildén 1965, Wiens & Rotenberry 1986).

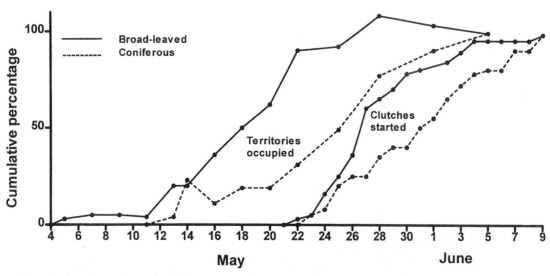

Figure 3.8   Settling and egg-laying dates of Pied Flycatchers *Ficedula hypoleuca* in broad-leaved and coniferous forest in central Sweden. In the preferred broad-leaved areas, food-supply increased earlier in spring, the birds settled earlier and at greater density, and annual breeding success (in terms of young per female) was higher than in coniferous areas. Redrawn from Lundberg *et al.* 1981.

The use of such areas declined over time, however, because existing occupants died or left, but no new ones settled.

Turning now to wintering areas, on one estuary, Oystercatchers accumulated first on the two most preferred mussel beds where feeding rates were highest (Goss-Custard *et al.* 1982, 1984). But as more birds arrived and interactions increased, birds increasingly occupied the less favoured mussel beds. Later arriving adults displaced immatures already present on the favoured beds. Moreover, as mussel stocks declined on the favoured beds, more birds left, but in reverse order of their dominance rank, to feed on less preferred beds or in other habitats.

Similarly, in Florida, wintering American Kestrels *Falco sparverius* arriving from the north occupied habitats in decreasing order of foraging potential. Early arriving birds (mostly females and juveniles) acquired most of the good places. They were not displaced by later arrivals which therefore took the less good places (in contrast to Oystercatchers) (Smallwood 1988). Hence, as in breeding areas, the best places were occupied first.

The second type of evidence for sequential habitat fill comes from annual variations in the occupation of different territories or habitats. Over a period of years, as the model in Figure 3.7 requires, numbers remain more stable in one habitat than in another. The interpretation has been that one habitat is preferred and fills to near maximum capacity every year, while the other takes up the surplus which varies in number from year to year, depending on the total population (the 'buffer effect'). This pattern was apparent among Great Tits *Parus major* in an area of Holland, where breeding density was higher and more stable from year to year in strips of broad-leaved woodland than in adjacent pine forest (Kluijver & Tinbergen 1953) **(Figure 3.9)**. Survival was not measured separately in the two types of wood but breeding success was lower in pine, which was therefore considered a less good habitat (Lack 1966).

The same pattern was also apparent in Chaffinches *Fringilla coelebs* breeding in the same area (Glas 1960), and in several other species elsewhere, including Yellow-hammer *Emberiza citrinella* (better breeding success in favoured habitats, O'Connor

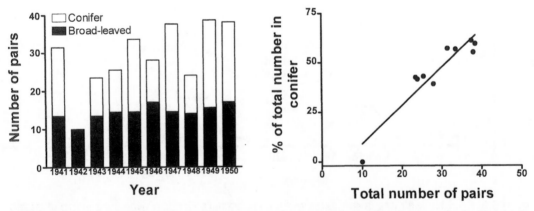

Figure 3.9   Annual fluctuations in the breeding numbers of Great Tits *Parus major* in 25 ha of good broad-leaved habitat and adjoining 100 ha of poor conifer habitat. Numbers were more stable in the good (= preferred) habitat (left), and as total numbers rose, an increasing proportion used the poor habitat (right). From Kluijver & Tinbergen (1953).

1980), Dickcissel *Spiza americana* (more polygyny in favoured habitat, Zimmerman 1982) and Red-winged Blackbird *Agelaius phoeniceus* (more polygyny in favoured habitat, Clark & Weatherhead 1987). Moreover, in Kirtland's Warbler *Dendroica kirtlandii* in Michigan, males greatly outnumbered females and occurred in greater proportion in secondary habitats, where they remained unpaired (Probst & Hayes 1987). In this same species then, in the same years, the sexes were distributed differently between primary and secondary habitats, but each according to its numbers. Density-dependent habitat use has been noted in several other species, but without the additional information on whether preferred habitats offered better survival or reproductive prospects (e.g. Pinowski 1967, O'Connor & Fuller 1985).

Several known behaviours help to ensure more consistent occupancy of good habitats. Firstly, in each year first-time settlers fill gaps in good habitats in preference to poor habitats (as above). Secondly, surviving adults of many species show greater year-to-year site-fidelity in good habitats than in poor ones, with more individuals moving from poor to good habitats (or territories) than vice versa, regardless of their previous breeding success (e.g. Newton 1986, Bollinger & Gavin 1989, Hatchwell *et al.* 1996). Thirdly, the greater tendency of adults to change territories after a breeding failure than after a success might also cause more birds to abandon poor habitats each year than good ones (Delius 1965, Harvey *et al.* 1979, Newton & Marquiss 1982, Dow & Fredga 1983).

The third type of evidence for sequential habitat fill has come from species that increased in numbers over a period of years, and in the process occupied an increasing range of habitats or areas. For conservation purposes, Seychelles Warblers *Acrocephalus sechellensis* were introduced from one island (Cousin) to another (Aride), where spatial variation in habitat had been assessed previously from measurements of the insect food supply (Komdeur *et al.* 1995). The first warblers to be released established territories in the best areas (with most insects) and, as their numbers grew over the next few years, birds spread to progressively poorer areas (with fewer insects) (**Figure 3.10**). The same phenomenon has been noted in several other species as their numbers recovered from the effects of a hard winter (see Williamson 1969 for Wren *Troglodytes troglodytes* and O'Connor 1985, 1986 for Mistle Thrush *Turdus viscivorus*) and in other species as they colonised new areas (von Haartman 1971).

In western Europe, several species of geese have increased in recent decades, and at the same time have occupied an increasing number of habitats, in both breeding and wintering ranges. Barnacle Geese *Branta leucopsis* on Svalbard, when first discovered nested entirely on cliffs and rocky slopes in mountain areas, but as their numbers grew, they established new colonies on offshore islands (Prestrud *et al.* 1989). Similarly, in wintering areas, as Dark-bellied Brent Geese *Branta b. bernicla* increased in numbers over a period of years, they occupied intertidal and salt marsh habitat first, and after reaching a certain population level began to use nearby farmland (Ebbinge 1992). In recent years, Brent Geese have shown two main habitat shifts within each winter, feeding first on intertidal algal beds, next on salt marsh, and then on farmland. These moves were associated with the successive depletion (and other loss) of their different food-sources (Vickery *et al.* 1995). As the population increased over a period of years, depletion occurred more rapidly, and habitat shifts occurred progressively earlier in the winter. At one site in Norfolk, southern England, in the 1950s when only 245–450 geese fed there, they moved from mud

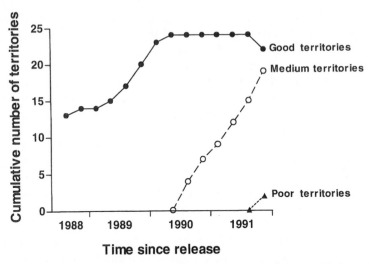

Figure 3.10   Sequential occupation of habitat by an expanding population of Seychelles Warblers *Acrocephalus sechellensis* on Aride Island. ● – high quality habitat, ○ – medium quality habitat, ▲ – low quality habitat. Quality of territories was measured by food-supply (leaf-dwelling insects). From Komdeur *et al.* (1995).

flats to salt marsh in late February, but by the 1990s when numbers had increased to 4500, they switched from mud flat to salt marsh in October–November. In the 1950s, no birds fed on farmland, but by the 1990s, they were spending more than four months there. Such changes in the feeding habitats of birds have sometimes been attributed to cultural learning, but in this instance goose numbers began to increase and use up existing food-supplies before the change to new feeding areas occurred. The implication of such changes is that while we can define the actual range of habitats occupied by bird species at any one time, it is difficult to define the full potential range that might be used if numbers increased further. To put it another way, some bird species may be able to occupy more habitats than they currently use.

Patterns involving the long-term sequential fill of different areas have been described for Grey Plover *Pluvialis squatarola* wintering in different parts of Britain, as their total numbers grew (Moser 1988), and for Cormorants *Phalacrocorax carbo* wintering in different parts of Switzerland (Suter 1995). In the latter, the process was stepwise, and each category of habitat experienced a rapid build-up in numbers, followed by stabilisation before the next type of habitat was occupied, and so on.

These various examples show that sequential occupation of habitats or areas, in order good to poor, may occur on various temporal scales, from a few days to many years, and at various spatial scales, from different territories in the same area to different parts of a geographical range. It can also occur in both breeding and wintering areas. In all the studies quoted, habitat quality was assessed in some way, but in only a small proportion was the pattern of habitat occupation shown to be linked with territoriality or other competitive interactions; in most studies the mechanism was not examined. Moreover, the patterns shown were no more than consistent with the model in Figure 3.7, and other explanations could not be excluded. However, experimental evidence for density-dependent shifts in habitat occupancy, and for the role of dominance behaviour in limiting numbers in particular habitat areas, is discussed in Chapters 4 and 5. None-the-less, the whole concept of sequential habitat occupation, while consistent with observations on many species, is best regarded as often being the most plausible explanation. Sometimes the long-term expansion of a species into new areas could be due to changes in the areas themselves, making them more acceptable as habitats.

Sequential habitat occupancy and decline in breeding performance

As stressed already, the spread of birds to poorer habitats as their numbers rise can cause their mean breeding performance to decline, even though the birds in good sites continue to do well. The greater the spatial variation in habitat, the stronger the density-dependent decline in the mean performance is likely to be. Just such an effect, consequent upon variation in the measured quality of nest-sites, became apparent among Shags *Phalacrocorax aristotelis* on the Farne Islands, northeast England, through a natural experiment (Potts *et al.* 1980). All nest-sites were placed in one of five categories, good to poor, according to several objective criteria, which included the degree of protection they afforded against rain and high seas. In 1968 and 1975, natural events killed a large proportion of the Shag population, thus releasing large numbers of nest-sites. On each occasion, remaining birds moved into sites previously classed as good, and thereby increased their nest success (which was more than twice as good in the best sites as in the worst ones). As the population was then entirely accommodated in good sites, the average breeding success of all colony members was high. On average, young Shags increased their production of young by 71% and older ones by 39%. But as the colony recovered and increasing proportions of birds had to use poorer sites, average nest success declined. This process thus gave a clear density-dependent response to rising density, associated with variation in the quality of nest-sites. Other examples of the same phenomenon, based on variable nest-sites, include Peregrine *Falco peregrinus* (Mearns & Newton 1988), Kittiwake *Rissa tridactyla* (Coulson 1968), Puffin *Fratercula arctica* (Birkhead & Furness 1985), Guillemot *Uria aalge* (Harris *et al.* 1997) and Great Shearwater *Puffinus gravis* (Rowan 1965).

Variation in territory or habitat quality can act in the same way, slowing the average breeding rate in the population as numbers rise. In the Spanish Imperial Eagle *Aquila heliaca*, breeding success declined as numbers recovered following cessation of human persecution. The effect occurred because the remnant population occupied a limited area of good habitat (with high prey densities), and as numbers expanded, new colonists settled in poorer areas (with fewer prey), where nest success was lower. Over the whole period, no significant decline in breeding success occurred in the better territories (Ferrer 1993). Other examples of species showing decline in mean reproductive rate, linked to greater occupancy of poor habitat at high population levels, include Mistle Thrush *Turdus viscivorus* (O'Connor 1986), Nuthatch *Sitta europaea* (Nilsson 1987), Lapwing *Vanellus vanellus* (Galbraith 1988) and Grey Partridge *Perdix perdix* (Panek 1997).

Studies of other species have shown that, over a period of years, territory sites varied in frequency of use. Some sites were used every year, others in most years, and yet others only in occasional years. Among Sparrowhawks *Accipiter nisus* in south Scotland, those territories that were used most often also showed the highest

Table 3.1   Mean contribution to future nesting population made by Sparrowhawk *Accipiter nisus* breeding attempts in different grades of nesting habitat. Some nesting places acted as 'sinks' and others as 'sources' (see text). From Newton 1991.

Grade of nesting place[1]	Mean number of young females raised per nest[2]	Mean number of females that survived to nest themselves[3]	Balance between mean annual production and mean annual mortality[4]
1	0.60	0.18	Negative
2	0.79	0.24	Negative
3	0.80	0.24	Negative
4	1.06	0.32	Even
5	1.34	0.40	Positive

[1]Calculated from the number of years the territory was occupied during a 15-year period. Grade 1 occupied in 1–3 years (not necessarily successive years), Grade 2 in 4–6 years, Grade 3 in 7–9 years, Grade 4 in 10–12 years and Grade 5 in 13–15 years.
[2]Calculated as half the mean number of young raised per nesting attempt, on the knowledge of an equal sex ratio among fledglings.
[3]Calculated as 30% of the numbers reared, on the knowledge that 28% of young females survived to lay eggs, and a further 2% nested without laying eggs.
[4]The annual mortality of adult females was estimated by different methods at 29–36% (mean 32%), so that to replace itself each breeding female must produce an average of 0.32 female breeders per year.

nest success, in terms of the number of young fledged per nesting attempt (**Table 3.1**). The highest grade territories produced more than enough young per year to offset annual adult mortality, so acted as 'sources', whereas the lowest grade territories produced insufficient young so acted as 'sinks' (see next section) (Newton 1991). In the area as a whole, the breeding population was in balance, and remained approximately constant over a 20-year period. Other species showed similar relationships, with known poorer territories occupied mainly in years of high population (for Tengmalm's Owl *Aegolius funereus*, see Korpimäki 1988; for Jay *Garrulus glandarius*, see Andrén 1990; for Magpie *Pica pica*, see Møller 1982; for Eastern Kingbird *Tyrannus tyrannus*, see Blancher & Robertson 1985; for Wheatear *Oenanthe oenanthe*, see Brooke 1979; for Nuthatch *Sitta europaea*, see Nilsson 1987, Matthysen 1990). This could have the effect of lowering the mean reproductive rate in years of high breeding numbers. In some species adult mortality was also higher in the less used territories, as shown for Nuthatch (Matthysen 1990), but in most species mortality was hard to separate from emigration, as birds were more likely to abandon poor territories than good ones.

SOURCES AND SINKS

The source–sink concept is applicable at various spatial scales from different territories in the same locality, through different habitats within a region, to different parts of a geographical range. In effect, it emphasises the notion that species populations are unlikely to be self-sustaining in all the areas where they are found. If the per capita surplus production in source areas greatly exceeds the per capita deficit in sink areas, a small amount of source habitat could ensure the continued occupancy of a large amount of sink (Pulliam 1988). Examples of apparent source and sink areas have been described now for many bird species, including

Sparrowhawks *Accipiter nisus* and others in different territories (as just mentioned), Dotterels *Charadrius morinellus* on different mountaintops (Thompson & Whitfield 1993), and Barn Owls *Tyto alba* in different landscapes (de Bruijn 1994).

Not all examples are water-tight, however, for some researchers have simply measured reproduction in different habitat areas, and assumed that mortality was the same in each. This assumption may not always have been justified, and to identify reliably source–sink areas, both reproduction and mortality should be measured within each area. Even then it may be difficult to distinguish sources from sinks if reproduction or mortality within habitat patches is density-dependent (Watkinson & Sutherland 1995). Potentially viable populations may appear to be non-viable because, at the time of measurement, occupants are crowded sufficiently to depress reproduction below the level needed to sustain numbers. The area thus appears as a sink when, with a lowering of density, it could become a source. Recognising genuine sinks from 'pseudo-sinks' is unlikely to be always straight-forward. From a true sink all birds would disappear if immigration stopped, but in a pseudo-sink they would only decline to a lower density.

While birds seem generally to prefer those habitats where they do best, this may not always be so, especially where conditions are recent or changing. The word 'trap' has been used for habitats, usually man-made, that the bird perceives as favourable, but that in reality are unsuitable, by virtue of some feature to which the bird is unresponsive. In other words, a trap is a sink that birds take as a source. Examples include areas of farm crops which attract ground-nesting birds whose nests are later destroyed by machinery.

OTHER RESPONSES TO HABITAT SATURATION

Polygyny

Where habitats are heterogeneous, individual males may differ greatly in the quality of territory they hold. Females may then prefer to mate polygynously with an already paired male in good habitat than monogamously with a bachelor male in poor habitat (Orians 1969). Polygyny may thus increase nesting density in rich habitat, while leaving unpaired males in poorer habitat. The finding that polygyny is more frequent in food-rich habitats is based on observational evidence in several species, notably Dickcissel *Spiza americana* (Zimmerman 1982), and on experimental evidence in others, such as Red-winged Blackbird *Agelaius phoeniceus*, in which the provision of extra food on territories led males to acquire more females (Wimberger 1988).

Serial use of habitat

In some bird species, two sets of occupants might use the same nesting area in the same year. In the Storm Petrel *Oceanodroma castro* in the Galapagos Islands, Harris (1969) found two separate populations, sharing nesting holes and breeding annually, one in May–June and the other in November–December. Non-breeders and failed breeders also remained faithful to their annual cycles over four seasons. In effect, such a system doubles the amount of nesting habitat available to a population, and thus the number of individuals that can breed. The same phenomenon has been recorded in various ducks and waders, in which individuals move away with their young after hatching, leaving the same nesting areas free for use by others of their species (Chapter 4). But it can occur only where one set of birds finishes with the nesting habitat in time to allow another set to breed in the same season, and where the occupants have sufficient food and alternative habitat in which they can live when not nesting.

Cooperative breeding

In so-called cooperative breeders, a group territory is defended by a dominant pair and their offspring from previous years, some of which may help to raise the current young. Offspring thereby gain living space for up to several years, but help with territory defence and chick care. Such systems occur in some resident species which seem to occupy their habitats to saturation, leaving no space for a separate non-breeding sector (Emlen 1978, Stacey & Koenig 1990). Well-studied examples include the Florida Scrub Jay *Aphelocoma coerulescens* and the Arabian Babbler *Turdoides squamiceps* (Woolfenden & Fitzpatrick 1984, Zahavi 1989). The parents of such species gain by allowing their offspring to stay until they can acquire territories for themselves, on the rare opportunities that arise. Thus, with the help received, the parents can raise more young than otherwise to independence, and these young are then assumed to survive better by staying with their parents until an opportune time to leave than by dispersing immediately into a habitat already saturated with territory-owners. Experimental evidence for this explanation of family territories was provided by the Seychelles Warbler *Acrocephalus sechellensis* which, when introduced to a new island, lived in pairs until all available habitat became occupied. Then, as numbers continued to rise, groups began to form, as offspring remained on their parents' territories (Komdeur 1992).

NON-TERRITORIAL ADULTS

In some bird species, non-territorial 'floaters' live a mainly secretive and solitary existence in and around the territories of breeders, and are therefore not much in evidence. This has been found, for example, in many songbirds and raptors (Kendeigh 1941, Delius 1965, Smith 1978, Newton 1979). In other species, the non-breeders live mainly in areas not defended by breeders, often in groups or flocks, and are much more in evidence. This is the case, for example, in various crows, swans and shorebirds (Carrick 1963, Harris 1970, Holmes 1970, Patterson 1980). In general, however, non-territorial birds are much more difficult to count than territorial breeders, and have been relatively little studied.

One species in which the total non-territorial component has been fairly precisely estimated is the Goshawk *Accipiter gentilis* on the Swedish island of Gotland, where the numbers of nests found were stable from year to year (Kenward *et al.* 1991). Among a large sample of radio-tagged hawks in spring, females greatly outnumbered males, at 177 and 108, respectively, per 1000 km². Of these birds, about 68 males and females subsequently bred, which left 40 males (37% of total) and 109 females (62%) which remained in the area as non-breeders. Hence, in this population, non-nesting females outnumbered breeding ones by 1.6 to one. No birds of either sex bred in their first year, and even some of three years and older were non-breeders. This contrasted with more persecuted (and depleted) Goshawk populations elsewhere, where up to a third of nesting pairs contained first-year birds.

Estimates of the proportions of non-territorial non-breeders present in some other bird populations in the breeding season are given in **Table 3.2**. The fact that in most species males outnumbered females among non-breeders was presumably a consequence of a surplus of males in the overall population, coupled with a monogamous breeding system. The surplus of males could itself be attributed to the greater vulnerability of females to a number of mortality causes, including starvation (associated with male dominance) and predation on the nest (Chapter 9). Like the Goshawk just mentioned, the Red-billed Gull *Larus novaehollandiae* of New Zealand was unusual in having an obvious surplus of females in the overall population, and this was also reflected in the non-breeding contingent (Mills 1989).

In most species, non-territorial birds are generally younger than territorial breeders, but often show a marked hierarchy, with the more dominant ones acquiring territories before the others (e.g. Smith 1978, Birkhead *et al.* 1986, Smith & Arcese 1989). Because individuals are more likely to acquire a territory as they get older, the proportion of territorial breeders increases through successive age-groups. Within particular age-groups, breeders have sometimes been found to be bigger in body size or some other respect than non-breeders (e.g. see Alisauskas 1987 for American Coot *Fulica americana*). In one study, only those Carrion Crows *Corvus corone* that grew to sufficient size succeeded in obtaining a territory, while smaller individuals remained non-territorial (Richner 1989). All these findings presumably reflect the intensity of competition for territories.

Whatever their distribution, the numbers of non-territorial, non-breeding birds might be limited in either of two ways. If the habitat available for non-breeders was limited, their numbers could be constrained by competition for resources, as described above. In this case, mortality among non-breeders would be density-dependent, and their numbers would become attuned to the extent and carrying capacity of their habitats. Alternatively, if habitat for non-breeders were so vast that they could never fill it, their numbers would be set instead at whatever level the annual inputs (from reproduction and exit from the breeding sector) matched the annual losses (from mortality and recruitment to the breeding sector). Because the input from reproduction is inevitably limited by the numbers and productivity of breeders, the input of young to the non-breeding sector must also be limited. Such a situation was suggested for Peregrine Falcons *Falco peregrinus* in which breeders are confined to areas with suitable nest-sites (usually cliffs), while non-breeders can also live in the much more extensive areas that lack nest-sites but hold sufficient prey (Hunt 1988, Newton 1988b). It could also hold in many other bird species, including other raptors, shorebirds and seabirds for which nesting areas are limited compared with potential feeding areas.

A theoretical maximum ratio of non-breeders to breeders was calculated by Brown (1969) for a range of bird species with different reproductive and mortality rates. In the most extreme examples, non-breeders could outnumber breeders by two or more times, a situation actually found in certain species (Carrick 1963, Kenward *et al.* 1991). Moreover, in some long-lived species, with a large ratio of non-breeders to breeders, it is usual for some individuals to wait several years beyond achieving physiological maturity or adult plumage before they breed. Thus

Table 3.2 Percentage of non-territorial non-breeders reported in various bird populations in the breeding season. + indicates which sexes were represented, when not counted separately.

Species	Males	Females	Both sexes	Reference
Australian Magpie	79	62	72	Carrick 1963
Gymnorhina tibicen				
Magpie	+	+	40	Birkhead *et al.* 1986
Pica pica	+	+	20	Baeyens 1981
Rufous-collared Sparrow	+	+	50	Smith 1978
Zonotrichia capensis				
Song Sparrow	4–8	0	2–4	Tompa 1964
Melospiza melodia				
Silvereye	+	+	15	Catterall *et al.* 1982
Zosterops lateralis				
House Wren	15(20)[1]	12(25)[1]	+	Kendeigh 1941
Troglodytes troglodytes				
Blackbird	19	8	14	Ribaut 1964
Turdus merula				
Tree Swallow	+	23–27	+	Stutchbury & Robertson 1985
Tachycineta bicolor				
Skylark	+	+	10	Delius 1965
Alauda arvensis				
Great Horned Owl	+	+	0–53[2]	Rohner 1996
Bubo virginianus				
Cassin's Auklet	+	+	50	Manuwal 1974
Ptychoramphus aleuticus				
Red-billed Gull	34	59	49	Mills 1989
Larus novaehollandiae				
Great Skua	+	+	46	Klomp & Furness 1992
Catharacta skua				
African Black Oystercatcher	+	+	60	Summers & Cooper 1977
Haematopus moquini				
Oystercatcher	+	+	46	Ens *et al.* 1995
Haematopus ostralegus				
American Coot	29	5	19	Alisauskas 1987
Fulica americana				
Ring-necked Pheasant	11	0	6	Burger 1966
Phasianus colchicus				
Sparrowhawk	+	45	+	Newton 1985
Accipiter nisus				
Goshawk	37	62	52	Kenward *et al.* 1991
Accipiter gentilis				
Mountain Duck	+	+	44	Riggert 1977
Tadorna tadornoides				
Mute Swan	+	+	69	Jenkins *et al.* 1976
Cygnus olor	+	+	58	Brown & Brown 1993
	+	+	64	Meek 1993

In some studies, birds that held territories but did not breed or attract a mate are counted as 'breeders'.

In most species, males predominate, especially because of a male-biased sex ratio in the total population.

[1]Figures refer to first period of breeding (and second period of breeding).

[2]Figures from different years.

although some Oystercatchers breed at two years of age, in at least one study other individuals waited an average of ten further years before they obtained a nesting territory (Ens *et al.* 1995). If such intense competition for nesting opportunities were general among Oystercatchers, a bird would gain little or nothing by moving elsewhere.

MEASUREMENT OF CARRYING CAPACITY

The concept of carrying capacity encapsulates the notion that, in any area of habitat, resources must ultimately limit the numbers of birds that can live there. It is a tricky concept, however, because carrying capacity depends not only on features of the habitat, such as cover, food or nest-sites, but also on bird behaviour. As an obvious illustration, more birds could occupy a given area by flocking than by defending large individual territories. Moreover, on a fixed food-supply, many birds might occupy an area for a short period, or fewer for a longer period. Carrying capacity would then be better measured as 'bird days' than as maximum bird numbers. So how can one tell whether a particular area, occupied progressively within a season or over a period of years, is indeed filled to capacity? It is not enough for numbers to have stabilised; they should have stabilised despite larger numbers of potential settlers available. One way to gain this information is to count birds over a period of time as they accumulate, and then plot bird numbers in preferred places against total numbers over a wider area that includes poorer places. This has been done within a season for several coastal species, including Green-winged Teal *Anas crecca* (Zwarts 1976), Knot *Calidris canutus* (Goss-Custard 1977a), Redshank *Tringa totanus* (Goss-Custard 1977b) and Oystercatcher *Haematopus ostralegus* (Goss-Custard 1977a, Zwarts & Drent 1981, Goss-Custard *et al.* 1982). In all these species, densities reached a plateau in preferred places as the total population over a wider area continued to rise (as in **Figures 3.1** or **3.7**). The increase in numbers in preferred places could thus be assumed to have stopped because of some site-specific density-dependent effect, and not because the supply of birds had dried up, and the plateau level could be taken as the area's carrying capacity at that time, on whatever social system prevailed (Goss-Custard 1985). This is not to imply that the carrying capacity of any area is fixed; it may vary during the course of a year, or from one year to the next, depending on food-supply and other conditions.

The same procedure can be used to estimate the mean carrying capacity of preferred habitats from counts over a wider area, made over a period of years, as overall numbers rise or fluctuate. For example, for the Great Tits *Parus major* in **Figure 3.9**, the carrying capacity of the broad-leaved woodland on the prevailing territorial system could be estimated at around 14 pairs. Numbers remained at around this level for several years, despite big fluctuations in the numbers of Great Tits in the wider area that included poorer coniferous habitat. However, much better evidence for populations at carrying capacity has come from experiments in which all the birds on sizeable study areas were shot. These same areas were then recolonised, either the same year or a later year, to the same level (see Chapters 4 and 5).

Another concept of carrying capacity, which takes account of changing resource levels, concerns the 'sustainable population size', which is the maximum number of individuals that can use a site over a defined period (Sutherland 1996a). Many birds of high latitudes rely on non-renewing food-supplies during winter, or eat renewing supplies faster than they can be replenished. With more consumers, depletion rates are greater and intake rates can be expected to fall more rapidly. Once their individual intake rates fall below the threshold needed for survival, birds must leave or die **(Figure 3.11)**. In other words, the higher the starting population, the greater the proportion affected by shortage. The maximum sustainable population for any locality could thus be defined as the maximum number that could live there for a given time without depleting the food-supply below the threshold for survival (a threshold that varies somewhat between individuals). The notion of carrying capacity must therefore be treated with care, and whatever definition is put on it, the concept is best applied to limited areas and to limited times of year.

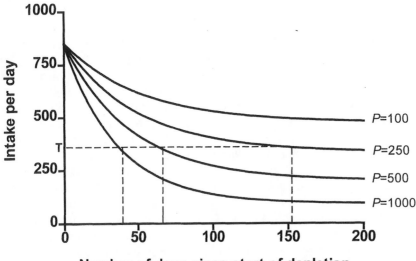

Figure 3.11   Change in intake rate over time resulting from depletion. The different lines are for different numbers of consumers (P). T is the threshold intake necessary for survival, and the dotted lines show the periods that the habitat can sustain different numbers of consumers. On the model depicted, for a season of 150 days, 100 consumers would have excess food, 250 consumers would just survive, but 500 or 1000 would deplete their resources so quickly after about 35 and 65 days, respectively, that many would starve or need to leave. Modified from Sutherland & Anderson 1993.

DISPARATE CARRYING CAPACITIES OF BREEDING AND WINTERING HABITATS

While some bird species live in the same areas year-round, perhaps even in the same territories, others become more widely distributed between the breeding and non-breeding seasons, and yet others spend the breeding and non-breeding seasons in geographically separate areas. In addition, many species eat different types of food in the breeding and non-breeding seasons, which may have different levels of availability. For both these reasons, therefore, the total carrying capacities of the breeding and non-breeding habitats of particular populations need not necessarily correspond, so that, as birds return to nesting areas each year and compete for territories or nest-sites, two scenarios are possible (**Figure 3.12**).

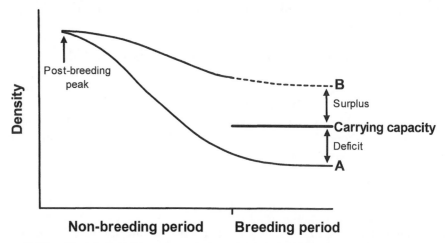

Figure 3.12   Model showing seasonal changes in total bird numbers in relation to the carrying capacity of nesting habitat (thick line). In the lower curve (A), numbers left at the end of the non-breeding season are fewer than the nesting habitat could support, and in the upper curve (B) numbers are more than the nesting habitat could support, leading to a surplus of non-territorial non-breeders. In (A) breeding numbers are limited by conditions in the habitat occupied in the non-breeding season, and in (B) by conditions in the habitat occupied in the breeding season.

(1) Too few birds are left at the end of the non-breeding season to occupy all the nesting habitat fully, so that practically all individuals of appropriate age and condition can breed. In this case, breeding numbers would be limited by whatever factors operate in the non-breeding season, and alleviation of the factors that influence the extent or carrying capacity of non-breeding habitats would be needed before breeding numbers could rise. For convenience, the breeding numbers of such populations could be described as 'winter limited'.

(2) More birds are left at the end of the non-breeding season than the available nesting habitat will support, producing a surplus of non-territorial non-breeders. In this case, alleviation of the factors that influence the extent or carrying capacity of the nesting habitat would be needed before breeding numbers could rise. For convenience, the breeding numbers of such populations could be described as 'summer limited'. The same holds for populations in which birds remain on their nesting territories year-round, yet at the start of breeding still show a non-territorial surplus of mature adults.

In later chapters, examples are given of some species in which breeding density appeared to be winter limited and of others in which it appeared to be summer limited. However, because both the extent and carrying capacities of habitats vary from year to year, and from area to area, we can expect that the same species might be winter limited in some years or areas and summer limited in other years or areas. The same patterns could hold at the end of the breeding season, as birds return to their non-breeding habitats. At that time, the total numbers of adults and young might be insufficient to occupy non-breeding habitats to the full, leading to good survival through the non-breeding season, or the total numbers may exceed the carrying capacity of non-breeding habitats, leading to poorer survival.

ESTIMATING THE EFFECTS OF HABITAT LOSS

Most bird species are associated with a particular type of habitat: with forest, grassland, desert or marsh, as the case may be. Their numbers in any one patch of habitat may vary directly with the size of that patch (examples in **Figure 3.13** and Prater 1981), and their total numbers in a region might depend largely on the overall area of all the habitat patches available (**Figure 1.4**). This is true only as a generalisation, however, and the fact that, in any bird population, not all habitats and areas may necessarily be fully occupied at any one time, makes the effects of piecemeal habitat loss on bird populations hard to assess. This is especially true where habitat varies in quality: the loss of patches of high carrying capacity would clearly have more effect on overall numbers than would the loss of similar-sized patches of low carrying capacity. Likewise, for resident birds in the longer term, loss of 'source' habitat would have more effect on overall numbers than would loss of the same amount of 'sink' habitat.

Under the continuing threat of land reclamation, attempts have been made to estimate the effects of estuarine habitat loss on wintering shorebirds (Goss-Custard *et al.* 1995a, b). The first problem is to find whether bird numbers in the estuary

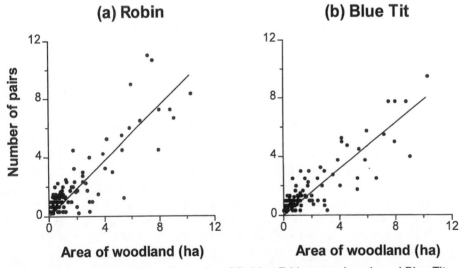

Figure 3.13   Numbers of breeding pairs of Robins *Erithacus rubecula* and Blue Tits *Parus caeruleus* in woods of different size, in southeast England. In both species, local population size was linearly related to the area of habitat available. Based on data collected by P. Bellamy and S. Hinsley.

concerned are at the maximum possible. If not, the reduced area resulting from development might still accommodate the same number of birds but at higher density. But if the estuary was already filled to capacity, there could then be knock-on consequences for bird numbers on this and other estuaries, as displaced birds redistributed themselves, intensifying competition within the same or other estuaries. Since all parts of a wintering range are unlikely to be filled to capacity, little can be said about the effect at wider spatial scales.

Reduced numbers of waders on estuaries where major reclamations have occurred suggests that such estuaries were indeed occupied close to capacity. In the Tees estuary, northeast England, a 61% loss of mud flat from 360 ha to 140 ha over three years was accompanied by local declines of 32–88% of different shorebird species counted in autumn, even though the national populations of all these species increased in this time (**Figure 3.14**). Further changes in numbers occurred over the next 20 years, with some species decreasing further and others increasing, as sediments changed and affected prey supplies (Evans 1979, 1997). In the Danish Waddensea, 1100 ha were reclaimed in 1980, and of the 12 wader species found there, eight declined by up to 85%, and the most numerous duck species declined by 60% (Laursen *et al.* 1983). Similarly, the extent of decline of Dunlin *Calidris alpina* on different British estuaries was proportional to the extent of encroachment by the plant *Spartina anglica* which covered the feeding areas (Goss-Custard & Moser 1988). This led to a decline in the overall British Dunlin population. In some other studies, however, birds displaced from one estuary were accommodated nearby, causing little or no decline in regional numbers (Lambeck *et al.* 1989).

On any estuary, reclamation usually occurs nearest the shore, on areas that waders use at the start and end of the tidal cycle. This reduces the feeding time, as well as the feeding area, for all individuals. But because habitat areas are not uniform, some species would be expected to be affected more than others, depending on which parts of a habitat area were lost. It is not necessarily peak bird numbers that are reduced, but the average numbers over a winter (or 'bird days'). Loss of birds following habitat destruction could involve deaths, as well as movements. Noticeable on-site mortality occurred among California Gulls *Larus californicus* after a 65% loss of nesting habitat from one year to the next, caused by rising water levels. This resulted in a 56% loss of gulls, and a 26% increase in their density in the remaining nesting habitat. Deaths were attributed to increased territorial strife (Jehl 1989).

For resident birds, in which breeding and wintering areas are the same, population declines should be in direct proportion to habitat loss, if habitat were uniform and fully occupied. For migrant birds the situation is more complicated because they occupy separate areas in winter and summer, and habitat loss may occur in one or both (**Figure 3.15**). The actual population change following loss of breeding or wintering habitat will depend on the relative strengths of density-dependence in the two areas (Sutherland 1996b). In the wintering area the strength of density-dependence is measured by the per capita rate of increase of mortality with increasing population size (or decreasing area) (slope d). In the breeding area, the strength of density-dependence is measured by the per capita rate of decrease in reproduction with increasing population size (or decreasing area) (slope b). If slope b > slope d, loss of breeding habitat would have most impact on overall population size, and if $d > b$, loss of wintering habitat would have most impact. If the two density-dependent relationships were known, the effect of loss of habitat (or food-supply) on

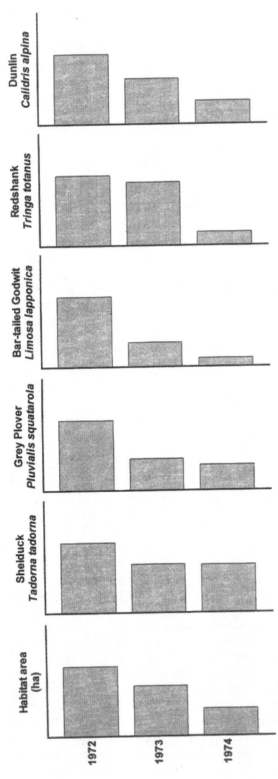

Figure 3.14 Effects of land reclamation on autumn bird numbers at Seal Sands, Teesmouth, northeast England. During the three-year period, the area of mud flat was reduced from 360 ha to 140 ha, and maximum feeding time per tidal cycle fell from 12.5 hours to 8.5 hours, because of longer inundation. Bird numbers are shown in relation to the 1972 totals (=100). All five species declined in numbers locally, despite increases in the national totals over the same period. From Evans 1979.

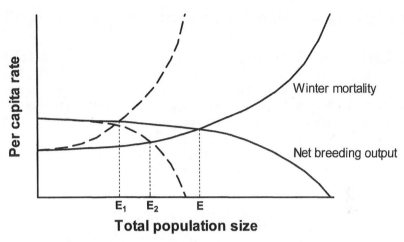

Figure 3.15   Model of relationship between per capita mortality and reproduction (continuous lines), and equilibrium population size (*E*). Per capita winter mortality increases, and per capita reproduction decreases, with increasing population size. The equilibrium population size is where the two lines intersect. The model shows how loss of habitat (or food-supply) results in population decline. If wintering habitat (or food-supply) is reduced by 50%, this results in a displacement of the relationship between mortality and total population size in direct proportion to the degree of habitat loss (dashed line). The equilibrium population size is reduced accordingly (*E*₁). A 50% loss of breeding habitat (or food-supply) similarly results in a displacement of the relationship between net breeding output and total population size (dashed line) and a reduced equilibrium population (*E*₂). In this example, 50% loss of breeding habitat results in a smaller reduction in equilibrium population size than does 50% loss of wintering habitat, because density-dependence is stronger in winter than in summer. Modified from Sutherland 1996b.

overall population size could be calculated as $b/(b+d)$ for the breeding area, or as $d/(b+d)$ for the wintering area.

If, in an extreme case, all the density-dependence occurs in winter habitat (say), with no density-dependence in breeding habitat (which at current population levels is present in excess, as in **Figure 3.12A**), then loss of winter habitat would cause a matching reduction in population size. In contrast, loss of breeding habitat would have no effect up to the point where breeding became density-dependent.

These considerations lead to a number of conclusions about the effects of habitat loss on equilibrium population sizes:

(1) knowledge of density-dependence within just the wintering or breeding area cannot be used to predict precisely the effects of habitat or food loss in either, for it is the ratio of density-dependence in the two areas that is important;

(2) unless there is no density-dependence acting during one of the seasons, a loss of habitat or food supply in either summer or winter areas will result in population decline;

(3) the consequence of habitat or food loss is greatest for the season in which density-dependence is strongest (winter in **Figure 3.15**); and

(4) if the effect of habitat loss in one season is small, it must be great within the other season.

These points hold only for populations that are near the limit of what their habitats (or food-supplies) will support and not for populations that are well below this limit, for example through human persecution. It should be stressed, moreover, that **Figure 3.14** is merely a model of how migrant population levels might be regulated. There can be few species for which enough information is available to test the model, or to judge the form of the density-dependent relationship over a wide range of densities at both seasons, though attempts have been made for the Oystercatcher *Haematopus ostralegus* (Goss-Custard *et al.* 1995a, Sutherland 1996b).

CONCLUDING REMARKS

In conclusion, a large volume of evidence supports the general idea of sequential habitat occupancy in birds, in order of good to poor, and its potential role in population regulation. As yet, the argument must necessarily be constructed from studies on a wide range of species, at different stages in the process of progressive habitat fill, and for no species has the whole process from colonisation to equilibrium been fully documented. Nevertheless, existing studies are sufficient to demonstrate different stages of the process, and the role of behavioural mechanisms in population regulation: in translating the effects of changes in resources and other habitat components into changes in bird densities, whether from one place to another or from one time to another.

In most of the studies described above, food-supply seemed to have a major influence on habitat quality and bird behaviour. If foraging birds were frequently attacked by predators, however, this would not only directly lower their survival chances, but would also reduce the value of those areas as feeding sites, because the need for greater alertness and continual evasive action would reduce feeding rates (Chapter 13). The presence of parasites can also reduce the value of otherwise good habitats, and birds clearly respond by avoiding those types of parasites (mostly ectoparasites) that they can detect (Chapter 10). Moreover, if predation or parasitism were density-dependent, this would be a further factor reducing the mean survival or breeding performance of individuals as their density rose. The salient points are that (1) the ability to exploit the food in any particular area does not necessarily depend on food-supply alone, but also on other factors that affect feeding rate, and (2) the benefits of a good food-supply can be negated by other factors that reduce feeding, breeding and survival rates. Not surprisingly, birds respond to prevailing conditions accordingly, an aspect discussed further in Chapter 13.

In the past, it was thought that birds normally filled their habitats to the full, like a hand fills a glove, and that in all habitat birds could normally be expected to maintain their numbers. Nowadays ideas have changed, however, with the realisation that populations have a spatial structure, and do not necessarily extend at any one time to all habitable areas, even good ones. In some populations, many individuals might live in areas where they could not maintain their numbers without continual immigration from better habitat elsewhere.

In the scientific literature, good habitats are not always defined on fitness criteria (such as breeding and survival rates), but rather according to the densities of individuals found there (in turn dependent partly on food and other resources). High density and high individual performance may often go together, but this is not always the case (van Horne 1983). Birds are sometimes found breeding at low density in habitats where their individual performance is high, and at high density where their individual performance is low. Shelducks *Tadorna tadorna*, for example, which feed on coastal mud flats, breed either at high density where large expanses of mud are available, or at low density where mud flats are small and fragmented. Yet in high-density areas breeding success is much poorer than in low-density areas, at 0.4–0.32 versus 0.72–1.22 young per territorial adult per year (Pienkowski & Evans 1982). In this species, therefore, habitats with high density were not the best as judged from individual breeding success (for other examples, see van Horne 1983). In other words, densities alone can sometimes be misleading indicators of habitat quality.

SUMMARY

The number of birds that can live in any area is limited, often in relation to resource levels, so that other potential settlers are excluded. Competitive behavioural mechanisms of limitation include territoriality and other interactions which act, ultimately in relation to resource levels, to influence density through effects on the feeding rates, survival and reproductive prospects of individuals. Birds excluded from prime areas may be forced to occupy poorer ones where, through lower food-supply or other disadvantage, they have lower survival and reproduction. Thus, as overall population size continues to rise, the mean per capita survival and reproductive rates of the overall population are likely to decline. This decline is likely to accelerate rapidly as the non-breeding component grows. In this way, variation in the extent and quality in habitat, coupled with competitive interactions, can provide a powerful density-dependent check on continuing population growth, leading to an equilibrium in population size, ultimately influenced by resource levels. As yet, not all components of this model have been properly tested, and for the most part the model has been pieced together from studies on a range of different species.

Song Thrush *Turdus philomelos* singing on territory.

Chapter 4
Territorial Behaviour and Density Limitation

Territorial behaviour is one of the most obvious manifestations of dominance. It is so ubiquitous a mode of social organisation, and in birds such a conspicuous activity, that it hardly needs description. In its most extreme form, a territory is seen as a defended area, of fixed location and exclusive use, with clear boundaries separating it from the territories of neighbours, and proclaimed with distinctive displays and vocalisations. In less extreme form, territories may overlap between neighbours, with a core area defended more vigorously than the rest. In even less extreme form, overlap between neighbours is almost complete and defence is barely noticeable. The term 'home range' is often used for dispersion systems of this type. But whatever the extent of defence, exclusive use or overlap, such dispersion systems lead to a fairly even distribution of individuals through suitable habitat. For present purposes, therefore, we can adopt a rather loose definition of territorial behaviour, as any form of spacing behaviour that involves site-specific dominance and that produces a dispersion pattern that, within the area occupied, is more regular than random (Newton 1992). Avoidance, as well as aggression, may be involved in producing such regular spacing (Jenkins 1961).

The benefits of territorial behaviour for the individual include the assurance of a resource, such as food, mate or nest-site, or some combination of these. They also include any increased protection against predation and disease that spacing out or familiarity with the area confers. In the breeding season territorial behaviour might also give increased security of parentage: for the male by reducing the chance of cuckoldry, and for both sexes by reducing the chance of dump laying by other females. Some of these benefits are immediate, while others accrue over the longer term. By excluding other members of its species, the owner of a feeding territory maintains exclusive or prior rights over the good foraging spots, and capitalises on its knowledge of where they are. It can also more efficiently exploit self-renewing resources, such as nectar in flowers (sunbirds) or insects washed on to river banks (wagtails) by timing its visits at appropriate intervals (Davies & Houston 1983).

Territories may be held by single individuals, by pairs, or by groups of individuals. Depending on circumstances, some species hold territories only in the breeding season, while others hold territories throughout the year for most of their lives. Many migrant species hold separate territories in both breeding and wintering areas, and some also set up temporary territories at migratory stop-over points (Gass 1979, Bibby & Green 1980). In some migrant species, territories are held by pairs in summer, when male and female act as partners, but by single birds in winter, when male and female act as rivals: examples include the Wheatear *Oenanthe oenanthe* and Reed Warbler *Acrocephalus scirpaceus* in the Old World, and the Northern Waterthrush *Seiurus noveboracensis* and Wood Pewee *Contopus virens* in the New World. Some species nest in colonies, defending small areas round their nests, but hold larger territories in their feeding areas, an example being the Grey Heron *Ardea cinerea* (Marion 1989).

Some of the most curious territorial systems involve some neotropical forest birds, which form group territories containing individuals of several species. Membership of these groups adheres to a strict quota of one family per species (Munn & Terborgh 1979, Gradwohl & Greenberg 1981). Once a bird becomes an established group member, it remains with the group for years (Munn 1985, Greenberg & Gradwohl 1986). Each day, the birds assemble to forage together. All group members recognise the same territorial boundaries, so that in effect the group, rather than its component pairs, holds the territory. When adjacent groups happen to meet, there is a sudden burst of vocal activity, as each pair postures and displays to its counterpart in the neighbouring group. The main species comprising such groups include insectivorous ant-shrikes, ant-wrens and ant-vireos. For half the year, these birds are joined by some migrant species from North America, including Blue-winged Warbler *Vermivora pinus*, Golden-winged Warbler *Vermivora chrysoptera* and Canada Warbler *Wilsonia canadensis*, which occur as pairs or single birds in each group (Greenberg & Gradwohl 1980, Greenberg 1985, 1987). The important implication of this system, as yet not fully explored, is that different species have similar densities, occurring up to the same maximum numbers in any stretch of forest. It is as though the social structure imposes a similar upper limit on each species, regardless of its individual food-supplies. The advantages of the system for individuals are assumed to be the same as for mixed flocks elsewhere, namely enhanced predator detection and foraging efficiency.

DENSITY LIMITATION

Whatever the benefits to territory owners, territorial behaviour provides a mechanism by which local densities might be limited, either by forcing some individuals into poorer habitats or by imposing on them a non-territorial lifestyle which carries greater mortality risks and poor reproductive prospects (Chapter 2). The role of territorial behaviour in limiting bird densities is supported primarily by circumstantial evidence from a wide range of species, including (1) observations of territory defence; (2) the division of habitat areas into territories; (3) a sudden change in local population level at the time of territory establishment; (4) the existence of non-territorial, non-breeding birds, including mature adults; (5) the rapid replacement of lost mates or territory owners from among non-territorial birds at various times of year; or (6) observed attempts by non-territorial adults to obtain territories by displacing residents (e.g. Kendeigh 1941, Snow 1958, Carrick 1963, Delius 1965, Rowan 1966, Watson & Jenkins 1968, Smith 1978, Newton 1979, Drent 1984, Birkhead *et al.* 1986, Smith & Arcese 1989). It is also supported by a wealth of experimental evidence, with which the rest of this chapter is concerned (Newton 1992).

REMOVAL EXPERIMENTS

Of the many experiments done to test the role of territorial behaviour in limiting bird densities, the most frequent type is the 'removal experiment', in which territorial residents are removed from a settled population. If the removed birds are then quickly replaced by new individuals (which are not merely neighbours expanding or changing their territories), it can be inferred that, prior to removal, the residents had prevented potential settlers from taking up territories. A removal experiment thus tests both for (1) the existence of density limitation, and (2) the presence in the vicinity of surplus individuals able to take a territory when a chance arises. Without a surplus, territorial density may still be limiting, but it is not possible to demonstrate it convincingly. Replacements should occur within the same season; if they do not occur until the next year, they could simply be part of the normal recruitment process and not the result of a surplus. If the newcomers from the surplus breed in the same year, the experiment further confirms that these birds were at the time mature and capable of breeding. Moreover, it is important in monogamous species that removal and replacement should involve both sexes; otherwise the experiment may reveal no more than an unequal sex ratio, with a surplus of one sex over the other.

Some removal experiments have involved taking out only a small number of selected individuals, scattered through a settled population, while others have involved the large-scale depopulation of extensive tracts of habitat. In some studies, replacement birds have been left in their newly-acquired territories, so it was possible to check whether they went on to breed. In other studies, replacement birds were shot as they appeared, one after another, in order to gain more idea of the number of potential recruits available. Hence, the conclusions that could be drawn from different removal studies have varied with the experimental format. In the rest of this section, the experimental evidence for density limitation is examined in relation to habitat variation, considering different groups of birds in turn (**Table 4.1**). Readers who want to skip the detail from individual experiments can pass straight to the Discussion.

Table 4.1   Summary of findings from removal experiments done to test the role of territorial behaviour in limiting bird breeding or wintering densities[1].

| | Number of species | Number of studies | Number with replacement of | | Number of studies in which one or more replacements were known to come from a non-territorial sector[3] |
			One sex[2]	Both sexes	
Songbirds					
Autumn–winter	7	11	3	6	3
Spring	23	34	21	8	5
Grouse					
Autumn–winter	2	2	1	1	1
Spring	7	12	6	5	5
Raptors					
Autumn–winter	1	1	0	0	0
Spring	3	3	0	3	0
Waterfowl, Spring	5	5	1	3	1
Waders, Spring	2	3	0	2	1
Seabirds, Spring	3	3	0	3	2
Overall	53	74	32	31	18

[1]Based on review by Newton (1992) in which further details and original references may be found, with additional details from Smith (1978), Knapton (1979), Fonstad (1984), Payne & Groschupf (1984), Gauthier & Smith (1987), Desrochers et al. (1988), and Stutchbury (1994).
[2]In some experiments, only one sex was removed, whereas in others both were removed but only one was replaced.
[3]The figures are minimal because they depended on replacements coming from known (marked) non-territorial birds.

Passerines

The earliest experiments on passerines were performed long ago in a 16-ha tract of conifer forest in Maine during a spruce budworm outbreak (Hensley & Cope 1951, Stewart & Aldrich 1951). The territorial males of some 40 species of birds were first counted and then, after nesting had started, as many as possible were shot in a three-week period. The total number shot was about twice the original estimate of birds present, implying considerable rapid replacement. In the following year, almost the same number of birds was back on the plot as in the previous year (154 compared with 148), and again the authors shot about twice this number within three weeks. This was an overall figure, and whereas most of the resident bird species showed little or no replacement within a season, for the summer visitors the total number shot formed up to five times the number initially present, depending on species.

Among many species, then, there were evidently more potential settlers than original breeders. Because the removals were conducted well after the migration season, the authors concluded that the newcomers were drawn from a non-territorial surplus, which was present in the general area and from which individuals moved into the study plot continually as gaps were created. In most species, however, including the ten most numerous ones, only males were replaced. There was thus no evidence in most species for a surplus of females, or that any females were being denied an opportunity to breed through habitat saturation. In a smaller scale one-year experiment in northern Finland, which involved eight migrant species, the numbers of birds removed also exceeded the number of initial settlers, but again it was not convincingly shown that both sexes were replaced (Mönkkönen 1990).

In a similar experiment in southern England, Edwards (1977) removed individuals of several resident territorial species (both sexes) from an area of farmland, reducing the number of territories in Blackbird *Turdus merula* from 14 to 3, in Song Thrush *T. philomelos* from 5 to 0, in Dunnock *Prunella modularis* from 13 to 2 and in Chaffinch *Fringilla coelebs* from 19 to 11. Re-invasion by Blackbirds (both sexes) was rapid, but stopped after 17 days when the population reached 57% of its former level; re-invasion by Song Thrush and Dunnock was slower and also incomplete; and by Chaffinch was zero. On a nearby control plot, by contrast, numbers remained stable throughout the season. These findings were consistent with the idea of density limitation in territorial birds, and with the existence of a surplus of both sexes in three of the four species; however, because the birds were not marked, the status of the replacements prior to removal remained uncertain, as in the studies above.

Turning now to experiments on individual species, in an area of western Canada where Song Sparrows *Melospiza melodia* held year-round territories, Knapton & Krebs (1974) did two types of experiment: (1) removing all the territory holders in an area at the same time, and (2) removing territory holders one at a time at 3–4-day intervals. Both types of removal were done in October, coinciding with an autumn peak of territorial activity, and again in March, coinciding with a spring peak. At both seasons all the empty territories were rapidly filled by first-year birds that had previously been non-territorial, revealing that the density of territorial Song Sparrows had been limited at both seasons. In the two simultaneous removal experiments, territory size decreased significantly after resettlement, and more birds than previously occupied the same area. But in the successive removal and control areas there was no such change. This implied that, in this species, settlement patterns could influence mean territory size, with individuals taking smaller territories when many settled simultaneously than when they settled gradually over a long period. Other observations on Song Sparrows in a different area revealed a surplus of males in the breeding season, but not of females (Smith *et al.* 1991).

The various species of titmice (Paridae), found in temperate and boreal woodland, also show peaks of territorial activity in both autumn and spring. In the Great Tit *Parus major*, autumn territorial behaviour does not necessarily lead to the immediate exclusion of subordinate non-territorial birds which, for a time, continue to live alongside the territorial birds. None-the-less, when territorial birds were removed from particular areas in autumn and winter, previously subordinate juveniles rapidly established territories in the same places (Drent 1984). As in the Song Sparrow, higher territorial densities were achieved when several individuals settled simultaneously than when they settled gradually over a longer period.

Territorial Great Tits were more likely to exclude others from their areas in spring than in autumn, but removal studies in spring have given variable results. In some oft-quoted experiments, Krebs (1971) removed in spring 14 Great Tit pairs from their territories in woodland. The vacated ground was rapidly re-occupied, partly by neighbouring pairs expanding their territories but mainly by new pairs which abandoned their territories in nearby hedgerows and moved into the wood. The vacated hedgerow territories were not subsequently re-filled. Because woodland was a superior habitat, which offered better breeding success than hedgerows, the pairs that changed territories had moved to better habitat when given the chance. These findings implied that territorial density was limited in the best habitat (woodland), causing some birds to settle in suboptimal habitat (hedgerow), but gave no evidence for a non-territorial surplus unable to breed. The birds that moved were yearlings which had not previously nested. Adults that held territories in the hedgerows did not move. This was consistent with previous findings that Great Tits that had bred before usually stayed in the same place, even in poor habitat.

Experiments on Great Tits in a nearby wood in a later year showed that, after hedgerow birds had moved to replace removed woodland birds, the hedgerows were re-occupied by new birds, assumed to come from a non-territorial surplus (Krebs 1977). Another spring removal, in Belgium, resulted in replacements of both sexes (Lambrechts & Dhondt 1988), as did another in primaeval forest in Poland, where densities were only one-tenth of those in western Europe (Wesolowski et al. 1987). Evidently, surplus Great Tits are present in some springs but not in others, and in low density as well as in high density populations.

Certain other species of tits form group territories in winter. Each group usually consists of several obvious pairs, and occupies a discrete range which overlaps little with neighbouring ranges. Group members consist of 'regulars' which stay in the same group, and lower ranking 'floaters' (mostly juveniles) which wander widely between groups (Ekman et al. 1981, Smith 1984). Any regular that disappears from a group is rapidly replaced by the highest ranking floater of the same sex (Smith 1984). In spring, these groups break up and the component pairs establish individual territories mainly within the area previously occupied by the group (Samson & Lewis 1979). Removal experiments have been done on several species in autumn, winter and spring. Five neighbouring Willow Tit P. montanus groups were removed in autumn from a forest in south Sweden (Ekman et al. 1981). Within a month the area was repopulated by a similar number of birds which arranged themselves into five new groups, occupying roughly the same areas as their predecessors. Three groups that were removed in winter were also replaced, but this time by birds from other groups rather than by floaters which had largely disappeared by then. Similar experiments in the same area with Crested Tits P. cristatus, in which no floaters were apparent, led to only partial replacement, again by birds from other groups. In both species all the newcomers that could be aged were juveniles. The authors concluded that, within the study area, both the numbers and sizes of groups were regulated. Any floaters that occurred as a result of territorial (dominance) behaviour in autumn soon disappeared through death, emigration or replacement of group members that died. In line with these findings, Willow Tits in a different area removed from groups in January were not replaced (Hogstad 1989), and nor were territorial Willow Tit and Crested Tit pairs removed in spring (Cederholm & Ekman 1976). In all these tit species, therefore, whether they formed pair territories or group territories, territorial behaviour in autumn could create a non-territorial surplus which disappeared gradually over the ensuing weeks or months.

These and other experiments on songbirds in **Table 4.1** may be summarised as follows. Of 11 experiments on seven species in autumn or winter, nine led to replacements of one or both sexes, but in only three species (Song Sparrow, Great Tit, Willow Tit) were the replacements known for certain to come from a non-territorial sector. A slightly earlier removal might also have provided evidence for a non-territorial sector in other species. Of 34 experiments on 23 species in spring, 29 led to replacement of one or both sexes following removal, but in only five (Song Sparrow, Rufous-collared Sparrow *Zonotrichia capensis*, Great Tit, Black-capped Chickadee *Parus atricapillus* and Red-winged Blackbird *Agelaius phoeniceus*) were replacements known for certain to come from a non-territorial sector. In most species, the origin of the replacements was unknown but on grounds of age and other circumstantial evidence, they were thought to derive from a non-territorial sector (Newton 1992). These various experiments included both resident and migrant species, on breeding or wintering areas.

In most of the species concerned, pairs held sizeable territories which provided both nesting and feeding requirements. However, in species that use special nest-sites, such as tree-holes, breeding density in some areas may be limited, not by shortage of foraging habitat, but by shortage of nest-sites. This has been shown repeatedly when a striking increase in breeding numbers has followed the provision of nest-boxes in areas where natural sites were previously sparse or lacking (Chapter 8). In such species, the nest-cavity is clearly the underlying limiting resource, and aggressive territorial behaviour the proximate mechanism securing ownership of a site in the face of competition. As in the defence of larger territories, defence of scarce nest-sites leads to the exclusion of subordinate individuals which must move on and breed elsewhere or remain as non-breeders that year.

In some passerine species, such as the Great Reed Warbler *Acrocephalus arundinaceus*, males and females occupy different habitats in their wintering areas (Nisbet & Medway 1972). This may be because the sexes prefer different habitats, or because the dominant males occupy the best habitats, leading the females to settle in less good ones. In the Yucatán Peninsula of Mexico, most male Hooded Warblers *Wilsonia citrina* defended territories in primary forest, whereas most females lived in scrub on abandoned fields (Lynch *et al.* 1985). To test whether females were excluded from forest by males, males were removed from habitat boundaries where the two sexes came into contact (Morton *et al.* 1987). No females expanded into the space left in the forest by the removed males, so in this species, the difference in distribution between the sexes may have occurred through preference rather than through dominance relationships. In the American Redstart *Setophaga ruticilla*, in contrast, removed males were replaced largely by females, implying that in this species males had excluded females from mutually acceptable wintering habitat (Marra *et al.* 1993).

In conclusion, the main points to emerge from studies on passerines, besides the frequent demonstration in autumn or spring of density limitation, include the evidence of a surplus of both sexes (but especially males) in many species in spring, and the importance of synchrony in settlement patterns to final density. Synchrony in settlement could be expected in many migrant populations, in which individual males often arrive on their breeding areas on the same day or within the space of a few days.

Grouse

Because of their importance as quarry species, grouse have been subject to many detailed studies, including several experimental manipulations. Many species are territorial, and their densities fluctuate greatly from year to year, often in regular (cyclic) patterns.

The most thorough studies concern the Red Grouse *Lagopus l. scoticus*, in which pairs occupy territories on heather moorland (Jenkins *et al.* 1963, Watson & Jenkins 1968, Watson 1985, Watson & Moss 1990). In studies in eastern Scotland, the autumn (October) population after shooting typically contained about 50% territorial pairs which survived well and bred in the following year, and 50% non-territorial birds which survived poorly and did not breed (Chapter 2). During the autumn and winter, the non-territorial birds were gradually driven away by territorial birds, and virtually all had died or emigrated before the following spring (many were found dead locally). Of 119 territorial birds of both sexes shot experimentally at various times of year, 111 were replaced, most by known non-territorial birds but others by territorial birds which moved up to 1 km to take the vacated ground (Watson & Jenkins 1968). Replacement occurred in years of both low and high territorial density, but in general was most rapid in autumn and least rapid or non-existent in spring, by when most surplus birds had died.

In some areas, all territorial Red Grouse were shot to create large empty spaces, and in other areas only selected individuals were shot. Yet in most cases the replacement birds took territories of similar size, and settled at similar densities, to their predecessors on the same ground. In control areas, by contrast, containing large numbers of birds, only three changes of occupant occurred during the period of the experiments: one bird moved to occupy a vacancy on an experimental area, and the other two disappeared and were themselves replaced. The conclusions were that densities were limited by territorial behaviour, leading to a large surplus of both sexes in the autumn, most of which had disappeared by the following spring. The lack of replacement in spring was later confirmed for females in another area (Hudson 1990). Unlike the situation in some songbirds, the degree of synchrony in settlement seemed not to influence density, for replacement birds settled to the same density as their predecessors in the same area.

These findings were broadly confirmed in various other species of grouse, including Willow Ptarmigan *Lagopus lagopus* and Rock Ptarmigan *Lagopus mutus*, and some forest species, such as Blue Grouse *Dendragapus obscurus* and Ruffed Grouse *Bonasa umbellus* (**Table 4.1**, review Newton 1992). But whereas in resident species, notably Red Grouse, territorial behaviour in autumn limited the subsequent breeding density, in migrant species territorial behaviour in spring was paramount. Although only one study, performed late in the spring, provided no evidence for a surplus (**Table 4.1**), it was clear that surplus birds were not present at all times in all populations. Some studies included only male removals, but evidence for a surplus of both sexes was found in all the ptarmigan (*Lagopus*) species (Watson 1965, Hannon 1983, Mossop 1985). Forest grouse also showed another interesting point: in areas where replacements occurred, not all non-territorial males of breeding age that were known to be present took up territorial vacancies that were created. Evidently some factor other than exclusion prevented these birds from taking territories (for Ruffed Grouse *Bonasa umbellus*, see Dorney & Kabat 1960, Fischer & Keith 1974; for Blue Grouse *Dendragapus obscurus*, see Lewis & Zwickel 1980; for Spruce Grouse *Dendragapus canadensis*, see Szuba & Bendell 1988). In most studies of forest grouse, only males were removed.

Raptors

Some raptor species show extreme population stability, with densities of territorial pairs varying by no more than 15% of the mean level over long periods of years (Newton 1979). Often pairs are regularly spaced through the habitat, occupying the same sites year after year, despite a turnover of occupants. These facts imply that densities are firmly limited. Other species, which have restricted diets based on cyclic prey, fluctuate in density from year to year, in step with their prey (Chapter 7). In many raptor species, non-territorial birds are not much in evidence, but rapid replacement of lost mates is commonplace, having been recorded incidentally in at least 26 species in 19 genera (Newton 1979).

Controlled removal experiments on at least three raptor species have also revealed a surplus of both sexes, mostly in the younger age-groups **(Table 4.1)**. Moreover, in two studies, involving the Sparrowhawk *Accipiter nisus* and American Kestrel *Falco sparverius*, replacements occurred only on good territories and not on poor ones (with territories classed as good or poor on previous occupancy and nest success; Bowman & Bird 1986, Newton & Marquiss 1991). In one experiment, replacement Sparrowhawks took good territories made available by removals, even though they had not previously taken up vacant territory sites in poor habitat. Also, when given the opportunity to breed, replacement birds bred much less well than other first-time settlers. This was particularly true of males. Findings from raptors thus underline the importance of both site features and bird features in influencing whether settlement, defence and breeding occur.

Waterfowl

Because most waterfowl are heavily hunted, it is hard to tell what their natural fluctuations might be, but in some regions many duck species show long-term fluctuations in numbers in response to periods of wet and drought (Chapter 11). Waterfowl show various types of spacing behaviour and some species have a large non-breeding sector. Some duck species defend distinct pair territories in winter quarters and in spring breeding areas, in addition to nest-sites and summer brood-rearing areas, all of which may be occupied for a short time in spatially separated areas. Moreover, where the occupation of particular areas is short compared to the potential breeding season, different pairs may defend and occupy the same territory sites in succession; in one study some sites were used in turn by up to three Shoveller *Anas clypeata* pairs per year, with each male deserting the site after his female had laid (Seymour 1974). At any one time, density is apparently limited, but with more than one set of occupants in each breeding season (Chapter 3). At the other extreme, certain swans defend large all-purpose territories throughout the breeding season or throughout the year, with non-breeders occurring in flocks elsewhere.

Experimental studies on three species of shelducks *(Tadorna)* showed that breeding densities were limited by territorial behaviour on the feeding areas (Young 1970, Riggert 1977, Williams 1979). Observational evidence on other species suggested that limitation could occur with respect to nest-sites (in hole-nesting species), while limitation of numbers and duckling production could also occur in brood-rearing areas (Raitasuo 1964, Jones & Leopold 1967, Pienkowski & Evans 1982, Savard & Smith 1991). Considering the potential for limitation on any of the different types of territories held by ducks at different times of year, and the occurrence of sequential occupancy, ducks would repay more study.

Waders

In the breeding season, waders typically hold sizeable territories, obtaining their food from the territory or elsewhere. Different patterns may occur in the same species, depending on the relative distributions of suitable nesting and feeding habitat (Ens *et al.* 1995). Sequential occupancy, with different pairs using the same areas in succession in the same breeding season, has been recorded in at least one species, the Golden Plover *Pluvialis apricaria*, in which two (or occasionally three) pairs occupied the same areas each year (Parr 1979): once a pair had left a territory with their chicks, another pair moved in. In many wader species, breeding densities are normally fairly stable from year to year, often with a non-breeding sector.

The most thorough experiments were on Oystercatchers *Haematopus ostralegus* nesting on Skokholm Island, Wales (Harris 1970). On 13 territories from which single individuals were removed, the remaining partner (male or female) rapidly remated. However, four territories from which both partners were removed were taken over by neighbouring pairs which moved to use the better habitat thereby released. The removals led to a great deal of movement and mate change, but most replacement birds were unmarked strangers to the area, although at least one (marked) bird came from a known flock of non-breeders. The replacement birds bred that year. The conclusions were that spring breeding density, which had been stable over several years, was limited by territorial behaviour, leading to a surplus of mature birds of both sexes, a finding confirmed in other studies of Oystercatchers (Ens *et al.*, 1995). Other removal experiments on waders, again leading to replacements, concerned Dunlin *Calidris alpina* (Holmes 1966).

Colonial seabirds

In effect a colony is a cluster of small territories, all centred on nest-sites, the defence of which can lead to the exclusion of surplus birds and to the build-up of a non-territorial non-breeding sector just as effectively as any other territorial system where space is short. In many colonial seabirds, non-breeding adults may be seen in and around the colonies, and the failure of some individuals to breed has been attributed to their inability to secure a nest-site within the colony limits. This is readily understandable where a colony occupies the entire space available, but is puzzling where there is room for expansion. At many colonies, vacant sites are available on the edges, and over a period of years the colony may indeed spread into such sites. At any one time, however, the centre of a colony seems more attractive to potential settlers than the edges, perhaps in some species because of greater exposure to weather or greater egg and chick predation at the edges (Kruuk 1964, Patterson 1965, Tenaza

1971, Gaston & Elliot 1996, Yorio & Quintana 1997). The difference in nest success between centre and edge can be substantial: some 21–30% of eggs in the centre of a Sooty Tern *Sterna fuscata* colony were lost, compared with 87–92% at the edge, the difference due mainly to predation (Feare 1976b). In the cliff-nesting Kittiwake *Rissa tridactyla*, at a colony with no obvious predation, previous non-breeders nested in the centre of the colony when extra ledges were created experimentally, while at the same time unoccupied ledges remained available on the periphery (Coulson 1968, Porter & Coulson 1987). A similar phenomenon has been noted in colonial songbirds (Orians 1961), and is analogous to the situation in some non-colonial species discussed above, in that some (mostly young) individuals took up territories in favoured places when they were given the chance, but had hitherto ignored less favoured places which lay empty. The implication is that, in these species, some individuals may prefer to wait until a favoured site becomes available within the colony than to settle at an earlier age in what seems to be a less favoured site on the edge.

Some of the most telling removal experiments concern Cassin's Auklet *Ptycho-ramphus aleuticus*, a small hole-nesting seabird which breeds in colonies on islands off western North America (Manuwal 1974). At a colony in California, estimated to contain 105 000 pairs, the birds remained year-round, and nested either in rock crevices or in self-made burrows at densities up to about one per square metre. The pairs defended a small area around the burrow entrance, driving away all other birds that landed. Non-territorial floaters were estimated to form 50% of the total island population, and about 80% of these birds had not previously bred. From a total of 60 burrows over two years, Manuwal (1974) removed 97 occupants after laying had started, including both members of 37 pairs. He also removed the replacements as they appeared, recording a total of 154 in the two seasons. In each year he left the later arrivals, many of which bred but with lower success than the rest of the population. Most of the newcomers were young birds (1–3 years old) which had not bred previously, and some also returned the following year. The conclusion was that colony size was limited by the aggressive behaviour of its members, in relation to available nest-sites, leading to a substantial non-breeding sector of both sexes, which included some sexually mature birds capable of breeding. As in some other seabirds, however, while parts of the island appeared saturated with nests, other parts did not. The avoidance of certain areas again revealed that some birds were more willing to wait for an existing site in a preferred area than to dig a new burrow in a less favoured area.

These and other experiments on gulls and skuas reaffirm that, in species that defend little more than the nest, aggression can be limiting where space is short or nest-sites are of variable quality (Young 1972, Pierotti 1980). Hence, insofar as can be judged from the small number of relevant studies, territorial behaviour could play just as central a role in limiting local nesting density in seabirds as in more dispersed landbirds. The main problems that require more investigation concern the apparent unattractiveness for some species of colony edges, and why some birds of breeding age remain as non-breeders rather than settle on vacant ground at the edges (e.g. Coulson 1968). There may be a trade-off for individuals, with the choice between setting up a new territory in the current year with an inexperienced partner on the colony edge (which for some colonies could have greater exposure and predation risks) or remaining as a non-breeder on the chance of acquiring by replacement an established territory and experienced mate in the more protected centre. Because most seabirds typically show great site-fidelity, for most individuals any choice could represent a lifetime commitment. Even where populations are expanding, reluctance to settle on the colony edge could both slow the growth of existing colonies and present a behavioural obstacle to the establishment of new ones. It seems, however, that not all species show it.

DISCUSSION

At face value, up to 63 of 74 experiments on 53 different bird species have provided evidence for density limitation by territorial behaviour and for the existence locally of potential replacement birds of one or both sexes, either in autumn, spring or both **(Table 4.1)**. In some species that were territorial year-round, a non-territorial sector seemed to be a permanent feature of the populations studied, whereas in other species a surplus was present in some seasons, some years or some areas, but not in others. Where non-territorial birds changed greatly in numbers during the course of the year, removals only a few weeks apart could give quite different results, with replacements of removed territorial birds occurring in one month but not in the next (Watson 1965, Rippin & Boag 1974, Hogstad 1989). Moreover, in some species, evidence for a non-territorial sector was found in years and in areas of low density, as well as in years and in areas of high density (e.g. Great Tit *Parus major*, Wesolowski *et al.* 1987; Red Grouse *Lagopus l. scoticus*, Watson & Jenkins 1968; Spruce Grouse *Dendragapus canadensis*, Szuba & Bendell 1988; Blue Grouse *D. obscurus*, Bendell *et al.* 1972, Zwickel 1972; Kestrel, Village 1990). This would be expected if territory sizes varied between years and areas in relation to food-supplies or other local factors.

Replacement of removed territorial birds was seen to occur 'within minutes' in Red-winged Blackbirds *Agelaius phoeniceus* (Orians 1961, Eckert & Weatherhead 1987), within hours in some other species (Krebs 1977, Hogstad 1989), but took up to a week or more in others (Edwards 1977, Newton & Marquiss 1991, Stutchbury 1994). Speed of replacement may have depended partly on the proximity of potential recruits, as well as on the attractiveness of the site. Replacement birds were usually younger than the birds they replaced and, as shown by a few studies, were usually drawn from among the most dominant individuals in the non-territorial sector (Knapton & Krebs 1974, Drent 1984, Smith & Arcese 1989, see also Chapter 3). On nesting territories, many replacement birds bred within a few weeks after moving in, confirming that they were sexually mature.

In studies in which removals involved a small number of selected individuals scattered through a network of territories, it is perhaps not surprising that individuals were usually replaced on a one-to-one basis, so that local densities remained unchanged. Newcomers were fitting into a pre-existing territorial framework, often pairing with bereaved residents. In contrast, experiments in which several neighbours were removed simultaneously to create a large area of vacant habitat gave two main types of result. In one type, the area was occupied to about the same level as previously (for Red Grouse, see Watson & Jenkins 1968; for Great Tit, see Krebs 1977; for Wren *Troglodytes troglodytes*, see Wesolowski 1980). The possibility that this was a chance result, dependent on the number of replacement birds that happened to be available, is remote; it could in any case be eliminated in studies with more than one set of removals-replacements in the same year, or with surplus birds still seen to be present after replacement was complete, as in the studies just mentioned. The most likely explanation of replacement to the same density is that the habitat itself had a major influence on the number of birds that settled there. In the second type of result, the vacant area was occupied to a higher density than previously, associated with simultaneous settlement (Knapton & Krebs 1974, Drent 1984). This implied that, as well as habitat, bird behaviour (settlement pattern) had some influence on the final density (Knapton & Krebs 1974). The fact that, in the same species, density did not increase in other areas where removals were made at the same time but singly and successively, supports this view. It seems that birds already settled in an area were better able to keep out others than birds in the process of settling at the same time as others, perhaps because of the defence benefits gained from prior residence (Chapter 2). A third type of result, where a vacant area was occupied to a lower density than previously (for Blackbird *Turdus merula* and others, see Edwards 1977), might have resulted merely from a shortage of candidates, rather than from an aspect of habitat or bird behaviour.

The experiments described above showed clearly that territorial behaviour could *limit* densities in particular areas or years, but they did not confirm that territorial behaviour could *regulate* densities over a period of years. This could only happen if, as the number of potential settlers rose, an increasing proportion was excluded, as in **Figure 3.1**. Such a pattern was seen in several species in which both territorial and non-territorial individuals could be counted over several years of varying total numbers. Examples include Great Tit *Parus major*, Nuthatch *Sitta europaea*, Tawny Owl *Strix aluco*, Great Horned Owl *Bubo virginianus*, Sparrowhawk *Accipiter nisus*, Oystercatcher *Haematopus ostralegus* and others (**Figure 4.1**; Southern 1970, Dhondt 1971, Kluijver 1971, Tinbergen *et al.* 1985, Nilsson 1987, Newton 1991, Goss-Custard *et al.* 1995a, Rohner 1996). In all these species, a greater proportion of individuals

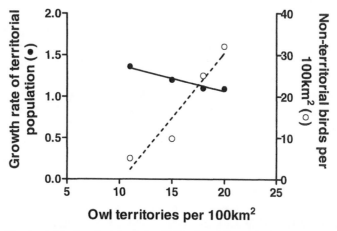

Figure 4.1   Territorial behaviour and the limitation of population growth in Great Horned Owls *Bubo virginianus*. Growth rates of the territorial population (●) declined as the numbers of territories in the study area increased. Numbers of non-territorial birds (○) increased as territories became more densely packed, revealing density-dependent exclusion of a surplus. Both findings fit the pattern modelled in Figure 3.1. From Rohner 1994.

failed to get territories in years when the overall population was high. Similar patterns were evident in the settlement of various species on their wintering grounds, as discussed in Chapter 3, although these did not necessarily involve territorial behaviour.

Mechanisms of density limitation

In a system in which territorialism leads to expulsion of practically all other individuals from the defended area (as reported in Red Grouse *Lagopus l. scoticus*, Watson & Jenkins 1968), the mechanism of density limitation is clear cut. The number of birds in the area is limited to those that hold territories. In many species, however, notably passerines, non-territorial individuals live inconspicuously in the same areas as the territorial ones, within and between the territories. So while territorial behaviour limits territorial (breeding) density, it appears not directly to limit total density. Moreover, where territorialism leads merely to dominance without expulsion, so that subordinates live openly within the territories of dominants, disappearing over a period of months (as reported in the Great Tit *Parus major* in autumn, Drent 1984), territorialism would seem to provide no more effective a density-limiting mechanism than any other expression of social hierarchy. As explained in Chapter 2, the 'strength' of territorial behaviour, and its consequences for other individuals, could depend partly on the cost–benefit ratio of the behaviour at the time.

Several resident territorial species, which were subjected to removal experiments, conformed to the same system of density regulation, depicted in **Figure 4.2**. Autumn territorial activity led to the separation of dominant territorial and subordinate non-territorial individuals. Not only did this process limit territory densities at the time, but it also set a ceiling on such densities for the ensuing months, in some cases for the whole year. As time passed, individuals were continually lost through mortality or emigration, but usually at a lower rate from the territorial than from the non-territorial sector. However, as long as non-territorial birds were present in the vicinity, they acted as replacements, so that territorial density was maintained. How long the reserve of non-territorial birds lasted depended on their initial numbers and on their subsequent rate of attrition. Once they had gone, further losses in the territorial sector were not made good, and the numbers of territorial birds declined. On this system, the density of territorial birds in spring could only be the same as, or lower than, their density in autumn. Spring densities thus depended primarily on autumn densities and secondarily on the extent of intervening loss through death or emigration. This situation held, for example, in Red Grouse *Lagopus l. scoticus* in eastern Scotland (Watson & Jenkins 1968) and in Kestrels *Falco tinnunculus* in southern England (Village 1990).

The situation differed where some individuals left in autumn for alternative wintering areas from which they could return in spring. Competition for territories was then renewed in spring and, if conditions permitted, densities could reach the same or even higher levels than in the preceding autumn. This situation held, for example, in Kestrels in Scotland, in which some birds migrated south for the winter (Village 1990), and in Great Tits in Holland in which some birds left the breeding woods for the winter, returning in spring (Drent 1984). Some of the birds that settled in spring may have originated from different areas in the preceding autumn, so that in any given area the number settling in spring need bear no consistent relationship

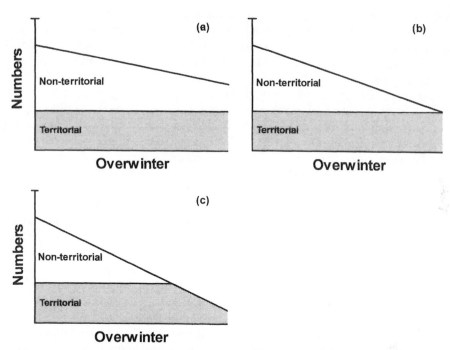

Figure 4.2   Model of density limitation in resident species that show autumn and spring (or year-round) territorial behaviour. (a) Overall mortality slight: sufficient to reduce numbers of non-territorial birds over winter, leaving a surplus in spring. Mortality is not restricted to territorial birds, but as non-territorial birds rapidly replace dead territorial ones, territorial density is maintained. Experimental removal of territorial birds at any time from autumn to spring would lead to replacement. (b) Overall mortality moderate: sufficient to remove all non-territorial birds but not to reduce territorial density. Experimental removal of territorial birds would lead to replacement over most of the period, but not at the end, in spring. (c) Overall mortality heavy: sufficient to remove all non-territorial birds, and to reduce the density of territorial ones. Experimental removal of territorial birds would lead to replacement in the first part of the period, but not in the later part.

to the number present in the preceding autumn. The same could hold for populations in which all individuals left to winter elsewhere.

A third situation, in which density limitation by territorial behaviour occurred only in spring, held in species such as the Chaffinch *Fringilla coelebs* which in some areas flocked in winter and were territorial only while breeding. In such species the extent of any apparent surplus would depend mainly on the numbers of potential breeders left in spring (relative to the carrying capacity of nesting habitat), after overwinter losses from whatever cause had taken their toll **(Figure 3.12)**.

Territorial behaviour and annual fluctuations in density

By its nature, territoriality tends to stabilise densities, reducing the year-to-year variation. Yet many territorial species, including many songbirds, fluctuate greatly in breeding density from year to year, their numbers sometimes doubling from one year to the next. If an area of habitat is incompletely occupied, with vacant areas and no exclusion of a surplus, we can assume that there are insufficient birds to fill it. But often the habitat may appear completely occupied year after year, with no vacant patches, and with the occupants holding territories of widely different sizes in different years. This situation could arise in several ways.

Firstly, potential occupants may vary in numbers from year to year, but simply divide the available habitat between them, seldom reaching a level that would lead to the exclusion of a surplus. Secondly, if territory sizes vary in relation to food-supplies (as shown in many studies, Chapter 3), and if food-supplies themselves vary from year to year, then so might territory sizes and resulting densities. Thirdly, if the number of contestants varies from year to year, territories might be more compressed in some years than in others, with a surplus excluded at a range of higher densities (as in **Figure 3.1**, see Zimmerman 1971, Myers *et al.* 1979, Ewald *et al.* 1980, Village 1983). Moreover, the number and persistence of contestants might themselves vary with the local food-supply, and contribute to the observed link between food-supply, territory size and density. Fourthly, the degree of synchrony in settlement might vary from year to year, giving higher densities when many birds settle at the same time than when they settle gradually over a longer period. Synchrony in settlement is most likely in short-lived songbirds which suffer a high but variable annual mortality. It is less likely in long-lived species in which few gaps in the territorial system arise in any one year, and in which individuals are usually replaced on a one-to-one basis. This latter system is likely to result in the same territorial pattern being maintained year after year, acting to stabilise density in the long term, despite some annual fluctuation in food-supply and other conditions. Adjustment to long term-changes in food-supply is likely to occur only slowly over a period of years (Newton 1979, Watson 1996). Fifthly, in some species, such as Red-legged Partridge *Alectoris rufa* and Red Grouse, territory sizes may be influenced by the relatedness of neighbours. This is because individuals of such species relinquish ground more readily to close kin than to genetically more distant individuals (Green 1983, Watson *et al.* 1994). Sons often settle next to fathers, and brothers to one another. Other things being equal, average density might then correlate with the average relatedness of neighbours, which itself is likely to vary between years. As yet, however, such a kin-based system has been described only in the two species mentioned, and in many other species close kin seldom settle next to one another (Greenwood 1980, Newton 1986).

In these various ways, territorial behaviour might contribute to fluctuation in annual densities. They help to dispel the belief that territoriality invariably acts to stabilise densities, and also to explain the observation that, in any given area, territorial behaviour can be limiting, and lead to exclusion of surplus, over a range of densities. In addition to promoting annual variations in density, the same mechanisms might promote area differences.

Other findings from removal experiments

Removal experiments have revealed several other points relevant to population limitation, and to the models of density regulation discussed in Chapter 3.

The role of site quality. In certain species, some males and females of breeding age remained non-territorial, even in the presence of vacant territory sites. This implies that factors other than the availability of a territory precluded territorialism and reproduction. One factor often involved was quality of territorial site, with 'good' territories defined by high occupancy or high breeding success in previous years. In some studies, replacements following removals occurred on good territories but seldom or not at all on poor ones, and some replacement birds took up good territories, even though they had not previously occupied vacant poor sites (for Cassin's Auklet *Ptychoramphus aleuticus*, see Manuwal 1974; for Blue Grouse *Dendragapus obscurus*, see Lewis & Zwickel 1980; for Kittiwake *Rissa tridactyla*, see Porter & Coulson 1987; for Sparrowhawk *Accipiter nisus*, see Newton & Marquiss 1991). Evidently some individuals would settle and breed if good sites were available, but not if the only vacant sites were poor.

Two speculations on this apparent choosiness have been offered. Firstly, in species such as the Sparrowhawk which obtain food from their territories, ability to breed depends partly on the food content of the territory and partly on the hunting skills of the bird. Hence, whereas a skilled bird might be able to breed on a good or a poor territory, an unskilled bird might manage to breed only on a good territory. There would then be little point in an unskilled bird occupying a poor territory, and it might have to wait until a better site became available or until its hunting skill improved (Newton & Marquiss 1991). The second speculation applies to long-lived species, such as the Oystercatcher *Haematopus ostralegus*, in which few territorial vacancies arise in any one year. When a bird eventually gets a territory, therefore, it must normally stay there for the rest of its life. Where territories vary in quality, a bird might thus face a 'career decision' on whether to accept a poor territory now, or wait another year or more in case a better territory becomes available (Ens *et al.* 1995). It is easy to imagine that a long-lived bird might raise more young in its lifetime by occupying a good territory for a few years than a poor one for many years. The decision of when to settle cannot be separated from the decision of where to settle. Both are dependent on the frequency with which territorial vacancies arise, on the quality of the sites concerned, and on the number of non-territorial contenders. Ens *et al.* (1995) wrote of Oystercatchers 'queuing' for territories, and Smith (1978) described a similar situation in the Rufous-collared Sparrow *Zonotrichia capensis*.

Territory moves. In some removal studies, some replacement birds simply moved from poorer territories. This was best documented in the Great Tit (from hedges to woods, Krebs 1971), but also occurred in Red Grouse (where a move provided the prospect of a larger territory, Watson & Jenkins 1968), Willow Ptarmigan *Lagopus lagopus* (to an area with earlier snow melt and more luxuriant vegetation, Blom & Myrberget 1978), Blue Grouse (to a favoured display site, Lewis & Zwickel 1980) and Oystercatcher (to incorporate areas with better feeding, Harris 1970). These observations indicate that occupancy of at least the best areas was limited by territorial behaviour, and that some established birds in poor areas would move to better areas that became available. As mentioned in Chapter 3, however, birds that had already bred in a site often remained there (= site-fidelity), foregoing a better opportunity elsewhere.

Poor quality of replacements. If replacements were drawn from a non-territorial sector consisting mainly of birds in younger age-groups, as shown in many species (e.g. Smith 1978, Birkhead *et al.* 1986), they would be expected to be subordinate, and perhaps also to breed less well, than existing territorial birds. In at least five studies in which some removed birds were held captive and later released, such birds mostly managed to displace their replacements to regain their original territories and mates, either in the same year or the next (for Red Grouse, see Watson & Jenkins 1968; for Oystercatcher, see Harris 1970; for Rufous-collared Sparrow, see Smith 1978; for Spruce Grouse *Dendragapus canadensis*, see Szuba & Bendell 1988; for Kestrel, see Village 1990). This implies that the original territory owners remained dominant to the birds that replaced them. Among Great Tits, however, the longer that replacement pairs had held a territory, the more difficult they became to evict, and after ten days in captivity only 10% of removed pairs regained their territories after release (Krebs1982).

In several species, replacement birds bred significantly less well than the rest of the territorial population, their younger mean age perhaps contributing to their poorer performance (e.g. Manuwal 1974, Village 1990). In raptors the male provides the food to the incubating female and has more influence on nest success than the female (Newton 1979). In line with this difference in roles, pairs in which the female was replaced (and the original male remained) showed little or no reduction in success compared with control pairs, while pairs in which the male was replaced (and the original female remained) showed a marked reduction in success (Kestrel, Village 1990; Sparrowhawk, Newton & Marquiss 1991). These various findings clearly indicate that, although some replacement birds were capable of breeding, their success was often much lower than that of established birds.

Sex differences. In monogamous species, notably songbirds, the higher replacement rate of males was held to reflect a surplus of males in the total population (Stewart & Aldrich 1951). But in other species, such as Red-winged Blackbird *Agelaius phoeniceus* and Blue Grouse, the apparent surplus of males was at least partly attributable to their polygynous mating system, which led to longer-deferred first breeding in males, and hence to the exclusion of more young males than young females from the breeding sector.

Effects of removal on recruitment and age of first breeding. One consequence of density limitation by territorial behaviour is that the number of young individuals that can get territories in any one year depends partly on the number of territorial adults remaining from the previous year. The greater the density of previous settlers, the fewer the gaps for new recruits (Newton 1991; Chapter 5). Because removals are tantamount to increased predation, they give some indication of the demographic response in bird populations to increased mortality of territory holders. Where full replacement occurs, extra mortality at that level would have no impact on the size of the territorial (potential breeding) population. Because in many species, non-territorial birds are younger, on average, than territorial ones, replacement would result in the recruitment of younger birds and a lowering of the age of first breeding, perhaps with an associated reduction in average breeding success.

CONCLUDING REMARKS

In general, removal experiments have provided substantial support for the models of density limitation in relation to habitat, presented in Chapter 3. They have shown unequivocally that territorial behaviour can limit local density. They have in addition provided insight into the influence of site quality and bird quality on occupancy, on the circumstances behind movements by existing territory owners (usually from poor to better territorities), on the ages, status and performance of replacement birds, on sexual differences in behaviour, and on the demographic consequences of increased mortality of territory owners. They have shown further that bird behaviour and population structure can influence density, in addition to any underlying control imposed by habitat and food supply.

SUMMARY

Experiments on a wide range of bird species confirmed that territorial behaviour could limit bird densities, and that non-territorial individuals were able to take territories in the same season when territory owners were removed. Some replacement birds, although younger on average than territorial birds, were sexually mature and able to breed on their newly acquired territories.

In certain species, non-territorial birds of breeding age were present in the general area despite the existence of vacant territorial sites, and replacements were observed on good territories but not on poor ones. The implications were that site quality influenced whether settlement, defence and breeding occurred, and that some individuals had more stringent site requirements than others. To some seabird species, sites in the centre of a colony were more attractive than sites on the edges.

Observations on several species showed that, as the number of potential settlers rose, an increasing proportion was excluded by territorial behaviour. The fact that the territorial densities of many bird species vary greatly from year to year, or from place to place, is not inconsistent with the limitation of density by territorial behaviour.

Grey Partridges *Perdix perdix* on territory.

Chapter 5
Density-dependence

Where habitats are not disturbed, most bird species remain relatively stable in abundance over long periods of years. Their breeding numbers may fluctuate from year to year, but between limits that are restricted compared with what their reproductive and mortality rates would allow (Lack 1954). This implies that their breeding numbers are regulated in some way, by factors that act to curb the rate of increase as these numbers rise, and to curb the rate of decrease as numbers fall. Without the operation of stabilising 'density-dependent' mechanisms, populations could increase or decrease without bounds. Following a change in environment, some populations may indeed increase or decrease for many years, an increase leading to a new and higher mean level, and a decrease to a lower level or extinction.

Density-dependence in any demographic parameter, whether births or deaths, immigration or emigration, can be brought about by competition for resources, such as food, nest-sites or territorial space, as explained in previous chapters. It can also be brought about by natural enemies if predators kill an increasing proportion of prey individuals as prey density rises, or if parasites infect a greater proportion of host individuals as host density rises (see Chapters 9 and 10 for examples). However, natural enemies do not always affect populations in a stabilising density-dependent manner. In some circumstances they can act in a manner that correlates with density but does not promote stability (see below), or in a manner that is independent of density. Other limiting factors almost always act in a density-independent manner, extreme weather and pesticide poisoning being examples. In understanding population changes, therefore, it is useful to separate the stabilising from the destabilising processes. For while only the former can act to regulate numbers, the latter can still have big effects on numbers **(Figure 1.8)**. The size and fluctuations of any population are determined by all the factors that impinge on it, whether they are influenced by density or not.

In order to detect density-dependence, we have to examine changes in numbers, birth or death rates in relation to changes in population density. This is less easy than it seems, and usually requires long series of records obtained over many years. One common problem is that density-independent factors may cause such big changes in numbers that they mask the effects of density-dependent factors. Another problem is that the resources over which birds compete can change greatly in abundance from year to year. The density at which competition sets in is then likely also to vary from year to year, so that density-dependence operates against a continually changing resource base. Hence, unless densities can be expressed as numbers per unit of relevant resource rather than as numbers per unit of area (as is usual), any density-dependence can be easily missed (Dempster 1983). There are also statistical problems in detecting density-dependence, mentioned below.

To understand the effect of any density-dependent process on a population, it is necessary to find the exact form and strength of the relationship. For example, over the range of densities observed, mortality may increase with density, but not sufficiently to prevent numbers from rising. It can then only slow the rate of increase, and is 'under-compensatory'. Alternatively, mortality may increase so much in response to a rise in density that it causes numbers to fall, and is 'over-compensatory'. The third possibility is that an increase in mortality compensates exactly for the rise in density, giving a constant number of survivors. In this situation, all the losses falling on the population at that stage exactly match the numbers that have been gained and the population remains stable in the long term. Moreover, a response that is under-compensatory at low densities may become compensatory or over-compensatory at higher densities, the overall relationship between mortality and density appearing curvilinear. The same holds for fecundity, immigration or emigration.

The two forms of regulation, by competition for resources or by natural enemies, could produce somewhat different effects (Pollard 1981). If resources were plentiful, competition may be absent at low densities and take effect only at high densities, imposing a 'ceiling' on eventual population size. Numbers cannot exceed the ceiling (except very temporarily) and regulation is brought about primarily by a relaxation of competition as numbers fall below the ceiling. In this model, if the percentage mortality (say) is plotted against density, the relationship is curvilinear, with the curve steepening rapidly towards the ceiling density. In contrast, predators and parasites could act at both low and high prey densities, giving a more linear relationship between percentage mortality and density. Instead of a ceiling density set by resources, there would then be an 'equilibrium' density, which could be well below the potential ceiling. Numbers could fluctuate both above and below the equilibrium, but in both cases would tend to return there under the action of predation or parasitism. These two regulatory processes, involving a ceiling or an equilibrium, are conceptually different, even though their effects are sometimes the same. In practice, they may also be difficult to separate if resources fluctuate from year to year to give a movable ceiling which, over a period of years, is indistinguishable from an equilibrium level.

In the rest of this chapter, information on density-dependence in bird populations is reviewed, beginning with stabilising density-dependence. Three types of evidence are considered, resulting from annual counts, life-cycle studies, and experiments in which the numbers in an area were changed artificially and the subsequent response of the population was monitored.

ANNUAL COUNTS

Fluctuating populations

For some bird species annual counts of breeding numbers in defined areas have been used retrospectively in attempts to detect density-dependence. If numbers are regulated, the argument goes, this should be apparent from the fluctuations, as the extent and direction of change from year to year should be related to the numbers themselves. In general, years of lowest population should be followed by the greatest proportional increases, whereas years of highest population should be followed by the greatest declines. If this were so, numbers would tend to return each year close to the mean level. If, instead, density-independent factors had prevailed, the extent and direction of change should have been independent of population level. An example for the Sparrowhawk *Accipiter nisus*, implying strong density-dependence, is given in **Figure 5.1**.

This simple pattern can therefore indicate whether annual counts fluctuate in a manner expected of a regulated population. Paradoxically, however, as the method depends on annual fluctuations, it is hard to apply to the most steady populations – those that are most tightly regulated. A second problem is that the pattern depicted in **Figure 5.1** could also be found in a non-regulated population that fluctuated at random. In any set of random numbers, such as simulated annual counts, extreme values would usually be followed by values nearer the mean, giving a pattern similar to that in **Figure 5.1**. A more serious statistical problem, however, arises from the lack of independence in data points. Because each change in numbers (numbers in year B/numbers in year A) is contingent on starting numbers (in year A), these two measures are not wholly independent of one another. This invalidates the use of simple statistical tests based on ordinary regression theory. Several procedures have been proposed to circumvent this problem, of which one of the most reliable at present is the randomisation procedure of Pollard *et al.* (1987).[1]

---

[1]*This method entails comparing the actual sequence of year-to-year changes with a large number of randomly generated sequences, which are based on the fluctuations observed in the actual sequence. A significant difference between the observed and predicted patterns will still not 'prove' density-dependence, but it will cast serious doubt on the underlying process being density-independent. By implication, then, it will indicate whether the annual counts conform to a pattern expected of a regulated population, as in the counts of Sparrowhawks in* **Figure 5.1**. *This method is effective even if the counts show a sustained upward or downward trend.*

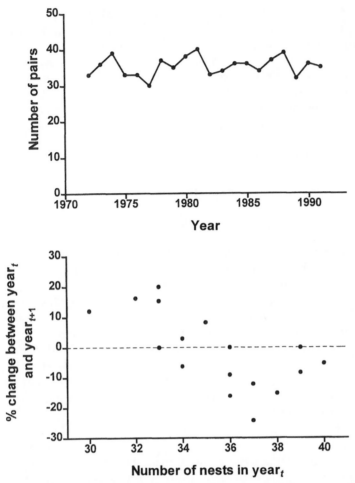

Figure 5.1   Evidence for density-dependence among annual counts of Sparrowhawk *Accipiter nisus* nests, south Scotland, 1972–91. In general, years of highest population were followed by the greatest declines, and years of lowest population were followed by the greatest increases, as expected in a regulated population. A randomisation test devised by Pollard *et al.* (1987) indicated that the pattern differed significantly from that expected of an unregulated population. Extended from Newton & Marquiss 1986.

In addition to the Sparrowhawks in **Figure 5.1**, evidence of density-dependence from annual counts was found in six out of ten duck species studied over 26 years on the Canadian prairies (Vickery & Nudds 1984), in Golden Plover *Pluvialis apricaria* studied over 24 years in Britain (Yalden & Pearce-Higgins 1997), and in various songbirds studied over periods of 22–27 years in Britain (Greenwood & Baillie 1991, Newton et al. 1998). In some of the latter, the evidence for density-dependence became stronger once the disturbing effects of winter weather or long-term trends were allowed for in the statistical analysis (Holyoak & Baillie 1996, Rothery *et al.* 1997, Newton *et al.* 1998).

Increasing populations

When a species establishes itself in a new area, following introduction or natural colonisation, its numbers usually grow rapidly for several years and then level off, roughly following a logistic pattern. The initial growth phase confirms the capacity for rapid increase, while the levelling off implies that the capacity for increase has been checked by factors that act most strongly at high density. In nature, many populations appear to follow this general pattern, but often with fluctuations, because numbers are also influenced by density-independent factors. The examples in **Figure 5.2** refer to both breeding and wintering populations.

In the absence of immigration, the rate of population growth that can be achieved by a species depends on its intrinsic rate of increase, which is higher in small fast-breeding (*r*-selected) species than in large, slow-breeding (*K*-selected) ones (Chapter

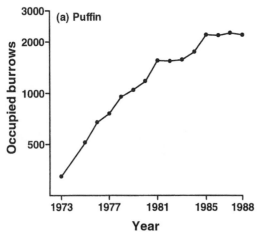

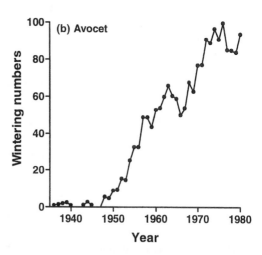

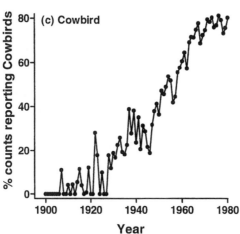

Figure 5.2 Patterns of numerical increase in various bird populations following colonisation of a new area. In all species, numbers increased for several years and then levelled off, though in two species the annual fluctuations were substantial.

(a) Puffins *Fratercula arctica* on part of the Isle of May, eastern Scotland. Redrawn from Harris & Wanless 1991.

(b) Avocets *Recurvirostra avosetta* at a wintering area in southwest England. Redrawn from Prater 1981.

(c) Brown-headed Cowbirds *Molothrus ater* in the southeastern United States. Points show the proportion of Christmas Bird Counts on which the species was recorded 1900–80. Redrawn from Brittingham & Temple 1983.

1). In practice, immigration often contributes to local increases to give much faster rates of growth. For example, for many decades the world Gannet *Morus bassanus* population grew consistently at about 3% per annum, a rate that matched the known reproductive and mortality rates. Some long-established colonies grew at rates less than this, but some newly-formed ones grew for a time at rates up to eight times higher than this through immigration (Nelson 1978). Similarly, among Kittiwakes *Rissa tridactyla*, small colonies in Britain showed faster growth rates during 1959–69 than larger ones **(Figure 5.3)**. The interpretation was that smaller colonies provided less competitive conditions, so attracted more recruits (Coulson 1983). Whatever the rates of local or global growth, however, they can eventually be expected to decline to nil, as numbers stabilise at a level influenced by prevailing conditions.

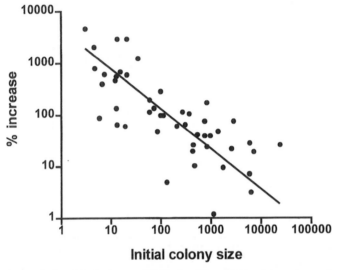

Figure 5.3   The relationship between Kittiwake *Rissa tridactyla* colony size and proportional increase between 1959 and 1969. The smallest colonies showed the greatest rates of increase, implying density-dependent growth. From Coulson 1983.

Other evidence for density-dependence in counts comes from natural events, such as hard winters, which suddenly reduce numbers well below the usual level. In most such species, numbers recover within a few years, but the pattern of recovery, in which the rate of increase slows progressively as numbers return to normal, is consistent with the operation of regulating factors that operate more strongly at higher densities. Such patterns are evident in the annual numbers of Grey Heron *Ardea cinerea* nests counted in Britain, which showed not only rapid recoveries after hard winters, but also relative stability at other times **(Figure 1.2)**. They are also apparent in Wrens *Troglodytes troglodytes* and other species recovering after hard winters or other catastrophes, as well as in species exposed to a sudden increase in food or habitat (Newton *et al.* 1997).

Not all expanding species can be expected to show a smooth transition from increase to plateau. Some might increase to an unsustainable peak, then decline to some lower level around which they fluctuate. We can expect this pattern in species that at first face a superabundant food-supply, increase to the point at which they over-eat their food, and then decline to a lower level that the food-supply can support long term. This is most likely to occur in species that limit their own food-supply, such as some geese feeding on natural marsh plants (Chapter 7). The best documented examples of overgrazing, followed by population crashes, concern deer and other herbivores (Caughley 1970).

LIFE-CYCLE STUDIES

Fluctuating populations

Analyses of annual counts can at best indicate only the general form of any density-dependent mechanisms that might be in operation. Analyses of life-table data, in which counts are made at different stages from egg to adult over a period of years, can also indicate at which stages such mechanisms operate. Any demographic measure, from clutch size to adult mortality, can be examined in relation to density, but the more stages of the life-cycle are examined, the more likely it is that all the density-dependent processes operating on a population will be detected (assuming that measurement errors are small).

The method of key-factor analysis has often been used to expose density-dependence in life table data. The method entails counting, over a period of years in the same area, the numbers present at successive stages, say from egg to adult, and then checking whether the additions or losses at any one stage in different years correlate with the numbers at the start of that stage (or with breeding numbers). The changes at particular stages are expressed as k-values (k for killing power), calculated as log N (initial population) – log S (survivors). A positive correlation between a k-value and log N could indicate a density-dependent relationship (Varley & Gradwell 1960, Dempster 1975).[2]

---

[2]*The use of k-values avoids some statistical distortions involved in the use of percentages, and unlike percentages, k-values can be added and subtracted. In an ideal data set, the way in which k varies with log N indicates the precise nature of the density-dependence. A slope (b-value) of zero suggests density independence, whereas slopes of less than 1, exactly 1 or more than 1 suggest, respectively, situations of under-compensating, compensating and over-compensating density-dependence. Sometimes the relationship between k and N is curvilinear, so that at low densities mortality is density-independent and then, as densities rise, mortality becomes progressively under-compensating, compensating and then over-compensating.*

Exact compensation (b=1) could reflect pure contest competition, because there is a constant number of winners (survivors), regardless of the number of contenders (Dempster 1975). After perturbations the population would then return to its equilibrium level within one year. At the other extreme is pure scramble competition, which is the most intense form of over-compensating density-dependence, in which all competing individuals are so adversely affected that few survive. This would be indicated by a b value close to infinity (a vertical line). The most common situation in birds is that competition is contest-like, with varying degrees of under-compensation (that is, b values between 0 and 1). Plotting k against log N for different stages of the life-cycle is thus an informative way of depicting the effects of density-dependent factors on numbers at different stages. Variations in the slope of the curve (b) give some idea of the way in which losses change with density.

The main statistical problem with the key-factor procedure is the lack of independence in the variables used in correlation analysis, as the losses at any one stage are partly dependent on the numbers at the start of that stage, as mentioned above. Various ways of overcoming this problem have been proposed, but they tend to be conservative, providing statistical evidence for only the most obvious of the linear density-dependent relationships in operation. They are much less suitable for determining non-linear relationships.

The same method can of course be applied to fecundity, where k is measured as failure to achieve the maximum possible reproductive rate. The so-called 'key-factor' in such analyses is whatever form of loss (k-value) accounts for most of the variation in total annual loss (K, calculated as the sum of the individual k-values from successive stages of the life cycle). The key factor is thus defined as whatever k-value over a period of years shows the strongest relationship with K. The more forms of mortality that are included in the measured k, the greater will be the likelihood of its being correlated with K (total loss). This is simply a result of including a large slice of K in the one k. Like any other k-value, the key factor may or may not be density-dependent.

Of the 22 key-factor studies of birds listed in **Table 5.1**, all revealed apparent density-dependence in one or more demographic measures. In some species, periods of density-dependent loss occurred in both breeding and non-breeding seasons, while in others they were concentrated in one or other season. Overwinter loss, which emerged as density-dependent in several species, was a 'catch-all' category, calculated as the difference between numbers at the end of one breeding season and the start of the next. Not only did it cover much of the year, but it included immediate post-fledging losses, autumn and winter losses (whether mortality or emigration), as well as the failure of living birds to recruit to the counted breeding population. In few studies was it possible to pinpoint which of these types of loss provided the density-dependent component (see **Table 5.1**).

In some species, the density-dependent factor was shown to act in a compensatory manner, offsetting the effects of earlier events. Thus, in the Tawny Owl *Strix aluco*, seasons of good breeding were followed by high overwinter loss, while seasons of poor breeding were followed by low overwinter loss, while in the Red Grouse *Lagopus l. scoticus*, seasons of high shooting mortality were followed by low natural mortality, and vice versa **(Figures 5.4 and 5.5)**. These effects were thus strongly regulatory, acting swiftly within years to reduce the extent of both increases and decreases in breeding numbers. The same was earlier found for Grey Partridges *Perdix perdix* (Jenkins 1961).

The timing of the major density-dependent events in the annual cycle can influence our perception of the degree of stability in a population. For convenience, the degree of stability is usually judged from the year-to-year fluctuations in breeding numbers, and if counts of breeders usually follow a period of strong density-dependent adjustment, the counts at that time would vary less than they might otherwise do. In contrast, if the counts of breeders followed a period of variable density-independent loss, they could show great year-to-year fluctuation. For example, in the Avocet *Recurvirostra avosetta*, the population at two sites in southern England increased and then levelled off. In the level phase, overwinter loss was almost perfectly compensatory, the numbers lost over winter matching the numbers previously gained from breeding (Hill 1988). In this species overwinter loss was clearly responsible for stabilising the breeding population. On the other hand, in the Woodpigeon *Columba palumbus* egg predation was strongly density-dependent, but was incapable of stabilising the breeding population, because it was quickly offset by replacement laying and followed by a large density-independent winter mortality (Murton & Westwood 1977). In general, populations that are strongly regulated after a major period of density-independent loss will show greater stability in breeding numbers than those in which the reverse occurs. Species that are strongly territorial while breeding tend to fall in the former category.

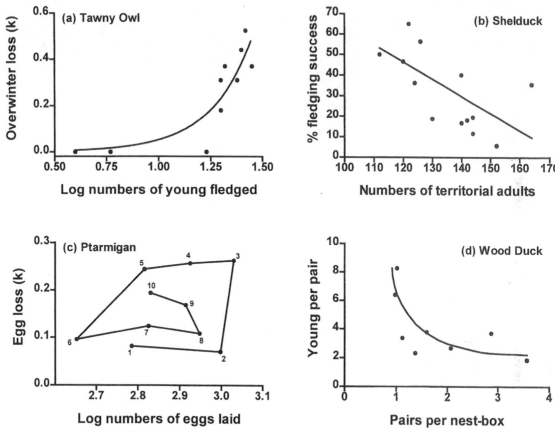

Figure 5.4   Some examples of density-dependent relationships in birds. In each example, each point refers to the value from a particular year.

(a) Relationship between overwinter loss and fledgling production in the Tawny Owl *Strix aluco*, expressed as *k*-values. Loss becomes density-dependent only above a certain density, probably through competition for territories. From Dempster 1975, based on Southern 1970.

(b) Relationship between duckling survival and territorial density in the Shelduck *Tadorna tadorna*. The linear pattern indicates that losses (mainly through predation) are density-dependent over all densities. From Patterson *et al.* 1983.

(c) Relationship between egg loss (through predation by Stoats *Mustela erminea*) and breeding density in the Rock Ptarmigan *Lagopus mutus*, showing the typical circular pattern of a delayed density-dependent response, formed by joining the *k*-values from successive years. From Dempster 1975, based on Weeden & Theberge 1972.

(d) Relationship between density and breeding success among Wood Ducks *Aix sponsa* competing for nest-sites. Failures were due to females fighting and laying in the same nests, giving rise to desertions and reduced hatching success. From Jones & Leopold 1967.

In addition to the key-factor analyses, other studies in **Table 5.1** detected density-dependence at one particular stage in the life-cycle that happened to have been examined. Such studies were useful in adding to the list of known regulatory mechanisms, but they did not necessarily reveal the full story for any of the species concerned. The example from Wood Duck *Aix sponsa* in **Figure 5.4**, in which egg

Table 5.1 Direct density-dependence in bird populations as revealed by studies that have examined production or losses at different stages of the annual cycle over a period of years. Studies using the method of key-factor analysis are marked with an asterisk. + = density-dependent: − = density-independent; P = predation or parasitism, S = shooting; C = competition for food or space, associated with territorial or other aggressive interactions; D = dispersal; I = interference by other birds; K = the key factor, accounting for most of the annual variation in losses.

Species	Area (Years)	DEMOGRAPHIC VARIABLES										Source
		Failure to breed	Clutch-size	Nest-failure	Chick Mortality	Young raised per female	Autumn loss	Winter loss	Autumn and winter loss	Adult survival	Recruitment to territorial population	
Great Tit *Parus major*[1]	Finland (16)									+(C)	+(C)	Orell 1989
	Netherlands (39)					+[2]		+				Kluijver 1951
	Belgium (10)		+	+(C)		+[2]						Dhondt et al. 1992
	England (17)	−	+		−	+						Lack 1966
	England (21)		+	+(P)					−	−		Krebs 1970*, Dunn 1977
	England (20)	−					−(K)	+	+			McCleery & Perrins 1985*
Blue Tit *Parus caeruleus*	Belgium (10)		+							+		Dhondt et al. 1992
Coal Tit *Parus ater*	England (8)		+									Lack 1966
Willow Tit *Parus montanus*	Sweden (6)	−	+		−		+[3]	+	+(D)	+(K)[4]		Ekman et al. 1981 & Ekman 1984*
Goldcrest *Regulus regulus*	Sweden (8)										+(C)	Nilsson 1986
Nuthatch *Sitta europaea*	Sweden (12)			+(I)		+					+(C)	Nilsson 1987
Wren *Troglodytes troglodytes*	England (20)								+			Peach et al. 1995b
Song Sparrow *Melodia melospiza*	British Columbia (5)	−	+	+(P)		+					+(C)	Arcese et al. 1992
Tree Sparrow *Passer montanus*	Various sites and periods			+						−		Summers-Smith 1996
Pied Flycatcher *Ficedula hypoleuca*	England (17)		+	+					+(K)			Stenning et al. 1988*
	Finland (7)		+	−	−	+[5]						Tompa 1967
	Finland (13)		−	−	−	+			−			Virolainen 1984*
	Finland (16)		−	−	−				+(K)			Järvinen 1987*
Sedge Warbler *Acrocephalus schoenobaenus*	England (25)		−	−	−	−			+(K)			Baillie & Peach 1992*
Blackcap *Sylvia atricapilla*	England (25)		−	−	−				+(K)			Baillie & Peach 1992*
Whitethroat *Sylvia communis*	England (20)		−	+	−				+(K)			Baillie & Peach 1992*
Willow Warbler *Phylloscopus trochilus*	England (25)		−	−	+				+(K)			Baillie & Peach 1992*
Redstart *Phoenicurus phoenicurus*	Finland (14)		−	−	−				+(K)			Järvinen 1987*
Song Thrush *Turdus philomelos*	Britain (21)								+			Baillie 1990
Fieldfare *Turdus pilaris*	Norway (15)			+(P)								Hogstad 1995
Barn Swallow *Hirundo rustica*	Denmark (19)		+	−	+[3]				+(K)			Møller 1989a*, Baillie & Peach 1992
Ground Finch *Geospiza fortis*	Galapagos (8)									+(C)		Gibbs & Grant 1987a
Cactus Finch *Geospiza scandens*	Galapagos (8)									+(C)		Gibbs & Grant 1987a
Woodpigeon *Columba palumbus*	England (12)		−	+(P)			−	−				Murton & Westwood 1977*, Dempster 1975*
Coot *Fulica atra*	Netherlands (21)									+(C)[7]		Cavé & Visser 1985

Species	Location	Parameters (density-dependence)	Reference
Tawny Owl *Strix aluco*	England (11)	−(K), −, +, +(C)	Southern 1970*, Smith 1973*
Little Owl *Athene noctua*	Germany (11)	−, −, +(C)	Exo 1992*
Partridge *Perdix perdix*	England (11)	−, +(K)[6], −, +(S,C)	Blank et al. 1967*
	England (various sites & periods)	+(P), +(P), +(S, C, D)	Potts 1986*
	Poland (7)	+	Panek 1997
Bobwhite Quail *Colinus virginianus*	Wisconsin (15)	+(D), +(D)	Lack 1954 from Errington 1946
Spruce Grouse *Dendragapus canadensis*	Illinois (24)		Roseberry 1979
	Alberta (10)	+, +, +	Boag et al. 1979
Blue Grouse *Dendragapus obscurus*	British Columbia (9)	+	Zwickel 1980
Willow Grouse *Lagopus lagopus*	Norway (10)	+(P), +(P)	Myrberget 1972
Red Grouse *Lagopus l. scoticus*	Ireland (4)	+(P), +(D), +(D)	Watson & O'Hare 1979
Black Grouse *Tetrao tetrix*	Scotland (18)	+(D), +(D)	Watson et al. 1984
Hazel Grouse *Bonasa bonasia*	France (11)	+, +	Ellison 1991
Capercaillie *Tetrao urogallus*	France (12)	+, +	Ellison 1991
	Scotland (9)	+(D), +(D)	Moss & Oswald 1985
Sparrowhawk *Accipiter nisus*	Scotland (15)	−, −, −, +(C), +(C)	Newton 1988a*, 1992
Mallard *Anas platyrhynchos*	England (16)	+(P), +(C), +(P), +(S), +	Hill 1984*
	North America (26)	+, +	Kaminski & Gluesing 1987
Shoveller *Anas clypeata*	Latvia (19)	+(S), +(S)	Mihelsons et al. 1985
Tufted Duck *Aythya fuligula*	Latvia (19)		Mihelsons et al. 1985
Blue-winged Teal *Anas discors*	Iowa (13)	+(P), +(P)	Weller 1979
Wood Duck *Aix sponsa*	California (8)	+(C), +(C)	Jones & Leopold 1967
	New York (7)	+(C)	Haramis & Thompson 1985
Shelduck *Tadorna tadorna*	Scotland (17)	+(I), +(I), +(I), −, −	Patterson et al. 1983
Barnacle Goose *Branta leucopsis*	Svalbard (13)	+(C), +(C), +(C)	Owen et al. 1986
			Owen & Black 1991
Avocet *Recurvirostra avosetta*	England (40)	−(K), +, +, +[8]	Hill 1988*
Oystercatcher *Haematopus ostralegus*	Wales & England (20)	−, +, +	Goss-Custard et al. 1995a

[1] Another parameter found in Great Tit to decline with increase in density was egg weight (Perrins & McCleery 1994)

[2] Proportion of pairs that raised second broods also decreased with increasing density

[3] Inverse density-dependence

[4] Summer survival both adults and yearlings; winter survival yearlings only density-dependent

[5] Based on re-analysis by Alatalo & Lundberg (1984)

[6] In areas where predators were controlled

[7] Delayed density-dependence

[8] Strictly the loss between post-breeding numbers one year and breeding numbers two years later, because most Avocets do not return to breeding areas until they are nearly 2 years old

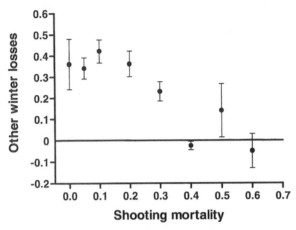

Figure 5.5   Relationship between losses through shooting and subsequent natural losses in Red Grouse *Lagopus l. scoticus*, expressed as *k*-values. Note that in some instances winter losses appear negative, the result of immigration of individuals into the study area after shooting stopped. From Hudson 1992.

mortality increased with breeding numbers, was unusual among bird studies in that density was calculated as numbers per unit of resource, in this case nest-boxes (Jones & Leopold 1967). In this example, the relationship between density and loss was curvilinear, implying that competition became progressively more marked as density rose. Otherwise, the main factors identified in the above studies as acting in a stabilising density-dependent manner included predation on eggs or chicks, and competition for food or territories.

In species in which breeding density remains fairly stable from year to year and is limited by territorial behaviour, the number of new breeders recruited in any one year is largely dependent on the number of vacancies created by the loss of previous breeders. This is illustrated in **Figure 5.6** for a Sparrowhawk population in which breeding numbers fluctuated by no more than 15% of the mean over a 15-year period (Newton 1991). One consequence of limited recruitment is that surplus birds die or emigrate or accumulate as a non-territorial sector, which in many species occupy areas not used for breeding (Chapter 4). Such a system of regulated recruitment might be expected in any territorial species, and has been demonstrated in various songbirds, grouse and others (Watson & Moss 1970, Boag *et al.* 1979, Zwickel 1980, Nilsson 1987). It would also be expected in species limited by shortage of nest-sites (Chapter 8). In most of the life-cycle studies discussed above, only the breeders were counted, and failure to recruit was included as a component of 'overwinter loss', so could not be examined separately.

Territorial behaviour, coupled with spatial variation in habitat, can underlie some other density-dependent responses. In years of low density, birds may occupy mainly good patches of habitat, but as their numbers grow, some may spread increasingly into poor patches where they survive or breed less well. This could cause the mean survival or breeding performance of the population to decline as numbers rise, even though the birds in good places continue to do well (Chapter 3). This habitat effect accounted for all the density-dependent reduction in mean clutch size in one study of

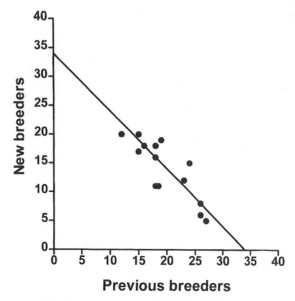

Figure 5.6 Density-dependent recruitment to a breeding population of the European Sparrowhawk *Accipiter nisus*, south Scotland, 1975–89. The breeding population remained fairly stable during this period, and the numbers of new breeders recruited each year (*y*) were inversely related to the numbers of established breeders remaining from the previous year (*x*). (Regression relationship: *y* = 31.7 – 0.90*x*, *r* = 0.69, *N* = 14, P<0.003.) Population stability was itself a consequence of a fairly stable territorial system, in which the landscape was apparently occupied to a similar level each year. From Newton 1991.

Blue Tits *Parus caeruleus* (Dhondt *et al.* 1992), and for part of a decline in the breeding success of Nuthatches *Sitta europaea* at high density (Nilsson 1987). Other examples, involving nest-sites and territories of variable quality, are given in Chapter 3.

   In some species, a build-up of non-territorial non-breeders can also have adverse effects on the performance of breeders. As the numbers of Golden Eagles *Aquila chrysaetos* in the Swiss Alps increased over several decades (in response to reduced human persecution), intrusions by single adults into the territories of breeding pairs led to fights which became a major cause of adult mortality and of breeding failure (Haller 1996). Intrusions increased the frequency of non-breeding and of hatching failure, as females spent more time off the nest during incubation. Average breeding success therefore declined as numbers recovered, and breeding performance was poorest in those pairs that experienced the highest rates of intrusion **(Figure 5.7)**.

Increasing populations

For some bird species that were seen to increase to a plateau level, attempts were made to identify the demographic variables involved in slowing population growth, whether changes in reproduction or mortality, immigration or emigration. Among wild geese that winter in western Europe, several species have increased greatly during the last 40–50 years, apparently mainly in response to reduced shooting (Ebbinge 1985, Owen *et al.* 1986, Madsen 1991). As their numbers grew, certain

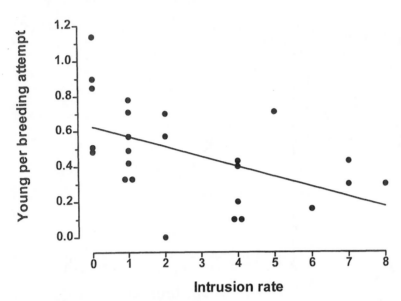

Figure 5.7 Mean number of young raised per year in individual Golden Eagle *Aquila chrysaetos* territories during 1985–94 in relation to the number of intruding single eagles seen in those territories. Pairs on territories subject to greatest intrusion produced fewest young. From Haller 1996.

populations showed reductions in the proportion of young present in wintering flocks, implying a density-dependent reduction in breeding rate. This occurred partly because fewer birds of breeding age produced young, and partly because broods were smaller. The pattern was most apparent among the Svalbard Barnacle Geese *Branta leucopsis* and Icelandic Greylag Geese *Anser anser* that winter in Britain **(Figure 5.8)**. In the Barnacles, natural adult mortality rates also increased over the

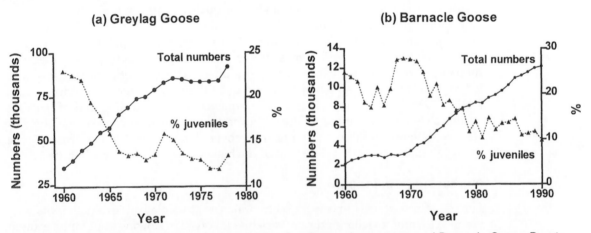

Figure 5.8 Reduced production of young by Greylag Geese *Anser anser* and Barnacle Geese *Branta leucopsis*, as their numbers have grown. The populations concerned breed in Iceland (Greylag) and Svalbard (Barnacle) and winter in Britain, where the counts were made. The graphs show 5-year moving mean values. Adapted from Owen *et al.* 1986.

years from 2% per year at a population level of 5000 birds to 8–10% per year at 10 000 birds. This increase was attributed to competition for food on breeding areas resulting in insufficient fat deposition for autumn migration, so that increasing proportions of birds died on the journey (Owen & Black 1991). Similarly, as a colony of Snow Geese *Anser caerulescens* in the Canadian Arctic grew from 2000 to 8000 breeding pairs over a 15-year period, various demographic measures declined, including mean clutch size (from 4.2 to 3.5), mean brood size at hatch (from 0.9–1.0 to 0.6–0.8), gosling growth and survival to 6 weeks (65% to 30–40%), and first-year survival (from 60% to 30%) (Cooch *et al.* 1989, Francis *et al.* 1992, Williams *et al.* 1993). All these changes were attributed to competition on the breeding grounds, associated with decline in the biomass of salt marsh food-plants from 50 g per m^2 to 25 g per m^2, resulting from increased grazing pressure. Other expanding goose populations have shown no such demographic changes, so perhaps they have not yet reached a level at which competition is important (Owen *et al.* 1986, Fox *et al.* 1989, Summers & Underhill 1991).

Gradual changes in demographic variables were also noted in some birds-of-prey, whose numbers were reduced by organochlorine pesticide poisoning in the 1950s, and then recovered after organochlorine use was stopped. Compared with a stable population of Sparrowhawks, a recovering population showed a similar reproductive rate, but higher adult survival, a higher immigration to emigration ratio and a lower mean age of first-breeding (Wyllie & Newton 1991). Moreover, as the recovering population grew, all these demographic measures converged progressively to match those in the stable population. The fact that it was mainly breeding performance that changed in the geese, and survival and movements in the Sparrowhawk, could be because competition occurred mainly during the breeding season in the geese, but outside the breeding season in the raptor. Similar changes in demographic variables were also noted in various seabird species during colony growth or decline (Coulson & Wooller 1976, Coulson *et al.* 1982, de Wit & Spaans 1984, Harris & Wanless 1991).

While in theory any demographic variable might change as a population grows, we cannot conclude for certain that it is changing in response to density as such. It might change over the years for reasons unconnected with density, but still act to curb the rate of population growth. Another deficiency in some such studies is that, although declines in reproduction coincided with increase in numbers, no account was taken of the area occupied, which may have expanded as numbers grew. This gave uncertainty on how density (as opposed to numbers) changed over the period concerned. As with any observational data, therefore, such studies give suggestive, but not conclusive, evidence for density-dependence in particular demographic measures.

Conclusions from studies on life-cycles

Despite the problems, important points have emerged from published studies:

(1) almost any aspect of avian demography can vary in a density-dependent manner;

(2) although bird populations can persist only by virtue of density-dependent factors, other factors uninfluenced by density can have marked effects on numbers;

(3) even if a given factor operates in a directly density-dependent manner, it may not stabilise numbers if its effect is small or counteracted by other (density-independent) factors;

(4) a density-dependent factor which in itself would not stabilise a population may do so in conjunction with other density-dependent factors;

(5) whether a given factor appears as density-dependent in an analysis depends on whether it is expressed in relation to a segment of the population or in relation to the total population.

Like annual count studies, life-cycle studies can at best only indicate likely density-dependent relationships. It is always possible that, over the study period, density-independent losses by chance conformed to a pattern expected of density-dependent losses. The risk of such chance effects is of course lessened the longer the study can be continued, but the evidence for density-dependence can never be as strong from observational studies as from experiments in which densities are deliberately altered, and the ensuing demographic response of the population monitored.

EXPERIMENTAL MANIPULATIONS OF DENSITY

Adjustments of densities

Experimental adjustments of breeding density provide an additional approach to the study of density-dependence. For birds that use nest-boxes, the density of breeding pairs can be readily manipulated. At least three such experiments have been conducted with Pied Flycatchers *Ficedula hypoleuca* in northern Europe (Tompa 1967, Alatalo & Lundberg 1984, Virolainen 1984) and one with Collared Flycatchers *Ficedula albicollis* in central Europe (Török & Tóth 1988). One study reported evidence of reduced clutch sizes at high densities (Virolainen 1984), and all four reported reduced fledging success (Tompa reported no such effect, but his conclusion was reversed by a re-analysis by Alatalo & Lundberg 1984). In no study did differences in laying dates occur between areas, or in the age-structure or frequency of polygyny among breeders. The poorer performance at high density was therefore attributed to enhanced competition for food (Alatalo & Lundberg 1984). Both these flycatchers are migrants that winter in Africa, where additional regulating processes might operate (Stenning *et al.* 1988).

Some of the experiments on flycatchers compared breeding in only two plots of different density. It was uncertain, therefore, whether demography differed because of density or because of some other site-related factor. Ideally, experiments of this type should involve more replication, with several plots at each density in similar habitat, together with changes of densities within plots as well as between them. Only in this way can density effects be separated from possible plot effects on breeding.

Reductions of densities

Density-dependence is also revealed by another kind of experiment. As every pest controller knows, when birds are removed from an area, newcomers soon begin to appear, and in a short time numbers may return to normal. It is not the fact that birds move into vacant habitat that is surprising, or that recolonisation can sometimes occur with a matter of days or weeks. What is interesting is that numbers often return to about the same level as before. This in itself provides a compelling indication that densities are regulated in relation to some feature of the local habitat. There are many instances among birds-of-prey, which have long been systematically persecuted in parts of Europe. For example, Rowan (1921–22) reported how four pairs of Merlins *Falco columbarius* were shot from the same area of moorland each year for more than 20 years. During this time no adult survived there and no young was reared, yet every year four new pairs were back on the same sites.

Some researchers have conducted large-scale removals on an experimental basis, checking the extent of re-occupation the same year or the following year. The speed with which an area is re-occupied depends on the availability of potential recruits in the vicinity. Sometimes full re-occupation may occur within days or weeks, but at other times only after a breeding season, when young birds become available to fill the gaps. Removal experiments with same-year replacements were discussed in Chapter 4, and we can now turn to large-scale removals in which numbers were re-checked the following year.

Some pioneering work on songbird populations in Maine was especially revealing (Hensley & Cope 1951). In one year the initial population in a 16-ha forest tract was estimated at 148 territorial males of various species, all of which were shot, together with a similar number of same-year replacements (Chapter 4). In the next spring the size and composition of the restored population was almost identical, with each species present at similar numbers and locations as the year before, to give a total of 154 territorial males (see **Table 5.2** for the ten most abundant species). The populations in two control plots were also stated as similar between years, but no figures were given. These findings provided strong circumstantial evidence that songbird densities were limited at about the same level in successive years in relation to local habitat.

Table 5.2   Comparisons of bird populations in a 16-ha tract of spruce-fir forest in successive years. In the first-year all the birds present in spring were shot, followed by their same-year replacements. The next year the same area had been re-occupied to a similar overall level as before by the same species. Only the data for the ten most numerous species are given below. From Stewart & Aldrich (1951) and Hensley & Cope (1951).

Species	Numbers of territorial males in	
	1949	1950
Bay-breasted Warbler *Dendroica castanea*	35	34
Magnolia Warbler *Dendroica magnolia*	24	19
Cape May Warbler *Dendroica tigrina*	12	12
Blackburnian Warbler *Dendroica fusca*	9	13
Myrtle Warbler *Dendroica coronata*	9	10
Olive-backed Thrush *Hylocichla ustulata*	8	14
Slate-coloured Junco *Junco hyemalis*	5	8
Golden-crowned Kinglet *Regulus satrapa*	4	9
Nashville Warbler *Vermivora ruficapilla*	4	5
Yellow-bellied Flycatcher *Empidonax flaviventris*	4	4
All species together*	148	154

*Including scarcer ones not listed

Later work on other species gave similar results. In British Columbia, all male and female Blue Grouse *Dendragapus obscurus* were shot from the same area in four successive years (Bendell & Elliott 1967). Each year the area was re-occupied by both sexes to about the same level as previously. Other experiments on this species involved populations in two areas showing different densities and different long-term trends. One population was dense and increasing, whereas the other was sparse and decreasing, but both gave similar findings (Bendell *et al.* 1972, Zwickel 1972). Each study area was divided into an experimental plot, in which virtually all birds were shot, and a control plot in which birds were merely counted. The following spring, the experimental plots in both areas had been re-occupied to a roughly similar level as in the previous year. The main difference was that the new occupants contained a larger proportion of yearlings than the previous ones **(Table 5.3)**. On the control plots numbers were also similar between years, but the proportion of yearlings showed no great increase in the second year. It was concluded, therefore, that the new settlers in the experimental plots were immigrants, and that many of them would probably have died if vacancies had not been created for them by the removals. Recruitment to the breeding population was therefore density-dependent with respect to already established birds (see also Zwickel 1980).

Table 5.3 Findings from two large-scale removal experiments on Blue Grouse *Dendragapus obscurus* in British Columbia. In each of two study areas, virtually all birds present were shot in one breeding season from a large experimental plot, while in a nearby control plot birds were merely counted. The following breeding season birds were counted on both experimental and control plots. Figures show the numbers of birds present or removed (and the percentage of yearlings). Experimental plots were re-occupied to about the same level as in the previous year, but by a greater proportion of yearlings (see text). From Bendell *et al.* (1972) and Zwickel (1972).

| | Study area A Population dense, increasing in the long term | | | | Study area B Population sparse, decreasing in the long term | | | |
| | Experimental plot | | Control plot | | Experimental plot | | Control plot | |
	Males	Females	Males	Females	Males	Females	Males	Females
1970	110(35)	110(34)	107(30)	123(38)	64(34)	43(30)	48(23)	46(36)
Removed	102	102	0	0	64	43	0	0
1971	100(79)	120(75)	124(30)	129(51)	46(94)	47(92)	31(37)	45(56)

The experiments conducted so far have shown only that numbers can be restored to the same level in successive years or have changed in a manner similar to those in a control area. They still fall short of the ideal, however, because the number of potential settlers was not known. For a convincing demonstration of regulation, rather than mere limitation, it must be shown that the population settles to a similar density each year, despite wide variations in the numbers of potential settlers. For most birds there seems little chance of getting this type of information, especially as settlers could be drawn from large and unknown areas. Where removals were made repeatedly over a number of years, however, as in Blue Grouse (Zwickel 1980), it is unlikely that the numbers of contenders were the same each year. Despite the problems, experiments provide a useful way forward in the study of density-dependence, because they circumvent some of the statistical problems of analysis, and can be done in a shorter period than is required for time-series or life-table studies.

Demographic responses to reduced density

Recovery of breeding numbers after a cull could result from increased reproduction or survival, or in the ratio of immigration to emigration, or from a lowering in the age of first breeding. In one species or another, changes in response to a culling programme have been noted in all these demographic parameters **(Table 5.4)**. In the Blue Grouse discussed above, it was the recruitment of yearlings that increased, probably in association with improved first-year survival.

Table 5.4   Demographic changes observed following culling programmes or other catastrophic population declines, + = response; − = no response. No symbol indicates lack of measurement. All possible demographic responses to reduced density were recorded in one species or another.

	Increased reproduction (young per pair per year)	Improved survival of juveniles	Improved survival of adults	Increased immigration/ emigration ratio	Reduced age of first-breeding	Source
Great Tit *Parus major*		+	+			Kluijver 1966 Tinbergen *et al.* 1985 Verhulst 1992
Willow Tit *Parus montanus*		+	+	+		Ekman *et al.* 1981
Feral Pigeon *Columba livia*	+		+	+		Kautz & Malecki 1990
Sparrowhawk *Accipiter nisus*	−		+	+	+	Wyllie & Newton 1991
Herring Gull *Larus argentatus*	+			+	+	Coulson *et al.* 1982
Blue Grouse *Dendragapus obscurus*		+		+	+	Bendell *et al.* 1972, Zwickel 1972

During the normal year-to-year fluctuations in a Great Tit *Parus major* breeding population, clutch size and nest success varied in a density-dependent manner **(Table 5.1)**. In years of low population, this resulted in a higher production of young per pair. Moreover, after an artificial reduction in density, the annual survival of full-grown birds also increased. On the Dutch island of Vleiland, Great Tit pairs produced an average of 11 young per year, only about 27% of adults survived from one year to the next, and only about 6% of young were later found breeding there. For four years, Kluijver (1966) removed 60% of the young, so that each pair raised an average of four young per year. In these years, the annual survival of adults rose to about 56% and as many as 22% of the remaining young subsequently bred locally. This was sufficient to compensate for the extra loss of young, and breeding numbers continued to fluctuate around the same level as before. This experiment thus gave a clear indication that the survival or emigration rates of both adults and juveniles were density-dependent.

In the Netherlands, Great Tits often raise two broods per year. In two successive years Kluijver (1966) removed 90% of first-brood young just before they could leave the nest. In the following years, 16 and 18% of the second-brood young were found breeding in the study area, compared with an average of 5% in other years. Evidently, the numbers of first-brood young influenced the numbers of second-brood young that stayed and bred in the area. The removal also increased the proportion of remaining first-brood young that subsequently bred in the area (Verhulst 1992). Other experiments that revealed improved fecundity, survival or immigration in response to reductions in density involved Willow Tits *Parus montanus* (Ekman *et al.* 1981), feral pigeons *Columba livia* (Kautz & Malecki 1990) and Herring Gulls *Larus argentatus* (Coulson *et al.* 1982).

OTHER DENSITY-DEPENDENT RESPONSES

So far in this chapter the emphasis has been on those processes that operate in a direct density-dependent manner and act to stabilise densities. Other types of density-dependent processes act in a way that destabilises densities, of which the following have been identified in some bird populations.

Inverse density-dependence

So-called 'inverse density-dependence' occurs when the per capita mortality decreases, rather than increases with population size (or when per capita reproduction increases with population size). This situation occurs, for example, where predators remove constant numbers of prey each year from a varying population. They would thus remove a smaller, rather than a larger, proportion at high prey densities. An example is given in Chapter 9. Such predation thereby accentuates fluctuations in prey numbers, and does not promote stability.

Delayed density-dependence

The term delayed density-dependence is used when the additions or losses in a population correlate, not with present numbers, but with previous numbers in an earlier year or generation. Suppose that the numbers of a bird species first rise steadily for several years, and then fall steadily for several years, and suppose that these changes are due to changes in the death rate. Then throughout each phase of increase the death rate is low, both at the start when numbers are low and later when they are high. Likewise, during each phase of decrease the death rate is high, both at the start when numbers are high and later when they are low. Hence, if the mortality at each stage is plotted against the ensuing initial numbers at that stage, no relationship between the two is at first apparent, since low death rates occur with both low and high numbers, as do high death rates. However, if the points are joined for successive years, a regular pattern is found in which the points, running anticlockwise, form a rough circle **(Figure 5.4c)**.

This type of pattern is characteristic of predator–prey (or parasite–host) oscillations, in which the numbers of the prey (or hosts) rise steadily for several generations and then fall, while those of the predator (or parasite) do the same but lagging behind, a situation first explored mathematically by Lotka (1925) and Volterra (1926). When the prey is abundant the predator increases, which eventually leads the prey to decrease, which in turn leads the predator to decrease, after which the prey can increase again and the cycle is repeated. Under these circumstances, the death rate of the prey does not vary directly with its own numbers at the time, but depends on the relative numbers of the predator. The latter have been determined by the numbers of the prey in the past, hence the term 'delayed density-dependent'

(Varley 1947). Such relationships involve a time lag of at least one year or generation, but time lags as such do not necessarily always produce cycles (Royama 1992). Annual count data are usually tested statistically for evidence of cycles using autocorrelation techniques (Brown & Rothery 1993).

Some rodent and grouse species show fairly constant densities in the south of their range, but increasingly marked and regular fluctuations towards the north, with gaps of several years between the peaks (Chapters 7 and 9). This geographical change is associated with a switch from predominantly direct density-dependence in the south to delayed density-dependence in the north, possibly associated with a change in the type of predation (Hanski *et al.* 1991, Chapter 9). Most examples of delayed density-dependence in birds come from cyclic grouse populations; they include predation on Hazel Grouse *Bonasa bonasia* by Goshawks *Accipiter gentilis* (Lindén & Wikman 1983), and parasitism of Red Grouse *Lagopus l. scoticus* by strongyle worms (Chapter 10).

Again, however, when a population fluctuation involves mortality or reproductive changes of the type shown in **Figure 5.4c**, these changes are not necessarily due to a predator or parasite, or to any other factor acting with a time lag. They could result from variable density-independent losses which by chance follow a cyclic pattern. This may be why such patterns have sometimes appeared inexplicably in non-cyclic species, such as Coot *Fulica atra* (in adult mortality) (Cavé & Visser 1985).

SPATIAL DENSITY-DEPENDENCE

Because biologists have been concerned primarily with explaining year-to-year fluctuations in numbers, most attempts to detect density-dependence have involved the analysis of year-to-year demographic changes within limited areas. But we should also expect that density-dependent processes would operate in a spatial context, with high reproduction, low mortality and net immigration in areas where numbers are low, and the opposite conditions in areas where numbers are high. Such processes would tend each year to even out densities between areas in which other conditions were similar. The fact that many birds can make local movements at almost any time of year makes for a swift response, and could lead to reduced fluctuations in local numbers and in the overall wider population.

Some of the same problems that beset the measurement of density-dependence over a series of years can also apply in comparisons between a series of areas. In particular, much of the demographic variation observed between areas can result, not from variation in density *per se*, but from variation in resources or some other aspect of habitat. Some areas might hold more food than others, and for this reason alone might have higher densities of birds, showing higher reproductive and survival rates. So again, unless densities can be expressed in relation to resources rather than in relation to area of habitat, there is little hope of examining spatial density-dependence effectively, when it results from competition. As bird populations rise, it is common for increasing proportions of individuals to be pushed by competition into poorer habitat, where survival and breeding success are lower (Chapter 3). In this way some of the spatial density-dependence in populations is likely to be translated into the temporal density-dependence necessary to regulate populations over time.

Despite the problems, some clear examples of spatial density-dependence have emerged. Two examples, showing overwinter loss in different areas, are shown in **Figure 5.9**. In both studies, those areas that started with the highest densities of birds in September or October lost the greatest proportion over winter. The role of movements was shown by the fact that areas that started with the lowest densities gained birds over winter, rather than losing them.

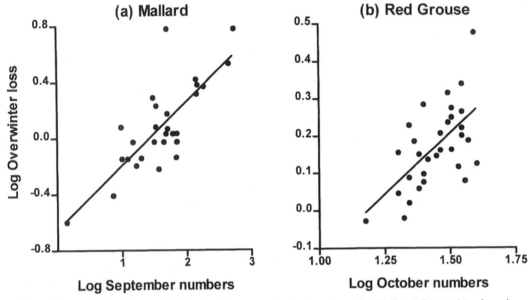

Figure 5.9   Spatial density-dependence in overwinter losses of (a) Mallard *Anas platyrhynchos* and (b) Red Grouse *Lagopus l. scoticus* in different areas. The greater the density in September or October, the greater the loss by the following March. Mallard numbers on particular waters and Red Grouse numbers on particular 0.5 km² areas of heather moor. Negative values indicate immigration. Significance of relationships: Mallard, $b = 0.46$, $r^2 = 0.62$, $N = 28$, $P<0.001$, Red Grouse, $b = 0.66$, $r^2 = 0.38$, $N = 31$, $P<0.001$. Calculated from Hill 1983 and Redpath & Thirgood 1997.

Another manifestation of spatial density-dependence is the greater predation sometimes seen on nests at high densities **(Figure 5.10a)**. This has been found by observation on the natural nests of songbirds, ducks and others (Chapter 9; Krebs 1971, Fretwell 1972, Fleskes & Klaas 1991). It has also been shown by experiments which involved laying out artificial nests at different densities in similar cover, and finding what proportion had their eggs removed by predators (Tinbergen *et al.* 1967, Göransson *et al.* 1975, Sugden & Beyersbergen 1986, Esler & Grand 1993). Predation on artificial nests may not be the same as on natural nests, but in all such experiments, predation emerged as density-dependent. The main observed predators were Crow *Corvus* species which, after finding an egg, increased their search in the immediate vicinity, enhancing the risk for other eggs nearby.

In colonial birds, spatial density-dependence is evident in interactions both between and within colonies. In some seabirds, colony sizes appear to be related to the numbers of conspecifics at neighbouring colonies **(Figure 5.11)**. Four species in

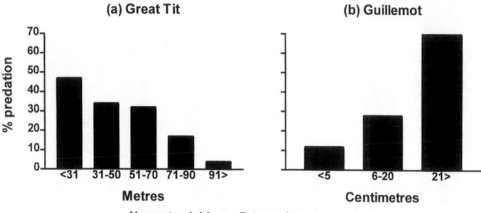

Figure 5.10   Spatial density-dependence in nest predation.
(a) Direct density-dependence. The closer together were Great Tit *Parus major* nests, the more likely they were to be preyed upon by Weasels *Mustela nivalis*. From Krebs 1971.
(b) Inverse density-dependence. The closer together were colonial Guillemots *Uria aalge*, the better they were able to protect their eggs against predatory gulls. From Birkhead 1977.

Britain bred in larger numbers at colonies that were not near to other large colonies whose members could have competed for the same local food-supplies (Furness & Birkhead 1984). An influence of food-supplies on colony sizes was also suggested by other observations. Colonies on small offshore islands in Alaska were, on average, larger than colonies on promontories, which were, on average, larger than those on linear coastlines (Birkhead & Furness 1985). This would be expected if numbers were related to food availability near to the colony, because birds on offshore islands could forage in an arc of 360° of sea, twice as much as is available to birds on linear coasts.

Reproductive performance in some seabird colonies has been found to decline with colony size. Negative relationships between mean chick fledging weights and colony sizes were found among nine colonies of Brünnich's Guillemots *Uria lomvia* and among six colonies of Puffins *Fratercula arctica* (Gaston 1985), and elsewhere between various aspects of breeding performance and colony size in Brünnich's Guillemots, Common Guillemots *Uria aalge* and Kittiwakes *Rissa tridactyla* (Hunt *et al.* 1986). Because they fledge at lower weights, chicks from larger colonies may also be subject to higher post-fledging mortality. These various findings support, but do not prove, the idea of population regulation by competition for food-supplies near the colonies (Ashmole 1963).

Turning to the situation within colonies, in some species the spatial relationship between nest density and success is complicated. On the one hand, nesting in colonies often confers some protection against predators of eggs and chicks because the colony members can join together to attack marauders (Kruuk 1964). Where communal defence is the main factor operating, it can lead to inverse density-dependence within the colonies, with better breeding success at higher densities or

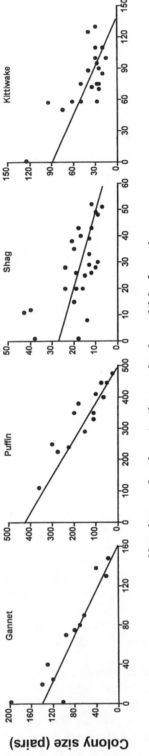

Figure 5.11 Relationships between sizes of long established seabird colonies in Britain and total number of pairs at other colonies within the foraging range of the species concerned. Colony sizes expressed as square root values, and foraging ranges taken as 100 km for Gannet *Morus bassanus*, 150 km for Puffin *Fratercula arctica*, 30 km for Shag *Phalacrocorax aristotelis* and 40 km for Kittiwake *Rissa tridactyla*. From Furness & Birkhead 1984.

in larger colonies (Kruuk 1964, Birkhead 1977, Andersson & Wiklund 1978). This was found among Guillemots *Uria aalge* subjected to egg-predation by gulls **(Figure 5.10b)**, and as indicated above, it has no regulatory role. In some colonial gull species, however, the gulls themselves prey upon the eggs and chicks of neighbours. Such cannibalism is often the main cause of individual breeding failure, and tends to be greatest in the densest parts of colonies (Fordham 1964). Among roof-nesting Herring Gulls *Larus argentatus*, pairs nesting in groups produced fewer young than pairs nesting singly on isolated sites (Monaghan 1979). The same was found among Glaucous-winged Gulls *Larus glaucescens* in which the difference was again due to greater cannibalism among the group-nesters (Vermeer *et al.* 1988). Within a single ground-nesting Herring Gull colony, however, the greatest success occurred among pairs nesting at the most common density, with success declining at both lower and higher densities (Parsons 1976). This was taken to illustrate the compromise in a colonial species between the conflicting advantages of clustering and spacing out.

Another manifestation of spatial density-dependence in birds occurs within broods at times of competition for food. In many species, the per capita mortality of nestlings increases with increasing size of broods. In a study of the Alpine Swift *Apus melba*, for example, the individual chicks in broods of one showed a 3% mortality on average, while those in broods of two showed a 13% mortality, those in broods of three showed a 21% mortality and those in broods of four a 40% mortality (Lack 1954). Similar results have been found in a wide range of other birds, from tits to crows and raptors. They are readily understood in terms of competition among brood-mates for the food provided by parents. Larger broods often receive fewer meals per chick, resulting in greater mortality and, in the survivors, lower weights at fledging (Lack 1954, 1966). These various observations thus indicate that spatial density-dependence can operate in birds at a range of spatial scales, from within individual nests to large geographical regions.

CONCLUDING REMARKS

Because for many years density-dependence proved so hard to detect in bird populations (Lack 1954, 1966), it seems worth examining with hindsight some of the obstacles. One problem was mentioned at the start, namely that density has usually been measured on a 'per area' basis. Yet the competition that can generate density-dependence is seldom over 'area' as such, but over the food and other resources that the area contains. In several species in **Table 5.1**, winter food-supplies (mostly tree-seeds) varied greatly in abundance from year to year, which is likely to have altered the density of birds at which competition set in. If the food-supply could have been measured precisely, and densities expressed as per unit of food, rather than as per unit of area, more density-dependent relationships might have emerged.

Similarly, in some species, breeding numbers fluctuated from year to year according to the severity of the intervening winters, a relationship that emerged in analyses as 'density-independent'. Yet a major effect of severe winter weather for many bird species is to make part of the usual food-supply unavailable, which can greatly enhance competition between individuals. So again, if numbers could have been expressed in terms of the food available during hard weather, losses could well have emerged on analysis as density-dependent. These problems with varying resource levels are likely to have led to under-estimation of the frequency and strength of density-dependence in many bird populations. Those species in **Table 5.1**, such as the Mallard *Anas platyrhynchos*, in which density-dependence was apparent at all major stages of the life-cycle, may be those in which area alone provided an adequate and consistent basis for density measurement.

A second problem in assessing density-dependence is that long-term upward or downward trends in population size may mask it. Populations that were relatively stable in the long-term would be expected to show strong density-dependent responses, but those undergoing long-term increase or decrease might at best show weak responses. Analyses that allow for long-term trends are therefore more likely to expose density-dependence than those that do not (Holyoak & Baillie 1996). Similarly, density-dependence may not show within the range of densities observed during a study, but might well emerge if numbers increased to even higher levels, and competition intensified. This problem was encountered by Ekman *et al.* (1981), who observed no density-dependence in the survival of young Willow Tits *Parus montanus* at normal densities, but found that survival was increased when densities were reduced experimentally.

A third problem results from the practical difficulties in most bird species of estimating the numbers of non-breeders. In most studies, the losses at each stage of the life-cycle were examined in relation to the numbers of breeders or to the numbers of eggs laid or young hatched or fledged, these being the only components of the population that could be counted. If the total adult population could have been counted, instead of just the breeders, conclusions could have differed. In many of the species that were studied, breeding numbers were limited by territorial behaviour, and surplus birds accumulated as a non-breeding non-territorial component. In this situation, as total numbers grew, an increasing proportion of the population would become non-reproductive. The measurement of breeding performance against total numbers rather than against breeding numbers alone would not only steepen existing density-dependent relationships, but would reveal all other aspects of reproduction as density-dependent.

The omission of non-territorial birds from counts in spring also affects the findings on overwinter loss. In some studies, this loss measured against the previous breeding population appeared as density-dependent. This was an inevitable consequence of a variable reproductive rate coupled with a relatively stable breeding density. The same loss might not have emerged as density-dependent if it could have been measured against the total population of territorial and non-territorial birds. In short, the conclusion of some studies in **Table 5.1**, that breeding season losses were density-independent and overwinter losses density-dependent, might have been completely reversed if non-territorial birds had been included in the spring counts.

A fourth problem centres on the amount of detail that can be collected. The various losses that occur at any one stage of a life-cycle, such as the nestling period, may have to be combined as one measure. But in practice they may be due to several factors acting simultaneously, some of which may have so large a density-independent effect that they mask the influence of density-dependent factors. Even if different factors act sequentially, their effects may not be separable if they fall within the same operationally-defined stage of the life-cycle. In an early analysis for Great Tit *Parus major*, overwinter loss (between the end of one breeding season and the start of the next) emerged as density-independent (Krebs 1970). But in a later analysis, when this period was split into two parts, 'autumn loss' emerged as density-independent and 'winter loss' as density-dependent (McCleery & Perrins 1985). The effect of the former was so great that, in the combined periods, density-dependence could not be detected. Clearly, the opportunities for detecting density-dependence increase with the number of stages into which the life-cycle can be split, and the number of mortalities that can be distinguished.

A fifth problem stems from the influence of other species competing for the same resource or supporting the same predators and parasites. Where several species exploit the same food-supply, it is the total density of all species that might affect the response of any one species, and not merely its own density. Similarly, where several species nest in the same place, such as a hedgerow or meadow, it may be the total density of all species together that influences the searching behaviour of predators, and not simply the density of any one species (Chapter 9). These problems relate to questions of competition (Chapter12), but may well reduce the chances of detecting true density relationships within species. Finally, to understand the effects of density-dependence in a population, it is not enough merely to detect it. To be useful in mathematical models, the exact strength and form of the relationship must be known.

SUMMARY

Population regulation can occur only when one or more density-dependent processes act on the rates of births or deaths, immigration or emigration. Such processes include competition for food or other resources, predation and parasitism. Whereas direct density-dependent processes can help to stabilise numbers, delayed density-dependent processes can lead to oscillations, inverse density-dependent processes tend to accentuate fluctuations and density-independent processes can cause big changes in numbers regardless of density. Abundance is determined by the combined effects of all the limiting factors that impinge on a population, whether they act in a manner which is dependent or independent of density.

By analyses of annual counts, the form of the density-dependent process has been determined in various songbirds, ducks, game birds, raptors and others. By life-cycle analyses, it has been determined in at least 63 studies of 43 species, affecting clutch size, nest success, and the survival or emigration of juveniles or adults (or net overwinter loss). Many other studies, in which only one demographic parameter was examined, have also suggested density-dependence.

The familiar pattern in species colonising new areas, of rapid increase in numbers followed by levelling off (with fluctuations), provides further evidence for density-dependence. In some increasing species, the main demographic change leading to a slowing in the rate of population growth has been identified.

Experiments in which densities were deliberately altered have provided some striking evidence for density regulation over a period of years, and for density-dependent change in demographic variables.

Difficulties in detecting density-dependence stem from (1) studying populations too low for density effects to occur, (2) studies over too short a time period to detect it, (3) failure to include non-breeders in density measures, (4) inappropriate measures of density (for example, density per unit of area rather than per unit of resource), and (5) density effects involving other species, as well as the species of interest.

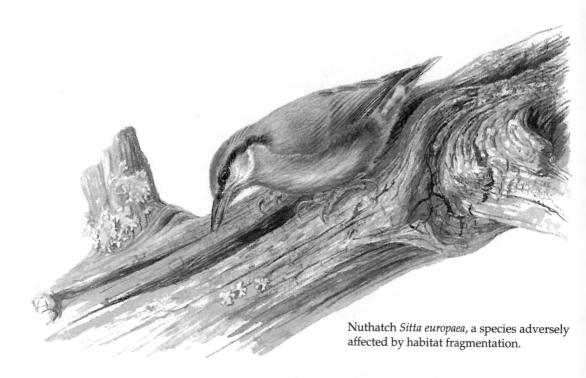

Nuthatch *Sitta europaea*, a species adversely affected by habitat fragmentation.

Chapter 6
Habitat Fragments and Metapopulations

It is not just the amount and quality of habitat that influence bird population levels, but also its distribution in the landscape. Some habitats, such as wetlands and mountaintops, have always been patchily distributed, but other once more extensive habitats, such as forest or grassland, have been fragmented by human activities. Over much of the world, therefore, most bird species now live in patchily distributed habitats. Moreover, it is becoming increasingly apparent that the size and spacing of habitat patches can greatly affect the regional occurrence of birds and other organisms. Imagine a pristine landscape covered with forest, and that over a period of years this forest is progressively felled, and replaced with farmland and towns. The felling does not start at one end and work to the other, but begins as scattered openings which gradually expand and coalesce. The original, almost continuous, forest becomes broken into patches which, as felling continues, become ever smaller and more widely spaced. This process of forest fragmentation thus involves a gradual reduction in the total area, while remaining patches become progressively smaller and further apart. More of the forest becomes 'edge' in character, because it borders open land, and relatively little remains as forest 'interior'.

Species characteristic of open land or forest edge are likely to benefit from this process, as their habitats become increasingly available. So are generalist species, such as Crows *Corvus corone*, which thrive in mixed landscapes, but may be scarce or absent in continuous forest or continuous open land (Andrén 1992). In contrast, species of the forest interior can be expected to decline. In the early stages of forest loss, their decline is likely to occur more or less in proportion to the habitat lost. But in the later stages, as some of the remaining patches become too small or far removed from others to be permanently occupied, other processes begin to act, and overall population decline comes to exceed that expected solely from loss of habitat. This is a result mainly of two simple facts: firstly, smaller patches of habitat usually support smaller populations which, in their year-to-year fluctuations, are more likely to die out than larger ones; and secondly, habitat patches that are far removed from others, relative to the dispersal powers of the species concerned, are less likely to receive immigrants from other patches, and so are less likely to be recolonised after a local extinction. In other words, regardless of limiting factors acting within habitat patches, the spatial configuration of those patches within a landscape can influence the persistence of populations, their overall distribution and abundance.

The level of fragmentation at which such processes come into play can be expected to vary between landscapes according to the way in which remaining habitat is distributed, and between species according to their dispersal and other features. However, from a review of published studies on birds and mammals, Andrén (1994) concluded that overall population levels were lower than expected from the amount of habitat present whenever that habitat formed less than 10–30% of the total landscape area. The evidence was that individuals in such highly fragmented landscapes were not distributed according to habitat, but were absent from many apparently suitable areas. For any given degree of fragmentation, overall populations can stabilise, but at levels below that which would occur if all suitable habitat patches were occupied at the same time. Appreciation of these facts has stimulated increasing interest in 'metapopulation dynamics' (Hanski & Gilpin 1991).

METAPOPULATIONS

The term metapopulation is applied to any population composed of a number of discrete and partly independent subpopulations that live in separate areas but are linked by dispersal. The subpopulations normally occupy discrete patches of breeding habitat separated by different habitat which is unsuitable for breeding. For a species of woodland bird, for example, the metapopulation may be distributed among different woods, set in an open landscape, but with individuals able to move between the woods. In a metapopulation, as originally conceived by Levins (1969, 1970), not all the habitat patches may be occupied at any one time, and the pattern of occupancy changes continually, because of local extinctions and recolonisations. All the habitat patches remain continuously favourable, but their subpopulations are assumed to have a finite lifetime, so that the entire metapopulation persists in a shifting pattern of patch occupancy. If the proportion of patches occupied remains constant, the metapopulation is in equilibrium. Whenever local recolonisations exceed local extinctions, the metapopulation expands to occupy more patches. But when local extinctions exceed recolonisations, the metapopulation contracts, to nil if the process continues.

In the first mathematical models of metapopulations, habitat patches were for simplicity assumed to be of similar size and spacing. But since then, more realistic models have been developed, in which patches vary in size and spacing, influencing local extinction and recolonisation prospects (e.g. Hanski 1994). The term 'mainland–island' metapopulation has come to be used for situations where dispersal from one or more extinction-resistant subpopulations (resistant by virtue of their large size) maintain smaller and more extinction-prone subpopulations nearby (Harrison 1991). The latter are extinction-prone primarily because of their small size, but if they happen to live in poor (sink) habitat, they are even more dependent on immigration.

By focusing on the processes of recolonisation and local extinction, metapopulation theory highlights the facts that, in any landscape, not all suitable habitat patches are likely to be occupied at any one time, and that the sizes and spacing of habitat fragments could be crucial to species persistence. Fragment size is important because of its influence on subpopulation size and local extinction risk, and fragment spacing is important because of its influence on recolonisation prospects. The interaction of the effects of area and isolation means that a large average patch area could compensate for a high degree of isolation, and alternatively that small inter-patch distances could compensate for small average area. The theory emphasises the crucial role that dispersal plays, both in recolonisation and in bolstering declining local subpopulations. Among birds, we might expect that it would be more applicable to resident species than to migrants which regularly move over long distances.

Metapopulation theory has some important implications for population persistence. Firstly, the risk of local extinction varies not only with fragment size, but also with the proportion of habitat patches occupied. This is because the immigration rate to any one patch is likely to increase with the proportion of other patches occupied, and immigration reduces the risk of local extinction. Other things being equal, then, a species that is found in only a small proportion of available habitat patches is at greater risk of regional extinction than is one that is found in a large proportion. Secondly, even though some habitat patches are vacant at any one time, if the number of available patches is reduced, a species could disappear from that region. This is because the rates of local extinctions may remain the same, but the rates of immigration may fall if the number of source areas for immigrants declines or if the remaining patches are too scattered for dispersing individuals to find or reach. Conversely, the provision of extra habitat may increase the numbers of a patchily distributed species, even when some existing habitat is vacant under present conditions. These points have practical implications in conservation, in that species currently occupying only a small proportion of the suitable habitat patches available to them may none-the-less decline if some patches are removed, or they may increase if extra patches are added (examples are given below).

Another important consequence of a metapopulation structure is that, beyond a certain threshold of habitat fragmentation, we can expect that species will decline to regional extinction. This is because remaining habitat patches are too small to allow subpopulations to persist long term, and too far apart to facilitate recolonisation. As a once-continuous habitat becomes increasingly fragmented, therefore, we can envisage for each species two critical thresholds. The first occurs when the species becomes scarcer than expected from the total area of habitat remaining, but can still persist long term in equilibrium with the habitat, occupying only a proportion of suitable patches at any one time. The second occurs when its long-term persistence becomes impossible, and it declines to regional extinction. Both thresholds are likely to vary between landscapes, according to the distribution of remaining habitat, and also between species, according to their natural densities and dispersal powers.

Metapopulation theory is one aspect of avian ecology in which the ideas and mathematical models have advanced more rapidly than relevant data have been collected, and it is as yet uncertain how well current theory fits the field situation. For although in much of the world most bird habitats are now highly fragmented, many bird species may move so freely between fragments that they are better considered as a single population than as a metapopulation. The metapopulation concept is valid only in so far as dispersal between fragments is sufficiently reduced to cause subpopulations to function to a large extent independently of one another. Clearly, there is no sharp distinction, and what we probably have in nature is a continuum of variation between (1) fragmented but inter-connected subpopulations functioning as a single unit, through (2) mainland–island situations, in which populations in small habitat patches are maintained by continual immigration from larger ones, on to (3) largely isolated subpopulations functioning largely independently of one another but still linked by dispersal, to (4) totally isolated subpopulations functioning wholly independently of one another. Moreover, a landscape that enables one species, with good dispersive powers, to function as a single population, could cause others to function as more or less isolated subpopulations, and yet others to decline to extinction. For conservation purposes, it is helpful to know for any given landscape where on this continuum particular species lie. One way to answer this question for birds would be with extensive ringing or radio-tracking programmes to assess the extent of interchange between subpopulations occupying different habitat patches. In practice, however, it has been most successfully addressed by recording the presence or absence of particular species in different habitat patches each year for several years. The resulting information has then been used to calculate for particular species:

(1) the proportions of different habitat patches occupied each year,
(2) changes in the occupancy of patches between years, and
(3) frequencies of occupancy, local extinction and recolonisation in relation to patch size and spacing.

This information can then be used to assess how close to a metapopulation structure particular species are in particular landscapes, and to predict the effects on overall population levels of further habitat change.

FIELD EVIDENCE FOR METAPOPULATIONS

Most relevant research has been on woodland birds. For a range of different song-bird species, presence/absence was recorded over several years in 151 different woods in an arable area of southeast England (Hinsley *et al.* 1995a, 1996, Bellamy *et. al.* 1996). These woods were mainly broad-leaved and similar in character, but varied in size (from 0.2 ha to 30 ha) and position with respect to other woods. From the data collected, the probability of finding particular species in woods of different sizes could be calculated **(Figure 6.1)**. In particular years, some species were widespread and found commonly in woods of all sizes, while others were absent from small woods and found with increasing frequency in larger woods. Often, closely related species showed markedly different patterns. In most (but not all) species, the pattern of woodland occupancy differed significantly from that expected on random placement, if species of a given abundance had simply occupied different-sized woods in numbers proportional to area. In practice, most species were absent from small woods more often than expected by chance.

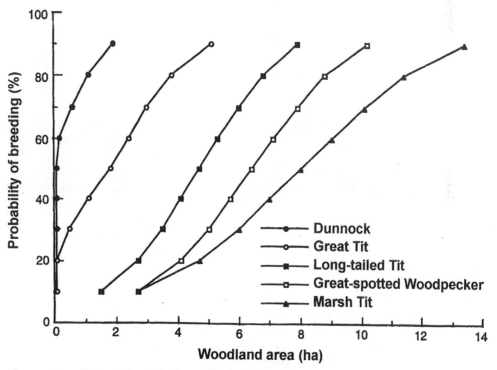

Figure 6.1   Probability of finding different bird species breeding in woods of different sizes, southeast England. Details for Dunnock *Prunella modularis*, Great Tit *Parus major*, Long-tailed Tit *Aegithalos caudatus*, Great-spotted Woodpecker *Dendrocopus major* and Marsh Tit *Parus palustris*. From Hinsley *et al.* 1995a.

To what extent could these species patterns of occupancy, related to woodland size, have arisen from the year-to-year patterns of local extinction and recolonisation? From one year to the next, breeding species were more likely to disappear from small woods than from large ones, while after a period of absence, the same species were more likely to recolonise large woods than small ones (see **Figure 6.2** for Wren *Troglodytes troglodytes* and Blackbird *Turdus merula*). For most species, in the landscape concerned, woodland area had much more influence than isolation on patterns of both local extinction and local recolonisation. This was probably because most of the bird species themselves were so mobile, relative to the distances between

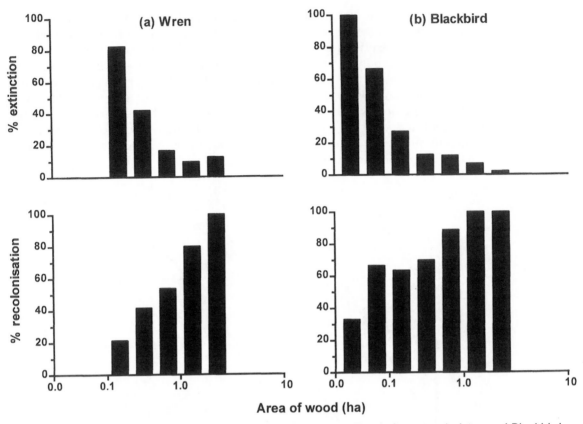

Figure 6.2   Local extinction and recolonisation among Wrens *Troglodytes troglodytes* and Blackbirds *Turdus merula* in relation to woodland area. From Newton 1995, based on data from P. Bellamy and S. Hinsley.

woods, that in this landscape isolation was of little importance to them. However, the distributions of a few species, notably Nuthatch *Sitta europaea* (see below), Marsh Tit *Parus palustris* and Long-tailed Tit *Aegithalos caudatus*, seemed to be influenced by isolation, as well as by woodland size.

The correlation between woodland size and species disappearance arose, as expected, because smaller woods had small populations, more prone to local extinctions (strictly local disappearances in which movements as well as mortality could play a role). Both local extinctions and recolonisations were species-specific features which depended on the movement patterns of the species concerned, as well as on their annual mortality rates. In the Wren, in the years concerned, a minimum of six pairs per wood was needed to ensure continuity of occupation from one year to the next, but in the Blackbird in the same years, three pairs were enough. It was clear from the movements of ringed birds that extinction may often have been prevented by immigration boosting the size of small populations (the 'rescue effect' of Brown & Kodric-Brown 1977). In some species, immigration may have been lower in the more isolated woods, which meant that extinction rates would have been influenced by distance as well as by area, in the same way that recolonisation was influenced by area, as well as by distance.

In this landscape, therefore, in most bird species examined, local extinctions declined with increasing woodland size, while recolonisations increased. This gave greater year-to-year continuity to the species populations of larger woods, and hence larger species numbers at any one time in larger woods (a familiar finding in many studies, Fuller 1982, Ford 1987, Loyn 1987, Newmark 1991, Willson *et al.* 1994, Newton 1995). There was nothing surprising in any of these findings, but they served to confirm that the chance processes of local extinction and recolonisation did indeed operate (or even prevailed) in this landscape, in ways that were related to woodland area and population size. The species concerned were all resident, present in the same general area year-round.

The findings from these English woods paralleled those obtained earlier from offshore islands, in which the main consistent predictor of local extinction was small population size (Diamond 1984, Diamond & Pimm 1993, Rosenzweig & Clark 1994). Other factors of apparent relevance on islands included:

(1) density (lower in larger species, increasing annual extinction risk),
(2) degree of year-to-year stability in population size (greater fluctuations increasing annual extinction risk),
(3) individual lifespan (longer in larger species, reducing annual extinction risk),
(4) intrinsic rate of increase (faster in smaller species, reducing annual extinction risk), and
(5) mobility and immigration chances (longer dispersal distances in larger species, reducing annual extinction risk) (Pimm *et al.* 1988, Diamond & Pimm 1993).

Local extinction risk on islands thus depended on species ecology and life-history, as well as on population size. The same patterns can be expected to hold in habitat patches, but have not yet been fully explored.

Studies in North America, where much larger forest tracts still occur, revealed other patterns. During a five-year period, Robbins *et al.* (1989) made counts at 469 points in forest fragments in Maryland and adjacent states. These fragments ranged in size from 0.3 ha to more than 3200 ha. Most were sampled by a single point count, in which an observer stood at a randomly-selected location for a 20-minute period on three occasions between late May and early July, and recorded all bird species seen or heard. So regardless of forest size, the sampling effort was the same. Yet many more species were detected per location in larger woods than in small ones.

Three main patterns were found in individual species **(Figure 6.3, Table 6.1)**. Some species, such as the Wood Pewee *Contopus virens*, were equally likely to be detected in woods of all sizes; they appeared to be uniformly distributed, regardless of woodland size. Other species, such as the House Wren *Troglodytes aedon*, favoured edges and so were more likely to be detected by counts in small woods than in large ones. For yet other species, such as the Red-eyed Vireo *Vireo olivaceus*, the probability of detection increased with woodland size. Several species were detected only in woods exceeding 100 ha in area, including the Black-throated Blue Warbler *Dendroica caerulescens*, Canada Warbler *Wilsonia canadensis*, Black-and-white Warbler *Mniotilta varia* and Cerulean Warbler *Dendroica cerulea*. Similar findings have emerged from other studies, and have raised the intriguing question of why certain species, in which individual pairs can obtain all their needs from a territory of around 1–3 ha, seem restricted to habitat patches two or more orders of magnitude larger than their minimum area needs. So marked is this phenomenon that some

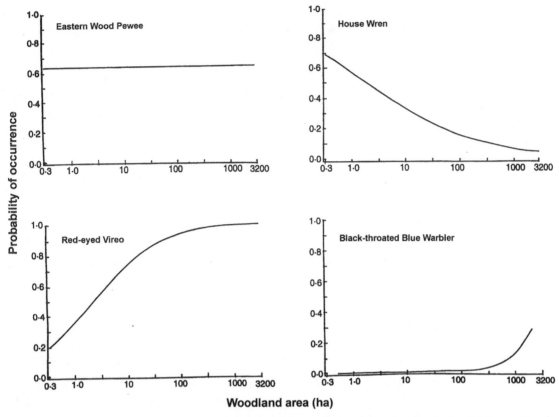

Figure 6.3 Probability of finding four bird species at a single randomly selected point in woods of different sizes in eastern North America. Details for Eastern Wood Pewee *Contopus virens*, House Wren *Troglodytes aedon*, Red-eyed Vireo *Vireo olivaceus* and Black-throated Blue Warbler *Dendroica caerulescens*. From Robbins *et al.* 1989.

ornithologists distinguish such species by the special name of 'forest-interior birds'. One explanation of the phenomenon is that such species favour particular rare habitats within the forest which are not normally found in small areas. If this were so, it would be their special habitat needs that restrict them to large areas, and not a need for large areas as such.

Forest-interior specialists do not occur in Britain, perhaps because if they ever did occur here, they disappeared through extreme forest fragmentation long before ornithologists could record them. Such species still occur in the larger forests of continental Europe, however, as exemplified by the White-backed Woodpecker *Dendrocopus leucotos* and Three-toed Woodpecker *Picoides tridactylus*, both of which seek particular types of habitat within extensive forest (Carlson & Aulén 1990).

Restriction to large forest areas is shown even more strikingly in the Kokako *Callaeas cinerea* of New Zealand, which is now found only in forests exceeding 1000 ha in area (Ogle 1987). There is no obvious reason for this restriction: at known densities a forest of this size should still support a population of 250 individuals. In this and other species that have gone from smaller forests in recent times, existing patterns could result from recent fragmentation, with species distributions settling

Table 6.1   Relationship between probability of detection at a randomly-selected point within forest and forest area (range 0.3 ha–3200 ha) for 38 bird species in eastern North America. From Robbins *et al.* 1989.

Probability of detection with increase in forest area	Number of species	Examples
Constant	4	Eastern Wood Pewee *Contopus virens* Yellow-billed Cuckoo *Coccyzus americanus* Ruby-throated Hummingbird *Archilochus colubris*
Decreases	8	Gray Catbird *Dumetella carolinensis* House Wren *Troglodytes aedon* American Robin *Turdus migratorius*
Increases	26	Red-eyed Vireo *Vireo olivaceus* Scarlet Tanager *Piranga olivacea* Ovenbird *Seiurus aurocapillus*

The following species (with minimum area of forest in which they were detected) were found only in forests exceeding 100 ha in area: Black-throated Blue Warbler *Dendroica caerulescens* (1120 ha), Cerulean Warbler *Dendroica cerulea* (138 ha), Black-and-white Warbler *Mniotilta varia* (203 ha) and Canada Warbler *Wilsonia canadensis* (187 ha).

to a lower equilibrium level, or they could result from general population declines, nothing to do with fragmentation, but manifest in the smaller patches (with lower numbers) first.

In some North American studies, effects of isolation on the woodland avifauna were also apparent, with woods far from others holding fewer species than expected from their size (Lynch & Whigham 1984, Robbins *et al.* 1989). The relative importance of patch area and isolation to species occurrence is likely to vary between studies, because both depend on the distribution of woodland in the landscapes concerned, on the particular bird species represented, and probably also on the way that isolation is measured: whether by the distance to the nearest wood or by the proportion of woodland in the surrounding area.

The restriction of certain species to large habitat patches is not confined to forest birds. In a survey of 63 bogs in Sweden, some bird species with small individual area needs (up to a few ha) were virtually absent from bogs less than 25 ha in area (Boström & Nilsson 1983). With an increase in bog size, increasing numbers of bog species were found. Because many species that favour large areas of habitat, whether woods or bogs, are migrants which occupy their nesting areas only in summer, their absence from smaller habitat patches is unlikely to be due entirely to extinctions of small populations, but to involve a need (or preference) for larger habitat areas. The specific habitat needs of certain species may be met only in large areas, as mentioned above, while other species seem actively to avoid edges (examples are given below). Edge-avoidance has been attributed to competitors or to generalist predators and parasites (cowbirds) which penetrate the habitat from surrounding areas, and render a wide peripheral zone unattractive.

We therefore have several possible explanations of why certain species are found mainly or only in large habitat areas that greatly exceed the area needs of individual pairs:

(1) Random placement, in which on statistical grounds the probability of occurrence of any species increases with habitat area (an explanation eliminated for most of

the species discussed above). On this basis, rare species would be almost entirely restricted to large habitat areas.

(2) The stochastic processes of area-related extinction or area-related recolonisation, both of which help to ensure greater continuity to the populations of large habitat patches, and hence greater species numbers at one time in large patches.

(3) Area-related habitat diversity, such that the special needs of particular species are more likely to be met in large than in small habitat areas.

(4) Preference for large habitat areas, in some species linked with edge-avoidance.

The first of these explanations is based entirely on chance processes, the second and third include both chance and behavioural elements, and the fourth is entirely behavioural, based on habitat preference. The four explanations are not mutually exclusive, and different explanations could hold to varying degrees in different species. Better knowledge of the needs of particular species may clarify the picture, but in any case the manifestations of chance processes are inevitably greatest in the smallest and most isolated habitat patches.

To sum up, in the distribution patterns of bird species among habitat patches of different sizes, the chance processes of local extinction and recolonisation become increasingly important the more fragmented the habitat. Over and above these effects, the specific habitat and area needs of individual species are also involved and must be met before any species is found in any particular habitat patch. Certain species with small individual area needs are restricted, for as yet unknown reasons, to very large habitat patches. The implications of these findings for conservation are obvious, namely the more that habitats are fragmented, the more likely are species with special habitat or area needs to disappear, and the more an element of chance is introduced into the long-term persistence of all species. This in turn will increase the likelihood of regional extinction and eventually of total extinction. If these points are apparent in present-day landscapes for birds, which include some of the most mobile organisms, it would not be surprising if they held in more extreme form in less mobile organisms.

Studies of the Nuthatch and others

In some species, findings from field studies have been examined specifically in the light of metapopulation theory. In a study of the Nuthatch *Sitta europaea* in the Netherlands, occupancy over six years varied from 14% in small (<1 ha) isolated woods to 100% in large (>10 ha) less isolated woods, where isolation was defined by the amount of woodland in the surrounding area (Opdam *et al.* 1995). Hence, not all woods were occupied at one time, and over a period of years local extinctions and recolonisations occurred, giving some changes in the woods that were occupied. Local extinction rates were related to the size and quality of the woods, and colonisation rates varied according to the number of Nuthatches in the surrounding area, so that the most isolated woods were least likely to be recolonised. The occurrence of this species in different woods could therefore be modelled explicitly as a metapopulation (Verboom *et al.* 1991). Application of the same model to a sparsely-wooded arable landscape in southeast England, where climate and other conditions seemed favourable, showed that Nuthatches were much scarcer than predicted from the area of suitable habitat available, and that their scarcity could be attributed to the effects of isolation. The woods concerned were inadequate to support a sizeable breeding population long term, and were too far removed from other suitable woods for recolonisation to be successful (Bellamy *et al.* 1998). In this study, then, a mathematical model developed from a metapopulation study in one region was used to explain the scarcity of Nuthatches in another region, accounting for an otherwise puzzling gap in the distribution across southern England.

Similar patterns have been found in birds of other habitats. In the Netherlands, Black Grouse *Tetrao tetrix* were distributed according to the size and spacing of heathland, with recolonisations seemingly not occurring over distances greater than 10 km. Among various marshland birds, some species were often absent from areas under 10 ha in size, and the presence of individual species was influenced by the size of the marsh and by the distance to the next marsh, as well as by the amount of marshland in the surrounding landscape (Foppen 1994, in Opdam *et al.* 1995). In each case, mathematical models of metapopulations, which incorporated data from actual landscapes, simulated patterns of bird distribution that closely resembled the patterns found in nature. Moreover, the regional status of species has been observed to change in predictable ways as habitat areas were removed or added to landscapes by human action. With increased planting of woodland in The Netherlands, certain species spread to existing woods that before were considered too small or isolated. They included Nuthatch, Marsh Tit *Parus palustris*, Hawfinch *Coccothraustes coccothraustes* and Lesser Spotted Woodpecker *Dendrocopos minor*. In the same way that a reduction in the number of habitat patches in a region could lead to population declines greater than expected from the area of habitat removed, the addition of more patches to existing ones could result in regional population increases greater than expected from the area of habitat added. New habitat could facilitate the more continuous occupancy of existing habitat, largely because the associated increase in immigration reduces local extinctions. It seems that many small resident bird species behave in modern landscapes in ways that are consistent with metapopulation theory. The same has been found for various butterflies and other insects (Hanski *et al.* 1995).

Birds of old forest

Among woodland birds, some of the most fragmented populations are now found in species dependent on forest more than 80 years old, for this is an age that few managed forests are allowed to reach. Such species include many that depend on large dead trees for nesting or foraging, such as certain woodpeckers and other hole-nesters (Carlson & Aulén 1990). The Red-cockaded Woodpecker *Dendrocopos borealis* in the southeastern United States was once abundant and widespread, but is now limited to small and isolated populations, mostly declining from continuing habitat loss and fragmentation (Reed 1990, Walters 1991). The birds excavate nesting and roosting cavities in the heart wood of live pine trees. Only large trees more than 80 years old are suitable, and these are the prime targets of logging companies. The species breeds co-operatively, and group territories are centred on clusters of nest and roost cavities. Although the species is generally thought to be fairly sedentary, long-distance dispersal (up to 90 km) has been recorded (Walters *et al.* 1988), so in a metapopulation context, clusters of cavities could be considered as habitat patches. The provision of artificial cavities (i.e. increasing the number of suitable patches) might increase the species' survival chances (Walters 1991).

The potential end result of increasing isolation of remaining subpopulations is regional extinction. Such an event was recorded for the last population of the

Middle-spotted Woodpecker *Dendrocopos medius* in Sweden (Pettersson 1985). Clearance of old oak woodland divided the population into several isolated remnants, but numbers remained fairly stable at 15–20 pairs for more than 20 years, before declining to nil around 1983. The habitat remained suitable, and proximate factors in the final demise included unusually heavy mortality caused by severe winter weather in three of the last four years, together with reduced fecundity, possibly due to inbreeding (the proportion of non-viable eggs increased as the population declined, Pettersson 1985). The longest recorded dispersal distance of a ringed Middle-spotted Woodpecker was 14 km and the species was absent from suitable habitat more than 9 km from an existing breeding locality. The closest other population was at least 90 km from this Swedish population.

In western North America, the Northern Spotted Owl *Strix occidentalis caurina* is dependent on old growth forest, but lives at low density, with home ranges covering from several hundred to several thousand hectares (Thomas *et al.* 1990). This species is assumed to have once occurred continuously through extensive tracts of mature forest, but is now restricted to clusters of old growth patches set in a matrix of younger forest and clear-fells created by logging. It is likely to survive in the long term only if sufficient old growth patches are preserved in a distribution pattern that ensures they are accessible to dispersing owls. The species has been studied well enough to produce the detailed information needed for credible mathematical models. Such models have predicted a sharp threshold in the degree of fragmentation that would cause a previously viable population to decline and that this owl would disappear if the patches of old growth forest were reduced to less than 20% over a large region (Lande 1988b, Lamberson *et al.* 1992). The development of management plans for the conservation of this owl provides a good example of the applied use of metapopulation theory (Shaffer 1985, Thomas *et al.* 1990, Lande 1991, Lamberson *et al.* 1992), but it remains to be seen to what extent the plans will be enacted by the forest industry and in any event whether the owl will survive in the long term.

The use of spatially explicit models linked with landscape maps that capture the heterogeneity of real landscapes can be used to suggest likely responses of birds and other organisms to different proposed land management options. They can be validated to some extent from patterns of distribution and occupancy from regions not used to develop the models, and then used to predict the responses of bird populations to different land-use scenarios (Opdam *et al.* 1995). Such models have already been used in explaining current distributions of some species (such as the Nuthatch above), and in conservation planning for others, including Northern Spotted Owl (above) and Bachman's Sparrow *Aimophila aestivalis* in the southeastern United States (Pulliam *et al.* 1995). Metapopulation theory has thus become a popular paradigm in conservation biology.

MAINLAND–ISLAND METAPOPULATIONS

Many bird species probably have a mainland–island type of population structure somewhere in their ranges, where small subpopulations are maintained mainly by immigration from large ones. In some species, geographical distributions tail off gradually, rather than stopping abruptly, leading to a mainland–island structure near the edge of the range. The width of the band within which tailing-off occurs, in relation to the species' dispersal ability, will determine whether colonists from the 'mainland' or from other 'islands' predominate (Harrison 1991). One study concerned Spruce Grouse *Dendragapus canadensis* in patches of coniferous forest at the southern boundary of their range in the Adirondack Mountains of the United States (Fritz 1979). Small forest patches were less likely to be occupied than large ones and, when inhabited, had small, extinction-prone populations. Suitable, but unoccupied, forest patches were significantly further from a source of colonists than were occupied patches, implying that for such remote patches immigration was insufficient to outweigh local extinction. The grouse in the study were not marked and so the relative contributions of the 'mainland' and other 'island' populations to immigration could not be assessed.

Sometimes small populations become isolated from the main range, so that the chances of immigration are much reduced. In Britain, for example, the Cirl Bunting *Emberiza cirlus* was once widespread across the south, but has now become confined to a small part of the southwest, separated by several hundred kilometres from the next nearest population in Europe. It is inconceivable that this isolated population receives immigrants on a regular basis. Other species, such as the Dartford Warbler *Sylvia undata* in Britain, show similar isolation, making their persistence wholly dependent on local events. Such species are thus especially vulnerable to events that lower their numbers, because they cannot be rescued by immigrants from elsewhere. The Dartford Warbler almost disappeared completely from Britain after two successive hard winters (Chapter 11).

OTHER FACTORS AFFECTING THE PERSISTENCE OF METAPOPULATIONS

Other factors, besides the size and spacing of habitat patches, are likely to influence the persistence of metapopulations.

Fluctuations in subpopulations. In any metapopulation, the different subpopulations, although linked by dispersal, function largely independently of one another. However, if they are all influenced by the same widespread environmental factor, such as weather, they might still fluctuate in synchrony. This synchrony could greatly increase the chances of regional extinction because subpopulations might crash at the same time, behaving like a single population. In contrast, if they fluctuated independently, out of phase with one another, the chances of regional extinction would be greatly reduced, because vacant patches could be continually recolonised from occupied ones. In effect, then, the degree of correlation in the fluctuations of different subpopulations could greatly influence the survival of a metapopulation (see Lahaye *et al.* (1994) for exploration of this aspect in the Northern Spotted Owl).

Sources and sinks. The source–sink concept acknowledges that habitat varies in quality, and that in good patches numbers can grow, whereas in poor ones they can only decline unless maintained by immigration (Chapter 3). This does not mean, however, that sink habitat contributes nothing to the persistence of a metapopulation.

For a single isolated source population, all emigrants would be lost if no other habitat were available. In the presence of a sink, however, at least some emigrants and their offspring may persist for a time, and they or their offspring might later return to the source. The lifetime of a source population could therefore be prolonged by the existence of a nearby sink (Howe *et al.* 1991), the overall population (and the associated gene pool) could be larger, and the chance of regional extinction thereby reduced. Ultimately, the persistence of any metapopulation depends not only on a balance between local extinctions and recolonisations, but also on a balance between source and sink habitat.

In practice, small habitat patches may often prove to be sinks. Predation rates on nest contents are often greater near the edges of habitat patches than nearer the centre, and birds in small patches suffer greater overall predation rates than do those in large patches (Chapter 9). The effect of this is to increase the vulnerability of subpopulations in small patches further than expected from their small size alone. In North America, nest parasitism by Brown-headed Cowbirds *Molothrus ater* acts in the same way (Chapter 10), because these birds of open land will penetrate forest by more than 100 m in their search for suitable hosts. Through increased predation and parasitism, forest fragmentation has been suggested as a factor in the widespread declines of some migrant songbird species in parts of eastern North America, although the evidence is equivocal (Chapter 9; Terborgh 1989, Askins *et al.* 1990, Maurer & Heywood 1993). Another problem of small forest patches is the greater exposure to wind, which gives colder and more drying conditions than occur in larger, more sheltered patches. At high latitudes, the increased chill factor in winter might well reduce the survival chances of small birds in small woods, but I know of no relevant studies.

Dispersal and recolonisation. If the amount of emigration from occupied habitat patches is density dependent, per capita dispersal from such patches increases with population growth within the patches, but if dispersal occurs regardless of density, then per capita emigration does not change as the population grows, and is an inevitable form of loss from occupied patches even at low population levels. It is also a source of colonists for other patches at all population levels. In any metapopulation, therefore, density-dependent dispersal could reduce the chance of local extinctions, while density-independent dispersal could raise the chance of recolonisation.

Dispersal between patches depends partly on the intervening terrain, and may be facilitated by habitat corridors linking patches (as hedgerows might link woods). This holds whether or not the corridor itself provides permanent habitat. Dispersal rates between patches linked by corridors may then be greater than between patches lacking corridors (Saunders & Hobbs 1991). Connectors between patches need not be continuous to enhance dispersal. Small habitat patches could act as stepping stones and facilitate dispersal between distant sites. Again this could hold whether or not the small habitat fragments themselves were suitable for long-term occupation.

For some woodland bird species, hedgerows provide extra habitat, while for others they act merely as pathways for movement between woods. Yet other woodland species are reluctant to cross open land, with or without hedges, and are poor colonists. In southeast England, the presence in particular woods of certain species, such as Long-tailed Tit *Aegithalos caudatus*, Robin *Erithacus rubecula* and Bullfinch *Pyrrhula pyrrhula*, was positively associated with the amount of hedgerow in the surrounding landscape (Hinsley *et al.* 1995b). So hedges may have been important in providing extra habitat or in the movement of these species between woods. In contrast, the presence of Treecreepers *Certhia familiaris* in particular woods was negatively related to the distance to the nearest wood, but was uninfluenced by hedges. This suggested that Treecreepers were reluctant to cross large gaps and did not routinely use hedges either as permanent habitat or as corridors between woods. At the other extreme, although Chaffinches *Fringilla coelebs* and Great-spotted Woodpeckers *Dendrocopus major* were more commonly found in woods in well wooded than in sparsely wooded areas, they were not influenced by hedges, suggesting that they readily crossed gaps between woods. Landscapes clearly vary in their permeability for different species and, while corridors or stepping stones of habitat may facilitate the movements of some species, they have no obvious effect on others. Whatever the species, however, the process of dispersal probably involves risks, and losses may be greater among individuals traversing unsuitable than suitable habitat. For example, forest birds that depend on cover for escape could be especially vulnerable to avian predators when crosssing open land or water. For such species, then, the more fragmented the habitat, the greater the risks from gap-crossing.

The role of intervening habitat was apparent in the occupation of natural forest patches by Hazel Grouse *Bonasa bonasia* in Sweden. In farmland, the species was seldom encountered in forest patches that were more than 100 m from continuous forest, but in managed conifer plantation the threshold was about 2000 m. Hazel Grouse evidently dispersed more readily through forest plantation than through open farmland in seeking remaining areas of suitable forest habitat (Swenson 1995).

The way that potential colonists behave on arrival, and whether they are attracted or repelled by conspecifics, could also influence colonisation patterns. Some species may be attracted by the presence of conspecifics and remain if they find other individuals, but move on if they do not (Alatalo *et al.* 1982, Shields *et al.* 1988, Ray *et al.* 1991). Conspecific attraction will tend to reduce the number of patches occupied at equilibrium, but should enhance persistence within patches, and hence of the whole metapopulation. On the other hand, presence of conspecifics to full capacity may prevent colonists from settling, and force them to move elsewhere. This process could promote the occupation of more patches, as mentioned above.

Finding suitable habitat. As habitats become extremely fragmented, with large distances between patches, another problem could arise, namely that birds that disperse from existing patches might have difficulty in finding new patches and mates. This problem could be heightened by the fact that, in many bird species, one sex disperses further between natal and subsequent breeding sites than the other (Greenwood 1980). Mathematical models that restrict the capacity of territorial species to locate vacant territories and mates indicate that a substantial proportion of habitat can become vacant even where many surplus birds of both sexes are available to occupy it (see Lande 1987, 1988b, for Northern Spotted Owl), and this was shown empirically in the Nuthatch (Matthysen & Currie 1996).

Finding suitable habitat patches and mates is likely to be particularly difficult for long-distance migrants in which many young do not return precisely to their natal area. Where the species' breeding range contains large tracts of continuous habitat, this will be of little consequence. But for a scarce species inhabiting scattered habitat fragments in a wide geographical range, weak natal philopatry could make it difficult for individuals to find both habitat and mates, leading inevitably to overall population decline.

The North American Kirtland's Warbler *Dendroica kirtlandii* is a long-distance migrant which breeds in temporarily suitable patches of young pine forest (Chapter 1). Once more widespread, the whole population has now become confined to an area spanning 50×70 km in Michigan, but even here birds have occupied less than half the apparently suitable habitat in any one year (38% in 1980, Ryel 1981), and at least 15% of singing males failed to obtain mates (Probst 1986). Although the population may be limited mainly by other factors, the difficulty in locating habitat patches and mates on the breeding range could contribute to the rarity of the species (Mayfield 1983). Apparently suitable habitat covers only 1–2% of the total breeding range, and isolated males have established territories more than 500 km outside the main area (Ryel 1981). At least some such birds were in their first year, having been ringed the year before in the main area. For more widely distributed species, long-distance dispersal presents no particular problem, because the majority of long-distance settlers still remain within the area where they are likely to find habitat and mates.

Difficulties of dispersal clearly do not account for all situations where a species is found in only a proportion of the habitat patches that seem suitable. Firstly, our perception of what is suitable habitat may be wrong, and areas of vacant habitat may in fact be unsuitable. In this case the species may be occupying all the suitable habitat. Secondly, only a proportion of the apparently suitable habitat may be fit for occupation in any one year, but the species must occupy different patches in different years, and could not persist indefinitely in any one patch. An example is the seed-eating Crossbill *Loxia curvirostra*, which requires conifer forest, but must usually change areas each year because of the sporadic cropping patterns of its food-plants. Thirdly, a species may be held at such a low level by other limiting factors, say by heavy mortality in the wintering areas, that it is never numerous enough to occupy all available patches of breeding habitat. Clearly, occupation of only a proportion of habitat patches can occur for a variety of reasons, and does not in itself imply a metapopulation structure.

HABITAT FRAGMENTATION AND REGIONAL EXTINCTIONS

As forest or other habitat becomes increasingly fragmented, an increasing proportion of the species originally present in the region would be expected to disappear, for reasons given above. Baseline information is available from very few forest areas before deforestation began, so that the process of species loss has seldom been studied directly. However, in an area of cloud forest in the western Andes of Colombia, bird surveys were made in 1911, 1959 and 1989–90 (Kattan *et al.* 1994). Mainly between 1911 and 1959, the total original forest area was reduced by more than 50%, to leave patches interspersed with farms and houses. Of 128 forest bird species present in the area in 1911, 24 had apparently gone by 1959 and another 16 by 1989–90, giving a total reduction of 31% in species numbers. Missing species were still present in the nearest tract of unbroken forest, about 10 km away.

Other studies, aimed to assess the number of local or regional extinctions, have compared the species numbers in particular remnant forest patches with those presumed to have been present in the region before the surrounding forest was felled (assessed from the nearest similar but larger forest). For example, three areas of forest on Sao Paulo plateau in southeastern Brazil, covering 1400, 250 and 21 ha, respectively, had been isolated for more than 100 years by deforestation of the surrounding land (Willis 1979, 1980). In 1975–78, the three fragments were found to contain 202, 146 and 93 bird species respectively. The original forest of the plateau was estimated to have contained about 230 species, so the largest tract could have lost about 28 (12%) species, the medium-sized tract 84 (37%) species and the smallest 137 (60%) species. The large tract contained virtually all the species held by the other two, and the medium-sized tract contained virtually all the species held by the small tract. This implied that the losses occurred in a sequence related to fragment size, and that certain species were more likely to disappear from small tracts. The losses were minimum estimates, firstly because the fragments contained at least 20 edge species that were included in the counts but which were probably absent from the original continuous forest. Secondly, collections of birds made about 15 years earlier in the largest fragment contained ten other species that had apparently gone by 1975–78 when the counts were made. This implied that not all losses occurred immediately, and hinted that further losses, resulting from reduced area, might occur in future.

These and similar studies elsewhere have generally confirmed that (1) the smaller the forest fragment surveyed, the greater the proportion of original forest species that were missing, and (2) loss of species was not necessarily immediate, but could occur over a period of years or even decades [for the neotropics see Leck (1979), Terborgh (1975) and Karr (1982); and for other regions, see Diamond *et al.* (1987), Saunders (1989) and Newmark (1991)]. They have also revealed the types of species most likely to disappear, namely (1) those that live naturally at low density, and so are generally scarce, such as many predators, (2) those that live at higher density, but during a year must range over large areas to obtain their food, such as frugivorous cotingids, toucans and parrots, and (3) those that are reluctant to cross open country, such as some insectivorous antbirds and others (Terborgh & Winter 1980). For the most part, then, the bird communities of remnant forest patches form nested series, with the impoverished communities of small patches consisting of non-random subsets of the species found in larger patches.

Such studies might over-estimate the numbers of species lost, because a small forest patch might not start out with all the species of the larger tract in which it originally lay. Also, in most such studies, the current avifauna was assessed by counts in 1–3 successive years. If counts had been continued for more years, more species might have been found, either because they had been missed earlier or because they had moved in. However, these concerns would not alter the general findings that (1) small fragments held fewer species than large ones and (2) certain species were absent from small fragments and found only in large ones. The main unresolved problem with such studies, in the absence of other information, is knowing how many species went because of habitat loss or change as such and how many because of other human impact.

In an experimental study, rainforest fragments of different sizes (1 ha, 10 ha, 100 ha and 1000 ha) were left near Manaus in Brazil, as surrounding areas were cleared to create pasture (Lovejoy *et al.* 1986, Bierregaard 1990, Bierregaard *et al.* 1992). As expected, species numbers declined most rapidly in the smallest patches, and least rapidly in the largest, but because changes were still occurring, it was not possible to assess their full extent, either in the individual patches or in the wider region. Surprisingly, however, only limited influx of 'edge' species occurred, although forest birds themselves soon declined near the edges. At 10 m from a large man-made clearing, 38% fewer birds were caught in mist nets than 50 m into the forest, and 60% fewer than at 1 km into the forest. Species numbers were also affected, with 28 species caught at 10 m, 47 at 50 m and 50 in deep forest. Even 50 m from the forest edge certain species were rare. If this distance was taken as the limit to which edge effects extend, then all of a 1-ha, and 19% of a 10-ha square reserve, was 'edge'. Hence, rather than an enhancement of species numbers at the edge, as is often observed in temperate woodland, in these Amazonian forests there was a decline, but this may have been because the edge in the Amazonian patches had not had time to develop a shrub layer.

CONTINUOUS POPULATIONS IN FRAGMENTED LANDSCAPES

Habitat fragmentation occurs at different spatial scales, affecting populations or individuals (Haila 1990). At the scale of individual territories, habitat fragmentation need not result in the sub-division of a population, only in a lowering of overall density. Some species can persist in landscapes in which habitat fragments are smaller than their home ranges, if individual birds can make use of several neighbouring fragments. This happens where fragments are close enough together that the costs of travelling between them can be borne, but it usually results in much bigger ranges than in continuous habitat. For example, the Tawny Owl *Strix aluco* depends on trees for roosting, nesting and hunting perches. It is usually associated with woodland, but can persist in open landscapes, with scattered clumps of trees. Compared with Tawny Owls in large woods, whose entire territories lay within the wood, Tawny Owls in nearby farmland had territories 5–6 times larger, embracing several small woods, and hunted both within the woods and in the intervening open land (Redpath 1995). In continuous woodland, no territory was smaller than 9.2 ha, but in open landscape owls survived with as little as 0.7 ha of woodland within their larger home ranges. Reproductive rates were similar in both woodland and open land, but mortality was higher in open land, and owl numbers there may have been maintained partly by continuing immigration **(Table 6.2)**. For this species, then, woodland fragmentation did not result in the isolation of subpopulations but in a reduction in habitat quality and territorial density, with no break in the spatial continuity of the population. In some areas the same was true for Northern Spotted Owls, which had 20–30% of old forest in their territories that also included younger forest (Forsman *et al.* 1984).

Throughout their range, Capercaillies *Tetrao urogallus* are found in old conifer forest. In Sweden, they have survived some degree of fragmentation by increasing their home ranges to incorporate several forest patches, but in districts where old forest formed less than 30% of the landscape, small patches were abandoned. Predation of eggs and young was higher in fragmented than in continuous forest, and overwinter survival was lower (Rolstad & Wegge 1987, 1989, Gjerde & Wegge 1989).

Many other European forest species with sizeable home ranges can persist in open

Table 6.2   Territory size and performance of Tawny Owls *Strix aluco* in continuous and highly fragmented woodland. Sample sizes in parentheses. Modified from Redpath (1995) with help from the author.

	Continuous woodland	Open landscape with small scattered woods
Mean territory size	27 ha (5)	152 ha (7)
Mean area of woodland within range	14.8 ha (10)	3.9 ha (20)
Turnover, as % of territories occupied in a 3-year period	87% (39)	43% (28)
Young per pair per year ($\pm$ SE)	0.61 ± 0.13 (109)	0.37 ± 0.10 (51)

landscapes by using several woodland fragments. Some, such as the Tawny Owl, make use of intervening habitat, but others, such as the Black Woodpecker *Dryocopus martius*, do not. In a fragmented landscape in central Sweden, Black Woodpeckers held territories about three times as large as in more continuous forest, but encompassing the same total area of trees. In this species, no difference in breeding success was found between continuous and fragmented habitat (Tjernberg *et al.* 1993).

At the scale of the individual territory, as well as at the larger scale of the population, species evidently vary in the degree of habitat fragmentation that they can withstand. This in turn depends on features of the species itself, such as its mobility, anti-predator behaviour and diet, as well as on features of the intervening habitat. However, simply because a species is found in fragmented habitat does not mean that it is secure there in the long term: its presence may depend on continuing immigration from a large expanse of other habitat nearby. To understand how a species is affected by habitat fragmentation, it is important to measure, not only occurrence, but also breeding success and turnover in different types of landscape. What is clear is that, for woodland species that readily cross open country, minimum territory sizes are not necessarily the same as the minimum size of habitat fragment in which the species occurs.

CONCLUDING REMARKS

Metapopulation structure has important implications for population persistence. It implies that bird populations can be limited not only by the obvious factors of food, predators and parasites, but also by the chance factors imposed by their own dispersal abilities in relation to the spatial configuration of habitats. Over and above any effect of habitat quality, the fragmentation of habitat can reduce populations to a level lower than expected from the total amount of habitat remaining. This fact is well documented by recent research. Although under continued habitat fragmentation, local extinctions could come to outrun recolonisations, leading to regional extinction, I know of no well-documented examples from birds. The problem is that bird populations that are declining for reasons other than habitat fragmentation could disappear progressively from fragments in a manner similar to that expected under disrupted dispersal (see also Simberloff 1994). It is therefore difficult to attribute regional extinction unequivocally to destruction of a proportion of habitat patches. However, this type of extinction is commonplace in other contexts, such as vaccination programmes in which the immunization of only a proportion of potential hosts (equivalent to removal of habitat patches) has proved sufficient to eradicate some viral and bacterial pathogens (May 1994).

Metapopulation theory is appealing because it attempts to explain the observed distribution of particular species among habitat patches in terms of only two processes of demonstrable importance. It has therefore stimulated much new research, and encouraged the study of aspects of bird ecology previously ignored. However, it is clearly an oversimplification of what happens in nature. The precise habitat types available in a habitat patch are important, as are features of the species themselves (such as the densities at which they can live and their dispersal powers), and of the surrounding landscape, particularly whether it contains corridors or stepping stones that might facilitate dispersal. Clearly, habitat fragments cannot realistically be viewed in the same way as islands in the sea (a frequent analogy). Birds that live in habitat fragments can be influenced by other species from the surrounding land, and some may themselves depend partly on surrounding land for food. Greater appreciation of the influence of the matrix landscape on the fauna of habitat patches has helped to shift the emphasis in conservation thinking from reserves alone to the overall landscape.

The timescale on which local extinctions and recolonisations occur would be expected to vary between different metapopulations. The recognition of any meta-population is therefore dependent on turnover occurring within the study period (seldom longer than three years), and many apparently isolated subpopulations may prove to be linked by dispersal in the longer term. Thus both spatial and temporal scales are likely to influence metapopulation processes. In addition, some studies were made in landscapes from which some habitat had been recently lost, so their bird populations might not have reached equilibrium with respect to the remaining habitat, as in the Brazilian studies mentioned above. Some effects of habitat fragmentation become apparent rapidly, but others take much longer, depending partly on the intervening habitats.

SUMMARY

Some bird species live in naturally patchily-distributed habitats, while others live in once-extensive habitats that have been fragmented by human activity. In both groups, long-term persistence depends partly on dispersal between patches. As habitat fragments become smaller and more isolated from one another, chances of local extinction rise and chances of recolonisation fall. Such processes could lead to lower equilibrium populations than expected from the total area of habitat available, and if continued, could lead to widespread population decline and eventual extinction, even in the presence of some remaining habitat. The fragmentation of habitat can thus cause bigger declines in populations than expected merely from the proportion of habitat lost, and has been proposed as a major factor in the declines and regional extinctions of some forest and other bird species. Within any given landscape, depending on their dispersal powers, species differ in the degree of population fragmentation they can tolerate without declining. In some species, large habitat patches offer better survival and reproductive prospects than small ones, partly because of greater predation on nest-contents in small patches.

Part Two
Natural Limiting Factors

Barn Owl *Tyto alba* with prey.

Chapter 7
Food-supply

Few ornithologists will question the importance of food-supply in influencing bird numbers, but its precise effects are seldom easy to quantify. One problem is that shortages are not always shown in an obvious way, such as starvation (loss of body condition to the point of death). As discussed in previous chapters, territorial and other interactions between individuals can operate to adjust bird densities to local food-supplies, causing hungry birds to move elsewhere, where they may survive or die from a variety of causes, not necessarily including starvation. Some birds exposed to food-shortage may thus succumb to predation or disease, which conceals the underlying cause of their deaths. Secondly, food shortage may reduce population size through lowering breeding rates, again not necessarily entailing the starvation of full-grown birds. This type of effect may be hard to detect because of the time lag between the food-shortage and the resultant decline in breeding numbers. In some long-lived species, individuals do not normally breed until they are several years old, so it may take several years before the effects of poor breeding are reflected in poor recruitment. Thirdly, because food-supplies fluctuate, shortages are seldom continuous, but fall in particular seasons or years, and sometimes last for no more than a few days or weeks at a time. Moreover, if numbers are reduced greatly during a period of shortage (as during hard weather), a species may then experience several years until the next shortage. So unless a study coincided with a shortage, the observer may underestimate the importance of food-supply. As a further complication, the effects of food-shortage on any population may be accentuated by predators, which constrain the birds to feed at particular places or particular times, and by parasites which directly or indirectly take part of the host's nutrient intake for themselves (Chapter 13).

The term 'food-shortage' has been used for food that is inadequate in quantity, quality or availability. Whatever the abundance of food in the environment, if it is for some reason unavailable, a bird may starve. Many intertidal invertebrates retreat deeper in the mud in cold weather, putting them beyond the reach of shorebirds, while many soil-dwelling ones move deeper during drought. They illustrate the general points that only a proportion of potential food-items may be accessible to birds at any one time, and that this proportion may change through time, in response to other conditions. Secondly, if food is plentiful but of low nutrient content or digestibility, a bird may be unable to process enough each day to maintain its weight. This is particularly true of herbivores eating fibrous vegetation, and shortage may arise partly from crop capacity, which limits the amount that can be stored for overnight digestion, or from gut morphology which limits the processing rate (Kenward & Sibly 1977). In general among birds, species-differences in gut structure correspond with dietary differences, and in species whose diet changes seasonally, gut morphology may adapt accordingly. This is true of small birds, such as Bearded Tits *Panurus biarmicus*, which switch from insects to seeds between summer and winter, and large birds, such as grouse, which switch between different kinds of plant material (Spitzer 1972, Savory & Gentle 1976, Al-Joborae 1980, Kehoe & Ankney 1985, Barnes & Thomas 1987). The need for change in gut structure can make some such species vulnerable at times of rapid shifts in diet, as when snowfall forces birds suddenly from one food-type to another. Captive American Robins *Turdus migratorius* and Starlings *Sturnus vulgaris* took several days to adjust to a switch from fruit to insects or vice versa (Levey & Karasov 1989), while Woodpigeons *Columba palumbus* took 10–15 days to acclimatise to a diet of brassica leaves (Kenward & Sibly 1978). This latter change entailed a marked loss of weight.

Many bird species cushion the effects of food-shortage by storage, either of body fat or of food itself. In small birds, fat stores are accumulated mainly for use overnight, and few species can last more than a day or two without feeding. Larger birds, with their relatively lower metabolic demands, can store relatively more energy per unit body weight. As a result, larger birds can rely on internal reserves for longer than small ones, more than two weeks in large waterfowl (Chapter 11). Many birds store some body fat and protein in preparation for breeding, extreme examples being Arctic-nesting geese in which the females produce and incubate their eggs almost entirely on reserves accumulated in wintering or migration areas (Chapter 11). For some weeks, then, they are protected against the need to feed on their breeding grounds. Most migratory birds accumulate body fat for the journey, the amount depending on the duration of unbroken flight (reviews, Berthold 1993, Biebach 1996).

Species that store food-items may retrieve them hours, days or even months later, depending partly on the type of food. Some meat-eaters cache prey for consumption later in the day or in the following days, while some resident seed-eaters may store food in autumn to last through the winter, or even year-round. Long-term storage is frequent among various tits, nuthatches and woodpeckers (mainly *Melanerpes*), reaching its extreme in certain jays and nutcrackers. In some such species, stored food can provide more than 80% of the winter diet, and can also feed the adults into the next breeding season, influencing the numbers of young raised (Hannon *et al.* 1987, Källander & Smith 1990). In good seed years, huge amounts of seed can be stored. Individuals of several nutcracker species, for example, can cache 20 000–100 000 pine seeds each autumn, some 2–5 times their total winter energy needs (Vander Wall 1988, Lanner 1996). Crucial aspects of hoarding behaviour are that individuals should guard their stores or should hide them (to reduce pilfering) and remember where they are. Some species store items individually, each in a separate place, and in experiments have shown incredible feats of memory (Krebs *et al.* 1996). But none of this behaviour alters the dependence of such birds on their food-supplies: it simply enables them to use their foods more effectively, taking advantage of temporary gluts to lessen later shortages. It does not prevent them being limited by shortages in years of poor seed-crops (see Hannon *et al.* 1987 for Acorn Woodpecker *Melanerpes formicivorus*).

Yet other birds can cushion themselves against temporary food-shortages by becoming torpid, achieving an inactive state in which body temperature and heart, respiratory and metabolic rates are greatly depressed, thereby conserving energy. In some small species, this happens in some degree every night, but it is especially marked in some small hummingbirds, such as Anna's Hummingbird *Calypte anna* of western North America (Bartholomew *et al.* 1957). Some other species, such as the Common Swift *Apus apus*, can remain torpid for several days at a time when cold weather renders their insect prey inactive (Koskimies 1950), and at least one species, the Common Poorwill *Phalaenoptilus nuttallii* can remain torpid for several weeks, in a state akin to mammalian hibernation (Jaeger 1949, Bartholomew *et al.* 1957). These adaptations help the species concerned to last through difficult periods, whether hours, days or weeks, but they cannot eliminate altogether the risk of starvation.

EFFECTS OF FOOD-SHORTAGE ON INDIVIDUALS

That food-shortage can affect individuals directly, through causing breeding failure or starvation, is beyond question. Most information on the effects of food on bird breeding comes from correlations between productivity (young raised per pair) and food-supply, either year to year or place to place (Martin 1987). Bird species that experience marked annual variations in food-supply, such as some rodent-eating raptors, show marked annual variations in breeding success (review Newton 1979). In one ten-year study of Rough-legged Buzzards *Buteo lagopus*, for example, mean production varied from nil in poor rodent years to four young per pair in the best rodent year (Hagen 1969). In other raptor species, annual production of young varied three-fold in Hen Harrier *Circus cyaneus* and Gyr Falcon *Falco rusticolus*, up to four-fold in Kestrel *Falco tinnunculus*, up to five-fold in Common Buzzard *Buteo buteo* and Ferruginous Hawk *Buteo regalis*, and up to six-fold in Goshawk *Accipiter gentilis*, all in association with annual fluctuations in prey densities (Newton 1979). Similar correlations between annual food-supply and breeding success have been documented in various owls, in insectivorous warblers, hirundines and gallinaceous birds (insectivorous as chicks), in seed-eating finches, nectivorous hummingbirds and many others (**Figure 7.1**; review Martin 1987, also Bryant 1975, Rands 1985, Baines 1996 and later examples).

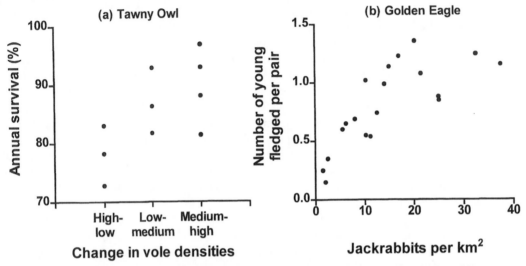

Figure 7.1. Examples of relationship between food-supply and annual survival and between food-supply and breeding success. (a) Tawny Owl *Strix aluco* in a forest in England where the Field Voles *Microtus agrestis* that formed the main prey fluctuated on a three-year cycle, so were increasing or decreasing in different years (from Petty 1992); (b) Golden Eagle *Aquila chrysaetos* in sagebrush desert in Idaho, where the main prey species was the Black-tailed Jackrabbit *Lepus californicus*. From Steenhof *et al.* 1997.

Moreover, some other bird species, which normally show little annual variation in breeding success, have occasional catastrophic years, after a sudden reduction in food-supply. Such events are especially marked in seabirds, with almost complete breeding failure recorded among various species after fish stocks collapsed (for Kittiwake *Rissa tridactyla* in Alaska and Britain, see Hatch 1987 and Hamer *et al.* 1993; for Puffin *Fratercula arctica* in Norway, see Anker-Nilsson 1987; for Arctic Tern *Sterna paradisaea* in Britain, see Phillips *et al.* 1996; for Shag *Phalacrocorax aristotelis* in Britain and Norway, see Aebischer 1986 and Vader *et al.* 1990; for various other species, see Chapter 13).

Further indication of the role of food-supply in breeding is provided by cooperative breeders in which pairs that receive help with chick feeding raise more young per year than those that do not (Brown *et al.* 1978, Emlen 1978, Stacey & Koenig 1990). It is also provided by some polygynous species, such as the Hen Harrier *Circus cyaneus*, in which primary females raise more young per year than secondary ones because primary females get more help from the male (Balfour & Cadbury 1979).

Depending on the stage at which they occur, effects of food-shortages can be manifest through non-laying, small clutches, egg-desertion, poor chick growth and survival, and (in multi-brooded species) in reduced numbers of nesting attempts per season, as well as in loss of weight and reduced survival prospects in adults. In small birds that can raise more than one brood in succession, much of the variation in annual productivity is caused by variations in the mean number of broods raised per year. With year-round access to protein-rich poultry food, some Tree Sparrows *Passer montanus* in Malaysia and Moorhens *Gallinula chloropus* in South Africa bred year-round (Siegfried & Frost 1975, Wong 1983).

Three types of field experiments have served to confirm that food-supply can influence the production of young. One type has involved the manipulation of food-supplies, most often the provision of extra food. In most such experiments, feeding was stopped before hatch but in seven out of eight altricial species where extra food was provided during the nestling period, survival and production of young improved **(Table 7.1)**. Increases were most marked in experiments done when natural food-supplies were low or when densities (and competition for food) were high (Smith *et al.* 1980, Newton & Marquiss 1981, Dijkstra *et al.* 1982). In these circumstances, food-provision usually raised production to the levels seen in the best natural food-conditions (Boutin 1990). Among fed birds, increases in mean production up to four-fold or five-fold were recorded, compared with control birds that received no extra food (A.J. Hansen 1987, Arcese & Smith 1988). In these and other species, food-provision also altered other aspects of performance, in a way likely to improve post-fledging survival, for example by resulting in earlier laying or heavier fledglings. In some species extra food acted partly to reduce rates of predation and parasitism, because fed birds spent more time at the nest, and could thereby better protect their eggs and young (Yom-Tov 1974, Högstedt 1981).

Different experiments on the same species have sometimes given conflicting results, with birds responding to extra food in one experiment but not in another, or responding in different ways in different experiments (see various studies on Song Sparrow *Melospiza melodia* and Magpie *Pica pica*, **Table 7.1**). The failure of food to influence reproduction, as observed in some experiments, was variously attributed to:

(1) reproduction being limited by factors other than food,
(2) natural food-supply being so abundant, or population density so low, that additional food had no effect,
(3) the food provided being inadequate in some way, or
(4) food being given at an inappropriate stage of breeding (say at the chick stage when limitation occurred at the egg stage).

In Tengmalm's Owl *Aegolius funereus*, fed pairs produced more fledglings even though feeding was stopped after egg-laying. This was because fed pairs produced more eggs, implying that productivity was limited at that stage rather than later in the cycle (Korpimäki 1989). The same held in Burrowing Owl *Athene cunicularia* (Wellicome 1993).

In other experiments, insecticides were used to reduce bird food-supplies. Following forest spraying with a bacterial insecticide, production of young by Black-throated Blue Warblers *Dendroica caerulescens* was reduced in a year of moderate caterpillar abundance, but not in two years of low caterpillar abundance, when breeding was poor anyway (Rodenhouse & Holmes 1992). In other experiments, insecticide use failed to affect bird breeding success in years of very high caterpillar

Table 7.1 Effects of experimental supplementation of food-supply on the breeding performance of birds. I = incubation, L = laying, N = nestling stage, PL = pre-laying, PF = post-fledging, T = throughout, − = no response, + = positive response, blank = not reported, * = positive response through polygyny.

Species	Stage food provided	Earlier laying	Larger clutches	Hatching success	Increased Chick weight	Increased Chick survival	More breeding attempts	Higher post-fledging survival	More young per pair	Source
Amakihi *Hemignathus virens*	T		+						+	van Riper 1984
Song Sparrow *Melospiza melodia*	T	+	+	+	+	+	+	−	+	Arcese & Smith 1988
	PL	+	−			−	−		−	Smith et al. 1980
Great Tit *Parus major*	PL	+								Källander 1974
Crested Tit *Parus cristatus*	T	+	−		+	−	+		+	Bromssen & Jansson 1980
Willow Tit *Parus montanus*	T	+	−		−	−	+		+	Bromssen & Jansson 1980
Red-winged Blackbird *Agelaius phoeniceus*	T	+	−	−		−			*	Ewald & Rohwer 1982
	T	+	−				−		*	Wimberger 1988
Seychelles Magpie Robin *Copsychus sechellarum*	T	+	−	+	+	+	−	−	+	Komdeur 1996
Pied Flycatcher *Ficedula hypoleuca*	N							+		Verhulst 1994
Cactus Wren *Campylorhynchus brunneicapillus*	N,PF		+		+	+	+	+	+	Simons & Martin 1990
Dunnock *Prunellus modularis*	T	+	−		−	−	+		*	Davies & Lundberg 1984, 1985
Jackdaw *Corvus monedula*	T	+	+	−	+	+			+	Soler & Soler 1996
Carrion Crow *Corvus corone*	T	+	−	+	−	+			+	Yom-Tov 1974
	N				+	+			+	Richner 1992
Magpie *Pica pica*	PL,L,I,N	+	−	+	−	+		−	+	Dhindsa & Boag 1990
	T	+	+	+	(+)	+	+		+	Högstedt 1981
	T	+	−			+	(+)		+	Hochachka & Boag 1987
	PL,L	+	−			−			−	Knight 1988
Common Starling *Sturnus vulgaris*	N			+	+	+	+		+	Crossner 1977
Burrowing Owl *Athene cunicularia*	PL,L,I	+	+	−			+		+	Wellicome 1993
Tengmalm's Owl *Aegolius funereus*	PL,L	+	+	+	−				+	Korpimäki 1989
Puffin *Fratercula arctica*	N				+		−			Harris 1978
Herring Gull *Larus argentatus*	PL,L	−	+		−	−				Bolton et al. 1992
Oystercatcher *Haematopus ostralegus*	N				+					Ens et al. 1992
American Coot *Fulica americana*	PL,L	+	+							Arnold 1994
Coot *Fulica atra*	N				+	+		−	+	Brinkhof & Cavé 1997
Red Grouse *Lagopus l. scoticus*	T†								+	Watson et al. 1984
	T†								+	Watson & O'Hare 1979
Kestrel *Falco tinnunculus*	PL,L	+	+							Dijkstra et al. 1982
Sparrowhawk *Accipiter nisus*	PL,L	+	+		−		−		−	Newton & Marquiss 1981
Bald Eagle *Haliaeetus leucocephalus*	I,N				+	+			+	Hansen 1987
Great Egret *Casmerodius albus*	N				+	+			+	Mock et al. 1987

†Through fertilizer application.

abundance when even massive reduction in caterpillar numbers still left plenty for birds[1]. Ponds experimentally sprayed with pesticide (as in aerial forest spraying), produced significantly fewer invertebrates, with consequent effects on the survival of Black Duck *Anas rubripes* and Mallard *A. platyrhynchos* ducklings (Hunter *et al.* 1984; Chapter 15). Another relevant study involved Herring Gulls *Larus argentatus* in which reduction in food-supply, resulting from the closure of a rubbish dump, was followed by smaller clutches and a 50% reduction in production of young (Pons & Migot 1995). Other experiments involving gallinaceous birds are reported below.

The second type of experiment, involving adding extra chicks to nests of altricial species to find whether parents could raise larger than normal broods, has given variable results. In 11 out of 40 brood-enlargement experiments reviewed by Dijkstra *et al.* (1990), enlarged broods suffered greater mortality and yielded fewer fledglings than control broods. These findings implied that the parents were already operating close to the food-limit, and were unable (or unwilling) to collect more food per day. In the remaining 29 experiments, however, enlarged broods produced more fledglings, on average, than control ones (although in some such studies the young were underweight). These experiments implied that for those birds food at the nestling stage was not limiting. Again, however, they did not exclude the possibilities that food might have been limiting at a different stage of the breeding cycle (laying or post-fledging stages), or that adults that raised extra young might themselves have suffered reduced future survival or reproduction (shown in 8 out of 12 species investigated, Dijkstra *et al.* 1990).

The third type of experiment involved the removal (or natural loss) of the male after hatch from species in which both sexes normally help to raise the young. From experiments on 14 such species reviewed by Martin (1987), 13 recorded increased chick starvation and reduced overall production of young. This was consistent with observations on some polygynous species, mentioned above.

These various types of study confirm that food-supply is important in limiting reproductive output in at least some types of bird. This limitation arises because adult birds are constrained by their food-gathering abilities in the numbers of eggs they can produce or (in altricial species) in the number of young they can feed. Individual adults may be unwilling to increase their daily work activity above a certain level if this would harm their future survival and reproductive effort, so that food gathering depends on some interplay between food availability, parental efficiency and effort. In precocial species, the chicks themselves cannot obtain enough when food is scarce.

Most direct information on proximate mortality causes in full-grown birds comes from autopsies of birds found dead, among which starved birds can be recognised by their low weights, absence of body fat, and emaciated muscle tissue **(Table 7.2)**. Because autopsy samples depend on intact carcasses found by people, they give an unrepresentative cross-section of mortality causes. Particular problems are that birds consumed by predators cannot be examined, and many starving birds are almost certainly removed by predators before they can die from weight loss, so that both predation and starvation victims are under-recorded in carcass samples. None-the-less, in some species starvation accounted for a substantial proportion of

---

[1]*The many other experiments on the effects of forest spraying on bird breeding were of limited value in the present context because of probable direct chemical effects on the birds.*

Table 7.2 Causes of mortality as determined by post-mortem analyses of birds found dead. + = less than 1%.

Species	Region (years)	Numbers examined	Diagnosed causes of death (%)							Source
			Starvation[1]	Predation[2]	Disease[3]	Accident/trauma[4]	Poisoning[5]	Shot or trapped[6]	Other[7]	
Various species[a]	Britain (1952–59)	1,000	?		28[b]	33	10		21	Jennings 1961
Various species[a]	Britain (1954–62)	2,044			155[b]	84	205		139[c]	Keymer et al. 1962
Various species[a]	Britain (1960–62)	583			12	26[d]	25		37	Macdonald 1962,1963
Various garden songbirds	USA (1989)	152	18		4	61	8	3	5	Okoniewski & Novesky 1993
Long-eared Owl / *Asio otus*	Britain (1963–95)	123								Wyllie et al. 1996
Great Horned Owl / *Bubo virginianus*	USA (1975–93)	132	32		4	33	8	9	13	Franson & Little 1996
Tawny Owl / *Strix aluco*	Britain (<1978)	172	18						82[e]	Hirons et al. 1979
Barn Owl / *Tyto alba*	Britain (1963–89)	627	20	+	3	54	9	2	11	Newton et al. 1991
Pheasant / *Phasianus colchicus*	Wisconsin (1955–58)	226		42		1		29	28[f]	Burger 1964
Bald Eagle / *Haliaeetus leucocephalus*	USA (1966–81)	692	6		8	31	11	38	15	Reichel et al. 1984[8]
Red-tailed Hawk / *Buteo jamaicensis*	USA (1975–92)	163	20		13	18	15	1		Franson et al. 1996
Sparrowhawk / *Accipiter nisus*	Britain (1962–80)	341	9		4	48	14	11	14	Newton et al. 1982
Kestrel / *Falco tinnunculus*	Britain (1962–80)	616	29		11	23	13	4	20	Newton et al. 1982
Kestrel / *Falco tinnunculus*	Britain (1966–80)	92	48		10	5	23	14		Keymer et al. 1981
Various raptors	USA (1972–77)	850			8	34		38	15	Redig 1978
Mute Swan	England (1979–85)	260			10	24	46	5	15	Birkhead 1982, Sears 1988
Cygnus olar	Scotland (1980–86)	147			17	21	14	9	39	Spray & Milne 1988
Whooper Swan / *Cygnus cygnus*	Scotland (1981–86)	57			2	28	47	11	12	Spray & Milne 1988

[1] Emaciated and no other obvious cause. [2] Includes birds scavenged. [3] Includes infectious diseases and other disorders; not all diseases, notably viruses, were checked for in all studies. [4] Includes collision, road casualties, drowning, electrocution. [5] Mostly by pesticides, also lead in some species, and entirely lead in swans. [6] Includes oiled birds. [7] Includes undetermined. [8] The last of 7 papers published during 1969-84 from which the results here are taken. [a] Most commonly examined birds were thrushes, finches, tits, starlings, sparrows, pigeons and gulls. [b] Includes 85 large parasites, 70 infections, excluding viral. [c] Called adverse environmental factors, including oiling, degenerative diseases and unknown. [d] Includes some predation. [e] No other cause of death examined other than starvation. [f] Shock, died within 3 days of release.

recorded deaths – up to 35% of 616 Kestrels *Falco tinnunculus* and 40% of 627 Barn Owls *Tyto alba* found dead in Britain (Newton *et al.* 1982, 1991). In other species, starvation appeared rare, accounting, for example, for only 6% of 692 Bald Eagles *Haliaeetus leucocephalus* found dead in the United States, and none of 407 Mute Swans *Cygnus olor* found dead in Britain (Reichel *et al.* 1969, Birkhead 1982, Sears 1988, Spray & Milne 1988), although heavy losses of Mute Swans to starvation were seen in other areas (Andersen-Harild & Bacon 1989, Meek 1993).

Other information on mortality causes has come from samples of radio-tagged birds, in which each individual was followed until its death. Most such studies have involved gallinaceous birds, large enough to carry a transmitter **(Table 7.3)**. Nearly always, starvation was rare or unrecorded, but this might have been because starving birds fell victim to predators, so were included with predation victims **(Table 7.3)**. Other information on bird mortality causes has come from recoveries of ringed birds reported dead, but here one is further restricted to forms of death easily recognised by the finder, without the benefit of autopsies. In summary, all these sources of information, biased as they may be, confirm that birds can starve to death, but suggest that the proportion relative to other deaths varies greatly between species.

Deaths of adult birds from starvation are usually restricted to the non-breeding season, when food is scarcest. In many species, it is only at times of massive food-shortage, such as during intense cold or drought, that starving birds are much in evidence (Chapter 11). Most examples of weather-induced starvation involve small songbirds, waders and waterfowl, and occasionally seabirds (Chapter 11). Among large concentrations of birds, the death toll can be spectacular. Examples include the 14 000 Eiders *Somateria mollissima* that starved off Denmark in the winter of 1981–82, or the 18 000 waterbirds (including 14 000 diving ducks) that died on the Dutch Waddensea in March 1986 (Wrånes 1988, Suter & van Eerden 1992). Such instances underline the role of weather in influencing food-availability, an aspect discussed in detail in Chapter 11.

Further evidence for the influence of food-supply on the survival of full-grown birds has come from correlations, over several years, between mean annual (or overwinter) survival and food-supply (Perrins 1979, van Balen 1980, Nilsson 1982, Gibbs & Grant 1987a, Hannon *et al.* 1987, Taylor 1994). In various species, mean survival rates were higher in years of plenty than in years of famine (e.g. **Figure 7.1**), with juveniles usually affected more than adults. Other evidence has come from experiments in which extra food was provided by the researcher. The survival rates of local birds receiving extra food were then compared with those of other birds elsewhere, which had only natural foods to eat. In such experiments, improved survival (especially of juveniles) was evident in all six species in which it was measured, being up to twice as high in fed birds as in unfed ones **(Table 7.4)**. The species concerned were all short-lived, with relatively low annual survival rates, and in some, part of the improvement in survival was again attributed to reduced predation, because fed birds were able to spend more time on the alert or in safe sites (Jansson *et al.* 1981).

*   *   *

The fact that some birds die of starvation, or that survival and reproductive rates can be influenced by food-supply, does not on its own mean that food-supply affects

Table 7.3 Causes of mortality in various bird species as determined for radio-marked individuals. + = Less than 1%.

Species	Location (year)	Number of birds		Causes of death (%)						Source
		Tagged	Died	Starvation	Predation	Disease[1]	Accident[2]	Shot[3]	Others and unknown	
Ruffed Grouse *Bonasa umbellus*	Wisconsin (1982–88)	381	222	0	67	1	4	28	0	Small et al. 1991
Red Grouse *Lagopus l. scoticus*	Scotland (1994–96)	348	207	0	89[4]	0	+	3	7	Redpath & Thirgood 1997
Turkey *Meleagris gallopavo*	Wisconsin (1988–94)	224	132	8[5]	71	+	+	7	10	Wright et al. 1996
Ring-necked Pheasant *Phasianus colchicus*	Minnesota (1967)	74	60	2	92		2		4	Hessler et al. 1970
	Wisconsin (1968–71)	105	–	2[6]	72	0	9	7	10	Dumke & Pils 1973
	Colorado (1979–81)	101	47	0	100	0	0	0	0	Snyder 1985
Goshawk *Accipiter gentilis*	Sweden (1980–87)	352	67	37	5	5	17	36	0	Kenward et al. 1993
Bald Eagle *Haliaeetus leucocephalus*	Alaska (1989–92)	159	34	31	(50)[7]	0	0	6	13	Bowman et al. 1995
American Kestrel *Falco sparverius*	California (1992)	18	5	0	80	0	0	0	20	Yamamoto et al. 1993
Northern Spotted Owl *Strix occidentalis caurina*	California (1987–89)	80	26	31	23	8	0	0	38	Paton et al. 1991
	Oregon-Washington (1975–90)	?	46	7	33	0	4	2	54	Foster et al. 1992

[1]Includes parasites and pathogens. [2]Mostly collisions. [3]Includes wounded and trapped. [4]62 killed by raptors, 21 by mammalian carnivores. [5]All in one of 5 winters, with prolonged snow. [6]Attributed to extreme weather. [7]Aggressive encounters with other eagles.

Table 7.4 Effects of artificial food provision on the overwinter survival of some bird species.

Species	Age groups	Region	(a) Survival without extra food (%)	(b) Survival with extra food (%)	Degree of improvement (b/a)	Source
Great Tit	Adults	Netherlands	26	50	2.0†	van Balen 1980
Parus major	Juveniles		3.4*	7.6*	2.2†	
Crested Tit		Sweden	39	76	1.2	Jansson *et al.* 1981
Parus cristatus						
Willow Tit		Sweden	45	82	1.8†	Jansson *et al.* 1981
Parus montanus						
Black-capped Chickadee		Wisconsin	37	69	1.9†	Brittingham & Temple 1988
Parus atricapillus	Adults	Alberta	69	65	0.9	Desrochers *et al.* 1988
	Juveniles		64	77	1.2	
Song Sparrow	Adults	British	82	77	0.9	Smith *et al.* 1980
Melospiza melodia	Juveniles	Columbia	56	76	1.4†	

* Proportion of birds hatched in the study area that subsequently bred there.
† Statistically significant, at least at P<0.05.

local breeding densities. Indeed, in some feeding experiments, improvement in local overwinter survival did not lead to higher local breeding density, but merely to greater emigration before nesting began (Brittingham & Temple 1988, Desrochers *et al.* 1988). Alternatively, losses from food-shortage might be compensated by reduced subsequent losses from other causes, giving no apparent effects on population levels. In particular, poor breeding might be offset by improved overwinter survival, or poor overwinter survival might be offset by better breeding (Chapter 5). The evidence that food-supply can in some circumstances have a major influence on breeding bird densities is discussed in the sections below.

CORRELATIONS BETWEEN BIRD NUMBERS AND FOOD-SUPPLIES

Correlations between the densities of birds and their food-supplies provide only circumstantial evidence for a causal relationship, yet they are widespread in nature. They occur at various spatial scales from individual territories to large geographical regions, and at temporal scales from a few minutes (as in seabirds exploiting surface fish) to periods of years. The examples in **Table 7.5** include various songbirds, raptors, shorebirds, seabirds and others, in all of which individuals were found at greatest density when or where their food was most plentiful. Such correlations can be divided into three main types.

(1) Short-term associations between bird numbers and food-supplies. Such associations are apparent mainly in species that exploit sporadic food-sources, abundant at different places at different times.
(2) Long-term changes in bird numbers, associated with long-term changes in food-supplies. Some such changes occur naturally, but many result from human activities, notably land-use and fishery practices.

Table 7.5 Evidence for the influence of food-supply on bird numbers, based on temporal and spatial correlations between bird densities and food-supplies.

Species	Food-supply	Spatial scale[1]	Breeding or non-breeding[2]	Region	Source
A. Variations from year to year in the same area					
Various sparrows	Seeds	3	NB	Arizona	Pulliam & Parker 1979, Dunning & Brown 1982
Ground finches *Geospiza fortis* and *G. scandens*	Seeds	2	B	Galapagos	Grant 1986, Gibbs & Grant 1987b
Brambling *Fringilla montifringilla*	Caterpillars *Epirrata*	1	B	Sweden	Enemar et al. 1984
Siskin *Carduelis spinus*	Spruce *Picea* seeds	2	B	Finland	Haapanen 1965
	Spruce *Picea* seeds	2	B	England	Petty et al. 1995
Redpoll *Carduelis flammea*	Birch *Betula* seeds	1	B	Sweden	Enemar et al. 1984
Common Crossbill *Loxia curvirostra*	Spruce *Picea* seeds	2	B	Finland	Reinikainen 1937
	Spruce *Picea* seeds	2		England	Petty et al. 1995
Bullfinch *Pyrrhula pyrrhula*	Seeds	2	NB	England	Newton 1972
Various seed-eaters	Seeds	4	NB	Several continents	Schluter & Repasky 1991
Nuthatch *Sitta europea*	Beech *Fagus* seeds	1	B	Sweden	Nilsson 1987
	Hazel *Corylus* seeds	1	B	Sweden	Enoksson 1990
Various warblers	Caterpillars*	2	B	New Hampshire	Holmes et al. 1991, Rodenhouse & Holmes 1992
American Redstart *Setophaga ruticilla*	Caterpillars*	2	B	New Hampshire	Sherry & Holmes 1991
Pied Flycatcher *Ficedula hypoleuca*	Caterpillars*	2	B	Finland	Virolainen 1984
Coal Tit *Parus ater*	Arthropods	2	NB	England	Gibb 1960
Rufous Hummingbird *Selasphorus rufus*	Nectar	1	NB	California	Gass et al. 1976
Great Horned Owl *Bubo virginianus*	Snowshoe Hares *Lepus*	3	B	Yukon	Rohner 1995
	Snowshoe Hares *Lepus*	3	B	Saskatchewan	Houston & Francis 1995
Long-eared Owl *Asio otus*	Voles *Microtus*	2	B	Finland	Korpimäki & Norrdahl 1991
Short-eared Owl *Asio flammeus*	Lemmings	3	B	Alaska	Pitelka et al. 1955
	Voles *Microtus*	2	B	Scotland	Village 1987
	Voles *Microtus*	2	B	Finland	Korpimäki & Norrdahl 1991
Snowy Owl *Nyctea scandiaca*	Lemmings	2	B	Alaska	Pitelka et al. 1955
Tengmalm's Owl *Aegolius funereus*	Voles *Microtus*	2	B	Finland	Korpimäki 1985
Barn Owl *Tyto alba*	Voles *Microtus*	2	B	Scotland	Taylor 1994
Woodpigeon *Columba palumbus*	Farm crops	2	NB	England	Inglis et al. 1990

Species		Food	Status	Location	Reference
Swift Tern *Sterna bergii*	3	Anchovies *Engraulis*	B	South Africa	Crawford & Dyer 1995
Arctic Tern *Sterna paradisaea*	2	Sandeels *Ammodytes*	B	Scotland	Monaghan & Zonfrillo 1986
Various terns	2	Herrings *Clupea*	B	Scotland	Monaghan & Zonfrillo 1986
Pomarine Skua *Stercorarius pomarinus*	3	Lemmings	B	Alaska	Pitelka *et al.* 1955
Goshawk *Accipiter gentilis*	3	Snowshoe Hares *Lepus*	B	Yukon	Doyle & Smith 1994
	2	Hazel Grouse *Bonasa*	B	Finland	Wikman & Lindén 1981
Hen Harrier *Circus cyaneus*	2	Voles *Microtus*	B	Wisconsin	Hamerström 1969
Peregrine *Falco peregrinus*	3	Partly Lemmings	B	NW Territories	Court *et al.* 1988
Rough-legged Buzzard *Buteo lagopus*	2	Lemmings	B	NW Territories	Court *et al.* 1988
	2	Lemmings	B	Norway	Hagen 1969
	2	Lemmings	B	Sweden	Enemar *et al.* 1984
Ferruginous Hawk *Buteo regalis*	2	Jackrabbits *Lepus*	B	Utah	Woffinden & Murphy 1977
Black-shouldered Kite *Elanus caeruleus*	2	Microtine rodents	B,NB	South Africa	Mendelsohn 1983
Gyr Falcon *Falco rusticolus*	3	Ptarmigan *Lagopus*	B	Alaska	Swartz *et al.* 1974
	3	Ptarmigan *Lagopus*	B	Alaska	Platt 1977
Kestrel *Falco tinnunculus*	2	Voles *Microtus*	B	Scotland	Village 1990
	2	Voles *Microtus*	B	Netherlands	Cavé 1968
	4	Voles *Microtus*	B	Britain	Snow 1968
	2	Voles *Microtus*	B	Finland	Korpimäki & Norrdahl 1991
	2	Voles *Microtus*	B	Norway	Hagen 1969
	2	Voles *Microtus*	B	Switzerland	Rockenbauch 1968
Various ducks	2	Chironomids*	B	Iceland	Gardarsson & Einarsson 1994
Cape Gannet *Sula capensis*	3	Anchovies *Engraulis*	B	Southwest Africa	Crawford & Dyer 1995
Cape Cormorant *Phalacrocorax capensis*	3	Anchovies *Engraulis*	B	Southwest Africa	Crawford & Dyer 1995
Jackass Penguin *Spheniscus demersus*	3	Anchovies *Engraulis*	B	Southwest Africa	Crawford & Dyer 1995

B. Variations from place to place in the same time period

Species		Food	Status	Location	Reference
Ovenbird *Seiurus aurocapillus*	1	Invertebrates	B	Ontario	Stenger 1958
	1	Invertebrates	B	Tennessee	Smith & Shugart 1987
Silvereye *Zosterops lateralis*	2	Figs *Ficus*	B	Australia	Catterall *et al.* 1982
Wren *Troglodytes troglodytes*	3	Insects	B	Britain & Iceland	Cody & Cody 1972
Dipper *Cinclus cinclus*	3	Invertebrates	B	Britain	Ormerod *et al.* 1985, Vickery 1991
Guillemots *Uria aalge* and *U. lomvia*	3	Mainly Capelin *Mallotus*	NB	Barents Sea	Erikstäd *et al.* 1990
Guillemot *Uria aalge*	2	Capelin *Mallotus*	NB	Newfoundland	Piatt 1990
Puffin *Fratercula arctica*	2	Capelin *Mallotus*	NB	Newfoundland	Piatt 1990
Laughing Gull *Larus atricilla*	2	Beetles larvae	NB	New York	Buckley & McCarthy 1994

Table 7.5 Continued.

Species	Food-supply	Spatial scale[1]	Breeding or non-breeding[2]	Region	Source
Oystercatcher *Haematopus ostralegus*	Mussels *Cerastoderma*	2	NB	England	Yates et al. 1993
Grey Plover *Pluvialis squatarola*	Invertebrates	2	NB	England	Yates et al. 1993
Curlew *Numenius arquata*	Invertebrates	2	NB	England	Yates et al. 1993
Bar-tailed Godwit *Limosa lapponica*	Invertebrates	2	NB	England	Yates et al. 1993
Redshank *Tringa totanus*	Invertebrates	2	NB	England	Yates et al. 1993
	Shrimps *Corophium*	2	NB	England	Goss-Custard et al. 1970b
Knot *Calidris canutus*	Molluscs	2	NB	England	Yates et al. 1993
Dunlin *Calidris alpina*	Chironomids	3	B	Alaska	Holmes 1970
	Invertebrates	2	NB	England	Yates et al. 1993
Common Crane *Grus grus*	Oak *Quercus* seeds	3	NB	Spain	Bautista et al. 1995
Red Grouse *Lagopus l. scoticus*	Heather *Calluna* shoots	3	B	Scotland	Miller et al. 1966, Moss 1969
Sparrowhawk *Accipiter nisus*	Birds	3	B	Britain	Newton et al. 1986
Buzzard *Buteo buteo*	Rabbits *Oryctolagus*	3	B	Britain	Newton 1979
	Rabbits *Oryctolagus*	2	B	Scotland	Graham et al. 1995
Golden Eagle *Aquila chrysaetos*	Carrion	3	NB	Scotland	Watson et al. 1992
Peregrine *Falco peregrinus*	Birds	3	B	Britain	Ratcliffe 1969
Shelduck *Tadorna tadorna*	Invertebrates	2	NB	England	Yates et al. 1993
Cormorant *Phalacrocorax carbo*	Fish	3	NB	Switzerland	Suter 1995
Cape Petrel *Daption capense*	Krill *Euphausia*	4	NB	Southern Ocean	Heinemann et al. 1989
Adelie Penguin *Pygoscelis adeliae*	Krill *Euphausia*	4	NB	Southern Ocean	Heinemann et al. 1989
Various seabirds	Krill *Euphausia*	3	NB	Southern Ocean	Veit et al. 1993
Various seabirds	Krill *Euphausia*	3	NB	Southern Ocean	Obst 1985
Various seabirds	Various prey types	3	B	Southeast Arabia	Bailey 1966

C. Sudden or long-term changes in food-supply

Species	Food-supply	Spatial scale[1]	Breeding or non-breeding[2]	Region	Source
Common Guillemot *Uria aalge*	Capelin *Mallotus*	3	NB	Barents Sea	Vader et al. 1990
Puffin *Fratercula arctica*	Herrings *Clupea*	2	B	Norway	Anker-Nilssen 1987
Arctic Skua *Stercorarius parasiticus*	Sand-eels *Ammodytes*	2	B	Scotland	Phillips et al. 1996
Great Skua *Catharacta skua*	Sand-eels *Ammodytes*	2	B	Scotland	Klomp & Furness 1992
Dunlin *Calidris alpina*	Invertebrates	4	NB	Britain	Goss-Custard & Moser 1988
Various heron species	Crayfish *Procambarus*	3	B	Louisiana	Fleury & Sherry 1995
Buzzard *Buteo buteo*	Rabbits *Oryctolagus*	2	B	England	Dare 1961
Various seaducks	Invertebrates	2	NB	Scotland	Campbell 1984

Various ducks	Mussels *Dreissena*	2	NB	Lake Eyrie	Wormington & Leach 1992
Mute Swan *Cygnus olor* and others	Pondweed *Elodea*	2	NB	Orkney, Scotland	Meek 1993
Cape Gannet *Morus capensis*	Pilchards *Sardinops*	3	B	South Africa	Crawford & Shelton 1978
Shag *Phalacrocorax aristotelis*	Fish	2	B	Scotland	Aebischer 1986
Cape Cormorant *Phalacrocorax capensis*	Pilchards *Sardinops*	3	B	South Africa	Crawford & Shelton 1978
Jackass Penguin *Spheniscus demersus*	Pilchards *Sardinops*	3	B	South Africa	Crawford et al. 1990
Various seabirds	Sand-eels *Ammodytes*	2	B	Shetland,Scotland	Furness & Barrett 1991, Monaghan et al. 1989
Various seabirds	Anchovies *Engraulis*	3	B	Peru	Schaefer 1970, Idyll 1973

[1] Spatial scale: 1=fine scale, metres to hundreds of metres (one territory to another); 2=coarse scale, 1-100 km (landscape); 3=meso scale, 100-1,000 km (regional); 4=macro, >1,000 km (continental, oceanic).

[2] B indicates breeding density, NB indicates non-breeding, measured in the non-breeding season or away from breeding areas in the breeding season (e.g. pelagic seabirds at sea).

*Correlation with a 1-year lag.

(3) Consistent regional differences in bird numbers, associated with regional differences in food-supplies. Some such variations are associated with regional variations in underlying rock or soil type, which influence biological productivity, including bird densities.

Short-term changes

In the short term, birds adjust numerically to changes in food-supplies in two different ways. The first is in their settling patterns, moving into areas where food becomes plentiful and leaving areas where food becomes scarce. In this way, birds can respond to changes in food-supply more or less immediately, so that their densities can match food-supplies at the time. The second response is through changes in local survival or reproductive success, as described above. This type of response produces a lag, so that high densities follow good food-supplies and low densities follow poor supplies. The two responses (simultaneous and delayed) are not mutually exclusive, as many bird species respond to a local increase in food-supply both by immigration and by enhanced survival and breeding success. Moreover, on the same food-supply, some species respond primarily by immigration, while others respond primarily through reproduction and survival, as later examples show.

Seed-eaters. Some of the most careful documentation of population change in relation to seed-supply concerns the ground finches *Geospiza fortis* and *G. scandens* on the Galapagos Islands (Grant 1986, Gibbs & Grant 1987a,b). In this arid environment, seed production depended on rainfall which fluctuated greatly from year to year (0–1359 mm over 15 years). Over a period of years, finch numbers on Daphne Island varied with the seed-supply, with most mortalities occurring during the dry season shortages that occurred in all but the wettest years. Reproduction also varied with rainfall: in a year with little rain, little or no breeding occurred, but in 1982–83 when rain fell over a long period, both seeds and caterpillars were plentiful; finch breeding occurred over eight months and some individuals raised up to eight broods in succession. Some young raised near the start of the season bred successfully before its end. By January 1984, the population had increased four-fold and young birds comprised more than 75% of the total, but numbers crashed again in the next dry season shortage. Comparing different islands in the same year, finch biomass was closely correlated with seed biomass at the end of the dry season. Thus both temporal and spatial variation in finch mortality, reproduction and numbers were all related to food-supply. The correlations were so close that other factors known to affect breeding and mortality, namely predation and disease, appeared to have little influence on overall numbers. In these birds, therefore, the case for food limiting numbers was about as compelling as could be expected on correlative evidence.

In some continental seed-eaters, which depend on variable seed-crops, similar events occur, but they are harder to document fully, because the main response is through movement which can occur on a huge spatial scale (**Table 7.5**; Pulliam & Parker 1979, Dunning & Brown 1982, Pulliam 1985). In an unusually widescale study, Schluter & Repasky (1991) measured the densities of seed-eating birds and their food-supplies outside the breeding season in similar snow-free habitats on different continents. Half the variation in finch densities across sites could be explained by seed abundance, and at any one site, numbers fluctuated from season to season in accordance with seed-supplies. Finch densities on different continents were roughly similar when seed-supplies were the same, despite the different types of finches present. It seemed that finches were food-limited in all the sites examined. Nevertheless, average densities on mainlands were only about one-fourth the densities found on the Galapagos Islands, a difference the authors attributed mainly to the paucity of competing seed-eaters (such as ants and rodents) on the islands. This broad-scale study confirmed the pattern seen in more local studies of finches where food-supply was measured (Pulliam & Parker 1979, Lack 1987, Schluter 1988), and in regional studies that used rainfall as a surrogate measure of seed-supply (Dunning & Brown 1982).

Other birds whose local densities fluctuate greatly from year to year are the northern finches that depend on tree species whose seed-crops vary from year to year (Newton 1972). Cropping depends partly on the natural rhythm of the trees themselves, and partly on the weather. In any one area, trees of different species tend to fruit in phase with one another, giving an enormous profusion of tree fruits in certain years and practically none in others. The trees in widely separated areas may be on different cropping regimes, partly because of regional variations in weather, so that good crops in some areas may coincide with poor crops in others. Usually, each productive area extends over thousands or millions of square kilometres, and is separated from the next by terrain that, in that year, is almost devoid of seeds. In some years, moreover, the productive patches are plentiful and widespread, and in others few and far between, so that the total production of seeds over a continent varies greatly from year to year.

The birds that feed on such seeds, such as Siskins *Carduelis spinus* and Redpolls *C. flammea*, generally concentrate wherever their food is plentiful at the time, and ringing has confirmed that individuals may breed or winter in widely separated areas in different years (Newton 1972). In this way, their numbers in particular localities fluctuate greatly from year to year in parallel with local seed-crops, and may range between total absence in years when appropriate tree-seeds are lacking, to thousands of birds per square kilometre in years when such seeds are plentiful. Some of the earliest data from a single area were collected by Reinikainen (1937) in mid Finland. He travelled the same route by ski each Sunday in March for 11 years, counted the Crossbills *Loxia curvirostra* he met on his journeys and estimated the cone crops of spruce and pine. The numbers of breeding Crossbills seen each year were strongly correlated with the size of the spruce crop (though not with the pine crop), the highest number of birds being 20 times the lowest, with an increase of this order occurring from one year to the next **(Figure 7.2)**. Similar relationships between local numbers and local food supplies have been found for Siskins, Bramblings *Fringilla montifringilla* and other boreal seed-eaters elsewhere (Haapanen 1965, Newton 1972, Petty *et al.* 1995).

Some tree-seed feeders, notably Crossbills, remain in the boreal forest through most years, leaving southward only in years of widespread crop failure. Other species, notably Siskin and Redpoll, seem to move south each autumn, but only until they find areas rich in food (Svärdson 1957). In consequence, the distance travelled by the bulk of the migrants varies from year to year, according to where the crops are good, and only when the migrants are exceptionally numerous or their food is generally scarce, do they reach the furthest parts of their wintering range, as an

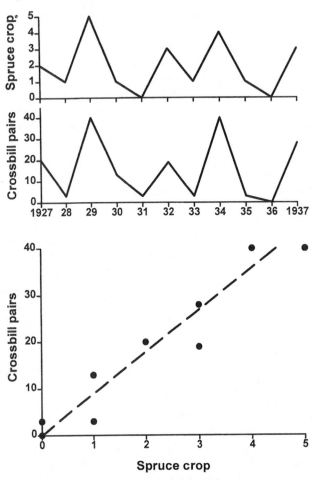

Figure 7.2   Relationship between the population density of the Crossbill *Loxia curvirostra* and the cone crop of spruce *Picea abies*. Crossbills in number of pairs per 120-km transect; Spruce crop classified in five categories. From Reinikainen 1937.

irruption. Over much of North America, conifer crops tend to fluctuate biennially, and it is in the alternate years of poor crops that seed-eaters migrate furthest south. At least eight species of boreal seed-eating birds tend to irrupt together in response to a widespread synchrony in seed-crops (Bock & Lepthien 1972, 1976). Even more remarkable, crop failures and irruptions are often synchronised between much of North America and much of Eurasia, as in 1949, 1955, 1959, 1961, 1963, 1965, 1968, 1969, 1971 and presumably also in more recent years. The synchrony is in turn attributed to circumboreal climatic events bringing tree-fruiting on both continents into phase. On the other hand, crop-failures and bird movements from montane areas south of the boreal region mostly occur in different years (Bock & Lepthien 1976). In the marked fluctuations of their local breeding and wintering populations, boreal tree-seed specialists contrast with some other seed-eaters, which feed from many kinds of herbaceous plants and (except in arid areas) have more stable food-supplies from year to year (Newton 1972).

For obvious reasons, movement emerges as a much more significant factor in continental areas than on small islands (although inter-island movement was recorded in Galapagos), enabling certain species to respond rapidly to local or regional changes in food supplies. For reproduction, some seed-eater species depend on insects, so that whatever the abundance of seeds, they can breed only when insects are available. In the Galapagos finches, caterpillars were plentiful over a long period in the wettest year, and helped to permit prolonged breeding (Gibbs & Grant 1987b). However, many other seed-eaters, including many emberizids, may be unable fully to exploit abundant seed-crops in summer, because of insect shortages. In such species, apart from movement, the main demographic response to improved local seed-supply is enhanced overwinter survival (e.g. Great Tit *Parus major*, **Figure 7.3**; Nuthatch *Sitta europaea*, Nilsson 1982; Acorn Woodpecker *Melanerpes formicivorus*, Hannon *et al*. 1987). In such 'winter-limited' species, we should not necessarily expect breeding densities to correlate with summer food-supplies (in most years following the lower curve (A) in **Figure 3.12**, see Enemar *et al*. 1984 for examples).

Yet other seed-eaters, notably Crossbills and some other cardueline finches, can if necessary feed their young on seeds alone (Newton 1972). These species can take greater advantage of abundant seeds, breeding over a longer period in years when crops are good. Some seed-eaters may even raise successive broods in the same

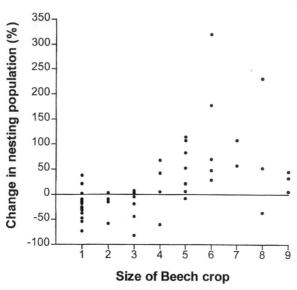

Figure 7.3 Relationship between annual changes in Great Tit *Parus major* breeding density and the size of the Beech *Fagus silvatica* seed crop in the intervening winters. Over a longer period of years with no Beech crop the mean annual population change was –8.9% (16 years), the mean annual adult survival was 47.5% (14 years) and the proportion of local young that survived to breed in the study area was 9.0% (15 years). For years with a Beech crop, the equivalent figures were +37.2% (10 years), 57.9% (8 years) and 15.6% (8 years). Differences between non-beech and beech years were significant (*P*=0.004, *P*=0.021, *P*=0.009, respectively). From C.M. Perrins 1979 and unpublished.

breeding season in localities up to several hundred kilometres apart, as they exploit temporary seed-crops in different places at different times (see Peiponen 1962 for Redpoll, Jaeger *et al*. 1986 for African Quelea Finch *Quelea quelea*).

Raptors and other predators. Further examples of short-term correlations between bird numbers and food-supply are provided by predatory birds that depend on cyclic prey. Two main systems are recognised: (1) an approximately 3–5-year cycle of small rodents on the northern tundras, boreal forests and temperate grasslands; and (2) an approximately ten-year cycle of Snowshoe Hares *Lepus americanus* in the boreal forests of North America (Elton 1942, Lack 1954, Keith 1963). The numbers of certain grouse species also fluctuate cyclically, in some regions in parallel with the rodent cycle, and in others with the longer hare cycle (Chapter 9). The populations of these various prey animals do not reach a peak simultaneously over their whole range, but the peak may be synchronised over tens or many thousands of square kilometres. In Fennoscandia, the periodicity of vole cycles tends to increase northwards from about three years between peaks at 60°N to about five years at 70°N. The amplitude of the cycles also increases northwards from barely discernible cycles south of 60°N to marked fluctuations at 70°N, where peak densities typically exceed the troughs by more than 100-fold (Hanski *et al*. 1991). The role that food plays in the cycles of these mammalian prey species is uncertain, but there is little doubt about the predators, all of which tend to breed most densely and prolifically in conditions of peak prey numbers **(Table 7.5)**.

Annual fluctuations in the densities of vole-eaters can be substantial. In an area of western Finland, for example, over an 11-year period, numbers of Short-eared Owls *Asio flammeus* varied between 0 and 49 pairs, numbers of Long-eared Owls *Asio otus* between 0 and 19 pairs, and Kestrels *Falco tinnunculus* between 2 and 46 pairs, all in accordance with spring densities of *Microtus* voles (Korpimäki & Norrdahl 1989, 1991). All these raptors were summer visitors to the area concerned, and settled according to vole densities at the time.

Predators of voles that respond rapidly by immigration include the three just mentioned that migrate south for the winter, and others, such as the Hawk Owl *Surnia ulula* and Tengmalm's Owl *Aegolius funereus*, that move around within the boreal zone, settling most densely each year wherever voles are plentiful (Korpimäki 1985). In some regions of tundra, the same holds for the Snowy Owl *Nyctea scandiaca*, Short-eared Owl *Asio flammeus* and Rough-legged Buzzard *Buteo lagopus*. Other vole predators, such as the Ural Owl *Strix uralensis* and Tawny Owl *S. aluco*, are residents which turn to other prey when voles are scarce (Saurola 1989, Petty 1992). Because they remain in the same areas year-round, they respond to a vole peak mainly by improved reproduction and survival, reaching peak numbers after the peak in food-supply. The same is true for the Barn Owl *Tyto alba* in Britain **(Figure 7.4)**.

The principal avian predators of Snowshoe Hares, namely Goshawk *Accipiter gentilis* and Great Horned Owl *Bubo virginianus*, are also mainly resident, and show a delayed response, reaching peak numbers after the hare peak. Like the resident vole predators, they respond to a rise in prey numbers by improved reproduction and survival, and to a crash in prey numbers by greatly reduced reproduction and survival, and also by massive emigration, appearing as irruptions far to the south of their usual breeding areas (Mueller & Berger 1967, Houston & Francis 1995, Rohner 1995, 1996).

Widening the comparison to include other raptorial birds, three levels of response to annual prey numbers can be recognised (Newton 1979). Those populations that are subject to the most marked prey cycles show big local fluctuations in densities and breeding rates (e.g. Kestrels in boreal regions); those subject to less marked prey cycles show fairly stable densities, but big fluctuations in breeding rates (e.g. Buzzards *B. buteo* and Tawny Owls *Strix aluco* in temperate regions); while those with stable prey populations show stable densities and fairly stable breeding rates (e.g. Sparrowhawks *Accipiter nisus* in temperate regions). Much depends on how varied the diet is, and whether alternative prey are available when favoured prey are scarce. The more diverse the diet, the less the chance of all prey-species being scarce at the same time. Raptors that have fairly stable food-supplies show some of the most extreme stability in breeding population recorded in birds, with breeding numbers varying by no more than 15% of the mean over several decades (Newton 1979). This level of stability has been recorded in a wide range of species, from Sparrowhawks and Peregrines *Falco peregrinus* to Golden Eagles *Aquila chrysaetos* (Ratcliffe 1980, Newton 1986, Watson 1996). Moreover, the same species may fluctuate numerically in one region, but not in another, depending on the stability of the local prey supply.

Insect-eaters. Some bird species eat insects or other arthropods year round, and others only in summer, turning to plant material in winter. An abundance of insects in summer can lead to improved breeding success, and in some species to increased breeding density in the following year. Some migrant warblers, which breed in the forests of eastern North America, responded to an abundance of caterpillars by raising two broods per year instead of one (Holmes *et al.* 1991, Rodenhouse & Holmes 1992). Over a period of years, and in different forest plots, warbler breeding numbers fluctuated in parallel with caterpillar biomass in the previous summer. The main species concerned were Black-throated Blue Warbler *Dendroica caerulescens*, Black-throated Green Warbler *D. virens* and Red-eyed Vireo *Vireo olivaceus*, but many other species were also affected (Holmes & Sherry 1988, Sherry & Holmes 1991). A similar response has been noted in Pied Flycatchers *Ficedula hypoleuca* in northern Europe **(Figure 1.6)**.

Most outbreaks of defoliating caterpillars in eastern North American hardwood forests are sporadic, occurring in one location in one year and somewhere else the next. The probability that any particular stand will experience an outbreak in any given year is low. Every now and then, however, one North American caterpillar species, the Spruce Budworm *Choristoneura fumiferana*, increases over a period of years to reach plague proportions over wide areas, causing extensive defoliation. Several bird species show strong numerical responses to this insect (Kendeigh 1947, Morris *et al.* 1958, Crawford & Jennings 1989). In one study, Bay-breasted Warblers *Dendroica castanea* increased in abundance from 2.5 pairs per 10 ha in uninfested stands to 300 pairs per 10 ha during an outbreak, Blackburnian Warblers *D. fusca* increased from 25–30 pairs to 100–125 pairs per 10 ha, and Tennessee Warblers *Vermivora peregrina* from 0 to 125 pairs per 10 ha (Morris *et al.* 1958). Because in each outbreak the numbers of the birds rose over several years in parallel with the caterpillars, it was hard to tell how much the increase was due to immigration and how much to the high local breeding success. However, as warblers reached their peak numbers one or more years later than the caterpillars, breeding success was probably involved.

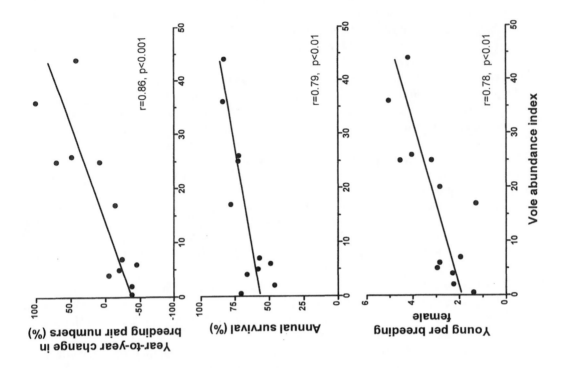

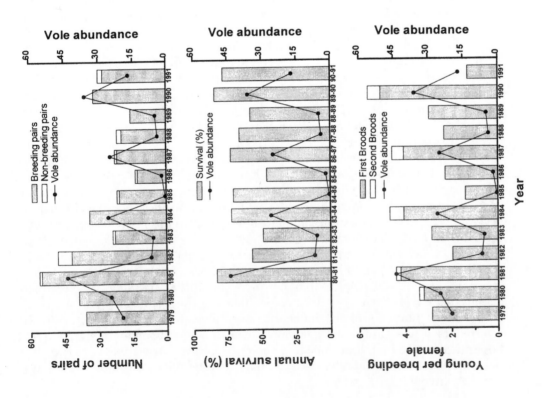

In some other bird populations, too, year-to-year fluctuations in breeding numbers have been linked to food-related fluctuations in previous breeding success. Among various duck species at Lake Myvatn in Iceland, production of young was correlated with the abundance in the previous year of chironomids and other insects which formed the food of ducklings. The mean number of fledglings produced per female in any one year influenced the subsequent change in spring population density, as found in Wigeon *Anas penelope*, Tufted Duck *Aythya fuligula*, Greater Scaup *Aythya marila*, Common Scoter *Melanitta nigra* and Harlequin Duck *Histrionicus histrionicus* (Gardarsson & Einarsson 1994). In only one species, Barrow's Goldeneye *Bucephala islandica*, did this relationship not hold.

*   *   *

Temporal correlations between bird numbers and food-supply in particular localities are sometimes impressive, with changes in numbers keeping strict parallel with changes in food, or keeping in step but a year or more behind. Such correlations are consistent with the idea of food limiting numbers, but they do not prove it because of the possibility that other influential factors may have changed along with food. Nor do they tell us if populations were as high as they might have been if food alone were limiting, so one cannot exclude an influence of other limiting factors. The fact that, on a similar food-supply, finch densities on mainlands were one-fourth the level on islands suggests that other factors, such as predation and competition, were involved. Spatial correlations between bird numbers and food-supplies might also have other explanations: for example, that birds already limited by some other factor then distribute themselves according to food-supply, or again that birds are limited in each area by some unknown factor that varies in parallel with food. As an example, relationships between caterpillar abundance and subsequent breeding density, described above for North American birds, have also been noted in some European species, but the interpretation differed (Tomialojc & Wesolowski 1990). Good breeding in the European study was attributed to reduced predation on nest contents in years of high caterpillar numbers, as some predators switched to feeding on caterpillars instead of on the contents of bird nests. In this area, then, year-to-year

Figure 7.4   Numbers, survival and breeding success of Barn Owls *Tyto alba* in relation to abundance of Field Voles *Microtus agrestis* in an area of south Scotland. In this population, immigration/emigration was slight and year-to-year changes in numbers were due mainly to local changes in survival and breeding success. From Taylor 1994.
(a)   Numbers of breeding pairs in different years in relation to an index of vole abundance. Owl numbers were higher in the peak vole years, but were also high in the years after the peaks. However, the percentage change in the number of breeding pairs varied between years in relation to spring vole abundance. The numbers of owls showed a rapid tracking of the vole population (variation 14–94 pairs between years).
(b)   Overwinter survival of adults in relation to an index of vole abundance. Survival was highest in the good vole years (variation 20–56% between years).
(c)   Number of young produced per breeding female in relation to an index of vole abundance. Production was highest in the good vole years (variation 1.3–5.6 young between years).

variations in breeding success were related mainly to predation, rather than to the frequency of second broods, as in the North American warblers, in which case different mechanisms were involved in producing the same correlation.

It is also possible that bird numbers are limited by food in food-rich areas, but not in poor ones. Many bird species occupy good habitats in preference, and when these are full, spread to poorer habitats (Chapter 3). This means that, in the best habitats, numbers may be proximally limited by social behaviour (perhaps in relation to food), whereas in poorer habitats, the numbers depend largely on overspill; they may vary more from year to year than in the best habitats, yet seldom (if ever) reach saturation. An example of a study showing that density was limited in relation to food in both high and low density areas was that of Holmes (1970) on Dunlin *Calidris alpina* in Alaska.

In the same way that birds have preferred habitats, some species also occupy parts of their breeding and wintering ranges every year, and other parts less often, perhaps only in years of high population or extreme food-shortage. If food were limiting, we might again expect correlations between food-supply and numbers in the core areas but not in the marginal areas which depend on 'overspill'. This phenomenon is evident among the various seed-eaters, discussed above, in which year-to-year correlations between local densities and seed stocks were most evident within the northern 'regular' range. They would not necessarily be expected in the most southerly wintering areas, which the birds reach only in occasional years. In some years, as observations confirm, seeds are plentiful in southern areas, but no wintering finches arrive to exploit them. And when they do, their numbers in different areas do not necessarily correlate with seed supplies if even small crops are enough to satisfy the birds present (Pulliam & Dunning 1987).

Long-term changes

Within particular regions, long-term or stepwise changes in bird populations from one level to another often coincide with long-term or stepwise changes in their food-supply. To begin with an extreme example, near Edinburgh city in southeast Scotland, every year some 16 000 seaducks wintered offshore, feeding on the rich invertebrate life that was in turn fed by raw sewage from the large human population of the region. After the installation of a new sewage processing plant in 1978, the invertebrates declined, and in two years the duck population had fallen to about 3000 (Campbell 1984).

Over much of Europe, many farmland bird species have declined in numbers over the past 30 years, some species in Britain by more than 90% (Tucker & Heath 1994, Fuller *et al.* 1995). For some species, the main likely causes are the massive reductions in food-supplies that have resulted from agricultural intensification (Table 7.6). Seed-eaters are affected by loss of wild food-plants (through herbicide use), and by the change from spring-sown to autumn-sown cereals, with the consequent loss of winter stubbles and their associated cereal and weed seeds (Chapter 15). Invertebrate feeders are affected by insecticide use (and resulting reduction in arthropod populations), by herbicide use (and resulting loss of insect food plants), by the conversion of pasture to arable (and associated reduction in earthworm and other soil invertebrate populations), and by land drainage (which dried the top soil, making surface-dwelling invertebrates less active, and soil-

Table 7.6 Causes of recent population declines in some bird species of European farmland that have been studied in detail.

Species	Crucial food-supply	Cause of reduction in food-supply	Mechanism of population decline	Source
Tree Sparrow *Passer montanus*	Winter seeds	Reduction of seed-supply through destruction of food-plants by herbicide use, depletion of seed bank in soil, and loss of winter stubbles.	Probably increased mortality of full-grown birds.	Summers-Smith 1996
Linnet *Carduelis cannabina*	Small weed seeds	Loss of seed-supply through destruction of food-plants by herbicide use, depletion of seedbank in soil, and loss of winter stubbles.	Probably increased mortality of full-grown birds.	Newton 1986, 1995 O'Connor & Shrub 1986
Cirl Bunting *Emberiza cirlus*	Winter seeds Summer insects	Loss of winter stubble fields, herbicide and insecticide use, destruction of old pastures which provided grasshoppers and other insects.	Increased winter mortality[1], and reduced breeding success.	Evans & Smith 1994
Corn Bunting *Miliaria calandra*	Winter seeds	Loss of winter stubble fields, herbicide use.	Increased winter loss[2]	Donald & Forrest 1995
Starling *Sturnus vulgaris*	Soil invertebrates Animal feed	Reduction in invertebrate food-supply associated with conversion of pasture to autumn-sown cereals and drainage of remaining pasture; redesign of poultry and pig units to exclude birds.	Increased mortality of nestlings and full-grown birds.	Feare 1994 Tianen *et al.* 1989
Skylark *Alauda arvensis*	Small weed seeds Arthropods	Reduction of arthropod supply through insecticide use and of seed supply through herbicide use and loss of winter stubble.	Inadequate chick production; probably also increased adult mortality.	Schläpfer 1988 Jenny 1990
Lapwing *Vanellus vanellus*	Soil invertebrates	Land-drainage, resulting in drying and hardening of top soil, making invertebrates less available, and giving access to machinery earlier in the year.	Inadequate chick production	Beintema & Muskens 1987 Peach *et al.* 1994
Partridge *Perdix perdix*	Insects for chicks	Destruction of insect food-plants by herbicide use; direct destruction of insects by insecticide use.	Inadequate chick survival[1]	Potts 1986 Rands 1985
Lesser Kestrel *Falco naumanni*	Grasshoppers and other insects	Reduction in insect abundance through pesticide use.	Inadequate chick survival	Hiraldo *et al.* 1996

[1] Supported by experimental evidence.

[2] Additional suggested cause: greater nest destruction through earlier grass-cutting for silage.

In addition to the above species, two other declining farmland birds that have been studied include the Corncrake *Crex crex* and Stone Curlew *Burhinus oedicnemus*, in both of which farm mechanisation was found to be the primary cause. Tractor-drawn grass-cutting equipment greatly increases the mortality of young and adult Corncrakes, causing population decline (Norris 1945, Green 1995), while spring harrowing of arable land destroys the eggs of Stone Curlews, causing population decline (Green 1988).

dwelling ones less available, especially in late summer). The quality of evidence varies between the species listed: in some, such as the Grey Partridge *Perdix perdix*, the conclusions from field observations have been well tested by experiments (Potts & Aebischer 1991, Chapter 15), but in others, the evidence as yet is little more than surmise.

In contrast, for much of the 20th century in both Europe and North America, many seabird species have undergone marked population expansions, which in some species were attributed to increased food-supplies. For certain species, the new food sources were largely land-based, provided by sewage outfalls and rubbish dumps, while for others they were largely or entirely sea-based, provided mainly by offal and other discards thrown overboard from fishing vessels and from factory ships. The spectacular expansion of the Fulmar *Fulmarus glacialis*, which in Britain (excluding St Kilda) increased from 24 pairs in 1878 to nearly 600 000 pairs in 1990, was attributed by Fisher (1966) to growth of the fishing industry and the huge supplies of offal food that followed, first from whaling and then from fish trawling. According to Furness *et al.* (1992), fishing boats in the waters around Britain may discharge 95 000 tonnes of offal and 135 000 tonnes of whitefish discards every year, enough to feed two million scavenging seabirds. Although Fulmars take most of this material, other beneficiaries include the Gannet *Morus bassanus*, Great Skua *Stercorarius skua*, Great Black-backed Gull *Larus marinus*, and (in smaller numbers) Lesser Black-backed Gull *L. fuscus*, Herring Gull *L. argentatus* and Kittiwake *Rissa tridactyla*. All these species have shown huge increases in population sizes during the 20th century (Cramp & Simmons 1983, Lloyd *et al.* 1991). In recent years, however, some seabird populations have begun to decline, after fish stocks collapsed, often associated with human overfishing **(Table 7.5)**. Similar declines have occurred periodically elsewhere in the world, linked either with overfishing (Chapter 13) or with natural events, such as changes in sea currents which have reduced local food-supplies (Chapter 11).

For some seabird species, attempts have been made to find which demographic variables were involved in the declines. In certain species, after the collapse of fish stocks, declines in breeding numbers were attributed primarily to low recruitment after several years of poor breeding (for Puffin *Fratercula arctica* in Norway, see Anker-Nilssen 1987; for Arctic Tern *Sterna paradisaea* in Shetland, see Monaghan *et al.* 1989). In other species, declines were attributed to increased overwinter mortality of adults (for Guillemot *Uria aalge* in the Barents Sea, see Vader *et al.* 1990; for Galapagos Penguin *Spheniscus mendiculus* in Galapagos, see Anderson 1989; for Guillemot and Shag *Phalacrocorax aristotelis* in Shetland, see Furness & Barrett 1991). In yet other species, declines were attributed to a combination of poor breeding and increased mortality (for Great Skua in Shetland, see Klomp & Furness 1992). The differences depended largely on the extent to which alternative foods were available in summer or winter, and as expected, declines in breeding numbers were most rapid when they involved increased adult mortality.

Turning now to some other examples of long-term change, recovery in the numbers of some endangered bird species has followed the provision of additional food. The restoration of the Trumpeter Swan *Cygnus buccinator* population in Montana/Idaho was an early example where winter food-shortage limited numbers until grain was provided. The population then increased from about 30 birds in 1936 to about 600 birds in the mid-1950s (Cade & Temple 1995). Other examples concern various crane species that had been reduced to low numbers through loss of winter habitat and food-supplies. Winter feeding programmes benefited populations of the White-naped Crane *Grus vipio* (which increased at Izumi, Japan, from 45 birds in 1959 to 732 in 1976), Hooded Crane *G. monacha* (increased at Izumi from 357 birds in 1959 to nearly 3000 in 1976), and Japanese Red-Crowned Crane *G. japonicus* (increased in Hokkaido from 33 birds in 1952 to 220 by 1976) (Cade & Temple 1995). In all these examples, however, food-provision formed part of a wider management package (including legal protection), so that it is hard to separate the role of extra food from other factors that might have changed at the same time.

Many other examples of long-term or sudden changes in bird populations and their food can be found in ornithological journals, though mostly at the anecdotal level. Such studies suggest that food limited numbers at different levels in different periods, and that sudden declines in numbers were caused by sudden declines in food; in some cases, other factors, such as disease and predation, were excluded as being important. Such events form natural experiments demonstrating the effects of food-supply, but they lack controls, and often one cannot exclude the possibility that some unknown limiting factor changed at the same time. Among seabirds, the increase since the 19th century could well have been due to food (which for some species is known to have increased), but it is hard to eliminate another frequent explanation, that some such species have been recovering from human persecution in the past. Probably both factors were involved, their relative importance varying between species and regions.

Differences between areas of different productivity

For geological or other reasons, some areas of land and sea are more productive than others, and support more life, including birds. In some studies, variations in bird densities have been related, not to food directly, but to some supposed index of food, such as soil productivity. Data of this type were collected for forest birds in Finland, where there are three main forest-forming trees, namely birch *Betula*, spruce *Picea* and pine *Pinus* (Palmgren 1930, von Haartman 1971). On similar soils in mature forest, songbirds were most numerous in birch, less numerous in spruce, and scarcest in pine, while in mixed spruce/birch woods they were most numerous of all. Most of the birds concerned were insectivores, whose densities were paralleled by similar variations in insect abundance. For any one forest-type, however, songbirds were more numerous in forests on good soil than in forests on poor soil **(Figure 7.5)**. This was again attributed to food-supply, for several aspects of productivity were better on good ground, such as the growth and fruiting of trees, insects on the foliage, and earthworms in the soil below. Many of the birds concerned were summer visitors to Finland, and of those that stayed the winter, some moved outside the woods. Hence, the variations in bird breeding numbers must have been due, not to birds surviving better over winter in some woods than in others, but to their settling patterns in spring. Similar trends were noted in songbirds elsewhere in Europe, whether or not they wintered locally (Moss 1978).

The studies above were concerned with the entire songbird fauna, all species together. But correlations between bird density and soil fertility have been noted within species, comparing densities in different areas. Among game birds, Red

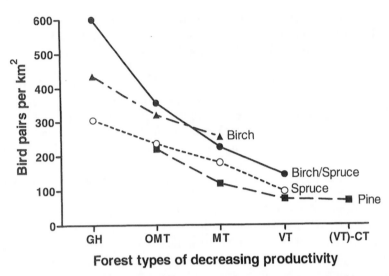

Figure 7.5   Densities of breeding birds in Finnish forests in relation to forest type and dominant tree-species. GH – grass-herb; OMT – *Oxalis-Myrtillus*; MT – *Myrtillus*; VT – *Vaccinium*; CT – *Calluna*. Modified from von Haartman 1971.

Grouse *Lagopus l. scoticus* were more numerous on land overlying basic than acidic rock, associated with greater nitrogen and phosphorus content in the food plant *Calluna vulgaris* on basic rock (Miller *et al.* 1966, Moss 1969). Among raptors, Sparrowhawk *Accipiter nisus* densities in 12 parts of Britain were strongly correlated with soil productivity and elevation, and, in areas where counts were made, with the abundance of small bird prey **(Figure 7.6)**. Within woodland, different hawk pairs nested only 0.5 km apart in the richest prey areas and more than 2.0 km apart in the poorest. Similar relationships were found for Peregrines *Falco peregrinus*, which spaced themselves at 2.6 to 10.3 km apart in different regions of Britain, according to soil productivity and associated prey densities (though prey were not measured directly) (Ratcliffe 1969). Higher Peregrine densities than any in Britain were found on the Queen Charlotte Islands, off western Canada, where the mean distance between about 20 pairs was 1.6 km, in this case linked with massive concentrations of seabirds (Beebe 1960). This population later dropped to about six pairs, following a decline in prey numbers (Nelson & Myres 1975).

In Hwange National Park, Zimbabwe, Tawny Eagles *Aquila rapax* nested an average of 3.5 km apart on basalt, which supported an abundance of plant and animal life, but 59 km apart on Kalahari sand, which was much less productive (Hustler & Howells 1989). No difference in breeding success occurred between areas. An opposite relationship with soil fertility occurred among Ravens *Corvus corax* in northern Britain, in which breeding densities were greater on base-poor (granite) areas than on sedimentary ones, and greater at higher than at lower elevations (Marquiss *et al.* 1978). However, these relationship arose because Ravens fed largely on carrion, and dead sheep were most frequent on poorer and higher areas.

Similar relationships hold in water birds, comparing nutrient-rich (eutrophic) lakes with nutrient-poor (oligotrophic and dystrophic) ones. **Table 7.7** illustrates this

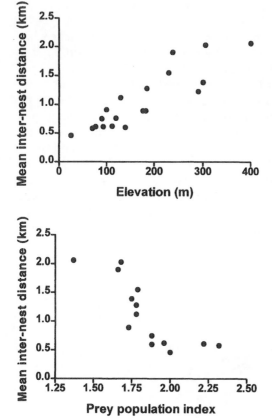

Figure 7.6 Mean nearest-neighbour distances of Sparrowhawk *Accipiter nisus* nesting territories in continuous nesting habitat shown in relation to altitude above sea-level, land-productivity and food-supply in different parts of Britain. Nearest-neighbour distances widen with rise in altitude or fall in land productivity and food-supply. From Newton *et al.* 1986.

point, again based on Finnish studies collated by von Haartman (1971). The greater productivity of eutrophic lakes was reflected in higher densities of invertebrates and fish, and in turn of birds. Other factors were involved, however, for the vegetation was better developed in and around eutrophic lakes and offered more nest-sites; also bird densities varied partly with the lake depth, with the length of shore in relation to water surface, and with the presence or absence of islands. In this case, as for the woodland birds discussed above, the density differences resulted from the settling patterns of breeding birds, not from their survival outside the breeding

Table 7.7   Densities of birds and other animals in different types of lakes in Finland. From von Haartman 1971.

	Lake types of decreasing productivity		
	Eutrophic	Mesotrophic	Oligotrophic and dystrophic
Bottom fauna (individuals per m²)	2000 – 10 000	?	0 – 1000
Fish production (kg per ha)	ca 10	2 – 3	2 – 3
Birds (pairs per km²)*	11 – 130	21 – 27	1 – 39

*Including ducks (Anatidae), grebes (Podicipedidae), divers (Gaviidae), rails (Rallidae), and Common Tern *Sterna hirundo.*

season. The lakes in question supported no bird-life in winter when they were frozen. The species of birds involved also differed between the different types of lake, with plant-feeders commonest on the eutrophic ones. On Swedish lakes, bird densities were linked specifically with phosphorus content, the main limiting nutrient, as well as with edge vegetation (Nilsson & Nilsson 1978).

As in terrestrial birds, correlations between nutrient status and density are evident in individual aquatic species, comparing areas. For example, in the Dipper *Cinclus cinclus* in different parts of Britain, territories along streams were longer, and densities lower, where the pH of the water was low **(Figure 7.7)**. This was in turn associated with decreases in suitable food-items, such as mayfly nymphs and caddis larvae (Ormerod *et al.* 1985, Vickery 1991). In addition, historical data from one Welsh river showed that Dipper numbers declined by 80–90% during a period of increasing acidity, when the invertebrate food-supply declined. In contrast, another stream-nesting bird, the Grey Wagtail *Motacilla cinerea*, showed no link with stream acidity, but its food was of terrestrial rather than aquatic origin, and included flies, caterpillars and spiders, unaffected by stream pH.

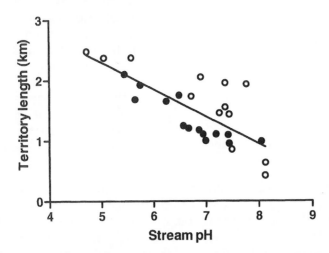

Figure 7.7   Territory length of Dippers *Cinclus cinclus* in relation to mean stream pH in Wales (open symbols) and Scotland (filled symbols). Redrawn from Vickery & Ormerod (1991).

Among shorebirds wintering on coastal flats, densities were related to the types of sediments present, with fewest birds on sand and most on mud which held the greatest densities of invertebrate prey (Yates *et al.* 1993). On a single large estuary, spatial variations in the densities of each of eight shorebird species were related to the densities of their prey species, which were in turn related to substrate type (mainly particle size) and to inundation time **(Figure 7.8)**.

Large concentrations of seabirds are often associated with oceanographic features, such as upwellings and current convergences. These features promote water mixing and nutrient cycles, which in turn influence the abundance of plankton and other marine life on which seabirds depend (Bailey 1966, Jehl 1974, Brown *et al.* 1975, Brown 1979, Pocklington 1979, Griffiths *et al.* 1982, Gould 1983, Hunt & Schneider

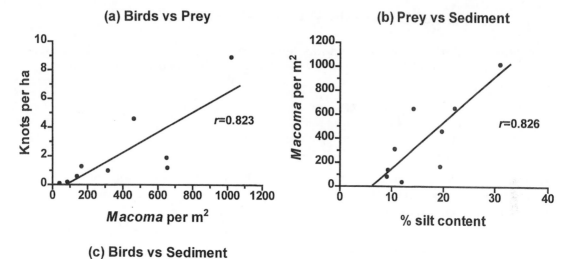

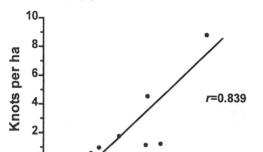

Figure 7.8   Relationship between density of Knots *Calidris canutus*, the molluscan prey *Macoma balthica* and the sediment type on the Wash, southeast England. From Yates *et al.* (1993).

1987). Some of the largest seabird concentrations in tropical seas, which are generally poor in birds, occur at upwellings (such as near southeast Arabia and northwest Africa) or where cold nutrient-rich currents penetrate warm nutrient-poor seas (such as the Peruvian coast and the Galapagos Islands). On a somewhat smaller scale, seabirds associate with other features that promote concentrations of food, such as ice edges (Bradstreet 1979), coral reefs and algal rafts (Haney 1986), and at even smaller scales, of tens or hundreds of metres, they associate with fish shoals or other local food sources (Brown 1980, Obst 1985). Hence, partly through the influence of external physical features, correlations between seabird numbers and food-supplies are evident at a range of different spatial scales.

In the absence of human intervention, such geophysical and chemical attributes of the environment promote consistent regional differences in biological productivity, affecting almost the entire biota. Within limits, they give a rough indication of what would be expected in particular areas, in the absence of detailed counts and surveys. They also indicate the changes in bird and other wildlife populations that are likely to occur under certain types of human impact, such as eutrophication and acidification (Chapter 15). Indirectly, they provide further evidence for the influence of food-supply on bird densities.

MEASUREMENTS ON THE BIRDS THEMSELVES

Apart from studies in which starvation was demonstrated, at least four types of data from the birds themselves have been used as evidence that food can limit bird numbers:

(1) low weights and poor body condition;
(2) low daily food-intakes and feeding rates;
(3) large proportion of available time spent feeding; and
(4) fighting over food.

Each type of evidence by itself is no more than suggestive, but becomes useful when added to other information. There are also pitfalls in the use and interpretation of such data, as discussed below.

Weights and body composition

In theory, data on weights and body composition through the year could serve the same purpose as data on starvation, in indicating periods when feeding is difficult. But they can also be misleading. In the north temperate zone, small birds are generally heavier with more fat and protein in their bodies in winter (when food is scarce) than in summer (when food is plentiful) (King & Farner 1966, Newton 1972). This has been attributed to the need for greater reserves in winter to survive the long cold nights and the greater unpredictability of food-supplies then. Winter weights also vary in relation to dominance status and predation risk (Chapter 13). In the milder months, weights fluctuate according to the stage of breeding, in females increasing before egg-laying and in both sexes declining during the nestling period (Newton 1972, 1979, Bryant 1975). Evidently factors other than available food influence body weights, and some of the lowest weights occur at times of plenty. Among small birds, even the heaviest individuals may die within a day or two if food becomes scarce. So starving birds are not present for long and, if they form only a small percentage of a population at one time, they may be hard to detect. With small birds, therefore, the scarcity of underweight individuals in trapped samples does not necessarily indicate that food is plentiful or that starvation is rare.

Among large birds, however, underweight individuals survive for longer, and are therefore more liable to be sampled. Birds as large as geese can survive for weeks on reduced rations (Newton 1977; Chapter 11), so all the birds in a population might be underweight at one time, yet recover later. The interpretation of weight data should therefore depend on the type of bird involved, and the stage in the annual cycle; at best it can only indicate periods of food-shortage and the individuals most affected.

Feeding rates and food-intake

Sometimes measurements of food-intake indicate that some birds are obtaining less food per day than they need to maintain their weight, so that if this continues such birds will starve. At certain seasons, Ward (1965) found that some of the *Quelea* collected as they arrived at a roost had begun the night with an insufficient reserve of crop food and body fat to last till morning. He concluded that such birds would die, and used changes in the proportions of under-fed birds in roost samples obtained through the year to indicate periods of food-shortage.

Measurements on feeding rate in the field have also been used to indicate periods of difficulty. As food density declines, so does the rate of intake, until a bird can no longer obtain its daily needs in the time available (Chapter 3). Not all individuals are affected similarly: some feed less efficiently than others, because they are either inexperienced or subordinate. As individuals become affected over a period of time, they either starve or move away, so that local numbers are progressively reduced by food-shortage. This kind of evidence – produced for Woodpigeons *Columba palumbus* feeding on farm crops and for various shorebirds feeding on coastal invertebrates – was discussed in Chapter 3. At best it can indicate the proportions of birds in difficulties at different seasons in the locality concerned, and the age, sex and social status of the individuals involved. Whether bird numbers are reduced over a wider area depends on whether the emigrants find alternative feeding places or whether they die.

Proportion of time spent feeding

Data of this type, collected through the year, have been used to define the seasons when birds spend most time feeding, and thus by implication when they have greatest difficulty in meeting their needs. If non-breeding birds spend practically the whole day feeding, the argument goes, they must be near the food limit (Lack 1954). When breeding, many birds forage for long periods to collect food for their young, so while the parents themselves may not be at risk from starvation, their chicks may be.

Outside the breeding season, such data have been obtained either by watching particular individuals for long periods (Gibb 1956), or by frequent scanning of flocks and each time counting the proportion of birds that were feeding (Goss-Custard *et al.* 1977a). Only by the first method can individual birds be compared, whereas the second method gives average values for a population, provided that feeding and non-feeding birds are equally visible. Most temperate zone species that have been studied have been found to spend more time feeding in winter than in autumn or spring. In mid-winter, this proportion rose to 90% or more of the daylight hours in various tit species (Gibb 1954, 1960), and to 95% or more in Woodpigeons and various waders (Murton *et al.* 1966, Goss-Custard *et al.* 1977a). For Rooks *Corvus frugilegus*, however, food was scarcest in mid-summer (after breeding) when birds spent more than 90% (equivalent to 15 hours) of the daylight feeding (Feare 1972). For all these species, the seasonal change in feeding routine paralleled seasonal changes in measured food-stocks.

Such data on their own do not prove food-shortage, because they give no indication whether the birds succeed or fail in their attempts to get enough. Some individuals may be merely sampling different food patches or involved in caching activities. Moreover, in some kinds of birds, a tendency to feed for only a short time each day does not necessarily indicate that all is well. Herbivorous birds might fill their crops in a short time, but may then be unable to digest the food well enough or fast enough to maintain body weight. In some sit-and-wait raptors, hunting periods may be hard to define, and food might be obtained in only a few minutes of intense activity, yet not frequently enough to keep the bird alive. Hence, information on feeding periods is of use in some species, but not in others, and the results will normally do no more than pinpoint the difficult periods.

Fighting over food

Birds often fight over food items or feeding places, and measurements of such fighting through the year have been used to indicate difficult periods. In some cases, the study was broadened to include other kinds of interactions that lower feeding rates, such as birds avoiding or paying more attention to one another. Several studies showed that birds interacted while feeding more in winter than in spring or autumn, or more as their density rose, with the harmful effects usually falling more heavily on the young or other subordinate individuals (Gibb 1954, 1960, Goss-Custard *et al.* 1982; Chapter 3).

Caution is again needed with this type of information. For one thing, fighting may be more related to the distribution or type of food than to its total amount. In Rooks, much more fighting occurred during periods of snow, when birds were forced to feed close together in limited places, than in summer, when food was scarcer but more scattered, and when most mortality occurred (Patterson 1970). Fighting may have influenced density in particular places, but had no bearing on limiting total numbers if alternative places were available, or if different individuals fed at different times. Likewise, the incidence of food-robbing may be more related to the size of food-morsels than to other factors. Food-robbing was rare when tits took small items that could be swallowed quickly, but increased greatly when they took large items, which had to be manipulated before being swallowed, thus giving more opportunity for food-robbing to occur (Gibb 1954, 1960). The same effects were seen in various shorebirds (e.g. Goss-Custard *et al.* 1982). In some seed-eaters, aggression decreased when food became scarce, because feeding took priority (Caraco 1979). The reciprocal result was obtained by supplementing the natural food of wintering Yellow-eyed Juncos *Junco phaeonotus*, which then spent less time feeding and more in fighting. Hence, again there are difficulties with this kind of information, and each case should be assessed carefully.

It has also been argued that the very occurrence of fighting for food can be taken as evidence that food is limiting, for there would otherwise be no point in the birds wasting time and energy (Lack 1954). This argument is probably invalid, however, for it would be advantageous for an individual to fight whenever it could obtain food more easily that way than by seeking it elsewhere; that is, whenever the benefit outweighed the cost. Some fighting over food may also serve to establish a peck-order, from which the winners may benefit in future.

*   *   *

In conclusion, all these various measures on the birds themselves have proved useful in indicating the times of year when feeding was difficult, and in certain studies the proportions and types of birds that were worst affected. All are more useful for certain species than for others, and in any case must be interpreted with care and along with other data, ideally on seasonal changes in numbers. Such data are also more valuable if they are obtained after predicting them from some previous independent data suggesting food-shortage, rather than being collected along with a variety of observations all made at the same time, and interpreted retrospectively.

MEASUREMENTS ON THE FOOD-SUPPLY

Short-term effects on food-supplies

In further attempts to assess the effects of food-supply on bird numbers, some biologists have measured the rate of decline in food-stocks over a period of weeks or months, as birds fed. Many birds of high latitudes have a more or less fixed amount of food to last them through the winter, with little or no replenishment of stocks until the spring. This applies, for example, to seed-eaters and to many northern insectivores. For some species, it has proved possible to measure the standing crop of food, and the extent to which it is depleted through the winter. If the birds remove a 'considerable proportion', the argument goes, they are more likely to be up against a food limit than if they remove only a small proportion.

In several species, food-stocks were depleted by more than 50% during winter, and in other species by more than 90% (Lack 1954, 1966). Wintering insectivores often removed a large proportion of invertebrates in woodland (Soloman *et al.* 1976, Norberg 1978), as did shorebirds feeding from coastal mud (Goss-Custard 1980, Evans & Dugan 1984, Piersma 1987). Geese and Wigeon *Anas penelope* removed 66–91% of the biomass of *Zostera* during winter (Madsen 1988); Barnacle Geese *Branta leucopsis* removed up to 82% of the yield of grasslands (Percival & Houston 1992); and various waterfowl have removed almost all the grain remaining in unharvested fields (Baldassarre & Bolen 1994). Seabird communities have been estimated to consume some 5–64% of the annual pelagic prey production in different areas (Montevecchi 1993). Many other examples of substantial depletion of winter food-stocks can be found in the ornithological literature.

In some species, depletion of food-supplies varied between years or between areas, indicating years or areas when food limitation was most likely. For Bullfinches *Pyrrhula pyrrhula* in an English wood, practically all the seeds from favoured food-plants disappeared during winter in years of poor seed production, but only a small proportion in years of good production (Newton 1972), and the same was true for various sparrows in Arizona (Pulliam & Parker 1979). If these seed-eaters were food limited, therefore, this was probably only in the years with poor crops. As for area differences, wintering Eiders *Somateria mollissima* removed 48–69% of the biomass of mussels *Mytilis edulis* in the Gulf of St Lawrence, compared with only 13% in the Dutch Waddensea and 6% in the Baltic (Guillemette *et al.* 1996). Eiders were thus more likely to be food limited in the former than in the latter areas.

Results from winter studies often contrast strikingly with those on other species from summer, when plants are growing and animal food-species are more abundant and reproducing. In most data in the literature, birds took less than 10% of potential food in the breeding season, but occasionally up to 50% (for examples, see Lack 1954, 1966). Many summer data were collected during insect plagues, however, when it was not surprising that insectivorous birds took a negligible proportion. Spraying infested forests with insecticide in summer has often had no effect on bird breeding, the insects remaining being sufficient for the birds' needs (see above), but the situation may differ in years of insect scarcity. Herbivores can also be affected by shortage in summer, as shown for Barnacle Geese *Branta leucopsis* in Svalbard (Prop *et al.* 1984), and for other species elsewhere (see below), in all of which depletion of food-supplies limited breeding output.

Measurements on food-stocks, like those on the birds themselves, can usually do no more than define the periods and localities of food-scarcity. There are also problems in measuring certain kinds of food accurately, and with the interpretation of data. In some cases the total stock of food was estimated, but no indication was given of what proportion was available to the birds – a proportion that may have varied with conditions. Secondly, while for some birds all known foods were measured, in others only favoured or particular foods were measured. It was always possible that when favoured (or known) foods were finished the birds could turn to alternatives, perhaps previously unknown to the observer. Thirdly, it was not usually known how much food the birds had to remove before they got into difficulties. In one species the removal of 90% of food might cause no obvious mortality, while in another the removal of only 10% might cause a lethal reduction in intake rate.

Herbivores, such as grouse, often appear surrounded by food, but feed selectively on the most nutritious items (Watson & Moss 1979). Such birds normally remove a tiny fraction of the food available (about 2% in Red Grouse *Lagopus l. scoticus*, Savory 1978), if food is defined as the total quantity of leaves, shoots, buds and fruits present. However, preferred plants often show obvious overgrazing, while other plants nearby remain untouched (e.g. Seiskari 1962). If only these preferred plants provide food adequate for survival, then overgrazing of these could often occur. This raises the related point that plants might respond to heavy grazing by reducing their nutrient content or by increasing the chemicals that make the plant unpalatable or indigestible, thus reducing their food value even further (Schultz 1964, Haukioja & Hakala 1975). Moreover, for some herbivores, the efficient exploitation of a particular food requires changes in behaviour, gut anatomy and gut flora (see above).

While the proportion of food removed often gives no indication of whether food is limiting, this statistic may sometimes be useful in showing how close birds are to the food-limit. If it is found, for instance, that birds already remove more than 90% of their food, and that no apparent alternatives are available, then a big increase in bird numbers locally is clearly not possible. In many cases, however, the proportion of food removed may be less than 20%, in which case other information is needed in order to predict whether a further increase in numbers would be possible. It would also be useful to know whether any decline in food-stocks was due to the birds themselves, or to something else (other animals or weather). If it was due to the birds, then larger numbers would lower food density further, increase competition, and cause more losses. But if it were due to something else, bird numbers may have relatively little influence on trends in food-supply.

In conclusion, measurements on the food-supply can normally do no more than measurements on the birds themselves, and indicate likely periods of difficulty. Again, they are more use on some species than on others, and special care is needed with species that feed selectively on vegetation. Before they can give a clue to the effects of food-supply on numbers, periodic counts of birds are also needed, for example: (1) if within one year a conspicuous drop in food-stocks coincided with a conspicuous drop in bird numbers or a low point in the annual population cycle (for *Quelea*, see Ward 1965), or (2) if annual variations in winter food-stocks correlated with annual variations in losses of birds (for Coal Tit *Parus ater* see Gibb 1960, for Great Tit *Parus major* see Perrins 1979). Such evidence would still be correlative, however, and could be strengthened by experiment.

Longer-term effects on food-supplies

Most measurements of the effects of birds on their food-supplies have involved assessments in particular years, as described above, but there can also be longer-term effects if the amount of food removed in one year affects the amount available in the next. Herbivores in particular can gradually destroy their food-supply over a period of years so that the carrying capacity of their habitats is progressively reduced. Such effects have been documented for Arctic-nesting geese, notably Snow Geese *Anser caerulescens* in northern Canada (Kerbes *et al.* 1990). The colony at La Perouse Bay in Manitoba built up from 2000 to 8000 pairs over a 15-year period. As the colony grew, the geese gradually removed their favoured food-plants (sedges), allowing mosses and other less palatable plant species to spread, and in some areas reducing the ground to bare mud. As this happened, the geese had to walk progressively further from the nesting colony to find suitable food, and the growth and survival rates of young geese declined. Another major colony at the McConnel River collapsed some years earlier, again associated with destruction of favoured vegetation. Perhaps this is a common process, with goose colonies periodically becoming established in new areas, building up over two or more decades until the vegetation is degraded, and then declining. If so, we would then expect to find at any one time colonies at various stages of rise and fall, together with evidence of past colonies at sites no longer suitable. As yet, however, it is unknown how long the vegetation takes to recover or whether it ever does recover.

Geese can also damage the vegetation in their marshland wintering areas, to their own detriment. This is familiar in Snow Geese on some east coast refuges in North America, where the destruction of dense salt marsh vegetation gives rise to areas of bare mud known as 'eat outs'. Because the geese have alternative feeding areas, their numbers are as yet unaffected (Hindman & Ferrigno 1990). In these areas, however, the geese are in effect pushing back the vegetation succession, so may actually help to maintain the habitat in a marshy state for longer.

EXPERIMENTAL MANIPULATIONS OF FOOD-SUPPLIES

Most of the evidence discussed so far on the importance of food-supply is of two types: (1) an association in space or time between changes in bird numbers and changes in food-supplies; or (2) a coincidence in one area between periods of population decline and apparent food-scarcity, as revealed by measurements on the birds themselves or on their food **(Table 7.8)**. Such correlations may arise because the change in food is (1) sufficient to account for that particular population change, (2) necessary to account for all population changes, (3) a non-causal factor associated with the important causal factor, or (4) a completely spurious association, in which change in food is in no way linked directly or indirectly with change in numbers. One way to identify causal relationships is by experimental alteration of the food-supply, monitoring the response of the population against an appropriate control **(Table 7.9)**. The ideal experimental species eats foods that can be easily provided (or removed), lives at high density and remains in the same locality year-round, so that the effects of food manipulation on its subsequent numbers can be readily assessed. These three constraints exclude most bird species as easy subjects for experiment,

Table 7.8 Evidence for population limitation by food-shortage in various bird-species studied in detail. Modified from Newton 1980.

Rook *Corvus frugilegus* in northeast Scotland (Feare 1972, Feare *et al.* 1974)	Population at lowest density, and spread over largest area, in mid-summer. At this time food-stocks were minimal, because of a reduction in the numbers of fields in which birds could feed, an absence of grain, and the disappearance of large invertebrates (earth-worms and tipulid larvae) from the soil surface. Associated with high mortality (especially of juveniles) and low weights, and with lower feeding rates (150 cals per day) and longer feeding periods (90% of daylight, or 15 hours) than at any other season. Another potential food-shortage occurred during periods of deep snow in winter, when birds competed for space at localised feeding sites; however, the birds then spent only 30% (3 hours) of the active day feeding, and obtained 240 cals per day.
Red-billed Dioch *Quelea quelea* in Nigeria (Ward 1965)	Period of sudden population decline (due partly to emigration) coincided each year with temporary shortage of food, when grass seeds germinated at the onset of the rains. Starved and underfed birds were most prevalent then.
Galapagos Finches, mainly *G. fortis* and *G. scandens* (Grant 1986)	Marked spatial and temporal correlation between finch densities and seed-supply; high mortality at times of seed-scarcity, accompanied by selection for change in beak size.
Great Tit *Parus major* in southern England and the Netherlands (Perrins 1979, van Balen 1980).	Increase in breeding density after winters of good beech mast crops and known high overwinter survival; decreases after winters with poor crops and known low overwinter survival. Increases in breeding density after winters of experimental food provision.
Coal Tit *Parus ater* in southeast England (Gibb 1960)	Over five years, a close correlation was found between winter bird density and winter food-stocks (arthropods on foliage): survival from October to March varied greatly from year to year in relation to measured food-stocks. During this period, the birds ate around 50% of several main prey species, and in mid-winter spent more than 90% of the day feeding.
Woodpigeon *Columba palumbus* in southeast England (Murton *et al.* 1964, 1966, Inglis *et al.* 1990).	(1) Period of population decline in winter coincided with depletion of grain, then clover stocks. The lowest bird densities, low weights, low feeding rates, and most starving birds occurred in late winter, when clover stocks were minimal. Temporary food-shortages occurred during periods of deep snow, when the birds were concentrated on localised brassica sites. (2) Winter numbers declined steeply from the late 1960s, following the change from spring-sown to autumn-sown cereals and ploughing of clover leys and pasture, which reduced the winter food-supply, but increased from the late 1970s following the introduction of an alternative food-crop, oilseed rape.

Table 7.9 Conclusions permissible from experiments involving manipulation of food-supply. From Newton 1980.

Experiment	Response	
	Density increases	Density stays the same or decreases
Food increased	1 Food is a sufficient or necessary explanation	2 (a) Increased food is not sufficient to prevent a decline (b) Reduced food is not necessary to cause a decline
Food decreased	3 (a) Increased food is not necessary to allow an increase. (b) Decreased food is not sufficient to prevent an increase.	4 Food is a sufficient or necessary explanation.

In this context, 'sufficient' is used when a change in food is one of several factors that might cause a change in density, and 'necessary' is used when a change in food is invariably associated with a change in density.
The conclusions in boxes 1 and 4 of this table are tentative because it remains possible that the next experiment could show the opposite outcome, but the conclusions in boxes 2 and 3 can be regarded as valid so long as statistical and other methodological criteria are met.

and in any case care must be taken to ensure than the food provided is nutritionally appropriate. Most published experiments concern tits (Paridae), which were given seeds in winter as a supplement to natural food-supplies. Subsequent breeding numbers in the same area were then measured from the numbers of nests in nest-boxes, which were supplied in excess.

Experimental design

The simplest type of experiment involved a single area in which bird numbers were monitored for some years, then food was added continually and bird numbers were monitored again for several further years, giving a before-and-after comparison (grade 1 in **Table 7.10**). The weakness of this procedure is that, without a control, one cannot be certain that a change in bird numbers was due to food-provision and not to some other unknown factor which changed at the same time. The second type of experiment involved two areas studied simultaneously, with food supplied in one area but not in the other (grade 2). If bird numbers increased more (or decreased less) in the treatment area than in the control area, this was taken as evidence that food-supply influenced numbers. Ideally, the two areas

Table 7.10   Effects of winter food provision on the breeding density of various birds.

Species	Region	Grade of experiment (years)[1]	Increase in density (proportionate increase)	Source
Great Tit *Parus major*	Germany	3 (1)	Yes (up to 1.4)	Berndt & Frantzen 1964
	Finland	1 (3)	Yes (1.6)	von Haartman 1973
	England	2 (1)	No	Krebs 1971
	Netherlands	2 (5)	Yes (1.4)	van Balen 1980
	Sweden	3 (2)	Yes (1.4-1.9)[2]	Källander 1981
Blue Tit *P. caeruleus*	Germany	3 (1)	No	Berndt & Frantzen 1964
	England	2 (1)	Yes (1.4)	Krebs 1971
	Sweden	3 (2)	No	Källander 1981
Willow Tit *P. montanus*	Sweden	2 (1)	Yes (2.2)	Jansson *et al.* 1981
Crested Tit *P. cristatus*	Sweden	2 (1)	Yes (2.2)	Jansson *et al.* 1981
Coal Tit *P. ater*	Scotland	2 (1)	No	Deadman 1973
Black-capped Chickadee	Pennsylvania	1 (1)	Yes (1.3)	Samson & Lewis 1979
P. atricapillus	Wisconsin	3 (3)	No	Brittingham & Temple 1988
	Alberta	2 (2)	No	Desrochers *et al.* 1988
Tufted Titmouse *P. bicolor*	Pennsylvania	1 (1)	No	Samson & Lewis 1979
Nuthatch *Sitta europaea*	Sweden	3 (2)	Yes (1.6)	Enoksson & Nilsson 1983
	Sweden	3 (3)	Yes (1.9)	Enoksson 1990
Song Sparrow *Melospiza melodia*	British Columbia	2 (1)	Yes (1.4)	Smith *et al.* 1980
Carrion Crow *Corvus corone*	Scotland	1 (1)	No	Yom-Tov 1974
Magpie *Pica pica*	Washington	1 (1)	Yes (2.0)	Knight 1988
	Sweden	3 (2)	No	Högstedt 1981
	Alberta	2 (2)	No	Dhindsa & Boag 1990
Jackdaw *Corvus monedula*	Spain	3 (1)	Yes (1.4-1.8)	Soler & Soler 1996
Red Grouse *Lagopus l. scoticus*	Scotland	2 (1)	Yes (up to 2.4)[3]	Watson *et al.* 1984
	Ireland	2 (4)	Yes (1.2)	Watson & O'Hare 1979
Black Grouse *Tetrao tetrix*	Sweden	3 (4)	No	Willebrand 1988

[1] Grade of experiment: 1= before-and-after study. 2= separate treatment and control areas. 3= reversal of treatments or other replication.
[2] Response in one year only.
[3] Response during increase phase of population growth but not decline.

should be far enough apart that food-addition in the one area does not benefit birds in the other. The third type of experiment involved some replication: either by reversing the treatments, so that after a time the experimental area became the control and vice versa, or by the simultaneous use of several experimental and several control areas ideally chosen at random (grade 3). Such replication strengthens the findings because it increases the likelihood that any response observed in the study species resulted from the treatment and not from some other feature peculiar to the area or time period. Among the food-provision experiments summarised in **Table 7.10**, five involved simple before-and-after comparisons in the same area (grade 1), 11 involved a simultaneous control area (grade 2), while ten involved reversal of treatments or replication, with several experimental and control areas (grade 3).

Experimental findings

The most substantial feeding experiment yet undertaken, covering several successive winters, involved Great Tits *Parus major* in the Netherlands (van Balen 1980). It was based in two areas of similar woodland 7 km apart, in one of which extra food was provided while the other acted as a control. Before winter feeding began, Great Tit breeding numbers in the two areas were similar, and fluctuated more or less in parallel from year to year **(Figure 7.9)**. After feeding began, breeding densities in the two areas diverged and no longer fluctuated in parallel. Over several years, nest numbers in the experimental area averaged 40% higher than those in the control area. The impact of artificial feeding varied between years depending on the Beech *Fagus silvatica* crop, which was the most variable major component in the natural food-supply. Comparing numbers in the experimental area in the years before and after food-provision, it seemed that extra feeding doubled the population in poor Beech years, but made little difference in good Beech years **(Figure 7.9)**. This contrast provided further evidence that winter food-supplies influenced local breeding densities, with shortages most marked in years of poor Beech crops. On this evidence, the Great Tits in this experiment fitted the lower curve (A) in **Figure 3.12** in that variations in winter feeding conditions caused variations in breeding density.

Further north in Europe, Great Tit breeding densities were greatly influenced by winter severity, with the lowest densities following the coldest winters **(Figure 7.10)**.

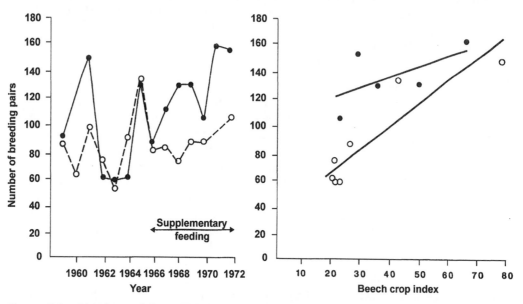

Figure 7.9   Numbers of Great Tits *Parus major* breeding in two areas in the Netherlands of similar size and habitat, in one of which extra winter food (hemp and sunflower seeds) was provided from 1967. After this the two populations diverged and no longer fluctuated in parallel. Filled symbols – experimental plot, open symbols – control plot. This experiment showed the role of winter food-supply in influencing local breeding density. From van Balen (1980).

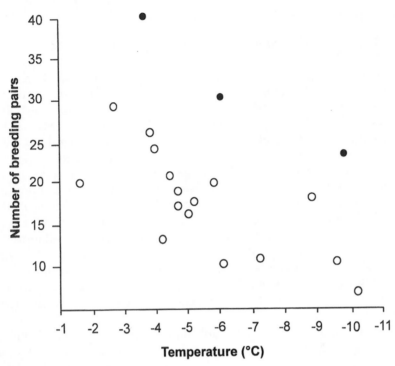

Figure 7.10 Relationship between Great Tit *Parus major* breeding numbers and mean temperature of the preceding winters in an area of Finland. In the winters 1969–71, extra food was provided. Filled symbols – pair numbers after winter of food provision, open symbols – pair numbers after other winters. From von Haartman 1973.

When food was provided artificially in three winters, the same relationship held with weather but breeding densities were correspondingly higher (von Haartman 1973). Other experiments on tits, covering 1–2 winters, gave variable results. In some studies, two tit species had access to the supplementary food, and if each of these is counted as a separate experiment, then 8 out of 15 winter feeding experiments led to obvious increases (1.3–2.4 fold) in subsequent breeding density, compared with control areas **(Table 7.10)**. The same species responded to food provision in one area but not in another, or in one year but not in another; and in the same area and year, sometimes one species responded while another did not (e.g. Krebs 1971, Källander 1981). Where the birds were ringed, increases in breeding density could be shown to result from some combination of increased local survival and increased immigration, both affecting mainly first-year birds (Krebs 1971, van Balen 1980, Jansson *et al.* 1981, Enoksson 1990). Similarly, in the Song Sparrow *Melospiza melodia* on Mandarte Island, British Columbia, seed-provision in one winter led to increased breeding density through improved survival of juveniles (Smith *et al.* 1980). Like the Great Tits in the previous paragraph, all the species that responded to winter food-provision by increased breeding density conformed to the lower curve (A) in **Figure 3.11**.

Two other winter feeding experiments on Black-capped Chickadees *Parus atricapillus* led to increased winter bird densities but had no effect on subsequent local breeding densities (Brittingham & Temple 1988, Desrochers *et al.* 1988). This was because some of the extra birds produced by winter feeding left the area at the end of winter. The experiments had the same effect as a garden feeder but on a larger scale. The implication was that some birds were surplus to the carrying capacity of the local nesting habitat, in which case the findings were consistent with the upper curve (B) in **Figure 3.12**. Another winter feeding experiment in a broad-leaved wood in Ohio led to greatly increased winter densities of at least six bird species, compared with the same wood in the previous year and with a nearby control wood in which no extra food was provided (Berner & Grubb 1985). Compared with the control wood, densities were 1.5–2.8 times higher in different species, but no measures were made of subsequent breeding densities. The same occurred after seed-provision in a Hungarian wood, where densities of four species were 2.5–10.4 times higher than in a control wood (Székely *et al.* 1989). In contrast, seed provision led to no increase in the winter densities of Chipping Sparrows *Spizella passerina* and other seed-eaters in Arizona in two years of moderate natural seed production. This was in line with earlier observational evidence indicating that food was limiting for these species only in years of poor seed production (Pulliam & Dunning 1987).

Supplementary feeding was associated with an increase in breeding density in one study of Magpies *Pica pica*, but not in two others, nor in a study of Carrion Crows *Corvus corone* (Yom-Tov 1974, Högstedt 1981, Dhindsa & Boag 1990). In each case, the feeding period was short (2 weeks to several months) compared to the time that birds hold territories. One might not have expected a sudden increase in density in species in which individuals are long-lived and keep the same territories for years, despite some fluctuation in natural food-supply. The result might have been different if food was provided over several years, or if existing birds had been removed to enable different ones to settle. Shorter-lived species have contracted their territories in response to food-provision (for Nuthatch, see Enoksson & Nilsson 1983, Enoksson 1990; for Dunnock *Prunella modularis*, see Davies & Lundberg 1984), or have expanded their territories in response to food-removal (for Rufous Hummingbird *Selasphorus rufus*, see Hixon *et al.* 1983; for Rufous-sided Towhee *Pipilo erythrophthalmus*, see Wasserman 1983). However, not all such studies were concerned with breeding density (Chapter 3).

In Red Grouse *Lagopus l. scoticus*, fertilising areas of heather (the food-plant) promoted an increase in breeding density one year later over the previous year, as well as over that in control areas nearby (Miller *et al.* 1970). This occurred when experiments were started in years with low or moderate densities, but in a later trial, fertilising an area at high grouse density failed to halt a big decline (Watson & Moss 1979). The Red Grouse in the area concerned underwent marked fluctuations, and these experiments suggested that the birds could respond to improved nutrition during a period of increase but not during a period of decline (Watson *et al.* 1984). Similar results were found in attempts to prevent the cyclic declines in small rodents by food-provision (Boutin 1990). In another area, however, where Red Grouse did not fluctuate cyclically, fertilising an area of moorland led to a 20% increase in breeding density, compared with an unfertilised control area (Watson & O'Hare 1979).

Artificial feeding of Black Grouse *Tetrao tetrix* in winter started in Finland as a widespread management procedure, in an attempt to stop their numbers declining. The birds were provided with oats at lekking sites throughout the winter. Although the birds took the food, the long-term population decline continued, as did shorter-term cycles in numbers (Marjakangas 1987). The same procedure was evaluated in Sweden, where Black Grouse were fed at two leks for four years, while those at three other leks in the same area were left as controls. No differences in body weights or survival were noted between birds at the two types of lek (Willebrand 1988). This last experiment was in an area of abundant natural winter food (birch), but the conclusion from both studies was that Black Grouse numbers were not limited by winter food. This was in line with other studies which pointed to deterioration in habitat and summer insect supplies as causing long-term decline, through poor chick production (Baines 1996), with predation also accounting for heavy losses in some years (Angelstam *et al.* 1984, Marcström *et al.* 1988).

Turning to manipulation of summer food-supplies, experiments involving insecticides have been conducted as an adjunct to forest-insect control. The chemicals most often used included DDT, carbaryl and fenitrothion, applied in a wide range of dosages and spraying regimes (Chapter 15). In some studies, no effects of spraying were apparent on local bird breeding densities, while in others the reported declines may have been due more to direct poisoning of birds by the pesticides than to reductions in their food-supplies (Pearce & Peakall 1977, Peakall & Bart 1983, Spray *et al.* 1987). However, in some studies, the chemical used was diflubenzuron, a growth regulator that kills larval insects but not birds (DeReede 1982, Stribling & Smith 1987, Cooper *et al.* 1990). In these studies, no significant differences in the numbers of various common species, or in their combined numbers, were found between treated and untreated plots. Evidently, in studies involving diflubenzuron, the breeding densities of insectivorous birds were not affected by reductions in larval insect numbers part way through the breeding season. However, spraying was invariably done in years of abundant caterpillars, so that even after a heavy kill, enough may have survived to feed the birds. If spraying had been done in a year with few caterpillars the results may have differed, but in any case alternative foods were often available. In the Grey Partridge *Perdix perdix* in Europe, widespread declines in numbers have been attributed to poor chick survival, in turn resulting from pesticide-induced reductions in insect numbers (Potts 1986). Experiments on farmland, omitting pesticide use, resulted in more insects, better chick production, and greater subsequent breeding density (Rands 1985). Similar improvements in chick survival also occurred in Red-legged Partridge *Alectoris rufa* and Ring-necked Pheasant *Phasianus colchicus* (Green 1984, Hill 1985). These experiments are discussed further in Chapter 15.

In summary, winter food-supplementation has often led to an increase of local breeding density, and had most effect when the natural food-supply was poor. Where mechanisms were examined, extra winter food altered the rates of survival and immigration/emigration. Increases in subsequent breeding density up to 2.4-fold were recorded, and these were evident after only one winter of experimental feeding. Failure to achieve an increase could be attributed to other factors limiting breeding numbers (including natural food being superabundant at the time), or to limitations in density imposed by local nesting habitats, so that any rise in winter density was not passed on to spring.

Experiments provide strong evidence that food can limit bird numbers, but they are still open to the possibility that some additional unknown factor changed in parallel with food, and caused the change in numbers. They are an improvement over observing natural changes, however, because in experiments it is the observer who brings about the change. Such experiments need careful fore-thought, including a knowledge of the bird's annual cycle of numbers and behaviour, so that the change in food-supply can be made at an appropriate time. Finally, even if a clear-cut result is obtained, allowing the conclusion that change in food changed numbers, it is not of course safe to assume that all changes in numbers are due to changes in food. Food may be sufficient for some changes, but not necessary to explain them all.

SUMMARY

Food-supplies can affect bird breeding densities by influencing the survival of full-grown birds or their production of young, or through influencing immigration/emigration. Observational evidence on these points from a wide range of species has been confirmed by experimental evidence on a narrower range.

For any given population, food may be limiting only at certain times of year or in certain years. It may sometimes involve deaths from starvation, but not always. This is especially so (1) if social behaviour regulates local density in relation to food-supply, and excluded birds succumb to other mortality, such as parasitism or predation; or (2) if food-shortage lowers breeding and recruitment rates, and thus population density, without increasing starvation rates in full-grown birds.

Circumstantial evidence that food can limit bird numbers comes from (1) spatial variations in bird densities that correlate with spatial variations in food-supplies; (2) year-to-year changes in local densities that correlate with local year-to-year changes in food-supplies; and (3) sudden or long-term changes in bird numbers that accompany sudden or long-term changes in food-supplies. Other circumstantial evidence arises when a drop in numbers coincides with periods of starvation, low weights, low feeding rates, long feeding times, increased fighting, or marked depletion of food-stocks.

In 15 studies, experimental provision of extra winter food was followed by an increase in subsequent breeding density (compared with earlier years or to control areas). In 11 other studies, this did not happen, evidently because breeding densities were limited by factors other than winter food, including (in some species) features of the nesting habitat. In summer studies, reduction of caterpillar numbers (by insecticide use) caused no reduction in the numbers of adult insectivorous birds that could be attributed to the decline in food-supply, but in some gallinaceous birds it reduced the production of young, which in turn lowered subsequent breeding numbers.

Osprey *Pandion haliaetus* nest on pylon.

Chapter 8
Nest-sites

While food-supply could potentially limit the numbers of all birds, in some species breeding density is often held at a level lower than food would permit by shortage of some other resource. Limitation by acceptable nest-sites is evident mainly in species that use special places, such as tree-cavities or cliff ledges. The evidence is partly correlative, in that variations in breeding density, from place to place or year to year, often parallel similar variations in nest-site availability; and breeding pairs are generally absent from areas that lack nest-sites but that seem suitable in other respects (non-breeders may live there). More convincingly, however, experimental manipulations of nest-site numbers have often led to corresponding changes in local breeding densities.

Most of the evidence concerns tree-hole nesters, because of their willingness to accept nest-boxes, but also includes species that use underground boxes (burrow-substitutes), buildings (cliff-substitutes), pylons (tree-substitutes) and rafts (island-substitutes). In each case, one has to be sure that the acceptance of new sites has allowed extra pairs to breed, and has not merely entailed shifts of existing pairs from other nest-sites. Experimental demonstrations of nest-site shortages provide some of the clearest and most unequivocal examples of resource limitation in birds. They also show the action of successive limiting factors, for once the shortage of nest-sites has been alleviated, numbers usually increase to a new higher level, at which they may be assumed to be limited by some other factor. In addition, appropriate manipulations of limited nest-sites have shown clearly the effects of inter-specific competition: of how one species, by taking most of the available sites, can affect the numbers of another (Chapter 12).

Where nest-sites are scarce, the process of density limitation depends on defence of the nest-site, or of the territory containing it, by the pair in occupation, so that other pairs are excluded. In species that defend little more than the nest-site itself, such as the Starling *Sturnus vulgaris* or Pied Flycatcher *Ficedula hypoleuca*, breeding density can have a fixed ceiling, set by the number of sites and regardless of the numbers of contenders. In other species, such as tits, each pair defends a large feeding territory containing up to several potential sites. In these species, then, it is not shortage of sites as such that limits density, but shortage of sites outside the territories of existing pairs. In such species, territories are often compressible to some extent, giving a more flexible ceiling to density, sensitive to the numbers of contenders (Chapter 3), and in some species polygyny can increase nest numbers above one per territory. Whatever the social system, however, nest-sites can clearly constrain breeding numbers and output in a density-dependent manner. The more that the numbers of potential breeders exceed the number of sites available, the smaller the proportion of birds that can obtain one and breed. Moreover, where shortage of nest-sites limits breeding density, it must eventually also limit total population size: where the output of young is limited, no population can increase beyond a certain level (Chapter 3).

SPECIES THAT NEST IN TREE-CAVITIES

Throughout the world, many forest bird species depend on tree-cavities for nesting. In Europe and North America, where most studies have been done, 5% and 4%, respectively, of all bird species are obligate hole-nesters, while in southern Africa and Australia equivalent figures are 6% and 11% **(Table 8.1)**. In all these regions other species use tree-holes, but are not tied to them. Most obligate hole-nesters are in the orders Passeriformes (songbirds), Piciformes (woodpeckers), Apodiformes (swifts), Coraciiformes (rollers), Strigiformes (owls), Psittaciformes (parrots) and Anseriformes (waterfowl). Certain species, such as woodpeckers, can excavate their own holes, but most species (the 'secondary cavity nesters') cannot, so have to rely on existing holes, produced earlier by woodpeckers (which usually make new holes each year), by insects (especially termites) or by fungal decay. It is these non-excavators that are most often short of sites, and with which this section is mainly concerned. Although dependent on existing holes, some non-excavators can enlarge or otherwise modify holes to suit their needs, especially in soft wood.

Formation and density of tree-cavities

In any tract of forest, the numbers of usable cavities increase with the age of the trees, which influences the amount of dead wood in which most holes are formed. By

Table 8.1   Numbers (%) of bird species in different parts of the world that use tree-cavities for nesting. From Saunders *et al.* 1982; Voous 1960; Witherby *et al.* 1939.

	Numbers of species	Numbers known to use cavities	Numbers known to be obligate cavity nesters
Europe			
Passerines	169	28 (17)	9 (5)
Non-passerines	250	32 (13)	13 (5)
Total	419	60 (14)	22 (5)
North America			
Passerines	292	18 (10)[1]	13 (4)
Non-passerines	192	30 (10)	11 (4)
Total	484	48 (10)	24 (4)
Southern Africa			
Passerines	333	20 (6)[2]	13 (4)[2]
Non-passerines	310	35 (11)	23 (7)
Total	643	55 (9)	36 (6)
Australia			
Passerines	297	23 (8)	7 (2)
Non-passerines	234	71 (30)	50 (21)
Total	531	94 (18)	57 (11)

In the regions shown, cavity-nesting birds occur in the following orders: Anseriformes (waterfowl), Falconiformes (falcons), Psittaciformes (parrots), Strigiformes (owls), Caprimulgiformes (nighthawks, one obligate hole nester in Australia), Apodiformes (swifts), Trogoniformes (trogons, one obligate hole-nester in southern Africa), Coraciiformes (rollers), Piciformes (woodpeckers) and Passeriformes (songbirds).

[1] Eight species are primary excavators able to penetrate hard wood.
[2] One species is a primary excavator.

providing foraging sites, dead wood in many regions also favours woodpeckers which further increase the numbers of holes available for other species. In undisturbed forest, therefore, the abundance and diversity of hole-nesting birds increase as the forest ages (Haapanen 1965, Mannan *et al.* 1980). Eventually, the number of cavities is likely to reach an equilibrium **(Figure 8.1)**, where the rate of annual gain (through decay, woodpecker or insect activities) roughly matches the rate of loss (through deterioration, natural sealing and tree fall) (Sedgwick & Knopf 1992).

In the absence of human intervention, any long-established forest can be expected to contain some large dead trees, each of which may stand for several decades, gradually losing limbs and softening, as it decays and finally falls (Cline *et al.* 1980). Large standing dead trees (called 'snags') are an important source of nest-sites for cavity-nesting birds. In general, the bigger the tree when it dies, the longer it will remain standing, the bigger the range of cavity sizes it can eventually provide, and the greater the range of bird species that can make use of it. In a Douglas Fir *Pseudotsuga menziessii* forest in Oregon, most hole-nesters were found in snags more than 60 cm across at chest height and more than 15 m tall. These snags usually had broken tops, few or no branches, decayed sapwood and heartwood and incomplete bark cover. They occurred mainly in unmanaged forests more than 110 years old (Mannan *et al.* 1980). Most species nested in the thicker snags of those available, perhaps because

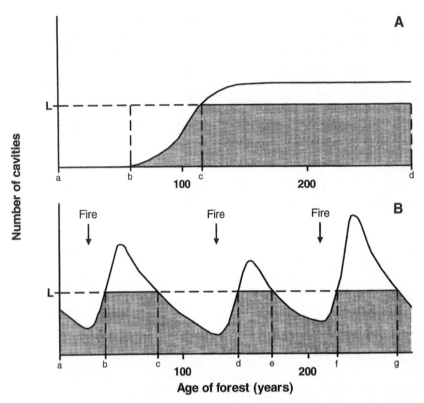

Figure 8.1   Hypothetical models of the changing numbers of cavities and cavity-nesting birds during natural forest succession. (A) from establishment to maturity in the absence of fire or other catastrophes; and (B) an established forest under the influence of periodic fires. The timescale is set roughly for forest in north temperate regions, but would vary with tree species. On the vertical scale, L indicates the level below which hole-nesting birds are limited in breeding density by shortage of nest-sites, and above which they are limited by other factors.

In A, cavity numbers increase progressively as large trees age and die, and then stabilise as large tree mortality rates stabilise. At equilibrium the annual additions of new cavities (arising mainly through decay and woodpecker activities) approximately match the annual losses (mainly through branch and tree fall). In period a–b cavities are absent, b–c increasing, and c–d present in excess. So in period a–b cavity-nesting birds are absent as breeders, in b–c they are present but limited at the level that nest-sites would permit. Managed forests are normally felled before stage c, and some before stage b.

In B, many large trees are periodically killed by fire, and then fall progressively over several decades as new forest develops around them. Years later, another fire sweeps through and the process is repeated. In each cycle, cavity numbers reach a peak some years after the fire, as trees rot, and then decline as trees fall. Cavity-nesting birds are limited by shortage of cavities in periods a–b, c–d and e–f, but not in periods b–c, d–e and f–g when cavities are surplus to needs. The exact pattern depends on the proportion of trees killed by each fire, on the intervals between fires and on other factors. If fires are too frequent, they prevent any trees from becoming big enough to provide usable cavities.

such snags were less likely to fall, as well as offering larger, better insulated cavities, further from the ground, and hence more secure from predators.

In old forests it is not unusual to find large snags at densities exceeding ten per hectare. This number increases periodically if the forest suffers from fire or other catastrophes which kill many trees, and then declines progressively as dead trees fall and rot (Figure 8.1). In much of the world, however, these natural processes are now rare, as more and more forests are managed for timber, a process that typically involves the shortening of rotation lengths (= tree life-spans), the suppression of fires and the routine removal of snags and other dead wood.

Other cavities develop in the trunks of living trees where branches become detached. Such holes also increase in numbers as the forest ages. Broad-leaved trees provide more holes of this type than do conifers. This is because when a limb detaches from a hardwood tree, protective gum-filled cells form only in the living sapwood, round the edges of the wound, for a time leaving the heartwood exposed to fungal and insect attack. But when the same process occurs in a conifer, resin released from the living tissue impregnates the heartwood, protecting it against decay (Peace 1962). Hence, where snags have been removed, many hole-nesting birds in conifer forests rely on the occasional broad-leaved tree (Haapanen 1965). Small numbers of cavities also develop in living trees by the action of lightning, frost cracks and other trunk wounds, or by the breakage of tops.

In natural old forest, containing large snags, tree-holes may exceed 40 per ha (McComb & Noble 1981), but in managed forest, with no snags, holes resulting from branch scars typically reach only 6–7 (occasionally up to 15) per ha in mature broad-leaved areas, and less than one per ha in conifers (van Balen et al. 1982, Waters et al. 1990). These figures refer to all cavities, and large ones suitable for ducks may be far fewer, usually less than one per ha, but occasionally up to five (Bellrose et al. 1964, Saunders et al. 1982, Souliere 1988). Moreover, not all holes of appropriate size for a given species are necessarily available. Some might be damp, infested with parasites or stuffed with nest material from previous years, while others might be within the territory of a conspecific. In managed woods under 50 years of age, tree-cavities of any kind are almost non-existent, unless old trees remain from the previous crop.

Observers familiar only with nest-box studies in the managed woods of western Europe can have little conception of the abundance and diversity of hole-nesting birds found in more natural forests, where dead and dying trees are not removed. In some regions woodpeckers can form a substantial part of the local avifauna, and hole-nesters in general can outnumber all other birds. Moreover, species differ markedly in the types of site they use: in the preferred sizes of cavity and entrance hole, in whether the cavity is high or low within the tree, on the trunk or side branches, and in whether it is exposed or secluded, and near or far from feeding areas (Edington & Edington 1972, van Balen et al. 1982, Saunders et al. 1982). Some excavating species, such as certain woodpeckers, can cope with hard wood, while others, such as some nuthatches, need softer, more decayed wood (Raphael & White 1984). Overlap is considerable, however, because species are often constrained by the types of site available, or by the presence of competitors which exclude them from some of the sites they might otherwise use (Chapter 12). For example, in one Michigan area where Starlings were absent, Mountain Bluebirds *Sialia currucoides* used cavities with a wide range of entrance holes, but in another area where Starlings were present, Bluebirds were largely restricted to sites with entrance holes small enough to exclude Starlings (Pinkowski 1976). Similarly, in Sweden four hole-nesting species all preferred higher over lower holes when they had a choice, the lower having greater predation rates. On average, however, they occupied decreasing nest heights, in order of the decreasing dominance status of each species, with the Starling taking the highest holes, followed by the Nuthatch *Sitta europaea* and Blue Tit *Parus caeruleus*, and the Marsh Tit *Parus palustris* the lowest (Nilsson 1984). None-the-less, particular species generally show considerable year-to-year consistency in the types of site they use (van Balen *et al.* 1982).

Differences between species in preferred natural sites are reflected in preferences for boxes with different dimensions, hole sizes and positions in the tree, and surrounding habitat. Again, however, overlap occurs, and over a period of years the same box may be used by different species (Löhrl 1970, 1977, Nilsson 1975, Munro & Rounds 1985). In both natural and artificial sites, therefore, the potential for competition is great. The size of the entrance hole is important because it influences whether competitors (and predators) can get. in. Each species is safest with the smallest hole through which it can squeeze to reach the cavity beyond. For any one species, moreover, the size of the cavity itself has been found to influence clutch size and the number of young raised (see references in van Balen *et al.* 1982). The fact that predation rates are often higher in natural holes than in nest-boxes provides further indication that hole-nesters must often make do with what is available rather than with what is optimal.

Circumstantial evidence that shortage of tree-cavities can limit bird breeding densities

Among conifer forests in the Sierra Nevada of California, correlations were apparent between the densities of hole-nesting birds and the densities of snags, with most birds on the plots with the most snags **(Figure 8.2)**. The relationship was not linear, however, and beyond a density of about seven snags per ha, bird density levelled

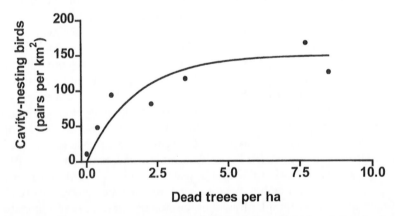

Figure 8.2   Breeding density of cavity-nesting birds (all species) in relation to density of standing dead trees (>38 cm diameter at breast-height) on seven conifer plots in California. The data imply that, below about five dead trees per ha, cavity nesters were limited in density by shortage of nest-sites. Above that level, bird densities stabilised and were apparently limited by other factors. Modified from Raphael & White 1984.

off. Presumably, at such high snag density other limiting factors, such as food-supply, came into play. In the same study the density of hole-nesting birds declined by 77% after snags were removed from a burned plot, while other birds declined by only 6% (Raphael & White 1984).

Several other studies, in various geographical locations, showed that the numbers and diversity of cavity-nesting birds increased with the density of snags available (e.g. Haapanen 1965, Mannan *et al.* 1980, Zarnowitz & Manuwal 1985, Snyder *et al.* 1987). Yet other studies showed that hole-nesting birds (but not other birds) were scarcer in managed forests that lacked snags than in unmanaged forests (Hildén 1965, Nilsson 1979), or that the densities of hole-nesting birds declined following snag removal (Scott 1979, Raphael & White 1984). Whereas some species used snags only for nest-sites, others used them also for roost sites, and yet others (notably woodpeckers) as foraging sites. For the latter, then, snag removal meant more than the loss of nest-sites.

Other arguments put forward to support the view that cavity-sites might often be limiting include:

(1) the occurrence of fighting for holes, and of evictions, both between and within species;
(2) successive use of the same hole by different pairs in the same season; and
(3) shifts in territory boundaries to incorporate nest-holes in territories otherwise lacking them (recorded in Great Tit *Parus major*, van Balen *et al.* 1982, East & Perrins 1988).

However, all these events are open to other explanations, and do not in themselves prove that nest-sites limited breeding density. Only experimental manipulations of nest-sites can provide reliable answers.

Experimental evidence that shortage of tree-cavities can limit bird breeding densities

In the managed woods of western Europe, studies of birds in nest-boxes have become popular with biologists, because boxes can be designed to attract the desired species and, once in place, they can provide large samples of easily-found nests in accessible and standardised sites. Moreover, where natural sites are scarce or lacking, it is possible to get the entire local population of a target species into boxes, and as some species also roost in boxes, individuals can be easily caught and examined both inside and outside the breeding season. Not surprisingly, then, nest-boxes have formed the basis of many long-term population studies of birds, and some box-nesting species, such as the Great Tit *Parus major* and Pied Flycatcher *Ficedula hypoleuca*, are now among the best known birds in the world. But despite the popularity of nest-box studies, relatively few observers have monitored the effect of nest-site provision on breeding density, by before-and-after comparisons. Fewer still have monitored density changes in the treatment plot against those in a similar 'control' plot with no boxes, and even fewer have replicated the experiment on several plots. The findings from some 46 studies on 27 species are summarised in **Table 8.2**. Almost all provided evidence that nest-sites limited breeding density.

Care is needed in interpreting the findings of nest-box studies. Many species of birds, not limited by shortage of nest-sites, will readily accept well designed artificial

Table 8.2 Limitation of breeding density among birds that nest in tree-cavities, as shown by the experimental addition or removal of nest-sites. Sites were added by providing nest-boxes or subtracted by blocking natural holes or removing dead snags.

Species	Area	Habitat	Addition (+) or subtraction (−) of nest-sites	Grade of experiment[1]	Increase (+) or decrease (−) in breeding density[2]	Source
Starling *Sturnus vulgaris*	New Zealand	Farmland	+	1	+	Coleman 1974
House Sparrow *Passer domesticus*	New Zealand	Farmland	+	1	+	Coleman 1974
Tree Sparrow *Passer montanus*	England	Farmland	+	1	+	Boyd 1932
	Germany	Farmland	+	1	+	Creutz 1949
	Germany	Pine plantation	+	1	+	Dornbusch 1973
	Germany	Woodland	+	1	+	Gauhl 1984
	Germany	Woodland	+	1	+	Löhrl 1978
	Germany	Woodland	+	1	+	Schönfeld & Brauer 1972
Western Bluebird *Sialia mexicana*	Arizona	Pine forest	+	3	+	Brawn & Balda 1988
Mountain Bluebird *Sialia currucoides*	Manitoba	Parkland and meadows	+	1	+	Cutforth 1968
	California	Pine-fir forest	−	1	−	Raphael & White 1984
	Manitoba	Mainly farmland	+	1	+	Miller 1970
Eastern Bluebird *Sialia sialis*	Manitoba	Mainly farmland	+	1	+	Miller 1970
Redstart *Phoenicurus phoenicurus*	Finland	Birch forest	+	1	+	Järvinen 1978
Pygmy Nuthatch *Sitta pygmaea*	Arizona	Pine forest	+	3	+	Brawn & Balda 1988
Mountain Chickadee *Parus gambeli*	California	Conifer forest	+	1	+	Dahlsten & Copper 1979
Great Tit *Parus major*	Britain	Broad-leaved woodland	−	1	−	East & Perrins 1988
	Japan	Second-growth forest	+	1	+	Higuchi 1978
Varied Tit *Parus varius*	Japan	Second-growth forest	+	1	+	Higuchi 1978
Blue Tit *Parus caeruleus*	Britain	Broad-leaved woodland	−	1	−	East & Perrins 1988
Pied Flycatcher *Ficedula hypoleuca*	Britain	Deciduous woodland	+	1	+	Currie & Bamford 1982
	Finland	Birch forest	+	1	+	Järvinen 1978
	Northern Sweden	Birch forest	+	2	+	Enemar & Sjöstrand 1972
	Southern Sweden	Broad-leaved woodland	+	2	+	Enemar & Sjöstrand 1972
	Southern Sweden	Broad-leaved woodland	+	2	+	Enemar & Sjöstrand 1972
Collared Flycatcher *Ficedula albicollis*	Sweden	Broad-leaved woodland	+	3	+	Gustafsson 1988
Violet-green Swallow *Tachycineta thalassina*	Arizona	Pine forest	+	3	+	Brawn & Balda 1988
Tree Swallow *Tachycineta bicolor*	Ontario	Mixed, rural	+	1	+	Holroyd 1975
Purple Martin *Progne subis*	Ontario	Mixed, rural	+	1	+	Holroyd 1975
Ash-throated Flycatcher *Myiarchus cinerascens*	Arizona	Riparian woodland	+ and −	2	+ and −	Brush 1983
Green-rumped Parrotlet *Forpus passerinus*	Venezuela	Farmland	+	2	+	Beissinger & Bucher 1992

Species	Location	Habitat		Grade[1]		Reference
Little Owl *Athene noctua*	Germany	Farmland	+	1	+	Exo 1992
Barn Owl *Tyto alba*	Scotland	Forest openings	+	1	+	Petty et al. 1994
European Kestrel *Falco tinnunculus*	Netherlands	Grassland	+	1	+	Cavé 1968
	Britain	Grassland	+	1	+	Village 1983
American Kestrel *Falco sparverius*	Wisconsin	Farmland	+	1	+	Hamerstrom et al. 1973
Goldeneye *Bucephala clangula*	Scotland	Forest and lakes	+	1	+	Dennis & Dow 1984
	Sweden	Forest and lakes	+	1	+	Sirén 1951
	Finland	Forest and lakes	+	1	+	Eriksson 1982
	Canada	Forest and lakes	+	1	+	Johnson 1967
Barrow's Goldeneye *Bucephala islandica*	British Columbia	Forest and lakes	+	1	+	Savard 1988
Wood Duck *Aix sponsa*		Forest and lakes	+	1	+	McLaughlin & Grice 1952
	New York	Forest and lakes	+	1	+	Haramis & Thompson 1985
	California	Marsh	+	1	+	Jones & Leopold 1967

[1] Grade 1 = before-and-after comparison in the same area; grade 2 = simultaneous comparison between experimental and control area; grade 3 = reversal of treatments between experimental and control areas, or replication of experimental and control areas.

[2] Studies with no response include Bufflehead *Bucephala albeola* (Gauthier & Smith 1987) and European Kestrel *Falco tinnunculus* in one area (Village 1983).

sites in preference to inferior natural sites. For example, Tawny Owls *Strix aluco* in a spruce plantation in northern England used a range of natural nest-sites, including cavities in old broad-leaved trees, crags and ground sites, man-made structures and stick nests of other species (Petty 1992, Petty *et al.* 1994). But after nest-boxes were erected, 83% of the population of more than 40 pairs switched to using boxes from the first year that boxes were available and all pairs had switched by the fourth year. Thereafter, for the remaining 12 years of study, no nests were recorded in natural sites, so complete had been the adoption of boxes. Yet no indication was found that the observed changes in Tawny Owl breeding numbers had anything to do with box provision. So in any such experiment with nest-boxes, one must check to what extent increased box use is due to:

(1) an increase in the density of territorial pairs,
(2) existing territorial pairs lacking sites being provided with an opportunity to nest, or
(3) existing territorial pairs shifting from natural sites.

Only in situation (1) would shortage of nest-sites limit pair density, whereas in (2) it limits the proportion of pairs that nest, and in (3) it limits neither pair numbers nor the proportion that nest. Insofar as could be judged from the information provided, all the experiments listed in **Table 8.2** in which a response occurred involved a change in the densities of pairs (situation 1).

Some researchers have examined the response of all the cavity-nesting species in an area to nest-site provision or removal. For example, in Ponderosa Pine *Pinus ponderosa* forest in Arizona, Brawn & Balda (1988) studied hole-nesting birds on five 8-ha plots of different structure. On three plots they counted all hole-nesting birds for four years, then added nest-boxes, and continued the counts for another four years. The other two plots were left as controls, and no boxes were provided **(Figure 8.3)**. The degree to which sites were limiting varied between species and between plots, depending partly on the number of dead snags present. Overall breeding densities (all hole-nesters combined) increased significantly after box provision on the two plots with fewest natural sites. Increases were not apparent in the first year after box provision, but occurred progressively thereafter, giving in the fourth year overall densities 230% and 760% higher than previously. On these plots, more than 90% of the hole-nests found were in boxes. Three species responded strongly, so were most clearly limited by nest-site availability before boxes were installed, namely the Violet-green Swallow *Tachycineta thalassina*, Pygmy Nuthatch *Sitta pygmaea* and Western Bluebird *Sialia mexicana*. The last species increased from 20 pairs per km² before box provision to 78–88 pairs per km² after box provision. Three other hole-nesters did not respond significantly, despite a surplus of boxes, so were assumed to have been limited by other factors. In contrast, on the two control plots where no boxes were provided, no significant density changes occurred. The same was true in the third experimental plot, which had the most natural sites, and where only 30% of hole-nesters used boxes. On this plot, therefore, cavities were already surplus to needs, and the occurrence of some box-nests was attributed to birds switching from natural sites.

In another experiment in riparian woodland in Arizona, the numbers of hole-nesters declined after cavities were blocked on a plot with many cavities, but increased after nest-boxes were provided on a plot with few natural cavities (Brush 1983). In contrast, breeding numbers remained stable in an unmanipulated plot

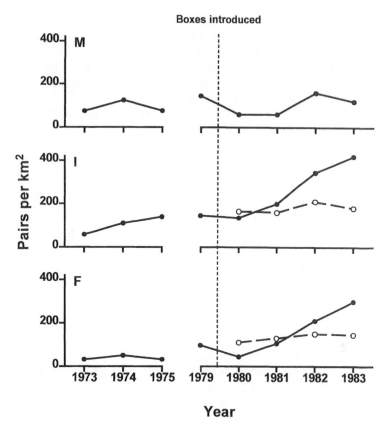

Figure 8.3   The effect of nest-box provision on the breeding densities of hole-nesting birds in Ponderosa Pine forest in Arizona. Some 60 boxes were provided in each of three 8-ha plots, which had many (M), intermediate numbers (I) or few (F) natural cavities, but not in two 8-ha control plots (dashed lines). Nest-box provision resulted in increased bird densities in two of the experimental plots, but not in the third plot, which had the most natural nest-sites. No significant density changes occurred in the control plots where no boxes were provided. The conclusion was that the densities of cavity-nesting birds were limited by shortage of nest-sites, except in the plot with most natural sites. Modified from Brawn & Balda 1988.

where natural cavities were available throughout. In this study, the species most affected by nest-site manipulation was the Ash-throated Flycatcher *Myiarchus cinerascens*. Differential response between species was also noted in another experiment, involving two tit species in an English wood. The removal of nest-boxes caused a decline in the numbers of Great Tits, but not of Blue Tits, which moved to natural sites (East & Perrins 1988).

   In young woodland in Europe, where natural sites were scarce, nest-box provision has led to massive increases in the breeding densities of many hole-nesting songbirds, including Pied Flycatcher *Ficedula hypoleuca*, Collared Flycatcher *F. albicollis*, Redstart *Phoenicurus phoenicurus* and various tit *Parus* species, and in some localities also Tree Sparrow *Passer montanus* and Starling *Sturnus vulgaris* (**Table 8.2**).

The two flycatchers showed some of the strongest responses. These migrant species arrive after resident hole-nesters have started to breed, and are unable to displace residents from holes. They can normally breed only in woods with many natural sites, and since they are not strongly territorial (defending only the nest-hole), they have often increased spectacularly in numbers after nest-boxes were installed. For example, in Finland Pied Flycatcher densities equivalent to 2000 pairs per km^2 were achieved, greater than all other birds together (von Haartman 1971). Without the boxes, few if any flycatchers could have bred.

In North America, the three bluebird (*Sialia*) species have declined greatly since 1900, supposedly because of the felling of old trees, coupled with competition for remaining cavities with introduced Starlings and House Sparrows *Passer domesticus*. Special-interest groups, such as 'Bluebirds Unlimited' and the 'North American Bluebird Society', were formed to encourage nest-box provision. These schemes have led to greatly increased bluebird densities, especially where boxes were designed and positioned so as to minimise their use by competing species (Cutforth 1968, Miller 1970, Backhouse 1986).

Turning now to non-passerine species, box provision has been followed by large increases in the numbers of various hole-nesting ducks, including Wood Duck *Aix sponsa* in several parts of North America (McLaughlin & Grice 1952, Jones & Leopold 1967, McGilvrey & Uhler 1971, Haramis & Thompson 1985), Goldeneye *Bucephala clangula* in northern Europe (Sirén 1951, Fredga & Dow 1984) and Barrow's Goldeneye *B. islandica* in western Canada (Savard 1988). In the eastern United States, Wood Duck numbers plummeted early in the 20th century, but have since returned in abundance, largely as a result of the thousands of nest-boxes that have replaced natural cavities lost to logging (Bellrose 1990). The same held for the Goldeneye in Scandinavia, where nest-boxes have been provided for several hundred years, initially to obtain eggs for human consumption. In the 1970s, there were an estimated 750 000 such boxes in Sweden alone. Typically, when boxes were first erected in a new area, Goldeneye numbers increased slowly over a period of years, and then levelled out, often at 0.3–0.5 pairs per km of shoreline. This was another indication that, once the shortage of nest-sites had been rectified, other factors limited numbers at a higher level (Eriksson 1982, Fredga & Dow 1984).

Among raptors, European Kestrels *Falco tinnunculus* increased in one year from fewer than 20 to more than 100 pairs when boxes were provided in a Dutch area with few natural sites (Cavé 1968). Similar results were obtained with European Kestrels in other areas (Village 1983), and also with American Kestrels *F. sparverius* (Hamerström *et al.* 1973). Various species of owls readily accept boxes, and densities of Little Owl *Athene noctua* and Barn Owl *Tyto alba* increased after box provision (Exo 1992, Petty *et al.* 1994). Among various parrots, scarcity of nest-cavities has been shown to limit breeding densities in some areas (Beissinger & Bucher 1992, Munn 1992), and the selective removal of large trees from tropical forests, together with the felling of nest trees to obtain parrot chicks for sale, may have reduced the sites available for some species (Beissinger & Bucher 1992).

Experimental demonstration of the presence of non-breeders

To test whether shortage of nest-sites prevents some birds from breeding altogether in certain years, leading to a non-breeding surplus, two types of experiment have been done. In one type, nest sites were installed late in the season, after other pairs had settled and started to nest. In normal circumstances, any birds that had not started by then were unlikely to nest anywhere that year, so occupation of late boxes could be taken as evidence of a non-breeding surplus. In one experiment, 20 out of 112 nest-boxes made available to Tree Swallows *Tachycineta bicolor* late in the season in Ontario were soon occupied by new pairs (Stutchbury & Robertson 1985). The newcomers included more yearlings than were present among the previously established birds, and because none was already ringed, they had not moved from other boxes nearby. By implication, these birds would not have bred that year if boxes had not been provided. A similar experiment with the same result involved European Kestrels *Falco tinnunculus* in artificial stick nests (Village 1990). In both these species, the increase in breeding pairs depended on the provision of extra nest-sites, yet similar vacant sites were already present within the territories of some established pairs, whose defence behaviour was seen to prevent access by other individuals. Hence, in both species, it was not shortage of nest-sites as such that prevented the establishment of additional breeding pairs, but the lack of such sites far enough from existing pairs. This deficiency excluded some birds from breeding.

In the second type of experiment, occupants of nest-boxes were removed well after nesting began, only to be promptly replaced by other individuals which then proceeded to nest. This again showed not only that nest sites were limiting, but also that other birds previously excluded from sites were available locally to take up the vacancies (Newton 1992). Such replacements have been documented following removals in the resident Starling *Sturnus vulgaris* (Heusmann & Bellville 1978) and House Sparrow *Passer domesticus* (Schifferli 1978, T.R. Anderson 1990) and in the migratory Pied Flycatcher *Ficedula hypoleuca* (males only, Alatalo *et al.* 1983) and Mountain Bluebird *Sialia currucoides* (Power 1975). Where removals were continued, several individuals sometimes occupied the same box in quick succession (for House Sparrow, see Nelson 1907; for Pied Flycatcher see von Haartman 1957). In all these species, in which individuals locate a hole and then defend it, the nest-site is evidently the limiting resource and defence the proximate mechanism securing ownership. In some other hole-nesting species, such as tits, which defend large feeding territories that include one or more nest-sites, replacements have also occurred following removals (Chapter 4); but in these species it is not always certain whether territories or nest-sites are limiting.

Conclusions on cavity-nesters

These various studies show that, over wide areas, the breeding densities of many non-excavating cavity-nesting birds are limited by shortage of nest-sites. This is not always the case in mature unmanaged woods, where natural holes are plentiful, but almost invariably the case in younger managed woods, as in much of western Europe. Nest-box provision has almost always increased the local breeding densities of one or more species of hole-nesting birds (as in nearly all the studies in **Table 8.2**). In several studies (notably on the Pied Flycatcher), numbers increased abruptly from the year after boxes were installed, implying that large numbers of potential occupants were available locally. Without the boxes these birds would presumably not have bred, but remained as part of a non-breeding contingent unable to breed through shortage of sites. Other studies indicated that, following box provision, numbers increased slowly over several years (up to 20 years in Goldeneye), implying that no more than a small surplus was present in any one year, and that reproduction or continued immigration contributed to the longer-term increase. Whatever the species and its rate of increase, numbers eventually levelled off, however many extra boxes were provided. The implication was that, once the shortage of nest-sites was rectified, another factor took over to limit numbers at a higher level. Evidence from various tits and raptors showed that this second factor was food-supply (Chapter 7). In such species, therefore, densities in different areas were limited by either nest-sites or food, whichever was in shortest supply (see also Newton 1979 for raptors).

The extent of the increase recorded following box provision varied greatly between studies, and sometimes the presence of boxes enabled species to breed in abundance in areas previously closed to them. In most studies, breeding density more than doubled after box provision and in some it increased more than 20-fold. In one of the most extreme examples, Tree Swallows increased 23-fold, from 7 to 163 pairs in a 160-ha area in New York State. This occurred over a 10-year period, as nest-boxes were added gradually (Swanson & Ryder 1979). In another instance, a colony of 150 pairs was established in 0.3 ha of meadow where none bred previously (Whittle 1926). The fact that nest success is often better in purpose-built boxes than in natural sites (Nilsson 1975, Møller 1989b) could help local numbers to build up rapidly after box provision.

SPECIES THAT NEST ON CLIFFS OR OTHER RESTRICTED SITES

Swallows and others

Some bird species that normally nest on cliffs will accept buildings. This has enabled them to exploit areas otherwise unavailable, and led to increases in both numbers and distribution. Striking examples are provided by swallows and other hirundines, which throughout much of the world have occupied buildings of one form or another, spreading with human settlement. Some such species, which formerly used cliffs or caves, have now become so widespread that it is hard to imagine how scarce and localised they must once have been. This is true of European Barn Swallows *Hirundo rustica* and House Martins *Delichon urbica*, which now rely almost exclusively on buildings for nest-sites. Some species have switched so completely to man-made structures that it is now hard to find them in natural sites anywhere.

The seven hirundine species that breed in North America vary in the extent to which they have adopted man-made structures, depending partly on their original nest-sites (Erskine 1979). The Purple Martin *Progne subis*, which nests naturally in tree-holes, has had the longest association with humans. Native Americans in the southeast hung gourds from trees near their villages or from poles in their corn fields. Martins accepted these sites and, as a result of their mobbing behaviour, kept predators away from the corn and foodstuffs hung up to dry. Early European colonists substituted ceramic jugs and wooden nest-boxes in attempts to attract martins (Jackson & Tate 1974). Over the years, martin houses have appeared in a wide range of styles, including multi-celled apartment blocks holding several pairs. In one extreme example near New York, some 250–300 pairs occupied a single giant house (Swanson 1980). It is inconceivable that so many pairs could have found natural sites locally. Nowadays in the eastern States the species breeds almost entirely in such man-made sites, but in the west it still uses mainly natural tree-cavities. The fact that the martins usually occupy boxes in the first year they are put up, increasing thereafter, again implies that 'surplus' birds are usually available to take up the sites (Jackson & Tate 1974).

As in Europe, the Barn Swallow *Hirundo rustica* in North America has taken to nesting in barns and sheds, or under bridges and house porches, and is still expanding into new areas following human settlement. Similarly, the Cliff Swallow *Hirundo pyrrhonota* spread rapidly in North America in the 1800s following its acceptance of buildings, although later it began to suffer from the expanding population of House Sparrows *Passer domesticus* which took over its nests (Bent 1942). One Cliff Swallow colony on a farm in Wisconsin increased from a single pair in 1904 to more than 2000 in 1942, helped by the addition of 'shelves' to the outside of barns and continuing sparrow control (Buss 1942).

The two North American species that use holes in earth banks, the Rough-winged Swallow *Stelgidopteryx serripennis* and Bank Swallow *Riparia riparia* (= Sand Martin), have spread as sand and gravel workings, and road and rail cuttings, have become increasingly available. The Rough-winged Swallow also nests in drainpipes and in holes in bridges, and has responded to the provision of tubes as artificial sites. When the birds arrived in spring in a Michigan area, few holes remained from the previous year, so most pairs had to wait until other holes were dug and abandoned by Bank Swallows. When Lunk (1962) installed nest-tubes, birds inspected some within an hour of completion and started nest-building within the first day. In all, 19 of 21 tubes were used in the first season. Lunk concluded that many Rough-winged Swallows were present in the area each spring 'waiting' for nest-sites to become available, and that without the extra sites only a proportion of pairs could have bred by the end of the season. So despite the presence of artificial substrates, shortage of nest-sites was still limiting the local breeding numbers of this species.

During the last 200 years, therefore, all these hirundine species, in accepting man-made sites, have become more abundant and widespread. The fact that the occupation of specific localities, and associated range expansions, closely followed the appearance of buildings or other man-made sites strongly implies that lack of nest-sites formerly excluded them. Almost certainly, range expansion has facilitated increased population size, and has not merely caused a similar total population to become more thinly spread. However, two other species, the Tree Swallow *Tachycineta bicolor* and Violet-green Swallow *Tachycineta thalassina*, which use mainly tree-holes, may be no more abundant now than formerly, because their acceptance of artificial sites is unlikely to have compensated fully for the loss of natural sites resulting from forest-clearance.

Like hirundines, some species of swifts that formerly used rock crevices or tree-holes for nesting have also taken to buildings, extending their range in consequence. The most obvious example from the Old World is the Common Swift *Apus apus*, and from the New World the Chimney Swift *Chaetura pelagica*, which formerly relied on large hollow trees but now uses chimneys, sticking its nest to the inside walls. Other passerines that have benefited in North America from buildings include the Eastern Phoebe *Sayornis phoebe* in the east and the Say's Phoebe *S. saya* in the west. These species are naturally associated with cliffs, nesting on ledges in caves or other recesses. But they will also nest in dark buildings, enabling them to spread greatly from their original habitat.

Raptors and others

Some cliff-nesting falcons have also taken to buildings, enabling them to breed in areas otherwise closed to them. Among small falcons, only kestrels have taken this opportunity, and several species over their whole range use anything from isolated ruins to churches and public buildings in city centres. The German name 'Turmfalke' (Tower Falcon) for *Falco tinnunculus* is testimony to this habit in central Europe. Nesting on buildings is one of the factors that have enabled this Kestrel to penetrate towns so successfully. In 1968 no fewer than 61 pairs were found within the city of Munich (one pair per 5 km²), of which 42 nested on buildings (Kurth 1970). In 1967, 142 pairs were found within about 30 km of St Paul's Cathedral in London, a density of one pair per 23 km² (Montier 1968).

Among the larger falcons, the Peregrine *Falco peregrinus* has used old buildings to a considerable extent, either isolated or in towns and cities. In Europe and North America, the cities used until 1970 showed a curiously scattered distribution, implying that the habit arose independently each time, and most cases persisted for no longer than a few years (Hickey & Anderson 1969, Newton 1979). However, since the 1970s, Peregrines have nested increasingly in North American cities. In an attempt to restore populations depleted by pesticide poisoning, captive young were deliberately raised on buildings and allowed to fly free, using the falconry technique of 'hacking back'. Many such young, which were identified from their rings, returned to nest in cities, and some attracted mates from outside cities. A survey in 1988 revealed 30–32 Peregrine pairs nesting in at least 24 cities and towns across North America and not only where releases had been made (Cade & Bird 1990). The fact that most cities now contain more tall buildings than formerly, as well as more pigeons as prey, may have helped the habit to spread.

Other large falcon species have occasionally used buildings and other man-made structures for nesting, as have various other raptors (Newton 1979). Ospreys *Pandion haliaetus* have long used artificial structures, including tree-substitutes such as pylons, and in recent years special platforms. During aerial surveys along the Atlantic coast of North America, platforms for nesting Ospreys were noted in every State from South Carolina to New York, many in use. In 1973, only 31% of the 1500 pairs around Chesapeake Bay were found on trees, the rest on duck blinds (29%), channel markers (22%), and other man-made structures (18%), including platforms (Henny *et al.* 1974). Because nests on man-made structures are often more secure than those on trees, nest success is usually higher, with 0.6 young produced per natural nest in Michigan, compared with 1.2 young per platform nest (Poole 1989). In southern Idaho and Oregon, raptors and Ravens *Corvus corax* began nesting on a 596-km segment of power line within one year of its construction. Within ten years, numbers had increased to 133 pairs, all in landscape lacking other nest-sites (Steenhof *et al.* 1993). By using man-made structures, all such species increased their nesting densities and distributions in places where natural sites were scarce or lacking. In some instances birds accepted new structures immediately, but in others only after many years or only in particular areas.

Some tree-nesting species have also benefited from the provision of nest-sites. An example is the Mississippi Kite *Ictinia mississippiensis* which now breeds on the Great Plains of America 'in hundreds, if not thousands', in places where trees have been planted (Parker 1974). Several other raptors of American grasslands were found to depend to some extent on planted trees or deserted buildings (Olendorff & Stoddart 1974), while in South Africa, Greater Kestrels *Falco rupicoloides* and Lanner Falcons *Falco biarmicus* have spread by using pylons, as have Black Sparrowhawks *Accipiter melanoleucus* by using exotic eucalypt or poplar plantations (Kemp & Kemp 1975). In none of these species is it likely that prey populations increased, and the spread could be attributed entirely to the provision of nesting places.

Relatively few cliff- or tree-nesting species have been subject to experiments. However, an increase in Prairie Falcon *Falco mexicanus* pairs from 7 to 11 followed the digging of suitable holes in riverine earth banks in Saskatchewan (Fyfe, in Cade 1974), and an increase in Oilbird *Steatornis caripensis* pairs from around 15 to more than 40 followed the construction of concrete ledges in a nesting cave in Trinidad (Cade & Temple 1995). Similarly, the placement of 12 small trees in an open area allowed eight pairs of Carrion Crows *Corvus corone* to settle where breeding was not previously possible, all the birds concerned coming from a known non-breeding flock (Charles 1972). These various examples indicate that limitation of breeding density by nest-site shortages is widespread among many types of bird that use special sites, whether cliffs, caves, burrows or scarce trees in open land.

Colonial seabirds and herons

While some bird species accepted buildings as soon as they were built, other species have only recently begun to do so, usually as their populations expanded and natural sites became saturated. Various gull species have in some areas established colonies on the roofs of town buildings, enabling a further rise in local breeding numbers. This has occurred, for example, among Herring Gulls *Larus argentatus* and Lesser Black-backed Gulls *L. fuscus* in parts of northwest Europe, and among Herring Gulls, Ring-billed Gulls *L. delawarensis* and Glaucous-winged Gulls *L. glaucescens* in parts of North America (Monaghan & Coulson 1977, Blokpoel & Smith 1988, Vermeer *et al.* 1988).

In some other seabirds too, the provision of extra nesting habitat has sometimes led to increased breeding numbers. Large nesting platforms, built for the collection of guano, were provided off Namibia during 1930–40. By the 1950s, they held about 500 000 pairs of Cape Cormorants *Phalacrocorax capensis*, expanding both numbers and distribution (Crawford & Shelton 1978, Crawford *et al.* 1992). Likewise, following the exclusion of mammalian predators by fencing, Peruvian seabirds (also valued for their guano) began to nest on headlands, increasing the country's total seabird population from 8 million to 20 million birds over a period of years, and increasing the rate of increase from 8 to 18% (Duffy 1983). Arguably, the increase might have occurred anyway from some other cause permitting population growth (Chapter 11), but because different islands and headlands were occupied in turn as they were predator-proofed, safe nesting habitat was probably influential. At several sites in Scotland, fences around coastal industrial and military sites have enabled large numbers of terns to breed safely in areas previously ignored, and in other areas terns have taken to nesting on rafts built specially for them, but whether these events led to increases in regional numbers (rather than to shifts from nearby areas) remained uncertain (Techlow 1983, Dunlop *et al.* 1991). In the Seychelles, tree-felling enabled a Sooty Tern *Sterna fuscata* colony to increase from 120 000 to 345 000 pairs in six years (Feare 1976b). In general among seabirds, limitation by nest-sites appears to be an exceptional rather than a widespread phenomenon, but wherever birds occupy newly-provided sites away from others, this could facilitate the use of new feeding areas and further growth in numbers.

A remarkable experiment with Snowy Egrets *Egretta thula* and related species was conducted at 'Bird City' on Avery Island, Louisiana, by McIlhenny (1934). In 1892, when he began, such species had been greatly reduced in numbers by commercial plume hunters. His management measures included:

(1) providing large nesting platforms about two metres above the surface of an artificial lake,
(2) fencing the lake against human disturbance,
(3) providing the birds with sticks for nest-building, and
(4) introducing alligators 'to deter children' and to control nest predators such as Raccoons *Procyon lotor*.

By 1912 the population of several egret and heron species on the platforms had increased to 22 204 nests. In this instance, it was the security of the nest-sites that was crucial, rather than shortage in the surrounding area.

Non-specialists

Although the argument for the role of safe nest-sites in limiting breeding density is most readily made for species that use specialised and rare sites, nest-sites could be important in a much wider range of species, including those that build nests in commonly-available trees and shrubs. In a study of the birds of various woodland areas in Arizona, species numbers were more closely correlated with the range and density of nesting sites available than with the range and density of foraging sites (Martin 1988). The implication was that safe nest-sites influenced the presence even of species that built their own nests in commonly available sites.

FACTORS CONSTRAINING THE TYPES OF SITE USED

Because shortage of nest-sites can limit the numbers of pairs that can breed, the question arises as to what constrains the types of site used. If species accepted a wider range of nest-sites, they would be less likely to be limited by shortages. Interspecific competition is one factor involved (Chapter 12). From the examples given earlier, particular species may use only part of the potential range of nest-sites available to them, because they are kept out of other sites by dominant competitors. In addition, predation pressure can influence the range of sites that particular species use, and the types of site that are acceptable can change if predation pressure alters. Some of the best documented examples concern raptors.

Over most of its range, the Peregrine *Falco peregrinus* nests only on high cliffs or on islets surrounded by water or bog. Such sites give some protection against mammalian predators, including humans. But on tundra and other areas where people are few, Peregrines will accept lesser crags, earth slopes or even low mounds or boulders on flat ground (Hickey & Anderson 1969, Kumari 1974). Generally speaking, 'the minimum height of cliff acceptable to birds varies inversely with the degree of wilderness' (Hickey 1942). Some easily accessible sites in the Canadian Arctic were abandoned after an increase in human population, and the same evidently happened in Britain before 1860, and also in the United States after settlement by Europeans (Hickey 1942, Fyfe 1969, Ratcliffe 1969). In recent years, Peregrines in Britain have occasionally nested on level ground or on slopes, but the habit has not spread, possibly because such nests usually fail. Likewise, Bald Eagles *Haliaeetus leucocephalus* use accessible nest-sites in the Aleutians only on islands where there are no Arctic Foxes *Alopex lagopus* (Sherrod *et al.* 1977), while Buzzards *Buteo buteo* and Kestrels *Falco tinnunculus* regularly nest on the ground on some Scottish islands where there are no wild mammalian predators (Newton 1979). In various parts of their range, Ospreys nest on the ground on islands with no mammalian predators, but they disappeared from a treeless island off California, following its colonisation by Coyotes *Canis latrans* (Kenyon 1947).

There are thus indications among raptors that the types of nest-site used by a given species are not fixed, and that minimum requirements (in terms of security) can alter in response to changes in disturbance and predation pressure. In this indirect sense, predators could restrict the breeding densities of raptors in areas where safe sites are scarce, for if it were not for their predators, the raptors might accept less safe sites, and thereby breed in greater numbers. Almost certainly, the same applies to many other birds (see also Chapter 13). Moreover, in the same way that some seabirds occupied certain sites after mammalian predators were excluded, other birds occupied particular sites only after parasites were experimentally removed (Chapter 10).

Some birds have changed their nest-sites in unexpected ways. One of the most curious examples is provided by Oystercatchers *Haematopus ostralegus* in the city of Aberdeen in Scotland. Normally these birds nest on shingle and feed on the seashore or in nearby fields, but in Aberdeen they have started to nest in safety on the flat roofs of buildings, and feed on the larger lawns and playing fields. The same is true for some other plovers elsewhere, including the Killdeer *Charadrius vociferus* in some eastern North American cities, the Little Ringed Plover *C. hiaticula* in parts of Europe and the Senegal Thick-knee *Burhinus senegalensis* in Egypt (Cramp 1983). In parts of North America, the Common Nighthawk *Chordeiles minor* nests on flat roofs, and hunts the moths that gather around street lamps. For all these species, roofs provide the most secure nesting substrates in areas where food is available. And with changes as striking as these, it is difficult to predict which species might colonise towns in future.

SUMMARY

The breeding densities of many bird species that require special nest-sites are in some areas limited by shortage of acceptable sites. This is evident from circumstantial evidence in which the numbers of breeding pairs in different areas correlate with the numbers of local nest-sites, or where changes in the numbers of nest-sites are followed by corresponding changes in the numbers of breeding pairs.

When nest-sites were provided experimentally, they were often occupied in the same year, leading to an immediate rise in breeding density. This implied that surplus birds were available in the vicinity and able to take them up. In future years, pair numbers often increased further but eventually they levelled off. This implied that, at this higher level, other limiting factors came into play. Evidently, different factors acted successively to limit breeding density at different levels.

By constraining breeding density, shortage of nest-sites not only prevents some birds from breeding, but must also limit the total numbers of breeders and non-breeders, because no population can increase beyond a certain point, when the production of young is limited. Both breeding numbers and output are thus limited in a density-dependent manner.

Although most bird species accept a fairly narrow range of nest-sites, in some species this can change under altered predation pressure. Acceptance of less safe sites, where mammalian predators are absent, enables pairs to occupy areas otherwise unavailable.

Sparrowhawk *Accipiter nisus* attacking
young Blue Tit *Parus caeruleus*.

Chapter 9
Predation

While the supplies of food and other resources provide the ultimate check on bird numbers, some populations might be held by predators below the level that resources would permit (Lack 1954, Martin 1991, Newton 1993). Although practically all birds are subject to predation, at least at the egg and chick stages, it is not easy to assess its effects on population levels. Even where predators kill a large proportion of their prey each year, they do not necessarily reduce breeding numbers. The numbers of many bird species can more than double each year through breeding so that, in a stable population, more than half the individuals present at the end of one breeding season must die before the next, if not from predation then from something else. For predators to reduce breeding numbers below the level that would otherwise occur, at least part of the mortality they inflict must be 'additive' to other mortality, and not simply 'compensatory', replacing other forms of death.

Compensation can occur only if other factors cause survival or reproductive rates to vary in a density-dependent manner. Such a situation would be expected in populations that are close to their resource limit, and experiencing strong intra-specific competition, whether for territories, food or nest-sites. These are the circumstances in which predation is most likely to be compensated by reduced losses from other causes, and so have least effect on breeding numbers. Conversely, populations that are well below their resource limit, and relatively free of intra-specific competition, are most likely to suffer further reduction from any predation (or parasitism) that they experience. In theory, any one species could switch from one state to the other, depending on the other factors that influence its numbers.

THE PREDATORS

Birds fall victim to a wide range of predators, mainly avian, mammalian and reptilian. In north temperate regions, where most studies have been made, the commonest mammalian predators include foxes, cats and mustelids, with the addition in North America of raccoons. These various species hunt mainly on the ground and vary in their ability to reach tree nests, some mustelids and raccoons being especially good at climbing. The commonest avian predators include corvids and gulls (on eggs and chicks) and raptors (on large young and adults). In warmer regions, monkeys and other mammals are frequent predators, as are snakes and other reptiles, especially on eggs and chicks, while some ant species also consume chicks. Typically, any given prey species may have a small number of important predator species, but suffer occasional attacks from a much larger number. For example, most predation on Ring-necked Pheasants *Phasianus colchicus* in many areas is from Red Foxes *Vulpes vulpes* and Goshawks *Accipiter gentilis*, but in Wisconsin alone more than 30 species of birds, mammals and reptiles have been reported as eating eggs, chicks or adults from time to time (Wagner *et al.* 1965). In addition, some bird species have their own special predators, as exemplified by penguins eaten by Leopard Seals *Hydrunga leptonyx*.

Most predators of birds are generalists, and take a wide variety of prey, switching from one prey species to another as opportunities and needs arise. Because the population levels of such predators can be influenced by the sum total of prey, they are often not greatly affected by shortage of any one type. Generalist predators do not usually take their various prey in strict relation to the numbers of those prey in the area, but take some (vulnerable) species more than expected from their numbers, and others less so (Figure 9.1). Thus, certain prey species may experience heavy predation not because of their own numbers, but because other prey species are sufficiently plentiful to attract many predators to the area, or because alternative prey suddenly become scarce (Tinbergen 1946, Kenward 1985, Keith & Rusch 1988). As an example, on the island of Foula (Shetland), the predation by Great Skuas *Catharacta skua* on other seabirds increased ten-fold over a period of years, as the major fish prey of the Skuas declined (Hamer *et al.* 1991).

Other predators are specialists, concentrating on one or a few main prey species. Specialists are rare among predators of birds, although in certain regions specialism may be forced upon generalist predators by lack of alternative prey. The dependence of Gyr Falcons *Falco rusticolus* on Ptarmigan *Lagopus mutus* over much of the Arctic provides an example. Understandably, specialists are highly susceptible to shortage of their main prey, so that the numbers of predator and prey are often closely inter-dependent. It is difficult for a specialist predator to eliminate its prey, for its own numbers fall before the prey is extinguished. Moreover, because of delays in response, with the predator taking time to increase in response to a rise in the prey population, closely coupled predator–prey populations tend to oscillate (see below).

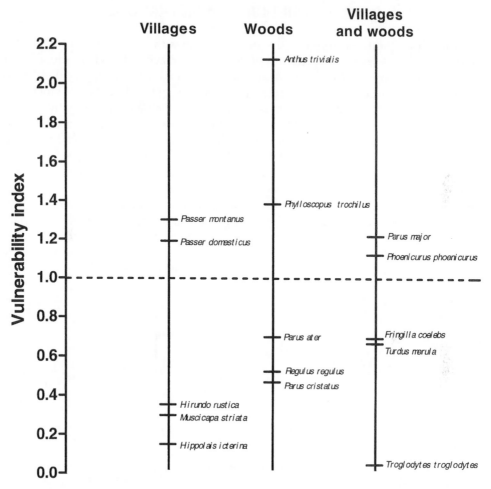

Figure 9.1. Different vulnerability of various avian prey species to predation by Sparrowhawks *Accipiter nisus*. An index of 1 (dotted line) indicates that a species was taken in exact proportion to its abundance in the area, more than 1 that it was taken more than expected from its numbers, and less than 1 that it was taken less than expected. Details from a study in the Netherlands by Tinbergen 1946.

Whether by generalists or specialists, predation is seldom distributed evenly through a prey population. It may be concentrated in particular localities, perhaps near where the predator itself is breeding, or where prey are unusually plentiful or vulnerable (Geer 1978, Erikstad *et al.* 1982). It may also be concentrated on particular age-groups, such as the young and inexperienced (Newton 1986, Bijlsma 1990) or on particular social classes, such as the non-territorial in a territorial species (Jenkins *et al.* 1964), or one sex more than the other (see below). The predation on a particular prey species also varies through time, in response to changes in the vulnerability of the prey, in the availability of alternative prey, or in the prey–predator ratio (Tinbergen 1946, Erlinge *et al.* 1983, Newton 1986).

SOME THEORETICAL CONSIDERATIONS

Predators can limit the numbers of their prey in either a regulatory (density-dependent) or non-regulatory (density-independent) manner. In density-dependent predation, the proportion of prey that are killed increases with rising prey density, so that eventually numbers stop increasing (or conversely, stop decreasing before they reach extinction). This provides one mechanism by which a prey population could be held below the level that resources would permit. Two responses of predators to changing prey populations can promote density-dependence: (1) the functional response, in which individual predators kill more of a given prey species per unit time as the numbers of that prey increase; and (2) the numerical response, in which predators increase in density (through immigration or reproduction) as the prey itself increases (Solomon 1949, Holling 1965). Either of these responses, or both together, can result in relatively heavier predation in conditions of high prey numbers than low. In practice, the exact forms of numerical and functional responses are hard to measure for predators of birds, because fluctuations in the abundance of alternative prey often distort the relationship **(Figure 9.2)**.

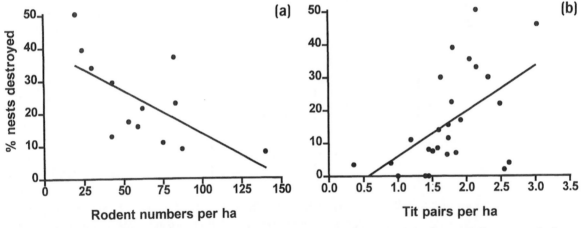

Figure 9.2   Predation by Weasels *Mustela nivalis* on the nest contents of tits (several *Parus* species).
(a) Relationship between percentage of nests preyed upon and the density of rodents, which formed the main prey of Weasels. The points cover 13 years, and indicate that nest losses were greater in years with lowest rodent densities.
(b) Relationship over 26 years between the percentage of nests preyed upon and the density of nests. In addition to the relationship with rodent density, predation was density-dependent with respect to tit density. Modified from Dunn 1977.

If predators remove a greater proportion of prey when prey numbers are low than when they are high, the relationship is 'inversely density-dependent' (Chapter 5). This could happen if, for example, the numbers of the predators were limited by some other factor (precluding a numerical response), and if the functional response was also weak. An example of inverse density-dependence in predation on Red Grouse *Lagopus l. scoticus* is shown in **Figure 9.3**. In density-independent predation, the numbers of prey removed each year by predators bear no consistent relationship

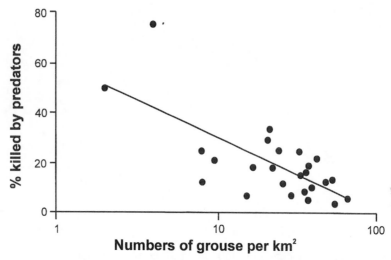

Figure 9.3   Inverse density-dependent predation on female Red Grouse *Lagopus l. scoticus*, mainly by Peregrines *Falco peregrinus*. The points refer to overwinter losses in different areas and years, and are shown in relation to autumn grouse densities. The greater the density of grouse, the smaller the proportion taken. This situation arose partly because Peregrine densities remained roughly constant (being determined by other factors), regardless of the densities of grouse. Modified from Hudson 1992.

to the numbers of prey present. It can occur, for example, when predation on a given prey species depends not on its own numbers but on the abundance of alternative prey, as described above. In one study, predation by Goshawks on Ring-necked Pheasants was much greater than expected, apparently because Goshawks were maintained at high numbers by a local abundance of Rabbits *Oryctolagus cuniculus* (Kenward 1985). In these circumstances, Goshawks removed 56% of hen Pheasants over one winter, a proportion the population could not sustain in the long term. They removed fewer cocks, however, which were larger and harder to kill. Density-independent predation can also occur when predators, whose numbers were determined elsewhere, suddenly move into an area (as during migration) causing a marked decline in prey numbers, before they themselves move on or die (see Keith & Rusch (1988) for effects of migrant Goshawks on Ruffed Grouse *Bonasa umbellus*). Other examples of predation reducing the numbers of secondary prey when primary prey are scarce are given below. In theory, predation that was either inversely density-dependent or density-independent could lead to a local extinction of prey.

Predation can switch from regulatory to non-regulatory with a change in prey density **(Figure 9.4)**. Under conditions of rising prey numbers, a point may come when predators can increase their kill rate no further. This happens because there are limits to both the amounts that individual predators can eat and to the numbers of predators that can live in a given area (for example because of interference or social intolerance). The numbers of prey that are taken then level off so that, if prey numbers but not predator numbers continue to rise, a progressively smaller proportion is taken. The relationship then switches from density-dependent (regulatory) to inversely density-dependent (non-regulatory) **(Figure 9.4)**. One

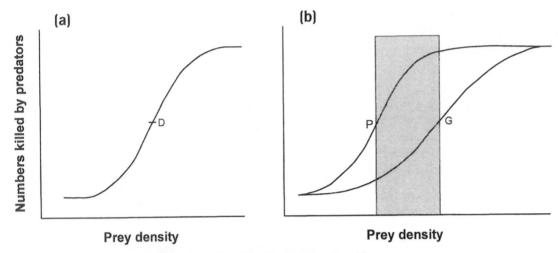

Figure 9.4   Models depicting predation at different prey densities.

(a) As prey density rises, the proportion taken by predators (by the combined action of numerical and functional responses) increases slowly to begin with, then more rapidly, before levelling off. The initial slow response occurs because at low density the prey are hard to find and the predators may be eating mainly alternative foods. The levelling off occurs because (in the numerical response) the numbers of predators eventually reach some socially imposed limit, while (in the functional response) individual predators become satiated or otherwise constrained in the amount they can eat. The combination of these two processes limits the numbers of prey taken, even though the numbers of prey available may continue to rise.

Up to point D, predation is density-dependent, involving the removal of an increasing proportion of prey as their numbers rise. This process acts to slow the rate of population growth and is therefore regulatory (stabilising). Beyond point D, the proportion taken declines progressively with further rise in prey density. Predation becomes inversely density-dependent and non-regulatory (destabilising), playing a progressively weakening role in limiting prey density. There is thus a dichotomy in the effect of predation, with density-dependence (regulation) at low prey densities and inverse density-dependence (non-regulation) at high prey densities.

In this way, some bird populations might be regulated by predation at low density, but limited (not regulated) by predation and other factors at high density (see text). However, most bird populations might not normally be expected to fluctuate over the full range of densities shown. The effect of predation on their population dynamics would therefore depend on the shape of the response curve within the range of densities experienced, in other words on the extent to which the predation rate accelerates or decelerates over the observed range of densities normally experienced. The whole curve covers the Type 3 response of Holling (1965) and from point D the curve covers the convex Type 2 response, while the central (linear) part covers the Type 1 response.

(b) The shape of the predatory response curve might be influenced by other factors, such as habitat structure or nest cover, which confer greater or less protection from predation. The two curves depict the response in habitat which is good (G) and poor (P) in this respect. In good habitat, at any given prey density, a smaller number of prey is removed and the density-dependence is less marked (less steep) than in the poor habitat. Over a certain range of density (shaded), predation may be density-dependent (stabilising) in good habitat, but inversely density-dependent (destabilising) in poor habitat.

consequence of this switch is that prey populations could have two equilibrium levels, a lower level at which they were regulated by predation, and an upper level at which they were merely limited by predation and other factors. A prey population might break free of regulation by predators if, for example, predation pressure lapsed temporarily, or if a sudden change in environmental conditions led to increased reproduction or immigration of prey. The salient feature of this situation is that, once the higher level has been reached and providing no other factors intervene, prey numbers should not return to their lower level once the original conditions return. Only then could we be sure that the prey, by increase alone, had escaped the regulating influence of predators (the so-called 'predation trap'). On the other hand, if numbers fell to near their original level, this might be because predators had re-asserted their regulatory effect, or because other factors (such as food-shortage) had intervened to reduce prey numbers. Escape from regulation by predation is the mechanism proposed to explain occasional outbreaks of some insect pests, which persist until other factors reduce their numbers again.

The role of predation in influencing prey numbers thus depends on the way in which predators operate. Only if the combination of functional and numerical responses provides an appropriate total response to changes in prey density can predators regulate prey numbers. Moreover, because predation is only one of several factors affecting prey populations, any regulatory effect it might have may be supported or counteracted by other factors.

EFFECTS OF PREDATION ON INDIVIDUALS

In many bird species, predation is the major cause of egg and chick losses, commonly accounting for around half of all nesting attempts and more than 80% of all nest failures (Nice 1937, Lack 1954, Ricklefs 1969, Martin 1991). But because many species can re-lay after a nest failure, such losses may have less effect on the annual production of young by individuals and by the population than these bald figures suggest. At the population level, the loss of say 50% of nests to predators often translates to a much smaller loss in eventual production, even though late broods may do less well than early ones. Moreover, the staggering of breeding attempts resulting from re-lays means that, in species that raise only one brood each year, the young appear over a longer period than otherwise and are thus not all equally vulnerable to infrequent destructive weather events, such as hailstorms.

Rates of predation on nest contents vary greatly between species, depending partly on the position of the nest (Lack 1954, Thiollay 1988). Losses tend to be heavy among species that nest on the ground (in which the sitting hen is sometimes taken too), somewhat less heavy among species that build open nests in shrubs and trees, and lighter still among species that use more protected sites, such as tree-cavities or cliff ledges. In addition, lower nests in trees and on cliffs tend to suffer more predation than higher ones, and in any one situation species that have open nests suffer more predation than those that build domed nests. All these findings would be expected from knowledge of the behaviour of predators. They are only tendencies, however, and in some areas different patterns occur, depending on local circumstances. An earlier finding that egg and chick predation tends to be higher in the tropics than elsewhere may be genuine, or it may be an artefact of the fact that many studies in the temperate zone were in secondary (human-modified) habitats that held reduced numbers of predators (Tomiałojc *et al.* 1984).

For any particular nesting situation, some sites seem to offer more protection than others. In species that hide their nests in ground vegetation, predation rates tend to be higher in sparse than in dense cover (Choate 1967, Chesness *et al.* 1968, Schrank 1972, Duebbert & Kantrud 1974, Newton & Kerbes 1974, Duebbert & Lokemoen 1980, Erikstad *et al.* 1982, Hines & Mitchell 1983, Rands 1986). Among Mallard *Anas platyrhynchos* in North Dakota, for example, the proportion of clutches that hatched varied from nil in the most open types of vegetation to more than 50% in the most dense (Cowardin *et al.* 1995). Similar differences were found in some shrub-nesting species, in some of which nest success also increased through the season as cover thickened (Newton 1972, Tuomenpuro 1991, Hatchwell *et al.* 1996). Certain species gain extra protection by nesting close to other species that drive away predators. For example, in Meadow Pipits *Anthus pratensis* and Yellow Wagtails *Motacilla flava* nest success was significantly higher among pairs nesting close to Lapwings *Vanellus vanellus* than among pairs nesting elsewhere (Eriksson & Götmark 1982). Other species gain protection by nesting near predatory birds or wasps (Lack 1968, Thiollay 1988). Taken together, these various findings imply that nesting areas (or sites) vary in quality, and there may therefore be strong competition for them. In some species, nest predation is also influenced by the nutritional state of the female, which may affect the amount of time she spends off the nest or the vigour of her defence against predators (Newton 1986). Because predation rates are so heavily influenced by features of the habitat, and of the birds themselves, as well as by the numbers and behaviour of predators present, they vary greatly in the same species between different areas and different years.

Far fewer studies have been made of predation on full-grown birds. Birds killed by predators are mostly consumed before they can be found by human observers, leaving only a few remains. They are therefore greatly under-represented among carcasses received for autopsy studies **(Table 7.2)**. Radio-tracking studies are of more value, and have implicated predation in the recorded deaths of many species **(Table 7.3)**. In a study of radio-tagged Black Grouse *Tetrao tetrix* in Sweden, predators accounted for almost all deaths recorded over three years, removing some 48–72% of all males in different years and 48–52% of all females (Willebrand 1988). In another study of Ring-necked Pheasants in Colorado, predators accounted for all the deaths recorded over three years (Snyder 1985). The validity of such findings depends, of course, on whether radio-tagged birds were representative of their population, and whether tagging itself influenced predation risk; moreover, they cannot reveal to what extent predation is predisposed by other factors, such as starvation. At the other extreme, in some large species, such as swans or eagles, the adults appear to be practically immune to predation, except from humans.

Other studies have involved estimating the numbers of birds killed by predators in a given time, and expressing these numbers as a proportion of birds in the area. For example, on an estuary in southeast Scotland, the overwinter losses of various shorebird species to raptors was estimated at 0–57% of the autumn populations (Cresswell & Whitfield 1994). In general, smaller species suffered greater predation than large ones, but Redshank *Tringa totanus* were especially vulnerable **(Table 9.1)**. Within species, juveniles were taken disproportionately more often than adults. Over the summer, Sparrowhawks removed some 3–14% of the breeding adults of all

Table 9.1 Mortality attributed to raptors of waders wintering on a Scottish estuary, September–March inclusive. Species arranged by increasing body size. From Cresswell & Whitfield 1994.

	Minimum percentage overwinter mortality		
	1989–90	1990–91	1991–92
Dunlin *Calidris alpina*	4.6	11.4	21.2
Ringed Plover *Charadrius hiaticula*	6.1	2.6	2.0
Purple Sandpiper *Calidris maritima*	0	6.3	12.5
Snipe *Gallinago gallinago*	28.6	33.3	25.0
Turnstone *Arenaria interpres*	28.6	9.5	11.6
Knot *Calidris canutus*	3.3	4.4	3.3
Redshank *Tringa totanus*	31.1	48.5	57.3
Greenshank *Tringa nebularia*	13.3	0	0
Golden Plover *Pluvialis apricaria*	0	0	0
Lapwing *Vanellus vanellus*	0	1.0	0
Grey Plover *Pluvialis squatarola*	7.9	2.5	5.1
Bar-tailed Godwit *Limosa lapponica*	0	0	0.8
Oystercatcher *Haematopus ostralegus*	0	0	0
Curlew *Numenius arquata*	0	0.3	0

Main predators included Sparrowhawk *Accipiter nisus*, Merlin *Falco columbarius* and Peregrine *Falco peregrinus*.

prey species together in different areas, and some 0–47% of the adults of individual species in particular study plots (Tinbergen 1946, Opdam 1978, Solonen 1997).

As with predation on nest contents, predation on adults can be influenced by habitat structure, and by the hunger or nutritional state of the bird. For example, in a study of radio-marked Hazel Grouse *Bonasa bonasia* in China, predation on adults was much greater in open mature forest than in thicker young (sapling) forests (Yue-Hua & Yu 1994). Similarly, Goshawk predation on Ring-necked Pheasants in Sweden increased with distance from cover (Kenward 1977). The role of hunger can be seen in many species which, when food is scarce, lower their guard and allow predators to approach closer than usual, an aspect discussed in detail in Chapter 13.

PREDATION IN THE ANNUAL CYCLE

Deaths from predation must reduce numbers at least temporarily, because some individuals die before they otherwise would. Inevitably, then, predators could influence the dynamics of prey populations, and could reduce their average level over the year as a whole. The question of most interest is whether the predation to which many species are subject throughout the year reduces their breeding numbers. In any seasonally-reproducing species, numbers are at their lowest point in the early stages of breeding, after most mortality of adults has occurred, but before any young are produced. If predation does not reduce breeding numbers, it has no effect on the numbers of eggs entering the next generation and is not limiting in the long term. It is thus irrelevant to the long-term persistence of the population at that level.

Game managers have a different perspective, because they aim to produce birds for shooting. They are thus concerned both with maintaining breeding stocks from year to year and with producing lots of birds for hunters to kill. Any natural predation or other mortality that lowers the production of young might reduce the numbers available for shooting in autumn, even though it may have no effect on subsequent breeding numbers. It is not contradictory, therefore, that predation might reduce post-breeding numbers in one year but have no impact on pre-breeding numbers in the next year. Similarly, predation might lower breeding numbers, but the reduction might be compensated by improved nesting success, so that big numbers still occur in autumn. In general, we can envisage three scenarios involving predation effects in the annual cycle of any seasonally-breeding bird species **(Figure 9.5)**:

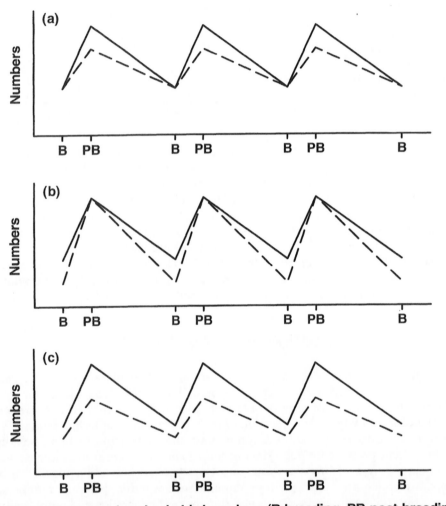

Successive annual cycles in bird numbers (B-breeding, PB-post-breeding)

Figure 9.5   Models showing possible effects of predation on the annual cycle of numbers in birds. Solid line – no predation effects, dashed line – with predation effects. For further details, see text.

(1) Predation reduces post-breeding numbers below what would otherwise occur, but not the subsequent breeding numbers. This scenario requires total compensation through overwinter survival.
(2) Predation reduces breeding, but not post-breeding, numbers below what would otherwise occur. This scenario requires total compensation through reproductive output.
(3) Predation reduces both breeding and post-breeding numbers below what would otherwise occur. In this scenario, compensation is insufficient to offset the full effects of predation, and throughout the year numbers remain below the level they might otherwise achieve.

As a general point, predation on eggs and chicks is much less likely to affect subsequent breeding numbers than is predation on the adults themselves. This is because, the earlier in life that losses are imposed, the greater the opportunity for these losses to be made good before the breeding stage, either by re-layings or by improved survival or immigration of remaining young. Similarly, losses of adults at the start of winter are less likely to reduce breeding numbers than are similar losses at the end, just before nesting begins.

EFFECTS OF PREDATION ON POPULATIONS

In this next section, examples are given where predators have been found to (1) have no obvious effect on the breeding numbers of their avian prey; (2) limit prey breeding numbers below what the habitat would otherwise support; (3) cause oscillations in prey numbers; and (4) exterminate their prey.

No effects on prey breeding numbers

In some bird species, predation might remove so few individuals that it has no discernible effects on breeding numbers. In other species, which are subject to substantial year-round predation, the losses involved are fully compensated by reductions in other losses. This happens when total mortality is density-dependent and 'pre-determined' by the extent to which numbers exceed the available territories or other resources.

This latter situation underlies Errington's (1946) view of a 'doomed surplus', that predators take only the number that would die anyway. He had earlier found that coveys of Bobwhite Quail *Colinus virginianus*, which were resident in winter 'territories' with poor cover, suffered more predation than those in areas with better cover (Errington & Hammerström 1936). He therefore concluded that lack of cover was the ultimate cause of death. Errington (1934, 1946) suggested that each habitat had a limited carrying capacity and that it was only when the numbers rose above this level that predation became limiting. You cannot study the effects of predation, he argued, by counting the numbers of prey killed. You have to determine the factors that predispose predation and make some individuals vulnerable and others not. His idea was that numbers were trimmed each year to a level that the habitat could sustain and that the excess succumbed to whatever mortality agents were available locally, whether predators, disease or starvation. This was an important point because it meant that, even though predation could be substantial, it might have no effect on breeding density.

Support for the notion of a doomed surplus came from a later study on Red Grouse *Lagopus l. scoticus* in eastern Scotland (Jenkins *et al.* 1964). These birds took territories in the autumn, but density on the heather moor habitat was limited, leading to exclusion of a non-territorial surplus, which could not breed. By tagging both territorial and non-territorial individuals, it was found that mortality was almost entirely restricted to the non-territorial birds, the remains of many of which were found in the area (Chapter 2). If a territorial bird died, its place was rapidly taken by a non-territorial individual, so that territorial (breeding) density was maintained through to the following spring, but by that time most of the non-territorial birds had disappeared. It thus seemed that predators and other mortality agents were merely removing the non-territorial surplus, and causing no reduction in breeding density. Although predators accounted for almost all the mortality, they were simply acting as executioners for individuals already condemned by the social system. In this area, however, the various predators of eggs and chicks were controlled by game-keepers, so that the post-breeding grouse population may have been higher as a result. One can imagine, if the post-breeding population were lower, the surplus might have been small or non-existent. In these circumstances, predators could well have removed some territorial birds, and reduced the breeding density, a situation subsequently described for Red Grouse in another area where predators were commoner (Hudson 1992).

In many other bird species, territorial behaviour, in association with limited breeding habitat or nest-sites, leads to the exclusion of a surplus of non-breeders (Newton 1992). Any territorial breeder that dies is then rapidly replaced from the non-territorial sector (Chapter 4). This process could thus account for a lack of detectable impact of predation on the breeding densities of many birds. Only if overwinter mortality is exceptionally heavy are numbers likely to be reduced below the level at which territoriality would be limiting. This situation again requires that habitats have limited carrying capacity and that territorial or other social interactions limit breeding densities. It is less certain to what extent the same process applies to species more tolerant of crowding, such as various ducks.

Another situation in which predation might be compensatory, rather than additive, is if sick or starving prey are taken in numbers greater than expected from their proportion in the population. Such selection is most evident in species that are normally difficult for the predator to catch, or among individuals caught after a tiring chase rather than quickly by surprise (Kenward 1978, Temple 1987). In one study, many of the Woodpigeons *Columba palumbus* caught by Goshawks *Accipiter gentilis* were thin, with 28% of victims already starved beyond the point of recovery (Kenward 1977). In such cases, the impact of predation on the population is clearly less than expected from the numbers taken.

A case study: Sparrowhawk predation on songbirds. The Sparrowhawk *Accipiter nisus* is one of the commonest raptors in Europe, nesting in woodland and hunting small birds, in both woods and open country. It can remove a large proportion of prey individuals each year, yet seems to cause no obvious depression of breeding densities in most species (Tinbergen 1946, Perrins & Geer 1980, Newton 1986, McCleery & Perrins 1991). This was shown incidentally over much of Europe around 1960, when Sparrowhawks were eliminated from large areas through the use of organochlorine pesticides, recovering and recolonising years later when organochlorine use was reduced (Chapter 15; Newton 1986). Yet when Sparrowhawks declined, no great upsurge in songbird populations occurred. This was apparent, for example, in a single 16-ha oak wood in southeast England, where all breeding birds were counted every year from 1949 to 1979 (Newton *et al.* 1997). These counts covered both the decline and the recovery periods of the Sparrowhawk, which bred in the wood and in the surrounding area up to 1959 and again from 1973. This gave three periods for comparison: 1949–59 (hawks present), 1960–72 (hawks absent) and 1973–79 (hawks present). Thirteen songbird species were sufficiently numerous for analysis. In nine such species, significant variations occurred between the mean counts for the three successive periods. In seven species, this was attributed to a sustained increase over the whole period, and in two species to a sustained decrease. These trends coincided with changes in the internal structure of the wood. No species was present in significantly greater numbers when Sparrowhawks were absent from the wood and at their lowest numbers in the surrounding area **(Figure 9.6)**. These data therefore gave no evidence that Sparrowhawks had any negative impact on the breeding densities of woodland songbirds. No other predator in Britain could take on the role of the Sparrowhawk, so other mortality agents among small birds must have achieved greater importance in the years when hawks were absent.

This general finding was paralleled by a more detailed study of Sparrowhawk predation on titmice nesting in the 320-ha Wytham Wood near Oxford in southern England (Perrins & Geer 1980). Most tits bred in nest-boxes, and had been studied for many years, with adults and young ringed each year. In all, nine hawk nests were studied intensively during 1976–79, out of the total hawk population in the wood of 6–8 pairs per year. The following points emerged:

(1) Within 60m of each successful hawk nest, both the occupancy and the success of nest-boxes were reduced, compared with boxes further away. This was attributed to the hawks eating more of the breeding tits near their own nests than elsewhere. No such effects were noted around hawk nests that failed soon after eggs had been laid, nor around one successful hawk nest on the edge of the wood whose owners hunted chiefly in farmland. The results were consistent between years, even though the locations of the hawk nests changed. The effect of this local predation of breeding tits on the subsequent tit population of the whole wood was immeasurably small.

(2) The hawks ate many fledgling tits, removing an estimated 18–34% of young Great Tits *Parus major* in the different years, and 18–27% of young Blue Tits *Parus caeruleus*. None-the-less, an average of more than six young tits per pair (including failed pairs) were raised in these years. Only about one chick per pair was needed to replace the annual losses of adults in order to maintain the population at a constant level, so many young tits must have died from predation or from other mortality causes at other times of year.

(3) Both the Great Tit and the Blue Tit breeding populations fluctuated greatly from year to year, but in general the numbers of nest-boxes occupied each year were not conspicuously higher in the period when hawks were absent than when they were present **(Figure 9.7)**. These results thus gave no evidence that Sparrowhawks, despite their predation, depressed the tit breeding densities (see also McCleery & Perrins 1991). Indeed, Blue Tits showed a long-term increase over the whole study period.

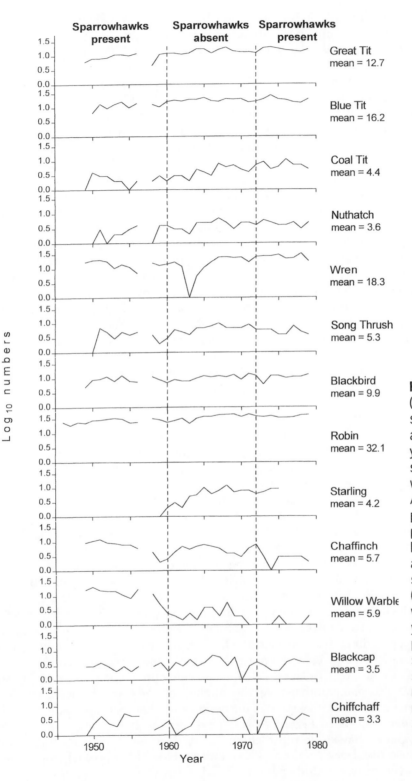

Figure 9.6 Counts (log values) of various songbird species in an oak wood each year during three successive periods when Sparrowhawks *Accipiter nisus* were present, absent and present, respectively. Mean values show average numbers of singing males (untransformed values) present each year over the whole period. None of the species shown was significantly most numerous in the middle period when Sparrowhawks were absent. From Newton *et al.* 1997.

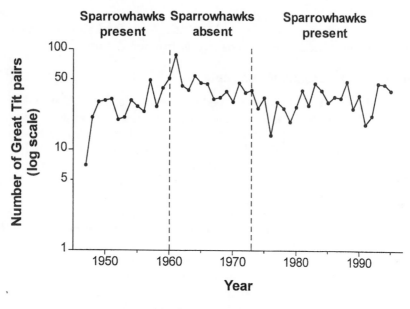

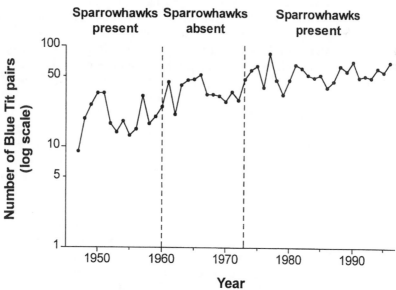

Figure 9.7 Numbers of Great Tits *Parus major* and Blue Tits *P. caeruleus* nesting in Marley Wood, Wytham, Oxford, during a 50-year period. Excluding the first year of study (when nest-boxes were first provided), nesting numbers were not conspicuously higher in years when Sparrowhawks were absent than in years when they were present. Based on data provided by C.M. Perrins (see Newton & Perrins 1997).

These various findings could probably be generalised to many other Sparrow-hawk prey species, in that the main effects of predation were to (1) change the seasonal pattern of mortality; (2) reduce the size of the post-breeding peak in the population; and (3) change the main agents of death; all without causing any noticeable decline in annual breeding numbers (Newton 1986). In practice, this meant that, instead of dying mainly in winter from food-shortage for example, potential prey birds died at all seasons largely from predation.[1]

Lowering of prey breeding numbers

The ability of predators to hold prey breeding density well below the level that could occur in the absence of predation is exemplified by research on the Grey Partridge *Perdix perdix* on English farmland (Potts *et al.* 1980, Potts 1986). The nests of this species are often located along hedgerows or other field margins which predators can easily search. The eggs are favoured by Crows *Corvus corone* and Magpies *Pica pica*, while the tight-sitting hens are often killed by Red Foxes *Vulpes vulpes*, Stoats *Mustela erminea* and house cats.

Field studies and computer simulations indicated how Partridge breeding densities could be regulated under different levels of predation (Potts *et al.* 1980, Potts 1986, Aebischer 1991). In areas where predators were present at near-normal densities, predation on Partridge nests was strongly density-dependent: the greater the density of nests, the greater the proportion of eggs and females that were destroyed, and the lower the breeding success **(Figure 9.8)**. However, the form of this relationship varied between areas, depending on the amount of nesting cover present. Where cover was plentiful (around 8 km of hedgerow per km^2 of farmland), the predation rate increased only slowly with a rise in nest-density, but where cover was sparse (say <4 km of hedgerow per km^2) the predation rate increased steeply, perhaps because, with so little cover, the predators had less to search. At levels of hedgerow cover greater than 8 km per km^2, Partridges gained no additional protection and there was no further change in the relationship between predation rate and nest-density. Computer-simulated population trends, based on field data, indicated that density-dependent nest predation was capable on its own of stabilising Partridge breeding densities at relatively low levels. The equilibrium level varied with the precise form of the density-dependent relationship, in turn dependent partly on the amount of available nesting cover (see also Chapter 13). As in the Bobwhite Quail studied by Errington (1946), the habitat influenced the population level through predation.

Over the years, the experience of game-keepers was that in any given habitat a considerable increase in Partridge numbers (both post-breeding and breeding) could be achieved by reducing the numbers of relevant predators. Where this occurred, the density-dependence in nest failures was almost removed, and nest success was consistently high **(Figure 9.8)**. This led to a big improvement in post-breeding density, and also in spring breeding density. Under such reduced predation, breeding densities were generally higher than under normal predation, but were again related to habitat quality, as reflected in the amount of nesting cover.

---

[1]*It may not be safe to generalise these findings to all Sparrowhawk prey species, however, or to the same prey in different habitats. In Britain as a whole, at least two prey species, namely Bullfinch Pyrrhula pyrrhula and Tree Sparrow Passer montanus, did reach temporarily much higher levels in the period when Sparrowhawk numbers were reduced (Newton 1967b, Marchant et al. 1990, Summers-Smith 1996), and the same was true for Blackbirds Turdus merula on Bardsay Island, off the Welsh coast (Loxton & Silcocks 1997). The latter offers somewhat unusual habitat for Blackbirds, with little cover.*

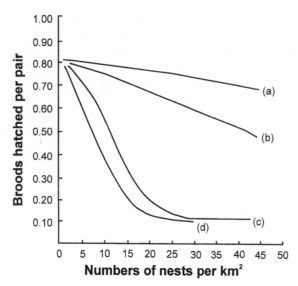

Figure 9.8 Density-dependent nest predation in Grey Partridges *Perdix perdix* in different areas.

Mean number of broods hatched per pair in relation to nest density, nesting cover and predator numbers in: (a) Norfolk, England – good nesting cover, predators destroyed; (b) Sussex, England – less good nesting cover, predators destroyed; (c) Sussex, England – same area, different period, predators not destroyed; (d) Various North American areas – poor nesting cover, predators not destroyed. Lines calculated by logistic regression analyses. Redrawn from Potts 1980.

When spring numbers exceeded the 'carrying capacity' of the habitat, the surplus was removed by emigration, and when spring numbers were deficient they could be made good by immigration (Jenkins 1961). Partridges thus re-distributed themselves each spring, so that breeding densities were matched to habitat quality. The spring population, prior to the re-distribution, had usually been influenced by winter shooting, leaving a surplus (relative to habitat) in some areas and a deficit in others. There were thus two natural density-dependent processes in the annual cycle of Partridges that could regulate numbers: (1) predation on nest contents which in the presence of predators gave an equilibrium density lower than the habitat could otherwise support, and (2) pre-breeding re-distribution, which was the main regulating influence where predators were controlled, and could to some extent mitigate the effects of high nest predation where predators were not controlled. These conclusions were based on comparative field studies in different areas, coupled with simulation modelling. They were subsequently supported by a predator-removal experiment, described below.

Nest predation has been found to increase with nest-density in other ground-nesting game birds. To judge from the slope of the relationship between nest loss and nest-density, the Red-legged Partridge *Alectoris rufa* lost about four times as many nests as the Grey Partridge at equivalent density in the same habitat, while the Ring-necked Pheasant lost about three times as many (Potts 1980). Similarly, in Errington's (1946) study, for any given density the Bobwhite Quail lost about twice as many nests as the Grey Partridge in the same region. Hence, in these species too, both post-breeding and breeding density might be reduced by predation below what would otherwise occur, and this was supported by comparative studies on Red-legged Partridges in areas with and without predator control (Table 9.2). But these relationships did not exclude the possibility that predation was also affected by habitat, especially by the amount of nesting cover, which could influence the equilibrium population level in the presence or absence of predators. Other examples of density-dependent predation on nest contents are given in Chapter 5. Moreover, where different species nest in the same type of site in the same habitat, predators may respond to their combined density, so that they all suffer greater predation than when nesting alone (Blancher & Robertson 1985). This is akin to the situation in which rare but vulnerable prey species could be eliminated through predation by the presence of common but less vulnerable ones (see above).

Table 9.2 Densities and hatching success of Red-legged Partridges *Alectoris rufa* in areas with and without predator control, based on the pooled data from several different areas. In areas where predators were killed, nest success and breeding density were greater than in areas where predators were not killed. From Potts 1980.

	Areas where predators were not killed	Areas where predators were killed
Total area covered (km²)	269.6	123.6
Breeding density (pairs per km²)	2.87	4.96
Clutches hatched per pair	0.33	0.53

Oscillations in prey numbers

Many grouse species, especially in northern latitudes, undergo cycles of abundance in which successive peaks and troughs in numbers occur at fairly regular intervals (Elton 1942, Lack 1954, Keith 1963, Keith *et al.* 1977). In boreal North America, for example, Ruffed Grouse *Bonasa umbellus* and others fluctuate in approximate ten-year cycles, in parallel with similar cycles in Snowshoe Hare *Lepus americanus* numbers. In much of boreal Europe, Hazel Grouse *Bonasa bonasia* and others fluctuate in 3–5-year cycles, in parallel with the numbers of voles. In parts of Britain, Red Grouse *Lagopus l. scoticus* are also cyclic or quasi-cyclic, but the period between peaks varies between four and seven (even up to nine) years in different regions (Lawton 1990, Hudson 1992). Moreover, some species, such as Willow Ptarmigan *Lagopus lagopus*, may follow a 3–5-year cycle in parts of Europe and a ten-year cycle in parts of North America (Watson & Moss 1979, Bergerud *et al.* 1985).

The cycle length refers to the regular intervals between peaks, but the size of the peaks often varies greatly from one cycle to the next (Figure 1.2). Knowledge of cycles in grouse numbers is based mainly on autumn hunting records which are influenced greatly by the production of young. The fewer long-term data on spring densities reveal the same cycle, but the fluctuations are less marked (Watson & Moss 1979, Angelstam 1983). It would seem, therefore, that the periodic peaks in abundance of such species result from a combination of both high breeding density and high breeding success.

Predation is one of several hypotheses proposed to explain the cycles in grouse numbers (Watson & Moss 1979), with predators causing regular crashes in the prey, which are followed by recoveries. However, any tenable hypothesis must also explain the facts that: (1) grouse and mammal (hare or vole) cycles are often coupled (Elton 1942, Siivonen 1948, Hagen 1952, Lack 1954, Myrberget 1972, Keith 1974, Hörnfeldt 1978); (2) in the decline stage of the cycle, grouse are not always synchronous with the mammals but can be up to a year or so behind (Angelstam 1983, Hörnfeldt et al. 1986); and (3) the same grouse species may fluctuate on different cycles in different regions, or in yet other regions may show no obvious cycles (Erlinge et al. 1983, Angelstam et al. 1984, Lindén 1988).

The main known feature that invariably links different species on the same cycle in the same area is that they share the same predators (in some areas they may share some of the same foods but in other areas they do not). In seeking causes of the cycles, however, it is important to separate the primary prey species (voles or hares) from the secondary (scarcer) ones (grouse). The evidence for predators causing the cycle is much stronger for the secondary species than for the primary (mammalian) ones. The supposed mechanism is as follows: as the numbers of herbivorous mammals rise over a period of years, so do the numbers of the predators that depend on them. Then, when mammal numbers crash, the abundant predators switch their emphasis to grouse (eggs, chicks or adults), reducing their numbers, and eventually themselves declining through starvation, emigration and non-breeding (Hagen 1952, Lack 1954, Keith 1974). This then enables both mammal and grouse numbers to rise again to start another cycle, and the remaining predators to shift their emphasis back to mammals. In these circumstances, grouse breeding numbers can be said to be limited by predation, at least in the decline phase of the cycle. Another likely effect of avian predators is in helping to synchronise the fluctuations in their prey over wide areas, as their movements lead them to concentrate at any one time in localities with the highest prey densities. What causes the cycles in mammal numbers is a more controversial question, but it is not necessarily predation.

Some evidence favours this 'alternative prey' hypothesis, as applied to grouse. In northern Europe different species of grouse fluctuate in synchrony, and in parallel with rodent numbers (Hörnfeldt et al. 1986, Small et al. 1993, Ranta et al. 1995). Moreover, the predation on the different grouse species clearly varies with the stage of the vole cycle. Among Willow Ptarmigan Lagopus lagopus in Norway annual nest losses varied between 10 and 63% (Myrberget 1984), among Capercaillie Tetrao urogallus and Black Grouse Tetrao tetrix in Norway between 11 and 78% (Storaas et al. 1982) and among Black Grouse in Sweden between 7 and 78% (Angelstam 1983). In all three species, the lowest nest predation occurred in years of peaks in rodent numbers and the highest when rodents declined. This was consistent with the view that generalist predators, such as foxes, switched to eating grouse (the alternative prey) when rodents became scarce and back again when rodents recovered (Lack 1954, Angelstam et al. 1984). Appropriate changes in diet were recorded in the relevant predators, which took relatively fewer voles and more grouse after voles declined. Moreover, when Lindström et al. (1987) provided extra food for predators during a vole crash, they found a higher autumn ratio of young to old grouse in the experimental than in the control area. This helped to confirm that predation on grouse was reduced when alternative food was available for the predators.

The breeding of other avian prey is also affected by the stage of the rodent cycle.

Correlations between bird nest success and rodent densities have been described for various hole-nesting and open-nesting songbirds in forest **(Figure 9.2)**, in some species affecting subsequent breeding densities (Dunn 1977, Järvinen 1985, 1990, Orell 1989, Hogstad 1993, McCleery *et al.* 1996, Viestola *et al.* 1996). They have also been described for some waterfowl and waders on northern tundra (Pehrson 1976, Roselaar 1979, Summers & Underhill 1987). However, declines in the nest success and numbers of these various alternative prey species may also occur for climatic or other reasons in years when rodents do not decline, which sometimes masks an otherwise cyclic pattern (Anglestam *et al.* 1985). The pattern has been best documented in Brent Geese *Branta b. bernicla* and two shorebird species which nest on the Taimyr peninsula of Siberia and winter in western Europe (geese) or southern Africa (waders), where the proportion of young in wintering flocks has tended to fluctuate on a three-year cycle **(Figure 9.9)**. Years of high production, which occurred in all species together, coincided with peaks in Taimyr lemming numbers, while the years of low production coincided with lows in lemming numbers or with late cold springs, which also reduced breeding success (Summers & Underhill 1987). The same pattern was later shown among Grey Plovers *Pluviatilis squatarola* from the Taimyr Peninsula (Underhill & Summers 1990), and among White-fronted Geese *Anser albifrons* from further west along the Barents Sea coast (van Impe 1996). Both these species winter in western Europe where the counts were made. The data for these various tundra nesting birds thus seem to support the notion that predators, such as Arctic Foxes *Alopex lagopus* and Arctic Skuas *Stercorarius parasiticus*, switch to eggs and young of birds when their primary lemming prey are scarce, causing large variations in wintering bird numbers from year to year, and probably also in subsequent breeding numbers.

Cycles in the numbers of small rodents are most marked on the northern tundra, and become progressively less marked in forested areas to the south, especially where continuous forest is replaced by a mixture of forest and farmland. In southern Sweden, for example, rodent numbers stay fairly stable from year to year (Erlinge *et al.* 1984) . This non-cyclic pattern was ascribed to the continuous high predation rate from generalist predators, such as Red Fox *Vulpes vulpes* and Buzzard *Buteo buteo*, which subsisted mainly on other prey, but changed their diet to concentrate on whatever was plentiful at the time. In this way, they were always numerous enough to prevent outbreaks of voles, a situation that depended on the presence of a wide range of alternative prey. They could also switch immediately, giving a density-dependent, rather than a delayed response, to change in prey numbers. In this way they supposedly kept vole densities relatively low and relatively stable, and hence prevented both voles and grouse from achieving the periodically very high densities found further north.

In North America, hares and grouse are also linked by common predators, mainly Lynx *Lynx canadensis* and Red Fox, but also Goshawk *Accipiter gentilis* and Great Horned Owl *Bubo virginianus*. The differing behaviour of these predators can explain two otherwise puzzling questions, namely why grouse usually decline one year earlier in the north (Canada and Alaska) than further south (in the Lake States), and why the cycles of grouse persist in the Lake States in the absence of similar cycles in the local hares (Keith & Rusch 1988). In Canada and Alaska, where Snowshoe Hares are strongly cyclic, the Ruffed Grouse cycle is thought to be generated primarily by resident predators (mainly Lynx and Red Fox), which switch to grouse when hares

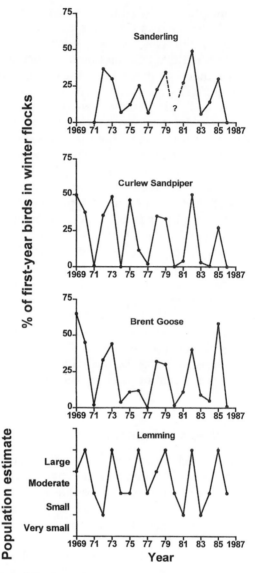

Figure 9.9 Proportion of first-year birds in wintering populations of various bird species that breed in the Arctic, in relation to numbers of lemmings on the breeding grounds in different years. High production of young in Sanderling *Calidris alba*, Curlew Sandpiper *Calidris ferruginea* and Brent Goose *Branta bernicla* occurred about every three years when lemming numbers were also high, providing alternative food for predators. From Summers & Underhill (1987).

decline. In the Lake States, by contrast, major surges of predation occur during periodic invasions of Goshawks and Great Horned Owls which leave the north when hares decline there. These birds greatly add to the usual predation on grouse, and cause grouse numbers to decline. The invasions of these predators usually occur in two successive autumns at 9–10-year intervals and are almost continent-wide

(Mueller *et al.* 1977). Demographically, summer survival of eggs and young grouse mainly determines the cyclic trend in Canadian populations, whereas overwinter survival is paramount in the Lake States (Keith & Rusch 1988). Although documentation is best for the Ruffed Grouse, other grouse species may be affected similarly, as may some other types of bird, including Whooping Crane *Grus americana* (Boyce & Miller 1985).

In both rodent-based and hare-based cycles, it is the temporary switch by predators from the main herbivorous prey to scarcer alternative prey (birds) that is held to be the chief cause of periodic reductions in grouse numbers. Just as the same grouse species, such as Willow Ptarmigan, may show ten-year cycles in North America and four-year cycles in northern Europe, the same predator species may show different cycles on different continents, depending on the role it plays. The Goshawk, which eats both hares and grouse, is apparently involved in driving the grouse cycle in North America but, because it does not eat voles, it can only follow the grouse cycle in Europe, which is driven mainly by mammalian predators.

The view that predators are responsible for certain cycles in grouse numbers thus hangs together quite well. But it is difficult to eliminate completely an alternative possibility, that the cycle is caused by some different factor, and the predators are simply removing a variable doomed surplus formed of birds that are destined to fail in their breeding or die for some other reason (Watson & Moss 1979). Moreover, Red Grouse in parts of Britain show cycles of abundance even though predators are kept at low density by game-keepers, disease being one likely cause (Chapter 10; Hudson 1992). The role of predators in the cyclic fluctuations of some grouse is thus not wholly proven, and predation cannot be an explanation that applies to all grouse populations.

Annihilation of prey: effects of introduced predators on island birds

Predatory mammals have been accidentally or deliberately introduced on many oceanic islands around the world, often with devastating effects on the local bird-life. Such case histories provide some of the most spectacular examples of predators eliminating their prey. They occur because both landbirds and seabirds on such islands have usually evolved in the absence of mammalian predators, often lack defensive behaviour and nest in accessible sites.

Alien predators thus present severe conservation problems. Of 127 species of birds known to have died out since 1600, about 92% were island forms, and alien predators are held responsible for at least 40% (Chapter 16; King 1985). Among seabirds, only the Guadalupe Storm Petrel *Oceanodroma macrodactyla* is thought to have been extinguished in this way, but many other seabirds have been eliminated from some of their breeding islands while remaining on others. The predators most frequently involved are rats *Rattus* and feral cats, but on some islands also Mongooses *Herpestes auropunctatus*, feral dogs, foxes (*Vulpes* and *Alopex*), Stoats *Mustela erminea* and other mustelids, monkeys (*Cercopithecus* and *Macaca*) and feral pigs.

With human help, rats have now reached more than 80% of the world's oceanic islands and island groups, though in many such groups, some islets still remain rat-free (Atkinson 1985). The three species involved are: (1) the Brown Rat *Rattus norvegicus* which is the largest (350–450 g) and preys mainly on ground-nesting and burrow-nesting birds and to a lesser extent on tree-nesters; (2) the Black Rat *R. rattus* which is smaller (100–180 g) but more agile, and is able to climb the highest trees; and (3) the Pacific Rat *R. exulans* which is smaller still (110–130 g) and also a good climber. All three species eat eggs, chicks and even full-grown birds up to their own weight, while Pacific Rats have also been seen to kill even larger species that lack defensive behaviour, for example adult Laysan Albatrosses *Diomedia immutabilis* on Kure Atoll, Hawaii, by eating holes in the backs of incubating birds (Kepler 1967). Together, these three rats are held responsible for about 56% of those avian extinctions attributed to predation (King 1985).

The impact of rats can be devastatingly rapid. For example, Black Rats reached Big South Cape Island, New Zealand, in about 1962. Within three years they had exterminated five species of native forest birds, including the last known population of the Bush Wren *Xenicus longipes* (Bell 1978). Similarly, following their establishment on Midway Island in the Pacific in 1943, the ground-dwelling Laysan Crake *Porzana palmeri* and Laysan Finch *Telespyza cantans* disappeared within about 18 months (Fischer & Baldwin 1946, Johnstone 1985).

Feral cats can take larger birds than rats, and are held responsible for 26% of the avian extinctions attributed to predators (King 1985). Again the effects can be rapid. In 1894 a lighthouse keeper took a pet cat to Stephen Island, New Zealand. Over some days the cat brought in 15 specimens of the previously unknown wren *Xenicus lyalli* which has not been seen since (Chapter 16). Feral cats on Socorro Island, Mexico, probably eliminated the endemic dove *Zenaida graysoni* from the wild and have driven the endemic mockingbird *Mimodes graysoni* to the brink of extinction. Both are ground nesters; tree-nesting birds on the same island have not been affected (Jehl & Parkes 1983). On islands where they have been studied, cats usually live at 4–14 individuals per km^2, but among vast seabird colonies they have reached more than 200 individuals per km^2 (Veitch 1985), killing more than one million birds per year on the Kerguelen Isles (Pascal 1980), where newly-introduced cats increased at more than 20% per year.

Where cats occur with potential mammalian prey, such as rats or rabbits, their predation on seabirds can be greater than where they occur alone. This is because, with alternative prey, the cats can maintain larger numbers through the off-season, when seabirds are largely absent. On Macquarie Island in the Southern Ocean, the ground-nesting parakeet *Cyanoramphus novae-zelandiae erythrotis* co-existed in good numbers with cats from about 1810 to 1880; but after the release of rabbits in 1879, the cats became much more numerous, and the parakeets were extinct by 1890 (Taylor 1979).

Although the crucial predators of island birds are usually mammalian, this is not always so. The western Pacific island of Guam once held 18 bird species, including 11 native forest species. In recent decades, forest birds have declined in wave-like manner, south to north, while savannah birds on the same island have maintained their numbers. The timing and pattern of the declines in forest birds coincided with the introduction (in the 1940s) and subsequent spread of the Brown Tree Snake *Boiga irregularis*, a mainly arboreal, nocturnal predator which eats both eggs and adult birds, as well as small mammals and reptiles (Savidge 1987). By the mid 1980s, at least seven formerly widespread forest bird species had not been seen for several years, and four others were critically endangered. This is the first time that a snake has been implicated.

The establishment of alien predators on an oceanic island has not inevitably led to extinctions of native birds. In some cases, it has led to mutual co-existence with no obvious decline in prey numbers, as is evident from the species that have remained widespread despite mammalian predation. In others, it has led to reduction of the prey to a new stable level, often restricted to places inaccessible to the predator (examples in Stonehouse 1962, Taylor 1979, Atkinson 1985). Only in a minority has it led to local extinction. However, the numbers of extinctions are likely to increase in future as predators continue to spread, and as already affected endemic species continue to decline.

To judge from known case-histories, several features of the predator–prey interaction influence the likelihood of extinction (Atkinson 1985). In the predators, the following seem important:

(1) body size, which affects the range of prey available and the ability to enter holes where nests occur;
(2) climbing skill, which affects the ability to reach nests on trees and cliffs; and
(3) abundance relative to the prey, in turn influenced by the availability of alternative foods at seasons when birds are least available.

In the prey, vulnerability is influenced by:

(1) body size relative to the predator;
(2) accessibility of nest-site;
(3) life-history features, such as inherent reproductive and mortality rates, duration of breeding cycle (when eggs and chicks are vulnerable), timing of breeding in relation to the seasonal cycle of predator numbers; and
(4) behaviour, including effectiveness of defence, whether eggs and chicks are left unguarded, and ability to fly.

The fact that alien predators have generally had a greater impact on the bird-life of temperate than of tropical islands may be because many of the latter have native rodents or land-crabs against which the local birds have evolved defences. These same defences may then give them some protection against new invaders (Atkinson 1985).

It is hard to assess in retrospect whether introduced predators caused all the extinctions attributed to them. For most species, the evidence is entirely circumstantial: rapid decline in numbers following the known release of a predator, accompanied by observations of predation. Whereas on some islands the establishment of a predator was the only obvious change, in others the effects of predators were confounded with other changes, such as forest clearance. However, in a few species evidence of decline has been accompanied by measurements which showed that prevailing predation rates were not sustainable – for example where breeding success of the prey species was almost zero year after year: e.g. the 2% nest success in Dark-rumped Petrels *Pterodroma phaeopysia* on Santa Cruz, Galapagos (Tomkins 1985), or the apparent complete nest failure of Bonin Petrels *Pterodroma hypoleuca* on Kure Atoll (Kepler 1967), of Freiras *Pterodroma madeira* on Madeira, and of Black Petrels *Procellaria parkinsoni* on Little Barrier Island, New Zealand (Collar & Andrew 1988). Moreover, on several islets the control of the predator by human efforts has reversed a population decline, or enabled a species previously eliminated from an island to thrive after reintroduction. For example, the Stitchbird *Notiomystis cincta* was once widespread on the North Island of New Zealand, but by 1890 fewer than

500 individuals remained only on Little Barrier Island (Griffin *et al.* 1988). After cats were eradicated in 1980, Stitchbird numbers increased to 3000, enough to colonise other islands. The North Island Saddleback *Philesturnus carunculatus* did not survive cat predation on Little Barrier or Cuvier Islands, but after cats were eradicated, Saddlebacks were reintroduced to both islands and now thrive (Veitch 1985). Similarly, on Baker Island, cats were exterminated in 1964 after they had greatly reduced the previously abundant seabird colonies. By 1978, large colonies of Sooty Terns *Sterna fuscata* and Lesser Frigatebirds *Fregata ariel* had become re-established, together with smaller numbers of other species (Forsell 1982).

It is not only the native faunas of oceanic islands that have suffered from introductions of predators. The same is true of many islands that lie close to continents and that once supported large seabird colonies. For example, mainly between 1900 and 1930, Arctic Foxes and Red Foxes were released by fur trappers on about 450 Alaskan islands, in each case with devastating consequences for ground-nesting birds (Bailey & Kaiser 1993). The Aleutian Canada Goose *Branta canadensis leucopareia* survived as a breeder on only a single island which escaped the fox releases. From the 1970s, however, when foxes were eliminated from some islands, populations of all surviving species increased, the most affected ones by 4–5-fold, as did reintroduced Canada Geese (Byrd *et al.* 1994).

Such case-histories from islands show what can happen when birds that have lived in the absence of mammalian predation are suddenly exposed to it, but they are of little relevance in assessing the effects of regular predation on mainland birds. Such devastating impacts, noted on island birds through human action, may occur naturally on continents from time to time as particular predators colonise new areas and encounter naive prey. The chances of witnessing them are small, however, and only the stable situations, in which predators and prey can co-exist, are likely to persist in the long term.

HABITAT–PREDATION INTERACTIONS

Some long-term declines in the numbers of mainland birds, in both Europe and North America, have been attributed partly to increased predation, itself associated with changes in landscape which have promoted an increase in predator abundance or in prey vulnerability. One idea is that reductions in the amount of nesting cover in farmland – for example through removal of hedges or other semi-natural areas – have led some nesting birds to concentrate in fewer areas, to the benefit of searching predators (as in the Grey Partridge *Perdix perdix*, discussed above). Similarly, reductions in the quality of cover – through overgrazing or conversion of semi-natural to crop vegetation – has made nests more exposed and easier for predators to find. Agricultural intensification could thus have caused many ground-nesting and shrub-nesting birds to become more vulnerable to predation. For the predators themselves, these changes are tantamount to an increase in food availability.

The most extensive studies concern the Mallard *Anas platyrhynchos* and other dabbling ducks on the North American prairies (Greenwood *et al.* 1987, 1995, Klett *et al.* 1988, Cowardin *et al.* 1995, Greenwood & Sovada 1996). An analysis of studies done during 1935–92 revealed that, over this period, nest success declined at a mean rate of 0.5% per year (Beauchamp *et al.* 1996a). Five common species showed similar rates of decline in success, but the two earliest breeders (Mallard and Pintail *Anas acuta*) which nested when cover was sparsest, showed the lowest success throughout (Beauchamp *et al.* 1996a). Declines in success were attributed to changes in habitat and predation communities. The increasing conversion of native grassland to cropland reduced the amount of good nesting cover, with nests in crops being more exposed to predation and to destruction by machinery. In the 1980s nest success in crops averaged less than the 15% needed to maintain a stable population. At this time, some common dabbling duck species could not maintain their former population levels where cropland exceeded about 56% of the landscape, yet over most of the prairies more than 70% of land was cultivated (Greenwood *et al.* 1995). The situation was exacerbated by agricultural policy which encouraged cultivation of areas once considered poor for crops, and by drought which enabled farmers to plough areas formerly too wet. At the same time, intensive control of Coyotes *Canis latrans* enabled Red Fox *Vulpes vulpes* numbers to increase. Both species eat ducks and duck eggs, but Foxes live at higher densities and are more serious predators of ducks (Greenwood *et al.* 1995). As a consequence of these changes, the overall productivity of Mallard and some other species fell below the level necessary to maintain numbers which therefore declined regionally.

The situation changed in the mid-1990s, when duck populations suddenly began to increase. This turn-around coincided with (1) the end of a long dry period, which had persisted through the 1980s, (2) the conversion, beginning in 1985, of millions of hectares of cropland to perennial grassland (providing better nesting cover), and (3) reduced Coyote control which led to an increase in Coyote numbers and an associated decline in Fox numbers (Greenwood & Sovada 1996). The whole episode illustrates the close interaction between human land management, nesting cover, predation and population trend, and showed that a bad situation could be reversed by changes in relevant conditions.

In some farmland areas, a different type of predation on ground-nesting birds is provided by farm animals or machinery which crush the nests. In a study of Lapwing *Vanellus vanellus* and other waders in Dutch meadows, the proportion of nests that hatched varied according to stocking density (lower success with more animals) and to mowing date (lower success with early mowing) **(Figure 9.10)**. Breeding success and population sizes of several species declined in recent decades because of improved land drainage, which reduced feeding opportunities and encouraged earlier grass growth. It also gave farmers earlier access with tractors, allowing grass rolling and fertiliser applications, which in turn led to earlier grass cutting (especially for silage) or to higher stocking densities (Beintema & Muskens 1987). The importance of nest crushing is that, for the most part, it is density-independent, so does not decline in proportion at low bird densities. It is therefore a very effective means of eliminating species from wide areas (see also Green (1988) for Stone Curlew *Burhinus oedicnemus*).

For the Lapwing, an analysis of British ringing results revealed no decline in the survival rates of full grown birds over the period 1930–90 (rather a slight increase since 1960). This implied that the widespread population decline was due to reduced production of young, as detailed local studies had shown. On prevailing mortality, Lapwings would need to produce 0.83–0.97 young per pair per year to maintain their numbers. This level of productivity was found in only eight out of 24 studies in different areas, those least subjected to agricultural changes (Peach *et al.*

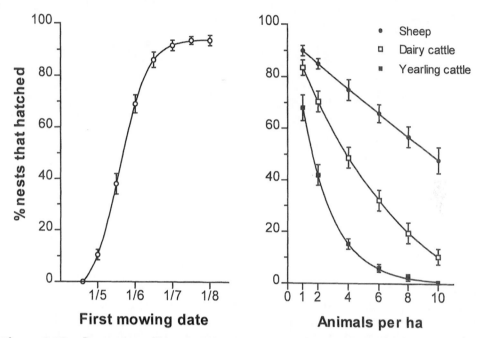

Figure 9.10   Proportion of Lapwing *Vanellus vanellus* clutches that hatched successfully in relation to stocking densities and mowing dates in Dutch meadowland. From Beintema & Muskens 1987.

1994). The most important change was the draining and re-seeding of former wet pastures, which reduced feeding opportunities and lowered nest success. Eggs and chicks were much more conspicuous to predators on the uniform green of a re-seeded area than against the more varied background of a rough undrained one (Baines 1990). The importance of a disruptive background to nest success was also evident among Whimbrel *Numenius phaeopus* in Manitoba (Skeel 1983).

Another landscape change thought to have promoted some bird population declines through enhanced predation is the growing fragmentation of natural and semi-natural habitat. Because of the patchy nature of modern landscapes, predators characteristic of only one habitat can more easily penetrate another, so that birds of most habitats could now be exposed to a greater range of predator species than previously. In addition, the fragmentation of natural habitats, through the development of farmland, has favoured generalist predators, such as crows and foxes (Andrén 1992). Not only do such species have a wider variety of prey available to them in human-modified than in natural landscapes, but they also get part of their food directly from humans, in the form of garbage and animal feed. They can thus achieve high densities, resulting in raised predation rates on vulnerable bird species, both in farmland and in the fragments of remaining semi-natural habitat. Several studies, with natural or artificial nests, have shown that predation is more frequent near forest edge and in small forest patches than in the interior of large forests (**Figure 9.11**; Gates & Gysel 1978, Andrén *et al.* 1985, Wilcove 1985, Small & Hunter 1988, Gibbs 1991, Møller 1991a, Sandström 1991, Andrén 1992), though this difference is not found everywhere (Yahner & Wright 1985). In fact, enhanced predation at

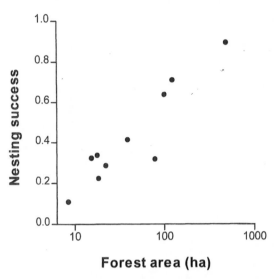

Figure 9.11   Proportion of Wood Thrush *Hylocichla mustelinus* nests that produced young in forest patches of different size, Pennsylvania ($r^2 = 0.86$, $P<0.001$). Nest success increased with size of forest fragment, mainly because of differential predation. Both avian and mammalian predators were more abundant in smaller fragments. From Hoover *et al.* 1995.

forest edges was apparent in ten of 14 studies involving artificial nests, and in four of seven involving natural nests, but seldom extended more than 50 m into the forest (Paton 1994). Overall predation rates also decreased with increasing area of forest patch in eight out of eight studies. Habitat fragmentation, together with enhanced nest predation, has thus been suggested (but not proven) as a factor in the widespread declines of some migrant songbirds in eastern North America (Chapter 5; Robbins 1980, Ambuel & Temple 1983, Wilcove 1985, Temple & Cary 1988, Terborgh 1989, Askins *et al.* 1990). The same points hold for grassland birds, among which nest success is also related to patch size and distance from edge (Ball *et al.* 1994).

MANIPULATION OF PREDATION

Predator-removal studies

Any supposed effect of predators on their prey can be shown most convincingly by experiment. If predators are limiting their prey, their removal from an area should be followed by an increase in prey numbers. As in food manipulations, several experimental designs have been used in studies of this type. Among the experiments summarised in **Table 9.3**, nine involved before-and-after comparisons in the same area, 12 involved a simultaneous control area where predators were left, while nine involved reversal of treatments or replication, with several experimental and control areas (for further discussion of experimental design, see Chapter 7).

One of the best of these studies was conducted on two forested islands off northern Sweden (Marcström *et al.* 1988). These islands were large enough to sustain populations of several mammalian predators, but were also joined to the mainland in winter by sea ice, enabling predators to move on and off. On one island, mammalian predators (mainly Red Foxes *Vulpes vulpes* and Martens *Martes martes*) were removed, while on the other they were left. After five years, the treatments were reversed for four further years, so that the experimental island then became the control and vice versa, giving nine years of study in all. The effects of predator removal were measured on four species of grouse, mainly Capercaillie *Tetrao urogallus* and Black Grouse *Tetrao tetrix*. Where predators were removed, more young grouse were produced, and subsequent breeding numbers were higher than where predators were left **(Table 9.4)**. The conclusion was that mammalian predation limited both breeding success and breeding density of the grouse species concerned.

On these islands, as on the mainland, the predators fed mainly on rodents, and only secondarily on game birds. Where predators were left, grouse breeding success was correlated with vole abundance, as most young grouse were produced in the peak vole years; but where predators were removed, no such relationship held. This finding was consistent with the view that predators turned more to grouse when voles were scarce, and confirmed that predation was mainly responsible for synchronising grouse productivity with vole abundance, as discussed above. However, the removal of Foxes and Martens had no significant effect on vole abundance itself during two four-year cycles, suggesting that these predators did not drive the vole cycle.

Of the 30 predator-removal studies summarised in **Table 9.3**, 13 involved galli-naceous birds (such as grouse and pheasants), 11 involved ducks (including two with simulated nests) and six involved other species. The parameters that were most commonly measured included nest success, post-breeding numbers (or ratio of large young to adults) and subsequent breeding numbers. In 27 studies in which nest success was measured, 23 showed an increase in success under predator-removal; in 17 studies in which post-breeding numbers were measured, 12 showed an increase; and of 17 studies in which breeding numbers were measured, ten showed an increase. Improved nest success was not always reflected in increased post-breeding numbers, and similarly, an increase in post-breeding numbers was not always reflected in increased subsequent breeding numbers. Overall, it seems that, in more than half the studies in **Table 9.3** that measured breeding density, this density was limited by predation. These studies conformed to model A and the others to model B in **Figure 3.12**.

Almost all the species studied in these experiments were ground-nesters, which as a group may have been more vulnerable to predation than other birds that nest in safer sites. In addition, the predators were all generalists and fed mainly on other prey (such as voles or rabbits), and whose numbers were therefore sustained mainly by these other prey. These are precisely the circumstances in which marked effects on vulnerable subsidiary prey species might be expected, because the predators are buffered against decline in their subsidiary prey by the availability of their main prey. Experiments that involved removal of only one predator species (whether crow or mammal) showed no significant effects, mainly because many nests were still found and robbed by other predators (Parker 1984, Parr 1993, Clark *et al.* 1995). In some studies, predation was influenced not only by the numbers and types of predators present, but also by the availability of alternative prey and by habitat features such as nesting cover.

Table 9.3 Effects on avian prey of various predator-removal studies: a summary of experimental findings. B-birds, M-mammals, ?-not measured. The extent of increase in post-breeding and breeding numbers was sometimes cumulative, increasing with the number of years the experiment continued.

Prey-species	Location	Predators removed	Grade of experiment[1] (years)	Increased nest success	Increased post-breeding numbers	Increased breeding numbers	Source
Gallinaceous birds							
Ruffed Grouse *Bonasa umbellus*	Connecticut Hill, New York	MB	3 (4)	Yes	No	No	Bump et al. 1947
	Valcour Island, New York	MB	1 (4)	Yes	No	No	Bump et al. 1947 Crissey & Darrow 1949
Ring-necked Pheasant *Phasianus colchicus*	Washington	MB	1 (4)	No	Yes	?	Einarsen 1950
	California	M	2 (4)	Yes	?	?	Hart et al. 1956
	S. Minnesota	MB	2 (5)	Yes	No	No	Chesness et al. 1968
	Denmark	Fox	2 (11)	?	Yes	?	Jensen 1970
	South Dakota	M	3 (7)	?	Yes	?	Trautman et al. 1974
Ring-necked Pheasant *Phasianus colchicus* Partridge *Perdix perdix* & Red-legged Partridge *Alectoris rufa*	England	MB	3 (1)	Yes	Yes	Yes	Brockless 1995
Ring-necked Pheasant *Phasianus colchicus* & Grey Partridge *Perdix perdix*	Germany	MB	2 (11)	?	Yes	?	Frank 1970
Wild Turkey *Meleagris gallapavo* & Northern Bobwhite *Colinus virginianus*	Texas	M	2 (2)	Yes	Yes	Yes	Beasom 1974
Grey Partridge *Perdix perdix*	England	MB	3 (6)	Yes	Yes	Yes	Tapper et al. 1996
Black Grouse *Tetrao tetrix* & Capercaillie *Tetrao urogallus*	Sweden	M	3 (9)	Yes	Yes	Yes	Marcström et al. 1988
Ptarmigan *Lagopus lagopus* & Black Grouse *Tetrao tetrix*	Sweden	Corvids	2 (4)	Yes	No	No	Parker 1984

	Region	Predator	Grade (n)[1]				Reference
Waterfowl							
Various ducks	North Dakota	Striped Skunk[4]	1 (2)	Yes	?	?	Kalmbach 1939
	Alberta	Striped Skunk[4]	1 (5)	No	?	?	Keith 1961
	Minnesota	MB	3 (6)	Yes	?	No	Balser et al. 1968
	South Dakota	M	2 (5)	Yes	?	Yes	Duebbert & Kantrud 1974
	South Dakota	M	2 (6)	Yes	?	Yes	Duebbert & Lokemoen 1980
	North Dakota	Striped Skunk[4]	3 (3)	Yes	?	?	Greenwood 1986
	Minnesota, Dakota	MB	3 (4)	Yes	?	?	Sargeant et al. 1995
	Saskatchewan	Crows	1 (3)	No	?	?	Clark et al. 1995
	North Dakota	M	3 (2)	Yes	Yes	No	Garrettson et al. 1996
Simulated ground nests	Manitoba	M	2 (1)	Yes	-	–	Lynch 1972
	North Dakota	M	2 (2)	Yes	–	–	Schrank 1972
Other birds							
Various songbirds	England	Corvids	1 (2)	Yes	?	?	Stoate & Szczur 1994
White-winged Dove *Zenaida asiatica*	Texas	Great-tailed Grackles[5]	2 (2)	Yes	?	Yes	Blankenship 1966
Various waders	Scotland	B	2 (4)	No	No	No	Parr 1993
Avocet *Recurvirostra avosetta*	England	Gulls	1 (8)	Yes	Yes	Yes[2]	Hill 1988
Various seabirds, waders and waterfowl	Alaskan Islands	Foxes	1 (20)	Yes	Yes	Yes	Byrd et al. 1994
Greater Sandhill Crane *Grus canadensis*	Oregon	MB	1 (7)	Yes	Yes	Yes[3]	Littlefield 1995

[1] Grade of experiments: 1=before-and-after study in same area, 2=separate experimental and control area, 3=replicated experiment, with several experimental and control areas or reversal of treatments in the same two areas.

[2] Increased breeding numbers progressively for 5 years, until population stabilised.

[3] Reversed a population decline, with pair numbers declining from 236 in 1975 to 168, then increasing to 230 by 1993 after predator removal from 1986.

[4] *Mephitis mephitis*

[5] *Quiscalus mexicanus*

Table 9.4   Effects of killing Red Foxes *Vulpes vulpes* and Martens *Martes martes* on the productivity and numbers of game birds in northern Sweden. Predator removal led to increased production of young and to increased breeding numbers. From Marcström *et al.* 1988.

	Predators preserved	Predators removed	Ratio
Productivity			
Brood-size in August	3.3 ± 0.1	5.5 ± 0.1	1:1.7
% hens with broods	59	77	1:1.3
Numbers of young per hen	1.9	2.2	1:2.2

Numbers in breeding season

	Under predator removal, % increase in counts of:	
	Capercaillie	Black Grouse
Lek counts	174	166
Transect counts	56	80

This table is based on the pooled data for different species and years from two forested islands (2350 and 1800 ha) in the northern Baltic, on one of which predators were removed, while on the other they were left. After 5 years, the treatments were reversed for 4 further years, giving 9 years in all. Broods were significantly larger (P<0.05) on the island with predator removal in 6 of the 9 years for Capercaillie *Tetrao urogallus*, in 6 years for Black Grouse *Tetrao tetrix*, in 5 years for Hazel Grouse *Bonasa bonasia* and in 2 years for Willow Ptarmigan *Lagopus lagopus*. The proportions of females with broods (all species) were significantly higher (P<0.01) in 8 of the 9 years on the island with predator removal. Increases in breeding populations with predator removal were calculated in relation to changes on the island where predators were preserved. The results for individual species were almost significant, but reached the 5% level when combined.

In all these experiments, increased breeding density was assumed to follow from the improvement in nest success and chick survival the previous year. However, in some experiments involving various duck species, breeding densities increased in the experimental area from the first year of predator removal, before any improvement in nest success could have occurred (Duebbert & Kantrud 1974, Duebbert & Lokemoen 1980). This implied that predator removal had influenced the settling patterns of ducks, with substantial immigration to the newly-created predator-free area. The fact that ducks can react quickly to a change in predator presence was evident in an observational study in North Dakota (Duebbert 1966). In two years, a 3-ha island in a lake held 78 and 121 nests, which showed high success. Then, in the next three years when water levels were lower and Raccoons *Procyon lotor* gained access, fewer than 20 nests were found and success was low. Gadwalls *Anas strepera* deserted the island at the start of the season. These examples, involving local variation in predation pressure, provided parallels with local variations in spring food-supply in influencing settling patterns (Chapter 7).

In all the experiments, the effects of predator removal were short lived, and when control stopped, predators soon moved back, and predation rates and population levels reverted to normal. Increases in breeding densities achieved under predator removal varied from nil to 2.6-fold, compared with the unmanipulated situation. The degree of increase depended partly on the number of years that predator removal was continued on the same area. In some studies, effects on breeding numbers were cumulative, at least up to three years (for example: 1.4× after one year and 2.6× after three years in Grey Partridge, Tapper *et al.* 1996). These effects were similar in magnitude to those found in food-provision experiments (Chapter 7), but trivial compared with those achieved with invertebrates in predator-removal experiments or in biological control programmes, in which density differences of 100-fold or more followed from the addition or removal of predators (Sih *et al.* 1985).

Only one experiment, involving the Avocet *Recurvirostra avosetta*, was continued until the increasing breeding population stabilised, about five years from the start of predator removal. Most of the other experiments were shorter, so could not be expected to show long-term effects, or whether breeding numbers increased to the maximum level possible under reduced predation. Nor could they show whether predation regulated, rather than merely limited, prey population density. If the experiments had been continued for longer, it is theoretically possible that the prey may have increased to the point at which predators could no longer reduce their numbers to former levels (release from the so-called predation trap, see above). This was the outcome of a predator-removal study on Rabbits *Oryctolagus cuniculus*, providing evidence for the existence of two equilibrium densities in the presence of predators, as mentioned earlier (Pech *et al.* 1992). It is also possible that, if habitat quality (food and cover) had been changed instead of predator numbers, a bigger and more lasting rise in density might have occurred.

In addition to experiments, occasional natural events that have affected predator numbers have revealed impacts on prey. For example, an epizootic of sarcoptic mange was prevalent among Red Foxes *Vulpes vulpes* in Sweden for about ten years from the late 1970s, substantially reducing their numbers. Breeding and post-breeding densities of several prey species increased, including Capercaillie *Tetrao urogallus*, Black Grouse *T. tetrix*. Hazel Grouse *Bonasa bonasia* and the hares *Lepus europaeus* and *L. timidus*, only to decline again when the disease subsided and Fox numbers recovered. During this period, fluctuations in these species also became de-coupled from those of voles. These events provided further evidence that Foxes can limit the numbers of gallinaceous birds, and that in Sweden they are important in conveying the 3–5-year cyclic fluctuation in voles to other species (Lindström *et al.* 1994). This event provided a contrast with the lack of prey response noted during the widespread decline of the Sparrowhawk *Accipiter nisus*, discussed above.

Predator-exclusion studies

The nest success of some bird species has been improved by providing safer sites. The breeding success of various tit (*Parus*) species in Wytham Wood, southern England, increased significantly after wooden boxes, which were accessible to Weasels *Mustela rivalis* and Great-spotted Woodpeckers *Dendrocopos major*, were replaced by concrete hanging boxes (Dunn 1977), leading to higher subsequent breeding densities (McCleery *et al.* 1996). Similarly, in North America, nest-boxes for Black-bellied Tree Ducks *Dendrocygna autumnalis* on poles fitted with sheet-metal predator guards showed 77% success, compared with 46% success in unprotected boxes and 44% in natural cavities (Bolen 1967). In this study, effects on subsequent breeding numbers were not recorded.

In other experiments, fences placed around individual nests or around large areas of nesting habitat resulted in improved nest success (Lokemoen *et al.* 1982, Green-wood *et al.* 1990, Beauchamp *et al.* 1996b). Fences could not exclude crows, however, and in some experiments mammals gained access, so some predation still occurred. No indication was given whether greater hatching success gave rise to greater breeding densities in later years. One tern colony increased in size the year after fencing, evidently from immigration to this safe site (Forster 1975). In fact, the most striking effects of predator exclusion involved seabirds (Chapter 8), but most examples involved the colonisation of predator-proofed sites and not simply an improvement in the nest success of birds already there.

Simulated predator-addition studies

Because predators deliberately released in an area might die or leave, few attempts have been made to increase predation by adding predators, and none concern birds (apart from the island situations described above). However, the effects of increased predation on young Great Tits *Parus major* were simulated experimentally on the Dutch island of Vleiland, where natural predation was negligible. Over four years, Kluijver (1966) reduced the production of young to about 40% of normal, yet observed no decline in subsequent breeding numbers. Immigration did not increase, but survival rates in the remaining birds (especially adults) improved to almost twice their previous value (Chapter 5). The implications were that individual survival was influenced by the numbers of conspecific competitors and that breeding numbers were limited by factors other than predation, at least up to a 60% loss of young.

In another experiment, Kautz & Malecki (1990) examined the effects of different culling regimes on Feral Pigeon *Columba livia* populations. In autumn 1981, they removed 0, 18, 33 and 39% of pigeons from four similar sized areas in different parts of New York City, and then monitored the response of the remaining birds. In the first three areas, population density was restored within a year, and in the high cull area within two years. This was achieved by compensatory improvements in the survival of eggs, chicks and adults (but not juveniles), and implied that an autumn cull of up to 35% could be sustained without causing long-term population decline.

These experiments were useful in indicating (1) the level of additional kill that could be imposed on populations at particular stages in the annual cycle without affecting breeding density, and (2) the compensatory changes in reproductive or survival rates that occurred to offset the impact of enhanced predation. Other similar experiments are discussed in Chapter 5.

OTHER ISSUES

Sex bias in predation

In many ground-nesting birds, the hen is often killed when the nest is found by a fox or other mammalian predator. Not surprisingly, females usually predominate among kills found at summer fox dens (Sargeant et al. 1984, **Table 9.5**). Such differential predation is thought to account in part for the usual predominance of males among adult game birds and ducks. In species examined, the sex ratio at hatching was about equal, but among increasing age-groups of adults, the ratio became progressively biased towards males (Lack 1954). In some duck populations, adult males outnumbered females as much as 3:1 (Lack 1954, Bellrose et al. 1961, **Table 9.6**). Although not necessarily due entirely to predation, these ratios give some idea of the maximum possible effect of female-biased mortality on breeding numbers, for

Table 9.5 Proportion of females among individuals of seven dabbling duck species found at Red Fox *Vulpes vulpes* dens in the Dakotas. The implication is that more females were killed, being more vulnerable on nests than the non-incubating males. Modified from Sargeant *et al.* 1984.

	Total ducks	% females
Mallard *Anas platyrhynchos*	239	81
Northern Pintail *Anas acuta*	312	73
Northern Shoveler *Anas clypeata*	103	90
Blue-winged Teal *Anas discors*	385	81
Gadwall *Anas strepera*	121	64
American Wigeon *Anas americana*	25	72
Green-winged Teal *Anas crecca*	34	74

without the distorted sex ratio there might have been many more potential nesting pairs. The same may hold to some extent in other ground-nesting bird species in which the female does most or all the incubation and is thus more vulnerable than the male.

Among some gallinaceous birds, predation away from the nest is also weighted towards females. This is apparent mainly in polygynous species in which males are bigger than females, and are therefore more difficult for some predators to tackle, but also because, in spring and summer, females are more active in food collection than males, and hence more vulnerable (Kenward *et al.* 1981, Widén 1987). Moreover, in some polgynous species, the more active breeding cocks are more vulnerable than younger, non-breeding ones. In one study of radio-tagged Black Grouse *Tetrao tetrix*, lekking adult cocks suffered 31% predation in spring and summer, whereas non-lekking cocks suffered none (Angelstam 1984). Hence, the behaviour of different sex- and age-groups at different times of year can clearly influence their susceptibility to predation.

Table 9.6 Adult sex ratio for some North American waterfowl. In all species studied, males outnumbered females. From Baldassarre & Bolen 1994, in which the original references may be found.

Species	% males
Black-bellied Whistling Duck *Dendrocygna autumnalis*	51.8
Lesser Snow Goose *Anser caerulescens*	52.2
Canada Goose *Branta canadensis*	53.5
Mallard *Anas platyrhynchos*	62.3
Green-winged Teal *Anas crecca*	56.6
Blue-winged Teal *Anas discors*	59.3
Northern Pintail *Anas acuta*	59.4
Canvasback *Aythya valisineria*	66.4
Redhead *Aythya americana*	60.2
Lesser Scaup *Aythya affinis*	70.0
Common Eider *Somateria mollissima*	51.0
Oldsquaw *Clangula hyemalis*	52.3
Hooded Merganser *Mergus cucullatus*	64.9
Ruddy Duck *Oxyura jamaicensis*	62.0
Ring-necked Duck *Aythya collaris*	77.0

Association between high nest success and high breeding density

Some species subject to heavy predation show a clear association, comparing areas, between high nest success and high breeding density. In some species, such as Grey Partridge and Black Grouse, this association could result from direct causal relationships, with nest success influencing subsequent breeding numbers, or from predation influencing both nest success and adult numbers (Potts 1986, Baines 1996). Another example in which high local breeding success contributed to high breeding density is provided by the Great Tit *Parus major* study in Wytham Wood mentioned above (McCleery *et al.* 1996). Prior to 1976, annual nest predation by Weasels *Mustela nivalis* averaged 30%, but after the nest-boxes were made predator-proof in 1976, nest predation fell to less than 5%. The resulting increase in production of young (3.4 per pair before 1976, 6.3 after), followed by increased local recruitment of these young, raised the size of the local breeding population. No obvious changes in other factors occurred across these two periods. The clear implication was that local production of young influenced the subsequent density of breeders. Other examples were given in **Table 1.1**.

In other species, the links between high density and high breeding success could result merely from birds being attracted to nest in areas where their chance of success is high. Among Woodpigeons *Columba palumbus* in Poland, nest densities were up to 450 times greater, and annual productivity 4–6 times greater, in town parks where predators were sparse, than in forests where predators were common. Food-supply could not account for these differences because pigeons from both nesting habitats fed alongside one another on local farmland in summer and in Iberian oak woods in winter (Tomiałojc 1980). In addition, after Crows *Corvus corone* settled in one park, Woodpigeon numbers and nest success declined, whereas after Crows and other corvids were removed from another park, Woodpigeon numbers and nest success increased. Tomiałojc argued that production of young had a direct effect on subsequent breeding density, through the return of young in later years. However, an alternative explanation was that predators influenced merely the distribution of nesting Woodpigeons which favoured localities where predation risk was low, while their total numbers were determined by some other factor.

Similar links between numbers and nest success occur among ducks and other birds which often nest at much higher density in safe places, such as islands in lakes, than on shorelines and other places accessible to mammalian predators (Duebbert 1966, Vermeer 1968, Newton & Campbell 1975, Hines & Mitchell 1983). Without a ringing–recapture programme, it is hard to tell to what extent local high densities in safe sites are due directly to the high success, survival and site-fidelity of birds that nest on islands, and to what extent to immigration of birds attracted by a safe site. In the first case, predation would influence breeding density directly through previous nest success, whereas in the latter it would influence breeding density through effects on settling patterns. Both mechanisms might often be involved, their relative importance varying between species or from site to site. In a study of Wood Ducks *Aix sponsa*, nest-boxes were made safe from predation so that nest success improved. Over the next 13 years, pair numbers increased from 10–15 to more than 90 (Bellrose *et al.* 1964). Integral parts of this increase, as shown by ringing, were not only the high production of young, but also the homing of the increased numbers of adults and young to the same area in successive years.

Regardless of whether local production of young directly or indirectly affects local breeding density, known behavioural mechanisms can ensure that more individuals attempt to settle each year in areas with high nest success. Some individuals might avoid areas where they somehow detect that predators are numerous, or they might react to a nest failure by moving elsewhere. Individuals of many bird species show greater site-fidelity after a breeding success than after a failure (Greenwood 1980), so over a period of years, in the absence of other constraints, their behaviour would tend to produce concentrations in the safest localities. Furthermore, birds nesting for the first time may be attracted to places where others are nesting, using this as a cue to habitat suitability (Alatalo *et al.* 1982, Ray *et al.* 1991). Thus spatial associations between high nest success and high densities in individual bird species could result from more than one mechanism:

(1) innate or learned attraction to safe sites;
(2) site-fidelity by young and adults, leading to greater build-up to the most productive sites;
(3) a greater tendency of adults to return to the same site after a nest success than after a failure; and
(4) greater attraction of new breeders to sites where others are nesting.

CONCLUDING REMARKS

Among animals in general, predator–prey relationships vary from closely-coupled interactions at one extreme, in which the numbers of prey and predators are inter-dependent, to neutral feeding links at the other extreme which are of no consequence for the population dynamics of either consumer or consumed. In between are prey-controlled dynamics in which the numbers of prey influence the numbers of predators but not vice versa, and predator-controlled dynamics in which the numbers of predators influence the numbers of prey but not vice versa.

To judge from the studies cited here, predation may reduce the post-breeding numbers of a large proportion of birds, but the breeding numbers of only a small proportion. Many species show evidence of limitation by food-supplies or nest-sites (Chapters 7 and 8), and among many territorial species that have been studied, removal experiments in spring have usually revealed the presence of 'surplus' birds unable to breed through lack of space (Chapter 4). Of the species that have been studied, ground-nesting ducks and game birds seem more often to be depressed in breeding numbers by predation than some other species, such as many songbirds. This may be because ground nests are vulnerable to a greater range of mammalian predators, and because some important predators (notably Fox) often kill the sitting hen as well.

In most of the world, the predator fauna has of course changed greatly under human influence. In some areas where studies have been done, predators may be scarcer than in natural habitats, leading to reduced effects on prey. In the managed woods of western Europe, nest predation on both open-nesting and hole-nesting birds is one-third as great as in the primeval forest at Bialoweiza in Poland (Tomiałojc *et al.* 1984). The breeding densities of many species are also higher in the woods of western Europe, but how much this is due to reduced predation is an open question. Tomiałojc *et al.* (1984) envisage that some bird populations in Europe exist in one of two states: either at low density limited by predation where predators occur in natural numbers, or at high density limited by food-supplies where predators occur in reduced numbers (fitting the model in **Figure 9.4**). They see this latter situation as holding over much of western Europe, reaching its extreme in towns, parks and gardens. However, predators are not necessarily scarcer in all human-modified landscapes, as indicated earlier. Several predators of birds and eggs spread across the North American prairies following human settlement and tree planting, so that such species as Raccoons and Crows are now common in areas formerly closed to them. Moreover, in both Europe and North America, the destruction of large predators, such as Wolves *Canis lupus*, may have allowed smaller ones, such as Red Foxes, to increase, with consequent greater effects on bird populations. Similarly, where large predators, such as Jaguars *Panthera onca*, have been eliminated from some neotropical areas, small omnivores and carnivores have increased, in at least one area becoming 4–10 times more abundant than before (Terborgh & Winter 1980). It is hard to imagine that the increase in small predators has had no effects on some prey populations.

SUMMARY

Many bird species suffer heavy predation at the egg and chick stages, and some also at the fledgling and adult stages. The effects of predation on any population depend on the extent to which predation is offset by compensatory reductions in other losses or by improved reproduction. Some bird populations can withstand heavy predation yet maintain their breeding numbers. This is apparent in some species whose breeding densities are demonstrably limited by other factors, often leading to a non-breeding surplus.

To reduce breeding or post-breeding numbers, at least part of the predation experienced must be additive to other losses. To judge from empirical evidence, predators might then (1) hold breeding density at a level well below what the habitat would permit, as in the Grey Partridge in some areas; (2) cause marked (often regular) fluctuations in breeding density, as in some northern game birds exposed to periodic heavy predation when their generalist predators experience a shortage of their main mammalian prey; or (3) cause decline to extinction, as in some island birds, exposed to introduced predators.

In recent decades, human-induced changes in landscapes and in predator communities may have led to increased rates of nest predation and in some species to population declines.

In 23 out of 27 experiments, the removal of corvid and mammalian predators usually led to improvements in the nest success of target species. In 12 out of 17 experiments this was followed by increased post-breeding numbers, and in 10 out of 17 experiments by increased breeding numbers (up to 2.6-fold). Most experiments involved ground-nesting game birds and ducks, and predation was often influenced by quality of nesting cover and availability of alternative prey.

In some bird species that have been studied, predation seems to play a minor role in the direct limitation of breeding numbers, at least in areas and time periods where studies have been made. In some ground-nesting gamebirds and ducks, however, predation can hold breeding numbers below the level that would otherwise occur, and females are more vulnerable than males, contributing to the uneven sex ratios usual in these birds.

Cliff Swallows *Hirundo pyrrhonota*.

Chapter 10
Parasites and Pathogens

Whereas predators may kill and consume their prey whole, parasites and pathogens usually live as populations within or on the living bodies of their hosts. They draw sustenance from their hosts, and have effects that vary from negligible to fatal. As with predation, there are conflicting views on the effects of parasitism. On one view, a parasite (or pathogen) tends to evolve in such a way as to become less harmful to its host with time, since it has a better chance of long-term survival if it does not destroy its habitat (except where transmission depends on the host's death). Likewise, the host tends to evolve resistance to the parasites and to any toxins they might produce. On this view, parasites would have minimal, if any, effect on their hosts. On the contrasting view, parasites are seen invariably to harm their hosts and in certain conditions may kill them. In practice, these two views probably represent opposite ends of a continuum of situations found in nature.

By definition, parasites must have some costs to their hosts, if only the reparation of tissue damage. These costs become important to individual hosts when they are sufficient to lower reproduction or chances of survival. They become important at the population level when they cause sufficient reproductive failure or mortality to reduce host breeding numbers below what would otherwise occur. Clearly, any parasite that affects reproduction or mortality has the potential to influence population levels. The questions of interest here are how often in nature parasites do indeed reduce the breeding numbers of their hosts and in what circumstances. Great progress in understanding parasite–host relationships has come in recent years as a result of the mathematical models developed by Anderson & May (1978, 1979). These models have done much to define the circumstances in which parasites of various kinds might influence host numbers, and the types of population response to be expected.

Throughout this chapter, I shall use the term 'disease' only to cover a clinically abnormal state resulting from parasite infection, and not to encompass the effects of other degenerative and metabolic disorders, nutrient deficiencies and toxic chemical effects. The important point about parasitic diseases is that they are infectious, passing 'horizontally' from one individual to another (via direct contact or vectors), or 'vertically' from parent to offspring. Because, in general, they are transmitted most efficiently at relatively high host population densities, they can act on host populations in a density-dependent or delayed density-dependent manner. They can thus contribute importantly to host population regulation, even if they cause only a small proportion of host deaths. Moreover, parasites can be sustained by a much smaller population of hosts than is needed to maintain a population of predators.

In assessing the impacts of parasitism on a population, the same question arises as with predation: whether the resulting deaths are additive to other mortality, or compensatory so that, in the absence of deaths from parasites, a similar number of deaths would occur each year from other causes. With disease, contributing causes are often apparent. For example, individuals weakened by starvation might become vulnerable to disease, as might those that are crowded. On the other hand, individuals that lose part of their daily food-intake to internal parasites may be more prone to the effects of food-shortage than other individuals that are free of parasites. Disease is thus increasingly seen as a response, not only to parasite infection, but to the overall condition of the host. This makes it hard to separate the effects of disease from the food-shortage or other environmental conditions that might favour it, and means that deaths from disease are often unlikely to be completely additive to other mortality. Interactions between disease and predation are also likely: disease may predispose certain individuals to predation, which may in turn reduce the spread of disease (Chapter 13).

These various interactions mean that the independent effects of many parasites on host populations can be properly assessed only by experiment. Only recently, however, have attempts been made to remove parasites from wild populations by use of chemicals, enabling effects on host populations to be examined. This has been done by use of pesticides which destroy the vectors (Kissam et al. 1975), by drugs or other chemicals which destroy the parasites (Hudson 1986, Chapman & George 1991), or by vaccination which reduces the impacts of the parasites on individual hosts (Hudson & Dobson 1991). The few such manipulations that have involved birds have been concerned more with the effects of parasites on individual performance than on population levels. However, natural experiments, involving the accidental introduction of a disease or disease-vector to a new area, have occurred from time to time, sometimes with devastating effects on the local bird-life (see below).

THE PARASITES

The great plethora of pathogens that affect birds and other animals fall naturally into two groups, namely microparasites (viruses, bacteria, fungi, protozoa) and macroparasites (helminth worms, arthropods) (Anderson & May 1979). The former are characterised by their small size, short life-cycles, and high rates of reproduction within individual hosts **(Table 10.1)**. With few exceptions, the duration of infection is short relative to the lifespan of the host. Some are transmitted by direct contact between host individuals, and others through vectors such as blood-sucking ticks or mosquitoes. Either way, the rate of spread of infection depends partly on the contact rate, which in turn is dependent on the host density (for directly transmitted pathogens) or on both vector and host density (for vector-transmitted ones). In those hosts that survive the initial onslaught, microparasites can promote some immunity to reinfection. Such recovered hosts can be recognised by the presence of appropriate antibodies in the blood, and in medical and veterinary practice the effects of microparasites can be reduced by vaccination. For microparasites, then, the host population can be sub-divided at any one time into distinct classes: susceptible, latent-infected, infectious, recovered-and-immune. Microparasitic diseases have caused many dramatic effects on populations, including myxomatosis (viral) in Rabbits *Oryctolagus cuniculus*, bubonic plague (bacterial) in humans, and chestnut blight (fungal) in American Chestnut *Castanea dentata*. Some of the best known

Table 10.1   Comparison in characteristics and usual mode of action between microparasites (viruses, bacteria, protozoa and fungi) and internal macroparasites (helminths*).

	Microparasites	Macroparasites
Characteristics		
Size	Invisible to naked eye	Usually visible to naked eye
Life cycle relative to host's	Short	Long
Population growth within host	Through reproduction	Through continuing uptake of infective stages
Immune response	Usually strong and long-lasting	Usually weak and short-lasting, or non-existent
Transmission	Through direct contact or vector, dependent mainly on density of hosts	Through food, body fluids or intermediate hosts, dependent on density of hosts and parasites
Mode of action		
Effect depends on	Initial infection	Total cumulative burden
Cause deaths	Often	Seldom
Epizootics	Frequent	Rare

*Helminths include platyhelminths (trematodes, tapeworms, flukes), acanthocephalans (spiny-headed worms) and nematodes (roundworms, hookworms).

microparasite diseases of birds include tuberculosis (bacterial), psittacosis (protozoal) and aspergillosis (fungal).

In contrast to microparasites, most macroparasites are visible to the naked eye **(Table 10.1)**. The internal ones (helminths) do not multiply directly within a host, but produce infective stages which usually pass out of the host before transmission to another host. They are relatively long lived, and usually produce only a limited immune response, so that macroparasite infections tend to be of a persistent nature, with hosts accumulating parasites through life or continually being reinfected. Transmission depends not only on the density of host individuals, but also on parasite density and other factors. Some parasite species require higher host densities than others before they can become established, so that the numbers of parasite species, as well as of individuals, can increase with a rise in host density (Dobson & May 1991).

Parasitic worms almost always have clumped distributions, being found in small numbers in most host individuals and in large numbers in a few. This is important because the effects of such parasites on individual hosts typically depend on the total burden of worms, and not simply on whether or not the host is infected (Anderson & May 1978, May & Anderson 1978). The death of only a few hosts (those with heavy burdens) can lead in turn to the loss of a large number of worms. This process tends to reduce the impact of macroparasites on host populations. Why certain individuals acquire heavy parasite loads, while others do not, may depend on behaviour, diet, body condition, age and genetic predisposition, as well as on chance. Some helminths tend to be concentrated in particular age-groups, perhaps the young or the very old, but the individual variation in total burdens is still great.

The variety of parasitic worms found in different bird species depends on the bird's diet (which influences infection risk), lifespan (which affects the time for accumulation), social behaviour (which influences transmission), migratory habits (which affects the number of sites visited), body size (larger having more) and habitat (aquatic species having more) (Gregory *et al.* 1991). Most waterbirds support a diverse community of such parasites, with large numbers of individuals per host **(Table 10.2)**. Some parasite species may be present in almost every individual in a population, while others are present in only a small proportion.

Macroparasitic associations are typically much more stable than microparasitic ones, owing to the persistent nature of the infections and the inability of the host to produce lasting immunity to reinfection. In some circumstances, however, especially if the free-living infective stages are long lived, helminth worms can induce recurrent epidemics that spread through a host population with devastating effects similar to those observed for microparasites (May & Anderson 1978, Hudson & Dobson 1991). Some examples are given below.

In contrast to macroparasitic worms, which live within the host body, parasitic arthropods mostly live on the outside. Such ectoparasites include various lice (Mallophaga), mites and ticks (Acarina), flies (Diptera), fleas (Siphonaptera) and bugs (Hemiptera). They mostly feed on blood, causing tissue damage and anaemia, as well as allergic responses and bacterial infections. In large numbers, they can thus weaken and sometimes kill their hosts. Most important, however, they act as vectors of microparasitic diseases. This is especially true of some ticks which can attach themselves to several hosts in succession. Some ectoparasites are transmitted from host to host at times of physical contact, but most can move from environment to

Table 10.2 Numbers of helminth parasites found in four species of grebes shot on lakes in Alberta. From Stock & Holmes 1987.

	Number of birds examined	Total numbers of parasite species	Numbers of parasite species per bird	Numbers of parasite individuals per bird
Western Grebe *Aechmophorus occidentalis*	20	16	4–10	112–1800
Red-necked Grebe *Podiceps grisegena*	33	23	4–14	326–10 459
Eared Grebe *Podiceps nigricollis*	31	26	2–15	231–33 169
Horned Grebe *Podiceps auritus*	7	14	4–8	212–10 943

host. As with internal macroparasites, therefore, transmission of ectoparasites depends on both parasite and host density and behaviour, as well as on other factors. Because the immune response is at best limited, birds suffer from permanent infection with ectoparasites or from continual reinfection. Ectoparasites themselves do not normally cause heavy mortality, except in the nestlings of some species (see below). Sometimes the same individual birds, examined on different occasions, are found to have similar parasite burdens each time. For example, ectoparasite loads were found to be repeatable traits on individual Barn Swallows *Hirundo rustica* (Møller 1991b); birds that had high infestations in one year had high infestations the next.

As with predators, it is also useful to distinguish specialist (host-specific) parasites from generalists, which can have more than one definitive host species. For specialists, the rate of spread is influenced by the density of the host, so that the infection rate can show density-dependence or delayed density-dependence, leading to predictable interactions between parasite and host populations. Because the two are coupled in a closed system, the host must usually attain a certain density (contact rate) before the parasite can persist and spread, and a disease tends to become self-limiting before it can annihilate a population. For generalist parasites, however, total numbers are influenced by all their host species, so that a particular vulnerable host might be affected more severely than expected from its numbers alone. In extreme cases, the numbers of the vulnerable host species may be kept at a low level through the persistence of a 'reservoir of infection' in alternative hosts living in the same area (rabies in various carnivores being an obvious example). The level of infection in the vulnerable host is not then density-dependent with respect to that specific host, but could be density-independent or inversely density-dependent. The latter situation arises because the ratio of reservoir host to susceptible host changes as the individuals of that susceptible host die. Whereas it is difficult for a specialist parasite to drive its host species to extinction (because the parasite population itself declines as host densities fall), a generalist can in theory extinguish one or more host species locally or globally so long as its population density is maintained by other, more robust, hosts. Examples are given below.

Certain host species are more vulnerable to particular generalist parasites than others, either because they are infected more easily or because they are more susceptible to the effects of an infection once acquired. In the initial infection, social organisation is important, colonial and flocking species being at greater risk than solitary ones. In the shallow lakes of Texas, the bacterial disease avian cholera has occurred in waterfowl mainly after summer droughts, when the wintering birds are unusually crowded. During an outbreak in January–March 1950, one bird in every 2500 died each day, up to 0.3% of the ducks being fatally affected at one time (Petrides & Bryant 1951). The smaller species, such as Green-winged Teal *Anas carolinensis*, Lesser Scaup *Aythya affinis* and Ruddy Duck *Oxyura jamaicensis*, suffered more than the larger ones. Teal, for instance, were 22 times more susceptible than Mallard *Anas platyrhynchos* **(Figure 10.1)**. Variation in susceptibility, once infected, has been shown in various Hawaiian birds with respect to avian malaria *Plasmodium relictum* **(Table 10.3)**, in different crane species with respect to equine encephalitis (Dein *et al.* 1986), in different pigeon species with respect to pigeon herpes virus (Snyder *et al.* 1985), in different grouse species with respect to the viral disease louping ill (Reid *et al.* 1980), and in various waterfowl species with respect to avian

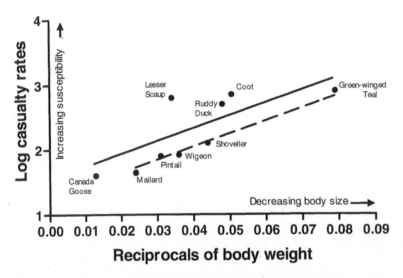

Figure 10.1   Relationship between mortality from avian cholera *Pasteurella multocida* and body weight among waterbirds at Muleshoe National Wildlife Refuge, Texas, 1950. Solid line – all species; dashed line – dabbling ducks only. Mortality ratings were calculated from observed deaths as a percentage of expected deaths, the latter estimated on the basis of equal vulnerability in all species. The proportion of fatalities increased in logarithmic ratio as body weight decreased. Species included Canada Goose *Branta canadensis*, Mallard *Anas platyrhynchos*, Green-winged Teal *Anas crecca*, American Wigeon *Anas americana*, Pintail *Anas acuta*, Shoveller *Anas clypeata*, Ruddy Duck *Oxyura jamaicensis*, Lesser Scaup *Aythya affinis* and American Coot *Fulica americana*. Modified from Petrides & Bryant 1951.

Table 10.3 Survival among various Hawaiian birds experimentally infected with avian malaria *Plasmodium relictum*. The Laysan Finch *Telespyza cantans* (endemic to the Pacific island of Laysan) had never been exposed to malaria, the endemic Hawaiian species had been exposed for less than 100 years, while the introduced species had had a long evolutionary association with the disease. All birds had the malarial parasites in their blood, but the only deaths occurred among endemic birds. From van Riper et al. 1986.

	Endemic					Introduced		
	Laysan Finch *Telespyza cantans*	Apapane *Himatione sanguinea*	Iiwi *Vestiaria coccinea*	Mauna Kea Amakihi *Hemignathus virens*	Mauna Loa Amakihi *Hemignathus virens*	Red-billed Leiothrix *Leiothrix lutea*	Japanese White-eye *Zosterops japonicus*	Canary *Serinus canaria*
Number infected	5	5	5	6	5	5	5	5
Number died	5	2	3	4	1	0	0	0

cholera (Rosen & Bischoff 1949, Petrides & Bryant 1951, Vaught *et al.* 1967), and viral duck plague (Spieker, cited in Wobeser 1981).

Variation in response to infection can also be seen between members of the same species: one race of Canada Geese *Branta canadensis* was more susceptible to avian cholera than another larger race on the same refuge (Vaught *et al.* 1967), and the pathogenicity of *Leucocytozoon simondi* within one subspecies of Canada Goose varied greatly between birds from two localities only 40 km apart, perhaps in association with other factors (Desser *et al.* 1978). There are many other examples of different susceptibility among zoo animals, and in extreme cases, the same pathogen may be wholly innocuous to one species but fatal to a closely-related one from another part of the world.

The impact of any pathogen on a population also depends on whether the pathogen influences mortality or fecundity. If individuals infected with a host-specific pathogen soon die, their swift disappearance from the population limits the spread of the pathogen. But if infected individuals do not die, but merely fail to reproduce, the pathogen can spread widely, lowering reproduction in enough individuals to cause population decline. An example of this process, involving a strongyle parasite in Red Grouse *Lagopus l. scoticus*, is given below.

Brood-parasites

Birds are also subject to another type of parasite. So called 'brood-parasites' are other species of birds (such as cuckoos and cowbirds) that lay their eggs in the nests of a limited number of host species. Most species of brood-parasites remove host eggs when they deposit their own eggs in host nests, and in some the resulting nestlings evict host young from the nest or grow more rapidly, diminishing the growth and survival of host young (Rothstein 1990, Payne 1997). In these ways, brood parasitism usually reduces the breeding rate of host species. Some brood-parasites specialise on a single host species, but others parasitise several or many. To be suitable, hosts must be of appropriate size, and have diets fit for the young parasite. Potential hosts vary in their susceptibility to parasitism, and some species have evolved anti-parasite behaviours, such as mobbing and ejection of strange eggs, which reduce the chance of victimisation.

Worldwide, about 103 bird species are inter-specific brood-parasites, about 1% of all bird species (Payne 1997). Besides the cuckoos (Cuculinae and Neomorphinae) and cowbirds (Icterinae), they include the Cuckoo-finch *Anomalospiza imberbis*, whydahs (Viduinae), honeyguides (Indicatoridae) and the Black-headed Duck *Heteronetta atricapilla*. The last-named is the only brood-parasite with precocial young, which live independent lives from the time they hatch in the nests of other waterbirds.

EFFECTS OF PARASITES ON INDIVIDUALS

Parasites can reduce the breeding success of birds, either by lowering the body condition of the adults, or by lowering the body condition and survival of the chicks. In most bird species, chick deaths from parasitism seem few or non-existent, and assume nowhere near the importance of predation. In certain species, however, deaths from parasitism are a major cause of reduced breeding output. Most studies have concerned ectoparasites (such as blood-sucking ticks, bugs and fly larvae), which can reach high infestations in species that use the same nest-sites year after year, such as many hole-nesters, cliff-nesters and colonial seabirds (Rothschild & Clay 1952). Some ectoparasites can survive for months or even years in sheltered nest-sites, breeding each time the sites are occupied, and building up substantial populations over a period of years. In one survey, hole-nesting Stock Doves *Columba oenas* were found to support 58 species of nest parasites at a mean density of 6573 individuals per decimetre of nest material, while the closely-related open-nesting Woodpigeon *C. palumbus* supported only one species at 20 individuals per decimetre of nest material (Nordberg 1936).

High parasitism is the price that many hole-nesters and colonial breeders pay for safety from nest predators. In some species, high ectoparasite burdens have been associated with increased nest desertions, and with reductions in mating success, clutch size, hatching success, nestling growth and survival, and in the number of broods per season. Ectoparasites have been found to be a major cause of nest desertion or chick mortality in Cliff Swallows *Hirundo pyrrhonota* and other hirundines, and in pelicans and various seabirds **(Table 10.4)**. In colonial species, they sometimes lead to nest desertions on a large scale. For example, some 5000 Sooty Tern *Sterna fuscata* pairs out of a colony of 40 000 pairs abandoned their eggs and chicks, apparently in response to an infestation of ticks *Ornithodoros capensis*. Many ticks were found in the deserted part of the colony but few or none where reproduction was normal. The ticks were still present in the following year in the same area, which the terns again avoided (Feare 1976a). Among 22 colonies of Kittiwakes *Rissa tridactyla* in Britain, all established for more than 30 years, those that were decreasing in size had a higher prevalence and density of ticks *Ixodes uriae* than those that were increasing (Boulinier & Danchin 1996). The authors suggested a cause-effect relationship: by influencing local reproductive success, ectoparasites could affect the recruitment of new breeders, and the site-fidelity of existing breeders, thus influencing local trends in numbers.

While colonial and other bird species, which use the same sites year after year, are often victims of wingless ticks and bugs, other bird species, which nest solitarily in different sites each time, are most often attacked by flying insects (flies) or by parasites that travel easily on their hosts. The most damaging are the blow-flies and muscid bot-flies whose maggots feed from live nestlings **(Table 10.4)**. Blood-feeders may reduce both haemacrit (% red cell volume) and haemoglobin, as well as stimulate an immune response via white blood cell production. Nestlings may be killed outright or fledge underweight which may reduce their subsequent survival.

The effects of parasites on breeding success have been assessed in both observational and experimental studies. In the rainforest of Puerto Rico, the hole-nesting Pearly-eyed Thrasher *Margarops fuscatus*, suffers from heavy infestations of the bot-fly *Philinus deceptivus* whose maggots feed from live nestlings. Arendt (1985) visited nests every 1–2 days, and in some he removed bot-flies from the chicks as soon as they appeared, leaving them at natural levels in others. In the 'cleaned' nests, some 99% of 74 nestlings survived, compared with 53% of 448 nestlings in natural nests. Among nestlings that died, 97% of mortality was due to botflies. Other studies have also found higher breeding success in unparasitised than in parasitised nests of the same species in the same area, while yet others have shown little or no effect of ectoparasitism on nestling survival despite heavy infestations **(Table 10.4)**.

Table 10.4 Effects of parasites on the breeding of various bird species.

Host species	Parasite species	Observation (O) or experiment (E)	Reduction in breeding performance	Location	Source
House Sparrow *Passer domesticus* and Tree Sparrow *P. montanus*	Micro-organisms *Escherichia coli, Candida* spp. and *Isospora* sp.	O	Hatching success (13–20%) Chick survival (24–30%)	Poland	Pinowski *et al.* 1988
House Wren *Troglodytes aedon*	Blow-fly *Protocalliphora hirundo*	O	Chick mass (8%)	Wyoming	Johnson *et al.* 1991
Great Tit *Parus major*	Hen Flea *Ceratophyllus gallinae*	E	Chick survival (68%) Brood size (24%) Fledging success (30%) Chick body mass (15%)	Switzerland	Richner *et al.* 1993
Chesnut-backed Chickadee *Parus rufescens* and Mountain Chickadee *P. gambeli*	Blow-fly *Protocalliphora* spp.	O	Chick survival (slight)	California	Gold & Dahlsten 1983
Pearly-eyed Thrasher *Margarops fuscatus*	Bot-fly *Philinus deceptivus*	E	Fledging success (46%) Chick body mass and growth (slight)	Puerto Rico	Arendt 1985
American Dipper *Cinclus mexicanus*	Blow-fly *Protocalliphora* spp. Fowl Mite *Ornithonyssus sylviarum*	O	Chick survival (slight)	California	Halstead 1988
Purple Martin *Progne subis*	Martin Mite *Dermanyssus prognephilus*	E	Fledging success (14%) Chick body mass (7%)	Kansas	Moss & Camin 1970
Tree Swallow *Tachycineta bicolor*	Blow-fly *Protocalliphora sialia*	O	Chick mass (no reduction) Chick survival (no reduction)	Ontario	Rogers *et al.* 1990
Cliff Swallow *Hirundo pyrrhonota*	Swallow Bug *Oeciacus vicarius*; Swallow Tick *Argas cooleyi*; Swallow Tick *Ornithodoros concanensis*	E	Fledging success (46%) Chick body mass (12%)	Texas	Chapman & George 1991
	Swallow Bug *Oeciacus vicarius* Swallow flea *Ceratophyllus celsus*	E	Nest success (up to 50%) Chick body mass (15%)	Nebraska	Brown & Brown 1986

Host	Effect	Parasite	Consequences	Location	Reference
Barn Swallow *Hirundo rustica*	E	Tropical Fowl Mite *Ornithonyssus bursa*	Delayed breeding[1] Incidence of second clutches (16%) Nest success (33%) Brood-size (19%) Chick body mass (5–6%)	Denmark	Møller 1990, 1993
	O	Blow-fly *Protocalliphora hirundo*	Nest success (17% in first and 57% in second clutches) Chick survival (21% in first and 61% in second nests) Seasonal production of young (55%)	New York State	Shields & Crook 1987
South African Swallow *Hirundo spilodera*	O	Tropical fowl mite *Ornithonyssus bursa*	Fledging success (34%)	South Africa	Burgerjon 1964
Starling *Sturnus vulgaris*	E	Red Fowl Mite *Dermanyssus gallinae*	Chick survival (15%)	South Africa	Royall 1966
	O	Tropical Fowl Mite *Ornithonyssus bursa*	No reduction in chick body mass or survival	New Zealand	Powlesland 1977
Hooded Crow *Corvus corone cornix*	O	Nematode *Syngamus trachea*	Fledging success (51%)	Sweden	Loman 1980
Brown Cacholote *Pseudoseisura lophotes* and Firewood Gatherer *Anumbius annumbi*	O	Bot-flies *Philornis pici* and *P. sequiji*	Chick survival (6%)	Argentina	Nores 1995
Prairie Falcon *Falco mexicanus*	O	Tick, probably *Ornithodoros concanensis*	Chick survival (65%)	Colorado	Webster 1944
Sharp-shinned Hawk *Accipiter striatus*	O	Bot-fly *Philornis* spp.	Chick survival (33%)	Puerto Rico	Delannoy & Cruz 1991
Goshawk *Accipiter gentilis*	O	Protozoan *Trichonomonas gallinae*	Chick survival (slight)	England	Cooper & Petty 1988
Red-tailed Hawk *Buteo jamaicensis*	O	Fly *Eusimulium clarum*	Chick survival (24%)	California	Fitch et al. 1946
Golden Eagle *Aquila chrysaetos*	O	Protozoan *Trichonomonas gallinae*	Chick survival (3%)	Idaho	Beecham & Kochert 1975

Table 10.4 Continued.

Host species	Parasite species	Observation (O) or experiment (E)	Reduction in breeding performance	Location	Source
Canada Goose *Branta canadensis*	Haematozoan *Leucocytozoon simondi*	O	Chick survival (16–87%)[2]	Michigan	Herman *et al.* 1975
Brown Pelican *Pelecanus occidentalis*	Tick *Ornithodoros capensis*	O	Total breeding failure through colony desertion	Texas	King *et al.* 1977
Sooty Tern *Sterna fuscata*	Tick *Ornithodoros capensis*	O	Nest desertion (12%)	Seychelles	Feare 1976b
Tengmalm's Owl *Aegolius funereus*	Haematozoan *Leucocytozoon ziemanni*	O	Clutch size (ca 20%)	Finland	Korpimäki *et al.* 1993
Ring-necked Pheasant *Phasianus colchicus*	Helminth gut parasites, mainly *Heterakis* sp.	E	Hatching success (>50%)	England	Woodburn 1995
Great Egret *Casmerodius albus*	Nematode *Eustrongylides ignotus*	O	Chick survival (16%) Chick weight (6–9%)	Florida	Spalding *et al.* 1994
Snowy Egret *Egretta thula* and others	Nematode *Eustrongylides ignotus*	O	Chick survival (up to 84% in different species)	Delaware	Weise *et al.* 1977

[1] Delayed breeding (for example through having to build a new nest rather than use an old one containing parasites) is often associated with reduced success, and reduction in chick mass is often associated with reduced post-fledging survival.

[2] Range of values in 13 different years.

Food-supply clearly affects the ability of nestlings to withstand parasitism, the effects of which my be greater in years of poor food-supply than in years of good supply (Moss & Camin 1970), or greater on the last-hatched (= least well fed) chicks among broods (Arendt 1985). Such effects were shown experimentally in the Eastern Bluebird *Sialia sialis*, in which Pinkowski (1977) restricted the food-intake of nestlings by use of neck collars for 90–150 minutes every second day. These broods accumulated significantly more blood-sucking blow-fly larvae (*Apaulina* sp.) than other broods that were not deprived of food, although in this study no increase in mortality was detected.

The effect of micro-organisms on nestling production was examined over two years among House Sparrows *Passer domesticus* and Tree Sparrows *P. montanus* nesting in boxes placed in suburban Warsaw (Pinowski *et al.* 1988). In the House Sparrow about 30% of eggs failed to hatch and in the Tree Sparrow about 21–25% in different areas. Some 30–50% of these eggs were infertile, but the remainder contained dead embryos, mostly heavily infected with bacterial pathogens, notably *Escherichia coli*. In addition, nestling mortality in both species was 24–30%. The authors collected dead and poorly-growing chicks, together with healthy chicks from the same boxes, and examined them for micro-organisms. In the House Sparrow, some 76% of dead and sick chicks were infected, compared with 11% of healthy chicks. The equivalent figures for the Tree Sparrow were 87% and 2%. The main organisms involved were *E. coli*, a yeast-like fungus *Candida* and a coccidial organism *Isospora*. The authors concluded that in both sparrow species microparasite infections were the main cause of egg failure and chick mortality.

As mentioned above, parasites can also affect bird reproduction through effects on adults. In Tengmalm's Owl *Aegolius funereus*, high levels of *Leucocytozoon* blood parasites in the females were associated with reduced clutch size, but only in years of poor food-supply; no such association was apparent in years of abundant food (Korpimäki *et al.* 1993). Following an outbreak among Mourning Doves *Zenaida macrura* of trichomoniasis (caused by a protozoan parasite *Trichonomonas*), many birds died, but some of those that recovered did not breed; the males examined had under-sized testes and the females showed no sign of egg development (Haugen 1952).

Ectoparasites and breeding success in swallows

Barn Swallows and other hirundines suffer greatly from ectoparasites, often having swarms of bugs, ticks and fleas in their nests. These parasites can overwinter in and around the old nest shells, and emerge to re-infest the birds on their return in spring. Over a few years, parasite numbers can build up and, in colonial species, can make the whole site uninhabitable for one or more years. Much work has been done on the Cliff Swallow *Hirundo pyrrhonota* which nests colonially on crags and buildings throughout much of North America, building nests of mud which last for several years. The birds tend to return to the same sites in successive years, but every few years they abandon certain colonies, apparently in response to ectoparasites, especially the Swallow Bug *Oeciacus vicarius*.

An important feature of parasite infestations, mentioned above, is that they can act in a density-dependent manner, as crowding facilitates transmission from one host to another. Among Cliff Swallows, the number of Swallow Bugs per nest and per nestling increased with size of colony and with density of nests within the colony **(Figure 10.2)**. The body mass of nestling swallows at 10 days of age declined significantly as the numbers of bugs per nestling increased and the effects of the parasites on breeding success increased with colony size **(Table 10.4;** Brown & Brown 1986). This provided clear evidence for density-dependence in both the rates and the effects of parasitism. Similar relationships between parasitism and colony size were also noted in Bank Swallows *Riparia riparia* infested with fleas *Ceratophyllus riparius* (Hoogland & Sherman 1976), and in Barn Swallows *Hirundo rustica* infested with blood-sucking blow-fly larvae *Protocalliphora hirundo* (Shields & Crook 1987) or with blood-sucking mites *Ornithonyssus bursa* (Møller 1990).

To find what role ectoparasites had on the growth and survival of Cliff Swallows, Chapman & George (1991) compared the parasite burdens, blood cytology, nestling growth and survival between colonies treated with a pesticide and other untreated colonies in the same area of Texas. Throughout the nestling period, two colonies were sprayed twice weekly with Dibrom®, a short-life acaricide which is non-toxic to birds. Another substance (Stickem) was smeared round the colony edges to reduce the immigration of bugs from outside. Five other untreated colonies acted as controls.

The parasite numbers, as monitored in traps, were significantly reduced in the

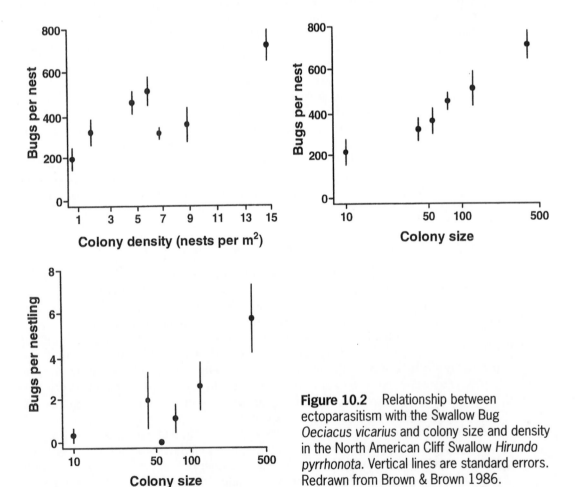

Figure 10.2   Relationship between ectoparasitism with the Swallow Bug *Oeciacus vicarius* and colony size and density in the North American Cliff Swallow *Hirundo pyrrhonota*. Vertical lines are standard errors. Redrawn from Brown & Brown 1986.

treated compared with the untreated colonies **(Table 10.5)**. Blood samples taken from 21-day nestlings had higher haemoglobin and haemacrit values in the treated colonies, giving one of the most direct measures of the influence of the parasites. In contrast, the levels of various leucocytes were higher in birds from untreated colonies, suggesting that they had suffered more from inflammation and bacterial infections. These blood differences were in turn associated with differences in hatching success and in the growth and survival of nestlings. Overall, some 91% of 260 eggs in sprayed colonies gave rise to fledglings, compared with only 24% of 755 eggs in untreated colonies. As expected, most nestling mortality was associated with severe anaemia and infection.

In a further experiment on Cliff Swallows in Nebraska, Brown & Brown (1986) treated half the nests in some colonies, killing the bugs, and left the others as untreated controls. They showed that Swallow Bugs *Oeciacus vicarius* lowered nestling mass and survivorship in large colonies where bug populations were high, but not in small colonies where bug populations were lower **(Table 10.6)**. In contrast, fleas (*Ceratophyllus celsus*) had no obvious effect on nestling body mass or survival, and their numbers showed no relationship with colony size or density. The effects of bug parasitism also appeared to influence post-fledging events, for the proportion of ringed young seen the next year was greater from treated (6.1%) than from untreated

Table 10.5 Effects of treating colonies of Cliff Swallows *Hirundo pyrrhonota* with an acaricide to reduce the numbers of ectoparasites. Asterisks indicate a significant difference between treated and untreated colonies: **P<0.01, ***P<0.001. Figures based on the pooled data from two treated and five untreated colonies. Most aspects of breeding performance, including production of young, were better at treated colonies, with reduced parasite numbers. From Chapman & George 1991.

	Treated		Untreated
Ectoparasites[1]			
Numbers caught per trap night	0.70	**	10.84
Blood characters			
Haemoglobin (g/100 ml)	14.6	**	10.8
% haematocrit[2]	51.5	**	47.3
Erythrocytes[3]	11 697	**	10 602
Leucocytes[3]	314	**	493
Lymphocytes[3]	105	**	213
Breeding success			
Number of eggs	260		755
Number (%) hatched	252 (97)	***	338 (45)
Number (%) fledged	236 (91)	***	182 (24)
Condition of young			
Weight at 23 days (g)	25.0	***	21.3
Primary feather length (mm)	49.3	***	46.1
Rectrix 5 length (mm)	42.2	***	39.8

[1]Main parasites included the Swallow Bug *Oeciacus vicarius* (Hemiptera), the Swallow Tick *Argas cooleyi* (Acari) and the tick *Ornithodoros concanensis* (Acari).
[2]% red cells per unit volume of blood.
[3]Numbers per mm[2] in a blood smear.

nests (3.5%). A later study revealed that young from highly parasitised broods were most likely to move to other colonies to breed than were young from lightly parasitised broods, which more often returned to the natal colony (Brown & Brown 1992). In the following spring, old nests that had been treated were occupied again, whereas other old nests were ignored, even though in some cases they were less than a metre from treated nests **(Table 10.7)**.

These experimental results imply that Cliff Swallows assess the degree of ectoparasite infestation in colonies early in the year, and quickly select the parasite-free nests. Since both bugs and fleas cluster conspicuously at the nest entrance in spring, the birds probably see them and avoid the infested nests. Birds that cannot find clean nests either build new ones or move elsewhere, both of which involve delay, and hence possible reduction in breeding output. The tendency of Cliff Swallows to build new nests more often (rather than re-using old ones) in large colonies than in small ones may be in response to heavier parasite infestations in large colonies. The synchrony in nesting, evident within colonies of this species, may also be a response

Table 10.6   Body mass and brood size at 10 days in fumigated (F) and non-fumigated (NF) Cliff Swallow *Hirundo pyrrhonota* nests in colonies of different size. The effects of parasitism were most marked in the larger colonies. Asterisks indicate a significant difference between fumigated and non-fumigated nests ($*P<0.05$, $***P<0.001$). No Swallow Bugs were seen on nestlings in fumigated nests, but up to 82 per chick were seen in non-fumigated nests. Chicks with five or more bugs on the 10th day were invariably feeble and probably soon died. In the largest colony, bugs accounted for a mean 3.4 g reduction of nestling weight and a 55% reduction in productivity. Brood sizes at hatching were the same in both classes of nests so the difference in brood sizes at fledging was due to differential nestling survival. From Brown & Brown 1986.

Colony size (number of active nests)	Nestling body mass (± s.e.)			Young per nest (± s.e.)		
	F		NF	F		NF
<10	22.0±0.9		20.3±1.1	2.3±0.5		1.1±0.6
43	23.9±0.5		22.1±0.8	2.9±0.3		3.4±0.2
56	24.0±0.5	*	21.5±1.1	1.4±0.3		0.9±0.3
75	23.8±0.3	***	21.1±0.5	2.7±0.3		2.5±0.3
125	24.2±0.3	***	21.0±0.7	2.4±0.1	***	1.4±0.1
345	23.7±0.2	***	20.3±0.3	3.1±0.1	***	1.4±0.2

Main parasites included the Swallow Bug *Oeciacus vicarius* and the flea *Ceratophyllus celsus*.

Table 10.7   Nest usage by Cliff Swallows *Hirundo pyrrhonota*, according to whether or not the nest had been fumigated to kill ectoparasites the previous summer. Results combined from two colonies. Old nests that had been fumigated were more often re-used. From Brown & Brown 1986.

	Fumigated	Non-fumigated
Old nests present	146	249
Number (%) used	125(86)	3(1)[1]

[1]Two of these nests were adjacent to the line dividing the fumigated and non-fumigated halves of a colony.
Main parasites included the Swallow Bug *Oeciacus vicarius* and the flea *Ceratophyllus celsus*.

to parasites, because any birds that nested late would face the onslaught of a much enlarged parasite population. In these and other swallows that raise more than one brood per year, parasites tend to build up during the course of the season, so that later broods are less successful than earlier ones (Shields & Crook 1987).

The longer-term impact of parasitism on a Cliff Swallow colony was shown experimentally in Wisconsin, where the normal shifting pattern of colony use was broken by management (Buss 1942). Every year the old nests were scraped from the barn, thus removing the parasites. This colony flourished without break for about 50 years and grew to a spectacular metropolis of 4000 pairs. When in 1951 the resident farmer failed to remove the old nests, the colony disbanded early, the birds deserting their nests en masse and leaving many half-grown young to starve (Emlen 1986). At the time, the whole surface of the nesting area was 'swarming with ticks and fleas'. The next year the swallows arrived in early April, and after making passes at the old parasite-ridden nests still present, they moved to a neighbouring farm. On 21 April, after nest-building had started on this other farm, Emlen scraped off all the old nests on one side of the old barn and sprayed part of the scraped area with insecticide. On his return a week later, the sprayed portion of the old colony had been densely reoccupied, while the unsprayed patch had not. This experiment demonstrated the effect that ectoparasites had on the tendency of birds to settle, and hence on colony stability and growth.

To judge from this and other studies, the presence of high parasite infestations can induce colony abandonment or discourage colony establishment at otherwise acceptable sites (Foster 1968, Brown & Brown 1986). In effect, then, in any one year parasitism reduces the number of colony sites available. Where potential sites are superabundant, this need have no effect on overall breeding numbers. But where potential colony sites are in short supply, parasitism could restrict the distribution and numbers of breeding pairs, by preventing in any one year the use of some areas that were otherwise suitable. By this indirect means, nest parasites could help to limit the breeding numbers of Cliff Swallows and other hirundines. This has not been demonstrated, however, so remains hypothetical.

Among Barn Swallows, mites also seemed to affect mating success. Males with high infestations more often remained unmated and had fewer extra-pair copulations than other males (Møller 1990, 1991b, 1993). The situation was complicated because males with many parasites had shorter tails than other males, and tail length was also related to mating success. The tail lengths of individual males changed from year to year according to parasite loads, and males that had high loads in one year showed a reduction in tail length the next. Whether females were responding to tail length or parasite loads, or using tail length as an index of parasite loads, was unclear. None-the-less, ectoparasite loads were clearly associated with reduced mating success in male Swallows and with reduced nesting success in both sexes. The findings did not reveal whether these effects on individual performance were reflected in reduced breeding numbers, so no effect on subsequent population level could be assumed.

Anti-parasite behaviour

In view of the effects of ectoparasites on nest success, it is not surprising that anti-parasite behaviour is well developed in birds (Møller et al. 1989). Infestations are reduced partly by the direct removal of parasites from the plumage and nest, and by routine preening, bathing and dust-bathing, or by anting in which the plumage is treated with formic acid (Simmons 1966). Nest sanitation also helps, including removal of faecal sacs, ejection of faeces over the nest rim and frequent renovation of nest material (Bucher 1988). The nest-lining used by some birds may reduce parasite loads, several favoured plant materials having repellent or pesticidal properties (Clark & Mason 1985). Some tropical birds nest near wasps and bees, which among other benefits, help to reduce the risk from parasitic diptera. Moreover, some species prone to nest parasitism by cuckoos or cowbirds have various behaviours, from mobbing to egg rejection, to reduce the risk.

Observers operating nest-box schemes have often noticed that birds prefer new boxes to old ones. As in the Cliff Swallows discussed above, parasite avoidance may be one reason for this preference. Fleas will jump on to almost any birds approaching an infested box, and Du Feu (1992) found that the presence of simulated fleas at the entrance of a nest-box reduced the frequency of prospecting visits by birds. In British Columbia, Tree Swallows *Tachycineta bicolor* preferred empty and clean boxes or those where the old material had been microwaved (to kill parasites) over those with old untouched material (Rendell & Verbeek 1996). While many hole-nesters have no choice but to re-use nest-sites, other birds generally build themselves a new nest in a different site for each attempt. This may be yet another anti-parasite strategy. Birds that get parasites from their food are sometimes able to recognise parasitised prey individuals and reject them. Thus Oystercatchers *Haematopus ostralegus* can distinguish *Macoma balthica* with high trematode loads (Hulscher 1982). Further, individuals of some bird species prefer to mate with unparasitised partners when given the choice. In addition to the swallows mentioned above, female pigeons *Columba livia* in captivity preferred louse-free males (Clayton 1990).

*   *   *

Studies on various bird species have thus demonstrated the effects of parasitism on breeding success, and by inference also on the size of the post-breeding population. They could not, however, reveal whether parasites helped to limit breeding numbers. Lowered production of young may have had no impact on subsequent breeding numbers if it was offset by improved survival of remaining young, as would be expected, for example, if breeding numbers were limited by resources.

The instances discussed are extreme examples, however, and in most birds that use different sites for each breeding attempt, deaths from ectoparasites are almost unknown. The importance of ectoparasites may have been underestimated in those well-studied species that use nest-boxes, because most observers clean out the old nests each year, and in the process reduce the parasite levels (Møller 1989b). This applies to many studies of tits, flycatchers and sparrows, but not to Starlings *Sturnus vulgaris* which remove the old nests themselves (Feare 1982). None-the-less, among birds as a whole, parasitism is clearly a much less widespread cause of breeding failure than is predation.

Adult mortality

Studies of adult mortality in birds have given similar results to those on breeding success, that is: in most species parasites and disease seem to be a minor cause of deaths, but in some species they are of over-riding importance (**Table 7.2**; Itamies *et al.* 1980, Dobson & May 1991). In waterfowl, for example, diseases probably account for the largest proportion of non-hunting deaths (Bellrose 1980). Particularly in North America, numerous epizootics have been recorded in which 25 000–100 000 birds have died, and smaller kills of 5000–10 000 birds are fairly commonplace. In some other bird species disease outbreaks appear every year, but do not necessarily kill every affected individual. One such disease is puffinosis, which appears every summer among Manx Shearwaters *Puffinus puffinus* nesting at certain Welsh colonies, chiefly in areas of dense ground vegetation which harbour ticks. This disease is caused by a 'virus-like' agent probably transmitted by ticks, and is evident from the blisters that appear on the feet, mainly in fledglings (Dane 1948, Harris 1965, Nuttall *et al.* 1982). Again, it is unknown whether such fledgling mortality affects subsequent breeding numbers.

Studies of diseases in full-grown birds typically rely on the pathological examination of carcasses. Some studies result from specific epizootics, and serve mainly to confirm the cause, while others result from more general surveys intended to assess the relative frequency of different mortality causes. The latter need to be treated with caution (**Table 7.2**, Chapter 7). For one thing, they usually depend on birds found dead by people, which are unlikely to represent a random cross-section of deaths. Secondly, the proportion of deaths attributable to parasites varies strikingly with the thoroughness of the examination procedure; many studies do not test for microparasites, especially viruses. Thirdly, if predators are present, they may remove many heavily parasitised individuals before they die, further reducing the perceived importance of parasites (for examples of predators selectively removing parasitised individuals from a population, both among birds and among other animals, see van Dobben 1952, Holmes & Bethel 1972, Vaughan & Coble 1975, Temple 1987). Fourthly, the fact that many macroparasites show markedly aggregated distributions means that the small proportion of heavily infected hosts could easily escape a sampling programme. Lastly, even if a large proportion of deaths is attributable to parasitic disease, this seldom tells us anything conclusive about the effects of disease on population levels, because of possible interactions with other mortality factors. The mere presence of parasites in a carcass does not necessarily imply that they were the primary cause of death.

Nor is it safe to infer much about the impact of macroparasites from the numbers found in individual hosts (Toft 1991). Because most parasites that kill their hosts also kill themselves, the more virulent the parasite, the lower the mean number of parasites expected per host, or the lower the proportion of hosts infected at one time. So the presence of few parasites or diseased individuals in a host population could mean that the parasites are a trivial cause of mortality, or that the parasites are so virulent that they are continually removing hosts (and themselves) from the population. It could also mean that transmission is low.

Experiments have proved more revealing than post-mortem studies about the effects of parasites on adult birds. For example, chewing lice (Mallophaga) eat the feathers of birds and can therefore affect insulation. They are normally kept at low numbers by preening, but can be removed completely by fumigation, enabling their effects to be assessed. Among feral pigeons in captivity, louse-infested birds had higher metabolic rates than cleaned ones, presumably in response to greater heat loss; and in the wild fumigated pigeons showed higher overwinter survival than sham-fumigated ones (Booth *et al.* 1993). In another experiment, survival of hen

Ring-necked Pheasants *Phasianus colchicus*, dosed with an anthelminthic, was about 10% higher in the three months (March–May) following treatment than that of untreated controls (Woodburn 1995). Treated hens also hatched more young. These birds contained various gut helminth parasites, most commonly *Heterakis*. For reliable assessment of the impact of parasites on population levels, however, additional information is required, as discussed in the following section.

EFFECTS OF PARASITES ON POPULATIONS

At the population level, parasites can show similar relationships with their hosts as predators with their prey, causing either:

(1) no obvious effects on breeding numbers,
(2) a permanent lowering of numbers below what would otherwise occur,
(3) regular fluctuations in numbers
(4) irregular fluctuations in numbers, associated with periodic epidemics, or
(5) declines to extinction.

The circumstances that might produce these various patterns have been elucidated by mathematical modelling (Anderson & May 1978, May & Anderson 1978, Hassell & Anderson 1989), and can be illustrated by particular case-histories from the field, as described below.

No obvious effects on host breeding numbers

This is probably a common situation, but the evidence is indirect: namely that all bird species that have been studied harbour some parasites, yet in many species deaths from parasitism or disease are almost unknown, in either chicks or adults. In some recent studies, for example, protozoan blood-parasites were found to have no obvious effects on the body condition, survival or breeding success of various passerines and raptors (Bennett *et al.* 1988, 1993, Ashford *et al.* 1990, Weatherhead & Bennett 1991, 1992, Tella *et al.* 1996), so at the levels found could not have affected their population levels. Possibly these parasites might not always be so benign, if the hosts were exposed to other debilitating influences at the same time, but the studies concerned produced no evidence for this view. Secondly, despite harbouring parasites, many bird species are known to be limited in breeding numbers primarily by other factors, such as food-supply, nest-sites or predation (Chapters 7–9; Newton 1980, 1993, 1994), often involving territorial behaviour which leads to the exclusion of a surplus of non-breeders (Chapter 4; Newton 1992). Any mortality they might suffer from parasites is seemingly trivial in the determination of breeding numbers. Parasites might cause some deaths which are inevitable for other reasons. Thirdly, many generalist parasites have probably been introduced into island avifaunas along with introduced birds, yet only rarely have they been implicated in population declines of their new hosts (see below).

Lowering of host breeding numbers

Mathematical models have revealed how a parasite–host interaction might lead to a stable equilibrium, in which the infection rate remains constant. If the parasite kills hosts in a density-dependent manner, the host population could be regulated at a level lower than might occur in the absence of the parasite. To my knowledge, no such stable, closely-coupled interactions between a specialist parasite and its host have yet been described from birds, but they may well exist. However, one example of population limitation by a generalist parasite involves the louping ill flavi-virus, which affects the central nervous system of Red Grouse *Lagopus l. scoticus* and other animals. The virus occurs patchily in the British uplands and is transmitted between hosts by the Sheep Tick *Ixodes ricinus*. Although the ticks may feed on a wide range of hosts, only sheep and Red Grouse are known to produce viraemias (virus in the blood) which exceed the threshold necessary for efficient transmission (Reid *et al.* 1978, Duncan *et al.* 1979). Thus louping ill is usually considered as a two-host one-vector system. The mortality that occurs in Red Grouse in areas where louping ill is endemic is sufficient to reduce stocks to low densities, but continuing infection of the grouse is dependent on the presence of sheep or other alternative hosts and ticks as vectors.

Although louping ill will kill adult grouse, its main effect is on chicks. In an experimental study of captive birds, Reid (1975) found that 78% of 37 experimentally infected juveniles developed clinical signs of weakness and died in six days. Comparative studies on wild grouse in northern England indicated that, in areas with ticks and louping ill, chick survival was about half the level found in areas with ticks and no louping ill, or in areas with neither ticks nor louping ill (Hudson & Dobson 1991). In areas where ticks were abundant, 84% of adult grouse had antibodies to louping ill, indicating previous infection (Reid *et al.* 1978). After vaccination, sheep no longer act as a reservoir host, which in theory provides a means of eradicating the disease from both sheep and grouse. This appears to have happened following a vaccination programme on several moors in northern England, but not following a similar programme on moors in Scotland (Hudson & Dobson 1991). The reason for the difference could be attributed to non-viraemic transmission between ticks that were feeding close together on the same host, in this case Mountain Hare *Lepus timidis*, which was absent from the English moors. The virus thus passed from vector to vector even though this host species did not become viraemic (Jones *et al.* 1997). Red Grouse may be susceptible to louping ill because they have not been exposed to this or similar viral disease during their evolution. Perhaps the virus has only recently been introduced (presumably in sheep) to their moorland habitat (Reid *et al.* 1980).

Cyclic fluctuations in host numbers

Both micro- and macroparasites can induce oscillations in host abundance, through acting in a delayed density-dependent manner. The best studied example concerns the Red Grouse as affected by the disease strongylosis, caused by a nematode worm *Trichostrongylus tenuis* (Lovat 1911, Hudson *et al.* 1992a,b). In its moorland habitat, the Red Grouse is the only host for this parasite. In some areas, the worms are found in the gut caeca of nearly every grouse examined, often causing little harm, but when present in large numbers weakening and sometimes killing the host. The eggs of the parasite pass out of the gut with the caecal faeces. The larvae hatch and, after some development, they climb to the tips of heather plants, which form the main food of Red Grouse. In this way they pass from one grouse to another. Active larvae can usually be found on the heather plants from June to September, when the worm burdens of grouse increase. By the end of the year the ingested larvae do not develop into the adult stage, but remain arrested within the grouse, developing into adults in the following spring (Shaw 1988). The adult worms themselves are long lived, probably as long as the grouse, so infection is cumulative. The infection intensifies year by year as the numbers of grouse rise, until eventually the population crashes, with a consequent reduction in the numbers of the parasite. Population cycles are thus thought to be produced by the effects of nematode parasitism acting mainly on the fecundity and survival of the grouse in a delayed density-dependent manner.

The impact of these parasites on the body weight, breeding and survival of grouse was examined in field experiments conducted over several years (Hudson 1986, Hudson et al. 1992b). The basic techniques were to catch grouse at random and treat half the birds with an anthelminthic to reduce worm burdens and the remaining birds with water as a control and then to monitor their subsequent performance. Overwinter weight gain was greater in treated birds than in control birds, as was subsequent clutch size, hatching success and chick survival (Table 10.8). Replication of the experiment in several other areas gave similar results, confirming that parasitism had a major influence on breeding success. Moreover, in the main study area, treated birds were significantly more likely to survive from spring and be shot in the following autumn, and less likely to be found dead in the interim, than were untreated birds. Grouse found dead on the study area in spring generally had high parasite burdens.

The experiment clearly showed that parasites reduced the productivity and survival of individual grouse. On its own, it could not show whether parasitism affected population levels, rather than merely influencing which individuals succumbed in a population ultimately limited by other factors. It none-the-less

Table 10.8 Effects of treating individual Red Grouse *Lagopus l. scoticus* with an anthelminthic drug to reduce nematode parasite numbers. Significance of difference between treated and untreated birds determined by one-tailed *t* test: *$P<0.05$, **$P<0.01$. Sample sizes in parentheses. Treated Grouse were heavier, and produced more young, than untreated ones. From Hudson 1986.

	Treated		Untreated
Male weight (g)[1]	784 (72)	*	680 (124)
Female weight (g)[1]	639 (100)	**	618 (57)
1982 Clutch size	8.3		7.9
Hatching success	97%		77%
Brood size[2]	5.5	*	3.6
1983 Clutch size	8.0	**	5.3
Hatching success	75%	*	38%
Brood size[2]	3.2	**	0.5
1984 Clutch size	8.8	*	7.6
Hatching success	96%		92%
Brood size[2]	7.4	**	4.9

[1]Five months after treatment
[2]At ten days

provided a crucial strand of evidence for the hypothesis of population limitation by parasites. In fact, this nematode–grouse system has the three conditions that in combination can generate population cycles, namely (1) parasite-induced reduction in host breeding success, (2) parasites distributed with a low degree of aggregation (almost random distribution) within the host population, and (3) time-delays in recruitment to the adult parasite population owing to larval arrestment (Anderson & May 1978). Given these conditions, mathematical models have shown that a simulated population would fluctuate with appropriate periodicity, with peaks every five years or so (Potts *et al.* 1984, Hudson & Dobson 1991, Dobson & Hudson 1992).

This experimental work also provided evidence for a link between parasitism and predation (Hudson 1986, Hudson & Dobson 1991, Hudson *et al.* 1992b). On several moors, grouse killed by predators had higher worm burdens than grouse that survived through the summer, to be shot in autumn. In addition, grouse that appeared to have died from the effects of parasites carried greater worm burdens than grouse killed by predators. During incubation, hen grouse are thought to produce less scent than usual, making it harder for mammalian predators to find their nests. The scent probably arises from the caecum which stops producing caecal faeces at that time. Wild hens that were dosed with an anthelminthic to reduce their worm burdens were found less often (15% of finds) by trained dogs hunting by scent than undosed hens (85% of finds). In contrast, for human observers hunting by sight, the figures were 45% and 55%. Hudson suggested that, compared with treated hens, heavily parasitised hens emitted more scent, making them more vulnerable to predation. If this was so, parasitism was the predisposing factor to death by predation. Parasites might also reduce the competitive ability of individual grouse, for birds without territories generally had higher worm burdens than those with territories (Jenkins *et al.* 1963).

Grouse disease may only have reached the importance it has because of predator control, which allows the grouse to achieve densities at which parasite transmission is enhanced. Thus, in recent studies, the presence of many heavily parasitised birds in the population was associated with a scarcity of predators. Mathematical modelling revealed how predators, by selectively removing the most infected grouse, could allow an increase in grouse density: they effectively blocked the regulatory role of the parasites, and stopped the cycle in grouse numbers (Hudson *et al.* 1992b). Even if predation merely reduced the post-breeding density of the grouse, it could greatly hamper the spread of the parasite. Low levels of predation were sufficient to achieve this effect in the model, but at higher levels, predation was itself enough to reduce grouse density to an even lower average level, despite the scarcity of parasites.

Population cycles involving strongylosis are known among Red Grouse chiefly on wetter, western moors. On drier moors parasite numbers are too low to kill many grouse. This has been attributed to poorer survival of the free-living stages of the parasite on drier moors, for in the laboratory their survival depends on humidity as well as on temperature. On drier moors, therefore, grouse numbers may cycle from some different cause or may simply fluctuate irregularly. One feature of parasite involvement is the marked crashes in numbers, so that the population typically falls from peak to trough in 1–2 years. Elsewhere the cycles are longer and more symmetrical, as both the rise and fall take place progressively over several years.

Clearly, although parasites might contribute to cycles in some areas, they are not responsible for cycles in all areas. It remains possible that their main effect is to induce crashes at high density, and thereby shorten a cycle that would still occur, but with longer periodicity, from different causes. This research on the grouse–nematode relationship is one of the few examples where the combination of field observations and experiments, coupled with the use of mathematical models, has indicated that pathogens can have a significant impact on the population dynamics of an avian host, in this case contributing to regular cycles in host abundance.

The Red Grouse is not the only gallinaceous bird that is susceptible to strongyle worms. Years ago, local outbreaks of *T. tenuis* were reported in Grey Partridges *Perdix perdix*, causing occasional population crashes, especially in wet years favourable to the free-living stages of the parasite. Their absence in recent decades has been attributed to the massive decline in Partridge densities, which are now well below the level necessary for effective transmission of the parasite (Smith 1913, Portal & Collinge 1932, Potts 1986). Low levels of infection with *T. tenuis* have also been found in Bobwhite Quail *Colinus virginianus* in North America (Moore *et al.* 1986). Another strongyle worm, *Syngamus trachea*, which causes 'gapes', is also frequent in Partridges, but is thought to cause only limited mortality (Potts 1986). A third species, *T. gallinae*, is found in various gallinaceous birds and doves, and is known to have killed several thousand Mourning Doves *Zenaida macroura* in the southeastern States in summer 1950 (Stabler & Herman 1951).

Notwithstanding the grouse example, cyclic changes in host–pathogen abundance are most likely to arise within microparasite–host associations. I know of no instances from birds, but examples from other animals include the two-year cycles in measles and three-year cycles in whooping cough that occurred in human populations prior to mass vaccination, and the recurrent rabies epidemics in European Red Fox *Vulpes vulpes* populations that occur every 3–5 years (Anderson *et al.* 1981).

Irregular fluctuations in host numbers

Irregular declines in the numbers of some birds are caused by periodic outbreaks of parasitic diseases, which kill large numbers of individuals in successive single events **(Table 10.9)**. Such epizootics have been recorded in a wide range of species, but particularly in waterfowl which are often crowded in large concentrations, facilitating transmission. The most important of such diseases include duck plague and avian cholera.

Duck plague results from a herpes virus infection, also called duck virus enteritis (DVE), which is found throughout the northern hemisphere, usually in domestic ducks. Strains of the virus vary in virulence, and not all waterfowl are equally susceptible. It attacks the vascular system, usually causing death within 14 days of infection (Wobeser 1981). The chief symptoms include haemorrhages throughout the body and diphtheritic membranes on the oesophagus and elsewhere. Infected birds excrete the virus, so transmission occurs through contact between individuals or with a contaminated environment. The virus can remain latent in carrier birds for up to four years before they begin to excrete it again at times of stress (Burgess *et al.* 1979). The first recorded epizootic in wild waterfowl occurred during the 1972–73 winter, when a single outbreak killed an estimated 42% of the 100 000 Mallard *Anas*

Table 10.9 Some examples of large-scale disease outbreaks in full-grown birds.

Disease	Agent	Host species	Locality (years)	Estimated numbers killed	Estimated population decline where known	Source
Bacterial						
Avian Cholera	*Pasteurella multocida*	Various waterfowl	Muleshoe Refuge, Texas (1956–57)	>60 000	–	Jensen & Williams 1964
		Various waterfowl	California (1965–66)	70 000	–	Rosen 1971
		Diving Ducks	Maryland (1970)	>4700	–	Locke et al. 1970
		Mainly Diving Ducks	Maryland & Virginia (1978)	31 000	–	Montgomery et al. 1979
		Mallard *Anas platyrhynchos*, Snow Geese *Chen caerulescens* and others	Missouri (1963–64)	7000	6% of Snow Geese	Vaught et al. 1967
		Various waterfowl	Texas (1947–48)	36 000	–	Petrides & Bryant 1951
		Various waterfowl	Texas (1949–50)	>4000	–	Petrides & Bryant 1951
		Various waterfowl	Various sites, Texas to Manitoba (1979–80)	>100 000	–	Brand 1984
		Eider *Somateria mollissima* females	Maine (1970–76)	633	–	Korschgen et al. 1978
		Eider *Somateria mollissima* females	Quebec (1964)	–	25%	Reed & Cousineau 1967
		Eider *Somateria mollissima* females	Quebec (1966)	–	17%	Reed & Cousineau 1967
		Eider *Somateria mollissima* females	Holland (1984)	80–100	–	Swennen & Smit 1991

Table 10.9 Continued.

Disease	Agent	Host species	Locality (years)	Estimated numbers killed	Estimated population decline where known	Source
Bacterial continued.						
Avian cholera	Pasteurella multocida	Cape Cormorants Phalacrocorax capensis	SW Africa (1991)	14 500	8%	Crawford et al. 1992
Fowl Cholera	Pasteurella sp.	Coots Fulica americana and others	California (1948–49)	Several thousands	–	Rosen & Bischoff 1949
Salmonellosis	Salmonella sp.	Mute Swan Cygnus olor	Switzerland (1964)	25	—	Bouvier & Hornung 1965
Avian diphtheria	Diphtheriae columbarum	Woodpigeon Columba palumbus	Southern England (1906–08)	>4000	–	Ticehurst (1908–11)
Streptococcal infection	Streptococcus zooepidemicus	Eared Grebes Podiceps nigricollis	Great Salt Lake, Utah (1977)	7500	–	Jensen 1979
Botulism	Clostridium botulinum	Various ducks	Great Salt Lake, Utah (1932)	250 000	–	Jensen & Williams 1964
		Various ducks	Western United States (1952)	4–5 million		Locke & Friend 1987
		Various ducks	California (1954–70)	1.5 million	–	Hunter et al. 1970
		Various waterfowl	Tulare Lake, California (1938–45, mainly 1940–41)	250 000	–	McLean 1946
		Various ducks	London, England (1968)	400	40%	Keymer et al. 1972

	Pathogen	Host	Location	Number	%	Reference
Protozoal						
Trichomononiasis	*Trichomonas gallinae*	Mourning Dove *Zenaida macroura*	Southeastern United States (1950–51)	Many thousands	–	Stabler & Herman 1951, Haugen 1952
Erysipelas	*Erysipelothrix rhusiopathie*	Eared Grebes *Podiceps nigricollis* and others	Great Salt Lake, Utah (1975)	5000	–	Jensen & Cotter 1976
Fungal						
Aspergillosis	*Aspergillus fumigatus*	Mallard *Anas platyrhynchos*	Colorado (1975–76)	Several hundreds	–	Adrian & Spraker 1978
		Wood Duck *Aix sponsa*	Illinois (1945)	>100	–	Bellrose et al. 1945
Viral						
Avian influenza	Avian influenza *A. virus*	Common Tern *Sterna hirundo*	Cape Province, South Africa	1300+	–	Rowan 1962
Duck plague (DVE)	*A herpesvirus*	Mallard *Anas platyrhynchos*	Lake Andes, South Dakota (1972–73)	42 000	42%	Friend & Pearson 1973
		Canada Geese *Branta canadensis*	Lake Andes, South Dakota (1972–73)	270	3%	Friend & Pearson 1973
Worms						
Acanthocephalan	*Polymorphus boschadis*	Eider *Somateria mollissima*	Finland (1931–38)	–	33%	Lampio 1946, Grenquist 1951
	Profilicollis botulus	Eider *Somateria mollissima*	Scotland (1955)	290	–	Garden et al. 1964
	Polymorphus botulus	Eider *Somateria mollissima*	Massachusetts (1958)	'Large numbers'	–	Clark et al. 1958

platyrhynchos wintering at the Lake Andes National Wildlife Refuge in South Dakota (Friend & Pearson 1973). Mortality peaked at 1000 birds per day in late January, when the population was concentrated on a limited area of open water. In contrast, only 3% of 9000 geese were affected, probably because they fed on land nearby.

Avian cholera (or pasteurellosis) is caused by the bacterium *Pasteurella multocida*. It is a highly infectious disease capable of killing birds within 4–6 hours after infection, though 4–9 days is more common. Typically, the birds die in good condition. External symptoms are virtually absent and the chief signs are small haemorrhages on the body surface and pinpoint lesions on the liver, heart and other organs. Transmission is primarily by inhalation or ingestion of the bacterium. The disease has been recorded in more than 100 bird species, but waterfowl are again its most obvious victims (Botzler 1991). More than 60 000 birds died at the Muleshoe Refuge in Texas during 1956–57 (Jensen & Williams 1964), and an estimated 70 000 died in California in 1965–66 (Rosen 1971). Even larger outbreaks, involving diving ducks, occurred in Chesapeake Bay in 1970 and 1978 (Montgomery *et al.* 1979). During an outbreak in California, affecting mainly American Coots *Fulica americana*, the disease spread to waterfowl and rodents, resulting in deaths of owls and Hen Harriers *Circus cyaneus* (Rosen & Morse 1972). Several severe epizootics of avian cholera have been recorded among breeding Eiders *Somateria mollissima*, especially among incubating females using stagnant pools for drinking. Two epidemics among a population of 4000 in Quebec killed 25% and 17% of birds, respectively (Reed & Cousineau 1967), while a similar epidemic on the Dutch island of Vlieland killed 80–100 females, having effects on nest numbers and age structure that were still apparent five years later (Swennen & Smit 1991).

In 1979–80, avian cholera occurred among waterfowl in North America at various points on the migration route from Hudson Bay in the north to the Texas coast in the south (Brand 1984). Several species of ducks and geese were affected, and the numbers killed at particular sites varied from a few individuals to an estimated 72 000. From the timing of outbreaks at different sites, the migrant birds seemed to have carried the disease from place to place along the migration route. Outbreaks occurred at some of the same sites in later years.

From a review of existing information on avian cholera, Botzler (1991) emphasised that: (1) mortality of single individuals (or small groups) is reported frequently, from a wide variety of locations; (2) epidemics occur only where waterfowl are concentrated, as on refuges; (3) epidemics typically begin in one host species (usually a common one), then spread to other species, using the same area; and (4) epidemics may be related to water quality (especially organic content) and other environmental conditions. Perhaps the biggest recorded incident involving species other than waterfowl killed 14 500 Cape Cormorants *Phalacrocorax capensis* off South Africa. The birds were short of food at the time (Crawford *et al.* 1992).

Epidemics involving macroparasites have also occurred among waterfowl. On the Baltic Islands of Finland, Eider Duck numbers increased between 1920 and 1930, but in the breeding season of 1931 many females and young died from infection with the acanthocephaline worm *Polymorphus boschadis*. The numbers of Eiders then rose again, reaching their former level by 1933–34. They were then reduced by a second epidemic in 1935, which continued less severely for the next three years, by which time breeding numbers were only two-thirds and in some places only one-third of what they had been (Lampio 1946, Grenquist 1951).

Waterfowl are also often infected with cestodes (tapeworms), as well as with trematodes which have molluscs as intermediate hosts. Of the 475 trematode species that had been identified in waterfowl, only 32 have been found to cause pathological changes or death (McDonald 1969). As an example, dabbling ducks have often died on the St. Lawrence river marshes in Canada during summer, as a result of infection with the trematode *Cyathocotyle bushiensis* (Gibson *et al.* 1972). Heavy burdens were acquired by the ducks feeding day after day on infected snails, resulting in severe caecal damage. Of the 264 species of cestodes found in waterfowl, none is normally fatal, except when infestation is so great as to block the gut (Grenquist *et al.* 1972, Persson *et al.* 1974), or when coupled with malnutrition or other debilitation, accounting in one instance for the deaths of 50 Mute Swans *Cygnus olor* (Jennings *et al.* 1961).

Many other epizootics among wild birds have been recorded, with examples from among game birds, pigeons, gulls, petrels and others **(Table 10.9)**. In many of these species, however, study of other limiting factors suggests that disease cannot be more than sporadically important. Similarly, in some species subject to long-term population studies, numbers suddenly and markedly declined, only to recover again in subsequent years (e.g. an unprecedented 70% decline in a Florida Scrub Jay *Aphelocoma coerulescens* population, Woolfenden & Fitzpatrick 1991). Although no disease was identified, an epidemic was suspected.

Our understanding of what precipitates irregular epizootics is poor, but not all follow the same pattern. Some result from an unexplained appearance of a particular disease agent, others from a disease agent that normally remains latent but that becomes fatal when the birds are exposed to some other stress such as food-shortage, or to crowding or other conditions that facilitate transmission. Thus some diseases that are not serious in the wild can kill many captive birds kept at high density in the same place for long periods. An example from poultry is Newcastle disease, caused by a paramyxovirus. Yet other epizootics have been attributed to a genetic change in a disease organism itself: to the evolution of a new virulent strain, or of a strain that can switch from one host species to another previously unsuited to it. Whatever their origin, most epizootics tend to be sporadic, affecting different localities in different years, but some have spread widely, with devastating impacts on populations over much of the range, myxomatosis in Rabbits *Oryctolagus cuniculus* providing an example.

Mass mortality caused by natural toxins. Other forms of mass mortality occur in birds from time to time, as a result of toxins produced by micro-organisms. The well known 'duck sickness' or botulism is not a parasitic disease, but a form of paralytic food-poisoning caused by a neuro-toxin from the bacillus *Clostridium botulinum*, mostly Type C strain (Kalmbach & Gunderson 1934, Sciple 1953). The anaerobic *Clostridium* is almost ubiquitous in nature, persisting in spore form for years; it grows well on organic matter in shallow stagnant waters or mud during warm weather, and occurs in most parts of the world. Deaths from botulism have been noted in about 70 bird species in 21 families, but mostly in waterfowl and gulls. The chief clinical feature is a flaccid paralysis which prevents the birds from flying. Often the wings droop and the head sags to one side (hence the term 'limber neck') and the eyelids may become encrusted. Birds die from respiratory failure or from drowning and may appear in good condition with no internal lesions (Locke & Friend 1987). Rows of carcasses, which coincide with receding water levels, often mark the aftermath of large-scale kills.

Year-to-year losses from botulism are highly variable but can be spectacular. An estimated million birds died from botulism at a lake in Oregon in 1925, 1–3 million at Great Salt Lake in Utah in 1929, and 250 000 at the northern end of Great Salt Lake in 1932, and 250 000 (25%) out of two million birds on Tulare Lake in California in 1938–41 **(Table 10.9)**. Many other outbreaks have involved smaller numbers of birds (Manuwal 1967, Smith *et al.* 1983), but totalled over wide areas or over a period of years they can be substantial, for example the 4–5 million waterfowl deaths attributed to botulism in the United States in 1952 (Smith 1975), or the 1.5 million in California alone during 1934–70 (Hunter *et al.* 1970). The organism multiplies, and the toxin concentrates, in the corpses of poisoned birds. These may then be eaten by raptors, which may in turn be affected, although Turkey Vultures *Cathartes aura* have shown unusual resistance (Kalmbach 1939). Deaths are not always concentrated but may sometimes be scattered over large areas wherever conditions are suitable: a survey during the hot summer of 1975 revealed more than 4000 birds (mostly gulls) that had died of botulism around Britain (Lloyd *et al.* 1976).

Although botulism is not infectious, its effects are exacerbated by droughts which cause waterbirds to concentrate in larger than usual numbers on remaining shallow wetlands. The birds become poisoned when they ingest the toxin-producing bacteria in the bodies of invertebrates that form their food, including the maggots living on rotting vegetation and carcasses. Once started, the dying waterfowl themselves become hosts for more maggots and perpetuate the epizootic. The outbreak comes to an end when temperatures cool, when birds leave the site or switch to other foods, when flies stop breeding, or when water levels stabilise (Wobeser 1981). The impact of botulism can be so substantial that after each severe outbreak the populations concerned could take several years to recover their numbers. Frequent outbreaks could thus cause populations to remain for much of the time below the level that would otherwise occur. Little wonder that, in North America, botulism has been said to constitute 'the greatest drain upon western waterfowl due to any single natural agency' (Kalmbach & Gunderson 1934), a view later echoed by Bellrose (1980).

Another type of mass mortality, attributed to a toxin, is the so-called paralytic shellfish poisoning, which occurs from time to time in seabirds in various parts of the world. The neuro-toxin concerned is produced by dinoflagellate species (*Gonyaulax*), whose presence in abundance turns the sea surface brown or red, giving rise to 'red tides'. The dinoflagellates are consumed by plankton-feeding fish or invertebrates, especially mussels, which concentrate the poison, passing it up the food chain to birds. The toxin acts by blocking the production of nerve impulses, leading to muscle paralysis and, at higher doses, to respiratory failure. Some species, notably Shag *Phalacrocorax aristotelis*, which get the toxin via contaminated fish, are much more sensitive than others, such as Eiders *Somateria mollissima*, which get the toxin directly from mussels. Some gulls can detect and avoid the toxin by regurgitating contaminated prey (Kvitek 1991). In a major incident off northeast England in summer 1968, so many Shags died that an island breeding population under study dropped by about 82%. The lesser mortality seen in other species was insufficient to cause marked reductions in their breeding numbers (Coulson *et al.* 1968). Another incident was recorded in the same area in 1975, by which year the Shags had recovered their numbers. This time the population was reduced by 63%, and again many deaths of other species were recorded too (Armstrong *et al.* 1978). Red tides have caused substantial mortality in Common Guillemots *Uria aalge* and divers *Gavia* spp. off the western coast of North America (McKernan & Schaffer 1942), and of Common Terns *Sterna hirundo* and other species off the eastern coast (Nisbet 1983). In addition, paralytic shellfish poisoning has been suggested as a factor limiting Black Oystercatcher *Haematopus moquini* populations in southern Africa, with one outbreak reducing numbers in several areas by about one half (Hockey & Cooper 1980). The causes of the dinoflagellate blooms are not well understood, but they are probably related to some combination of nutrient availability and water temperature. They may have been the cause of the Biblical statement 'and all the waters that were in the river were turned to blood' (Exodus 7:19–21).

Birds can be affected by other toxic marine algae. For example, in 1991 an outbreak of the diatom *Nitzchia occidentalis* in Monterey Bay, California, resulted in the accumulation of domoic acid in zooplankton. These were eaten by small fishes which then poisoned large numbers of fish-eating birds, especially Brown Pelicans *Pelecanus occidentalis* and Brandt's Cormorants *Phalacrocorax penicillatus*, along with some humans who ate contaminated shellfish (Fritz *et al.* 1992). Other large scale mortalities have been recorded in birds from time to time, but again without their cause being identified, one of the largest involving 150 000 Eared Grebes *Podiceps nigricollis* which died in the Salton Sea, California in 1992 (Jehl 1996).

Decline to extinction

Like other animals, birds may have little resistance to a pathogen with which they have had no previous contact. A disease may reach new areas as a result of natural spread of the parasite or its vectors, or through inadvertent introduction by humans. In this context, alien diseases may have contributed to the decline in the numbers of certain endemic landbirds on oceanic islands, restricting the distribution of some and annihilating others. The best documented examples are from Hawaii, where the release of infected non-native birds, together with mosquitoes (present since 1826), led to the establishment of avian malaria and avian pox (Warner 1968). These parasites are only mildly virulent to the introduced birds, with which the pathogens have had long evolutionary associations. But they often kill the indigenous Hawaiian birds, which have only recently been exposed to these diseases and have little or no natural resistance. The evidence is strongest for malaria *Plasmodium relictum capistranoae*, which, at low and medium elevations where mosquitoes occur, is readily transmitted from introduced to native birds (van Riper *et al.* 1986). This has led to a considerable reduction in the range and numbers of many endemic bird species, which are now mainly restricted to high or dry areas where mosquitoes are scarce or absent. In fact, Warner (1968) contends that the extinction of roughly half the indigenous landbirds of the Hawaiian Islands since their discovery by Europeans in 1778 was due mainly to introduced diseases. Parts of the islands below 1500 m, where mosquitoes are commonest, are now occupied almost entirely by resistant introduced bird species.

Deductions from field observations were supported by experiments in which various Hawaiian birds in captivity were inoculated with malaria. Many individuals of the endemic species died, whereas all the introduced ones survived (**Table 10.3**). All of the experimental birds had malarial parasites in their blood and those that died were found on necropsy to have extremely high levels, but no other obvious parasites or diseases. In a survey of the occurrence of malarial *Plasmodium* in various bird species caught in mist nets or found dead, the infection rate of different species ranged from 0 to 29% (van Riper *et al.* 1986). Interestingly, birds struck by cars had a significantly higher prevalence of malaria than did birds caught in mist nets. Debilitation resulting from the disease may thus have increased the likelihood of deaths from another cause. Similarly, more birds than expected had both pox and malaria than had one or other disease alone. This could have been because infection with one disease lowered resistance to another, or because such twin-infected birds had received greater attention from mosquitoes, which transmit both.

Although the evidence is largely correlative, it appears that both the elevational distributions of the native birds and their relative abundances in different (xeric and mesic) habitats are influenced to a major extent by the malarial parasite. Because the species differ in their susceptibility to both diseases, the extent of this influence varies among species. The introduced species, being largely resistant to both diseases, are distributed independently of them and reach greatest densities in the lowland mesic forests where food is most abundant. Levels of parasitism are not the only factor influencing distributional patterns, however, as habitat changes and introduced predators are also involved. Disease has been suggested as a contributory cause of extinctions of endemic birds on other oceanic islands, but without serious evidence.

OTHER CONSERVATION PROBLEMS

Unfamiliar parasites

One of the most significant, yet largely unexplored, impacts of humans on diseases in wild animals is likely to stem from the massive changes in animal distributions that have followed from human activity, whether from habitat changes or deliberate translocations of species. As on Hawaii, this has brought species in contact with new parasites, to some of which they have little or no resistance. Disease outbreaks have often occurred in domestic poultry, following attempts to establish flocks in areas where particular parasites were present. Epizootics of the protozoan blood parasite *Leucocytozoon simondi* occurred among domestic geese introduced to subarctic Quebec (Laird & Bennett 1970), while *L. smithi* affected turkeys introduced to South Carolina (Kissam *et al.* 1975).

Some of the most striking examples of the consequence of exposure to a new parasite concern the Brown-headed Cowbird *Molothrus ater* in North America. This brood-parasite lays its eggs in the nests of a wide range of passerine host species, thereby reducing their own production of young (Mayfield 1977). It is a bird of open country, originally confined to the grasslands of the mid-continent. But after deforestation, it has spread to occupy most of the continent (Mayfield 1977, Brittingham & Temple 1983). It thereby gained access to new host species from about 50 in pre-colonial times to more than 200 today (Mayfield 1965). Many of these new hosts have no innate defences, and will not desert or eject strange eggs, but raise the resulting young. Because Cowbirds use many hosts, moreover, they are not vulnerable to a population decline in any one, and could in theory parasitise favoured species to extinction, while maintained at high density by other common hosts. Extinction could occur if Cowbirds parasitised enough nests year after year to reduce the host reproductive raté below the threshold necessary to balance adult mortality (May & Robinson 1985).

During the 1960s, Cowbirds parasitised up to 70% of nests of the rare Kirtland's Warbler *Dendroica kirtlandii*, reducing the production of young warblers to less than one per nest (Mayfield 1983, Walkinshaw 1983). This was fewer than needed to maintain the population level and during 1961–1971 the warbler population declined by 60%. Then, with the start in 1972 of a campaign to remove Cowbirds, parasitism dropped below 5%, warblers fledged about 3 young per nest and their breeding numbers stabilised (Kelly & DeCapita 1982). Later they increased (trebled) when the habitat expanded as a result of a forest fire (Rothstein & Robinson 1994), but to maintain Kirtland's Warbler in the long term, a continuing programme of Cowbird control may prove necessary. Similarly, in the Least Bell's Vireo *Vireo v. belli*, one population in southern California fell to fewer than 20 pairs, with a 47% rate of parasitism in the early 1980s, but increased to at least 250 pairs after Cowbird removal was initiated in 1983 (Rothstein & Robinson 1994). Elsewhere in California, this vireo has disappeared altogether. On the other hand, the removal of Cowbirds from Mandarte Island off British Columbia led to an improvement in the nest success of Song Sparrows *Melospiza melodia*, but to no increase in subsequent breeding density which was determined by other factors (Smith 1981). Thus, of three experimental removals of Cowbirds, all three led to improved breeding success in the host species, but only one led to an unequivocal increase in breeding numbers.

Local studies of many other songbird species have revealed variable high rates of nest parasitism **(Figure 10.3)**. While these rates do not necessarily reveal the impact of parasitism on total seasonal success (because of some re-nesting attempts), some species seem no longer to be producing enough young to offset known adult mortality. In the White-crowned Sparrow *Zonotrichia leucophrys* in California, parasitism increased from 5% to nearly 50% of nests in 15 years, along with a rise in Cowbird numbers (Trail & Baptista 1993). In the Sierra Nevadas, local numbers of the heavily parasitised Warbling Vireo *V. gilvus* were inversely correlated with Cowbird abundance, suggesting a depressive effect of the brood-parasite on host population levels (Verner & Ritter 1983). Other species, which may have declined because of recent contact with this generalised brood-parasite, include the Golden-cheeked Warbler *Dendroica chrysoparia* in parts of Texas, and the Black-capped Vireo *Vireo atricapillus* in parts of Oklahoma (Grzybowski *et al.* 1986).

North American songbirds might soon be subject to another brood-parasite, namely the Shiny Cowbird *Molothrus bonariensis*, which is spreading northwards from South America via the West Indies. It is already established in Florida, and has appeared as far north as Maine (Post *et al.* 1993). In the West Indies, some local species are aggressive to the Cowbird and suffer little parasitism, but others lack defences and are highly vulnerable. On Puerto Rico, the Yellow-shouldered Blackbird *Agelaius xanthomus* suffered up to 94% nest parasitism and a massive population decline (Post & Wiley 1977, Cruz *et al.* 1985). In contrast, the closely-related Yellow-hooded Blackbird *A. icterocephalus* on Trinidad, where the cowbird has long been present, sustained only 45% parasitism. The two cowbirds provide some of the most telling recent examples of what can happen when a parasite meets new potential hosts, with little evolved resistance. As with all parasite–host associations, persistence of declining host species could depend on whether they can

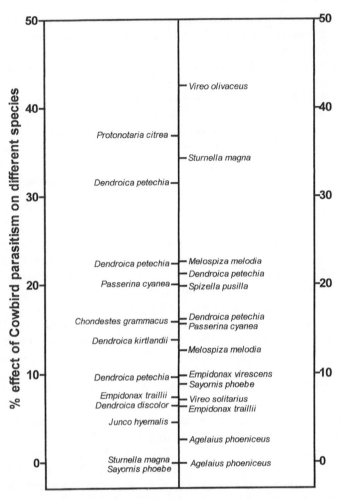

Figure 10.3 Percentage reduction in breeding success of various North American songbirds caused by the brood-parasite Brown-headed Cowbird *Molothrus ater*. Effect calculated as % nests parasitised x % difference in success between parasitised and unparasitised nests. Based on data assembled by Payne (1997), in which references to the original species studies may be found.

achieve some new lower level of stability (as could happen if parasitism were density-dependent), or alternatively evolve some measure of resistance before they die out. No other brood-parasites are known to have such devastating effects on other species, presumably because none has come into recent contact with new hosts.

Exposure to an unfamiliar disease has also caused mortality and population decline in some mammal populations. The most extensive epidemic ever recorded in humans was bubonic plague, caused by an insect vector-borne bacterium, *Yersinia pestis*, which killed nearly 20 million Europeans in the 14th century. Other epidemics killed millions of New World people in later centuries after their exposure to Old

World pathogens. The disease rinderpest, which is benign to its ancient cattle hosts, killed 95% of Wildebeest *Connochaetes taurinus* and Buffalo *Syncerus caffer* in East Africa in the 1890s (O'Brien & Evermann 1988); canine distemper killed about 70% of the last known population of Black-footed Ferrets *Mustela nigripes* in North America in the 1980s (Thorne & Williams 1988), and myxomatosis devastated European Rabbit *Oryctolagus cuniculus* numbers·in the 1950s, as did haemorrhagic viral disease in the 1980s and 1990s. With myxomatosis, the Rabbits are much less affected now than when the disease first arrived, implying that over time disease virulence and host resistance have co-evolved, lessening the impact of the disease. Some of these diseases had a major impact not only on their affected hosts but also on other organisms with which the hosts interacted, notably their food-plants and predators.

Diseases in Endangered Species

In widespread populations, disease often comes in a shifting pattern of local outbreaks, influenced partly by local conditions, including host densities. Particular localities might lose their populations temporarily, only to be subsequently recolonised by unaffected individuals from elsewhere. But as the distribution of populations becomes ever more reduced through human action, and ever more restricted to reserves, the chances of an epidemic wiping out an entire species become greater. The cautionary tale of the Black-footed Ferret, mentioned above, illustrates two epidemiological features that are common to endangered species. Firstly, because such species have small populations, they are themselves unlikely to sustain infections of virulent pathogens, acquiring them only from other more common host species. Secondly, because most individuals in the population are never exposed to a pathogen, there is little acquired immunity. In these circumstances, when an epidemic does occur, it can kill a large proportion of individuals. The reduction of any small population through disease, whether in the wild or in captivity, can have the additional effect of genetic impoverishment, with the risk that the population is less able to withstand future challenges, including the next disease. Increasingly, the question will arise whether we should aim to preserve endangered species in natural association with their parasites and pathogens, or whether we should try to keep them – like humans in developed countries – unnaturally free from infectious diseases (May 1988). The latter will almost certainly be necessary to get some rare species through the bottleneck of low numbers.

One frequent effect of bringing endangered species into zoos or reserves is to increase their population density, thereby increasing the efficiency of disease transmission. Another effect, especially in captive populations, is to bring species from different parts of the world together so that some are exposed to new diseases. This is a frequent cause of mortality in captive parrots, as susceptible and resistant carrier species come into contact. Many infectious agents have been implicated, including viruses, bacteria, fungi and protozoa (Gaskin 1989). In several other bird species, captive stocks maintained for conservation reasons have been heavily reduced in numbers and genetic variance by disease outbreaks, often contracted through contact with other more resistant species housed on the same site, examples including various cranes and the Mauritius Pink Pigeon *Columba mayeri* (Snyder *et al.* 1985). The deaths from mosquito-borne eastern equine encephalitis virus of seven out of 39 rare Whooping Cranes *Grus americana* housed at the Patuxent Wildlife Research Center in Maryland contributed to a decision to move half the flock elsewhere (Dein *et al.* 1986). Other dangers come from the release of captive individuals which can ferry novel pathogens back to wild populations that lack immunity (Wilson *et al.* 1994), or from the translocation of infected wild individuals from one locality to another, which can thereby extend the range of a pathogen.

In addition, the threat of new diseases is ever present. Soon after people began to domesticate wild animals, they unwittingly shared with them the mix of pathogens from which new potent diseases evolved. Smallpox has been traced back to cowpox, measles probably evolved from rinderpest or canine distemper, and influenza from ancient pig diseases. One way in which new viral diseases arise occurs when, within the same animal, two different viruses invade the same cell, and during propagation, exchange genetic material with one another in a process of 'reassortment'. They can then produce a new virus which, in some instances, might be more virulent than either parent form, or able to infect a new host resistant to both parent forms. Moreover, the production of different strains of virus by a mixture of mutations and selection can occur on a timescale that is faster than any co-evolutionary response in the host. On an alternative view, however, most emerging diseases are not new, just 'freed from obscurity by acts of man' (Morse 1990). Whatever their origin, it is hard to evaluate the threat posed by new or 'emerging' diseases (Morse 1991). What is certain is that the increasing scale of movement of people around the globe, and the associated trade in wildlife and wildlife products, will result in bird and other animal populations being increasingly brought into contact with new pathogens or old pathogens in new settings. Two of the current major killing diseases of waterfowl in North America, avian cholera and duck plague, may have originally come from domestic poultry imported to the continent. Both diseases were first identified in poultry in the 1880s, and in wild duck populations in 1944 (avian cholera) and 1979 (duck plague) (Baldassarre & Bolen 1994).

Other conservation problems have arisen from domestic animals acting as reservoirs of infection for more vulnerable wild species. A high mortality occurred in monkeys (*Presbytis entellus and Macaca radiata)* after cattle were introduced into forest in India (Boshell 1969). The cattle increased the numbers of a biting tick and hence the rate of infection of the flavi-virus causing Kysanur Forest Disease. On the reciprocal view, agriculturalists often view wildlife as a reservoir of infection for domestic stock; and a programme of Badger *Meles meles* control in parts of Britain was aimed at reducing the incidence of tuberculosis in cattle.

CONCLUDING REMARKS

With so few case-histories, it is impossible to guess at the full ecological significance of parasites in bird populations. Existing information would suggest that, in most bird species, parasites cause little mortality among adults or young and are much less prevalent as a major limiting factor for breeding numbers than are predation or food-shortage. Many species might experience the occasional epidemic, after which their numbers take several years to recover fully, and in waterfowl in some parts of North America epizootics might be so frequent as to assume the role of main limiting factor. These incidents are likely to increase in future, as human impacts on habitat force ever greater numbers of waterfowl to concentrate on refuges, or remain there for longer periods. The disease louping ill can depress Red Grouse *Lagopus l. scoticus* populations in some areas and strongylosis is involved in the cyclic fluctuations of Red Grouse numbers in other areas. The present level of both diseases is a result of human action in increasing the numbers of alternative hosts (louping ill), or in decreasing the numbers of predators (strongylosis). Moreover, while ectoparasites can reduce the breeding output of swallows and others, this does not necessarily reduce the subsequent breeding numbers, especially as such species are often limited by a shortage of nest-sites (Chapter 8).

Because disease is often associated with other stress factors, such as starvation or crowding, which might predispose it, much of the mortality it causes is not necessarily additive to other mortality. Moreover, the effects of one parasite species might well vary according to the number of other parasites the host contains, with multiple infections having greater impact than single infections, an aspect largely unexplored. The immune system of birds can also be suppressed by certain pollutants, increasing susceptibility to disease. Mallard experimentally dosed with PCBs, DDT, dieldrin or selenium showed reduced resistance to a duck hepatitis virus (Friend & Trainer 1970, Whiteley 1989), while Mallard exposed to petroleum oil had decreased resistance to the avian cholera bacterium *Pasteurella multocida*. In other experiments, many chemicals that occur in the modern environment have also reduced the resistance of laboratory animals to viral, bacterial or macroparasite infection. The list of implicated toxicants includes organophosphate, carbamate or organochlorine pesticides, halogenated aromatic hydrocarbons and heavy metals, such as arsenic, cadmium, lead, mercury and zinc (Whiteley 1989). Avian mortality from this source could rise as environments become increasingly contaminated (Chapter 15).

In this chapter, I have been concerned primarily with the effects of disease on bird numbers, but disease organisms (like predators) could also restrict bird distribution patterns, preventing certain species from occupying large areas of otherwise suitable habitat. We would normally have no way of knowing this, except in special circumstances, as exemplified on Hawaii. Disease organisms are also likely to have marked effects on the evolution and genetic make-up of host populations (Price 1980). Throughout history, many animal and plant species have been continually afflicted with devastating disease outbreaks, leading to selection for disease resistance, and in extreme cases which left few survivors, leading to marked reductions in genetic variance. The importance of maintaining genetic diversity with respect to disease defence is indicated by certain natural populations which have reduced genetic variability and apparent increased vulnerability to infectious disease (O'Brien & Evermann 1988).

Despite the sparsity of information on disease impacts on bird populations, the parallels with predation are striking. These include similar dynamic interactions between populations, the ability of generalist parasites, maintained by alternative hosts, to depress or even annihilate populations of scarcer but more susceptible host species, and the importance of heterogeneity in host exposure and susceptibility to infection in influencing the dynamics of host–parasite associations. As yet, however, much more uncertainty surrounds the effects of diseases than of predators on populations. This is in no small measure due to the greater difficulties of studying disease impacts.

As with predation, moreover, the best way to assess reliably the impact of parasites is to remove them in some way, and measure the response in the host population. Now that such work has started on wild populations by use of pesticides or drugs, the doors are open to further experimental work which should improve our understanding of disease impacts. There is an obvious need for experiments that examine the effects of parasites on populations and not merely on individual survival and breeding success. Those involving Cowbird removal from areas occupied by target songbird host species, or strongyle worm removal from Red Grouse, are a step in this direction (Smith 1981, Kelly & DeCapita 1982, Hudson *et al.* 1992b, Rothstein & Robinson 1994).

SUMMARY

In understanding interactions between parasites and their hosts, a useful distinction is between microparasites (viruses, bacteria, fungi and protozoa) and macroparasites (helminths and arthropod ectoparasites). Microparasites are responsible for most recorded epizootics which periodically cause large-scale reductions in bird numbers. Macroparasites occur more often as persistent infections which reduce breeding or survival rates, but in a few species cause sudden massive reductions in numbers. In some species, notably various swallows and others that nest in the same sites year after year, the build-up of ectoparasites can cause substantial reduction in breeding success and occasional desertion of colonies. Avian brood-parasites can also reduce the breeding success and subsequent breeding numbers of newly-exposed host species which have little or no natural resistance.

As with predators, parasites can have the following main effects on bird breeding densities, depending on circumstances:

(1) cause no obvious change (e.g. various species with certain blood-parasites);
(2) hold breeding density below the level that could otherwise occur (e.g. Red Grouse with louping ill);
(3) cause periodic massive reductions in numbers, either regular (e.g. Red Grouse with strongyle worms) or irregular (e.g. waterfowl with avian cholera); and
(4) cause decline to extinction (e.g. some Hawaiian bird species with avian pox and malaria).

Some major impacts have occurred when bird species, mainly through human action, have come into contact with generalist parasites to which they have no evolved resistance. Species can be driven to low densities or even extinction by parasites that are maintained at high densities by alternative, non-susceptible hosts. The role of parasites in bird population dynamics can change with changes in the density of (1) the host and parasite species themselves; (2) alternative host species, and (3) any necessary vector species, all of which influence the ease of transmission. They can also alter with changes in other environmental conditions, which affect infection and transmission rates. The impact of diseases on wild populations could increase in future, as a result of various human activities.

Waterfowl in snow storm.

Chapter 11
Weather

Climatic extremes can have major effects on bird numbers. Two main types of effect can be distinguished: firstly, the prolonged periods of regional severity, such as hard winters or droughts, which affect most bird species by reducing food availability; and secondly, the sudden episodic events, such as hurricanes or hail storms, which can kill many birds directly, as well as damaging their habitats and food-supplies. Longer-term effects of weather on bird populations are discussed in the context of climatic change in Chapter 15.

Unlike other limiting factors, extreme weather events, whether brief or prolonged, tend to occur unpredictably, but with certainty. They often (but not always) act in a density-independent manner, and regardless of initial population size, they can result in the deaths or breeding failure of a large proportion of individuals. The population size can then take up to several years to recover. Where such events occur in successive years, they can result in stepwise population declines, and in extreme cases to local extinction. In effect, much limitation of bird populations can take place on infrequent occasions, and during what might be considered as atypical conditions; and with long recovery periods, extreme weather events can have a big influence on mean population sizes.

In any one region, birds could in theory be influenced by any climatic extreme, whether of heat or cold, wet or dry, wind or fog. At high latitudes, however, the most influential weather factor is temperature, and marked declines in bird numbers often occur during hard winters. At lower latitudes, the dominant weather factor in many regions is rainfall, and marked declines in bird numbers are often associated with prolonged drought. Moreover, dry periods may increase the risk of fire, which can have pronounced effects on habitats, affecting bird populations for years to come.

Compared with most other types of animal, birds are well suited to resist the usual changes in temperature and rainfall. They are warm-blooded, well insulated, relatively waterproof and mobile. By their behaviour, they can avoid extremes, temporarily seeking shelter or flying long distances to avoid them, or can conserve heat by clumping and roosting together. They also have some behavioural control over thermoregulation – for example, by sleeking or fluffing their body feathers. For these reasons, birds seldom suffer immediate physical damage from weather extremes (but see later), and most effects are indirect, acting through habitats and food-supplies. Moreover, because each species is adapted to its normal environment, losses due to climate are caused mainly by weather that is unusual for the region concerned.

Direct effects result partly because, as homeotherms, birds must maintain a high and constant body temperature. Within a certain range of ambient temperatures (the thermoneutral zone), the cost to the bird of maintaining body temperature is minimal. But in air temperatures below this zone, birds must burn more energy to keep warm; and in temperatures above this zone, they must spend energy (and water) to keep cool by the evaporative chilling of moist tissues. Similarly, most birds can withstand some wetting of their feathers (some aquatic birds more than most), but torrential rain can soak their plumage, so that its insulating air-layer is lost and the bird may die of heat loss (Kennedy 1970). Young birds are most vulnerable to weather extremes before they can thermoregulate, when they mostly rely on their parents to keep them warm, cool or dry.

To some extent birds can protect themselves from transient weather extremes by the use of shelter, so as to keep dry and to minimise the metabolic costs of high or low temperatures. In cold weather, for example, they might (1) select warmer places within their habitat, (2) reduce energy-demanding activities not concerned with feeding (such as display), (3) increase the time spent feeding or, if the metabolic costs of feeding at low temperatures exceed the gains from food obtained, they may (4) cease feeding altogether and rest. The latter is often seen in larger birds, such as waterfowl, but can only be a short-term tactic, as it depends on using body reserves. When severe conditions persist, forcing birds to feed in the open, their survival and reproduction may be affected, so that populations decline. In addition, birds on migration may encounter thick fog which can hinder their navigation, or unfavourable winds which can slow their progress or blow them off course. Such conditions can cause heavy losses among migrants, especially over water, and such losses are sometimes manifest in reduced breeding densities (see below).

GENERAL WEATHER EFFECTS

The foraging success of some bird species is so much influenced by weather that annual changes in relevant climatic variables can promote annual changes in survival, breeding and population levels **(Figures 11.1** and **11.2)**. In certain species almost every aspect of performance seems to be influenced by weather. Among Peregrine Falcons *Falco peregrinus* in Australia, for example, the timing and duration of egg-laying, clutch size, nest success and breeding density were all negatively related to rainfall at the time, so that the total production of the population was lowest in the wettest years **(Figure 11.2)**. In this species, the effects were mediated, not so much through reduced hunting success, as through the wetting of their cliff-ledge nest-sites. Those pairs with sheltered nest-sites bred as well in wet years as in dry ones, whereas those with exposed sites often failed in wet years because the site flooded. So many pairs had poor sites that their failure affected markedly the mean performance of the population (Olsen & Olsen 1989). In other raptors, wet weather affected breeding by reducing the food-supply, suppressing hunting behaviour or prey availability. Negative relationships between breeding and rainfall over a number of years have been noted in a wide range of European species, including Kestrel *Falco tinnunculus*, Peregrine *Falco peregrinus*, Buzzard *Buteo buteo*, Honey Buzzard *Pernis apivorus*, Red Kite *Milvus milvus*, Sparrowhawk *Accipiter nisus*, Goshawk *Accipiter gentilis* and various harriers (Cavé 1968, Schipper 1979, Davis & Newton 1981, Newton 1986, Mearns & Newton 1988, Kostrzewa & Kostrzewa 1989, 1991, Village 1990).

Not all bird species in an area are affected in the same way by a particular weather extreme. In Europe, unusual wetness in summer benefits worm-eating species, such as Blackbird *Turdus merula* and Rook *Corvus frugilegus*, which probe for food more easily in damp ground (Dunnet & Patterson 1968), but the same weather can be disastrous for species such as Swifts *Apus apus*, that raise their young on adult insects. In this species, the food-shortage resulting from prolonged cold and wet can lead to delay in egg-laying and incubation, to egg desertion, small clutches, poor nestling growth and starvation, in some years in northern Europe leading to almost total reproductive failure (Koskimies 1950). The development of many invertebrates is temperature-dependent, with insects and others completing their life-cycles more rapidly in warm conditions. This affects the periods that foods, such as caterpillars, remain available for breeding birds, and hot summers can thereby lower the breeding success of certain species.

In winter at high latitudes, ground-feeding species are affected by snow lying for long periods, and twig-feeders mainly by glazed frosts. For most songbird species, prolonged snow cover seems to have more effect on numbers than low temperature and frost (Greenwood & Baillie 1991). Among shorebirds, some species are affected by the freezing of mud flats, which prevents them from probing for food, and others by strong winds which interfere with surface feeding (Dugan *et al.* 1981). Different responses to the same conditions provide one reason why different bird species in the same region do not necessarily fluctuate in synchrony from year to year. Moreover, many species may be more vulnerable to poor weather in years when their food is scarce, being better able to withstand extremes in years when food is plentiful (for Nuthatch *Sitta europaea* see Nilsson 1987, Matthysen 1989; various tits see Perrins 1979). Hence, the same species may be affected in some years of severe weather but not in others.

The influence of weather cannot always be judged by its effects at the time. In temperate regions aerial insect-feeders breed most successfully in warm dry summers, but if every summer were dry, the lack of rain would in time reduce their

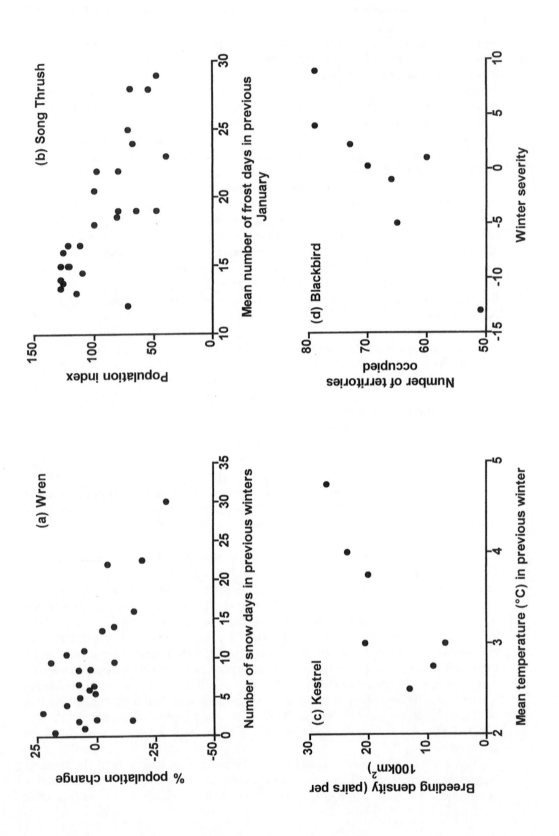

food-supplies, with eventual effects on nest success. Similarly, the small chicks of many wader and gallinaceous species survive best in dry weather, but they depend on rainfall in preceding months to provide their insect food (Moss 1986).

The Golden Eagles *Aquila chrysaetos* that nest in the deserts of Israel provide an example of the delayed effect of rainfall on breeding. Over a 17-year period, the mean number of young raised per pair in any one year was correlated with rainfall in the preceding year. High rainfall led to good plant growth, which in turn promoted good reproduction by hares and other herbivores which the eagles ate. No effect of rainfall was noted in two other areas where the food-supply remained more stable: one included irrigated farmland and in the other the eagles fed mainly on lizards whose numbers were less affected by rain (Bahat 1989). The same sort of effect was noted among Wedge-tailed Eagles *Aquila audax* in western Australia (Ridpath & Brooker 1986). Similarly, in the Western Kingbird *Tyrannus verticalis* in North America, annual variations in production of young were correlated with insect abundance, which was dependent on rainfall in the preceding year (Blancher & Robertson 1987). Weather can also affect birds in less obvious ways, such as by influencing the availability of mud which some species need for their nests.

PROLONGED SEVERE WEATHER

Although birds are continually affected by weather, it is the extreme events that cause the most marked population declines. Such events may last from less than an hour (certain storms) up to several years (prolonged droughts) but in general, the more extreme the event, the greater the range of species affected.

Hard winters

At high latitudes, the effects of hard winters on bird populations are plain for all to see. Typically, at the start of a cold snap, large-scale movements occur, as birds of

Figure 11.1   Examples of weather effects on the spring population levels of some resident birds. Each point refers to a different year.
(a) Relationship between the extent of population change between successive breeding seasons for Wrens *Troglodytes troglodytes* in woodlands and number of snow days in the intervening winter, England (r=0.71, P<0.01). From Greenwood & Baillie 1991.
(b) Relationship between population level of Song Thrushes *Turdus philomelos* on British farmland each spring and the number of freezing days in the preceding January, given as an average from 20 weather stations (r=0.75, P<0.001). From Baillie 1990.
(c) Relationship between breeding density of Kestrels *Falco tinnunculus* and the mean temperature of the preceding winter ($r = 0.51$, P<0.05). From Village 1990.
(d) Relationship between spring numbers of Blackbirds *Turdus merula* in part of the town of Lund, south Sweden, and the severity of the preceding winter weather (as measured by the deviation of the mean temperature during November–March from the long-term mean for the period) (r_s=0.86, P<0.01). From Karlsson & Källander 1977.

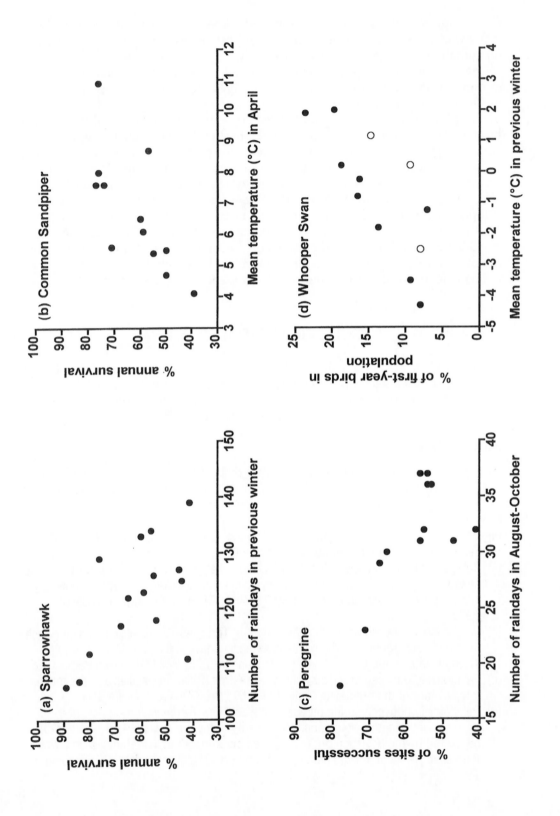

(a) Sparrowhawk — % annual survival vs Number of raindays in previous winter

(b) Common Sandpiper — % annual survival vs Mean temperature (°C) in April

(c) Peregrine — % of sites successful vs Number of raindays in August–October

(d) Whooper Swan — % of first-year birds in population vs Mean temperature (°C) in previous winter

several species leave for warmer climes, while others from even colder regions move in. As the cold continues, birds become increasingly concentrated wherever food remains available, and more than usual appear around houses and sea-coasts, preferring sheltered and other warmer sites. Many birds become tamer than usual, as they sacrifice safety for food; they forage for longer each day, and some normally nocturnal species, such as the Barn Owl *Tyto alba*, may hunt by day. Waterbirds are sometimes found frozen to the ice on ponds, and landbirds to the branches on which they have roosted; while others have ice on their feet or plumage. Many more birds than usual are found dead or dying, and by the following spring several resident species appear scarcer than usual (Jourdain & Witherby 1918, Ticehurst & Hartley 1948, Dobinson & Richards 1964, Cawthorne & Marchant 1980, Greenwood & Baillie 1991). Moreover, as weather severity increases with elevation, impacts on resident bird populations are generally greater on higher ground (for Barn Owl *Tyto alba* see **Figure 11.3**).

In western Europe, winters with unusually severe weather (above level 40 in **Figure 11.4**) occurred 14 times in the 90 years up to 1990 at intervals of 1–15 years, and four of these (1940, 1942, 1947, 1963) were extremely severe (above level 70 in **Figure 11.4**). These winters varied in the duration of frost and snow periods, in the lowest temperatures reached, in snow depths, in the numbers of glazed frosts, and in the extent to which saline habitats froze, all of which influenced the range of bird species affected. Not surprisingly, prolonged unbroken periods of frost or snow had more impact on bird populations than did several shorter periods, interspersed with mild spells. Mortality was especially evident in various songbirds, shorebirds and waterfowl, and hardly at all in some others, such as crows (see also Errington 1939, Trautman *et al.* 1939 for North America).

The ability of any species to withstand a cold snap depends partly on body size, as larger species have relatively more reserves, and need to devote a smaller proportion of their daily energy intake to keeping warm (because of their reduced

Figure 11.2   Examples of weather effects on bird survival and breeding success.
(a) *Winter weather and survival.* Relationship between annual survival of Sparrowhawks *Accipiter nisus* and winter (October–April) rainfall, south Scotland (logistic relationship: y (%) = 100 exp $(4.31-0.033r)/[1 + \exp(4.31-0.033r)]$, $P<0.05$). From Newton *et al.* 1993b.
(b) *Spring weather and survival.* Relationship between annual survival of Common Sandpiper *Actitis hypoleucos* and temperature in the latter half of April, soon after arrival in breeding areas, England, 1977–89 ($y = 0.074x + 0.40$, $r = 0.82$, $P<0.001$). Birds suffered higher mortality than usual in years when snowstorms occurred in the days following their arrival, so that breeding densities in these years were reduced. From Hollands & Yalden 1991.
(c) *Spring weather and breeding.* Relationship between breeding success of Peregrines *Falco peregrinus* and the number of raindays in the breeding season, Australia ($y = 101.60-1.42x$, $r = 0.76$, $P<0.01$). From Olsen & Olsen 1989.
(d) *Winter weather and breeding.* Percentage of first-winter birds among Whooper Swans *Cygnus cygnus* in south Sweden in January 1967–78 shown in relation to mean temperatures for the preceding December–March in south Sweden ($r = 0.76$, $b = 2.05$, $P <0.01$). Open circles denote winters after a cold spring in the Siberian breeding areas. From L. Nilsson 1979.

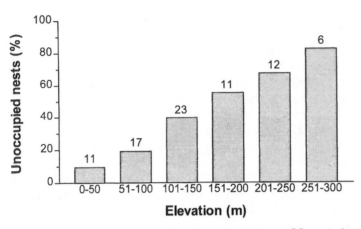

Figure 11.3   The loss of breeding Barn Owl *Tyto alba* pairs at 80 nest-sites in southern Scotland following the severe winter of 1978–79. Those at the highest elevations suffered the greatest losses. The numbers of nests are shown above the columns. From Taylor 1994.

surface area to volume ratio) (Blem 1990, Meininger *et al.* 1991). With increasing body size, therefore, the safety margin for survival without feeding increases **(Figure 11.5)**. In winter, small birds are unlikely to last more than a day without food, medium-sized species (of around 300 g) might last two or more days, and large geese and swans could last for two or more weeks (Boyd 1964). Although few birds lose all their food in hard weather, these figures give some idea of possible survival periods. We can therefore expect that short periods of hard weather will kill mainly

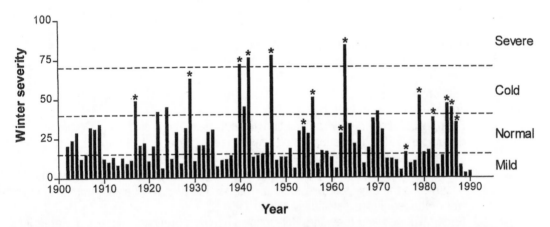

Figure 11.4   Winter severity in western Europe, as measured by temperature records from the Netherlands. Asterisks show winters in which large-scale bird mortalities were recorded. The mortality in 1976 occurred during a short cold snap in a generally mild winter. Winter severity (V) was calculated by the equation $V = 0.000\,275v^2 + 0.667ij + 1.111z$ where v = number of frost days (with minimum temperature below 0°C), ij = number of ice days (with maximum temperature below 0°C) and z = number of very cold days (with maximum temperature below −10°C). Redrawn from Meininger *et al.* 1991.

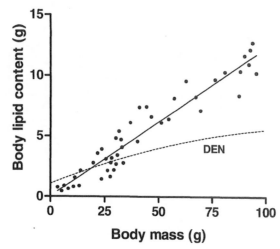

Figure 11.5 Mean lipid weight in relation to total body weight among wintering songbird species. Each dot refers to a different species. The dashed line shows the lipid equivalent of the daily energy need (DEN, assuming 1 g of lipid yields 37.7 kJ), and indicates that the safety margin of surplus fat increases with body size. In other words, larger species could survive for longer without feeding than small ones (some of which can survive some degree of reserve deficit by nocturnal torpidity). Modified from Blem 1990.

small birds, but that as the cold continues, larger species will form an increasing proportion of the casualties. To judge from birds found dead, small birds usually lose about a third of their body weight during starvation, while larger birds (thrushes to waterfowl) can lose up to a half, by which time practically all their body fat has been used, together with some body protein, as shown by emaciated pectoral muscles (Harris 1962, Hope Jones 1962, Marcström & Mascher 1979). In such cold conditions, birds die either from starvation (that is, when they have insufficient energy reserves to meet the demands of heat production) or from hypothermia (that is, when they cannot metabolise their reserves fast enough to meet these demands). Wind chill presumably contributes to hypothermia. When the weather improves, small birds can recover their body condition within a few days, much more rapidly than larger ones.

In line with these facts, small resident bird species generally show the largest population declines after hard winters (**Figure 11.6**; Graber & Graber 1979). Diet also has a major influence on survival prospects, because certain types of food become less available than others during cold and snow. In general, ground-feeders fare worse than tree-feeders, and, among tree-feeders, insect-eaters fare worse than seed-eaters. Thus, the most marked population declines recorded in Britain after hard winters were in small insectivorous birds, notably Wren *Troglodytes troglodytes*, Long-tailed Tit *Aegithalos caudatus*, Tree-creeper *Certhia familiaris*, Stonechat *Saxicola torquata*, Grey Wagtail *Motacilla cinerea* and Pied Wagtail *Motacilla alba,* and the localised Bearded Tit *Panurus biarmicus* and Dartford Warbler *Sylvia undata*. Two successive cold winters (1961–62 and 1962–63) reduced the numbers of Dartford Warblers, restricted to a few parts of southern England, by 98% (457 to 11 pairs, Bibby, cited in Gibbons *et al.* 1993).

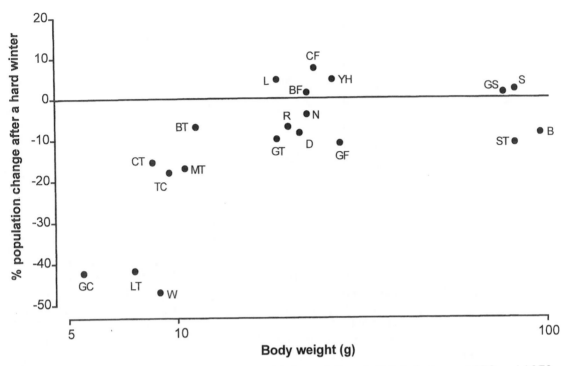

Figure 11.6   Percentage changes in woodland bird populations in Britain between 1978 and 1979 (a hard winter), shown in relation to the body weights of the species concerned. More species decreased than increased, and the most marked decreases occurred in three species of lowest body weight. Redrawn from Cawthorne & Marchant 1980.

Key: B – Blackbird *Turdus merula*, BF – Bullfinch *Pyrrhula pyrrhula*, BT – Blue Tit *Parus caeruleus*, CF – Chaffinch *Fringilla coelebs*, CT – Coal Tit *Parus ater*, D – Dunnock *Prunella modularis*, GC – Goldcrest *Regulus regulus*, GF – Greenfinch *Carduelis chloris*, GS – Great-spotted Woodpecker *Dendrocopos major*, GT – Great Tit *Parus major*, L – Linnet *Carduelis cannabina*, LT – Long-tailed Tit *Aegithalos caudatus*, MT – Marsh Tit *Parus palustris*, N – Nuthatch *Sitta europaea*, R – Robin *Erithacus rubecula*, S – Starling *Sturnus vulgaris*, ST – Song Thrush *Turdus philomelos*, TC – Treecreeper *Certhia familiaris*, W – Wren *Troglodytes troglodytes*, YH – Yellowhammer *Emberiza citrinella*.

It is not hard to imagine that, if the next winter had also been severe, this species could have been obliterated altogether from Britain. Larger ground-feeding species, such as Song Thrush and Lapwing, showed marked declines in some hard winters but not in all, while some aquatic species, such as Heron *Ardea cinerea*, Kingfisher *Alcedo atthis* and Moorhen *Gallinula chloropus*, declined in all severe winters. Among coastal waders, the Redshank *Tringa totanus* appeared particularly susceptible, and among waterfowl the Shelduck *Tadorna tadorna*.

Although the largest population declines were recorded among small birds, not surprisingly most of the carcasses found were from larger and more conspicuous species (Dobinson & Richards 1964). Such carcasses have repeatedly revealed that cold weather losses are not random through a population. Among shorebirds in southeast England, cold-weather mortality was weighted towards males (in at least three species), towards juveniles (in at least one) and towards smaller individuals (in five) (Clark *et al.* 1993). Females appeared most vulnerable in Shelduck and Wigeon *Anas penelope* in Britain (Harrison & Hudson 1964), and in males in Mallard *Anas platyrhynchos*, Ring-necked Pheasant *Phasianus colchicus*, Ruffed Grouse *Bonasa umbellus* and Wild Turkey *Meleagris gallopavo* in North America (Latham 1947, Jordan 1953). This situation in gallinaceous birds is surprising because males are much bigger than females and would have been expected to have survived better. Dead individuals were often found to be parasitised, injured or incapacitated in some other way (Trautman *et al.* 1939, Swennen & Duiven 1983). Even large Mute Swans *Cygnus olor* are not immune to starvation, as shown by the thousands found dead in Denmark in winters when ice covers their shallow-water feeding areas (Andersen-Harild 1981).

Weather movements, which begin at the start of a cold spell, occur mainly in ground-feeding birds, such as various thrushes, larks and plovers. They are apparent both from observations, and from subsequent ring recoveries, which occur in greater proportion than usual in the furthest parts of the wintering range (Holgersen 1958, Baillie *et al.* 1986). Although such birds may escape the worst of the weather, ringing has confirmed that they may still suffer higher mortality than usual. Not all individuals manage to find milder areas, and those that do are often present in great concentrations, increasing the pressure on local food-supplies. Their hunger and unfamiliarity with the area seem to make waterfowl and waders in these circumstances more vulnerable to human hunters, which increases the losses still further (McIlhenny 1940).

Not all species that overwinter at high latitudes suffer from hard winters. Those whose food remains available, such as some finches in years of heavy tree-seed crops, survive well while seeds remain available on the trees (Newton 1972, Nilsson 1987). Other species, which winter in the tropics and migrate to breed in northern regions, may benefit in spring from the shortage of local competitors. The large numbers of Pied Flycatchers *Ficedula hypoleuca* that bred in Britain in 1917 and 1947 were attributed to the unusual availability of nest-cavities in those years, occasioned by the scarcity of resident tits caused by preceding hard winters (Elkins 1983).

Heat and drought

In regions of low rainfall, such as much of western North America, central Asia, the Middle East, Africa and Australia, water is a major factor limiting plant growth. Annual variations in rainfall can cause marked variations in bird food-supplies, as well as in the amount of wetland available to water birds. In consequence, local bird populations usually increase during wet periods and decrease during dry ones, and several dry years in succession can devastate the populations of most species. In this way, through affecting food- and water-supplies, rainfall becomes the major ultimate determinant of bird numbers. During a drought, birds become increasingly concentrated around remaining water sources; many species (especially insect-eaters and nectar-eaters) experience increased mortality from food-shortage, while others stop breeding (Smith 1982). Compared with the effects of hard weather on northern birds, however, which in most regions affect only a proportion of species, several-year droughts in arid regions affect almost all species, even drought-adapted ones. In parts of Australia, many bird species die out or move out completely, and are absent from certain regions for years at a time, returning only when fresh rainfall re-creates suitable conditions (Serventy & Whittell 1976).

The effects of drought on the duck populations of the North American prairies have been studied through surveys conducted each May and July for many years by the United States Fish & Wildlife Service and the Canadian Wildlife Service. Known popularly as the 'duck factory', the Prairie Pothole Region of the Great Plains provides some of the best duck habitat in the world (Baldassarre & Bolen 1994). It covers about 777 000 km², extending from the grasslands in the south to the aspen parklands in the north, and contains more than 11 million hollows (potholes). In periods of high rainfall, every hollow is flooded and the landscape is dotted with ponds of various shapes and sizes, but in dry periods only the larger and deeper lakes remain. This gives enormous year-to-year fluctuations in the amount of wetland habitat (Figure 11.7). In a study in southern Saskatchewan, for example, water-filled hollows averaged 39 per km² in 1955 (a wet year) and less than 1.2 per km² in 1961 (a severe drought year) (Smith 1970). Marked annual variations in snow and rainfall thus produce a 'boom or bust' system in which abundant highly-productive waterfowl habitat occurs in about three years in ten.

Most waterfowl are present on the prairies only for breeding, spending the autumn and winter further south. When the prairies are well watered, many millions of ducks nest there, reaching more than 30 pairs per km² of land-water surface, and breeding well. But in drought years, fewer ducks nest; they are more densely concentrated on the remaining wetlands, and breed poorly: many females make no attempt to nest, and because nesting cover is more restricted than in wet

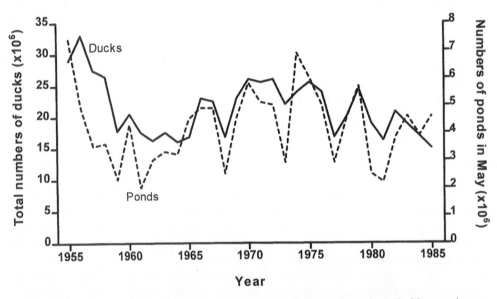

Figure 11.7 Relationship between duck numbers and pond numbers in May each year in the Prairie Pothole Region of North America. Linear regression relationship : $y = 12.56 + 2.11x$, $r^2 = 0.40$, where y = duck numbers and x = pond numbers. Modified from Batt *et al.* (1989), based on surveys by the United States Fish & Wildlife Service and the Canadian Wildlife Service.

years, those that do breed often lose their eggs to predators, and do not attempt to re-nest (see Rogers 1964 for Lesser Scaup *Aythya affinis*, Krapu *et al.* 1983 for Mallard *Anas platyrhynchos*, **Table 11.1**). Moreover, as ponds increasingly dry out in the intense heat of late summer, drought takes a heavy toll of the stranded flightless females and their young. Hence, through poor reproduction and enhanced mortality, high populations decline rapidly during droughts, only to increase again in the next wet period. The degree of annual variation can be illustrated from counts in an area of the Dakotas, where 557 duck broods were found in a wet year (1969) compared with 23 in a drought year (1961) (Duebbert & Frank 1984).

Table 11.1 Mallard *Anas platyrhynchos* breeding success in contrasting years on the prairies of North Dakota.

In the drought year, fewer hollows held water and, although smaller numbers of ducks occurred in the area, the numbers per remaining pond were higher. Ducks were mainly restricted to the larger, deeper ponds, and ranged over smaller areas. Many females made no attempt to breed, but left the area earlier than usual. Those that did breed laid slightly smaller clutches, and a smaller proportion than usual laid repeat clutches after failure of their first nests. From Krapu *et al.* 1983.

	Wet year (1976)	Drought year (1977)
Ponds per km²	10.8	1.5
Pairs per pond	0.6	3.0
Length of stay of females (days following capture ± SD)[2]	43.5 ± 16.5	16.4 ± 12.1
Home range per female	486 ha[1]	246 ha
Nesting attempts per radio-tagged female[2]	1.75	0.13
Numbers of nests found	135	93
Spread in laying dates (days)	85 (N = 135)	66 (N = 93)
Mean clutch size (± SD)	9.8 ± 1.5	9.1 ± 1.5
Re-nesting attempts	Many	Few

[1] 1973–75.

[2] For eight radio-tagged females only; all eight such females nested in 1976, but only one out of eight in 1977.

So important are the prairies to the continental duck population that, over a period of years, the overall ratio of young to old among ducks shot further south in winter (taken as a measure of breeding success) was correlated with the number of prairie ponds counted by aerial survey in July. Similarly, the total numbers of Coots *Fulica americana* shot in North America each year was dependent on prairie pond numbers in May **(Figure 1.4)**. In recent decades, however, agricultural policy has destroyed more than half of all potholes, so that even in wet periods waterfowl populations can no longer reach their former levels. Owing to a combination of prairie droughts and drainage, the total wintering population of ducks in North America declined substantially for more than 15 years (Boyd 1990).

The state of the prairies also influences the migration patterns of ducks. In years with abundant wetland habitat, most of the returning ducks breed on the prairies, and a relatively small proportion moves further north and northeast to breed in the more stable wetlands of the forests and tundras of northern Canada and eastern Siberia. But in years of prairie drought, a much greater proportion of birds move north and northeast, a pattern confirmed by aerial surveys and by ring recoveries (Henny 1973). The trend was particularly marked in Pintail *Anas acuta* **(Figure 11.8)**, but occurred to some degree in all dabbling species (Hansen & McKnight 1964, Smith 1970). In northern areas, breeding is less good than in good years on the prairies, which causes a further decline in overall numbers. Hence, duck numbers in the relatively constant northern wetlands depend not just on local conditions, but on those further south, being linked across millions of square kilometres. In addition, the numbers found in one year depend on the numbers in several previous years.

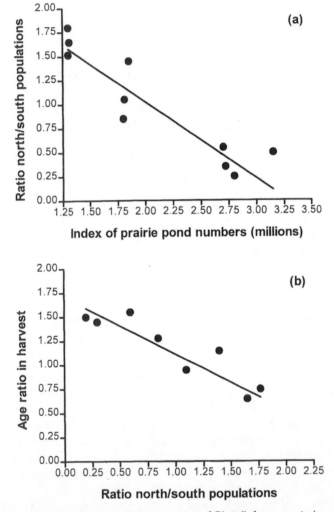

Figure 11.8   Distribution and breeding success of Pintail *Anas acuta* in relation to wetland conditions on the Canadian prairies. (a) Relationship between numbers of ponds and lakes on the prairies in May and the proportion of the total Pintail population found breeding north of the prairies, 1959–68 (r=0.91, P<0.001). (b) Relationship between the proportion of the total Pintail population in northern areas and overall breeding success (r=–0.92, P<0.001). Redrawn from Smith 1970.

Even greater fluctuations have been recorded among waterfowl in Australia, where the depth and extent of wetlands fluctuate even more widely. In some regions, following flood, wetlands may cover hundreds of square kilometres, but they soon dry up, leaving the land parched for many years. The waterfowl themselves move around, concentrating wherever floodlands are available. An example is the Australian Grey Teal *Anas gibberifrons*, which breeds annually in well-watered (mostly coastal) sites (Frith 1967). In years with good rainfall, numbers expand rapidly and, in exceptionally wet years, breeding occurs through much of inland Australia. However, in dry years, breeding is negligible, and when two or more drought years occur together, mortality increases and overall numbers decline. This species is thus greatly at the mercy of rainfall, fluctuating hugely over the years, both in numbers and in distribution. An even more extreme example is the nomadic Pink-eared Duck *Malacorhynchus membranaceous*, which can be present in a region in one year and then absent for several further years until conditions again become suitable (Frith 1967, Braithwaite 1976). The period of drying out after flooding can be especially important for ducks because it stimulates a massive rise in food-supply, notably chironomid flies which develop on the rotting vegetation. The fact that such pools often dry out keeps many of them free of fish which, if present, would remove much of this food.

As indicated above, it is not only waterfowl that are affected by droughts. In certain finches of the Galapagos Islands, rainfall controls the growth and the seeding of their food-plants, and hence their breeding, survival and population sizes (Grant 1986). Regular measurements of finch and seed numbers indicated that population limitation was by food-supply, through dry-season shortages that occurred in all but the wettest years (Chapter 7). Drought-related reductions in breeding success have been noted in many arid-land species in western North America, including insectivores, such as the Cactus Wren *Campylorhynchus brunneicapillus* (Marr & Raitt 1983), Sage Sparrow *Amphispiza belli* and Brewer's Sparrow *Spizella breweri* (Rotenberry & Wiens 1991). They have also been recorded in nectivores, such as the Long-tailed Hermit *Phaethornis superciliosus* in Costa Rica in which survival and numbers of adults were reduced by about one-third during a summer drought, after which 3–4 years were needed for recovery (Stiles 1992). In some dry parts of Australia, drought-induced gaps of 3–4 years sometimes separated successive breeding attempts in Wedge-tailed Eagle *Aquila audax* territories (Ridpath & Brooker 1986). Flamingos usually breed in huge concentrations and can suffer spectacular breeding failure when lakes evaporate. At Etosha Pan, in Namibia, for example, major successful breeding events occurred only three times in 40 years, and in some other years up to 100 000 flightless chicks starved as the pan dried (Simmons 1996). Significant relationships between breeding success, survival or population level and rainfall have been noted in several other species of arid regions elsewhere (Berry & Crowe 1985, Curry & Grant 1989, Zahavi 1989).

Correlations have also been found between survival, population level and rainfall in several migratory bird species which breed in Europe but spend the winter in sub-Saharan Africa. In the Sedge Warbler *Acrocephalus schoenobaenus*, for example, both the annual survival and the spring population levels as measured in Britain have been linked to annual rainfall in the western Sahel zone where this species winters **(Figure 11.9)**. The proposed mechanism is as follows: high rainfall in Africa → high food-supply in Africa → high survival in Africa → high breeding population in

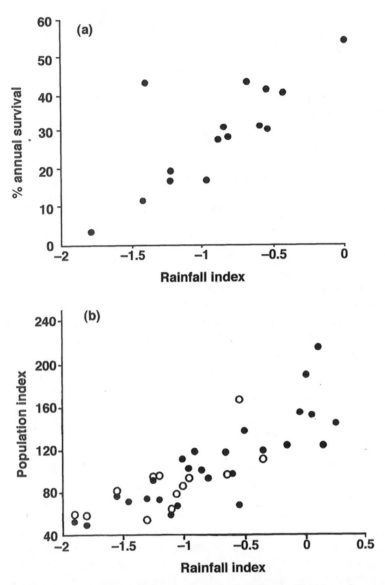

Figure 11.9 Relationship between (a) annual survival of Sedge Warblers *Acrocephalus schoenobaenus* at various sites in Britain or (b) annual spring population index in Britain, and preceding annual rainfall in the Sahel zone of Africa, where Sedge Warblers spend the winter. Symbols reflect different counting programmes: closed symbols – Common Bird Census, open symbols – Wetland Bird Survey. From Peach *et al*. 1991.

Europe. Likewise, the numbers of migrant herons counted at colonies in the Netherlands and France each year were correlated with water discharges through major rivers (reflecting rainfall) during the previous winter in west African wintering areas **(Figure 11.10)**. Overwinter survival of Purple Herons *Ardea purpurea* (Cavé 1983), White Storks *Ciconia ciconia* (Dallinga & Schoenmakers 1989, Kanyamibwa *et al*. 1990), Whitethroats *Sylvia communis* (Winstanley *et al*. 1974,

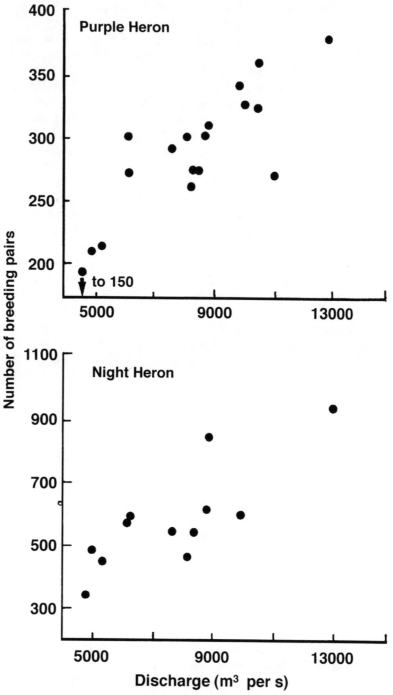

Figure 11.10   Relationship between the numbers of migrant herons nesting at colonies in western Europe and wetland conditions in wintering areas in west Africa during the preceding year. Purple Herons *Ardea purpurea* were counted at one major colony in the Netherlands, and Night Herons *Nycticorax nycticorax* at several colonies in southern France. Wetland conditions were measured as the maximum monthly discharges through the Senegal and Niger Rivers. Both relationships were statistically significant (on Kendall's rank correlation test, *P*<0.005). From den Held 1981.

Baillie & Peach 1992) and Sand Martins *Riparia riparia* (Bryant & Jones 1995) has shown similar correlations, as has the survival of Barn Swallows *Hirundo rustica* with March rainfall in southern Africa (Møller 1989a). In all these migratory species, climatic conditions several thousand kilometres south of the breeding areas affected the spring population levels seen in Europe.

The White Storks breeding in Europe show a migratory divide, those from western Europe crossing at Gibralter to winter in West Africa and those from eastern Europe crossing at the Bosphorus to winter in East Africa. Correspondingly, annual changes in the numbers of storks breeding in Alsace (France) during 1948–70 were correlated with annual rainfall in West Africa, while annual changes of storks breeding in parts of Germany during 1928–84 were correlated with rainfall in East Africa (Dallinga & Schoenmakers 1989). In years with high winter rainfall, storks also arrived earlier in their breeding areas and a higher proportion of pairs produced young. The supposed link with food-supply is supported by statistics on plagues of four important locust species on which storks are known to feed. On average, the arrival dates in Europe of storks wintering in West Africa have got later since the 19th century, but no such trend is evident for those wintering in East Africa. This was attributed to a worsening of food-conditions in the western wintering area, as a result of climatic changes and, more recently, locust control and desertification (Dallinga & Schoenmakers 1989).

The extreme conditions of some deserts, with sizzling temperatures, intense solar radiation and scarcity of surface water, present severe problems to many organisms, including both resident and migrant birds. In the Australian desert, 71 (60%) of 118 species studied by Fisher *et al.* (1972) were never seen to drink, or visited water on fewer than half of all days when temperatures exceeded 25°C. Some carnivores and insectivores could live without drinking water, getting all their needs from food, but other species, especially seed-eaters, had to drink daily, and were absent from areas lacking surface water. In drought, such species could die of thirst, while in extreme heat, they could die even in the presence of water if their evaporative cooling mechanism could not cope. During a heatwave in the interior of south and central Australia in 1932, when the temperature remained above 38°C (100°F) for over two months, a huge mortality occurred among Chestnut-eared Finches *Taeniopygia castonotus*, Budgerigars *Melopsittacus undulatus* and other small birds. Water tanks were coated with corpses, and vast numbers sought shade in railway stations and under trains; populations over hundreds of square kilometres were affected (Finlayson 1932).

EPISODIC WEATHER EVENTS

Depending on the region, storms of one sort or another can occur at any time of year, and hence can affect birds at any stage of the annual cycle. The unpredictability of storms means that their effects are hard to document. It is largely a matter of chance if they hit localities where bird counts were made beforehand, enabling their effects to be assessed by repeat counts. However, it has sometimes been possible to compare bird numbers after the event in affected and unaffected areas, as in some of the examples below. Consideration of the effects of storms on seabirds is left until later.

Spring snowstorms. Cold snaps, with heavy snow, sometimes occur at high latitudes after birds have started nesting. Mortality and breeding failures are greatest in small species, especially in migrant insectivores, recently returned from winter quarters, which are affected by lack of insects. Small species that use holes and other protected sites fare better, as do some large species (such as ducks) which can continue incubating through the snow. Birds at the laying or hatching stages are particularly vulnerable, as are those with small chicks (Bengston 1963, Bull & Dawson 1969, Pulliainen 1978, Ojanen 1979). Unseasonal cold and wet in tropical regions can affect not only local breeding species, but also migrants from higher latitudes that are wintering there, especially hirundines and other insectivores (Steyn & Brooke 1970).

Summer rainstorms. Storms of torrential rain, hail and wind, are fairly frequent in parts of the Canadian prairies in late summer, often flattening herbaceous vegetation over wide areas and killing many birds. On 14 and 18 July 1953, two such storms affected about 2500 km² of northern Alberta, killing an estimated 64 000–149 000 waterfowl, about 93% of the total in the area at the time (Smith & Webster 1955). Large but unrecorded numbers of other species also died, including songbirds, grouse and raptors. A similar storm occurred on 15 June 1977 in North Dakota. By comparing affected bird populations in this area with those in an adjoining area missed by the storm, Johnson (1975) concluded that the total bird population had been reduced by 45%, with estimated losses of 126 individuals per km². As the storm covered 390 km², the total losses were estimated at around 49 000 birds, excluding young. Such storms, although sporadic, could clearly have major temporary impacts on local bird populations (Bremer 1977, Higgins & Johnson 1978).

Elsewhere, rainstorms have killed many young birds, especially when they occurred soon after hatch, as exemplified by the 90% chick mortality recorded at a Herring Gull *Larus argentatus* colony in Newfoundland (Threlfall *et al.* 1974). The young of gallinaceous and other nidifugous birds seem especially vulnerable to rain soon after hatch. Their plumage becomes soaked from wet vegetation and they can then die of chilling. In addition, the flooding and wave action resulting from rainstorms has often destroyed the nests of aquatic birds over wide areas (Harris & Marshall 1957).

Tornadoes. At 3 am on 9 July 1940, a small town in western Iowa was struck by a tornado (McClure 1945). The next morning many birds were found dead. A month later, when counts were made, bird densities in this town averaged 1.3 per ha, compared with 30.5 per ha in a similar town nearby which escaped the storm. From this, it was inferred that the tornado destroyed more than 95% of the birds in its path. The least affected species were those that had roosted within the shelter of buildings or tree hollows.

Hurricanes. Hurricanes (or cyclones), with strong wind and lashing rain, are a feature of tropical regions, notably the Caribbean, typically affecting areas 300–600 km wide across their path. Within the few hours they take to pass, they can have devastating effects on habitats, smashing down trees or removing leaves, flowers and fruit (Tanner *et al.* 1991, Wunderle *et al.* 1992). The accumulated debris can later dry and burn, which leads to further damage (Lynch 1991). In coastal regions, the associated tidal surges can flood large areas, causing further loss of trees and other susceptible vegetation, and result in flushes of nutrients into coastal waters, while in inland areas rainfall can trigger landslides (Tanner *et al.* 1991).

The marked declines in bird populations that usually follow hurricanes are probably due more to structural damage to habitats and to loss of food-supplies than to direct mortality. In areas where birds had been censused before a hurricane, subsequent counts (up to several months later) revealed overall declines in numbers up to 60%, and declines in individual species up to 90% (Cheke 1975, Askins & Ewert 1991, Wauer & Wunderle 1992, Wunderle *et al.* 1992). The biggest declines were in species that ate tree-seeds, fruit and nectar, confirming loss of food-supply as a major causal factor. Several species were found to have increased in nearby less-damaged areas, suggesting that some of the declines in the most damaged areas were due to movements **(Table 11.2)**. Hurricanes that occur while birds are breeding inevitably cause massive nest losses (Tate 1972). Forests normally start to recover within a few weeks, as trees sprout new growth and seedlings germinate, enabling bird populations to recover (Askins & Ewert 1991). Where forest is extensively damaged, the avifauna may change, as species of forest-edge and scrub are favoured over those of mature forest, changing back over several years as the trees regrow (Lynch 1991).

Winter blizzards. The population of Song Sparrows *Melopiza melodia* on Mandarte Island, British Columbia, crashed from 103 breeding individuals in 1988 to 12 in

Table 11.2 Changes in bird numbers judged from counts nine months apart, five months before and four months after, the passage of Hurricane Gilbert. Details for ten habitats in Jamaica, affected by the hurricane, and two habitats in Puerto Rico, unaffected, as a control. Figures show the numbers of individuals per 10-minute point count (± s.e.). The most severe declines were in the most damaged upland areas on Jamaica; the increases in some lowland areas may have been due to movement from damaged upland areas. Significance: *$P<0.05$, **$P<0.01$, *** $P<0.001$, n.s. – not significant. From Wunderle *et al.* 1992.

	Before hurricane	After hurricane	Difference (%)	Significance of difference
Jamaica (hurricane)				
Montane cloud forest	3.3 ± 0.3	2.3 ± 0.4	−30	**
Montane pine plantation	5.0 ± 0.6	2.1 ± 0.3	−58	***
Montane coffee plantation	2.9 ± 0.3	1.5 ± 0.3	−48	***
Wet limestone forest	3.8 ± 0.6	5.0 ± 0.6	+32	*
Lowland coffee plantation	3.0 ± 0.4	2.5 ± 0.6	−17	n.s
Lowland secondary forest	4.8 ± 0.5	4.8 ± 0.4	0	n.s
Mangrove	4.4 ± 0.3	5.5 ± 0.4	+25	*
Lowland pasture	4.5 ± 0.6	4.3 ± 0.6	−4	n.s
Dry limestone ruinate	3.7 ± 0.4	3.4 ± 0.5	−8	n.s
Dry limestone forest	3.0 ± 0.4	3.5 ± 0.4	+17	n.s
Puerto Rico (no hurricane)				
Montane second growth forest	5.6 ± 0.3	6.1 ± 0.4	+9	n.s
Lowland pasture	4.8 ± 0.4	4.7 ± 0.3	−2	n.s

1989. This mortality was caused mainly by a single ice storm (Smith *et al.* 1996). Similarly, following a severe wind storm in Belgium in January 1990, about 62% of Crested Tits *Parus cristatus* in a study population died and many of the surviving ringed individuals were blown up to a few kilometres away from the area (Lens & Dhondt 1992). After a single winter storm, Carolina Chickadee *Parus carolinensis* numbers in an area of Illinois were reduced by 22% (Graber & Graber 1979).

Other extreme events. Some other weather effects are quite bizarre. For example, in 1928 during a thick fog, several hundred Tundra Swans *Cygnus columbianus* were swept over Niagara Falls, and more recently about 4000 Common Mergansers *Mergus merganser* and other ducks suffered the same fate (Baldassarre & Bolen 1994).

*   *   *

From these and other examples of catastrophic weather events, a number of generalisations emerge. Firstly, severe storms which occur sporadically, and usually affect some tens or hundreds of square kilometres at a time, can cause severe losses of a wide range of bird species, resulting in massive declines in local populations, sometimes of more than 90%. Secondly, when storms occur in the breeding season, they can also cause widespread breeding failure. Thirdly, wind storms can result in severe physical damage to habitats, which can take months or years to rectify, so that the total effects on the avifauna exceed those resulting from immediate mortality. Fourthly, although spatially restricted, such events can have disastrous consequences for the bird populations of small islands, and on a regional scale, their sporadic occurrence could be a continual factor in reducing local bird populations.

The frequency and significance of extreme weather events clearly vary with the size of area considered. We can be reasonably certain that a hurricane will occur somewhere in the Caribbean every year, much less certain that it will strike a particular island, and even less certain that it will hit a particular patch of forest. However, the overall significance of the event, if it does strike, varies inversely with the size of area considered. At the scale of the whole Caribbean the effect of a single hurricane is trivial, but at the scale of an individual forest it could be devastating. In effect, this means that, while localised effects on bird populations can be substantial, recovery is often likely to be rapid, helped by immigration from surrounding unaffected areas. The problem nowadays is that so many island species have been restricted to small areas by human action that, when severe storms strike, they can cause extinction (see Chapter 16 for examples).

Effects of fire

In some regions of savannah and grassland, fires occur nearly every year. They are responsible for holding the habitat in its current state, and hence for maintaining the present fauna, including bird species. In other regions, as in large parts of the boreal forest, wild fires occur at longer intervals of decades or centuries, usually killing the trees over wide areas and enabling new ones to grow. During forest regrowth, bird species composition usually changes completely, and both species numbers and overall bird density increase gradually as the habitat becomes structurally more complex. In the Sierra Nevada of California, for example, birds were counted for 25 years after a forest plot burned in 1960 (Raphael *et al.* 1987). As new forest grew and species numbers increased, ground- and shrub-nesting species gradually gave way to arboreal ones, as the new canopy developed. Cavity-nesters decreased over the first few years, because trees killed by the fire gradually collapsed. Over the same period, few changes occurred on a nearby unaltered plot of well-grown forest, which kept most of the same species throughout.

Because in forested regions different areas may burn in different years, fires are a major factor underlying habitat diversity on a landscape scale. They are thus responsible for maintaining a wide diversity of plant and animal life, and for enabling species of open and scrub habitats to persist in an otherwise forested region. In the boreal forests of Canada, an average of three million hectares of trees are burnt each year, mostly through lightning strikes (Freedman 1995). Although fires are natural events, human action has altered their frequency in many regions, as people often deliberately start fires in grassland and heathland areas (to favour grazing animals), and deliberately suppress them in managed forests, which makes their effects even more devastating when they do occur.

Weather and migration losses

Migrant landbirds caught by fog or storms over water must often be lost without trace. However, losses are occasionally recorded, as when large numbers of Quail *Coturnix coturnix* were found drowned during a spell of sea fog (Moreau 1927), when large numbers of dead swifts and martins were washed up on a beach in North Africa (Perrins *et al.* 1985), or when hundreds of Garden Warblers *Sylvia borin* and other species were washed ashore in Spain following a night of heavy rainstorms (Mead 1991). In the New World, a massive kill estimated at 40 000 migrants of 45 species occurred during a tornado and storm on 8 April 1993 on an island off Louisiana. The storm occurred at a time of day when large numbers of birds were arriving at the coast after crossing the sea at night (Wiedenfeld & Wiedenfeld 1995). Other mortality events in the Gulf of Mexico included one of 10 000 birds washed up on Padre Island, Texas, in May 1951 (James 1956), and another of 5000 birds on Galveston Island, Texas, in May 1974 (King 1976). In the former, half the casualties were of one species, the Magnolia Warbler *Dendroica magnolia*. After Hurricane Agnes killed many birds on spring migration, reductions in the numbers of Purple Martins *Progne subis* were evident over much of the breeding range (Tate 1972). Waterbirds probably suffer fewer losses on overseas flights than do landbirds, being more robust and able to rest on water when navigation or flying conditions become difficult (as shown for radio-tagged Whooper Swans *Cygnus cygnus* on their migration between Iceland and Britain, Pennycuick *et al.* 1996).

As indicated above, insectivorous migrants seem especially vulnerable to cold spells occuring during their northward migration. Many migrants turn back when they encounter cold weather (reversed migration), but this is not always possible, especially if they have exhausted their fat reserves. One such catastrophe affected newly arrived Cliff Swallows *Hirundo pyrrhonota* in Wisconsin when a sudden drop in temperature caused insects to lie dormant. The swallows did not fly during this period, but clung to their old nests, with one observer collecting 'a milk pail full' of the birds that had died (Buss 1942). Similar large losses among other newly-arrived migrant birds (mainly insectivores) were recorded during more recent cold snaps (Ligon 1968, Whitmore *et al.* 1977, Zumeta & Holmes 1978).

Among Common Sandpipers *Actitis hypoleucos*, which migrate from Africa to breed in Britain, annual survival fluctuated in one area according to the weather in April, when they arrived **(Figure 11.2)**. The mean survival over 13 years was 79%, but following late snowstorms in 1981 and 1989 survival fell to 39% and 50%, respectively, and breeding pairs from 21 to 14 and from 20 to 12. Recovery in population level was slow, with annual increments of only 1–2 pairs (Hollands & Yalden 1991). Similarly, in the spring of 1966, frost and snow on 11–17 April caused massive mortality of newly-arrived Lapwings *Vanellus vanellus* across southern Sweden and Finland, with reductions in populations of 30–90% in different regions (Vespäläinen 1968, Marcström & Mascher 1979).

As another example, in the spring of 1964 an estimated 100 000 King Eiders *Somateria spectabilis* – some 10% of the whole west Canadian population – died at migration staging areas in the Beaufort Sea. This mortality occurred when newly opened leads among sea ice re-froze, preventing the birds from feeding (Barry 1968). Such mass starvation events have been seen in this species several times this century. In a more recent event, birds found dead had lost about half of their weight, and many survivors had insufficient body reserves for breeding, so that the effects of the event extended beyond the mortality caused at the time (Fournier & Hines 1994).

Aseasonal weather can also affect birds on southward migration in autumn. In central Europe in September 1931, during an early cold snap, 'immense numbers of Swallows *Hirundo rustica* and House Martins *Delichon urbica*, and some Sand Martins *Riparia riparia*, numbed by the cold and unable to find food, sought shelter in stables, barns and houses where, in some cases, they congregated in great masses and were so inert that they could be picked up by hand' (Alexander 1933). The 89 000 individuals that were collected and transported by air for release in Italy 'were probably a small fraction of the total lost'. A similar event occurred in 1974 (Bruderer & Muff 1979).

Headwinds provide another hazard for migrants, because in effect they force the birds to fly for longer, depleting their energy reserves. No doubt many birds are lost at sea while fighting headwinds, while others die or suffer reduced breeding success after arrival. In Brent Geese *Branta bernicla*, which breed mainly on the strength of body reserves, productivity was noticeably reduced in years when the birds had to perform their migration against a strong headwind (Ebbinge 1989).

Many other examples could be given of large-scale losses associated with migration but, because losses that occur over sea or desert are usually impossible to detect, it is hard to tell whether they are of more than sporadic occurrence. Those that occur in spring, when numbers are near their seasonal low, are more likely to affect subsequent breeding densities than are losses that occur in autumn when numbers are near their highest. Although such losses represent a major cost of migration, for the migratory habit to persist we must assume that, in the long run, they are less than any losses that would be experienced if the birds stayed on their breeding areas all year round. For many species breeding areas are completely untenable in winter.

Weather, body reserves and the breeding of large Arctic-nesting waterfowl

Large waterfowl often arrive when their Arctic breeding areas are still snow-covered, and lay their eggs as soon as bare ground is exposed. This probably applies to all Arctic-nesting geese and swans, but three species have been studied in detail, namely the Snow Goose *Anser caerulescens* (Ankney & MacInnes 1978), Canada Goose *Branta canadensis* (Raveling 1979) and Dark-bellied Brent Goose *Branta b. bernicla* (Ebbinge 1992). In the accumulation of the necessary body reserves, the birds are influenced by weather conditions on wintering and staging areas, and in their subsequent breeding success by conditions in nesting areas (Davies & Cooke 1983).

When the geese arrive in the Arctic, food is largely unavailable, and in years when snow melt is late, the birds use more of their body reserves than usual for survival. This prevents some females from laying and reduces the clutch sizes of others, leading to a general reduction in breeding success (Barry 1962, Newton 1977, Davies & Cooke 1983). The females are solely responsible for incubation and if their reserves run out, they desert their nests to feed, leading to further reduction in breeding success.

Three environmental conditions may have led to heavy dependence on body reserves: (1) summers are so short at these latitudes compared to the period needed to breed that, if the birds arrived late enough to avoid snow altogether, in some years they could not raise their young before winter returned; (2) at the time the birds arrive and start nesting, food is largely unavailable; and (3) unattended nests are usually lost to predators if females take feeding breaks. It is these circumstances that could have led to a system in which females accumulate, south of the breeding areas, all the reserves necessary for migration followed by egg production and incubation (Ankney & MacInnes 1978). This system in turn makes breeding dependent on weather conditions on wintering and migration areas, as well as on breeding areas.

In some Arctic-nesting geese, marked annual variations in wintering numbers are associated partly with variations in the spring weather on breeding areas, through its effects on success. Among some populations of Brent Geese and Snow Geese, for example, the proportion of young in winter flocks has varied between less than 2% after late springs to more than 50% after early springs (Newton 1977). Similarly, the proportions of breeding Barnacle Geese *Branta leucopsis* that produced young in Spitzbergen were 60–80% in early seasons and 15–25% in late ones (Prop *et al.* 1984). Among Whooper Swans *Cygnus cygnus* wintering in Sweden, the proportion of young among winter flocks was correlated with the severity of weather both in the preceding winter (assumed to influence the body reserves of females), and in the preceding spring on the Siberian breeding grounds (assumed to influence the use of reserves) **(Figure 11.2d)**. Overall, 58% of the variance in breeding production between years was attributed to previous winter weather in Sweden and 32% to spring weather in Siberia. Effects of winter weather on the subsequent breeding of Mute Swans *Cygnus olor* have been noted in Denmark (Andersen-Harild 1981).

Similar relationships between spring weather and breeding success may hold for some species of ducks, such as the Eider *Somateria mollissima*, which also relies heavily on body reserves for egg production and incubation (Milne 1976). On Svalbard, female Eiders lost 46% of their weight between pre-laying and hatching, which included 81% of their body fat and 37% of body protein (Parker & Holm 1990). At high latitudes, many waders also fail to breed in late springs, returning to their wintering sites weeks or months ahead of normal (Tate 1972, Green *et al.* 1977).

OCEANIC CONDITIONS

Changes in sea currents

Oceanographic events, associated with shifts in currents, can have huge impacts on seabird populations by affecting their food supplies. The best known is the Southern Oscillation, which affects climate and currents over much of the world, but especially in the eastern Pacific, where it is known as the El Niño[1]. Off the coast of Peru the warm waters characteristic of tropical seas are replaced by a cold upwelling, the Humboldt current, which supports a great abundance of marine life, including seabirds. Three species, the Guanay Cormorant *Phalacrocorax bougainvillii*, the Peruvian Booby *Sula variegata* and the Peruvian Brown Pelican *Pelecanus occidentalis*, breed so abundantly that their guano is a commercial asset. In the poor warm waters to the north, other seabirds occur in the low numbers typical of tropical seas. In the El Niño years, warm waters come much further south (and north) than usual, and surround the Peruvian guano islands. Then the abundant plankton disappears, as does the anchovy *Engraulis ringens* which supports the seabirds, which in turn can die in millions from starvation (Vogt 1942, Hughes 1985). This has occurred more than 20 times in the last 100 years, but with more severity in some years than in others **(Figure 11.11)**. After each crash, seabird numbers recovered progressively year by year until the next crash. The biggest recorded decline occurred in 1957–58, when numbers dropped by 70% from 20 million to six million birds, after which numbers increased steadily to 17 million in 1962–64, when the current again shifted, causing another decline in fish and another crash in bird numbers. In the last 100 years at least, populations have thus alternated between precipitous declines and steady recoveries, not once reaching a level at which they stabilised. Associated declines in fish and bird numbers have also been observed in the central Pacific (on Christmas Island, Schreiber & Schreiber 1984) and to the north in California (Ainley & Lewis 1974). No evidence of unusual disease or predation has come at the time of the declines, but in recent years human overfishing of anchovies may have accentuated the problem, preventing or slowing the recovery in bird numbers.

---

[1]*The causes of the El Niño phenomenon are not well understood, but seem to begin with the failure of the easterly trade winds. These winds usually blow steadily and cause the surface waters of the Pacific to be pushed westwards, resulting in a bulge on the western edge of the ocean. When the winds fail, the bulge of warm water flows back eastwards along the equator. The warm water reaches the coasts of Central and South America by late September, and over a period of weeks raises the surface water temperatures in the eastern Pacific by as much as 10°C. During this time, warm water spreads north and south through the temperate zones, causing drastic changes in the weather patterns of both North and South America and beyond. As surface water temperatures rise, so does rainfall on adjacent land. While seabirds, which rely on productive cooler water, may experience high mortality and breeding failures, life on nearby land areas burgeons in response to greater rainfall, but in more distant areas it may suffer in response to droughts. Thus, the impact of the Southern Oscillation on northern and southern hemispheric circulation is associated with warm winters in western Europe and Siberia, droughts in eastern Asia, and Australia and high summer rainfall over central China, as well as high rainfall from Peru north to California (a reversal of the usual pattern), and with many other extreme weather events elsewhere (Schreiber 1994).*

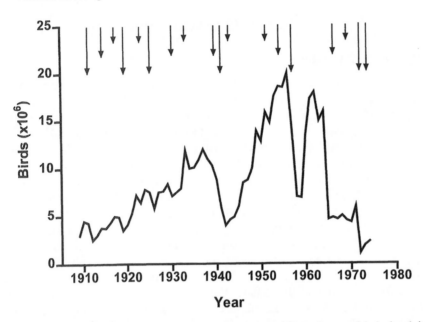

Figure 11.11   Fluctuations in the combined numbers of Peruvian seabirds (mainly *Phalacrocorax bougainvillii, Sula variegata* and *Pelecanus occidentalis*) in relation to El Niño events (current changes that result in food-shortage and mass starvation). The longest arrows indicate the strongest events and the shortest arrows the weakest events. Every event was followed by a decline in breeding numbers, but the biggest proportionate declines occurred when a severe event occurred at times of high bird numbers. At no time in the period shown did the bird populations stabilise, but alternated between sudden declines and subsequent recoveries. From information in Shannon *et al.* 1984. The large increase in the 1940s may also have been helped by the provision of extra nesting habitat (Duffy 1983).

Another concentration of seabirds, also exploited for their guano, occurs in association with the cold Benguela current off southwest Africa (Hutchinson 1950, Duffy *et al.* 1984, La Cock 1986). These birds are also affected by current shifts and their associated effects on food-supplies, which lead to mass mortality or breeding failure. The main species affected include the Cape Gannet *Morus capensis*, Cape Cormorant *Phalacrocorax capensis* and Jackass Penguin *Spheniscus demersus* (Crawford & Shelton 1978). During 1901–83, such events were recorded in 32 different years, at a mean interval of 2.6 years (SD = 2.5 years), mostly in the same years as an El Niño or in the year before an El Niño. They were often linked with red-tides caused by toxic algal blooms, and with droughts on distant parts of the southern African mainland (La Cock 1986).

In northwest Britain, over a 33-year period, a correlation emerged between the breeding performance of Kittiwakes *Rissa tridactyla*, the abundance of Herrings *Clupeus harengus*, zooplankton and phytoplankton, and the frequency of westerly winds (Aebischer *et al.* 1990). All declined at a similar rate from the 1950s to reach a trough around 1979–80, and then recovered somewhat (Aebischer *et al.* 1990). Although the mechanisms were unclear, the existence of such correlations again pointed to some underlying oceanographic–climatic influence, whose effects were felt through four trophic levels. Similarly, changes in the numbers of Adelie Penguins *Pygoscelis adeliae* in the Ross Sea over a ten-year period were correlated with climatic changes (Taylor & Wilson 1990). Year-to-year fluctuations in the breeding of some seabirds in other regions have been linked with sea temperatures (Boersma 1978, Murphy *et al.* 1986), and at high latitudes with the timing of ice break-up. There can be no doubt, then, that seabird populations in many regions are from time to time greatly influenced by the food-shortages caused by changes in currents and sea temperatures.

Seabird wrecks

Prolonged gales, which make feeding difficult, sometimes result in high mortality among seabirds. This is evident in so-called 'wrecks', when many birds are washed ashore or blown inland by onshore gales. Such events have been recorded in both northern and southern hemispheres, but affect some species more than others, depending partly on how close to land they feed.

In the 127 years from 1856, 14 wrecks of auks were recorded in Britain, mostly in late summer or autumn (Underwood & Stowe 1984). One occurred in September 1969, when over 12 000 seabirds (mostly Guillemots *Uria aalge*) came ashore in the Irish Sea. This event was attributed to a combination of stresses, including poor weather (and resulting food-shortage), moult and contamination with organo-chlorine pollutants, especially PCBs (Holdgate 1971). An even larger event, probably involving more than 100 000 Guillemots, occurred in Alaska in April 1970, after a five-day period of storms (Bailey & Davenport 1972). These birds weighed up to 25% less than normal, had no fat reserves or food in the gut, but no obvious disease or pollutant poisoning. This event was therefore attributed entirely to storm-induced starvation.

Similar events have been recorded in other species elsewhere, including Little Auks *Alle alle* in the North Atlantic (Murphy & Vogt 1933), prions on the western shores of South America, Streaked Shearwaters *Calonectris leucomelas* in Japan, Leach's Petrels *Oceanodroma leucorrhoa* in Britain and Japan, various diving petrels and Blue Petrels *Halobaena caerulea* in Australia and New Zealand, and various auks around Britain (Murphy 1936, Warham 1990). Such mass starvation events could be much more frequent among oceanic birds than those brought to the attention of land-based observers by unusual circumstances. Storm conditions could increase mortality by:

(1) displacing birds to areas with less food,
(2) making food more difficult to obtain,
(3) destroying food-supplies, or
(4) increasing heat and energy loss and battering.

It is curious, if such wrecks are primarily storm-induced, that they usually consist overwhelmingly of one species. This could be because only one species is present in the area at the time, or because predisposing factors, such as prior food-shortage, affect one species more than others. Although large numbers of birds are involved, it is often unclear which breeding populations they represent (in winter they could include migrants), and even whether they affect breeding numbers at all. The fact that seabirds usually have a large non-breeding component, from which new breeders could soon be recruited, may help to minimise the impact on breeding numbers. After a wreck on the east coast of Britain in February–March 1994, some 20 000–50 000 Guillemots *Uria aalge* and 3000–5000 Shags *Phalacrocorax aristotelis* were estimated to have died (Harris & Wanless 1996). This mortality had no detectable effect on Guillemot breeding numbers, but caused a massive decline in Shag breeding numbers. On the Isle of May, the return rate of ringed adult Shags dropped to 13% (compared with 75–82% in previous years) and the nesting population fell to the lowest level in 35 years. The difference in effect was because the mortality affected a smaller proportion of the Guillemots in the area, and mainly immatures, whereas in Shags it affected a larger proportion, and similar numbers of adults and immatures.

Hurricanes and other gales can also blow seabirds far outside their normal range, increasing their mortality rates. Most of the records of tropical seabirds in eastern North America occur after hurricanes. At such times, species such as Sooty Tern *Sterna fuscata* and Black Skimmer *Rynchops niger*, which are normally associated with the Caribbean, have appeared as far north as Canada (Godfrey 1966, Tuck 1968).

SELECTIVE MORTALITY RESULTING FROM WEATHER EVENTS

It will be clear by now that extreme weather, leading to bird deaths and population declines, does not affect a random cross-section of the birds in the area at the time. Some species are particularly vulnerable, and, within species, some sex- and age-groups are affected more than others. In certain species, moreover, individuals of a particular size and shape die in greater proportion than others, providing some striking examples of natural selection in action. Perhaps the first example was provided by Bumpus (1899) who found that some House Sparrows *Passer domesticus* that died in a severe storm differed in weight and measurements from those that survived (see also Brown & Rothery 1993). In recent years, other examples have come to light.

During a several-month drought on Daphne Island in the Galapagos archipelago, large individuals of the Ground Finch *Geospiza fortis*, which had large beaks, survived best because they were able to deal with the large and hard seeds that remained after most of the smaller and softer seeds had been eaten (Grant 1986). Through differential mortality, the mean bill size of the population was altered within a few months and, as bill size was shown to be inherited, this represented an evolutionary change by natural selection. Another result of the drought was that males survived better than females, giving a sex ratio of 5:1. At the next breeding season, females mated with the largest (= largest billed) of the available males, so that the same traits were further accentuated by sexual selection. In other years, advantages in small size were shown, so evolutionary changes away from the average tended to be short term. Once again, however, a single climatic event changed the morphology of the species.

More recently, it has emerged that the mean body size (as assessed by sternum length) of Sand Martins *Riparia riparia* nesting in Britain has varied over the years, according to rainfall in their African wintering areas (Bryant & Jones 1995). Drought years were followed by reductions in mean body size and wetter years by increases, another consequence of selective mortality, as shown from ringed birds.

Among Oystercatchers *Haematopus ostralegus* that died during a cold spell in the Netherlands, about 61% were handicapped by anatomical deviations (such as curved or crossed mandibles), a significantly higher proportion than in the population at large. These birds had survived normal circumstances, but succumbed at a time of stress. Of the remaining birds that died, 19% of those older than one year were still in moult, two months later than normal. While these abnormalities were not necessarily under genetic control, this incident provided another striking example of selective mortality (Swennen & Duiven 1983).

CONCLUDING REMARKS

Some of the extreme weather events described above lasted no more than a few hours, others lasted for several days or weeks, and yet others for several years. They also varied in their spatial impacts, with some short-lived storms affecting areas covering only a few tens or hundreds of square kilometres, while severe cold or drought sometimes affected large parts of a continent. But whatever their duration and geographical extent, extreme weather events have occasionally reduced the numbers of some bird species by more than 90%.

In species whose numbers are affected in this way, we can envisage three scenarios, depending on the frequency and severity of extreme events relative to the recovery rates of the populations concerned. Where such events occur at long intervals, but cause big reductions in numbers, the population may recover within a few years, so that for most of the time its numbers are limited at a higher level by other factors **(Figure 11.12a)**. This type of pattern is exemplified by the Grey Heron *Ardea cinerea*, whose numbers in Britain are reduced by hard winters. In the last 100 years, hard winters have mostly occurred at intervals much longer than the 1–3 years taken for Heron numbers to recover, so that the periodic reductions in numbers were interspersed by periods of relative stability **(Figure 1.2)**.

The second scenario is where extreme events occur at such frequent intervals and cause such heavy losses that the population can never recover fully from one event before it is reduced again by the next, giving the saw-tooth pattern of **Figure 11.12b**. The population is then limited primarily by successive catastrophes, and can never reach the level that it might achieve if such events were less frequent or absent. This type of pattern was shown by the several seabird-species affected by El Niño events off western South America **(Figure 11.11)**. Numbers alternate between single years of abrupt decline and several years of continuous increase. Because numbers can seldom, if ever, reach a level at which they would stabilise, they are presumably for most of the time well below the level that the local environment would otherwise support.

In the third scenario, severe events follow one another so rapidly and each time cause such heavy losses, that the species dies out from the area concerned **(Figure 11.12c)**. We cannot expect to observe this pattern frequently, but it may have been largely responsible for the final demise of the Middle-spotted Woodpecker *Dendrocopos medius* in Sweden (Pettersson 1985), and two successive hard winters almost eliminated the Dartford Warbler *Sylvia undata* from Britain in 1961–63 (see above). It could well apply to many resident species in the northern hemisphere, as their northern range boundaries move first north then south over periods of years,

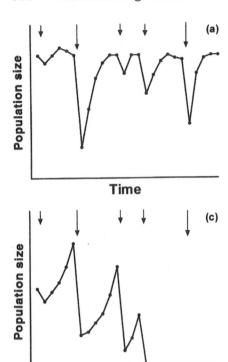

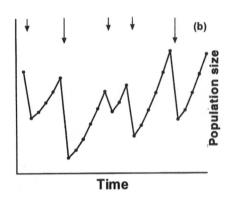

Figure 11.12   Model representing severe weather effects on populations. (a) Severe weather occurs mostly at long and irregular intervals, each time causing marked reductions in numbers from which the population recovers within a few years. For the rest of the time, the population remains fairly stable. (b) Severe weather events occur so frequently that, before numbers have recovered to a level at which they would stabilise, they are again reduced. The population shows a saw-tooth pattern in which occasional abrupt declines are separated by periods of continuous increase. The level reached by the population at any one time depends largely on the duration of the recovery periods between successive declines. (c) Severe weather events occur with such high frequency relative to the recovery rate of the population that they quickly lead to local extinction. Arrows showing the timing and severity of extreme weather events are the same in each diagram.

depending on the severity of the winters. Clearly, particular species might move from one pattern of fluctuation-limitation to another, according to changes in the frequency and severity of extreme events, or to changes in their recovery rates.

Although the timing of extreme climate events is largely unpredictable, small birds with short lifespans are unlikely to experience more than one in their lives, and several generations may pass without any. In the same area, individuals of larger, long-lived species can expect to experience one or more extreme events during their lives, and few if any generations can escape them altogether. It is perhaps not surprising, therefore, that when extreme events do occur, they weed out vulnerable individuals, and in short-lived species sometimes lead to measurable morphological change in the population.

Mortality resulting from sudden storms is almost certainly density-independent, for it occurs within a few hours, with little or no opportunity for competition between individuals (except in special circumstances where sheltered sites are limited). However, in more prolonged severe weather, which leads to food-shortages, the opportunity for competition during the event is much greater, so that initial density may influence the proportion of birds that die. This seems likely from the pattern of seabird fluctuations in **Figure 11.11**, in which higher populations were followed by relatively greater declines. In some studies, attempts were made to take account of density in assessing the effects of weather on winter survival or year-to-year population changes. In the Wren *Troglodytes troglodytes*, for example, 81% of the variation in year-to-year changes in breeding numbers in a wood in southern England could be explained by two factors, density in the first year and the duration of snow-lie (Newton *et al.* 1998). By implication, for any given density, more birds died in snowy than in mild winters, and for any given duration of snow-lie, a greater proportion of birds died in years of high than low densities. Similarly, among Song Thrushes *Turdus philomelos* in Britain, during the period 1962–76, 90% of the variance in year-to-year changes in breeding numbers could be explained by population density in the first year coupled with the number of freezing days in January. The relationship became less marked in subsequent years, when a long-term decline set in, caused by other factors (Baillie 1990).

SUMMARY

Sporadic storms of one sort or another can cause severe mortality of birds, and in several recorded cases have caused reductions of more than 90% in local population levels. Storms in the breeding season have also caused widespread breeding failure. Where habitat damage occurs, recovery of the bird population depends on the rate at which the habitat itself recovers, until which stage different species may be favoured. Similarly, hard winters in northern regions, droughts in arid regions, or shifts in ocean currents can cause massive declines in bird numbers, mainly through effects on food-supplies. Recovery from such events can take up to several years. Species exposed to frequent catastrophic mortality may alternate between periods of decline and recovery, but seldom (or never) reach a level at which their numbers would stabilise.

Heavy losses can occur among migrant birds that encounter fog or storms en route (especially over water) or are caught by aseasonal weather on their breeding areas soon after they arrive in spring or before they depart in autumn. For some species, such losses may represent a major but unavoidable cost of migration.

Severe weather events usually kill a non-random sample of the local avifauna. Some species are much more vulnerable than others (depending on body size and feeding habits), and within species particular sex- or age-groups are often more affected than others, as are particular body-size (or bill-size) classes. Because the latter features are genetically-influenced, such selective mortality provides some striking examples of natural selection in action.

Great Tit *Parus major* repelling Pied Flycatcher *Ficedula hypoleuca* from nest hole.

Chapter 12
Inter-specific Competition

Acting mainly through resources, inter-specific competition is yet another factor that could limit the distribution and abundance of birds. It is one of the principles of ecology that, where resources are limiting, species with identical needs cannot persist together indefinitely in the same area (Gause 1934). Invariably one would be better adapted or more efficient, and would out-compete and replace the other completely. Hence, different species of birds normally differ from one another in distribution, habitat or feeding ecology, or in more than one of these respects (Lack 1947, 1971).

Despite such specialisation, most bird species still share part of their food and other resources with other species, both closely related and more distant ones. If such species are limited by resources, then it follows that the numbers of one could influence the numbers of another. At the population level, competition can thus be defined as a reduction in the distribution or numbers of one or more species that results from their shared use of the same resource. It may involve depletion, where individuals of one species reduce the amount available to individuals of another species. Or it may involve interference, where individuals of one species, by aggressive or other means, reduce access to a resource by individuals of another species. In any pair of competing species, it is rare for both to suffer equally, and in any situation the effects of competition are usually 'asymmetric', falling more heavily on one species than on the other, though the winner in one situation may be the loser in another, as later examples show.

In general, the greater the overlap in the resource needs of different species, the greater the potential for competition between them. But because resources are not usually limiting at all times or in all places, the effects of competition on any given species are likely to vary from time to time and from place to place. When particular food-sources are especially abundant, they may be exploited by a wide range of species which at other times differ in diet. For example, when winged termites emerge from their nests en masse, they are eaten not only by small birds, but by a wide range of other species, including storks and eagles. When the feast is over, all these species revert to their usual different foods. The assumption is that, at high food densities, more species than usual can exploit the resource efficiently. In the presence of such superabundance, they may do so without competing. However, species can also take the same foods at times when little else is available, in which case, if those foods are scarce, competition could be strong. Measures of dietary overlap between species are therefore of limited value in assessing competition unless they also take account of food availability.

To demonstrate inter-specific competition beyond doubt, several criteria should be met, which can be approached in a stepwise fashion (Wiens 1989). As one proceeds through the list in **Table 12.1**, it becomes increasingly difficult to meet successive criteria, but the evidence gathered is progressively stronger. None of the criteria listed is sufficient on its own, but many researchers have inferred competition on criterion 1 alone, or on 1 and 2 together.

Table 12.1   Criteria that establish the occurrence of inter-specific competition with varying degrees of certainty. Modified from Wiens (1989).

Strength of evidence		Criteria
Weak	(1)	Observed patterns of ecological and geographical segregation should be consistent with those expected if competition were occurring.
	(2)	Species should overlap in the resources they use.
	(3)	Intra-specific competition for resources should occur in one or more of the species concerned.
	(4)	Use of a resource by one species should reduce the availability of that resource to another.
	(5a)	Individual effects: individuals of one species should be affected negatively in some way or another, which could reduce breeding or survival.*
	(5b)	Population effects: distribution or abundance of one species should be reduced by another.*
Strong	(6)	Alternative explanations of observed patterns should be ruled out.

*Negative effects on individuals do not necessarily translate to negative effects on breeding population levels (see text).

EFFECTS ON INDIVIDUALS

The effects of competition are best shown by experiments in which the numbers of one species are altered (by addition or removal), and then another is seen to respond as predicted. As in other field experiments (Chapter 7), the response is measured either as a before-and-after comparison or against an appropriate simultaneous control (Connell 1983, Schoener 1983). Individual responses are mostly measured as changes in either feeding rates or 'niche-use', or as a change in some demographic parameter, such as reproduction or survival, which could influence numbers **(Table 12.2)**. The first type of response merely indicates that the feeding rate or habitat use of one species might be constrained by the presence of another. If that species is resource-limited, its numbers might well also be affected, but this cannot be proved unless measured. The second type of response, involving reduced individual performance, gives the possibility of a lower population density, but only if the reduced performance of some individuals is not offset by improved performance or immigration of others.

Feeding rates and niche shifts

When different species feed together on the same food, individuals of a dominant species can greatly reduce the feeding rates of individuals of subordinate species. For example, many seabirds feed on discards from fishing boats. In the north Atlantic, Gannets *Morus bassanus* are dominant to all others. In one study, the feeding success of Herring Gulls *Larus argentatus* was found to decline in direct proportion to the number of Gannets present **(Figure 12.1)**. However, dominance relations can change with circumstance: Herring Gulls can dominate Lesser Black-

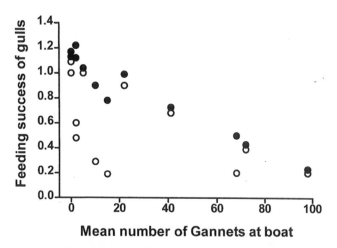

Figure 12.1   Mean feeding success of adult (closed circles) and immature (open circles) Herring Gulls *Larus argentatus* around fishing boats in relation to the number of Gannets *Morus bassanus* present.

The measure of feeding success for each gull takes account of the number of other gulls at the boat of the same age-group. The greater the number of Gannets present, the lower the feeding success of the gulls. Redrawn from Furness *et al.* 1992.

Table 12.2 Field experiments on inter-specific competition in which resources or bird densities were manipulated.

Species	Location (duration)	Variables manipulated	Response observed	Resource involved	Type of competition	Type of experiment*	Source
A. Experiments that led to niche shifts							
Hummingbirds *Lampornis clemenciae, Archilochus alexandri, Eugenes fulgens*	Arizona (1 week)	Food quality and density	Dominant species caused density-dependent habitat shift in subordinate species	Sugar solution in feeders	Interference	3	Pimm et al. 1985
Hummingbirds mainly *Phaethornis superciliosus*	Costa Rica (3 days)	Food availability	Bees excluded hummingbirds from flowers	Nectar	Interference and depletion	3	Gill et al. 1982
Juncos *Junco hyemalis* and Golden-crowned Sparrows *Zonotrichia atricapilla*	California (4 months)	Sparrow density reduced	Habitat shift by Juncos	Water source	Interference	1	Davis 1973
Tits *Parus* and Goldcrest *Regulus regulus*	Sweden (1 winter)	Willow and Crested Tit density reduced	Coal Tits and Goldcrests increased their foraging in the inner tree canopy	Arthropod food-items	Interference and depletion	3	Alatalo et al. 1985
Tits *Parus* and Goldcrest *Regulus regulus*	Sweden (1 winter)	Coal Tit and Goldcrest removed	Crested and Willow Tits shifted their foraging sites	Arthropod food items	Depletion	3	Alatalo et al. 1987
Carolina Chickadee *Parus carolinensis* and Tufted Titmouse *Parus bicolor*	Ohio (4 winters)	Titmouse density reduced	Chickadee expanded feeding niche	Food	Interference	3	Cimprich & Grubb 1994
Birds and rodents	Arizona (1 year)	Rodents excluded	More of seed-crop eaten by birds	Food	Depletion	2	Christensen & Whitham 1993
Birds and ants	Sweden (1 summer)	Access to trees by ants prevented	Insectivorous birds increased use of experimental trees	Food	Depletion	2	Haemig 1992
B. Experiments that led to improved performance							
Great Tit *Parus major* and Blue Tit *P. caeruleus*	England (1 summer)	Blue Tit young removed	Nestling Great Tits heavier	Food	Depletion	2	Minot 1981
House Sparrow *Passer domesticus* and Tree Sparrow *P. montanus*	Missouri (5 years)	Tree Sparrow density increased	House Sparrow fledging success decreased	Food	Depletion	1	Anderson 1978
Magpie *Pica pica* and Jackdaw *Corvus monedula*	Sweden (2 years)	Jackdaw density in Magpie territories increased	Magpies attacked Jackdaws, spent more time foraging and raised fewer young	Food?	Interference and depletion	3	Högstedt 1980

Species	Location (years)	Manipulation	Result	Resource	Mechanism	No.	Reference
Great Tit *Parus major* and Pied Flycatcher *Ficedula hypoleuca*	Sweden (2 years)	Both species' breeding density increased	Both species' success reduced	Food?	Depletion	1	Alerstam 1985
Collared Flycatcher *Ficedula albicollis* and Tits *Parus*	Sweden (1 summer)	Tit nesting densities increased	Flycatcher fledgling number and weight increased	Food	Depletion	2	Gustafsson 1987
Blue Tit *Parus caeruleus* and Great Tit *Parus major*	Hungary (3 years)	Tit breeding densities reduced	Blue Tit nestlings larger when Great Tit numbers reduced; Great Tit nestlings heavier when Blue Tit numbers reduced	Food	Depletion	2	Török 1987
Mallard *Anas platyrhynchos* and fish	England (5 years)	Fish removed	Mallard duckling survival increased	Food	Depletion	2	Wright & Phillips 1990

C. Experiments that led to increased density

Species	Location (years)	Manipulation	Result	Resource	Mechanism	No.	Reference
Various songbirds	Norway (8 years)	Densities of hole-nesters manipulated	Inverse relationship in 3 open-nesting species	Food?	Not known	2	Hogstad 1975
Various songbirds	Arizona (3 years)	Densities of hole nesters increased	No response in densities of open nesters	Food?	Not known	2	Brawn et al. 1987
Bark-foraging birds	Illinois (1 winter)	Dominant species removed	Other species increased in density	Foraging sites	Interference	2	Williams & Batzli 1979
Blue Tit *Parus caeruleus* and Great Tit *Parus major*	Belgium (2 years)	Great Tit excluded from winter roost-sites	Blue Tit nesting density increased	Winter roosts	Interference	3	Dhondt & Eyckerman 1980
Great Tit *Parus major* and Chaffinch *Fringilla coelebs*	Scotland (1 summer)	Chaffinch removed	Great Tit expanded into former Chaffinch territory	Territory	Interference	2	Reed 1982
Garden Warbler *Sylvia borin* and Blackcap *Sylvia atricapilla*	England (3 years)	Blackcaps removed	Garden Warbler increased and occupied former Blackcap territories	Habitat (territories)	Interference	1	Garcia 1983
American Redstart *Setophaga ruticilla* and Least Flycatcher *Empidonax minimus*	New Hampshire (3 years)	Flycatchers removed	Redstarts increased	Food	Interference	3	Sherry & Holmes 1988
Goldeneye *Bucephala clangula* and fish	Sweden (3 years)	Fish removed	Use of lake by Goldeneye increased	Food	Depletion	1	Ericksson 1979
Waterfowl and fish	England (5 years)	Fish removed	Use of lake by waterfowl increased	Food	Depletion	2	Phillips 1992
Collared Flycatcher *Ficedula albicollis* and tits *Parus*	Sweden (2 years)	Tit density reduced	Flycatcher density increased	Nest-sites	Interference	3	Gustafsson 1988

Table 12.2 Continued.

Species	Location (duration)	Variables manipulated	Response observed	Resource involved	Type of competition	Type of experiment*	Source
Blue Tit *Parus caeruleus* and Great Tit *Parus major*	Germany (6 years)	Access of larger Great Tits to nest boxes reduced	Blue Tit density increased	Nest-sites	Interference	1	Löhrl 1977
Blue Tit *Parus caeruleus* and Great Tit *Parus major*	Belgium (2 years)	Access of larger Great Tits to roost sites reduced	Blue Tit density increased	Roost-sites	Interference	3	Dhondt & Eyckerman 1980
Great Tit *Parus major* and Starling *Sturnus vulgaris*	Holland (1 summer)	Starling nesting density increased or decreased	Great Tit nesting density decreased or increased	Nest-sites	Interference	2	van Balen et al. 1982
Cahow *Pterodroma cahow* and White-tailed Tropicbird *Phaethon lepturus*	Bermuda (several years)	Access of Tropicbirds to nest-sites prevented	Cahow nesting densities increased	Nest-sites	Interference	1	Wingate, in Cade & Temple 1995

* 1 = Before-and-after comparisons; 2 = simultaneous experimental and control areas; 3 = reversed treatments or replications.

backed Gulls *L. fuscus* and Great Skuas *Stercorarius skua* at land-based feeding sites, but are less successful than both these species in competing around fishing boats at sea (Furness *et al.* 1992). Such interactions are often construed as competition, but they are not necessarily translated into reduced survival or reproduction, let alone reduced population level.

As another example of interference competition, Pink-footed Geese *Anser brachyrhynchus* and Barnacle Geese *Branta leucopsis* were studied on moulting areas in Greenland (Madsen & Mortensen 1987). Where they occurred separately, both species ate mainly grasses and sedges. Where together, both ate more moss which was less nutritious, but Barnacle Geese more than Pinkfeet. Where separate, both species spent 41–46% of the 24 hours grazing. Where together, Barnacle Geese spent 62% of the time grazing, while Pinkfeet seemed unaffected. Hence, in this area, the two species seemed to compete for favoured resources, the Barnacle Geese suffering more than the Pinkfeet. By aggressive interference, Pinkfeet kept Barnacles from the best feeding areas, but no effects on numbers were discerned.

Experiments on niche shifts have been conducted with mixed flocks of tits and other species. In conifer forests of northern Europe, dominant Willow Tits *Parus montanus* and Crested Tits *P. cristatus* fed mainly at favoured sites within trees, while the smaller Coal Tits *P. ater* and Goldcrests *Regulus regulus* were relegated mainly to the outer twigs, where the risk of predation was greater (Suhonen *et al.* 1992). When the dominant species were experimentally reduced in numbers in three forest plots, the subordinate species in these plots increased their foraging in the inner canopy in comparison with three control plots (Alatalo *et al.* 1985). This segregation was presumed to be due largely to interference, for the larger species often chased the smaller ones. But when subordinate species were experimentally removed, so that food-supplies in the outer parts of trees were less depleted, the dominant species spent significantly more time feeding there (Alatalo *et al.* 1987). So by an apparent combination of interference and depletion, the range of foraging sites used by each species was reduced by the presence of other species.

In other experiments, the food consumed by one species was measured in the presence or absence of another. When seed-eating mammals were denied access by metal barriers to certain Pinyon Pine *Pinus edulis* trees, a greater proportion of the seed-crop was consumed by birds, compared with unprotected trees (Christensen & Whitham 1993). Likewise, increased use of arboreal insects by birds occurred after insect-eating ants were excluded from certain trees (Haemig 1992). None of these experiments included an assessment of the effects of the expanded niche or food-supply on numbers, so they are of limited value in our present context. They serve mainly to confirm that particular species are constrained in their use of resources by individuals of other species, and can change continually in response to prevailing conditions.

Individual performance

Other experiments have explored the effects of inter-specific competition on the breeding or survival of individual birds **(Table 12.2)**. In southern Sweden, Magpies *Pica pica* and Jackdaws *Corvus monedula* fed their chicks the same sorts of food, and as chicks often starved, food seemed limiting. To test for competition, Högstedt (1980) encouraged Jackdaws to nest within certain Magpie territories by placing

Table 12.3   Breeding performance of Magpies *Pica pica* with and without Jackdaws *Corvus monedula* nesting in their territories.

Magpies and Jackdaws feed their chicks on similar foods, so could compete. Jackdaw nest-boxes were placed experimentally within the territories of some Magpie pairs, leaving others as controls. The next year the treatments were reversed. Magpies that had Jackdaws nesting in their territories produced fewer young than those without. Magpies often attacked Jackdaws, but Jackdaws did not attack Magpies. *$P<0.05$, ***$P<0.001$. From Högstedt 1980.

	Magpie territories		
	With Jackdaws		Without Jackdaws
Number of nests started	18		38
Mean weight of nestlings at 21–25 days (± SE)	182±7g	*	208±4g
Number of young per successful brood (± SE)	2.0±0.6	*	4.0±0.3
Number of young per breeding attempt (± SE)	0.33±0.20	***	1.68±0.34

No significant differences between the two groups were noted in laying dates, clutch sizes or weights, or in the proportions of clutches that reached the hatching stage.

three nest-boxes in each, leaving other Magpie territories without Jackdaws. In the second year, he moved the boxes, so that the same Magpie pairs acted as experimental and control pairs in different years, thus allowing for effects of individuals and territories on performance. Magpies with Jackdaws in their territories suffered significantly reduced breeding success, producing fewer young with lower fledging weights than those without **(Table 12.3)**. Högstedt attributed this difference both to increased starvation, because of food depletion in Magpie territories by Jackdaws, and to increased predation of nest contents by Hooded Crows *C. corone*. Increased predation occurred because the Magpies spent longer away from their nest searching for food and chasing Jackdaws, and their hungry young called more, attracting the attention of Crows.

Another experiment involved two related tit (*Parus*) species which nested in the same wood and ate many of the same foods (Minot 1981). The mean fledging weights of young Great Tits *P. major* in different parts of the wood were inversely correlated with the local density of Blue Tits *P. caeruleus*: Great Tits were heavier where Blue Tits were scarcest. So in one part of the wood, all young Blue Tits were removed at hatching, thus reducing the food demands of their parents. Young Great Tits in this area then fledged at higher weights than those elsewhere, increasing their subsequent survival chances. By implication, reduced pressure on a shared food-supply, occasioned by the removal of young Blue Tits, left more food for Great Tits, whose young were then better fed. Other studies elsewhere showed that Blue Tits reduced other aspects of Great Tit performance, including nestling and adult survival, and the proportion of pairs that raised second broods (Dhondt 1989).

In another experiment, when the densities of Great Tits and Blue Tits were experimentally reduced by more than 90%, Collared Flycatcher *Ficedula albicollis* pairs in the same areas raised more nestlings, and to greater weights, than those in control areas nearby. This held in both years of study, and in one year in one plot, the survival of adult flycatchers was also increased. Thus, those flycatchers that bred under conditions of reduced competition from tits were found to contribute more recruits to future breeding populations than did those breeding in control plots (Gustafsson 1987). The proposed mechanism was again differential access to food for the young, in both nestling and post-fledging periods. Other studies that led to improved breeding success in tits and flycatchers after reduction in the density of a competing species are listed in **Table 12.2**, but as in the above studies, effects on population levels were not examined.

As with any other adverse factor, declines in individual performance are not necessarily translated into declines in subsequent breeding numbers. The mortality or breeding failure of a certain proportion of individuals through competition might be compensated by improved survival of the remaining birds, so that breeding numbers are maintained. In the rest of this chapter, therefore, I shall concentrate on those forms of evidence that implicate inter-specific competition in limiting the distributions or breeding densities of birds.

EFFECTS ON POPULATIONS: LONG-TERM RESPONSES

Evidence for effects of competition on populations is of two main types **(Table 12.4)**. The first involves a short timescale, in which effects on the ecological niches, performance and numbers of individuals can be measured. These effects can be immediately reversed if conditions change, and are amenable to experiments of the kind discussed above. They give evidence of competition acting here-and-now. The second type involves a longer (evolutionary) timescale, and leads to effects on bird morphology (promoting species differences in ecology) and geographical range (promoting species differences in distribution). Because of the timescale involved, these evolutionary effects are not amenable to experimentation, and nor are they

Table 12.4   Types of circumstantial evidence that inter-specific competition can influence the distribution and abundance of species.

A. Short-term responses
1. Complementary density changes
   (a) In time
   (b) In space
2. Observed mechanisms of competition
   (a) For food
   (b) For nest-sites
   (c) For territories

B. Evolutionary and other long-term responses
3. Ecological segregation (in habitat or feeding behaviour)
4. Allopatric distributions
5. Expanding–retracting species pairs
6. Competitive release
7. Character displacement
8. Numerical release
9. Species replacements in time
10. Failure to establish

immediately reversible. They can only be inferred from existing patterns of ecology and distribution among related species, and so provide no more than circumstantial evidence of past competition. One partial exception, however, occurs when the establishment of a species in a new area (whether natural or human-assisted) is followed by reductions in the distributions or numbers of other species with similar ecology already present. This section is concerned with supposed long-term responses to competition, recognising that no clear-cut distinction exists between short-term and long-term impacts, and that given sufficient time, short-term functional responses could lead to long-term evolutionary ones.

Ecological segregation

In the absence of competitors, some species might well use a wider range of habitats and food sources, and thereby achieve greater population levels, than in the presence of competitors. Because of this possibility, Hutchinson (1957) distinguished between the *fundamental niche* of a species which embraces the full range of environmental conditions, habitats and food-sources that a species could potentially exploit in the absence of competitors, and the *realised niche* which reflects the narrower range of conditions, habitats and food-sources that it actually does exploit. The realised niche may differ from one region to another, depending on which competing species happen to be present.

The morphologies of related species living in the same area sometimes differ by constant ratios (Hutchinson 1959). Each species in a series might, for example, be 20% larger than the one below. This patterning of species has been attributed to competition: if they differed by less than 20%, they would be too similar in their needs and one would eliminate the other; conversely, if they differed by more than 20%, another species might become established, occupying a niche between the two. On this view, competition could in time lead to morphologies (and hence ecologies) that are 'patterned', with species differing by constant amounts, rather than by random amounts. In this way, the argument goes, competition between species could have played a central role in shaping community structure. As yet, however, no-one to my knowledge has examined species in a range of different communities to find how widespread such patterned morphologies are. Those described, such as the Galapagos finches (Grant & Schluter 1984), may be exceptions that have forced themselves upon the attention of ecologists, and might anyway have other explanations.

Some researchers have sought evidence for competition simply by documenting range or niche differences in the field (Criteria 1 in **Table 12.1**). Closely-related bird species in the same area invariably differ in some aspect of their ecology, as explained above, but this could have arisen in at least three ways. They might differ because of continuing competition, co-existing by virtue of niche differentiation, but retaining the ability to expand their realised to fundamental niches in the absence of competitors. The second interpretation invokes past competition, through which species have come to diverge in ecology, so that they no longer compete (called 'the ghost of competition past'). On the third explanation, species might differ not because of competition, past or present, but because of other features, evolved in their period of isolation from one another, which have made them different. On this view, related species might differ in ecology simply because they are different species, with different structure and morphology. On the basis of observational data alone, it is impossible to reject any of these interpretations, and all three could hold to varying degrees in one group or another.

Despite lack of proof for the principle of competitive exclusion, it has become widely accepted because of its inherent appeal, because of the weight of indirect evidence in its favour, and because of its firm theoretical base (notably in the equations of Lotka and Volterra). For the fact remains that, without some degree of ecological segregation between species, whatever its origin, such species could not co-exist in the long term if resources were limiting. They might co-exist if resources were limiting only for brief periods, with the effects of competition relaxing before any of the competing species could disappear completely from the region concerned. They might also co-exist if resources were never limiting, as when the numbers of potential competitors were kept by natural enemies below the level at which competition would occur.

Allopatric distributions

Similar species are often allopatric, occupying separate geographical ranges which abut or overlap to a small extent (called contiguous allopatry). The fact that the different species do not spread through one another's ranges has usually been attributed to competition, with one species favoured in one region and another in the adjoining region. As Lack (1971) put it, species inhabit separate but adjoining ranges where (1) they have such similar ecology that only one of them can persist in any one area, and (2) each is better adapted than the other to part of their combined ranges. Competition is the more likely explanation if the zone of contact between them does not correspond with an obvious change in habitat.

Between some species, such as the Black-capped Chickadee *Parus atricapillus* and Carolina Chickadee *Parus carolinensis* in North America, the boundary is abrupt, as their respective ranges are almost mutually exclusive, and where they meet they use different habitats (Snow 1954). Between most species, however, a much broader overlap zone occurs, in which species gradually replace each other in a region of similar habitat. In Europe, this situation is exemplified in varying degrees by the Chaffinch/Brambling (*Fringilla coelebs/F. montifringilla*), the Crested Tit/Siberian Tit (*Parus cristatus/P. cinctus*), the Curlew/Whimbrel (*Numenius arquata/N. phaeopus*), and the Red-necked Phalarope/Grey Phalarope (*Phalaropus lobatus/P. fulicarius*). In each pair, the first named species breeds mainly south of the other, and their zone of equal abundance has shifted northward during this century coinciding with climatic change. Such boundary changes thus implicate environmental factors in influencing the competitive balance between related species in similar habitats (Lack 1971).

Although two related species may occupy mainly different areas, supposedly through competition, this on its own does not explain why the species are found in the areas they are, and not the other way round. To explain this, a second difference is usually invoked, say in morphology or physiology, which might give one species the advantage in one area, and the second species in the other area. For example, Brünnich's Guillemot *Uria lomvia* occurs mainly to the north of the Common Guillemot *Uria aalge*, and is also slightly larger, so better adapted to a colder climate. The two species together conform to Bergman's rule (originally applied to subspecies rather than species), that northern forms are larger. A similar difference occurs in other species pairs on winter quarters, for example the Imperial Eagle *Aquila heliaca* and Steppe Eagle *A. rapax*, the Greater Spotted Eagle *A. clanga* and Lesser Spotted Eagle *A. pomarina*, and the Curlew and Whimbrel. In all three pairs, the larger (first-mentioned) species winters mainly north of the other. While this association does not explain which came first, the difference in distribution or the difference in body size, once evolved, the morphological difference may help to maintain the distributional one.

For some species, competition may not be 'direct', involving a single closely-related congener, but 'diffuse', involving several species, not all of which may be closely-related. Such diffuse competition is hard to detect, because we have no taxonomic or distributional clues as to who the competitors might be. But the idea is that the community of species present in an area divide up the feeding opportunities in such a way that an invader would have to displace several of them from parts of their niches if it were to persist. We can only surmise the existence of diffuse competition in special circumstances, in localities where whole constellations of species are missing, and range (or niche) expansions occur in species that have no clear single-species competitor. Examples are provided by the many species that occupy wider niches on islands, where potential competitors are fewer, than on mainlands (Lack 1971, MacArthur 1972), and by montane bird species, some of which occupy a wider range of elevation on those mountains that hold the fewest other species (Terborgh & Weske 1975).

Related bird species, which live side by side in the complex habitat of mainlands, often replace each other, in checkerboard fashion, in the simpler habitat of islands. Different species occur on different islands, but no more than one species occurs on any one island. The species thus have mutually exclusive, but interspersed distributions. An example is provided by two cuckoo-dove species *Macropygia mackinlayi* and *M. nigricollis*, which are found interspersed on various islands in the Bismarck Archipelago off New Guinea (Diamond 1975). Other examples occur in isolated mountain forests in the African savannahs, each of which holds only one species of woodpecker, but the several species involved co-exist in more extensive forest elsewhere (Moreau 1966).

Although such distributions imply competition, involving the local exclusion of one species by another, they do not explain why a particular species is present on certain islands (or mountains), and another species on others. The pattern could result from chance, depending on which species happened to become established first on each island; or it could result from subtle differences between islands, including the mix of competing species, which could tip the balance in favour of one species on certain islands and of another species on other islands. Whatever the explanation, checkerboard distributions imply that the presence of one species on an island, for whatever reason, will exclude another similar one, and thus restrict its range. In none of the studies involved, however, was habitat examined in detail, and alternative explanations could not be eliminated, including on some islands the possibility of human interference.

Expanding–retracting species pairs

As stronger evidence for the effects of competition in restricting bird distributions, one species can sometimes be seen in the process of expanding at the expense of another. Many examples involve closely-related species or subspecies which have come together in secondary contact, after a period of isolation in separate regions in which they diverged. On the North American prairies, at least 14 pairs of species or subspecies replace each other in woodland on the east and west sides (Rising 1983). In 11 of these pairs, hybridisation is known to occur at least occasionally along the contact zone. Some of the species may have come together only recently, as a result of human activities, such as tree-planting. Examples include the Baltimore Oriole *Icterus galbula* and Bullock's Oriole *I. bullockii,* the Rose-breasted Grosbeak *Pheucticus ludovicianus* and Black-headed Grosbeak *P. melanocephalus,* and the Indigo Bunting *Passerina cyanea* and Lazuli Bunting *P. amoena.* In several such pairs, one species has expanded in recent times at the expense of the other. For example, along the Niobrara River in Nebraska, the Indigo Bunting spread westward by about 200 km in 15 years, replacing the related Lazuli Bunting (Emlen *et al.* 1975). In this pair, hybridisation was infrequent, and the change was attributed primarily to competitive exclusion. Similarly, the Baltimore Oriole is replacing Bullock's Oriole, but in this case with a hybrid zone spreading west at about 10 km per year.

In conclusion, competition could be involved in limiting many bird ranges, but the evidence is circumstantial, and alternative explanations of the findings are hard to eliminate. Because the outcome of competition is likely to be influenced by other factors, we should expect that the boundaries between apparent competitors would move as climate and other conditions change.

Competitive release

Some potentially competing species occur in some areas together (sympatry) and in other areas alone (allopatry). Evidence for competition comes from niche contraction where the species occur together, or from niche expansion where they occur alone (known as competitive release). Many closely-related species of bird that occupy mainly different geographical ranges, diverge in habitat or feeding habits only in the overlap zone where they occur together (Lack 1971). For example, throughout its wide European range, the Chaffinch *Fringilla coelebs* breeds in both coniferous and broad-leaved woods. On the islands of Gran Canaria and Tenerife, however, it is replaced in pine forests by the Blue Chaffinch *F. teydea.* This is the only part of the range of *F. coelebs* where a second chaffinch species occurs and the only place within its range (including some other Canary Islands) where it does not breed in pine forest. Again, two species of nightingales breed in thickets in Europe, *Luscinia megarhynchos* in the west and southwest and *L. luscinia* in the east. They replace each other geographically, except in parts in eastern Europe where their ranges overlap. Here, and only here, *L. megarhynchos* is restricted to drier and *L. luscinia* to wetter habitats of the same general type. In these and other species pairs, the differences in ranges, and in habitats in the zone of overlap, have been attributed to competition. They again indicate how the presence of one species in an area might restrict the distribution and numbers of another through niche contraction.

Turning now to niche expansion, when a species is missing from a locality where it would be expected to occur, its place is sometimes taken by another species, found elsewhere in a different kind of habitat. Ground doves in the New Guinea region provide an example (Diamond 1975). On New Guinea itself, three similar species of ground dove occur in sequence from coast to inland: *Chalcophaps indica* in the coastal scrub, *C. stephani* in the light or second growth forest, and *Gallicolumba rufigula* in the rainforest. However, on the island of Bagabag, where *G. rufigula* is absent, *C. stephani* extends inland into rainforest, while on Karkar, Tolokiwi, New Britain and numerous other small islands where *C. indica* is also absent, *C. stephani* expands coastwards to occupy the whole habitat gradient. On Espiritu Santu, on the other hand, where *C. indica* is the only species present, it occupies all three habitats. Hence, it seems that *C. stephani* and *C. indica* occupy rainforest only where *G. rufigula* is absent, *C. stephani* occupies coastal scrub only where *C. indica* is absent, and *C. indica* occupies light forest only where *C. stephani* is absent. One possible explanation is that competitive exclusion and niche differentiation occur on the New Guinea mainland, and various degrees of competitive release on nearby islands. Similar patterns, also attributed to competitive release, have been described among montane forest birds in East Africa and South America (Moreau 1966, Terborgh & Weske 1975).

The House Sparrow *Passer domesticus* and Tree Sparrow *P. montanus* present a more complicated picture (Summers-Smith 1988). In the west of their joint Palaearctic range, the House Sparrow occupies towns and villages, in close commensal relationship with humans, while the Tree Sparrow occurs in the countryside away from buildings. But in southeast Asia, where the House Sparrow is absent, the Tree Sparrow occupies the towns and villages. In between, the separation is less clear cut: in some regions the Tree Sparrow occupies the towns and the House Sparrow the rural areas, while in other regions the situation is reversed. The general trend, however, is for a change from House Sparrow to Tree Sparrow in the towns, as one moves from west to east, as other conditions change. Moreover, in much of the Mediterranean region where both these species are absent, the Spanish Sparrow *P. hispaniolensis* takes over the urban niche. In eastern Sardinia, which the Tree Sparrow has recently colonised, it has displaced the Spanish Sparrow from built-up areas. In a sense, such examples provide natural experiments on the role of competition. But one can never be sure that some other relevant factor does not differ between areas, besides the presence or absence of a potential competitor, and accounts for the pattern.

Character displacement

In some species, it is not only the realised niche of a species that appears to alter from sympatry to allopatry, but also its morphology. Two species may differ most strongly in body or bill-structure, and hence in feeding habits, in the overlap zone. An example is provided by the two rock nuthatches, the eastern *Sitta tephronata* and the western *S. neumayer*, both of which frequent rocky hillsides. The two species differ in body and bill size only in Iran where they overlap, but are similar in size where they occur separately **(Figure 12.2)**.

This phenomenon, which is also found in other animals, was called 'character divergence' by Darwin (1859) and 'character displacement' by later authors (Brown & Wilson 1956). The divergence is regarded as an evolutionary response to competition (past or present) between the species where they overlap. However, it always remains possible that two species differ in the overlap zone for some completely different (unknown) reason. Also, there remains the question of why the overlap zone is apparently stable, and why the species do not diverge more widely, and spread through the whole of their mutual ranges. There seem to be two possible explanations for this lack of spread, one that food-sources are unusually diverse in

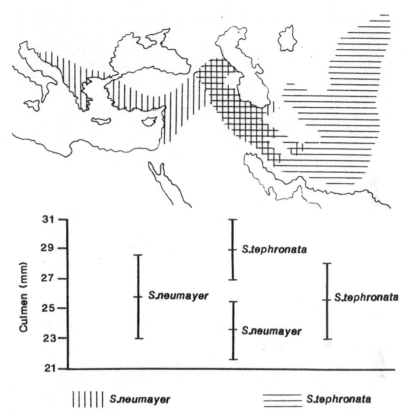

Figure 12.2 Geographical ranges (above) and beak-lengths (below, mean and range) of Western Rock Nuthatch *Sitta neumayer* and the Eastern Rock Nuthatch *S. tephronata*. The two species differ in bill size only where they overlap in Iran, and not elsewhere. From Vaurie 1951.

the overlap zone, so that only there can the niche be divided; the other that the species came into contact only in recent times and that ecological segregation is taking a long time to spread further. Good examples of character displacement are few, however, and not all supposed examples stand critical examination.

Examples of a related phenomenon, 'character release', are provided by the European tits. Particular species diverge from the usual morphology of their species only in small parts of their range where competitors are absent. Thus, on the Swedish island of Gotland, Coal Tits *Parus ater* are larger than on the mainland, associated with the absence from this island of the larger Crested Tit *P. cristatus* and Willow Tit *P. montanus* (Alatalo *et al.* 1986). The change in morphology coincides with a shift in foraging niche, as the Coal Tits of Gotland feed more on the inner parts of trees, while on the mainland they feed mostly on the outsides of trees and on needles. In other tit communities elsewhere in Europe, the absence of any one species is commonly associated with a morphological change in another, with small species becoming larger than elsewhere (Lack 1971, Dhondt 1989).

Because birds overlap in diet with a wide range of animals, not all character shifts need occur in response to other birds. On the Galapagos Islands *Geospiza* finches eat mainly seeds, but some also eat nectar in the dry season, a food-source they share with a large bee *Xylocopa darwini*. On islands where the bee is present, nectar comprised 4% of the diet of the finches *G. fuliginosa* and *G. fortis*, compared with 20% on islands where the bee was absent (Schluter 1986). In addition, the finches were smaller on islands without bees than on islands with bees. Because there were no obvious differences in the flower types between islands, Schluter (1986) suggested that the bee influenced nectar use by the birds, and that character displacement in finch body size occurred in response to the presence of the bees. Again, however, this plausible interpretation is no more than consistent with competition theory; it may have other explanations (such as other differences between areas) and is practically impossible to test.

Numerical release

Species limited by resources might be more numerous if they did not share those resources with other species, whether other birds or other animals. Consistent with this view, some species achieve greater densities in areas where competitors are absent than where they are present. Species on islands are typically fewer than on nearby mainland, but often occupy wider niches and achieve higher densities. On any given density of seeds on the ground, finches were four times more numerous on the Galapagos Islands than on various continental areas (Schluter & Repasky 1991). The authors attributed the difference to the lack of ants, rodents and other competing seed-eaters on the Galapagos. Similar differences have been found between large and small islands in the same region, if they held different numbers of potentially competing species. Among Wrens *Troglodytes troglodytes*, for example, territory sizes on various islands were inversely related to their per area food-contents, as found for many birds (Chapter 3). However, they were also related to the numbers of potential competitors present, being larger (for a given food density) on islands that had the most other species of small insectivorous birds (Cody & Cody 1972). Other examples have been found on other islands (Terborgh & Faaborg 1973).

The same phenomenon is sometimes apparent in patches of habitat. An example involving several species comes from the evergreen woodlands that occupy the southern slopes of the mountain ranges along the southern coast of Africa (Cody 1983). These woodlands have similar vegetation, but occur in smaller and more isolated patches from east to west. Bird species numbers also declined gradually in the same direction, from 45 in the extensive woodland in the east to 15 in the smallest and most isolated patch in the west. Yet despite this two-thirds decline in species numbers, estimated total bird densities remained roughly constant, as individual species became correspondingly more numerous **(Figure 12.3)**. Density compensation was complete in small insectivorous birds, partial in some other groups, and absent in yet others (such as pigeons). The implication was that some species at sites with fewer competitors reached higher densities through having access to resources that were taken by other species elsewhere. This was plausible, but again other explanations of the findings could not be excluded.

Species replacements in time

The invasion of an area by a new species is sometimes followed by population declines among those species already there that use the same resources. Forest clearance, initially to form roads, enabled non-forest birds to spread south through

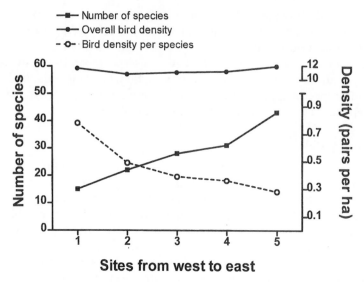

Figure 12.3   Density compensation in the bird species of Afro-montane woodland patches in southern Africa. Five count sites are ranked from the smallest and most isolated patch at the left through increasingly larger and less isolated patches on the right. Species numbers varied by 300% between patches but total bird density within the patches remained almost constant, varying only about 10%. The densities of individual species were greatest where fewest other species were present. From Cody 1983.

the Malyasian peninsula to reach Singapore (Ward 1968). As new invaders moved in, indigenous species with similar ecology declined or disappeared. In fact, the declines in eight out of ten species recorded during the 20th century were clearly associated with the arrival and establishment of similar species from elsewhere. Striking examples were provided by munias, with *Lonchura punctulata* replacing *L. maja*, and by the kingfishers, with *Halcyon chloris* replacing *H.smyrnensis*. Other examples of new invaders replacing earlier ones with similar ecology can be found among introduced Hawaiian birds (Moulton & Pimm 1986).

In other cases, an invader displaced an established species from only part of its niche, so that the two persisted side by side in different habitats. More than a century ago, the Rock Sparrow *Petronia petronia* bred in both towns and countryside of the central and western Canary Islands, but from the 19th century the Spanish Sparrow *P. hispaniolensis* established itself there, after which the Rock Sparrow disappeared from the towns, but not from the countryside (Cullen *et al.* 1952). Another example, involving the replacement of the Spanish Sparrow by the Tree Sparrow *P. montanus* in Sardinian towns, was mentioned earlier.

Other invaders have not eliminated competitors from any of their habitats, but merely reduced their densities. Thus, as the Starling *Sturnus vulgaris* spread over North America, it took over nest-cavities previously used by Eastern Bluebirds *Sialia sialis* and others whose densities declined in consequence (Chapter 8). The purported mechanism was later confirmed when the widespread provision of a surplus of nest-boxes enabled both Bluebird and Starling numbers to increase (Chapter 8).

Not surprisingly, most recorded instances of species replacements, whether complete or partial, are in man-made habitats. Even here, all that is observed is 'replacement' and that it is due to 'competitive displacement' is an inference. Replacement might be due to some other interaction, such as disease transmission, or simply to the continuing modification of habitats by human action, or to the independent loss and gain of similar species. However, it would be straining coincidence too far to suggest that in every such case the two events were unconnected.

Failure to establish

Species that successfully establish themselves in new areas attract attention, but there are many more that fail (Lack 1971). Every birdwatcher delights in the fact that occasional individuals of certain bird species, which normally live far away, periodically turn up as rarities. More than half the species on the British list are vagrants, which appear from time to time but do not remain to breed. Some such species appear in numbers every year, while others appear less often, in extreme cases perhaps once in several decades. The same has been noted in every well-studied region of the world, leaving no doubt that birds continually reach new areas. Clearly, dispersal is not the main obstacle to range extension. The main problem is the difficulty of establishment, and competition is one of several factors likely to be involved. The evidence is again circumstantial, namely that invading or introduced birds have more often established themselves in areas with impoverished avifaunas (such as some remote islands) and in areas of new man-made habitats (such as cultivated land) than in areas of natural climax vegetation **(Table 12.5)**. It is in the latter, with long-established species-rich communities, that any new invader is likely to meet the strongest competition.

One way or another, about 5% of all bird species have been introduced by human action elsewhere, many to more than one region (Long 1981, Lever 1987). Yet despite large numbers of releases, less than 30% of released species have managed to

Table 12.5   Habitat distribution of native and exotic birds (passerines and doves only) on three well studied Pacific Islands. In each island group, native bird species are mostly associated with native habitats, and exotic species mostly with new man-made habitats. From Case 1996.

	Percentage species found:		
	Only in native habitats	In both native and non-native habitats	Only in non-native habitats
Fiji-Viti Levu			
Native species (N=27)	52	44	4
Exotic species (N=8)	0	43	57
Hawaii-Kauai			
Native species (N=13)	100	0	0
Exotic species (N=17)	0	53	47
Societies-Tahiti			
Native species (N=5)	80	20	0
Exotic species (N=9)	0	22	78

establish themselves in a new area. Most species died out quickly, probably without reproducing. Others persisted for a few generations and then disappeared. Yet others established themselves only in a small area, but spread no further, and only a minority increased and spread over large areas. Of about 100 bird species introduced to North America, only six have become widespread, namely the House Sparrow *Passer domesticus*, Starling *Sturnus vulgaris*, Feral Pigeon *Columba livia*, Chukar *Alectoris chukar*, Ring-necked Pheasant *Phasianus colchicus* and Grey Partridge *Perdix perdix*. A few others have persisted only in particular localities, such as the Skylark *Alauda arvensis* on southern Vancouver Island, the Tree Sparrow *Passer montanus* in northern Missouri, and the Mute Swan *Cygnus olor* in some eastern states. More recently, the Monk Parakeet *Myiopsitta monachus* and the House Finch *Carpodacus mexicanus* have become widely naturalised, and a number of escaped cage-birds, notably parrots and doves, have become established in some large cities. All the rest quickly died out. Moreover, most successful species settled in the man-made habitats of farmland, parks and towns, and few successfully penetrated more natural habitats. Similarly, of at least 143 exotic bird species released in New Zealand, only 34 became established, again mainly in man-made habitats (Baker 1991). The same pattern is repeated in other parts of the world.

The success rate of bird introductions (defined as the percentage of attempted introductions that led to establishment) on tropical and sub-tropical islands declined steeply with the species richness of the extant native avifauna (Diamond & Case 1986). It varied from 90% for Ascension, with no native landbirds, to 60–80% for Lord Howe, Bermuda, Rodriguez and Seychelles, which had 3–17 native landbirds. In contrast, the success rate was nil on Borneo (with about 420 species) and New Guinea (with 513 species). The chance of an invader establishing itself thus seemed to vary inversely with the numbers of species already present. Secondly, the penetration of native forest by successfully introduced species also declined with species richness of the extant native avifauna. For example, on Viti Levu, the largest island of the Fiji archipelago (with 48 native species), the introduced species are strictly confined to human-modified open habitats. On Oahu in Hawaii, with ten native species, numerous introduced species are abundant in native forest, while Kauai is intermediate between Oahu and Viti Levu, both in species numbers (19) and penetration of forest by introduced birds (Diamond & Case 1986). However, these analyses took no account of island area, and the degree of human-induced habitat modification. Other studies revealed that a newcomer was less likely to establish itself if another species with similar morphology (and hence ecology) was already present (Moulton & Pimm 1983). All these findings would be expected if competition were important in hindering establishment.

None-the-less, competition has probably had less influence on the present avifauna of islands than previously thought. From a study of the birds on 71 island or island groups around the world, Case (1996) concluded the following:

(1) The number of exotic species gained by different islands was close to the number of native species lost through recent extinction. However, allowing for island area, both figures depended on the proportion of native habitat that had been replaced by man-made habitat: the numbers of native species declined with loss of natural habitat while the numbers of exotic species increased with expansion of man-made habitat.

(2) More than half of all known species extinctions on islands probably occurred before exotics were introduced. Hence, most extinctions of native birds could not have been caused by competition with newcomers, although some may have been. Similarly, the species lost from islands were often totally different from those gained. Hawaii, for example, lost many waterfowl, raptors, rails and forest nectar-feeders, but has gained mostly small seed-eaters and insect-eaters of open country. Hence, most exotics have probably not greatly benefited from the loss of native species, although again some may have. The most successful exotics on Hawaii, as elsewhere, are those that prosper in the new habitats provided by human activity.

(3) Other factors that influence the ability of introduced species to establish themselves include the number of introduction attempts made and numbers of birds released each time, the area of the island (more exotic species on bigger islands) and the remoteness of the island (more exotic species on remote islands that once had many endemic species).

(4) Exotics may find it more difficult to penetrate species-rich than species-poor communities, but this idea is hard to test on most islands because native species tend to remain in natural forest habitat, while most introduced species throughout their range prefer open habitats. There are thus inherent differences in the habitat preferences of most species of the two groups.

In conclusion, competition with native birds has probably not been a major factor in preventing the establishment of exotic birds on islands modified by human activity. This is partly because most of the species introduced, in contrast to native birds, prosper in man-made habitats, so with few exceptions contact between the two groups is limited. Competition could have played a bigger role in the replacement of one exotic species by another, and in preventing native species from colonising man-made habitats.

Discussion of evolutionary and long-term evidence

The exact mechanism by which one species might surrender part of its potential range, habitat or food-supply to another is poorly understood. In the long term, habitat differences between two species are probably brought about gradually by natural selection. Individuals of the one species might survive better in those places where the other is at a comparative disadvantage, and vice versa, so that in time each evolves a specific habitat preference. The same could be true for food segregation, where each species changes in bill and body structure to enable it to deal more effectively with its reduced spectrum of foods (see Grant (1986) for evidence of rapid evolutionary change in bill structure with change in food).

In this account, I have given a few examples to illustrate each point, but the same patterns of distributional and ecological segregation are repeated again and again in different birds and different regions around the world. Although competition is presumed to have played a major role in the development of such patterns, this is no more than a plausible inference. Most types of evidence considered above were generally consistent with the patterns expected from competition theory, and thus fulfilled criterion 1 in **Table 12.1**. Because most comparisons involved species of similar ecology (sharing a common resource), they also met criterion 2. In most studies, observed patterns in distribution, morphology, niche shifts and the like were taken as evidence for competition, but in the absence of any examination of criteria 4 and 5. The occurrence of intra-specific competition (criterion 3) was usually assumed implicitly and alternative explanations of the patterns (criterion 6) were seldom considered (Wiens 1989). This is hardly surprising, because testing alternative explanations for the patterns is not easy, and one can anyway never be sure that all possible explanations have been identified.

On the basis of such circumstantial evidence, competition has been portrayed as a major force in evolution, leading to the divergence of species and the adaptive radiation of whole groups of organisms (Lack 1971). It has been seen as important in shaping the distribution patterns of organisms, and the composition and structure of communities (MacArthur 1972, Wiens 1989). The most compelling examples involve pairs of similar species with overlapping resource needs. But it is likely that the effects on any one species result not just from a single major competitor, but from whole assemblages of species which share the same resources, predators or pathogens in some degree. The idea of diffuse competition has not been rigorously tested, however, and is still based primarily on inference. The fact that on islands species often reach greater densities than on mainlands where competing species are more numerous gives a strong indication that competition affects numbers. None-the-less, most of the evidence reviewed above on the importance of competition in influencing the distribution and abundance of birds is circumstantial, and is thus open to alternative explanations. But some of it is still compelling, and if we reject it merely because it cannot be checked by empirical tests, we risk greatly under-estimating the role of competition in nature. However, some situations represent natural experiments, and others can be manipulated by the researcher to provide more conclusive evidence on competition, as explained in the sections below.

EFFECTS ON POPULATIONS: SHORT-TERM RESPONSES

Complementary density changes

The evidence for competition reviewed above was based mainly on distribution patterns, involving spatial separation of similar species, or their replacement in the same area through time. For species that live together in the same habitat, evidence for competition has been sought by comparing their relative abundance levels. If changes in the numbers of individuals are determined mostly by inter-specific interactions within the community, numerous negative correlations should occur among species, with some increasing as others decline. But if the numbers of different species change independently of one another, no significant negative correlations should occur, except a few by chance.

This test was applied to the annual changes in a bird community of the natural forest at Bialoweiza, Poland, where counts of bird breeding numbers had been made each year during 1975–84 (Tomialojc & Wesolowski 1990). Within 325 pairwise comparisons among 26 species, only 42 (13%) statistically significant correlations were found. Thus the numbers of most species (87%) changed from year to year, independently of one another. Significant negative correlations, suggestive of competition, were extremely rare. Only seven (2%) emerged, and even these few could hardly reflect causal relationships, because the species concerned – for example

Woodpigeon *Columba palumbus* and Wren *Troglodytes troglodytes* or Siskin *Carduelis spinus* and Blackbird *Turdus merula* – were so different that they were unlikely to have been in competition. Even within the groups of congeneric species (two *Dendrocopus* woodpeckers, three *Ficedula* flycatchers, two *Phylloscopus* warblers, three *Parus* tits and two *Turdus* thrushes), no significant negative correlations emerged. Clearly, competitive relationships were not so prevalent in this community as to influence the year-to-year fluctuations of individual species.

On the other hand, the densities of two species pairs were positively correlated, tending to fluctuate in parallel over the years, namely Blue Tit/Great Tit and Blackbird/Song Thrush. Perhaps these species responded in similar fashion to the same environmental factors. Simultaneous fluctuation does not, of course, rule out the possibility of competition: it may simply mean that the effects of fluctuations in resources or other factors that affect both species simultaneously outweigh any negative effects of one species on the other. Competition is anyway likeliest in the years when resources are scarcest relative to bird numbers. Overall, however, the evidence for competition within this community, based on patterns of numerical fluctuation, was practically non-existent. Similar findings emerged from studies of other woodland bird communities elsewhere in Europe (Enemar *et al.* 1984, Virkhala 1991, Hogstad 1993, Morozov 1993), and in North America (Holmes *et al.* 1986, Brawn *et al.* 1987). They also emerged from studies of North American ducks, in which most dabbling and diving species fluctuated in synchrony from year to year, again suggesting common limiting factors rather than competition (Johnson 1996).

The same approach has been used to examine spatial relationships between species in the same expanse of habitat: for example, the relative densities of 14 species of Hawaiian forest birds at different sites, after allowing for effects of forest variation. The prevalent pattern here was of positive association between species, which probably resulted from mixed flocking and from the attraction of several species to local food-sources. However, some species showed consistent patterns of negative association. Wherever the introduced Japanese White-eye *Zosterops japonicus* was abundant, the native Iiwi *Vestiaria coccinea* and Elepaio *Chasiempis sandwichensis* were scarce (Mountainspring & Scott 1985). Also, negative correlations between population sizes over a 50-year period emerged between the introduced Japanese White-eye and the native Apapane *Himatione sanguinea* and Elepaio, and between the introduced Red-billed Leiothrix *Leiothrix lutea* which was declining, and the introduced White-rumped Shama *Copsychus malabaricus* which was increasing (Ralph 1991). In both studies, these correlative patterns were attributed to effects of competition, but in reality, they do little more than highlight likely candidates.

Some 30 years ago, bird communities were thought to be structured primarily by competition. They were viewed as interacting units that were generally stable and equilibrial (MacArthur 1959). It is now clear, however, that bird communities are far from stable, being composed of loose assemblages of species which, for the most part, fluctuate independently of one another. Inter-specific competition is only one of several factors that can influence the abundance of individual species, and its importance may vary from one species to another and from one community to another. It may well be more important in tropical forest than in the temperate and human-modified areas where most studies have been made.

Observed mechanisms of competition

In the examples discussed so far, competition at the population level was merely inferred from patterns of distribution or numerical fluctuation. In other situations, the actual mechanism of competition (whether depletion or interference) was observed in the field, when one species affected the distribution and numbers of another. To begin with depletion competition, Wigeon *Anas penelope* and Bean Geese *Anser fabalis* fed on the same area of grass on a nature reserve in southeast England (Sutherland & Allport 1994). Over a period of years, Wigeon increased in numbers and ate more grass, leaving less for the later arriving geese. Not surprisingly, the geese fed less on the reserve as the years went by. However, in this instance, the geese found alternative feeding areas nearby, so while depletion competition limited their use of the reserve, it did not reduce their overall numbers.

Sometimes one species indirectly removes the food of another: for example, by removing the habitat of potential prey organisms. Declines in the numbers of Capercaillie *Tetrao urogallus* and Black Grouse *T. tetrix*, which have occurred in Britain during recent decades, have been attributed partly to grazing pressure from increased numbers of sheep and deer (Baines *et al.* 1994, 1995, Baines 1996). These animals at high density shortened ground vegetation and removed most of the fresh growth. This in turn reduced the numbers of arthropods, especially caterpillars, which formed the food of young grouse chicks. The survival of chicks was thereby lowered, leading in turn to a decline in adult numbers. Comparison of vegetation height, insect densities and chick survival in several different areas showed that all these measures were inversely correlated with densities of sheep and deer. In the most heavily grazed areas, chick production by Black Grouse was much less than the 1.5–2.0 chicks per hen calculated as necessary to maintain the population level (Baines 1996). In this example, competition between herbivorous mammals and insects for the same fresh plant growth influenced the population trends of grouse. Indirectly, then, mammals and grouse were competitors. The same process involves other herbivorous birds, for example Mute Swans *Cygnus olor*, which, by eating aquatic vegetation, not only remove the food-supply of other plant-eating water-fowl, but also the habitat of many invertebrates eaten by ducks (Allin *et al.* 1987).

Turning now to interference competition, the House Wren *Troglodytes aedon* in eastern North America is expanding in distribution as the Bewick's Wren *Thryomanes bewickii* is contracting. Interference takes the form of one-sided vandalism, in that House Wrens throw out the eggs and chicks from Bewick's Wrens' nests. In a three-year study, House Wrens accounted for 81% of all nest failures in Bewick's Wrens, and victims suffered, on average, a two-thirds reduction in their annual production of young (Kennedy & White 1996). This may not have been the only interaction between the two species.

Sometimes birds in competition can be seen to fight over resources in short supply, such as nest-sites, larger species ousting smaller ones, whose breeding density is thereby lowered. This is evident among songbirds nesting in tree-cavities, among birds-of-prey nesting on cliffs and among seabirds on islands. For example, on islands off Peru, the Peruvian Booby *Sula variegata* and the Guanay Cormorant *Phalacrocorax bougainvillii* competed on a first come/first served basis to occupy nesting space before it was taken. Neither species could displace the other from nest-sites, once established (Duffy 1983). However, the Brown Pelican *Pelecanus occidentalis* dominated the other two and often took over their nests. But pelicans were confined to nesting on level ground, whereas the other two could also nest on gradients, so could persist in reduced numbers in the presence of pelicans. Similarly, in both Europe and North America, increases in the numbers of large gulls over the past 50 years caused local reductions in the numbers of smaller gulls and terns through use of nesting areas (see Nisbet 1978 for effects of Herring Gull *Larus argentatus* and other large gulls on the smaller Laughing Gull *Larus atricilla*). Various measures to reduce gull densities in particular areas have allowed terns to return and reversed the downward trend in their numbers (Kress 1983, Morris *et al*. 1992). Competition for breeding space has also occurred between seabirds and seals. Off southern Africa, Jackass Penguins *Spheniscus demersus* were eliminated from one island, and reduced in numbers on two others, when Fur Seals *Arctocephalus pusellus* occupied the nesting areas (Crawford *et al*. 1990). Like dispersed landbirds, then, certain colonial seabirds are in some areas limited in breeding numbers by shortage of places to nest, and in the face of this shortage, the numbers of dominant species can influence the numbers of others.

Another means by which one species can affect the distribution of another, at least on a local scale, is through inter-specific territoriality, in which individuals of one species oust the other from suitable breeding areas **(Table 12.6)**. In general, the competing species have similar feeding habits but occupy partly different areas or habitats, holding mutually exclusive territories where they overlap. Inter-specific territoriality thus reduces the area available to subordinate species, perhaps lowering their overall numbers. Alternatively, it might force individuals of the subordinate species into less suitable habitat, where breeding and survival could be lower (see Rice 1978 for Red-eyed Vireo *Vireo olivaceus*). Some species pairs show inter-specific territoriality in some areas but not in others. An example is the Chaffinch *Fringilla coelebs* and Great Tit *Parus major* which held mutually exclusive territories on a Scottish island, but overlapping territories on the more diverse habitat of nearby mainland (Reed 1982). Yet other species are aggressive to a range of other species, driving them from their territories; the reaction of territorial Mute Swans to smaller waterfowl provides an example (Allin *et al*. 1987). Other instances of inter-specific territoriality involve the short-term defence of specific food-sources, such as fruiting trees (Snow & Snow 1988).

In addition to the examples in **Table 12.6** overleaf, inter-specific territoriality was common among closely-related bird species distributed along a successional forest gradient in Amazonia (Robinson & Terborgh 1995). Species pairs in over 20 genera showed contiguous but non-overlapping territories, such that young forest was occupied by one member of the pair, and older forest by the other. By use of playback calls, ten species pairs showed evidence of interaction, mostly with the larger species being the aggressor (as judged by the birds' approach to the playback speaker). Such spatial segregation of congeneric bird species on habitat gradients (including elevational) may be commonly underpinned by inter-specific territoriality, with the larger species occupying the more productive end of the habitat gradient. In this way, inter-specific territoriality could reinforce a pre-existing habitat difference.

EXPERIMENTAL EVIDENCE FOR EFFECTS OF COMPETITION

Most experiments have measured competition by its effects on the feeding, breeding or survival of individuals, as discussed earlier, but some have checked for effects on population levels. It is this third type of response, measured as altered distribution or density, that is most relevant in the context of population limitation. Manipulations have occurred of resources (notably food and nest-sites) and of the densities of potential competitors. Of 14 such experiments listed in **Table 12.2**, 13 led to an increased density of the target species (though in two this was winter density, not breeding density).

Competition for food and feeding areas

One such experiment concerned several bark-foraging birds in deciduous woodlands of central Illinois (Williams & Batzli 1979). Aggressive Red-headed Woodpeckers *Melanerpes erythrocephalus*, which stored acorns in bark crevices and defended the surrounding area against all intruders, kept out other species. When the Red-headed Woodpeckers were removed from one 6-ha wood, Red-bellied Woodpeckers *M. carolinus*, and White-breasted Nuthatches *Sitta carolinensis*, which had been previously excluded, settled in this wood, but not the control wood where Red-headed Woodpeckers were left. The implication was that Red-headed Woodpeckers limited the local densities of the others through aggressive exclusion. In addition, Downy Woodpeckers *Picoides pubescens* foraged higher in the trees than previously in the experimental area, and higher than in the control area. They presumably therefore had access to a greater food-supply, but it was unclear whether their numbers were also affected.

Other manipulations have involved inter-specific territoriality. In an English wood, Blackcaps *Sylvia atricapilla* and Garden Warblers *S. borin* held mutually exclusive territories in those areas that were acceptable to both. Blackcaps arrived first in spring and established territories before Garden Warblers. When Garcia (1983) removed Blackcaps, some of these territories were rapidly taken over by Garden Warblers. After the removals ended, more Blackcaps settled in the area, some of which displaced established Garden Warblers. The implication was that Blackcaps could displace Garden Warblers from some locations, restricting the total area and the range of habitat that Garden Warblers could use, and thereby reducing their breeding densities. Other similar experiments are listed in **Table 12.2**.

As indicated already, birds compete not just with other bird species, but with other animals too. After the removal of fish from a 17-ha lake in southern England, submerged plant cover increased from less than 1% to 95% of the lake surface, while benthic invertebrates doubled in numbers (Giles 1992, Phillips 1992). Use of the lake by avian herbivores, such as Mute Swan, Coot *Fulica atra* and Gadwall *Anas strepera*, increased in association with greater weed growth, and use by Shoveller *Anas clypeata* and Pochard *Aythya ferina* in association with greater invertebrate numbers. Mallard duckling survival also improved (Wright & Phillips 1990). Such changes were not seen in other lakes nearby, whose fish populations were left intact. Similarly, in Sweden it had been observed that Goldeneyes *Bucephala clangula* preferred lakes without fish to those with fish, and that the diets of birds and fish overlapped. After fish were removed from one lake and prey species increased, more Goldeneyes were seen feeding there than formerly, and more than on an adjacent control lake (Eriksson 1979). So in both experiments, removal of fish resulted in increased food-supplies, and in turn in increased use of the lakes by waterfowl. Other studies of interactions between birds on the one hand and insects, reptiles or

Table 12.6 Examples of bird species that show inter-specific territoriality. As far as possible, dominant species in each example are listed first.

Species	Location	Source
Rufous Hummingbird Selasphorus rufus – Calliope Hummingbird Stellula calliope	Wyoming	Armitage 1955*
Lucifer Hummingbird Calothorax lucifer – Black-chinned Hummingbird Archilochus alexandri – Broad-tailed Hummingbird Selasphorus platycercus	Texas	Fox 1954*
White-eared Hummingbird Hylocharis leucotis – Magnificent Hummingbird Eugenes fulgens, Blue-throated Hummingbird Lampornis clemenciae, Broad-billed Hummingbird Cynanthus latirostris	Mexico	Moore 1939*
Costa's Hummingbird Calypte costae – Black-chinned Hummingbird Archilochus alexandri	Arizona	Bene 1952*
Anna's Hummingbird Calypte anna – Allen's Hummingbird Selasphorus sasin	California	Pitelka 1951*, Legg & Pitelka1956*
White-throated Sparrow Zonotrichia albicollis – Song Sparrow Melopiza melodia	Ontario	Kendeigh 1947*
Hammond's Flycatcher Empidonax hammondii – Dusky Flycatcher E. oberholseri – Grey Flycatcher E. wrightii	California	Johnson 1963*
Red-backed Shrike Lanius collurio – Masked Strike L. nubicus	Egypt	Simmons 1951*
Mourning Wheatear Oenanthe lugens – Pied Wheatear O. pleschanka	Egypt	Simmons 1951*
Blue Rock Thrush Monticola solitaria – White-tailed Wheatear Oenanthe leucopyga – Black-eared Wheatear Oenanthe hispanica – Northern Wheatear O. oenanthe	Aegean Is.	Stresemann 1950*
Pied Wheatear Oenanthe pleschanka – Northern Wheatear O. oenanthe	Ukraine	Stresemann 1950*
Mourning Wheatear Oenanthe lugens – Black-eared Wheatear O. hispanica – Desert Wheatear O. deserti	Egypt	Hartley 1950*
Hooded Wheatear Oenanthe monacha – White-tailed Wheatear O. leucopyga – Mourning Wheatear O. lugens	Egypt	Hartley 1949*
Eastern Meadowlark Sturnella magna – Western Meadowlark S. neglecta	Wisconsin	Lanyon 1956*
Black-winged Bishop Euplectes hordeaceus – Zanzibar Red Bishop E. nigroventris	Tanzania	Fuggles-Couchman 1943*
Eurasian Reed Warbler Acrocephalus scirpaceus – Sedge Warbler A. schoenobaenus	England	Brown & Davies 1949*
Red-winged Blackbird Agelaius phoeniceus – Common Grackle Quiscalus quiscula	Wisconsin	Nero 1956*
Philadelphia Vireo Vireo philadelphicus – Red-eyed Vireo V. olivaceus	Ontario	Rice 1978
Blackcap Sylvia atricapilla – Garden Warbler S. borin	England	Garcia 1983
Yellow-headed Blackbird Agelaius xanthocephalus – Red-winged Blackbird A. phoeniceus	Saskatchewan	Miller 1968
Nightingale Luscinia megarhynchos – Thrush-Nightingale L. luscinia	Europe	Lack 1971
Great-tailed Grackle Quiscalus mexicanus – Boat-tailed Grackle Q. major	Texas	Selander & Giller 1961*
Red-bellied Woodpecker Melanerpes carolinus – Golden-fronted Woodpecker M. aurifrons	Texas	Selander & Giller 1961*

Table 12.6 Continued.

Species	Location	Source
Common Ringed Plover *Charadrius hiaticula* – Little Ringed Plover *C. dubius* – Kentish Plover *C. alexandrinus*	England	Simmons 1956*
Grey Plover *Pluvialis squatarola* – American Golden Plover *P. dominica*	NW Territories	Drury 1961*
Pomarine Jaeger *Stercorarius pomarinus* – Parasitic Jaeger *S. parasiticus*	Alaska	Pitelka et al. 1955*
Golden Eagle *Aquila chrysaetos* – Bonelli's Eagle *Hieraaetus fasciatus*	France	Cheylan 1973
Common Buzzard *Buteo buteo* – Rough-legged Buzzard *Buteo lagopus*	Sweden	Sylven 1979
Red-tailed Hawk *Buteo jamaicensis* – Red-shouldered Hawk *Buteo lineatus*	Michigan	Craighead & Craighead 1956*

*References in Orians & Willson (1964), where other examples may be found. Not all the species shown are necessarily inter-specifically territorial throughout their mutual range.

mammals on the other have provided further support for the view that competition with other organisms can limit bird numbers (Wiens 1989).

Competition for nest-sites

Among cavity-nesters, competition for nest-sites is often strong, and larger species can usually evict smaller ones, except from holes that are too small for them to enter. Where nest-sites are scarce, the presence of one dominant species can limit the numbers of others. The population trends of competing species may then vary in inverse relation to one another, with the subordinate species increasing as the dominant declines. Most experiments have been done in forest-areas where natural nest-cavities are scarce. This has enabled the numbers of nest-sites for particular species to be manipulated by providing nest-boxes or by changing the sizes of their entrance holes.

In six forest plots in Sweden, almost all nest-boxes were occupied every year by one or other of three species. The numbers of all three fluctuated from year to year, but increases in the numbers of Great Tits *Parus major* and Blue Tits *P. caeruleus* were accompanied by decreases in the numbers of Collared Flycatchers *Ficedula albicollis*, and vice versa (Gustafsson 1988). The implications were that boxes were not limiting for the earlier nesting tits, but that the later arriving flycatchers were relegated to whatever boxes were still unused. The numbers of flycatchers on the plots thus depended on the numbers of tits. As an experiment, the numbers of tits in one plot were reduced in two successive years. This led to a large increase in the numbers of flycatchers, compared with those in a control plot **(Figure 12.4)**. As a further experiment in a Norwegian forest, Slagsvold (1979) blocked the entrances of all nest-boxes

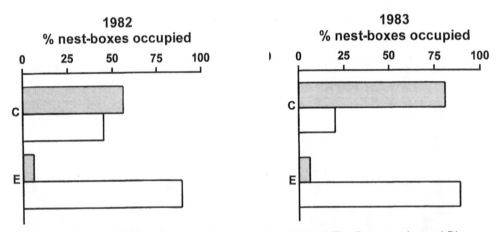

Figure 12.4   Proportion of nest-boxes occupied by Great Tits *Parus major* and Blue Tits *P. caeruleus* (shaded) and Collared Flycatchers *Ficedula albicollis* (open) in an experimental plot (E) where tit numbers were deliberately reduced and in a control plot (C) where tit numbers were not reduced. The densities of nest-boxes were the same in both plots, and the differences in species occupation between plots were statistically significant in both years (1982: $\chi^2 = 12.2$, $P<0.001$; 1983: $\chi^2 = 30.1$, $P<0.001$). Figures show number of pairs. Through competition for nest-sites, large numbers of tits limited the numbers of flycatchers that could nest. Redrawn from Gustafsson 1988.

except those already occupied by Great Tits. In consequence, later arriving Pied Flycatchers *Ficedula hypoleuca* were completely excluded from breeding there. Males fought unsuccessfully with Great Tits for possession of boxes, while females left the area. These experiments revealed how competition for nest-sites could limit the breeding densities of flycatchers, and how annual fluctuations in the breeding densities of flycatchers varied inversely with the numbers of tits.

In other studies, the numbers of Great Tits and Blue Tits were found to be inter-related. In a broad-leaved wood in southern England, the breeding densities of both species in different areas depended on densities of nest-boxes over the range 2–8 boxes per ha (Minot & Perrins 1986). Over this range, both species were most numerous in areas with most boxes. At the same time, the ratio of Blue to Great Tits increased with the density of boxes, with Blue Tits breeding in greater proportion in areas where boxes were most numerous **(Figure 12.5)**. Moreover, in areas with low densities of boxes, changes in the densities of the two species from year to year were inversely correlated, with Blue Tits increasing in years when Great Tits declined. In areas with high densities of boxes, however, the numbers of both species varied in parallel. By implication, where sites were limiting, the numbers of Great Tits (the dominant species) affected the numbers of Blue Tits, but where sites were not limiting no such interaction occurred, and the two species fluctuated in parallel in response to other factors.

An experiment in a German forest showed how strikingly the numbers of Great Tits and Blue Tits could be manipulated by changing the size of the holes on nest-boxes

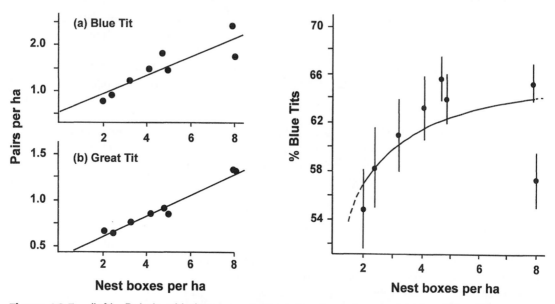

Figure 12.5   (left). Relationship between nest-box density and mean density of Blue Tits *Parus caeruleus* and Great Tits *P. major* in eight sections of Wytham Wood, Oxford. Regression line for Blue Tit: $y=0.53 + 0.20x$, $r = 0.89$, $P<0.001$; for Great Tit: $y=0.38 + 0.11x$, $r = 0.98$, $P<0.001$. (right). Effect of nest-box density on the mean (±SE) percentage of Blue Tits among the total numbers of both species nesting in eight sections of Wytham Wood, Oxford. Regression relationships: $y=0.67x / (x + 0.36)$, $r = 0.65$, $P<0.05$ (one-tailed test). From Minot & Perrins 1986.

(Löhrl 1977; Dhondt & Eyckerman 1980). Entrance holes of 32 mm diameter let in the larger Great Tits as well as Blue Tits, but holes of 26 mm diameter let in only Blue Tits. The numbers of one or both species were changed markedly over the years by appropriate manipulations, and Blue Tits bred in greater numbers in years when the dominant Great Tits were excluded by smaller hole sizes **(Figure 12.6)**.

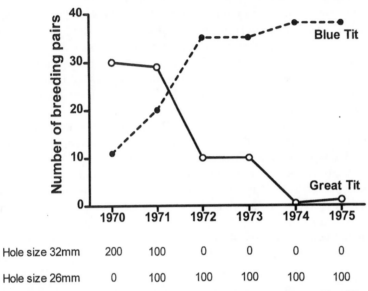

	1970	1971	1972	1973	1974	1975
Hole size 32mm	200	100	0	0	0	0
Hole size 26mm	0	100	100	100	100	100

Figure 12.6 Numbers of breeding Great Tits *Parus major* and Blue Tits *P. caeruleus* in a 10-ha wood, Germany, 1970–75, in relation to changes in the sizes of entrance-holes on nest-boxes. Holes of 32 mm would admit both species, but holes of 26 mm would admit only the smaller Blue Tit. Figures below the graph show the numbers of large-holed and small-holed nest-boxes present in each year. Redrawn from Dhondt & Eyckerman 1980, based on Löhrl (1977).

In some Belgian woods, Great Tits affected Blue Tits by preventing access to nest-boxes, not so much for breeding as for roosting in winter (Dhondt & Eyckerman 1980). Exclusion of Great Tits by reducing hole sizes led not only to a decrease in the numbers of Great Tits breeding, compared with control woods, but also to an increase in the numbers of Blue Tits. The same result was obtained in each replication, involving two different woods and two different years. It was attributed to competition, in which the numbers of Great Tits in the area influenced the numbers of Blue Tits. In this experiment two factors were changed at the same time, the numbers of suitable holes and the numbers of Great Tits; and because both species ate some of the same foods, it was hard to tell whether reduced competition for roost-sites or for food-supplies led to the increase in Blue Tit numbers. However, a later experiment, in which the numbers of holes for Blue Tits and for Great Tits were altered independently, confirmed that holes and not food-supplies were limiting for both species (Dhondt *et al.* 1991). This situation may not hold in all woods, because the numbers and sizes of natural holes vary from one wood to another. These findings may explain the results of other experiments, involving the provision of extra food, which in some studies led to an increase in the numbers of one or both tit species, but in other studies did not (Chapter 7). Whether numbers responded to food-provision would depend on whether they were limited in the area concerned by food-supplies, cavities or other factors.

From various observational and experimental evidence, then, the interactions between Great and Blue Tits are both asymmetric and reciprocal, depending on the types of competition involved. Where the species fight over localised resources, such as nest-sites, the larger Great Tit has the upper hand, and can reduce the density of Blue Tits (interference competition), but where the two species depend in summer on the same caterpillar supply, the smaller Blue Tit wins out, lowering the breeding success of Great Tits (depletion competition). This was shown by comparisons of Great Tit nest success in areas with differing densities of Blue Tits (see earlier), and occurred because the Blue Tits took smaller caterpillars, removing them before they could grow to the size favoured by Great Tits (**Table 12.2**, Dhondt 1989).

Great Tits themselves are not free from the effects of nest-site competition where Starlings *Sturnus vulgaris* occur (van Balen *et al.* 1982). In some Dutch woods, Great Tits nesting in natural holes seldom raised young because Starlings took over the holes before the tits had finished laying. Some of the tit pairs then remained in the area without re-nesting, while others moved elsewhere, so that density was lowered. As an experiment, entrance holes to Great Tit nests were narrowed to exclude Starlings, and most of the tits then bred successfully. As a reciprocal experiment, the entrances of seven nest-boxes occupied by Great Tits were widened, and within two days five had been taken over by Starlings. Hence, in woods where they were common, Starlings restricted Great Tits to breeding successfully mainly in cavities with entrance holes less than 35 mm across, only a proportion of the holes that Great Tits could otherwise use. Such competition could have resulted in a lowering of the breeding density of Great Tits. In North America, introduced Starlings compete for nest holes with a wide range of native birds, from Tree Swallows *Tachycineta bicolor* to Wood Ducks *Aix sponsa* (McGilvrey & Uhler 1971), while Starlings and House Sparrows *Passer domesticus* often use boxes intended for Purple Martins *Progne subis* (Jackson & Tate 1974). House Sparrows also take over the natural nests of Cliff Swallows *Hirundo pyrrhonota*, sometimes reducing their breeding numbers (Erskine & Temple 1970).

Although all experimental studies of competition for nest-sites have concerned bird–bird interactions, in many regions tree-cavities are used by various other animals, from bees to bats and squirrels. In Victoria (Australia), for example, 38% of 76 species of non-aquatic mammals use tree-holes, compared with 18% of birds (Cowley 1971). Competition with other animals could thus be an additional factor influencing the numbers of some cavity-nesting birds, or vice versa.

*   *   *

These various experiments confirmed that different bird species do indeed compete in ways that can be measured and that, through effects on survival or movements, the local density of one species can influence the density of another. Depending on circumstances, one species could exclude another completely from an area, or the two could co-exist using different parts of a resource. Evidence of inter-specific competition has also emerged from experiments on other organisms. About 90% of 148 studies reviewed by Schoener (1983), and 93% of 72 studies reviewed by Connell (1983), revealed effects of one species on the performance or densities of another.

However, these figures, which included few bird studies, did not necessarily reflect the true prevalence of inter-specific competition in nature. As with experiments on any other limiting factor, researchers were more likely to select species in which prior experience suggested competition, and studies showing an effect were probably more likely to be published than those showing no effect (Connell 1983). None-the-less, to judge from published studies, depletion and interference competition seem to be widespread among both plants and animals, and asymmetric competition seems commoner than symmetric. In these respects, then, birds are no different from other organisms.

APPARENT COMPETITION: SHARING OF NATURAL ENEMIES

As explained in earlier chapters, some species can be depressed in numbers by the action of a generalist predator, whose abundance in the area is in turn largely dependent on another prey species. Both species of prey are adversely affected by the predator, which is in turn beneficially affected by both species of prey. The two prey species can in a sense be regarded as competitors because, via the predator, the numbers of the one are influenced by the numbers of the other **(Figure 12.7)**. In effect, the outcome is the same as if the two species were competing for the same resource, but because no limiting resource is involved, the term 'apparent competition' was coined by Holt (1977, 1984).

The same holds where two or more host species are affected by the same parasite species. To illustrate the point, recall from Chapter 10 the viral disease, louping ill, which is prevalent in sheep in parts of Scotland, and is transmitted by a tick, not only

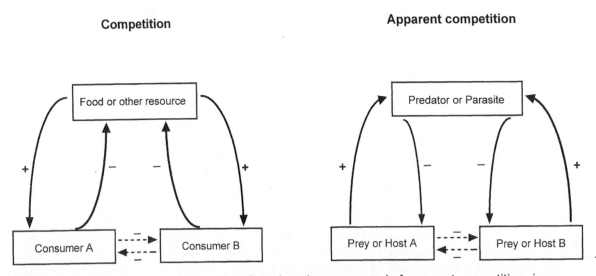

Figure 12.7 Mechanisms of competition via a shared resource, and of apparent competition via a shared predator or parasite. In terms of the effects of their interactions, two prey species being attacked by a common predator (apparent competition) are indistinguishable from two species consuming a common resource (competition). Solid lines depict direct interactions, and dotted lines indirect ones. From Holt 1984.

between sheep, but also to other animals in the same habitat, including Red Grouse *Lagopus l. scoticus*. Infections are often fatal to grouse, and can greatly reduce their numbers. However, the disease is maintained chiefly by the presence of a high density sheep population, which acts as a reservoir, and if the sheep are removed, the disease may disappear, for the grouse themselves live at too low a density for effective transmission. The sheep can thus be regarded as competitors to the grouse because, via the parasite, infected sheep can reduce the numbers of grouse. The same could be said for any pair of species that share the same pathogen, if one is more abundant or more tolerant of that pathogen than the other, and can therefore act as a reservoir for continuing infection.

In these ways, just as competition between two species can involve a resource, 'apparent competition' can involve a natural enemy. Both types involve a third party (a resource, parasite or predator), and the effect on the out-competed population is the same, namely depressed abundance or distribution. Among some species that appear to compete, experiments may be necessary to tell whether the interaction involves a resource in common, or a predator or parasite in common.

SUMMARY

Where resources are limiting, species must differ in their needs in order to co-exist. The occurence of competition between bird species has been inferred from observations that: (1) closely-related species often occupy mainly different geographical ranges, but with varying degrees of overlap; (2) in the areas of overlap, such species may diverge in habitat, or (3) in morphology and feeding behaviour, thus reducing the potential for competition; (4) species that occur together on mainlands often occur singly on islands, with different species on different islands but at higher densities than on mainlands; (5) colonisation of an area by a new species sometimes leads to the elimination of an ecologically similar one, or to a restriction in its habitat or food, as the two species divide resources. Although these various findings could be explained by competition, they provide only circumstantial evidence.

In experiments, removal of one species (or group of species) from an area has led to an expansion in the niche of another (or several others), to an improvement in its survival or breeding success, or to an increase in its numbers or distribution. The latter provide unequivocal evidence that the presence of one species in an area can depress the density of another that shares the same resource. In woodland where tree-cavities are scarce, the numbers of hole-nesting birds have been manipulated by the provision of nest-boxes with specific hole sizes. This has led to experimental demonstrations of how the numbers of one species can affect another, through competition for the nest-boxes themselves or for food.

Apparent competition occurs between species that share the same enemies, whether predators or parasites. It arises because predator abundance depends on total prey numbers, so if one prey species increases, it can lead to an increase in predator numbers, and hence to a decline in alternative prey. The same can hold where a parasite has more than one host. In essence, the effects are similar to more conventional forms of inter-specific competition, which arise from sharing the same limited resource.

Arctic Skua *Stercorarius parasiticus* chasing Arctic Tern *Sterna paradisaea*.

Chapter 13
Interactions Between Different Limiting Factors

In earlier chapters, the main limiting factors to bird numbers, namely resources and natural enemies, were discussed for the most part as though they acted separately on population sizes. Changes in the densities of many populations do indeed seem to be influenced by variations in one major limiting factor, as indicated by both observational and experimental evidence, but it is equally clear that different limiting factors can sometimes act together to influence numbers and that their effects are not always straightforwardly additive. Sometimes, one factor might enhance the effect of another, so that their combined impact on population levels is greater than the sum of their individual effects. At other times, one factor might reduce the effect of another, so that their combined impact is less than the sum of their separate effects. Although such interactions have been long recognised, the field evidence is still skimpy, but it is worth considering because it points the way to further research and experiment.

REVIEW OF EXPERIMENTAL AND OTHER EVIDENCE

Most of the published field experiments on factors limiting bird numbers have involved the manipulation of only one factor at a time. The findings from those experiments discussed in Chapters 7–12 are summarised in **Table 13.1**. Taken together, they have served to confirm that, within areas of suitable habitat, the main potential limiting factors – whether resources or natural enemies – have affected the breeding densities of one bird species or another. They have also confirmed that the same species can be limited in breeding density by different factors in different areas or in different years. It is clearly untenable to think that particular types of birds, or particular species, might be limited consistently throughout their range, or year after year, by the same factors. The provision of food or removal of predators in these experiments led in extreme cases to a doubling of breeding density compared with control areas, but provision of nest-sites often led to much bigger increases, up to 20-fold in the most extreme examples.

Table 13.1. Summary of experiments on the limitation of breeding density in birds. See Chapters 7–12 for further details.

	Grade of experiment[1]			Total	Number in which increase in breeding density occurred (%)
	1	2	3		
Food provision (winter)	5	11	10	26	15 (58)
Nest-site provision	37	5	4	46	44 (96)
Predator reduction[2]	4	7	6	17	10 (59)
Parasite reduction[3]	3	0	1	4	3 (75)
Competitor reduction	3	5	4	12	11 (92)
TOTALS	52	28	25	105	83 (79)

[1] Grade 1 = simple before-and-after comparison in the same area; Grade 2 = simultaneous comparison between experimental and control area; Grade 3 = reversal of treatments between experimental and control areas, or replication of experimental and control areas.
[2] A total of 30 predator-removal experiments included only 17 that examined the effects of predator removal on breeding density (as opposed to breeding success).
[3] Includes one experiment on Red Grouse and a strongyle parasite (Hudson *et al.* 1992b), one on Least Bell's Vireo *Vireo v. belli* and Brown-headed Cowbirds *Molothrus ater* (Rothstein & Robinson 1994), one of Kirtland's Warbler *Dendroica kirtlandii* and Brown-headed Cowbirds in which warbler density did not increase but stopped declining (Kelly & DeCapita 1982), and one on Song Sparrow *Melospiza melodia* and Brown-headed Cowbird (Smith 1981).

The fact that most of the experimental results were positive should not surprise us, because the experiments were usually done only after observational evidence had indicated what the main limiting factor might be. If species had been chosen at random for each type of manipulation, rather than on the basis of prior knowledge, far fewer positive results might have emerged. The findings cannot therefore be expected to reflect the relative importance of different factors in nature. Also, not all bird species lend themselves to small-scale field experiments, and, for practical reasons, most experiments have concerned species that occur at high densities. The need now is for experiments on birds that involve the simultaneous manipulation of more than one factor, as already done on some other kinds of animals. For example, in Snowshoe Hares *Lepus americanus* in Canada, the removal of predators from one area led to a two-fold rise in hare breeding densities over those in a control area; fertilising of vegetation in a second area led to a three-fold rise in density; while simultaneous predator removal and fertiliser application in a third area led to an 11-fold increase in density (Krebs *et al.* 1995). Hence, the effect of manipulating the two factors together was not simply the sum of effects of the two factors individually.

The same conclusion has emerged from mathematical models based on good field data. For example, the recent declines of Grey Partridge *Perdix perdix* numbers in western Europe have been attributed to three main factors: (1) shortage of insect food for the chicks, resulting from the use of herbicides in cereal crops; (2) shortage of suitable nesting cover for the hens, resulting from hedgerow removal; and (3) heavy predation on the eggs and sitting hens by corvids and predatory mammals (Potts 1980, 1986). Use of a mathematical model enabled the effect of management aimed at each of these individual factors to be assessed in terms of resulting population trend, together with the response of the population to different levels of human hunting (**Figure 13.1**; Aebischer 1991). In the current situation, the model showed that the stock, at 4.7 pairs per km^2 in spring, had little resilience to autumn hunting: a 20% harvest rate would approximately halve the spring stock, and a 50% harvest rate would drive it to extinction.

The first management step imposed in the model was to restore the insect food of chicks by abandoning the use of herbicides in the crop or in the crop headlands (Sotherton 1991). Under this regime, the model gave a new equilibrium level of 8.2 pairs per km^2 in the absence of hunting, nearly double the previous level. At similar harvest rates (20% and 50%), spring stock was again approximately halved, but the yield was up by 250%, and the population could withstand a harvest rate of more than 60% before going extinct.

The second step was to restore the amount of nesting cover. The prevailing hedges and grassy banks amounted to about 4 km per km^2. The effect of doubling this nesting cover to about 8 km per km^2 was again approximately to double the spring stock size, which increased to 16.3 pairs per km^2 with no harvesting. At similar harvest rates (20% and 50%), the spring stock was less than halved, but the yield doubled again, and the population could again withstand a harvest of more than 60% before going extinct.

The last step was to reduce the amount of egg and hen predation by control of predators. In this final stage of management, the unharvested spring stock rose to 51.6 pairs per km^2, over a ten-fold increase when compared with the level in the original unmanaged situation found on modern arable farmland. At the harvesting rate of 40%, which produced the maximum sustainable yield, the stock was still higher than under any other form (or lack) of management. In such optimal conditions, the population could sustain harvesting rates of more than 70% per autumn before being driven to extinction. Although the figures given were merely predicted by a model, and were therefore subject to any weakness in the model, they were in fact similar to figures found in nature where different management procedures had been applied, and to figures found in experiments where populations were monitored after changing the various conditions (Rands 1985, Tapper *et al.* 1996). As for Snowshoe Hares, the cumulative effect of altering several factors was greater than the effect of managing each factor separately. Thus, controlling the predators of Partridges

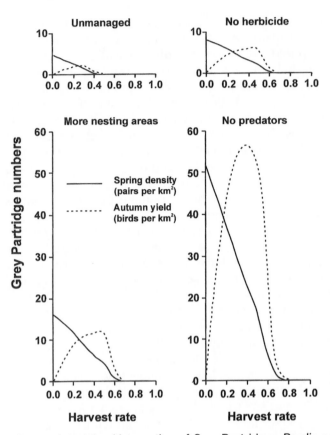

Figure 13.1   Effects of sustained harvesting of Grey Partridges *Perdix perdix* upon the spring stock size and yield as assessed by modelling. Harvest rates were set as a constant percentage of autumn stocks. Four different scenarios are considered, from worst to best: the current unmanaged situation under present-day arable agriculture (top left), the same without herbicide use (top right), the combination of no herbicides and more nesting cover (bottom left), and the same with predator control (bottom right). From Aebischer 1991, 1997.

without improving the habitat beforehand increased spring stock size from 4.7 to 13.4 pairs per km² without harvesting, compared with an increase from 16.3 to 51.6 after habitat improvement (Aebischer 1997). In both cases, the increase was approximately three-fold, but while predator control added only 9 pairs per km², habitat improvement followed by predator control added 35 pairs. The rest of this chapter considers other potential interactions between different limiting factors, some of which are additive and others counteractive in their effects on population levels.

INTERACTIONS INVOLVING DIFFERENT RESOURCES

For breeding, birds require both food and nest-sites, and the nesting densities of many species are limited by one or the other, whichever is in shortest supply. For some species certain regions may offer abundant food but few nest-sites, while other regions may offer plenty of nest-sites but little food (Newton 1979). Take the Peregrine Falcon *Falco peregrinus* in Britain, which for the most part nests on cliffs. It breeds commonly only in those parts of the country where cliffs are available, and its local densities there are influenced mainly by food-supply (Ratcliffe 1980). It is effectively absent from the rest of the country where cliffs (or quarries) are lacking, but potential prey are plentiful.

The restriction of some colonial species to patches of nesting habitat can lead to depletion of food-supply near the colonies, while food remains abundant further afield. Indeed, some seabird populations, it has been argued, may be regulated by such localised competition for food in the breeding season, restricted to the vicinity of the colonies and limiting production of young (Ashmole 1963). Regulation is thus achieved not by numbers of nest-sites alone, nor by food-supply alone, but by a combination of the two, with local concentrations of nest-sites leading to local concentrations of birds, leading in turn to localised food-shortage. Although individual birds might survive by feeding further afield where food is more plentiful, they could not reproduce there, through lack of nest-sites. This situation is likely to reach its extreme in some pelagic birds which, when not breeding, range over vast areas of open sea, but gather in very large numbers to breed at a small number of oceanic islands; however, it may also apply in some degree to more coastal seabird species with a greater range of available colony sites.

As yet, evidence for this means of population regulation, entailing an interaction between localised colony-sites and widespread food-supply, is entirely circumstantial: (1) colony size in some species is inversely related to the number of other colonies (and individuals) in the general area (Furness & Birkhead 1984; Chapter 5), and adjacent colonies often tend to be inversely related in size (Cairns 1989, 1992), suggesting density-dependent regulation of colony sizes; (2) in some species, breeding success and fledgling weights are inversely related to colony size, suggesting local competition for food (Birkhead & Furness 1985, Gaston 1985, Hunt *et al.* 1986); (3) the density of prey fish is lower near the colonies than further away (Birt *et al.* 1987), suggesting prey depletion; and (4) occasional mass starvation and breeding failure are observed at colonies when food is scarce in the vicinity, but known to be abundant beyond the colony-based foraging range of the birds (Chapter 8). We should not, of course, expect all colonies to be close to the food limit at any one time (Ainley *et al.* 1995).

Similar localised access to a widespread food-supply occurs in deserts where the need to drink restricts most bird species to feeding within flight range of a water source, either year-round or in the driest season. This restriction in turn may lead to depletion of local food-supplies, with effects on bird numbers, while in more distant areas food is present but out of reach. In western Australia, one of the major human impacts on resident bird species has been through the widespread provision of permanent water for sheep and cattle. This water has promoted increases in the numbers of many species that need to drink, while other species have declined, perhaps unable to compete (Saunders & Curry 1990). The actual foraging range varies between species depending on their flight capabilities and drinking frequencies. Other familiar examples of birds restricted by water to feed in certain areas are provided by geese and other waterfowl, which need the safety of water for roosting, and thereby feed only on land within easy flying range of their roost-sites.

Those species that can habitually commute further than others gain access to wider areas, as shown from comparison of the winter distributions of different goose species (Newton et al. 1974).

Likewise, in open land, some species may be restricted in distribution by their need for trees or cliffs for nesting and roosting, leaving intervening areas largely unexploited. Such focal point regulation of bird numbers, amid a widespread food-supply, may be the norm in many types of habitat. In effect, one patchy resource permits only patchy exploitation of a second more widespread resource. Examples were mentioned in earlier chapters where provision of nest-sites led to expansion of bird distributions, the implication being that food and other needs were adequate. These experiments, deliberate and incidental, provide the strongest evidence for interactions between different resources in influencing the distributions and numbers of birds. The examples described result from one essential resource being more restricted in distribution than another.

INTERACTIONS BETWEEN NEST-SITES, PREDATORS AND PARASITES

To protect their nests from predators, some bird species rely on cover, others on camouflage, and yet others on relatively inaccessible sites, such as high trees or cliffs, which give security at least against some mammalian predators. However, the same species sometimes use less safe sites in places that lack mammalian predators. On the Dutch island of Vleiland, with its 300 ha of pine plantations, some Woodpigeons *Columba palumbus*, Stock Doves *Columba oenas*, Kestrels *Falco tinnunculus*, Great Tits *Parus major* and Redstarts *Phoenicurus phoenicurus* nest on the ground (Spaans & Swennen 1968). Yet on the nearby mainland, where mammalian predators are numerous, all these species nest entirely in trees.

A review of nesting habits among various raptors revealed that:

(1) some geographical differences in the types of site used by particular species were linked with differences in local predation and disturbance pressures, with species accepting ground sites only where mammalian predators were absent;
(2) minimum requirements within an area changed with time as disturbance and predation changed;
(3) individuals occasionally attempted to nest in less safe sites but seldom succeeded; but
(4) if they did succeed the habit might then spread and lead to local expansion in distribution and density (Newton 1979).

The minimal level of security of nest-sites that any local population accepts may therefore be in continual process of adjustment to predation pressures (Chapter 8). In this indirect sense, predators are responsible for restricting breeding density in areas where safe sites are scarce, for if it were not for predators, many species might accept less safe sites, and thereby breed in greater numbers.

There is also some interaction between parasites and nest-sites. Many arthropod parasites depend for their transmission on birds using the same nest-sites in successive years: individual birds are reinfected each year from the legacy of eggs and larvae left from the previous year. Birds that can build their own nests in a fresh site each year escape this problem, but those with specialised site needs, such as tree-cavities or cliff-ledges, or which return year after year to the same colony sites, cannot (Brown & Brown 1986, Emlen 1986, Chapter 10). Individuals of many such species, when given the choice, change their sites from year to year, breaking the line of transmission. But where nest-sites are scarce, individuals may be forced to use the same sites in successive years, with immediate reinfection, or may be discouraged from breeding altogether. The availability of nest-sites could thus influence the general level of ectoparasitic infestation in bird populations although I know of no attempt to test this view. The important points are that scarcity of nest-sites could enhance parasite effects, and parasites could further enhance the nest-site shortage.

INTER-SPECIFIC INTERACTIONS

Species that eat the same type of food from the same places, and are limited by it, can be expected to be in competition, in that the numbers of one may affect the food-supply, and hence the numbers, of another (Chapter 12). The interaction is not always negative, however, as some species benefit from having others exploit the same resource (Charnov *et al.* 1976). Some vulture species find their food mainly by watching other vulture species (Houston 1988), and many other birds, from gulls to raptors, sometimes rob other species of their food (Brockman & Barnard 1979). Another type of feeding enhancement is seen in some neotropical forest birds which follow the trails of army ants to feed on the insects disturbed (Willis & Oniki 1978). It is also seen in the association of some seabirds with predatory fish, which drive small fish to the surface where they become available to the birds (Brown 1980).

In all these examples, the availability of food for one species need bear little relationship to food abundance, but is governed largely by other species exploiting the same resource. In most, the relationship appears markedly asymmetric, in that one species benefits greatly at the expense of the other, and may be largely dependent on the other for its livelihood. In theory, a decline in the numbers of the facilitating species could in turn lead to a decline in the beneficiary species, despite decreased overall predation on their mutual food supply. To my knowledge, this view has not properly been tested, but over a 21-year period, fluctuations in the numbers of Arctic Skuas *Stercorarius parasiticus* on the island of Foula (Scotland) paralleled similar fluctuations in their main food providers, Arctic Terns *Sterna paradisaea* (Phillips *et al.* 1996). While such kleptoparasitic species clearly benefit from robbing others, the costs to the victims are often substantial, and reflected in reduced production of young (as shown for Puffins *Fratercula arctica* robbed by Herring Gulls *Larus argentatus*, Nettleship 1975).

In the same way that some birds gain food through kleptoparasitism, others gain protection by nesting close to more aggressive species which are able to deter predators. Some species nest from choice close to Fieldfares *Turdus pilaris*, Lapwings *Vanellus vanellus*, terns, gulls or falcons, benefiting from their nest defence (Chapter 9). Most such relationships seem one-sided, in that one species gains protection from another, with no apparent cost, giving little in return. On the coast of Peru, Night-hawks *Chordeiles rupestris* formed mixed nesting colonies with more aggressive terns and skimmers. On all six occasions when terns and skimmers abandoned a beach, the remaining Nighthawk colony was destroyed by predators within two days (Groom 1992). The Nighthawks thus depended on the other species for successful breeding. Their presence was of no obvious disadvantage to the terns and skimmers, but also of no obvious benefit.

Sometimes the presence of one predator can facilitate predation by another, as when one species puts birds off their nests, exposing the contents to another. On Mandarte Island off British Columbia, predation by Northwestern Crows *Corvus caurinus* on eggs of cormorants and other seabirds was greater on days when Bald Eagles *Haliaeetus leucocephalus* were present than on days when they were absent (Verbeek 1982). The same was noted on Tatoosh Island off Washington State, where Bald Eagles scared Guillemots *Uria aalge* from their cliff-top nest-sites, so that the eggs were exposed to crows and gulls, in some years causing complete breeding failure. The provision of artificial plants as shelter caused Guillemots to stay on their nests as Eagles passed, and hence to avoid egg predation, as happened naturally in crevice-nesting Guillemots (Parrish & Paine 1996). In this example, habitat structure influenced the interaction between three predators and their prey. Human disturbance can also lead to increased egg and chick predation in the same way (see later).

Fisheries and seabirds

The complexities of interactions between species are shown by the effects of human fishing, which can lead to either increases or decreases in seabird numbers, depending on circumstances. Many species have increased in numbers because the discards thrown overboard have provided a massive increase in food supply. In particular, demersal fisheries bring to the surface large amounts of fish material that would otherwise remain inaccessible to birds. In the North Atlantic, the Fulmar *Fulmarus glacialis*, the Gannet *Morus bassanus*, various gulls and skuas have been the main beneficiaries of fishery waste (Fisher 1966, Burger & Cooper 1982, Furness *et al.* 1992). Further south, in the Benguela Current off southern Africa, seabird densities at sea increased by eight times in a 30-year period following the establishment of a demersal fishery (Abrams 1985). Furthermore, the removal for human consumption of large predatory fish has probably led to an increase in the abundance of small fish (such as sprats and sandeels) suitable as seabird food (Sherman *et al.* 1981, Furness 1982, Swatzman & Haar 1983). In northern waters, various alcid species are the main beneficiaries of this supply. Similarly, the removal of more than 95% of the large whales from Antarctic waters over the last century is thought to have enabled a massive increase in the krill *Euphausia superba* (as much as 150 million tonnes per year) which formed their food, leading in turn to the large increases in numbers of penguins and other krill-dependent seabirds recorded over the past 50 years (Laws 1977). This is simply a plausible explanation based on coincidence, however, so the increase in seabirds remains open to other explanations.

On the other hand, as the human exploitation of smaller fish has intensified, the food-supply of some seabirds has fallen, causing declines in their numbers. Human overfishing has been suggested as the cause of population crashes in several seabird populations. During the 1960s, marked declines in the breeding numbers of Cape Gannets *Morus capensis*, Cape Cormorants *Phalacrocorax capensis* and Jackass Penguins *Spheniscus demersus* off South Africa were associated with decline of Pilchard *Sardinops ocillata* and Anchovy *Engraulis capensis* stocks (Crawford & Shelton 1978, Crawford *et al.* 1983). Other examples include various seabird species and anchovies off Peru (Schaefer 1970), anchovies and Brown Pelicans *Pelecanus occidentalis* off southern California (Anderson & Gress 1984), various species and sandeels off Scotland (Heubeck 1989, Monaghan *et al.* 1989, Klomp & Furness 1992), and Guillemots *Uria aalge* and Capelin *Malotus villotus* off Norway (Vader *et al.* 1990). In some such species, breeding numbers fell by more than 70% within a few years. In other examples, collapse of fish stocks has been held responsible for mass breeding failure among seabirds, which if continued over several years could cause a decline in breeding numbers (for Kittiwake *Rissa tridactyla* in Alaska, see Hatch 1987; for various species around Britain, see Monaghan *et al.* 1989, Harris & Wanless 1990, Furness & Barrett 1991; for Puffin *Fratercula arctica* in Norway, see Anker-Nilsson 1987; for Shag *Phalacrocorax aristotelis* in Britain and Norway, see Aebischer 1986 and Vader *et al.* 1990). However, small short-lived fish species are highly susceptible to oceanographic changes, which have accompanied most major crashes in small fish stocks, so it is often hard to disentangle human from natural impacts in the declines of their populations.

These examples indicate the effects on populations of competition and food enhancement, and also show how the removal of certain organisms (such as whales) from an ecosystem could have repercussions through the whole system. Likewise, in terrestrial environments, the removal of large predators is thought to have allowed smaller ones to increase, with effects on some bird populations (Chapter 8). Examples include the effects of Wolves *Canis lupis* and Coyotes *Canis latrans* in reducing Red Fox *Vulpes vulpes* numbers to the benefit of ducks and other ground-nesting birds (Greenwood *et al.* 1995), and the effects of Peregrines *Falco peregrinus* and other raptors in reducing crow numbers, thereby enhancing the nest success of various birds (Paine *et al.* 1990).

INTERACTIONS BETWEEN FOOD-SUPPLIES, PREDATORS AND PARASITES

Food-shortage might increase the risk of predation, or predators might increase the risk of starvation. The same can be said of the interaction between food-supply and parasitism.

Food and predation

Compared with well-fed chicks, hungry ones are known to be more vulnerable to predation: altricial ones because they beg more, creating more noise (for Great Tit *Parus major*, see Perrins 1979; for Magpie *Pica pica*, see Redondo & Castro 1992), and precocial ones because they move around more in search of food, becoming more visible (for gallinaceous birds, see Green 1984, Warner 1984, Rands 1986; for water-fowl, see Hunter *et al.* 1984; for Oystercatcher *Haematopus ostralegus*, see Ens *et al.* 1995). In one study, predation by Herring Gulls *Larus argentatus* on Eider *Somateria mollissima* ducklings varied greatly from year to year, and in some years practically all ducklings were eaten during their first ten days of life (Swennen 1989). However, gull predation was heaviest when the small ducklings were already weakened by food-shortage, partly because they did not then respond appropriately to alarm signals from the adult females. Similar interactions may exist between the condition of young birds and their susceptibility to parasites (Chapter 10; Møller 1994).

Food-shortage can also enhance chick predation by causing adults to spend more time foraging rather than guarding their chicks. This has been noted in such diverse species as gulls, crows and raptors. Moreover, when supplementary food was placed near the nests of Magpies *Pica pica* and Carrion Crows *Corvus corone*, more young were raised, not because the young were better fed, but because they suffered less predation (Yom-Tov 1974, Högstedt 1981; Chapter 7). The extra food enabled the parents to spend more time near the nest, in better positions to protect their chicks against raids by other crows. Similar effects were shown in adult tits, where experimental food-provision led to improved overwinter survival, partly through reducing predation losses (Chapter 7). The extra food enabled birds to spend more time than otherwise in safer sites, and on vigilance behaviour (Jansson *et al.* 1981). In such cases, then, food-shortage and predation interact to influence nest success or survival.

Apart from causing direct mortality, predators can accentuate food-shortage in their prey when their repeated attacks discourage prey from feeding in places where they are at high risk, such as far from cover, or by disturbing them so often as to increase their flight costs and reduce the time they have for feeding. Once attacked by predators, many birds take time before they start to feed again, and then spend more time on vigilance. In flocks of small birds, return times after an attack can be as little as a few minutes (Morse 1973), but in other birds much longer. According to Errington (1939), Bobwhite Quail *Colinus virginianus* may remain in hiding without feeding for a day or more, following persistent attacks by hawks or human hunters. Thus, when exposed to predation or human hunting, birds typically spend more time on the alert, on movement and on hiding, leaving less for feeding (Krebs 1980, Paulus 1984).

The non-lethal effects of predators on bird numbers and distributions could in fact be greater than their lethal ones. Seabirds nest on islands partly to avoid mammalian predators: at present, many islands are made unusable for certain seabird species by the presence of introduced predators, but if such predators are removed, seabirds often recolonise (Chapter 9). Otherwise most adult seabirds seem almost immune from predation. Here is a situation then where predators, by restricting distribution, could well limit overall population levels, yet not a single adult in a secure site may die of predation. Expansion of populations could most effectively be achieved by removing predators from other islands, freeing them for use by seabirds. This happened in Peru where seabirds colonised islands and peninsulas after predators were excluded, increasing the total seabird population of the country (Duffy 1983). Situations where predators exclude prey from certain areas also occur naturally. In the Galapagos archipelago, for example, Red-footed Boobies *Sula sula* nest only on islands that lack Galapagos Hawks *Buteo galapagoensis* (Anderson 1991). Of 22 islands surveyed, no island had both species, seven had neither, ten had hawks but no Red-footed Boobies, while five had Red-footed Boobies but no hawks. This pattern was significantly different from that expected on independent distributions ($P=0.03$). The main factor influencing the distribution of hawks seemed to be the presence of *Tropidurus* lizards, their main prey.

By their mere presence, then, predators may limit the distribution of some bird species, which avoid areas of high risk and thus surrender part of their resource base, whether food or nest-sites. In addition, several bird species have been found to nest at lower density near to bird-of-prey nests than further away. One example concerns Kestrels *Falco tinnunculus* in western Finland, whose nesting locations in open farmland were manipulated by provision of nest-boxes (Suhonen *et al.* 1994). The Kestrels were summer visitors to the area, hunting small birds (as well as rodents), mostly within 1 km of their nests. Small migrant bird species that arrived after the Kestrels settled at lower density near Kestrel nests than further away, but no such difference occurred in small resident species that were already settled when Kestrels arrived, nor in larger species immune to Kestrels. Avoidance was especially marked in Skylarks *Alauda arvensis*. Similar avoidance of the vicinity of Merlin *F. columbarius* nests by songbirds was noted in Canada (Sodhi *et al.* 1990). These observations did not reveal whether any songbirds were prevented from breeding by the presence of predators or whether they simply settled elsewhere. They did indicate, however, that predators, by their mere presence, lowered the attractiveness of certain areas, and thus influenced local prey densities.

Many birds, in areas where predators occur, seldom forage more than a certain distance from cover, which can make food at greater distances unavailable to them. For example, in a study in Scotland, Ptarmigan *Lagopus mutus* rarely fed more than three metres from rocks or other cover, which rendered large areas of intervening ground unused (Rae 1995). In other species, the proportion of time spent on vigilance increased with distance from cover, which reduced food-intake rates and again rendered parts of the habitat unusable (Barnard 1979, 1980, Caraco *et al.* 1980, Elgar 1989). In this indirect way then, the interaction between habitat structure and predation pressure can limit the proportion of habitat that birds can exploit, and effectively limit their numbers below what the total potential food-supply would permit. It provides another example of focal point exploitation of widespread food supplies. Alternatively, competition for food near to cover may be intense, leading to depletion of local supplies, which may force birds to feed further from cover, increasing the proportion lost to predation. In situations like this, increases in numbers could in theory be achieved either by altering habitat (to increase cover), enhancing food-supply near to cover or reducing predator densities, or any combination of these.

Species differ in their reaction towards shrubby cover, depending partly on their escape behaviour. In an experiment, Cowie & Simons (1991) placed peanut-laden feeders at different distances from a hedgerow. Total food consumption at feeders close to cover was double that at feeders only 7.5 m away **(Table 13.2)**. House Sparrows *Passer domesticus* fed only near cover, Blue Tits *Parus caeruleus* mainly near cover, but Greenfinches *Carduelis chloris* fed frequently at greater distances, though still relying on cover for escape. In contrast, species of open habitats, such as shorelines and grassland, often employ aerial escape which is independent of cover. For such species, cover may represent a threat that could conceal predators. Such species increase their vigilance, and hence lower their food-intake, when feeding close to cover (Metcalfe 1984). Consistent with these findings, some grassland bird species responded markedly to minor experimental manipulations in the distribution of woody cover (Lima & Valone 1991). Following the addition of bushes to an open area, species with cover-dependent escape tactics, such as Chipping Sparrows *Spizella passerina* and Vesper Sparrows *Pooecetes gramineus*, increased in numbers, while cover-independent species, such as *Ammodramus* sparrows and Horned Larks *Eremophila alpestris*, declined. In other words, species abundances were influenced by their species-specific responses to cover.

Table 13.2   Frequency of use of bird-feeders (containing peanuts) placed different distances from cover. From Cowie & Simons 1991.

	Distance of feeder from cover (m)			
	0	2.5	5.0	7.5
% total food consumption	33	27	23	17
% birds at each distance:[1]				
Greenfinch *Carduelis chloris*	80	86	94	95
Blue Tit *Parus caeruleus*	12	10	3	4
House Sparrow *Passer domesticus*	4	0	0	0
Greenfinch:				
% time feeding[2]	72	66	66	63
% time vigilant[2]	12	18	20	27

[1] Other species accounted for the remaining birds seen.
[2] Other activities made up the remaining time, including movement and fighting.

The need for vigilance in the presence of predators is influenced by flocking, as well as by habitat structure. Many birds spend less time on vigilance when feeding in flocks than when feeding alone. Among Goldfinches *Carduelis carduelis*, the rate of individual energy gain increased markedly with flock size at least up to eight birds. Birds in flocks of eight or more ate up to 2.3 times more seeds per unit time than singletons. Usually the birds fed among tall herbs where visibility was limited, but when feeding in more open sites, such as freshly-mown fields, flock-feeding individuals increased their feeding rate even further, up to seven times the solitary feeding rate (Glück 1986). Relationships between scanning and group size have been noted in many other bird species (Barnard 1979, Benkman 1997), as have relationships between scanning and visibility (Metcalfe 1984). It is another way in which the risk of predation, interacting with habitat structure, influences feeding rate and hence starvation risk.

In any species, individuals must strike a balance between the risks of starvation and predation, and by implication, on any given food-supply, more birds might starve in the presence of predators than in their absence. Predators can also influence the quality of feeding areas over which birds compete, so that dominant species (or individuals) get the most secure sites and subordinates the least secure ones (Chapter 2). Such segregation was apparent among feeding flocks of tits in northern Europe, where the dominant Willow Tits *Parus montanus* and Crested Tits *P. cristatus* fed mainly on the insides of trees, forcing the subordinate Coal Tits *P. ater* and Goldcrests *Regulus regulus* to the outer twigs, where the risk of predation was greater. Over winter, much larger proportions of the subordinate than the dominant species were killed by Pygmy Owls *Glaucidium passerinum* (Suhonen *et al.* 1992). Within species, dominant adult Redshank *Tringa totanus* fed in winter in an area that offered low food density but greater security from predators, while juveniles were relegated to food-rich but risky areas (Cresswell 1994). A Redshank in a food-rich area could attain twice the feeding rate as one in a food-poor area, but was five times more likely to be killed by a hawk. This example illustrates a general point, that the quality of any feeding area is enhanced by food abundance but reduced by high predation risk.

In these ways, by both direct and indirect means, predators can have negative effects on prey populations. They might also have positive effects, however, as when predators reduce prey density to a level at which disease transmission is impaired or food competition is lowered. Imagine a population, such as that of a northern seed-eater, which has a fixed stock of food to last the winter. High numbers in autumn may cause the food-supply to be depleted quickly and few birds to survive the winter, but lower numbers in autumn (resulting from predation) could ensure that the food-supply lasted longer, so that more birds survived the winter than might otherwise have done so (Newton 1972, Murton *et al.* 1974). Autumn shooting of Rock Ptarmigan *Lagopus mutus* led to increased spring breeding densities and a high proportion of yearlings in local populations (McGowan 1975). In this instance, the increased spring densities were thought to be partly due to immigration of young birds tolerant of crowding. Further evidence that shooting may sometimes increase breeding densities relative to control areas comes from Red Grouse *Lagopus l. scoticus*, in which a population crash was curtailed experimentally by preventing a build-up to a peak density (Watson *et al.* 1988).

Similarly, if predators selectively removed diseased individuals as they appeared, they might help to suppress epidemics, and thus contribute to the maintenance of numbers. In this instance, they would also be taking birds destined to die from another cause, and so have no negative effect on numbers. An example of selective predation is provided by Red Grouse, in which nesting birds that were killed by predators contained more parasites than birds that were not (Hudson & Dobson 1991, Hudson *et al.* 1992a). Mathematical models indicated that small numbers of predators selectively removing heavily infected individuals could allow Red Grouse density to increase because the predators effectively reduced the regulatory role of the parasites. However, moderate to high levels of predation could itself suppress the grouse population (Chapter 10). Among other animals there are examples of parasites making infected prey more prone to predation either by weakening them or affecting their behaviour (Coombes 1996). In addition, predators or parasites might increase the numbers of some species by killing their competitors, thus making more resources available. Conversely, experiments on other organisms have shown that predator removal can sometimes lead to marked increases in the numbers of some prey species, but at the expense of others that share the same resources (Sih *et al.* 1985). The implication is that predation can influence community structure by depressing populations of competing species below resource limitation levels, allowing certain species to co-exist that would otherwise be incompatible. In all these situations, the negative effects of direct mortality from predation or parasitism are overridden by the benefits that facilitate larger populations.

Human disturbance

Human disturbance has the same effects as predator disturbance, disrupting feeding and other activities, and increasing energy expenditure from flight. It may also keep birds from potential feeding areas, or cause them to forage in areas where intake rates are lower or natural predation risks are greater. Any keen observer will have noticed that:

(1) some bird species are more sensitive to human presence than are others,
(2) individuals of most species can habituate to predictable and benign disturbance, yet remain sensitive to unpredictable or damaging disturbance (such as shooting),

(3) most species respond to people more than to other forms of disturbance, such as cars or aircraft,

(4) most quarry species are more sensitive (flying at greater distance from the source of threat) in seasons or in areas where they are shot at than in seasons and areas where they are protected, and

(5) human presence can prevent some species from using certain areas altogether.

Birds evidently respond continually to perceived predation risk, altering their behaviour, and hence the range of areas they can use.

Many studies concern waterfowl which, through some human activities, such as boating, may be totally excluded from otherwise usable habitat. Fishermen on banks and in boats were found to affect the number of ducks using part of a large reservoir in southern England (Cooke 1975). Pochard *Aythya ferina* and Coot *Fulica atra* almost entirely avoided the area when fishermen were present, but these and other species used the area heavily after mid-October when fishing stopped **(Figure 13.2)**. The same is seen in response to hunting, as some areas open to hunters remain virtually unused by waterfowl until after the end of the shooting season. Such distributional changes may be unimportant to the overall population level if they simply alter the dates at which food-supplies are exploited, but they could reduce the overall level if they cause birds to feed in other areas with greater predation risk or prevent them from fully exploiting potential food-supplies.

In many areas established as reserves, use by waterfowl throughout the winter increased greatly once shooting was stopped (Fox & Madsen 1997). In one study, disturbance effects of shooting were tested by setting up experimental reserves in two Danish coastal wetlands (Madsen 1995). Over a 5-year period, these reserves became important staging areas for coastal waterfowl, and the national totals of several species were increased, as more birds remained there. Hunted species increased the most, some 4–20-fold, while non-hunted species increased 2–5-fold. Furthermore, most quarry species stayed in the area for up to several months longer each winter than in earlier years. No declines in usage were noted in other areas still open to wildfowling, so the accumulation of birds in the reserves was attributed to the short-stopping of birds that would otherwise have migrated further south. One consequence of this build-up of birds was a greater depletion of the local food-supply, especially eelgrass *Zostera* **(Figure 13.3)**. In this and other studies, hunting disturbance emerged as a major factor influencing the winter distribution of water-fowl, on both local and regional scales. It was tantamount to local loss of habitat and food-supply, but was reversible. It seems that birds, like some other animals, select feeding areas by balancing the benefits of food abundance and predator avoidance. By implication, to obtain a given intake rate, higher food densities would be required at a disturbed site than at an undisturbed one. And of course, the risk of being killed is also greater at the disturbed site.

In a second study, the impacts of disturbance by farmers on spring fattening of Pink-footed Geese *Anser brachyrhynchus* were analysed. On undisturbed sites in northern Norway, the geese rapidly put on weight, whereas in disturbed sites they did not. Subsequently, geese that had fed in undisturbed sites produced more young per female than geese from disturbed sites (Madsen 1995).

Attempts were made to quantify the effects of disturbance on Pink-footed Geese *Anser brachyrhynchus* wintering in an area of southeast England where they fed

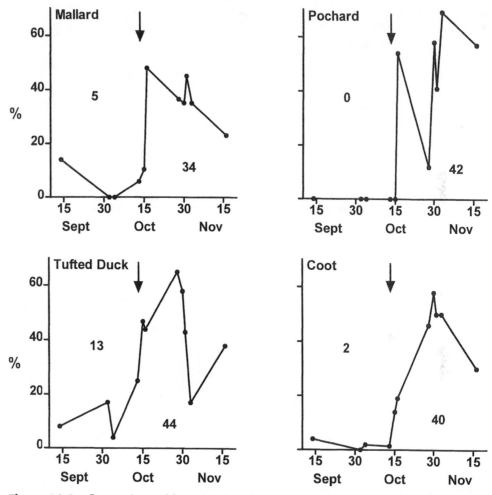

Figure 13.2   Proportions of four species of waterfowl in a sanctuary area of Grafham Water, southern England, in September–November 1974. The area was fished from the banks and from boats until 14 October (arrows). Numbers on each graph are the mean percentage of birds in the area before (left) and after (right) the end of the fishing season. Details for Mallard *Anas platyrhynchos*, Pochard *Aythya ferina*, Tufted Duck *Aythya fuligula* and Coot *Fulica atra*. Based on Cooke 1975.

mainly on the post-harvest remains of sugarbeet *Beta vulgaris* (Gill *et al.* 1996). The geese were often shot, so were wary, especially when feeding close to roads. By measuring the biomass of beet fragments before and after the geese had fed, it was discovered that a smaller percentage of beet had been eaten in fields near to roads than in fields further away **(Figure 13.4)**. The total number of 'bird-days' for each field depended both on its initial content of beet remains and on its distance from a road. If all fields were exploited by the geese to the level of the least disturbed fields, many more geese could have been sustained in the area over winter. The reduction in use of these feeding grounds caused by overwinter disturbance was quantified by translating the biomass of food not consumed into the number of birds that this food

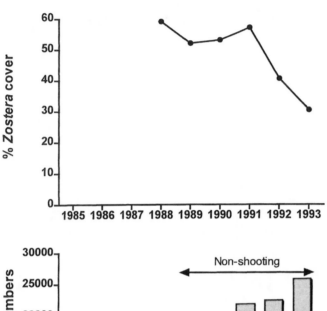

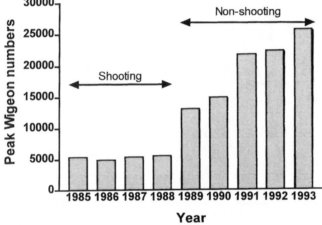

Figure 13.3 Development of Wigeon *Anas penelope* numbers after establishment of a non-hunting reserve in Denmark, and consequent effect on eelgrass (*Zostera*) food supplies. From Madsen 1995.

could have supported. Admittedly, if the geese were really hungry they might have become less wary, but they would then increase their risk of being shot.

Another effect of human disturbance is to facilitate natural predation. In one Finnish study, only about 40% of Velvet Scoter *Melanitta fusca* ducklings survived to three weeks old, the rest being eaten by gulls (Mikola *et al.* 1994). Predation was greatly influenced by disturbance from passing boats, which caused broods to scatter. Gull attacks were 3.5 times more frequent in disturbed than undisturbed situations, and broods disturbed more often than average ended up smaller than average. Increased predation on Eider *Somateria mollissima* ducklings in the presence of human disturbance has been noted in other studies (Åhlund & Götmark 1989, Keller 1991). Cannibalism is a major cause of egg and chick loss in Herring Gulls *Larus argentatus,* and in one study, hatching success on different islands was inversely related to human disturbance which caused adults to leave their eggs exposed to other gulls and to the sun which overheated embryos (Hunt 1972). In a review of

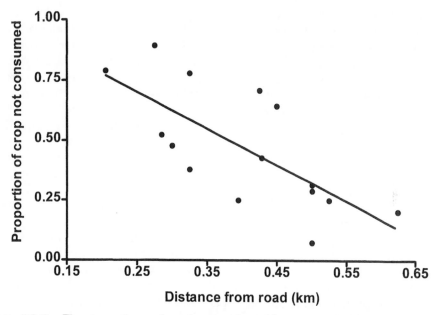

Figure 13.4   The percentage of waste sugar beet biomass eaten by Pink-footed Geese *Anser brachyrhynchus* in relation to the distance from a road. Each point refers to a separate field. Studies of this type help to quantify the effects of disturbance (= predator avoidance) on the ability of birds to exploit a food-supply, and to explore the likely consequences of changes in disturbance. Thus under greater disturbance, the geese might have avoided even more fields near to the road. If the slope in this Figure was thereby increased from the −1.5 shown to −3.0, this would reduce by 61% the numbers of geese that could be sustained (or more correctly, goose-days). Similarly, if hunting was reduced so that geese became less wary, reducing the slope to 0.6 (say), the carrying capacity of the area would be increased by 42%. From Gill *et al.* 1996.

relevant literature, Götmark (1992) found that half the studies that examined the effect of observer nest visits reported some reduction (by 40% on average) in nest success. The same has been seen with natural predators (see above), so although two predator species are exploiting the same resource, their effects on one another are not necessarily negative. Studies of the effects of human disturbance on wildlife provide useful surrogate studies of the non-lethal effects of predators. They are also of direct applied relevance to wildlife conservation, however, as human disturbance of rural areas continues to increase. As with predator disturbance, the critical question is whether human disturbance results in lower overall population sizes.

Winter fattening

From feeding during the day, small birds accumulate body fat for use overnight, typically carrying more fat in winter when nights are colder and longer (Newton 1969, Evans 1969, Biebach 1996). The greater its fat reserve, the longer a bird can resist starvation. However, one cost of carrying extra fat is reduced flight efficiency, which is assumed to increase predation risk, because slight differences in weight make big differences to the take-off speed and manoeuvrability of small birds

(Witter *et al.* 1994). During the course of a single day, in experimental conditions, a 7% increase in body weight (typical for small birds) caused a 30% increase in the time taken to reach cover (Metcalfe & Ure 1995). Another cost of fat storage is the extra energy needed to move around, and because fat birds need to feed more, they can spend less time in sites secure from predators. Fat reserves thus have costs as well as benefits (Lima 1986, McNamara & Houston 1990). If such reserves are viewed as a compromise between the risks of starvation and predation, several curious facts about winter fattening in small birds are more readily explained. These include the tendencies for fat reserves to increase with reducing daylengths, temperatures and predictability of food-supplies (found in many species; Biebach 1996), and for subordinate birds (which often have most difficulty in obtaining food) to carry more fat than dominant ones in the same area (found in Great Tits *Parus major*, Gosler 1996). Fattening can thus be viewed as a strategic response to prevailing conditions (Lima 1986).

Remarkably, during a 15-year period when Sparrowhawks *Accipiter nisus* were absent from southeast England (through pesticide poisoning), Great Tits *Parus major* in that region became heavier (fatter) **(Figure 13.5)**. When hawks recovered in numbers the tits became lighter again (Gosler *et al.* 1995). No such changes occurred in Great Tits in western districts where Sparrowhawks were present throughout, nor in a second species, the Wren *Troglodytes troglodytes*, which was less often killed by hawks. Again the implication was that tits regulated their own fat contents at some compromise level, depending on the relative risks of starvation and predation. In the presence of hawks, starvation chances were presumably higher if food-supply was suddenly reduced, as by snow. Hence, in assessing whether starvation or predation

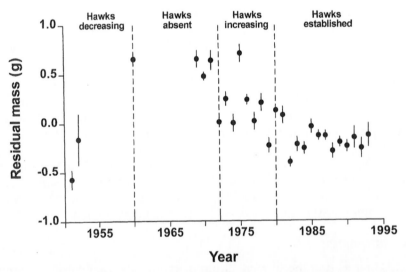

Figure 13.5   Changes in body mass (after correction by regression analysis for body size, time of day and ambient temperature) of Great Tits *Parus major* in Wytham Woods, Oxford, between 1951 and 1993. Points shown are means ± SE. Great Tits were significantly heavier during the years when Sparrowhawks *Accipiter nisus* were absent. From Gosler *et al.* 1995.

is the major cause of winter mortality in small birds, the distinction is blurred when one considers the population consequences of individual energy budgets. If food is limiting, many individuals may die from predation as a result of following the optimal strategy with respect to body reserves, or through feeding hard at the expense of reduced vigilance.

Some of the same points apply to migratory fattening, when birds lay down the reserves necessary to sustain them on long flights. The risk of predation could act in favour of rapid fattening immediately before the flight (as opposed to a slow build-up over a longer period) and of a reserve not much greater than needed for each leg of the journey (as opposed to a large safety margin).

Food and parasites

Not only are many internal parasites transmitted via food, once established within their hosts, they commandeer, directly or indirectly, part of the host's nutrient intake for themselves. Thus, it may be expected that a bird with parasites needs to eat more than one without in order to perform at the same level, or that, on the same amount of food, a bird with parasites would perform less well than one without. Not surprisingly, birds with many parasites often die selectively during periods of food-shortage, and reproduce less well than those with fewer (Chapters 7, 10 and 11). The implications are that (1) birds with high parasite loads may survive at times of plenty, but are among the first to succumb at times of scarcity, and (2) on any given food-supply, an unparasitised population could attain higher density than a heavily parasitised one.

The transmission of any parasite that is passed directly from host to host is dependent on the density of the hosts, which can in turn be influenced by other factors, such as predation and food-supply. In the past, a strongyle parasite caused frequent crashes in Grey Partridge *Perdix perdix* populations in Britain, but as the density of Partridges has declined in recent decades, through other causes, the parasite has ceased to be important (Potts 1986). The opposite trend occurred in Red Grouse, in which a strongyle parasite became important as a limiting factor in the late 19th century, once predator removal and habitat management had led to greatly increased grouse densities (Lovat 1911). Many other diseases, that have largely disappeared from bird populations or appeared only in recent years, could be associated with changes in host density brought about by changes in other limiting factors, or with changes in the abundance of alternative hosts. Because most parasites require both susceptible hosts and suitable conditions for transmission, it is easy to appreciate how changes in environmental conditions could enhance or suppress diseases.

Changes in the conditions in which waterfowl live have greatly increased the prevalence of disease, especially in North America (Friend 1992). Being gregarious for much of the year, waterfowl are predisposed to the transmission of infectious agents, especially outside the breeding season when they gather in largest numbers. Through human action (destruction of natural wetlands and creation of refuges), the birds are being increasingly concentrated in fewer places. Through artificial feeding, they are remaining in certain refuges for much longer each year than formerly, further increasing the opportunities for pathogens to spread.

Droughts also accentuate crowding, given the diminished base of wetland habitat.

Little wonder then that the first known outbreaks of two virulent diseases in waterfowl (avian cholera and duck viral enteritis) occurred on refuges (Friend 1992). Avian cholera has since become a major factor, periodically reducing the numbers of various waterfowl species in North America (Chapter 10). Moreover, the migratory habits of waterfowl mean that they can carry diseases from one place to another. In these ways, changes in the conditions that favour the development and spread of diseases have almost certainly influenced the proportion of waterfowl deaths from disease, and the importance of disease in limiting populations, compared with earlier times.

The anti-parasite behaviour of birds has been much less studied than their anti-predator behaviour. The situation is complicated in that, while some parasites, such as lice and cuckoos, are detectable, others, such as helminths obtained from food-items, are usually not. But like anti-predator behaviour, anti-parasite behaviour has costs, so trade-offs are presumably involved. In the same way that a bird mobbing a predator to drive it from the nest area risks being caught, one ejecting a cuckoo's egg risks removing one of its own instead. Similarly, large invertebrate prey items might give a shorebird high food-intake, but may be more heavily parasitised than small ones, and so may be rejected to reduce the risk of infection (Hulscher 1982). Protection from parasites is obtained, but only at the expense of reduced food-intake.

FOOD AND POLLUTANTS

Birds get pollutants mainly from their food, and the types of food eaten influence the levels acquired. In addition, however, the nutritional state of the bird at the time can influence the severity of some pollutant effects, so that birds in good condition are less likely to die than those in poor condition (Chapter 15). Sub-lethal effects on bird behaviour can also be important. Various chemicals at low levels of exposure have been shown to affect the learning ability of birds, their response to stimuli, their flight coordination, their nest-building and incubation, and their general activity, making them more or less active than normal (Peakall 1985). Although such studies have been confined to birds in captivity, the changes in behaviour induced could influence the ability of birds in the wild to obtain food or to avoid predators and other risks. In the reciprocal relationship, predatory birds are more likely to take affected prey behaving abnormally than normal prey, and thus increase their own risk (Hunt *et al.* 1992). By such indirect means, therefore, pesticide or pollutant exposure could influence the survival and reproductive prospects of birds in the wild through increasing their vulnerability to other limiting factors. Needless to add, this is an aspect that is extremely difficult to study outside the laboratory context. The effects of pollutants on community structure are discussed in Chapter 15.

OTHER INTERACTIVE FACTORS

All the types of interaction discussed above are in turn influenced by two other all pervasive factors, namely habitat and weather. The structure of the habitat, and the cover and shelter it provides, can have a great influence on densities of birds (including territory sizes), as well as on their feeding efficiency and predation risks.

For many bird species that both nest and forage within their territories, it is hard to evaluate the relative influences that cover and food-supply have on habitat selection (Martin & Roper 1988). In Grey Partridges *Perdix perdix* on English farmland, the amount and disposition of cover had a marked influence on territorial spacing regardless of food-supply, with pairs occupying larger areas where cover was sparse (Jenkins 1961). But for some other species in which nesting and feeding occur in different habitats, cover seems paramount in the selection of nesting habitat. The evidence for this view is largely indirect, that habitat preferences of particular species relate to cover rather than food-supply and that, through reduced nest predation, breeding success improves with quality of nesting cover (Martin & Roper 1988). An example is provided by the North American Clay-coloured Sparrow *Spizella pallida* in which individuals hold territories in low scrub and fly to feed in distant grassland. In this species, territory sizes were related to shrub density, with the smallest territories in areas of densest shrub cover and the largest in areas of sparsest cover (Knapton 1979).

The smaller the amount of nesting cover in a landscape, the easier it is for predators to find nests and the higher their foraging success. Increased predation on the eggs of several ground-nesting species in recent years has been attributed to destruction of cover by agricultural intensification (Chapter 9; for Grey Partridge, see Potts 1986; for prairie ducks, see Greenwood *et al.* 1987). Conversely, habitat structure can influence the ability of predators to exploit their prey. For some raptor species, the presence of cover or of suitable perches in open land can facilitate prey capture. To test the idea that shortage of hunting perches limited the use of open areas by raptors, Widén (1994) provided man-made perches on 11 clear-cut forest areas, while 11 other areas served as controls. Foraging raptors used experimental clear-cuts significantly more than control ones, and usage changed accordingly when the experimental and control areas were switched. By implication, perch provision was tantamount to increased food-supply. Likewise, Yosef & Grubb (1994) provided extra perch sites in an area occupied by 11 Loggerhead Shrike *Lanius ludovicianus* pairs. The birds responded by reducing the size of their territories which enabled six new pairs to settle in the same area. Again, the provision of hunting perches in open land amounted to increasing the food-supply. In contrast, no such density change occurred in areas occupied by 11 control pairs.

Another form of interaction between habitat and predation involves the long-term memory of predators. In a study of Tengmalm's Owls *Aegolius funereus*, Sonerud (1993) noticed that the risk of nest predation by Pine Martens *Martes martes* increased with the age of the nest-cavity (nest-box), and was higher in boxes where the previous nest was preyed upon than in boxes where the previous nest was successful. To test the view that this difference was due to Martens remembering the locations of boxes as likely sites for a meal, he relocated a randomly-selected half a sample of 12-year-old boxes, and then some years later relocated the other half. In both cases, predation on owl nests was reduced in the relocated boxes but unchanged in the unmoved ones (22% vs 83%). This supported the view that long-term memory for nest-sites played a role in predation by Pine Martens. The same phenomenon may explain why Black Woodpeckers *Dryocopus martius* that excavated a new cavity suffered lower nest predation than those that re-used an existing cavity from the previous year (Nilsson *et al.* 1991). Predator memory has also been used to explain the pattern of predation by Goshawks *Accipiter gentilis* on male Ruffed Grouse *Bonasa umbellus* taken from their display logs (Gullion & Marshall 1968). Those males that used logs not used by a predecessor lived on average four months longer than those that replaced earlier males on logs that had been used for several years running. Moreover, male grouse that shifted from one temporary log to another lived on average ten months longer than those that shifted from a temporary to a traditional log. Gullion & Marshall (1968) believed that this was because raptors learned the whereabouts of the traditional display logs and often waited in ambush when male grouse were using them. These various findings show that, when nest-sites and display-sites are in superabundant supply, so as to enable individual users to move around, predation may be lower than when such sites are less abundant, forcing users to re-use the same sites year after year. This is equivalent to the situation involving the build-up of arthropod parasites in frequently-used nest-sites (Chapter 10). It is not enough for such resources to be sufficient for users, they must be much more than sufficient if the rates of predation and parasitism are to be kept low.

Similar interactions may occur between weather and predation. In Arctic areas, in years of late snow melt, Arctic Foxes *Alopex lagopus* need only walk from one snow-free patch to another to find nests, instead of searching the entire terrain for more dispersed nests, as in years of early melt (Meltofte *et al.* 1981). Likewise, weather has a great influence on the abundance of food, as well as on its accessibility, which can in turn influence bird numbers, as illustrated by the effects of snow cover on ground-feeding birds (Chapter 10). And the incidence of disease is also greatly influenced by habitat and weather conditions which favour or impede it, the association of botulism with drought being an obvious example (Chapter 10).

CONCLUDING REMARKS

If resource limitation can be shown to exist, predation, disease and competition are often assumed to have little or no influence on numbers. The main argument of this chapter is to show that this view is too simplistic, and that a common and possibly prevailing pattern in nature is for joint limitation of population levels by both resources and natural enemies, themselves interacting with habitat structure and weather. While some types of interaction discussed above are well founded, however, others cannot be considered established until more documentation is available.

Birds are clearly up against different constraints at different stages of the annual cycle, depending on whether they are breeding, moulting, migrating or simply surviving. Changes from one activity to another alter the balance of risks, and the probability of dying from different causes. In some bird species, the very act of breeding, which normally occurs when food is plentiful, seems to increase the risks from predators and pathogens. Parents become vulnerable to predation because of the greater foraging and other activity that breeding entails, and females of ground-nesting species, in particular, are often plucked from the nest when incubating (Chapter 9). In addition, parasite infections often increase in birds that are breeding (Møller 1993, Norris *et al.* 1994). This may be partly because the endocrine changes associated with reproduction can lead to immuno-suppression, as can the greater energy demand involved, as shown experimentally in captive Zebra Finches *Taeniopygia guttata* (Deerenberg *et al.* 1997). In addition, some birds, when working to feed young, sacrifice their own maintenance activities, such as feather care, which enables ectoparasites to build up (Clayton 1991). These various risks provide costs to reproduction which have seldom been measured. Other risks result from the feather loss associated with moult and from the fattening associated with migration and winter survival.

The anti-predator behaviour of birds appears plastic, in that it can be varied by individuals according to the prevailing threat that predators pose (Lima & Dill 1990). The same may be true of anti-parasite behaviour. It would be surprising if this behavioural plasticity did not have effects on the sizes and dynamics of wild bird populations.

SUMMARY

Particular bird species can be limited primarily by different factors in different areas and at different times. However, different limiting factors often interact so that one can enhance or reduce the effects of another on population levels. Nest-sites, water, cover, predators or disease organisms can all restrict a species to only parts of an area where food is available, while predators and parasites can reduce rates of food-intake and body condition, thereby enhancing the effects of food-shortage. Alternatively, certain factors, such as increased food-supply or reduced predation, might enable species to increase to high densities, which facilitate the spread of disease, causing epidemics which would not occur at lower densities.

While certain species, through shared resources, predators or parasites, can indirectly reduce the numbers of their competitors, others can make more of a shared resource available to their competitors. Commercial fisheries have affected seabird numbers either negatively by removing food-supplies or positively by providing new food sources.

A frequent phenomenon in nature is 'focal point' regulation, where birds are concentrated in specific localities by their need for colonial nest-sites, cover or water. This can lead to heavy exploitation of the local food-supply, while further afield food remains abundant and largely unexploited. Examples occur among some desert birds concentrated by water holes and among pelagic seabirds concentrated by nesting islands.

Part Three
Human Impacts

Mallard *Anas platyrhynchos* descending towards decoys and hunter.

Chapter 14
Hunting and Pest Control

By killing birds, people can influence wild populations in the same ways as can natural predators, with effects on breeding numbers that vary from nil to annihilation. But because most human populations no longer depend on wild prey, human numbers do not decline in response to its over-exploitation. Aided by technology, humans now have unprecedented powers to reduce or eliminate bird and other wildlife species through over-kill – that is, culling at a rate that exceeds natural replenishment.

In some parts of the world, birds are still killed primarily for subsistence, and populations are sometimes over-exploited until they disappear. In the developed world, however, most killing of birds now occurs in the interests of recreational hunting or crop protection. The same principles apply in both, but hunting requires a sustained harvest, and crop protection a lowering of population level and consequent reduction of damage. A third aspect is the capture of wild birds for the pet trade, for which the ideal is again a sustained harvest. All three aspects are discussed in this chapter, leaving extinction through over-kill for Chapter 16.

HUNTING

The impact of hunting depends greatly on whether the quarry species is managed as a sustainable resource, such as resident game birds on some private estates, or whether it is hunted on a free-for-all system where anyone can kill what he wants. By manipulation of habitat, disease and natural predation, densities of some managed game species have been increased well above the level they would otherwise achieve, and, with careful control of hunting, have continued to provide large annual 'harvests' for long periods, on some estates for longer than 150 years. Where this system has failed, and game populations have declined, this can usually be attributed to land-use changes and poor habitat management (Potts 1986). In contrast, species hunted on a free-for-all basis have almost always declined, and many now have breeding populations well below the level that the habitat could support. The truth of this statement is evident in the fact that, whenever hunters gain unrestricted access to previously unexploited populations, these populations usually decline rapidly, recovering only if some measure of protection is later imposed. Examples involving the over-exploitation of whale and fish-stocks are commonplace, but the same holds for many hunted bird species which are shot without restraint. Migratory waterfowl populations in western Europe were once hunted on this system, but all have increased greatly in recent decades following the creation of reserves or the enactment of protective legislation (Monval & Pirot 1989, Kirby et al. 1995). The extent to which bird populations can be maintained in the face of such open hunting depends largely on the extent to which regulations, such as closed seasons and bag limits, can be introduced and enforced over wide areas. I shall return to these matters later.

Harvesting theory

In managed game species, the aim is to crop the population, removing some individuals each year and leaving others to reproduce and provide future harvests. The ideal is to take as many individuals from the population each year as is possible without causing a sustained decline in breeding numbers. In seasonally breeding species, this ideal can be achieved most effectively (1) if hunting is concentrated in the immediate post-breeding period, when numbers are at their highest, and (2) if hunting losses are then entirely compensated by reduced natural losses over winter or by improved reproduction the following spring, so that by the next hunting season numbers are fully restored. If this were so, the population could be harvested at a certain level indefinitely, causing no long-term decrease. Experience has shown, however, that this ideal is hard to achieve, and that it is easy to over-harvest a population, causing it to decline.

Sustained harvesting depends on the fact that depleted populations can quickly restore their size, through density-dependent changes in births or deaths. The maximum sustainable yield (MSY) can be defined as the highest harvesting rate that the population can match year after year with its own recruitment. The maximum yield at one time could be obtained by harvesting the whole population, but this could be done only once, and not sustained. Under continued over-exploitation, more individuals are removed than the population can stand, leading eventually to extinction. Under continued under-exploitation, fewer individuals are removed than the population can bear, and the 'crop' is smaller than possible, but can still be sustained long-term. The MSY, then, represents the balance between over-exploitation and under-exploitation. In theory, the MSY for any species could be calculated from knowledge of how birth and death rates respond to change in density; in practice, however, it is most readily found by trial and error, checking the effects of different levels of cull on the population trend. The MSY would also be expected to change with time, however, as other factors affecting the population changed.

Theories of harvesting often assume that population growth is logistic, with some maximum number of individuals when the population reaches the carrying capacity (K) for that habitat **(Figure 14.1)**. The sigmoid curve produced by the model is a simple description of density-dependent population growth in which, beyond a

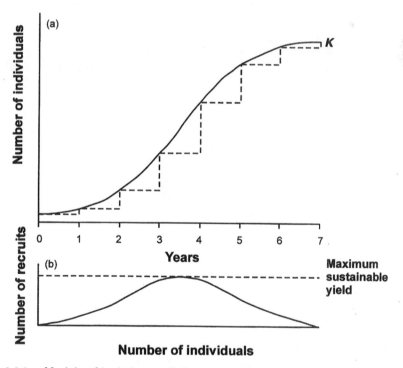

Figure 14.1 Models of logistic population growth and sustainable yield. (a) The growth of a population through time to level K set by the capacity of the area. Annual growth increments, shown by dotted lines, are greatest where the curve is steepest, around $0.5K$. (b) Size of sustained yield in relation to population size. The maximum sustainable yield – the maximum number that can be removed per year without causing long-term population decline – comes at around $0.5K$.

certain point, the growth rate of the population falls with further increase in population size. The growth rate is at its maximum (the intrinsic rate of increase) when the population is small, and falls to zero when the habitat is 'full'. To obtain the maximum annual harvest, the population must be at whatever level gives the highest annual numbers of new recruits. The harvest should be set so as to hold the population at that level (Caughley 1977). When harvesting is apportioned across all ages of animals, the greatest annual growth increment occurs where the logistic growth curve rises most steeply, which is near intermediate densities (or approximately ½K, **Figure 14.1**). This results in an 'n-shaped' sustained yield curve, which shows that the population size capable of yielding the greatest harvest in this imaginary population is near ½K.

There is an important lesson here. Although many managers of game are concerned with keeping their breeding stock as high as possible, this is a poor strategy for some populations because the highest sustainable harvest will be obtained when the population is below carrying capacity. This is especially the case when reproduction is strongly density-dependent. The point can be illustrated with some data for Ring-necked Pheasants *Phasianus colchicus*, a few individuals of which were released on an island off New York State **(Figure 14.2)**. As expected of any species in suitable but vacant habitat, Pheasant numbers increased rapidly and then more slowly as they approached the level that the island could support. Before they could reach this level, however, many of the Pheasants were shot. However, it was at intermediate levels of abundance that productivity was greatest, and at which the maximum sustainable harvest could have been taken – in this case by shooting enough individuals to keep the breeding stock at around 1400 **(Figure 14.2)**.

Maximum sustainable yields, estimated for a number of other hunted bird populations, have ranged up to 55% of autumn numbers **(Table 14.1)**. However, because each value depended on the reproductive and natural mortality rates prevailing at the time, it was specific to the population studied and could not be safely extrapolated to other populations of the same species elsewhere. In populations in which reproduction was density-dependent, the MSY could be achieved only by killing enough birds each year to hold breeding density below the level it would reach in the absence of shooting. For example, in one study of Bobwhite Quail *Colinus virginianus* in Illinois, the predicted MSY could be obtained by shooting 55% of autumn numbers. Subsequent breeding density was thereby reduced to 72% of its unharvested level, which gave the maximum production of young (Roseberry 1979). Within species, MSY values tended to be highest in populations that were managed, with predator control leading to enhanced production of young, and a high post-breeding density.

Harvesting strategies

The notion of MSY provides a useful conceptual framework, but the model in **Figure 14.1** is oversimplified. For one thing, it takes no account of an annual cycle that is split into a breeding period when numbers increase and a non-breeding period when numbers decline. In such a cycle, as indicated above, the timing of the harvest is crucial to its effect on the population. If shooting coincided with high post-breeding numbers, it would include many birds that would die anyway, and thus

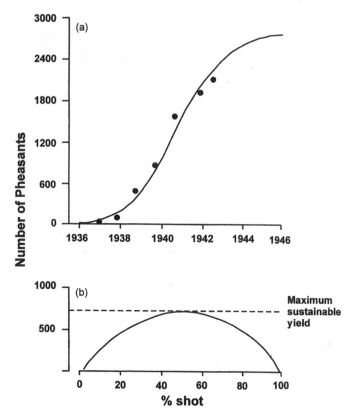

Figure 14.2   (a) Growth of a Ring-necked Pheasant *Phasianus colchicus* population on Protection Island following its introduction in 1937 (Einarsen 1945); (b) Recruitment curve calculated from the same data (modified from Begon *et al.* 1986). The maximum sustainable yield (about 700 birds per year) occurred with a breeding population of about 1400 birds. The aim of management for hunting should be to shoot enough Pheasants each winter to hold the breeding population at that level.

provide a high yield at no extra cost to the population. But if shooting occurred in spring, after most natural losses had already occurred, it would probably reduce the breeding stock excessively, giving low numbers for the next hunting season. As an example, the shooting of 40% of Rock Ptarmigan *Lagopus mutus* in autumn caused no obvious decline in breeding numbers compared with control areas, but the shooting of 40% of spring numbers led to a big decline in breeding numbers (McGowan 1975). It is for this reason that most regulated hunting of bird populations is set to occur in the immediate post-breeding period.

In addition, this simple model takes no account of annual variations in breeding or survival. With '*fixed quota*' harvests, it is easy to over-exploit populations because no allowance is made for occasional years when production might be low for natural reasons **(Table 14.2)**. To get round this problem, the harvest rate might be set at a level lower than the MSY, so as to maintain the population at a size greater than that required to give the MSY but thereby allowing for 'poor' years. Alternatively,

Table 14.1　Estimates of maximum sustainable yield in some hunted bird populations.

Species	Region	Percentage of post-breeding population that could be killed	Source
Mallard *Anas platyrhynchos*	Britain	29	Hill 1984
Red Grouse *Lagopus l. scoticus*	Scotland	>30[1]	Jenkins *et al.* 1963
Ruffed Grouse, males *Bonasa umbellus*	Minnesota	17–18	Ellison 1991
White-tailed Ptarmigan *Lagopus leucurus*	Colorado	30	Braun & Rogers 1971
Bobwhite Quail *Colinus virginianus*	Illinois	55[2]	Roseberry 1979
Grey Partridge *Perdix perdix*	Britain	30–45[1,2]	Potts 1986
Ring-necked Pheasant *Phasianus colchicus*, hens only	Britain	20	Robertson & Rosenberg 1988
Rock Ptarmigan *Lagopus mutus*	Alaska	>40[3]	McGowan 1975
Feral Pigeon *Columba livia*	New York City	35	Kautz & Malecki 1990

[1]The Red Grouse and Grey Partridge populations were 'managed', and benefited from predator control.
[2]Obtained only by shooting enough birds to hold breeding density below the level it would otherwise achieve.
[3]Surrounded by unshot areas, facilitating immigration.

numbers and breeding success might be monitored every year, and the annual harvest set accordingly. By manipulation of hunting seasons, numbers of hunting permits or individual bag limits, an attempt can be made to adjust the hunting toll each year to the stocks available at the time. For many years this was one objective of waterfowl management in North America (Nichols 1991). On such a *'variable quota'* system, the numbers killed may fluctuate greatly from year to year, but the risk of over-exploitation is reduced. The advantages and disadvantages of these and other harvesting systems are summarised in **Table 14.2**.

Harvesting strategies usually assume that populations are closed, and that any recovery from harvest occurs solely through changes in reproduction and survival within the population concerned. In widespread species, however, maximum sustainable yields can be higher if cropping is restricted to particular localities than if it is undertaken over large areas. Within small areas, immigration can help in the replenishment of depleted local stocks but, as a cull is expanded to wider areas, immigration becomes progressively reduced. As a related issue, another proven method of protecting hunted bird species involves the establishment of refuges in which hunting is barred. This reduces overall hunting pressure without the need to alter other regulations, and can help to sustain stocks, thereby supporting a larger

Table 14.2 Features of different harvesting systems.

Harvest system	Explanation	Advantages	Disadvantages
Free-for-all	Anyone can take what they want, with no controls.	No administration and no enforcement.	Insensitive to changing densities: usually leads to population decline and, in extreme cases, to extinction of stock.
Fixed quota	Same number taken each year.	Easy to administer, constant yield.	Insensitive to changing densities: strongly destabilising to population; can lead to population decline, needs enforcement.
Fixed effort	Same effort each year, whatever the take. *	Less destabilising than fixed quota.	Effort is difficult to control: yield is variable and fairly insensitive to changing densities; can lead to population decline.
Variable quota	Numbers taken vary in relation to population level.	Sensitive to changing densities; can ensure maximum sustainable harvest with no long-term population decline.	Costly: gives large fluctuations in the annual harvest, and in some years perhaps no harvest; requires annual measurements of numbers, and rigid enforcement.
Fixed percentage	Same proportion of population taken each year.	At sufficiently low level, can prevent population decline.	Costly: requires annual measurements of numbers, and, at sufficiently high level, can promote population decline. Needs enforcement.
Fixed escapement	Same numbers left to breed at the end of each year.	Sensitive to changing densities; gives maximum protection to breeding stock, preventing population decline.	Costly: requires annual measurement of numbers, gives large fluctuations in the annual harvest, and in some years no harvest. Needs enforcement.

*Could be achieved by allowing a fixed number of hunters access for a fixed number of days per year.

harvest in the longer term. Such refuges can work well for species that concentrate at high densities, such as breeding seabirds or wintering waterfowl.

A further assumption of simple harvesting models is that full-grown individuals are removed from the population at random. But in many bird species, certain sex- or age-classes are more easily killed than others, which can affect the population trend. More usefully, shooting can be directed to certain types of individuals, so as to minimise the effect on next year's harvest. For example, preferential harvesting of males in polygynous grouse species can substantially increase the MSY, if all females can still be fertilised and their individual reproductive rates are unaffected. With the same number of individuals, faster growth (and higher MSY) would then be gained with an excess of females over males, rather than with an equal sex ratio. In one study, shooting that pushed the adult sex ratio in Black Grouse *Tetrao tetrix* to 2:1 in favour of hens had no effect on the reproductive rate of hens (Ellison *et al.* 1988). In Ring-necked Pheasants, productivity and numbers were maintained with post-hunt ratios as distorted as ten females per male (Ball 1950, Allen 1954). Similarly, removal of 77% of Blue Grouse *Dendragapus obscurus* cocks in spring had no effect on the reproductive success of hens (Lewis 1984). This type of management is feasible in many polygynous game birds, because the larger, more colourful males can be readily distinguished by hunters. In many other species, however, such planned hunting is impracticable because the different age- and sex-groups look too alike.

In developing any harvesting strategy, it is important to allow for extra mortality associated with the harvest. The numbers of birds killed and retrieved are measurable, but form only part of the total. One additional source of loss is formed of birds that are hit and die later unrecorded. In North America, most estimates of this 'crippling loss' in waterfowl are around one-fifth of the number retrieved, but have ranged in different studies between 9 and 41% (Nieman *et al.* 1987). A second additional loss results from effects of the hunting methods used on both hunted and non-hunted species, remembering that some impacts (including lead poisoning) can continue for years into the future (Chapter 15). Allowance should also be made for other human-induced mortality. For instance, large numbers of gallinaceous and other birds may die from collisions with fences or overhead wires. From sample counts of carcasses under sections of power line that ran through suitable habitat, Bevanger (1995) estimated the annual loss in Norway at about 20 000 Capercaillie *Tetrao urogallus*, 26 000 Black Grouse *Tetrao tetrix* and 50 000 Willow Ptarmigan *Lagopus lagopus*. These numbers were equivalent to about 90%, 47% and 9% of the annual hunting harvest of these species. Unlike the harvest, however, these losses occurred year-round including spring, so their effects on population levels are likely to have been greater. Similarly, from a wintering population of 450 000 Eiders *Somateria mollissima* off Denmark, 140 000 (31%) were shot each year, but another 28 000 (6%) were killed by oil pollution, bringing the total human-caused loss to 37%, which was probably not sustainable (Joensen 1977).

Birds of most species are not normally hunted until they are full grown, after which all age-groups are acceptable as quarry. The above models are less useful when applied to other organisms, in which interest centres on only a portion of the population (for instance baby parrots or mature trees). For such species, other approaches are required which take these needs into account. Moreover, some other animals can withstand much higher levels of harvest than can birds. In some exploited fish stocks, for example, total mortalities exceeding five times the natural mortality are not uncommon, and may be sustained for decades (Shepherd & Cushing 1990). The regulating mechanisms in some fish populations are evidently extremely strong, and affect individual growth rates, as well as survival.

Additive and compensatory mortality

If mortality from hunting was entirely compensated by reduced natural loss, it would have no effect on the total annual death toll or on breeding numbers. But if hunting were partly or entirely additive to natural loss, it would raise the annual death toll and reduce breeding numbers. Above a certain threshold, of course, any extra mortality becomes additive, but up to that point birds could be shot sustainably **(Figure 14.3)**. Hunting can be offset by reduced natural loss only if the natural loss is density-dependent, as might occur if it were caused by competition for resources, predation or disease. Given an appropriate density-dependent response, species with the highest reproductive and mortality rates should be able to withstand the greatest proportionate loss to hunting.

Among Ruffed Grouse *Bonasa umbellus* in different parts of North America, the killing of up to 18% of cocks in autumn was entirely offset by reduced natural loss, giving a total annual mortality of 53%, but the killing of 25–39% was partly additive, giving a total annual mortality of up to 70% (Ellison 1991). The implication was that Ruffed Grouse in the areas concerned could not withstand hunting at the higher levels without declining. In contrast, among unmanaged White-tailed Ptarmigan *Lagopus leucurus* in Colorado and managed Red Grouse *Lagopus l. scoticus* in Scotland, shooting at 30% of the post-breeding numbers was entirely compensated by reduced natural loss, causing no depletion in breeding stocks (Jenkins *et al.* 1963,

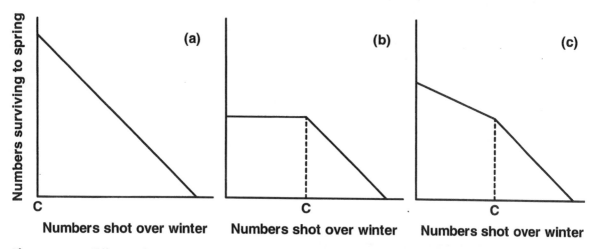

Figure 14.3 Effects of additive and compensatory hunting mortality on overwinter survival. (a) Hunting mortality totally additive to natural mortality, so that every bird that is shot reduces subsequent breeding numbers. (b) Hunting mortality totally compensatory up to a threshold level (C) beyond which it is totally additive. Up to point C, the numbers shot are entirely offset by reduced natural mortality, so that shooting up to that point causes no additional loss and no reduction in breeding numbers. (c) Hunting mortality partly compensatory up to a threshold level (C) beyond which it is totally additive. Shooting at all levels causes some reduction in breeding numbers, but greater beyond point C.

Braun & Rogers 1971). In fact, the Red Grouse population appeared under-exploited, and more birds could have been shot without affecting breeding numbers. Among Mallard *Anas platyrhynchos* in North America, prevailing hunting mortality was largely (but not entirely) offset by reductions in natural mortality **(Figure 14.4)**, and the same was found for Tufted Duck *Aythya fuligula* in Latvia (Mihelsons *et al.* 1985), for Bobwhite Quail *Colinus virginianus* in Illinois (Roseberry 1979), and for managed Grey Partridge *Perdix perdix* stocks in Britain (Jenkins 1961, Potts 1986). Again such findings might not hold in other populations of those species, subject to different breeding and natural mortality rates. In general, partial compensation by reduced natural loss appears to be the most frequent finding to emerge from studies of hunting losses on managed populations (Nichols *et al.* 1984, Nichols 1991). Some such populations were in decline, while in others the remaining hunting loss was compensated within the year by improved breeding success.

The fact that some bird populations increased after hunting was reduced implies that shooting at the previous level was at least partly additive. In several European goose populations, ringing revealed an improvement in annual survival rates after hunting declined (see Owen 1984 for Barnacle Goose *Branta leucopsis*). In North American Snow Geese *Chen caerulescens*, as hunting declined between 1970 and

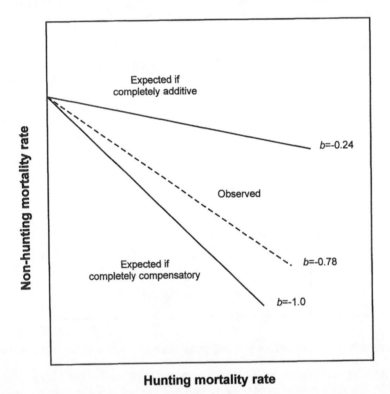

Figure 14.4   Observed mortality of Mallard *Anas platyrhynchos* in North America, estimated from ring recoveries, in relation to two alternative hypotheses of totally additive or totally compensatory mortality in response to shooting. The observed line fell between the two extremes, indicating that Mallard populations partly compensated for hunting mortality. Modified from Anderson & Burnham 1976.

1987, the annual survival of adults increased from 78 to 88%, but the annual survival of first-year birds decreased from 60 to 30%. Evidently young geese suffered more natural mortality. This was attributed to deteriorating feeding conditions on the breeding grounds as numbers grew, resulting in slower growth rates and reduced body size. Different processes influenced survival rates at different stages of the life-cycle, and this population could probably have tolerated greater hunting during 1970–87 without declining (Francis *et al.* 1992).

Studies on several species have shown that, up to a certain level, hunting itself could be density-dependent, removing a larger proportion of individuals from high-density populations (for Mallard *Anas platyrhynchos* see Schifferli 1979, for Grey Partridge *Perdix perdix* see **Figure 14.5**; Potts 1986). In other words, mortality from hunting was regulatory in its effects, and provided some safeguard against overhunting. With the hunting methods used, above a certain density of Grey Partridges in Britain, it became practically impossible to kill a larger proportion, and

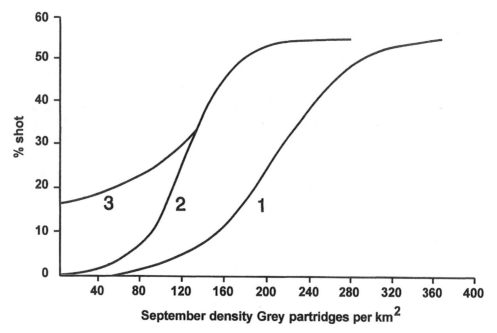

Figure 14.5   Density-dependent hunting mortality on Grey Partridges *Perdix perdix* on private estates in Britain. Lines were calculated from logistic equations showing the relationship between the numbers shot and the population density at the start of the shooting season in September. Line 1 refers to an estate well managed for Partridges; Line 2 refers to estates with less good (more open) habitat; and Line 3 refers to an estate where the main quarry was Red-legged Partridges *Alectoris rufa*, the shooting of which simultaneously increased the pressure on Grey Partridges at low density. The initial low response was caused by the practical difficulties of finding and driving birds at low densities, coupled with a reluctance to shoot many birds when stocks were low. The low response at high density was caused by the increased wariness of birds after they had been driven several times, coupled with a reluctance to overshoot. Modified from Potts 1980.

mortality from hunting became inversely density-dependent (non-regulatory). It then required a special effort, emigration or some natural catastrophe, to bring breeding numbers back down to a level at which productivity was maximised.

Multispecies problems

So far, I have assumed that species can be harvested individually, but more difficult problems arise if several species are harvested together, as in duck hunting in some areas, especially if the hunters cannot distinguish between species. The main problem is that species in the same situation differ in their chance of getting shot, so vulnerable species can be killed to the point of rarity, while others remain common enough to encourage continued shooting. At Loch Leven in southeast Scotland, ducks were shot as they passed over hunters in the dim light of dawn. This led to overshooting of Gadwall *Anas strepera* and Green-winged Teal *Anas crecca*, which were easily killed, and to undershooting of the more numerous Mallard which were more wary. Perhaps as a direct consequence, Gadwall and Green-winged Teal numbers were declining, while Mallard numbers were not (Allison *et al.* 1974).

In North America, waterfowl hunting often involves the use of wooden decoys to lure birds within gunshot. Three species of scoters were watched responding to decoys off the New Hampshire coast (Stott & Olson 1972). Black Scoters *Melanitta nigra* were extremely vulnerable to hunting in this way, Surf Scoters *M. perspicillata* were intermediate, while White-winged Scoters *M. fusca* were least vulnerable. In consequence, the three species were not shot in relation to their numbers in the area, which may have accounted for recent changes in their status. In another study, female Canvasbacks *Aythya valisineria* were more attracted to decoys than were males, so were more readily shot, further distorting an already uneven sex ratio (Olson 1965). The effect of hunting on this population was thus greater than if both sexes were harvested equally because it condemned more males than usual to non-breeding. Similar effects occur in other hunted species. On the other hand, protection of a favoured game species can allow other game species in the same habitat to increase. In Britain, for example, an increase in numbers of Red-legged Partridges *Alectoris rufa* followed partly from a decline in the hunting of Grey Partridges *Perdix perdix*. This was because, where the two species occurred together, the Red-legs were more easily shot than the Greys (Potts 1986). In other situations, where several species of similar vulnerability are taken simultaneously, the more K-selected species are affected most, because of their slower rates of recovery (Clark 1981). It is in fact difficult to regulate for r and K species together: if the fast-breeding r-species are harvested at the optimum rate, then K-species taken in the same proportion are likely to be driven to extinction. Alternatively, if harvest levels are set appropriate to K-species, then a large potential harvest of r-species is left untaken.

The human factor

In the real world, many populations are not harvested scientifically, but on the basis of tradition or greed, with no restraints applied. In areas where hunters are sparse relative to the numbers of birds available, or use primitive methods, unrestrained hunting may be acceptable, because it harvests well below the MSY. But nowadays, with modern weaponry, most widely-exploited bird populations are unlikely to persist unless harvesting is controlled in some way. If it is not, individual hunters tend to compete against one another to maximise their share of the harvest. Over-exploitation of bird and other animal populations usually results when, through lack of 'ownership', several interest groups have unrestricted access to the same population. No one group is prepared to limit its take, because this would merely leave more for the others. The system then operates on the basis of competing greed, what Hardin (1968)[1] called 'the tragedy of the commons'. It is this free-for-all which has most often led to the over-exploitation of fish and whale stocks in areas that are not owned or protected by any one interest group, and has also led to the over-exploitation of game birds and waterfowl on public lands, and of some seabirds on offshore islands. This is true even when a 'closed season' has been imposed.

The alternatives to an unmanaged free-for-all are of two types – what Hardin (1994) called privatism and socialism. In privatism, the resource is subdivided into many individual properties. Each owner is responsible for his patch, and if he manages sustainably, he will reap the long-term rewards. In socialism, the resource is treated as common property, and is managed by some central authority for the benefit of all. This is the basis of waterfowl management in North America, where government agencies are responsible for monitoring numbers, and for setting and enforcing sustainable bag limits. Resident bird populations can be managed successfully on the basis of privatism or socialism, but for migratory populations, socialism is the only viable option. This is because migratory populations are inevitably shared by different interest groups.

Management of resident populations

Bird species that remain in the same areas year-round can be managed and harvested by the same people, especially on private land protected from poaching. This provides a strong incentive for careful management of populations to provide a sustained yield, and also for habitat improvement to increase carrying capacity. The MSY model in **Figure 14.1** holds only for populations living naturally, exposed to normal predation and other risks. The situation can be altered greatly if predation and disease risks are reduced. Thus, if predation is the main limitation on reproduction, and relevant predators are removed, the density-dependence in reproductive rates can often be removed. The highest production is then achieved from high, rather than intermediate, population levels, giving a greater surplus for hunters to kill. The removal of nest predators can result in a 3–5-fold increase in the number of Grey Partridges that can be shot per unit area in Britain (Potts 1980). For more than 150 years, privatism has been the basis of management for resident game birds in Britain and other parts of Europe, where each estate manages and harvests its own stock. It could not work for migratory species that move on to be

[1]Hardin (1968) used the analogy of a publicly owned pasture in which sheep or cattle owners have free and unrestricted access to graze their animals. This system causes a degradation of the pasture by overgrazing, because individual farmers perceive that there are short-term economic benefits from having as many of their own livestock as possible grazing the pasture, causing an excessive aggregate use. Each herdsman gets the full benefit of adding to his herd, while the disbenefits are shared by all. One of Hardin's conclusions was that 'freedom in a commons brings ruin to all'. Such problems are avoided only if resources are managed in ways that pursue sustainability rather than growth and unabated exploitation.

over-exploited elsewhere, or for species that live in areas that cannot be adequately protected (such as the open sea).

Management of migratory populations

For populations that are migratory or live on publicly-owned lands, socialism is the only management option. Its main problem, in contrast to privatism, is that it is not self-policing, so without rigidly enforced controls, individual hunters compete against one another. Management by socialism therefore needs an effective and widespread enforcement system, as well as continued monitoring of stocks. In North America the hunters themselves pay for this system by means of a charge levied on hunting permits.

An unexpected problem emerged from the hunting of migratory waterfowl in which the open season was set at progressively later dates down the flyway to coincide in each area with the autumn passage. In Wood Ducks *Aix sponsa*, ring recoveries revealed that northern populations, which migrated the longest distances, suffered greater mortality (mostly from hunting) than southern ones (Nichols & Johnson 1990). This was because birds from northern populations encountered a succession of opening dates for hunting as they moved southward, resulting in an extended open season. The mortality in any one area was not excessive, but the cumulative total over the whole migratory route resulted in northern populations suffering high losses. Southern populations, in contrast, were exposed to only a single hunting season, during which they were buffered to some extent by the presence of northern birds. Under uniform hunting regulations, therefore, northern populations declined while southern ones thrived. The problem was resolved by re-adjustment of hunting seasons, shooting southern populations before northern ones arrived. The same phenomenon could be expected to occur in other migratory species, whether or not hunting was regulated.

For many migratory bird populations, which move from breeding areas where they are protected, to migration and wintering areas where they are not, the effects of hunting are especially hard to assess. For example, unrestricted recreational hunting in several countries in the Mediterranean region is said to remove a staggering 1000 million songbirds during each autumn migration (Magnin 1991, McCulloch *et al.* 1992). In situations like this, it is hard to judge the accuracy of the estimates, let alone the impact on north European breeding populations.

Commercial harvesting

Where the harvest of a species is of economic value, other complications arise. With any self-renewing resource, the temptation to over-exploit when prices are high will always be strong, even though this may jeopardise the very existence of the resource in the long run. The problem is accentuated if an illegal harvest develops on top of a legal one, and thrives because it can undercut the legal operators. As stocks decline and prices rise, the incentive to take more increases, so that the harvest becomes inversely density-dependent, leading in extreme cases to extinction. This process has occurred in some rare parrots taken for the pet trade: for example, Spix's Macaw *Cyanopsitta spixii*, which now survives only in captivity. Exploited populations of many organisms are continually being pushed to the brink of extinction, reaffirming that a sustained harvest depends on sound data, and on effective and enforceable regulations. The same pressures encourage the exploitation of additional species formerly safe from human interest.

In wild species that have very low rates of natural increase, such as some whales at less than 5% per year, it may be more profitable for the harvester to remove the entire stock at one time (causing extinction). This is because the resulting money, when carefully invested, will earn profit at a higher rate than will sustained harvesting of the stock (Clark 1981). The important thing is the population's intrinsic growth rate, r, in relation to the economist's inflation-corrected discount rate, δ (often taken to be around 5–7%). If $r > \delta$, then management for sustainable harvests makes economic sense, provided there is an effective 'sole owner'. In the opposite case, where $r < \delta$, the optimal harvesting strategy is to cash in the entire stock (or at least harvest to the point where the last remnant is too expensive to collect) and invest the realised capital where returns are higher. These purely economic considerations favour liquidating some species of parrots, large whales, tropical hardwoods and other slowly renewing resources. However abhorrent this idea may seem to most people, it is the basis on which much commercial exploitation of wild populations operates. In practice, it is only the fact that harvesting costs become prohibitive at low population levels that has saved certain large whales from extinction by a profit-driven industry.

The difficulties of enforcing regulations on commercially-exploited species are increased by the fact that, at least in the initial years, they entail a reduction in profits. There is also the problem of uncertainty over the future: whether the stock will actually recover, whether benefits of recovery will go to those who bear the cost of reducing the current harvest, and whether after recovery the demand for the resource will still exist. People understandably have more concern over the immediate future than over the distant future, especially if the benefits of reducing the current harvest can only accrue to future generations.

International harvesting

Even greater problems of regulation occur when the stock is shared between several nations, as are many migratory birds, or are even more international in nature, as are many marine animals. Any regulations require the existence of a central authority with power to impose them. If the stock is geographically restricted to a few states, effective management agreements have sometimes been enacted, as in the North American Migratory Birds Treaty, which involves Canada, the United States and Mexico. The simplest procedure consists of an agreed allocation of the harvest between states, a system that can work only if all states stick to the agreement. But if the stock is shared between a larger number of states, regulation becomes more difficult and (as in the case of some marine animals) is largely unresolved: the more parties become involved, the more difficult it becomes to maintain universal restraint.

A problem of a different kind is that managers, succumbing to bribes or to political expediency, often ignore the advice of their scientific advisers. The pattern is familiar in fisheries management. The stock is over-exploited and begins to decline. So scientists advise reducing the harvest. The exploiters object. Politicians and bureaucrats take the middle road. Reductions in harvest may be enforced but are insufficient, so the stock declines and everyone suffers (except perhaps for the politicians and bureaucrats).

Effectiveness of regulations

Over the years, many unprotected bird species have been over-hunted, so that their numbers have been reduced to a fraction of those that their habitats could support. In one region or another, this was true, for example, of various waterfowl, gallinaceous birds, shorebirds, cranes and seabirds killed as food, egrets and others killed for their feathers, and various parrots and other birds taken for the pet trade. Once harvests were controlled, however, most populations recovered rapidly, providing that suitable habitat was still available.

To begin with a specific example, in response to low populations, more restrictive hunting regulations were introduced during 1958–65 for Canvasback *Aythya valisineria* and Redhead Ducks *Aythya americana* in the United States. As shown by ring recoveries, this procedure reduced both the harvest and the total annual death toll. Mortality estimates fell from around 50% or more before the restrictions to 14–21% afterwards, allowing widescale population growth. This gave a clear demonstration of the effectiveness of reduced bag limits in a country where enforcement was possible. It benefited the entire populations of two wide-ranging species.

Almost all European waterfowl species have increased in numbers in recent decades since shooting was banned or reduced (Madsen 1987, Monval & Pirot 1989, Kirby *et al.* 1995). By 1990 most goose populations were between two and ten times larger than in 1960, and (where measured) mortality rates were lower (Owen & Black 1991). But perhaps the most impressive increase recorded in a goose population after shooting stopped concerned Ross Goose *Anser rossii* in western North America, which increased from 2000–3000 in the early 1950s to more than 188 000 in 1988 (Ryder & Alisauskas 1995). Similarly, the massive increases in some seabird populations during this century, in both Europe and North America, have been attributed, at least in part, to reduced hunting pressure. However, certain bird populations have not recovered after the virtual cessation of hunting, the Eskimo Curlew *Numenius borealis* in North America being an obvious example. The same is true for certain fish stocks after fishing was greatly reduced (Clark 1981). One possible reason is that such populations were reduced to a level at which they could be held down by other factors, such as predation, and require some special circumstance to free them again from this control (Chapter 9).

In many parts of the world, seabirds have long been harvested by local people for food, without causing long-term declines in population levels. Examples include Puffins *Fratercula arctica* on the Faeroe Islands, Great Shearwaters *Puffinus gravis* on Tristan da Cunha and Short-tailed Shearwaters *Puffinus tenuirostris* around Tasmania. In the latter, about 750 000 are taken every year from huge colonies and sold as 'mutton birds'. The population is large enough to withstand this level of take, and continues to flourish. Yet where competing interest groups have become involved, or commercial incentives have become paramount, over-harvesting has soon occurred, causing population declines. Examples include Cory's Shearwater *Calonectris diomedia* on the Salvage Islands near Madeira, and the Short-tailed Albatross *Diomedea albatrus* off Japan.

Some seabird populations continue to be heavily exploited, but in the absence of relevant information, it is hard to tell whether they are declining. An example is the Brünnich's Guillemot *Uria lomvia* which is shot in huge numbers off western Greenland and Newfoundland (being the second most shot bird species in Canada).

This is in addition to similarly large numbers killed in fishing nets. These wintering birds are drawn from a large sector of the Arctic, mostly from unstudied colonies, but precautionary bag limits were introduced in Newfoundland from 1993–94 in an attempt to cut the harvest by half (Gaston *et al.* 1994).

Other consequences of hunting

Side-effects of hunting can also affect populations. The constant disturbance caused by hunting not only increases the energy needs of birds, but also reduces the time available for feeding or the areas in which feeding can safely occur. Together, these effects amount to a reduction in food-supply (Chapter 13). Another side effect on waterfowl is the massive additional mortality that results from lead poisoning, caused when birds ingest spent gunshot (Chapter 15).

A further consequence of hunting, this time for fish, has been the accidental capture and death in recent decades of millions of seabirds. The two most lethal methods are drift netting and long-line fishing. Monofilament nylon drift (or gill) nets – called 'walls of death' by some people – came into widespread use from the 1960s. Being inconspicuous and durable, they are a particular hazard to diving seabirds of which more than 60 species worldwide have become frequent casualties, in addition to marine mammals and turtles. In the North Pacific alone, an estimated 48 000–64 000 km of this netting are set every night, more than enough to encircle the earth at the equator, and a further average of 40 km are lost or discarded every night, remaining a potential hazard for years (Manville 1991). Such nets are estimated to kill more than one million seabirds per year in the North Pacific alone, and more than enough seals and dolphins to account for the precipitous declines observed in Alaskan populations.

The following are some published estimates of the numbers of birds killed in particular areas at particular times:

- 875 000 birds per year in the north Pacific Japanese and Taiwanese squid fisheries, including 424 000 Sooty Shearwaters *Puffinus griseus* (Tennyson 1990);
- 250 000–750 000 birds per year in the north Pacific and Bering Sea Japanese salmon fishery, mainly Short-tailed Shearwaters *Puffinus tenuirostris* and Sooty Shearwaters *P. griseus* (King 1984);
- 100 000 Guillemots *Uria aalge* killed in 1985 in a cod fishery off Norway (Vader *et al.* 1990);
- 215 000–350 000 birds, mainly Brünnich's Guillemots *Uria lomvia*, killed annually in 1970–72 in salmon and cod fisheries off southwest Greenland and eastern Canada (Evans & Nettleship 1985);
- 17 500 birds per year in the Gulf of Gdansk, southern Baltic, during 1972–76 and 1986–90, about 10–20% of the total birds in the area, mainly Long-tailed Ducks *Clangula hyemalis* (8400), Velvet Scoter *Melanitta fusca* (1000) and Eider *Somateria mollissima* (1000) (Stempniewicz 1994).

The bird mortalities caused by drift nets are largely hidden from the public eye, and fishermen are understandably loathe to disclose figures, but it is even harder to assess the impact of this mortality on populations, except in localities where colonies have been monitored. A 53% decline in Guillemot *Uria aalge* numbers off central California from 229 000 in 1980–82 to 109 000 in 1986 was attributed primarily to intensive gill netting, compounded by oil spills and an El Niño event in 1982–83 (Takekawa *et al.* 1990). Individual colonies declined by 46–100%, depending on their distance from areas of high gill netting activity. More northern colonies, not affected by the netting, showed no such declines. Similarly, net casualties are thought to have accounted for a decline in Brünnich's Guillemot *Uria lomvia* numbers at two sites in North Norway from 250 000 in 1965 to 12 000 in 1989 (Strann *et al.* 1991), and for big declines in this and other species in eastern Canada and west Greenland (Piatt & Reddin 1984, Evans & Nettleship 1985). Many seabirds killed in the north Pacific are from species that breed in the southern hemisphere, mostly from colonies that have never been counted.

Long-line fishing for tuna developed mainly in the 1970s. Each boat tows one or two lines, each up to 100 km long and carrying up to 3000 baited hooks. It takes about five hours to set the line, and throughout the process, the last 100 m of line remain near the surface where the baits attract birds, some of which become hooked and drown. From on-board measures of catch frequency, and knowledge of the number of boats in operation, an estimated 38 000–173 000 birds were killed per year worldwide by the Japanese Bluefin Tuna *Thunnus thynnus* industry in 1981–86, including 10 000–50 000 per year in New Zealand waters alone (Brothers 1991). The main casualties were various albatross and large petrel species. In the Southern Ocean, an estimated 44 000 albatrosses were killed each year, of which at least 10 000 were Wandering Albatross *Diomedea exulans*.

The effect of long-line fishing on the Wandering Albatross has been studied in detail. In this species, as in other long-lived seabirds, only a slight increase in the naturally low mortality rate can tip the species into decline. Numerical declines have occurred at almost every colony studied, with annual rates of decline measured at 2.6%, 4.9% and 6% on three Crozet Islands and at 1.0% since 1961 on Bird Island, South Georgia (Weimerskirch & Jouventin 1987, Croxall *et al.* 1990). This fits with an observed halving in the size of some breeding colonies in the past 30 years. Mortality was heavier in females, apparently because they spend more time in heavily-fished waters. No reduction was observed in the mean annual reproductive output of breeders. At some colonies birds have begun to breed at an earlier age than formerly, in line with the reduction in density, but this has been insufficient to offset the rise in mortality. More recently, population declines have been detected in other oft-caught albatross species, notably *D. chrysostoma* and *D. melanophris* (Prince *et al.* 1994).

Commercial fishing operations also kill many waterfowl. For example, fishhooks on trotlines in the Laguna Madre of Texas could catch nearly 22 000 Redheads *Aythya americana* during a three-month period each winter, of which about 3200 died and another 7400 had poor survival prospects when released (McMahan & Fritz 1967). Together these formed less than 3% of the total 420 000 present. On Lake Okeechobee in Florida, commercial trotlines killed nearly 75 000 Lesser Scaup *Aythya affinis*, about 8% of those present during the winter of 1985–86 (Turnbull *et al.* 1986). Leghold traps set for muskrat and other aquatic furbearers also claim many waterfowl.

Birds for aviculture and falconry

Worldwide, the number of birds taken each year as pets is surprisingly large (Mulliken 1995). Most are taken in poor countries, on a free-for-all system, with financial rewards increasing progressively up the chain from trapper to pet dealer.

The only reliable data concern birds moved between countries, from records of international trade. Over 7 million birds were transported between countries per year in the 1970s, 80% of them songbirds (Inskipp & Gammell 1979). More than half entered the United States, western Europe and Japan, where demands were high. Since then, figures have fallen to an estimated 2–5 million birds per year (Mulliken 1995). All these figures exclude casualties that occur during capture and holding, as well as any illegal trade, on which aspects reliable figures are unobtainable. International trade accounts for only a part of the total, however, which has been estimated at more than 20 million birds per year. In one region or another, more than 20% of all bird species are taken as cage birds, at least some of which are declining as a result. The consequent rarity of some parrots (such as *Anodorhynchus hyacinthinus*) has raised their market value, putting even more pressure on remaining stocks. Many parrots are taken as chicks, which often entails destroying the nest-holes. Where such sites are scarce, this procedure can in turn reduce the numbers of parrots that can breed in future years. Moreover, the pet trade is now the main route through which birds become established in new areas, sometimes affecting the native birds through competition and disease transmission (Chapters 11 and 12). This could be by far its greatest impact.

Because of the declines in some wild parrot populations caused by avicultural demand, thought has been given to controlled harvesting (Beissinger & Bucher 1992). The situation differs from game hunting because the demand is for young parrots, which make the best pets. Hence, traditional MSY models do not apply, and the objective of parrot management should be to maximise the production of nestlings (rather than adults). This could be done by increasing the proportion of adult parrots that breed (by provision of food or nest-sites where these are limiting), by reduction of nest predation (by installing predator guards on nest-trees), or by the more labour-intensive methods of double-clutching (taking one clutch for artificial incubation, so that the pair re-lays) and reducing hatching asynchrony (by incubating the whole clutch from the same date, so that all eggs hatch together rather than at intervals, which improves nestling survival). Because the survival of wild nestlings is low, many more chicks than adults can be removed from a population without causing decline. This is especially so if subsequent survival is density-dependent, in which case the removal of a proportion of chicks can be compensated by improved survival of the remainder. By box provision alone, in an area where natural sites were limiting, Beissinger & Bucher (1992) estimated that 105–298 nestling Green-rumped Parrotlets *Forpus passerinus* could be taken per 100 boxes each year, without reducing the population below its former (natural) level. Therefore a parrot population would not have to be reduced well below K to maximise productivity, and yield could be maximised close to K. Moreover, the effect of harvesting could be monitored readily from the numbers of boxes occupied each year.

In some countries, raptors are taken in the same way as parrots, as nestlings, and then used for falconry or other purposes. The same principles apply as in parrots, and the largest breeding populations can be expected to provide the largest sustainable yields. Certainly, more birds could be taken at this age on a sustained basis than if the birds were trapped as independent free-flying juveniles or adults, as is the practice in the Middle East and elsewhere. The effect of live harvesting on raptor populations is especially hard to estimate, however, because substantial numbers of birds escape, or are turned loose, at a later date (Kenward 1974).

No species subject to controlled harvesting has yet become extinct (though some have come close), and some biologists maintain that in the long run many wild species (such as parrots) will need to be harvested on a regulated basis if they are to survive at all. In general, it may be easier to implement the sustainable use of wildlife for the captive market when:

(1) young age-classes with low survival probability in the wild are harvested,
(2) birds are marketed soon after harvesting, so that long holding periods are avoided, thus minimising suffering, losses and expenses in husbandry,
(3) the potential to increase productivity through management is high, so that sustainable harvests can be larger, and
(4) species are chosen that are resident in small areas, so that management can be done by individual land owners (under privatism), thus avoiding the competitive hunting of the same population by several interest groups.

One hidden value in exploiting wildlife on a sustained basis for profit is that it transmits an economic value to natural and semi-natural habitats, and thus promotes the conservation of other species too.

PEST CONTROL

The term 'pest' is usually applied to wild species that conflict with human interest. Certain bird species feed on crops, reducing the yield; while others eat domestic animals, fish or game, or creatures of special conservation concern. Yet other species by their roosting habits can foul buildings, drinking water or stored food; or by merely moving around can transmit diseases or collide with wires, windows and aircraft, causing damage. This section is concerned with control measures aimed to reduce the abundance of pest species, and not with other means of reducing damage (summarised in **Table 14.3**). The universal feature of long-standing pest species is that they can be culled heavily year after year without causing long-term reduction in their population levels, and often without achieving much reduction in damage. Other pest species are quickly obliterated, so soon disappear from the headlines.

The most important crop pests are various obligate seed- and fruit-eaters, including some finch-like birds (Fringillidae, Estrildidae, Ploceidae), pigeons (Columbidae) and parrots (Psittacidae), together with some facultative seed-eaters, such as starlings (Sturnidae) and icterids (Icteridae). All these birds can feed in dense flocks and accumulate in large numbers wherever food is temporarily plentiful. Most also have a high reproductive rate which enables them to recover rapidly from reductions in numbers and hence to withstand control measures in the long term. On a global scale, cereals are the most important food crops, and birds that regularly eat wheat, rice, millet, maize and sorghum often become serious pests. The species concerned feed naturally on wild seeds, and have benefited from the increased food supplies provided by agriculture. Many can live in extensive farm monocultures, obtaining most or all their needs from the cropping system, independently of natural habitats and food-supplies. Some such species have reached incredible numbers. The Red-winged Blackbird *Agelaius phoeniceus*, for example, is supposedly the most numerous bird in North America, with an April population estimated at 165 million, increasing by July to 350 million (Dolbeer 1990). This species, together

Table 14.3   Main methods of crop protection and pest control.

Methods	Comments
Crop protection	
1. Physical protection, as by netting.	Feasible only for small areas.
2. Protection by chemical repellant or scaring device.	Birds often habituate to scarers, which in time may become less effective. This is less so for natural scarers, such as falcons used to scare birds from airfields.
3. Crop resistance.	The breeding of varieties resistant to damage. A long-term solution, and in time the pest may overcome the resistance through genetic or behavioural change. Example: Droophead sunflowers are resistant to most birds except parrots.
4. Cultural control.	Use of cultural practices to reduce the risk of damage, as by changing crops, habitats or crop siting, including the planting of 'lure crops' to reduce damage elsewhere.
Pest control	
5. Shooting or other direct killing.	Specific to pest, but labour intensive.
6. Chemical pesticides.	Usually non-specific, so kill other animals. In insects, can lead to the development of resistance, so must be continually changed.
7. Chemical sterilants.	Effective for some organisms. Not yet in wide use in birds. Effect on population delayed.
8. Biological control.	Use of natural enemies (predators or disease agents) to reduce pest numbers. May be non-specific, with effects on desirable species.
Other methods	
9. Payment of compensation.	Expensive and open to abuse.

with three other grain-eaters – the Common Grackle *Quiscalus quiscula*, Brown-headed Cowbird *Molothrus ater* and Starling *Sturnus vulgaris* – are thought to achieve a combined winter population in North America of 500 million (Meanly & Royall 1976). When seed-eaters feed directly from the plant, grain yield is almost always reduced. In contrast, when growing crops are grazed (for example by geese), yield is not necessarily reduced because the plants can recover, although their growth and ripening may be delayed (sometimes with economic consequences).

Before embarking on any control programme aimed at population reduction, it is helpful to study the natural factors governing the reproductive and mortality rates of the pest and to define the level of kill that must be achieved to cause decline. At the same time, it is helpful to understand the animal's general biology in case there are weak links that can be exploited to reduce numbers. In the sections below, the control of particular bird pests is discussed in more detail, in order to illustrate the issues involved.

Woodpigeon

Originally a forest-dweller, this large pigeon has benefited from agricultural activities, and in western Europe now gets almost all its food from crop-plants. It feeds on newly-planted or ripening grains of wheat and barley, followed by ripe grain spilt at harvest. When grain is unavailable, it turns to rape or clover leaves, but in times of snow it eats high-standing brassica crops, causing serious damage (Murton *et al.* 1964).

For many years the killing of Woodpigeons *Columba palumbus* was encouraged in Britain by the payment of a subsidy on cartridges (Murton *et al.* 1974). The pigeons were killed like game birds, using double-barrel shotguns, usually by small numbers of people shooting the birds as they entered woodland roosts in the evening. One feature of this control method was that, regardless of flock sizes, it limited the numbers of birds that could be killed from each flock, because after two shots the gun had to be re-loaded, by which time the flock had passed. As the population rose, therefore, and flocks became larger, a steadily increasing number, but decreasing proportion, of birds was killed **(Figure 14.6)**. In other words, on the method of control used, shooting was inversely density-dependent, and above a certain population level could not control numbers.

Another problem with traditional pigeon control was that, although permitted year-round, most shooting occurred in the autumn and winter, a few weeks before the numbers were reduced naturally by large-scale starvation. It merely removed a surplus that was doomed to die from food-shortage in the ensuing weeks. Because no serious crop-damage occurred in this period, nothing was gained by control. Moreover, the natural survival was compensatory, so that seasons of heavy loss from

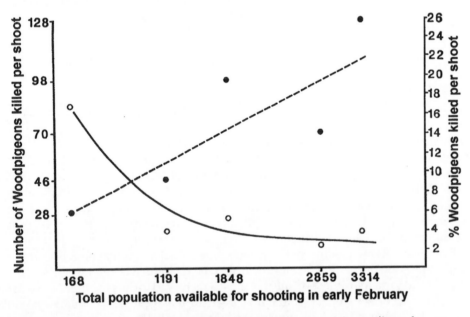

Figure 14.6   Inverse density-dependent relationship between the number of Woodpigeons *Columba palumbus* available for shooting and the percentage actually shot. The figures refer to the number of birds in a 40–ha area in southeast England in February and the numbers shot there per evening shoot in February and early March. Filled symbols refer to numbers shot and open symbols to percentage shot. From Murton *et al.* 1974.

shooting were followed by low natural loss, and vice versa. Clearly, this control system was not achieving its objective in reducing the Woodpigeon population, and in 1965 the subsidy on cartridges was withdrawn. Allowance was made for the continued subsidised shooting of birds over vulnerable crops, but more as a means of deterrence than control. No increase occurred in Woodpigeon numbers, which in fact declined in subsequent years in response to agricultural changes which reduced the food-supply (Inglis *et al.* 1990).

In Britain, the Rook *Corvus frugilegus* (which eats grain) is also traditionally 'controlled' by shooting young birds as they leave their nests in April, although without the help of a government subsidy. Again numbers are reduced only temporarily because a month later shooting is compensated by lower natural losses when population size is reduced by the seasonal decline in the availability of earthworms and other soil invertebrates (Murton 1971). This natural population reduction occurs a few weeks before the survivors begin to cause agricultural damage by feeding on ripening cereal grains. To reduce numbers further, it would be better to insert control measures between the periods of natural mortality and subsequent crop damage, but this involves more effort than shooting fledglings. So in this species, too, the traditional control is futile. In the absence of detailed knowledge of the biology of the pest, or of an effective means of population reduction, it is more sensible to concentrate remedial action at the sites of damage rather than attempting widescale population control. The latter is often impractical, as well as expensive.

Quelea

The African Red-billed Quelea *Quelea quelea* is one of the most numerous birds in the world, and probably the most important vertebrate pest of food-crops anywhere (Crook & Ward 1968, Bruggers & Elliott 1989). It occurs over most of Africa in semi-arid regions, and feeds on the seeds of grasses and cultivated cereals, notably sorghum, millet, rice and wheat. It gathers in huge communal roosts, numbering many millions of individuals, from which flocks spread over surrounding land to feed. It lives in regions where the short rainy season stimulates a flush of annual grasses, the seeds of which fall to the ground and last throughout the dry season. As seeds become exhausted, the birds become increasingly short of food, culminating when the rains begin and the remaining seeds germinate. The birds then move to areas where rain has been falling for several weeks, and new seeds have formed. None-the-less, during this time of food-shortage, considerable mortality occurs, and numbers seem to be naturally limited then (Ward 1965). Large-scale control operations have been conducted in most African countries for many years, often supported by international aid agencies. Control efforts were centred on the large roosts or breeding colonies where, by use of explosives or aerial spraying, many millions of birds could be killed in a single night.

Such spectacular kills were expected soon to reduce Quelea populations to levels at which crop-damage was no longer serious. Yet despite the international effort put into Quelea control, with hundreds of millions of birds killed every year, the species remained as common as ever. This was because, despite the extent of the control programme, it could still reach only a small proportion of roosts, some of which were in areas remote from effective control. And again, the birds' high natural rate of increase ensured that any losses imposed by control operations were quickly made good, so that population reduction could offer no more than temporary relief from damage. In addition, Quelea proved to be long-distance migrants on a large scale, presenting awesome logistical and political difficulties in achieving sufficient coverage. The dynamics of Quelea populations have so far proved sufficiently robust to withstand the levels of control that most countries can afford.

Because the range of Quelea in Africa is so great, because the birds move over large distances, and because opportunist breeding allows rapid increases, birds are always available to recolonise areas cleared at great expense. Under these circumstances, a policy of protecting important areas of crops, by destruction of nearby roosts and colonies while the crops are vulnerable, seems more sensible than large-scale attempts to reduce the population over wide areas. If no crops are at risk, the killing of Quelea is pointless. Quelea in West Africa did decline during the 1970s, but this was probably in response to the Sahelian drought which reduced their natural food-supplies, rather than to control operations.

*　　*　　*

Many attempts to control bird numbers, which have involved widescale and costly slaughter schemes, have proved ineffective (Ward 1979). Both apparent and real damage give rise to intense reaction on the part of the sufferers who see their livelihoods under threat, and a lobby is generated which pressurises the appropriate authority for action. If government departments are pressed to spend money to tackle the problem, they usually prefer high-profile control campaigns to scientific evaluations. Superficially this can seem a sensible approach, for tangible damage is witnessed and extrapolations to lost revenue are readily made. Yet often this leads to a waste of money, because the killing effort is too small relative to the population, or because efforts become concentrated at seasons when individuals are easiest to kill, rather than when numbers are low. Moreover, the control operations often bring other environmental problems, including damage to non-target species. For these and other reasons, control costs are often unrelated to damage costs, and may even exceed them. Formal economic analysis requires that the damage be quantified, usually in monetary terms, so that the potential benefits ensuing from control of the pest can be compared with the costs of control. Such costs may be social and environmental, as well as economic.

For these and other reasons, pest control practices in agriculture are in process of change. Slowly the emphasis is shifting from direct culling alone (including pesticide use) towards alternative techniques, such as biological control, or to combinations of several methods in integrated pest management programmes. In the interests of bird conservation, farmers in some areas are paid to tolerate losses of crops or stock. Such schemes have met with success, but must be sustained long term and are open to abuse. In the worst cases, farmers can blackmail conservation agencies into paying larger and larger sums. Inevitably money runs out, and all that has been achieved is the survival of a valued bird population for a few more years.

One of the most unusual control programmes ever mounted concerned the Tree Sparrow *Passer montanus* in China. When Chairman Mao proclaimed the sparrow an enemy of the state, a national campaign was launched, in which everyone was required to participate in a battle lasting three days. Those too old or infirm to play an active role had to keep the birds continuously in flight by banging tins and gongs, shouting and waving flags, and setting off fire crackers. (It had earlier been discovered that after two hours of flight, a sparrow drops exhausted and could be caught.) While no figures were produced on the total kill, nearly a million sparrows were killed in Beijing alone, and about three million in Shantung province. In the years that followed the campaign, rice production actually fell and the campaign was allowed to lapse, enabling sparrow numbers to recover (Summers-Smith 1996 from Chinese sources). The species is still harvested, however, as evidenced by the seizure in 1993 by Dutch customs officers of a cargo of two million frozen specimens en-route from China for human consumption in Italy (Summers-Smith 1996).

Raptors

In some countries, some raptor species are regarded as pests because they prey upon domestic stock or game, and were once killed at every opportunity. Some carrion-feeding species were (and still are) also killed incidentally in attempts to get rid of mammalian carnivores by use of poisoned meat baits. In an attempt to protect domestic stock, the killing of large raptors was officially encouraged in parts of Europe as early as the 16th century by payment of bounties (Newton 1979). This seems to have been sporadic, however, and to have had no marked or long-term effects on populations. It was with the rise in small game management in the 19th century that persecution reached its peak, and spread to the smaller species. Game shooting increased in popularity with the introduction of Ring-necked Pheasant *Phasianus colchicus* rearing, and increased yet further with the improvement in the shotgun, from muzzle loader to breech loader. The objective of total elimination was soon achieved for some raptor species over large areas.

Nowhere was the killing of raptors done on a more systematic basis than in Britain, mainly by game-keepers who were employed for the express purpose of 'vermin' control. The killing was widespread and occurred mainly in the breeding season, when pairs were present at known breeding sites. In time five species were eliminated from Britain completely as breeders, including Marsh Harrier *Circus aeruginosus* (by 1898), Honey Buzzard *Pernis apivorus* (1911), Goshawk *Accipiter gentilis* (1889), Osprey *Pandion haliaetus* (1908) and White-tailed Eagle *Haliaeetus albicilla* (1916) (Newton 1979). Several other previously widespread species became much restricted in range, the Buzzard *Buteo buteo* remaining only in some western districts, the Hen Harrier *Circus cyaneus* mainly on offshore islands, and the Red Kite *Milvus milvus* in a tiny area in central Wales, where game preservation did not take hold. The species least affected were the Merlin *Falco columbarius*, Kestrel *F. tinnunculus* and Sparrowhawk *Accipiter nisus*. These were the three smallest species with the highest breeding rates, and thus having the best ability to recover year after year from sustained killing. Comparing species, the most marked reductions were seen in species that had low and localised populations at the start, fed on carrion (which made them easy to poison), or had slow breeding rates, or a combination of these features. Lesser reductions were associated with large populations living partly away from game preserving areas, little or no carrion feeding and high breeding rates (Newton 1979).

In parts of Europe the payment of bounties for dead raptors often meant that records were kept of the totals killed. Using such statistics from various European countries, Bijleveld (1974) estimated that in the 20 years up to 1970, several million raptors were killed in Europe by game-bird hunters alone, with especially large numbers in France and Germany. The sheer magnitude of the figures led some people to doubt them, but they were repeated in similar order in region after region, and in each case feet or beak were required as proof of killing. However, the annual figures for many districts or estates often included many more raptors than could have lived there at one time, a testimony to the effects of movements or to the existence of neighbouring less disturbed populations, from which new recruits continually came.

Comparing the figures from the 20th century with those of the 19th, the main difference is in the reduced representation of eagles and other large species in some lists and their complete disappearance from others. That this was in some regions due to the culling itself is suggested by the large initial kills, followed by a swift decline, as the scheme continued. In other lists, however, the totals showed no obvious decline over many years, which suggests that in these areas the hunters were merely cropping the populations concerned, and causing no long-term decline. This is indicated in some official statistics from Austria, which show that between 1948 and 1968, premiums were paid annually on about 12 000–20 000 birds. Likewise, the cull of 6000 Goshawks *Accipiter gentilis* destroyed each year by Finland's 170 000 hunters was also thought to cause no long-term decline in the Goshawk breeding population, but most of the cull was of juveniles killed in the few months following breeding (Moilanen 1976, Saurola 1976).

Evidence for the impact of human predation on raptor populations has come, not only from the declines in populations that followed widespread persecution, but from the recoveries in numbers and distribution that have followed its reduction. In recent decades, the change in public attitudes, the abolition of bounty schemes and the enactment of protective legislation have enabled many populations to increase and spread, so that in parts of Europe some raptor species are probably more numerous now than at any time in the past 150 years. On a world scale, only one raptor species is known to have been eliminated completely by human persecution, the Guadalupe Caracara *Polyborus lutosus*, because it attacked young goats. Being an island form, its population would never have been large.

The most striking difference between the crop-pests such as Quelea, in which control operations have proved expensive and ineffective, and certain raptors in which control operations quickly obliterated populations over wide areas, was in the densities and life-histories of the species themselves. Significant crop-pests usually live at high density and have fast breeding rates and high natural mortality (*r*-strategists), so are able to recover quickly from any reduction in numbers. They are often flock-feeding, and range over large areas, found in one place one week and another the next. The raptors, in contrast, occur at only low densities, with slow breeding rates and low annual mortality (*K*-strategists), so are unable to recover quickly from reductions in numbers. Most are found consistently in the same territories year after year, nesting in known traditional sites where they can be killed in the breeding season. Moreover, carrion feeders can be culled at low cost simply by laying poisoned bait. Interestingly, the only crop-pests known to have been exterminated completely include parrots with life-history features more akin to raptors than to Quelea, the most striking example being the Carolina Parakeet *Conuropsis carolinensis* of the southeastern United States, shot to extinction because it ate grain (Chapter 16).

In conclusion, the success of any control operation in reducing the population levels of birds depends not only on the proportion of birds killed at one time, but also on the movement patterns and on the life-histories of the species concerned, whether they are long-lived and slow-breeding (*K*-selected) or short-lived and fast-breeding (*r*-selected). It depends also on whether the deaths from control operations increase the overall mortality or simply replace deaths from natural causes, so that the overall mortality remains the same. This in turn depends critically on the timing of control operations in relation to the main period of natural loss. As a further point, in many western countries, animal welfare and conservation issues are gaining influence, and restricting the range of species that can be controlled, as well as the methods that can be used.

SUMMARY

The effect of hunting depends greatly on whether the quarry species is managed as a renewable resource, with appropriate controls, or whether it is hunted on a free-for-all system with no effective restraints. Providing that habitat is suitably maintained, the former can provide sustained bags indefinitely, whereas the latter usually leads to population decline. Many populations of waterfowl, shorebirds, cranes, seabirds and others owe their present low status to over-hunting in the past (from which some are now recovering) or to continued over-hunting.

Sustained harvesting depends on the fact that, up to a point, depleted populations can quickly restore their size, through density-dependent changes in births and natural deaths. In some species with density dependent reproductive rates, the maximum sustainable yield (the highest harvesting rate that the population can match year after year with its own production) is theoretically achieved with a breeding population lower than would occur in the absence of hunting. Sustained harvesting is most readily achieved where the 'stock' is owned and managed by a single interest group. It is more difficult to achieve where (as in migrant waterfowl) the stock is exploited by several competing interest groups, a situation that almost always leads to over-exploitation unless harvest limits can be firmly enforced. Moreover, where several species are hunted simultaneously (again as in many waterfowl), the more vulnerable species can be shot to rarity, while others maintain their numbers.

Many seabirds are killed as human food, and even more as a by-catch in fishery operations, but the impact on breeding populations has seldom been assessed.

Worldwide, the most important bird pests of agriculture feed on cereal grains. They remain pests because for one reason or another they have proved hard to eradicate or to reduce in numbers permanently. Many have high reproductive rates, range over large areas and feed in dense flocks, accumulating rapidly wherever food is temporarily abundant. In contrast, control operations aimed at large raptors rapidly led to substantial population declines. These species were easily killed at all times of year (notably by use of poisoned baits), and their low reproductive rates prevented rapid recovery.

Peregrine *Falco peregrinus* with crushed thin-shelled eggs.

Chapter 15
Pesticides and Pollutants

Many toxic chemicals are now added continuously to the natural environment, either as pesticides, industrial effluents or combustion emissions. Some of these chemicals are man-made, while others are natural substances whose concentration in the biotic environment has been increased by human action. Some such chemicals have now become important agents of bird population declines, influencing distribution and abundance patterns on both local and widespread scales. The aim of this chapter is to discuss the main mechanisms of these population declines, using selected case-studies to illustrate the issues involved.

For present purposes, a pollutant can be defined as any substance that occurs in the environment at least in part through human action, and that, at the levels found, has deleterious effects on living organisms. This definition includes all pesticides. Birds absorb most pollutants through their food, but some also through their skin, by preening and inhalation. Ground-nesting and other birds may have pesticides sprayed directly onto their eggs, while these and other pollutants can be transferred to eggs from parental plumage. Adverse effects are significant in the present context only if they cause reductions in population levels.

Three main mechanisms of bird population declines have emerged. Firstly, some pollutants act directly by causing breeding failures or deaths. These forms of loss reduce breeding populations only if they add to natural losses, and are not offset by reduced losses from other causes. Secondly, other pollutants act indirectly on birds by reducing the food-supply (either for adults or for chicks), and thereby causing declines in numbers. Thirdly, other chemicals, notably atmospheric pollutants, can alter the physical or chemical structure of habitats, making them less suitable for certain species.

SOME GENERAL POINTS ABOUT PESTICIDES

If pesticides destroyed only the target pests and then quickly broke down to harmless by-products, problems from their use would be minimal. But most pesticides are non-specific and kill a wide range of organisms. Secondly, while some break down rapidly, others last for weeks, months or even years in animal bodies or in the physical environment. Thirdly, some pesticides accumulate in body fat and readily pass from prey to predator, causing secondary poisoning, or even pass along several steps in a food-chain, affecting animals far removed in trophic position from the target pest. Depending on their chemical properties, all pesticides lie somewhere within this three-feature spectrum of variation, with respect to specificity, persistence and propensity to accumulate in animal bodies. In addition, by contaminating air and water, some pesticides (like other pollutants) can reach areas and affect organisms far removed from points of application (witness the presence of dichlor-diphenyl-trichlor-ethane (DDT) residues in penguins and other Antarctic birds, George & Frear 1966). Other problems are caused by pesticide manufacture which, through accidents and discharges, has led to pollution of rivers, lakes and coastal areas, with loss of aquatic life. These problems are accentuated by many local accidents and abuses, excessive application, drift and careless disposal. Some of these same points apply to some other chemical pollutants, but because such chemicals were not designed as biocides, and have not been deliberately applied to large land areas, the problems they cause have been slower to appear.

Pesticide usage is increasing year by year: in terms of the area treated, the number of applications per year, and the variety of chemicals in use. By 1990, about 300 chemicals were in use worldwide as insecticides, 290 as herbicides, and 165 as fungicides and other pesticidal compounds, with a grand total of more than 3000 formulations (Freedman 1995). Between 1945 and 1989 their use increased by ten times in the United States, a country that accounts for about one-third of the global total (Pimental et al. 1992). For all these reasons, then, we can expect that the indirect impact of pesticides, via the depression of avian food-supplies, is likely to have increased progressively over the past 50 years, and will continue to increase into the future. Moreover, pesticides are now manufactured and marketed mainly by international companies, so that the same chemicals are used simultaneously in many parts of the world. This, together with the fact that some pesticides can be transported in the bodies of migratory animals, as well as in air or water currents, means that few parts of the globe are still entirely free from their effects.

The main types of pesticides that are applied against animal pests include the organochlorines, organophosphates, carbamates and pyrethroids **(Table 15.1)**. The non-specificity of such chemicals has been examined in experimental trials. One of the earliest involved the application of a commonly-used carbamate insecticide, carbaryl (trade name Sevin), to one of two adjacent one-acre (0.4 ha) patches of millet (Barrett 1968). Applied in mid-summer, the carbaryl remained toxic for only a few days, but in this time the numbers and biomass of arthropods declined almost to nil **(Figure 15.1)**. Their populations took seven weeks to recover in biomass, and longer to recover in numbers and diversity. The chemical is also likely to have affected other organisms not studied, including micro-arthropods, earthworms and other soil-dwellers, as evidenced by the build-up of leaf-litter. There was nothing unusual in this result, as similar effects were obtained in trials with other chemicals. Moreover, the several North American studies that examined the effects of aerial forest spraying with DDT recorded up to 90% declines in the overall insect population following a

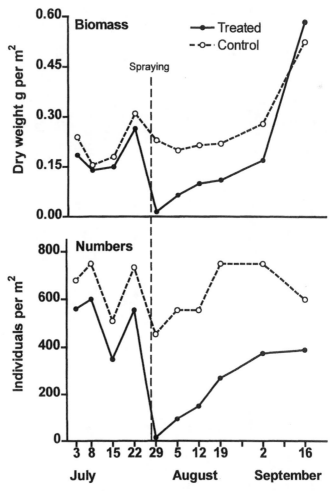

Figure 15.1   Impact of one application of the insecticide carbaryl (applied at 4.5 kg per ha) on the invertebrates in a patch of millet. From Barrett 1968.

Table 15.1 The main classes of pesticides in current use, their toxicity to birds, persistence and bio-accumulation properties. Mainly from Hudson et al. (1984), Hayes & Laws (1991) and Freedman (1995).

Classes of pesticides	Acute toxicity to birds	Persistence	Bio-accumulation	Examples
Organochlorine insecticides[1]				
DDT and relatives	Low	Very high	Very high	DDT, TDE, DDE, DDD methoxychlor
Lindane	Low	Moderate	Moderate	Lindane
Cyclodienes and relatives	High	High	Very high	Aldrin, dieldrin, endrin, heptachlor, chlordane, mirex toxaphene, chlordecone
Other organochlorines	Moderate	High	High	
Organophosphate insecticides[1,2]	Very high	Low-Moderate	Low-Moderate	Bromophos, carbophenothion, chlorfenvinphos, chlorpyrifos, diazinon, dimethoate, famfur, fenthion, fenithrothion, malathion, mevinphos, monocrotophos (azodrin), parathion, phosphamidon, triazophos
Carbamate insecticides[1,2]	High	Moderate	Very low	Aldicarb, aminocarb, carbaryl, carbofuran, carbosulphan, CPMC, methiocarb, pirimicarb
Natural organic insecticides[1]				
Pyrethroids	Very low	Low	Very low	Cypermethrin, deltamethrin, permethrin, tetramethrin
Other	Very low	Low	Very low	nicotene, nicotene sulphate, rotenone
Inorganic and organometals				
Inorganic metal salts	Low	High	Moderate	Mercuric chloride, mercurous chloride
Organometals	Moderate	High	High	ethyl mercury, methyl mercury, methoxyethyl, mercury, phenylmercury acetate
Chlorphenoxy herbicides	Low	Low	Very low	2, 4-D, 2, 4, 5-T, MCPA
Quarternary nitrogen herbicides	Moderate	High	Low	Diquat, paraquat
Azole fungicides	Low	Low	Low	Prochloraz

[1]Some of the compounds classed here as insecticides are also used against other invertebrate pests, such as nematodes and molluscs.
[2]These neurotoxicants inactivate the enzyme acetylcholinesterase which is responsible for the breakdown of acetylcholine, the neurotransmitter at cholinergic nerve endings and myoneural junctions. This enzyme inhibition allows the build-up of acetylcholine which results in continued stimulation and fatigue of cholinergic end organs and muscles, leading to tremors, convulsions and eventually death from respiratory failure. As most pesticides in these groups break down rapidly, in diagnosis cholinesterase levels in plasma or brain are usually measured instead.

single application, and a similar reduction of aquatic insects in forest streams (Adams *et al.* 1949, Hoffman *et al.* 1949, Hanson 1952, Crouter & Vernon 1959).

In small field trials, invertebrate populations usually recovered within two months of spraying, but recovery was likely to depend largely on immigration from surrounding areas. The chances of this happening are much less in modern extensively farmed landscapes where the treatment of almost every hectare leaves few refuges where vulnerable species can survive to produce colonists. In such areas, recovery is likely to depend mainly on reproduction by the small proportion of individuals that survive the treatment, and thus take much longer than where immigration is possible. Moreover, when arthropod populations recover from a pesticide application, certain species sometimes become more abundant than before and even previously rare species can increase to pest proportions. This can happen when the arthropod predators take longer to recover than their prey. This usually leads to further applications of pesticides – the so-called 'pesticide treadmill' – so that many invertebrate species become depressed over even longer periods. Different species vary greatly in their susceptibility to any particular pesticide, but in general the greater the application rate, the greater the range of organisms killed. The mode of application is also influential.

During the years of DDT use to control forest insects, application rates of 1.1 kg per ha were found to have minimal direct effects on bird populations, but higher rates caused population declines. For some bird species, almost total kill occurred at 5.7 kg per ha (Hotchkiss & Pough 1946, Adams *et al.* 1949, A.W.A. Brown 1978; **Figure 15.2**). Another organochlorine pesticide, toxaphene, applied at 1.7 kg per ha to control grasshoppers around wetlands in North Dakota, caused only small losses of waterbirds, but at 2.3 kg per ha in oil it resulted in a total kill of young ducks, coots, rails and terns (Hanson 1952). Likewise, a pesticide that kills no birds when sprayed as a liquid may kill large numbers if applied as granules which some species eat. And while spraying from the ground affects few areas outside the crop, spraying from the air can result in considerable drift, affecting wild areas interspersed among the crops and well beyond. Aerial spraying is usual in the widespread control programmes in non-crop habitats aimed against locusts, tsetse flies or mosquitoes.

The phenomenon of secondary poisoning by pesticides first became apparent in the USA in the late 1940s when DDT was applied to elm trees to control the insect vectors of Dutch elm disease (Barker 1958). The insecticide was picked up from leaves and soil by earthworms which were then eaten by American Robins *Turdus migratorius* with lethal consequences. The dead birds were found to contain 50–200 ppm (mg per kg) residues in the brain, a fatal dose obtained by eating about 100 worms. Most deaths occurred, not at the time of spraying, but in spring or after heavy rain, when worms were most available to birds. On one campus of Michigan State University, the Robin population was reduced within two years from 370 to a few individuals, and virtually no young were raised (Wallace *et al.* 1961). Similarly, the use of DDD over several years to control midges on Lake Clear, California, resulted in the decline of Western Grebe *Aechmophorus occidentalis* numbers from more than 1000 pairs in 1949 to 30 non-breeding adults in 1960 (Hunt & Bischoff 1960). This was again associated with food-chain contamination, as confirmed by chemical analyses of midges, fish and grebes found dead. Such secondary poisoning is particularly associated with DDT and other organochlorines, but has also been recorded from some other pesticides **(Table 15.2)**.

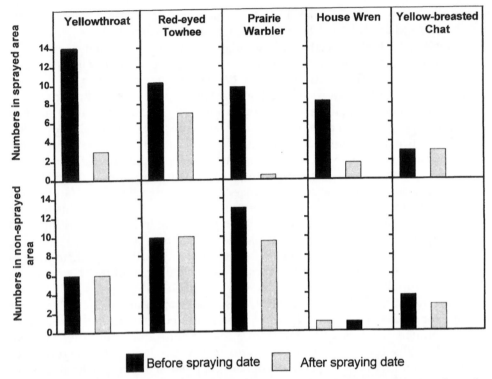

Figure 15.2   Impact of DDT sprayed at 5.7 kg per ha on the five commonest breeding bird species in an area of scrub in Maryland. Birds were counted in a 12-ha plot in the midst of a 36-ha sprayed area, and in another unsprayed area nearby. Of the five commonest species the Yellowthroat *Geothlypis trichas*, Prairie Warbler *Dendroica discolor* and House Wren *Troglodytes aedon* were reduced by 80% and the Red-eyed Towhee *Pipilo erythrophthalmus* by 35%, while the Yellow-breasted Chat *Icteria virens* showed no obvious change. The total decline in the five commonest species (which formed 77% of the original population) was 65%. Redrawn from Robbins & Stewart 1949.

A curious food-chain effect involved the use in the American west of the organophosphate famfur, poured on to cattle to protect them from warble flies *Hypoderma* spp. (Henny *et al.* 1985). For unknown reasons, Magpies *Pica pica* ate cattle hair, picking it from fences or directly from the animals, and thereby became poisoned. In one experimental area, the Magpie population declined by 67% in a two-month period after treatment of the local cattle, compared with 24% in another area where the cattle remained untreated. Moreover, the Magpies were sometimes eaten by Red-tailed Hawks *Buteo jamaicensis* which also died. The famfur remained active for up to three months after application. Magpie numbers in the western states declined between 1968 and 1979, coinciding with the widespread use of famfur, although other factors may also have been involved. In Canada, another organophosphate, fenthion, was used for the same purpose and also killed Magpies (Hanson & Howell 1981).

Because of their high fat-solubility and persistence, organochlorines readily pass up food chains. Increasing concentrations at successive trophic levels were evident

Table 15.2 Examples of bird mortality incidents resulting from secondary poisoning or food-chain effects involving pesticides.

Chemical	Application	Pathway to affected species	Area	Source
Organomercurials				
Methyl-mercury	Sown-seed treatment against fungi	Seeds – seed-eating birds – Goshawks *Accipiter gentilis*	Sweden	Borg et al. 1969
Organochlorines				
DDT	Nearby farm crops	Water – mud – crustacea – fish – ducks and gulls	Lake Michigan	Hickey et al. 1966
DDT	Elm trees against beetle vectors of Dutch elm disease	Leaves – earthworms – American Robins *Turdus migratorius*	Illinois	Barker 1958
DDD	Clear Lake Gnat *Chaoborus astictopus*	Water – fish – Western Grebes *Aechmophorus occidentalis*	California	Hunt & Bischoff 1960
Aldrin, dieldrin	Sown-seed treatment against insects	Seeds – seed-eating birds – Sparrowhawks *Accipiter nisus*	Britain	Newton 1979
Endrin	Voles	Leaves – voles – Barn Owls *Tyto alba* and others	Washington	Blus et al. 1989
Toxaphene	Rice against various pests	Water – fish – fish-eating birds	Texas	Andreasen 1985
Sodium pentachlorophenate	Rice against various pests	Water – Water-snails *Pomacea glauca* – Snail Kites *Rostrhamus sociabilis*	Surinam	Vermeer et al. 1974
Organophosphates				
Azodrin	Alfalfa against voles *Microtus guentheri*	Alfalfa – voles – raptors	Israel	Mendelssohn & Paz 1977
Famfur	Warble fly on cattle	Cow hair – Magpie *Pica pica* – Red-tailed Hawk *Buteo jamaicensis*	Western States	Henny et al. 1985
Carbamates				
Carbofuron	Soil pests	Vegetation and invertebrates – small birds and mammals – Red-shouldered Hawks *Buteo lineatus*	California	Balcomb 1983
Carbofuron	Beetseed treatment	Soil – earthworms – Buzzards *Buteo buteo*	Switzerland	Dietrich et al. 1995

in many (but not all) studies in which different kinds of organisms from the same place were examined (Newton 1979). Within areas, concentrations were lowest in plants (first trophic level), higher in herbivorous animals (second trophic level), and higher still in carnivores (third trophic level) and so on up the food-chain. It is not trophic level as such that is important, but the rates of exposure (or intake) which, because of accumulation, tend to be higher in carnivores. In addition, rates of accumulation were often greater in aquatic than in terrestrial systems because many aquatic animals absorb organochlorines through their gills as well as from their food. Fish rapidly pick up fat-soluble pollutants from water, and concentration factors of 1000 or 10 000 times between water and fish are not uncommon (Stickel 1975). At the other extreme, the bioaccumulation properties of carbamate pesticides are almost zero, although some can last for years in soils (A.W.A. Brown 1978).

DIRECT EFFECTS ON POPULATIONS

Organochlorine pesticides and predatory birds

Two direct effects of pesticides on bird population levels can be illustrated by different types of organochlorine compounds (Figure 15.3). This group of chemicals contains such well-known insecticides as DDT and the more toxic cyclodienes, such as aldrin, dieldrin and heptachlor (Table 15.1). DDT was first introduced into widespread agricultural use in the late 1940s, and the cyclodienes after 1955. For a time they were widely used throughout the 'developed' world, but during the 1970s and 1980s, they were banned progressively in one country after another, as their environmental effects became increasingly apparent. Now they are used without restriction mainly in tropical and sub-tropical areas.

Being fat-soluble and persistent, these chemicals accumulate in animal bodies, and pass from prey to predator. Three groups of birds have been particularly affected, namely (1) raptors, especially bird-eating and fish-eating species, such as Peregrine *Falco peregrinus*, Sparrowhawk *Accipiter nisus*, Osprey *Pandion haliaetus* and Bald Eagle *Haliaeetus leucocephalus*; (2) various other fish-eating birds, such as cormorants *Phalacrocorax* and pelicans *Pelecanus*; and (3) various seed-eating species, such as doves, geese and cranes, which eat newly-sown seeds of cereals and other plants that have been treated with organochlorines as protection against insect attack.

DDT is not especially toxic to birds, and very high exposures (as in the early forest spraying) are needed to kill birds outright. The main effects are on breeding. At sub-lethal level, DDE (the main metabolite of DDT) reduces the availability of calcium carbonate during eggshell formation so that the eggs are thin-shelled and break. Some thin-shelled eggs survive incubation, but the embryo dies from dehydration caused by excess water loss through a thinned shell. If reduction in the mean breeding rate of individuals is sufficiently marked, it leads to population decline, because recruitment can no longer match the usual mortality. The effects of DDT/DDE, which were initially deduced from field studies (Ratcliffe 1970), were subsequently confirmed by experiments on captive birds (Cooke 1973, Newton 1979, Risebrough 1986). All these effects were via the female, but DDT and its derivatives are also oestrogenic (mimicking the effects of the hormone oestrogen), and have been found to reduce sperm production in domestic fowl (Albert 1962).

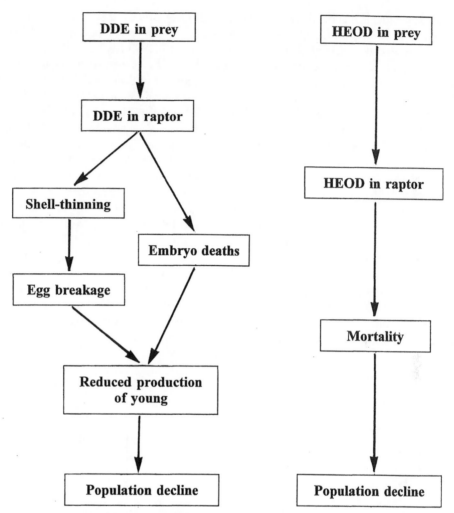

Figure 15.3   Modes of action of DDE (from the insecticide DDT) and HEOD (from aldrin and dieldrin) on raptor populations. From Newton 1986.

The cyclodienes, such as aldrin and dieldrin, are several hundred times more toxic to birds than is DDT (Hudson *et al.* 1984). These chemicals act mainly by killing birds outright, increasing mortality above the natural level sufficiently to cause rapid population declines **(Figure 15.3)**. In many countries, they were used chiefly as seed-treatments, so killed seed-eaters, as well as their predators.

The massive declines in the numbers of some bird-eating and fish-eating raptors, which occurred in Europe and North America in the 1960s, were thus attributed to the combined action of DDE reducing breeding rate and HEOD (from aldrin and dieldrin) increasing mortality rate. The relative importance of these mechanisms of population decline seems to have differed between regions, depending on the relative quantities of the different chemicals used. In much of North America, DDE-induced reproductive failure seems to have been paramount, but in western Europe HEOD-induced mortality was probably more important (Newton 1986). As the use of these chemicals has been reduced, shell-thickness, breeding success, survival and population levels of affected species have largely recovered (see **Figures 15.4–15.6** for Sparrowhawk; Cade *et al.* 1988, Crick & Ratcliffe 1995 for Peregrine). All these improvements were associated with reductions in the residues of organochlorine chemicals in eggs and tissues. In Britain, the recovery of Sparrowhawk numbers in different regions followed the decline in geometric mean HEOD residues in Sparrowhawk liver tissue to below 1 ppm in wet weight **(Figure 15.6)**.

An indication of the worldwide contamination of ecosystems with DDT can be gained from patterns of shell-thickness in the Peregrine Falcon, which breeds on all continents except Antarctica (Peakall & Kiff 1988). The greatest degree of shell-thinning (26%) was found in the eastern United States, from which Peregrines disappeared altogether within 20 years after the chemical came into widescale use (Cade *et al.* 1988). Marked shell-thinning also occurred in Peregrines nesting in Arctic North America and Eurasia (17–25% in different regions), reflecting the fact that these falcons and their prey migrated to winter in regions of DDT-use further south. The smallest levels of shell-thinning occurred in falcon populations that were resident in areas with no DDT use, and at the same time fed on prey species that were themselves year-round residents in these areas. An example was the Peregrine population of the Scottish Highlands, dependent mainly on Red Grouse *Lagopus l. scoticus* (Ratcliffe 1980). The falcons showed only 4% shell-thinning in that region, compared with more than 19% in most of the rest of Britain (Ratcliffe 1980). In most

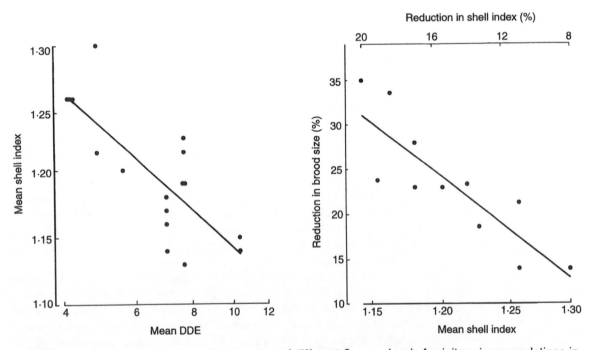

Figure 15.4   Shell index and breeding success of different Sparrowhawk *Accipiter nisus* populations in relation to geometric mean concentrations (µg per g) of DDE in eggs. Shell index measured as shell weight (mg)/shell length x breadth (mm). From Newton 1986.

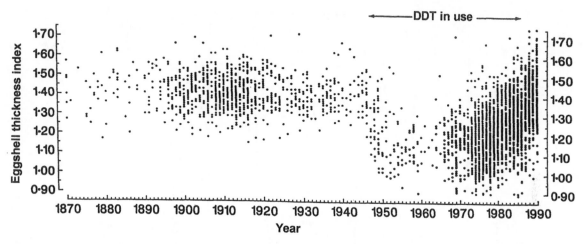

Figure 15.5   Shell-thickness index of British Sparrowhawks *Accipiter nisus*, 1870–1990. Shells became thin abruptly from 1947, following the widespread introduction of DDT in agriculture, and recovered from the 1970s, following progressive restrictions in the use of the chemical, which was banned altogether from 1986. Each spot represents the mean shell index of a clutch (or part-clutch), and more than 2000 clutches are represented from all regions of Britain. Based on shells housed in museum and private collections, and at Monks Wood Research Station. Shell index measured as shell weight (mg)/shell length x breadth (mm). Extended from Newton 1986.

of the populations studied, DDE residues in eggs were measured and mean levels were highest in populations showing most shell-thinning.

Comparisons of shell-thinning and population trend in different Peregrine populations from around the world showed that all populations with an average of less than 17% shell-thinning maintained their numbers, while all those with more than 17% declined, some to the point of extinction **(Figure 15.7)**. An average of 17% shell-thinning thus emerged as critical to population persistence, associated with an average of 15–20 ppm DDE in the wet weight of egg content. These are average figures applicable at the level of populations and do not apply rigidly to individual eggs. The same mean level of shell-thinning was also associated with population declines in other raptor species (Newton 1979).

Comparing different eggs from the same species, the relationship between the concentration of DDE in the egg content and the degree of shell-thinning invariably emerged as linear with DDE plotted on a logarithmic scale **(Figure 15.4)**. This was true of all the species so far examined. However, some families of birds were more sensitive to a given DDE level than were others, raptors and pelicans showing the greatest response and gulls and gallinaceous birds the least. Raptors thus emerged as particularly vulnerable to DDE, both because of their position high in food-chains and because of their high sensitivity (Newton 1979). The greatest degree of shell-thinning recorded in any species came from Brown Pelicans *Pelecanus occidentalis* in California, which were contaminated by the effluent from a DDT factory. On Anacapa Island off Los Angeles, only two chicks resulted from 1272 nesting attempts in 1969. Virtually all the eggs collapsed on laying, and the shells were on average 50% thinner than normal (Risebrough *et al.* 1971). This species is the most

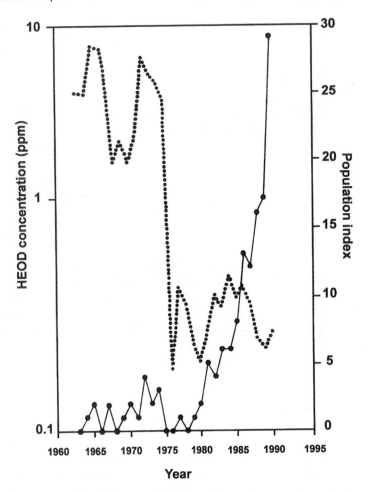

Figure 15.6 Trend in geometric mean HEOD levels (dotted line) in the livers of Sparrowhawks *Accipiter nisus* found dead in eastern England in relation to an index of population level in the same area (continuous line). HEOD is the chemical residue derived from the insecticides aldrin and dieldrin and the population index is based on the number of carcasses received for analysis. From Newton & Wyllie 1992.

sensitive to DDE of all those so far studied, 3 ppm in the fresh egg content being sufficient to cause nearly total breeding failure (Blus 1982).

The problems caused by organochlorines result partly from their extreme persistence, a quality that adds to their effectiveness as pesticides. The longevity of chemicals in any medium is usually measured by their 'half-life', the period taken for the concentration to fall by half. The half-life of DDE in soils has been variously calculated at between 12 years in some cultivated soils and 57 years in some uncultivated soils (Cooke & Stringer 1982, Buck *et al.* 1983). So even after the use of DDT is stopped, soil-dwelling organisms can remain a source of residue for some bird species for years to come. HEOD is much less persistent, but the estimated half-life in soil of 2.5 years is probably much too short for many areas (A.W.A.

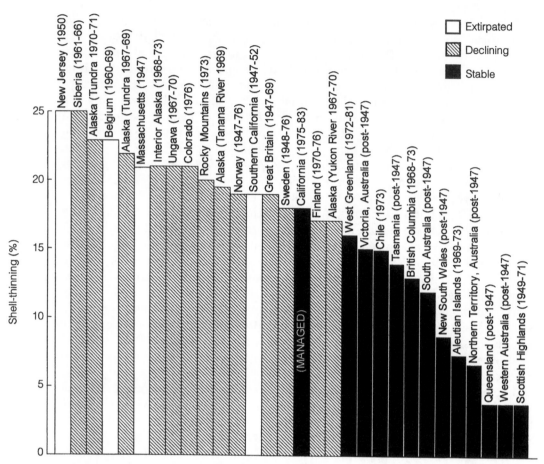

Figure 15.7 Shell-thinning and population trend in Peregrine Falcons *Falco peregrinus* in different parts of the world. All populations showing more than 17% shell-thinning (associated with a mean level of 15–20 ppm DDE in egg content) declined, some to the point of extinction. In the one exception, extra eggs and young were added by biologists to maintain numbers. The same level of shell-thinning was associated with population decline in other raptor species too (Newton 1979). Redrawn from Peakall & Kiff 1988.

Brown 1978). In the United States, the use of organochlorines was restricted in the 1970s, but birds continued to die from these chemicals for many more years. During 1986–90, 122 investigated mortality incidents were caused by chlordane and dieldrin in urban and suburban sites, involving 11 songbird and nine raptor species. In a summer survey of dead songbirds sent in by members of the public, organochlorine poisoning accounted for 17% of deaths. These incidents were attributed to the use of organochlorines 20 years earlier to control scarab beetles in turf. The chemicals were still present in both the surface soil and beetles from lawns (Okoniewski & Novesky 1993). In the tropics, the turnover of DDT and other organochlorines may be faster than in temperate areas, because higher temperatures favour faster degradation and volatilisation, through which more is exported elsewhere by wind (Berg 1995).

Organochlorines can disappear from animal bodies much more rapidly than from the physical environment, but again DDE lasts longer than HEOD. For example, in pigeons the half-life of DDE has been measured at 240 days, compared with 47 days for HEOD (Walker 1983). These rates vary between species, and with the condition of the individual. However, persistence in the body means that effects of organochlorines on individual birds can become manifest weeks or months after acquisition. Mortality is most likely to occur at times of fasting, as during hard weather or migration, or when body fat is metabolised, releasing organochlorines into circulation and enabling them to reach lethal levels in the nervous system (Bernard 1966). Thus, wintering Snow Geese *Anser caerulescens* initially survived contamination with organochlorine pesticides, but later succumbed during migration when fat mobilisation freed the accumulated residues. The geese were exposed in Texas but died in Missouri, more than 1000 km to the north (Babcock & Flickinger 1977). Delayed mortality also occurred among female Eider Ducks *Somateria mollissima* nesting on the Dutch Waddensea, where nest numbers declined by 77% between 1960 and 1968, mainly through deaths of incubating females (Swennen 1972). The birds became contaminated via their food (mussels) from organochlorines discharged into the River Rhine, but died mainly during incubation when they did not feed but depended on their body fat. Discharges were stopped in 1965, and within three years the Eider population began to recover. Likewise, in experiments, some Black Ducks *Anas rubripes*, which had been fed DDE with their food, were still laying thin-shelled eggs after two years on clean food (Longcore & Stendel 1977).

In general, birds have a less effective detoxification system for organochlorines than mammals have, which explains why birds are generally more contaminated than mammals from the same area, and why avian predators of birds, such as the Sparrowhawk and Peregrine, have been more affected by organochlorines than have avian predators of mammals, such as the Kestrel *Falco tinnunculus*. The bird-feeders also top longer food-chains than the mammal feeders do, giving more opportunities for residues to concentrate. The relationship between vulnerability and feeding habits was shown in many studies (Newton 1979), but perhaps most vividly in a forest area in the northwestern United States. This study involved a single experimental spraying of DDT (at 0.8 kg per ha) conducted in 1974, two years after the chemical had been banned from general use in the United States. In insectivorous American Kestrels *Falco sparverius*, residues in blood plasma peaked one year after the spraying, with a mean value of 0.78 ppm. In comparison, residues in three species of accipiters, ranked in ascending order of proportion of birds in the diet, were 2.6 times higher in Goshawk *Accipiter gentilis*, 3.8 times higher in Cooper's Hawk *A. cooperii* and 6.1 times higher in Sharp-shinned Hawk *A. striatus*. The last species had almost disappeared from the area two years after the spraying (Henny 1977).

Looking back, the organochlorine episode had an enormous impact on the public perception of pesticides. It highlighted for the first time the potentially severe environmental consequences that could result from the widescale use of chemicals, hitherto regarded as wholly beneficial. The effects of DDT were insidious: they were not apparent immediately, but only after several years, and could not have been predicted beforehand by the testing in operation at the time. They provided the first example (after nuclear fallout) of a genuine and well-documented global pollution problem, on a scale that we now take for granted. They led to wide debate on the value of circumstantial versus experimental evidence, for at first the case against the organochlorines was entirely correlational, a point exploited to the full by some elements of the agricultural and agrochemical lobbies. This in turn led to the suggestion, perhaps for the first time by non-ecologists, that the precautionary principle should be applied, and that no pesticide should be widely used until its potential effects were understood. Lastly, the events of more recent years have confirmed that, if remedial action is taken and if remnant populations still persist, such populations can and do respond by recovery. In other words, reducing pollution and pesticide use can rapidly bring measurable benefits.

Other pesticides and pollutants with direct effects on birds

Other chemicals noted for their persistence and accumulation properties include industrial polychlorinated biphenyls (PCBs) and alkyl-mercury pesticides. Although PCBs are oestrogenic, and have been suspected to affect mammalian reproduction, they have not been shown to have had major effects on bird populations despite their frequent presence in bird bodies. Alkyl-mercury pesticides were widely used as seed-dressings in the 1960s when they killed many seed-eating birds and raptors, contributing to population declines, notably in Sweden (Borg *et al.* 1969). Their replacement by other pesticides led to immediate reductions in the mercury levels of affected wildlife, especially raptors, and contributed (along with the banning of organochlorines) to population recoveries.

Other pesticides in common use include the carbamates, organophosphates and pyrethroids. Some chemicals in these groups are highly toxic to vertebrates and have caused large-scale mortalities **(Table 15.3)**. One of the most toxic is the carbamate insecticide, carbofuran, which is thought to kill at least two million birds per year in the United States, more than any other pesticide in current use (Mineau 1993). It is available in granular form, so can remain on the soil surface, to be eaten directly by birds as grit or food or to contaminate earthworms and other food organisms (Flickinger *et al.* 1986).

Field evidence for lethal effects of pesticides on birds has come mainly from carcasses found after spraying, with the cause of death confirmed by chemical analysis **(Table 15.3)**. Some studies also included counts of live birds before and after spraying, for comparison with counts in nearby untreated areas. Such studies mostly involved chemicals that were very toxic to birds, and often revealed local population reductions exceeding 80%, with total elimination of certain species from sprayed areas **(Figure 15.2, Table 15.4)**. Providing the spraying was not repeated, songbird populations usually recovered within 1–3 years, but other slower-breeding species took longer (A.W.A. Brown 1978). These studies were mainly in the 1950s and 1960s, since when most of the chemicals concerned have been replaced by less toxic ones, at least in Europe and North America, so that their overall direct impact on bird populations has been reduced (apart from carbofuran). Although many modern pesticides kill birds, sometimes in large numbers, they break down more quickly than organochlorines, so do not have such lasting or far reaching effects.

Whatever the chemical, large birds usually need bigger doses to kill them than do small birds, but birds from different families also differ greatly in their sensitivity, regardless of body size (Tucker & Crabtree 1970, Hill *et al.* 1975, Hudson *et al.* 1984).

Table 15.3 Examples of large-scale bird mortalities caused by pesticides.

Pesticide	Use	Location	Species affected (and corpses found)	Source
Organochlorines				
DDD	Against Gnats *Chaoborus astictopus*	California	Western Grebes *Aechmophorus occidentalis* (100)	Hunt & Bischoff 1960
Aldrin	Rice seed treatment against Rice Water Weevil *Lissorhoptrus oryzophilus*	Texas	Fulvous Whistling Duck *Dendrocygna bicolor* and other waterbirds, shorebirds and songbirds (192)	Flickinger & King 1972
Aldrin	Rice seed treatment against Rice Water Weevil *Lissorhoptrus oryzophilus*	Texas	Snow Geese *Anser caerulescens* (112)	Flickinger 1979
Aldrin and Dieldrin	Seed-treatment (mainly wheat), against various insect pests	Britain	Seed-eaters, including finches, pigeons and game birds (many thousands)	Cramp et al. 1962
Endrin	Against voles	Washington	California Quail *Lophortyx californicus*, raptors and others (194)	Blus et al. 1989
Sodium pentachlorophenate	Against water snails *Pomacea glauca*	Surinam	Snail Kites *Rostrhamus sociabilis* (50), also egrets, herons, jacanas	Vermeer et al. 1974
Toxaphene	Against goldfish	Big Bear Lake, California	Fish-eaters (ducks, terns, gulls, grebes, pelicans)	Rudd 1964
Organophosphates				
Azodrin (Monocrotophos)	Against Voles in alfalfa	Israel	Raptors (400)	Mendelssohn & Paz 1977

Azodrin (Monocrotophos)	Against grasshoppers	Argentina	Swainson's Hawks *Buteo swainsoni* (5000)	Goldstein *et al.* 1996
Carbophenothion	Seed-treatment (cereals)	Britain	Various goose species (several 100s)	Stanley & Bunyan 1979
Fenthion	Against mosquito larvae	North Dakota	453 warblers	Seabloom *et al.* 1973
Fensulfothion	Against pasture pests	New Zealand,	Mainly White-backed Magpie *Gymnorhina tibicen*, Black-backed Gull *Larus dominicanus* and Harrier Hawk *Circus approximans* (394)	Mills 1973
Parathion	Against aphids on cole crops	Britain	Various species	Cramp 1973
Parathion	Against cotton pests	Texas	Laughing Gulls *Larus atricilla* (216)	White *et al.* 1979
Phosphamidon	Against Spruce Budworm *Choristoneura fumiferana*	New Brunswick	An estimated three million songbirds killed in New Brunswick in 1975	Pearce & Peakall 1977
Carbamates				
Carbofuran	Rape seed treatment (granular application)	Saskatchewan	Many thousands of Lapland Longspurs *Calcarius lapponicus* killed	Mineau 1993
	Against turnip seed pests	British Columbia	Many thousands of Green-winged Teal *Anas crecca* killed	Mineau 1993
	Against alfalfa pests	California	American Wigeon *Anas americanus* (2450)	Mineau 1993
	Against alfalfa pests	Oklahoma	Canada Geese *Branta canadensis* (500)	Mineau 1993

Table 15.4 Direct immediate effects of pesticides on bird numbers.

Pesticide used	Pest	Area	Bird species affected	Numbers in sprayed area		Change (%)	Numbers in unsprayed area[3]		Change (%)	Source
				Pre-spray	Post-spray		Pre-spray	Post-spray		
Organochlorines										
DDT	Gypsy Moth *Lymantria dispar*	Pennsylvania	Various forest species, mostly songbirds[1]	128	2[4]	−98	112	108	−4	Hotchkiss & Pough 1946
DDT	Pine Beetle *Dendroctopus monticolae*	Wyoming	Various forest species, mostly songbirds[1,5]	61	21	−70	28	40	+43	Adams et al. 1949
DDT	Forest insects	Maryland	Various forest species, mostly songbirds			−65				Robbins & Stewart 1949
Aldrin	Japanese Beetle *Popillia japonica*	Illinois	Pheasant *Phasianus colchicus*			−25–50[2]				Labisky & Lutz 1967
Dieldrin or Heptachlor	Fire Ant *Solenopis saevissimi*	Alabama	Bobwhite Quail *Colinus virginianus*[1]	61	8	−87	76	74	−3	Clawson & Baker 1959
Organophosphates										
Phosphamidon	Spruce Budworm *Choristoneura fumiferana*	Montana	Various forest species, mostly songbirds	29.5 ± 12.4^6	6.8 ± 3.3^6	−77	46.8 ± 2.6^6	59.2 ± 11.6^6	+26	Finley 1965
Phosphamidon	Forest insects	Switzerland	Various forest species, mostly songbirds			−70				Schneider 1966
Famfur	Warble Fly *Hypoderma* sp. on Cattle	Western States	Magpie *Pica pica*[5]	29.0^7	9.7^7	−67	46.1^7	35.0^7	−34	Henny et al. 1985
Carbamates										
Carbofuran	Grasshoppers	Saskatchewan	Burrowing Owl *Athene cunicularia*			−54 (chicks only)				James & Fox 1987

[1]Carcasses found and analysed for chemical residues (organochlorines) or for cholinesterase levels (organophosphates and carbamates).
[2]Based on comparisons of sprayed and unsprayed areas, but figures not given. In addition to deaths of adults, subsequent breeding success was reduced.
[3]No spraying done but counts at same dates as in treatment area.
[4]One week later increased to 21.
[5]Two sprayings conducted.
[6]± standard errors.
[7]Per 40 ha.

Under experimental conditions, many chemicals in current use have also been found to cause embryo deaths and deformities when applied directly to eggs. The most toxic chemicals applied in this way include the herbicides paraquat and trifluralin (Hoffman & Albers 1984) and some organophosphate insecticides, such as parathion and diazinon (Hoffman & Eastin 1981, Hoffman 1990). Whether they affect wild bird populations through contamination of eggs remains an open question. Other potential and less obvious effects of certain pesticides include immuno-suppression which can render birds and other animals more susceptible to pathogens (Chapter 10).

Rodenticides

In some countries, chemicals used to control rodents in farm crops have had major effects on local bird populations. In part of Israel, in the 1950s, wintering raptor populations were almost obliterated in this way (Mendelssohn 1972). The main chemical used was thallium sulphate on grains which were also eaten by seed-eating birds, causing their deaths. The raptors died mainly through eating dead and dying rodents. In addition, an estimated wintering population of 100 000 Rooks *Corvus frugilegus* disappeared completely. Similar problems occurred in the Russian steppes in the 1950s, when zinc phosphide was used on grain baits against rodents, reducing the numbers of seed-eating birds (including Grey Partridge *Perdix perdix*, Great Bustard *Otis tarda* and Demoiselle Crane *Anthropoides virgo*) and of raptors (including Steppe Eagle *Aquila rapax*, Pallid Harrier *Circus macrourus* and Long-legged Buzzard *Buteo rufinus*) (Belik & Mikalevich 1995).

In some regions, rats and mice have become resistant to the anti-coagulant warfarin which has therefore been replaced by compounds such as difenacoum, bromadiolone and brodifacoum. These products act in the same way as warfarin, but are more toxic and more persistent, giving rise to secondary poisoning in rodent predators (Newton *et al.* 1994). Poisoned owls have been found in many areas, but the clearest evidence of population effects has come from the oil palm plantations of Malaysia. Until 1981, the rats that ate the palm fruits were controlled by a combination of warfarin-baits and Barn Owl *Tyto alba* predation. Then a switch to new rodenticides was followed by rapid declines in Barn Owl numbers. On one estate, the number fell from 20 pairs to two individuals over a 30-month period following the introduction of coumachlor and then brodifacoum (Duckett 1984). Several carcasses were found with signs of haemorrhaging, the most typical symptom of anti-coagulant poisoning.

Avicides

All the above impacts on birds occurred as incidental consequences of pesticide use. Sometimes, however, pesticides have been deliberately applied against birds. The largest-scale example is the use of the organophosphorous compounds, fenthion or parathion, to kill seed-eating Quelea in Africa (Chapter 14). The chemical is usually sprayed from the air over roosts or feeding flocks, and kills through inhalation or oral ingestion up to an estimated billion Quelea per year (Feare 1991). It also kills many other species in the area at the time, as does the aerial spraying of locusts.

Birds-of-prey are sometimes killed by use of poison baits, set for the birds themselves or for other predators. Several cases of population decline have been documented, involving eagles, vultures and others (Bijleveld 1974). They include the virtual elimination of the once-common Griffon Vultures *Gyps fulvus* from Romania and Bulgaria, linked with the widespread use of strychnine for Wolf *Canis lupus* control. Other poisons commonly used to kill raptors include the organophos-phorous compound phosdrin (or mevinphos), the narcotic alpha-chlorolose, and various other pesticides.

Mixed exposure

Birds are often exposed to combinations of pesticides, either because two or more have been applied together, or because the birds themselves move from one crop to another. For some chemicals the combined effect is much greater than expected from their individual toxicities. Such synergism happens where one chemical activates another or slows its detoxification. An example is provided by prochloraz fungicides which can accentuate the effects on birds of organophosphate insecticides (Johnson et al. 1994). So far this aspect has been studied only in laboratory conditions, but could presumably occur as a widespread phenomenon in nature.

Similarly, water bodies near to industrial sites may receive cocktails of chemical pollutants, together with run-off from farmland. Animals living there are exposed to so many pollutants that, when effects are observed, it is sometimes impossible to pinpoint the causal ones. For many years, gulls, terns and cormorants nesting around the Great Lakes showed reduced hatching success and chick deformities and, although a wide range of chemicals was identified in the eggs, the relative contributions of different chemicals to the problem remained obscure. However, organochlorines probably played a major role, especially as populations increased when organochlorine use ceased (Gilbertson 1974, Gilbertson et al. 1976, Gilbertson & Fox 1977, Hoffman et al. 1987, Kubiak et al. 1989).

General comments

Although the numbers of most bird species can recover quickly from direct pesticide impacts, if usage continues, local populations may remain permanently depressed or die out. The frequent use of a lethal but non-persistent pesticide can have consequences for bird populations just as severe as the less frequent use of a persistent one. During the past 50 years, many bird species over large areas have evidently been held by direct pesticide impacts well below the level that contemporary landscapes would otherwise support.

New pesticides are tested before acceptance in an attempt to eliminate at an early stage those of high toxicity to vertebrates. They are then further tested in field trials before they are approved for general use. However, laboratory tests are inevitably performed on a minority of captive species, while field trials are done over short periods in restricted areas, so new chemicals cannot be tested before release on all the species that could be exposed to them once they come into general use. Hence, we can expect that some chemicals will get into use that subsequently prove to affect certain non-target bird-species, in either typical or unforeseen circumstances. This happened with the organophosphate pesticide, carbophenothion, which, when first used as a seed-dressing, killed thousands of wild geese in Britain, in several different incidents (Stanley & Bunyan 1979).

There are other problems in translating laboratory findings to field situations. Knowing the dose of a chemical necessary to kill an individual bird, or a given proportion of a population as determined in laboratory tests, is helpful, but it cannot reveal what proportion of a population is likely to be killed in the wild, and to what extent those losses will be additive to natural ones. Like any other form of loss, mortality from pollutants may be offset to greater or lesser extent by reductions in any natural losses that are density-dependent. Experience in Britain with Sparrowhawk and Peregrine showed that production of young could be reduced to less than half its normal level without causing a decline in breeding numbers. This was because, with reduced competition, the survival of remaining young was higher, and with more territories than usual falling vacant, these young could begin to breed at an earlier mean age than usual. However, further reproductive failure could not be fully compensated, and the populations of both species declined, to nil in some regions. Using field experience of this type, it should be possible to calculate for other populations the degree of additive breeding failure or mortality that could occur from pollutants without causing population decline; in other words to calculate, in the conditions prevailing, the maximum sustainable yield, as described in Chapter 14. The next step is to relate this level of breeding failure or mortality to some threshold level of pollution that will cause it, as in the DDE response in the Peregrine or the HEOD response in the Sparrowhawk discussed above.

Lead poisoning in waterfowl

Lead poisoning of waterfowl results from the ingestion of spent gunshot. Like other birds that eat plant material, waterfowl habitually swallow small grit particles which remain in the gizzard and function in breaking down food. The gunshot is apparently taken as grit. It is then gradually eroded and absorbed, causing lead poisoning. The digestion of 1–3 lead shot (16 mg per kg of body weight) is considered sufficient to kill any species from duck to swan. The initial symptoms are fairly obvious. The bird may be reluctant to fly when approached, and develop drooping wings. Later, it is unable to lift its head, so the neck is laid along the back, giving a condition known as 'lumber neck'. The gut itself becomes paralysed, and peristalsis stops. The foregut may be full of food, yet the bird gradually starves to death, losing 30–50% body mass and producing green bile-stained droppings. On chemical analysis, elevated lead levels are found in blood, liver and kidneys, and after long exposure, also in bone.

Lead poisoning of waterfowl has increased in recent decades because of continued intensive shooting over wetlands, affecting birds over much of Europe and North America. It has been estimated to kill each year 2–3% of the autumn population of waterfowl in North America, some 2–3 million birds (Bellrose 1959, 1980). But because afflicted individuals can take several weeks to succumb, becoming gradually weaker, they often seek cover or are removed by predators, so that few carcasses remain as evidence (Zwank et al. 1985). In the early stages, lead-poisoned birds are also about 1.7 times more vulnerable to human hunters than are healthy birds (Bellrose 1959). In other words, the impact of lead poisoning is much greater than would be supposed from casual impression.

In two major surveys, more than 230 000 gizzards from waterfowl collected throughout the United States and Canada were examined for lead shot (Bellrose 1959, Sanderson & Bellrose 1986). Lead was found in 7–9% of gizzards. Allowing for under-recording (which is usual in manual examinations) and the erosion rates of shot in gizzards, these findings led to speculation that up to one-third of the continental duck population could consume lead in a given year, though not all consumers would be expected to die. Whatever the true figure, shooting indisputably caused three types of loss: shot birds killed immediately, shot birds wounded and not retrieved which died later, and unshot birds which died of lead ingestion. Further, because most deaths from lead poisoning occurred in late winter, they were not necessarily offset by reduced deaths from other causes, and could thus have lowered breeding numbers (Sanderson & Bellrose 1986). However, effects on breeding populations were not unequivocally demonstrated separately from more direct effects of shooting.

In the 1950s, Bellrose (1959) calculated that no less than 3000 tons of lead shot were deposited into the wetlands of North America every year; and based on samples from several areas, the numbers of spent pellets then averaged nearly 70 000 per ha. Because of species differences in feeding behaviour, however, not all species were equally vulnerable. Generally, bottom-feeders, such as Redheads *Aythya americana*, Canvasbacks *Aythya valisineria*, and Lesser Scaups *Aythya affinis* had the highest rates of ingestion (12–28% at any one time), whereas Mallards *Anas platyrhynchos*, American Black Ducks *Anas rubripes*, and Northern Pintails *Anas acuta*, were intermediate (7–12%), and most other ducks ranged from 1 to 3% (Bellrose 1959, Sanderson & Bellrose 1986). Regardless of intake, lead does not become toxic until it is eroded and absorbed, so that waterfowl feeding on hard foods such as corn experience greater exposure than those eating soft foods. Because of the problems, use of lead shot for wildfowling is now illegal in some countries, notably the United States, and steel shot is used instead. Among European countries, at least four have now banned lead shot, although the incidence of lead in European waterfowl is as high as, or higher than, in the United States (Pain 1991).

In Britain, a related problem arose from the use of lead sinkers on fishing lines. Over the years, many such sinkers were discarded by anglers, and from the 1970s became a major cause of mortality in Mute Swans *Cygnus olor*, causing marked population declines on several rivers (Birkhead & Perrins 1985, Sears 1988). The sinkers were found in the gizzards of dead swans, presumably again swallowed as grit. After the introduction of alternative materials to weight the lines, swan survival improved, and numbers began to recover.

Ingested lead has also killed some predatory birds which ate birds and mammals that had been shot but not retrieved. Such secondary poisoning may have contributed to the decline of the California Condor *Gymnogyps californianus*, but by the time the problem was recognised, the wild population was already down to fewer than ten individuals. Within two more years, three of these birds had died of lead poisoning (Wiemeyer *et al.* 1988, Snyder & Snyder 1989). Their stomachs contained worn fragments of lead bullets, used to kill deer. The birds were assumed to have fed on the deer and ingested the bullets accidentally. Unlike most other raptors, Condors do not regurgitate hard material in the form of pellets. Some other species, notably Bald Eagles *Haliaeetus leucocephalus*, feed heavily on shot-crippled waterfowl but mortalities from lead poisoning are uncommon (Pattee & Hennes 1983, Reichel *et al.* 1984), probably because eagles regurgitate most of the shot along with other undigested remains. Studies of lead-levels in the blood of other raptor species have revealed that occasional individuals have high levels (Pain *et al.* 1995), but it is impossible to estimate the scale of mortality, let alone its impact on populations. In addition, lead from mining and smelting waste poisoned many Tundra Swans *Cygnus columbianus* in northern Idaho (Blus *et al.* 1991), while organic lead from petrochemical effluent killed about 2400 waterbirds in one autumn on the Mersey estuary in Britain (Bull *et al.* 1983).

Oil spills and seabirds

In recent decades millions of seabirds have died from oil pollution. The most vulnerable species are the surface swimmers, such as auks, seaducks and grebes. Major incidents result from the wrecking of laden tankers and often kill many thousands of birds at a time, as the following examples show:

- Torrey Canyon, England, March 1967, spilling 117 thousand tonnes, resulted in 10 000 birds found dead and an estimated total kill of at least 30 000 birds, mainly Razorbills *Alca torda* and other alcids (Bourne *et al.* 1967, Bourne 1970);
- Amoco Cadiz, France, April 1978, spilling 233 thousand tonnes, resulted in 4600 birds found dead, and an estimated total kill of 15 000–20 000 birds, mainly Puffins *Fratercula arctica* and other alcids (Hope Jones *et al.* 1978, Conan 1982);
- Exxon Valdez, Prince William Sound, Alaska, March 1989, spilling 260 thousand barrels, resulted in 30 000 birds found dead, and an estimated total kill of 100 000–300 000 birds from 600 000 in the area at the time (Piatt *et al.* 1990); casualties included Common Murres *Ura aalge* (75%), other alcids (7%) and seaducks (5%);
- Sea Empress, Wales, February 1996, spilling 72 thousand tonnes, resulted in 7000 birds found dead, and an unknown total kill; casualties included mainly Common Scoters *Melanitta nigra* (Edwards 1996).

Much larger quantities of oil (0.8 million tonnes) were released by Iraqi forces during the Gulf War in January 1991, but the number of casualties found was about the same as in the Exxon Valdez disaster, in this instance mostly grebes and cormorants. The 500 km of affected coastline would have normally supported 100 000 waders, whose absence that spring was attributed to the oil-induced destruction of their invertebrate food-supply (Symens & Salamah 1993).

When seabirds swim into oil, their feathers become soiled and matted, allowing water to penetrate and causing loss of insulation and buoyancy. The birds soon die from hypothermia, drowning or starvation, or from ingesting oil removed during preening (Clarke 1984). Because oil floats on water, a relatively small volume can cover a large area. It can form an emulsion with water, called chocolate mousse. The volatile part gradually evaporates (faster at higher temperatures), leaving a sticky asphaltic residue, and eventually 'tarballs', which cause much less harm to birds.

In any incident, the numbers of birds killed depends not only on the numbers and species of birds in the area at the time, but also on the amount of oil spilled, how long it persists (which varies with sea temperature and clean-up operations), and where it drifts. The size of a spill thus tells little about its potential for damage. Near Norway in 1981, a small operational discharge of oily bilge washings killed an estimated 30 000 seabirds (more than in some major incidents), because the oil affected an area where birds were seasonally abundant (Kornberg 1981). In another instance, more than 16 500 oiled Magellanic Penguins *Spheniscus magellanicus* appeared on an Argentinian coast but no slick was seen (Fry 1992). Moreover, it is hard to estimate the total casualties from any incident, for while some oiled birds appear on nearby shores, where they can be counted, others appear on more distant shores or sink without trace, the proportions varying with wind and other conditions (Bibby & Lloyd 1977). There are also longer-term effects of oil on various invertebrates eaten by birds, which can take up to a few years to recover, or longer if toxic detergents and dispersants were used in the clean-up.

It is surprisingly difficult to assess the impact of large-scale local kills on seabird populations. This is partly because relevant pre-incident counts are seldom available, and in any case it is usually hard to tell from how wide a breeding area the casualties are drawn, or what proportion consist of breeders, as opposed to non-breeders. Incidents in the spring, such as the Exxon Valdez and Torrey Canyon, can reduce local nesting numbers (up to 89% after Torrey Canyon, **Table 15.5**), and one spill off South Africa was reported to have wiped out an entire colony of 8000 Jackass Penguins *Spheniscus demersus* (Ross 1971). Despite the low reproductive rates shown by some species, heavy losses of seabirds could be made good within a few years by immigration and by non-breeders starting to nest at an earlier age than usual. Incidents in winter could involve migrant birds drawn from wide (and usually unknown) breeding areas, so that the impact is borne by a wider range of populations. Whether in the breeding or non-breeding season, however, major oiling incidents increase the frequency of catastrophic mortalities to which many seabird populations are exposed naturally. Whether they tip populations into decline will depend on their frequency relative to the recovery powers of the population. Although repeated oil pollution seemed to have caused a decline of some seabird populations (including the virtual disappearance of auks from the English Channel), other potential causes of long-term decline could not be eliminated, and other similar populations elsewhere increased despite their exposure to frequent incidents (Clarke 1984).

Major incidents are estimated to account for only 10–15% of the oil that reaches the sea. Most comes from thousands of smaller incidents, accidental and deliberate, that occur at sea, both from shipping and from offshore platforms. These ongoing minor incidents provide a continuous source of seabird mortality, over and above the occasional major incidents. They occur especially in areas of heavy tanker traffic, so that oiled birds are nowadays continually found on shorelines. In western European waters, the numbers of seabirds routinely killed by oil has been estimated from beach surveys at tens of thousands per year, while the total numbers expected to die per year from all causes is in the hundreds of thousands (Dunnet 1982). It is

Table 15.5   Pairs of seabirds, Sept Isles, northern France, before and after the wrecking of the Torrey Canyon, listed in order of extent of decline. Declines were most evident in the most aquatic species, namely alcids and Shags. From Bourne 1970.

	Before	After	Reduction (%)
Razorbill *Alca torda*	450	50	89
Puffin *Fratercula arctica*	2500	400	84
Guillemot *Uria aalge*	270	50	81
Shag *Phalacrocorax aristotelis*	250	190	24
Gannet *Morus bassanus*	2600	2500	4
Herring Gull *Larus argentatus*	7210	7000	3
Lesser Black-backed Gull *Larus fuscus*	400	400	0
Great Black-backed Gull *Larus marinus*	52	52	0
Kittiwake *Rissa tridactyla*	30	30	0
Fulmar *Fulmarus glacialis*	20	20	0

not known to what extent this regular oil-induced mortality is additive to natural losses, and again, it is usually impossible even to estimate the numbers of birds at risk, or the full geographical provenance of the birds that die. However, this ongoing pollution has been considered a major threat to the seaduck populations of Europe, with tens of thousands found dead every year around the coasts of Denmark alone (Joensen 1977). In addition to incidents at sea, spills on land occur from well blow-outs and pipeline ruptures. These usually affect small areas, unless they reach a watercourse.

Much effort has been devoted to the rescue and cleaning of oiled birds, on the assumption that subsequent survival chances are improved. Based on North American ring recoveries, the median survival time of oiled Guillemots *Uria aalge* after cleaning, ringing and release was 6 days (range 0–919 days), compared with 216 days (range 1–9259 days) for unoiled birds after ringing. Equivalent figures for Western Grebes *Aechmophorus occidentalis* were 11 days (range 1–763 days) for cleaned oiled birds and 624 days (range 19–658 days) for unoiled ones, and for Velvet Scoters *Melanitta fusca*, the figures were 7 days (range 1–16 days) for cleaned oiled birds and 466 days (range 8–4939) for unoiled ones (Sharp 1996). With such low post-release survival, the cleaning of oiled birds seemed pointless, a common experience elsewhere (Clarke 1984). On the other hand, off southern Africa, some 37–84% of Jackass Penguins *Spheniscus demersus* appeared back on their breeding colonies after cleaning, so had evidently survived well (Morant *et al.* 1981). Much could depend on how much oil the birds had consumed while preening, and the extent of gastro-intestinal damage. The penguins excepted, however, severely oiled seabirds, whether cleaned or not, are most realistically regarded as lost to the population.

Birds that are not oiled sufficiently to die can sometimes show reduced breeding success (Trivelpierce *et al.* 1984, Fry *et al.* 1986). Levels of reproductive and other hormones were measured in Magellanic Penguins *Spheniscus magellanicus* that were partly covered with oil following a spill off the Patagonian coast. Certain hormones were at lower levels in oiled than in unoiled birds, and few of the pairs containing an oiled bird later established nests (Fowler *et al.* 1995). Other birds may pass oil and other hydrocarbons on to their eggs, reducing hatchability, partly through embryo suffocation. This has been repeatedly shown in various aquatic species when micro-litre amounts of crude or refined oils were applied experimentally to the surface of fertile eggs, either in laboratory or field conditions (Clarke 1984). The resulting mortality depended on both the dose and the stage of embryo development. Some teratogenic effects have occurred when oil was applied during the first few days of incubation (Eastin & Hoffman 1979). To my knowledge, this type of contamination has not yet resulted in massive breeding failure in any wild population, but it remains a threat.

Plastic and seabirds

The oceans now contain large quantities of plastic, particles of which are often found in the stomachs of seabirds. Two types are distinguished: user-plastics, which contain fragments of items such as polystyrene cups, plastic toys or synthetic netting; and plastic pellets which look like the raw material from which user-products are made. Mixtures of both types have been found at up to several thousand particles per square kilometre of surface in the Atlantic and Pacific Oceans, and are probably now universally present in sea waters (Carpenter & Smith 1972, Colton *et al.* 1974, Morris 1980).

Seabirds may swallow such particles in mistake for food, or get them from the stomachs of prey species. They are indigestible but can be eroded slowly. Their main effects are to reduce gizzard volume, theoretically to the point when the bird can no longer feed (Connors & Smith 1982). Species that habitually regurgitate pellets, such as gulls and penguins, can get rid of them, but shearwaters and petrels apparently cannot and it is these species in which the largest quantities are found (Furness 1985a). The effects on seabird populations are unknown, but as ever more plastic is used and discarded, increasing quantities are likely to be consumed.

Pollutants and life-history features

The direct impact that any pollutant has on a bird population is greatly influenced by the life-history features of the species concerned. Large species, which tend to be long-lived and slow-breeding (*K*-selected), are likely to withstand longer periods of reduced breeding success than are small, short-lived, fast-breeding (*r*-selected) ones. However, once reduced in numbers by a given percentage (through mortality or breeding failure), *K*-selected species would be expected to take longer to recover. Five years of complete reproductive failure would (on the usual 64% adult survival) reduce a Sparrowhawk *Accipiter nisus* population to 17% of its former level, a Peregrine *Falco peregrinus* population (90% adult survival) to 66% of its former level, and a Golden Eagle *Aquila chrysaetos* population (95% adult survival) to 81%. Because the populations of larger species would normally contain a substantial proportion of non-territorial non-breeders which can replace any territorial breeders that die, population decline in large species might be missed for several years if counts were restricted to breeders only. It is the resilience of long-lived species to reproductive failure that may have enabled Golden Eagles to maintain their breeding density in western Scotland through the 1960s, despite reproductive failures resulting from the use of organochlorines in sheep dips (Newton 1979).

INDIRECT EFFECTS ON POPULATIONS THROUGH FOOD-SHORTAGE

Pesticides and farmland birds

The second main mechanism of bird population declines is indirect, resulting from the removal by chemicals of a crucial food-supply. This is mainly because most pesticides kill a wide range of organisms, both pests and non-pests. Thus most insecticides kill many kinds of invertebrates; some fungicides kill invertebrates as well as removing the fungal food-supply of some insects; while herbicides remove the food-plants of other insects, as well as reducing the numbers of weed-seeds available for seed-eating birds. In many countries, the use of pesticides is now so ubiquitous, often with several applications per growing season, that invertebrate populations over huge areas could be held permanently at much reduced levels, not only in the crops themselves, but also in hedges and other adjoining habitats to which the chemicals are carried by wind.

An example of the indirect effects of spraying a carbamate insecticide on forest birds in New Jersey was given by Moulding (1976). The pesticide carbaryl was applied against the Gypsy Moth *Lymantria dispar*, and birds were counted on several dates before and after spraying. Compared with counts on unsprayed plots, those on sprayed plots revealed a gradual decline in bird numbers and species richness during the eight weeks following spraying **(Figure 15.8)**. By late July, overall densities were 55% lower than on the control plots. No further spraying was done,

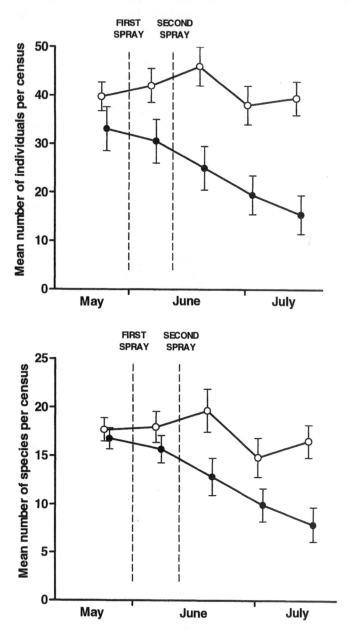

Figure 15.8   Effects of two sprays of the pesticide carbaryl on total bird numbers and bird species numbers in forest in New Jersey. Closed circles refer to sprayed plots and open circles to unsprayed plots. Redrawn from Moulding 1976.

but bird densities were still depressed on the sprayed plots the following spring, being 45% lower than in the same period the previous year. Canopy-feeding birds were reduced more than ground feeders. No dead birds were found, and the effect was attributed to emigration following massive reduction in the arthropod food-supply. Similarly, after spraying the organophosphates, fenithrothion and chlorpyrifos, to control locusts in Senegal, total bird numbers decreased on all treated plots. Some of the decrease was due to bird mortality (adults and chicks), but most was attributed to emigration following the sudden reduction in arthropod prey (Mullié & Keith 1993).

The best documented example of more widespread population effects through reduction of food-supply concerns the Grey Partridge *Perdix perdix*, whose widescale decline has been attributed largely to poor production of young, in turn due mainly to lack of insect food for chicks (Potts 1986, Potts & Aebischer 1995). This is because broad-leaved weed species, which form the food-plants of the relevant insects, are now absent from most modern cereal fields, as a result of herbicide use, and many other insects are killed directly by insecticide use **(Figure 15.9)**. The survival of young Partridges (up to 3 weeks old) was found to be directly correlated with the insect-supply in cereal fields. This proposed mechanism was confirmed by experimental field trials in which reduced herbicide use resulted in improved weed and insect populations, better chick survival, and greater Partridge densities (**Table 15.6**; Rands 1985). The decline in Partridge numbers is not restricted to Britain, where this research was done, but is general throughout the agricultural range of the species, with an estimated decline in the world population from 120 million in 1950 to ten million in 1990 (Potts 1991). Research findings led to the recommendation that farmers keen to conserve Partridges should leave an unsprayed strip of crop around the edges of fields, so-called 'conservation headlands'. This procedure enables

Table 15.6   Comparison between weed and insect populations, and game bird brood sizes in sprayed and unsprayed plots in cereal fields. Figures show mean values with standard errors. From Sotherton 1991.

	Plots with herbicide	Plots without herbicide	Significance of difference (P)[3]
Weeds[1]			
Number of weed species per 0.25m^2	2.1±0.5	6.8±1.0	<0.01
Total weed biomass	0.9±0.2	9.7±2.3	<0.001
Percentage weed cover	2.9±0.6	14.2±2.3	<0.01
Insects[1]			
Total game bird chick food-items per 0.5m^2	18.9±9.2	67.9±23.4	<0.001
Bird brood sizes[2]			
Grey Partridge *Perdix perdix*	7.5±0.8	10.0±0.6	<0.01
Pheasant *Phasianus colchicus*	3.2±0.5	6.9±0.5	<0.001

[1]Records from 1988.
[2]Records from 1984. In the Partridge, similar data for 1983, 1985 and 1986 showed significant differences in all but 1986, and in the Pheasant similar data for 1985 and 1986 showed significant differences in both years.
[3]Some analyses on transformed data.

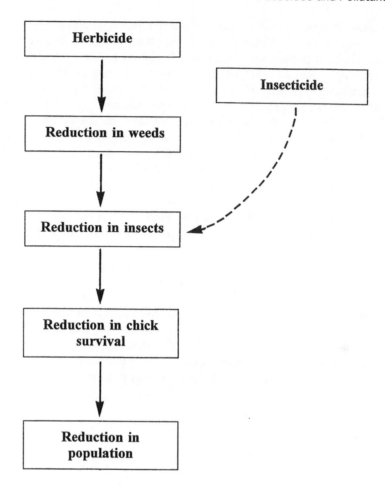

Figure 15.9 Proposed means by which herbicides have caused population decline in the Grey Partridge *Perdix perdix* in Britain. Insecticides may have had a minor additional effect. From Potts 1986.

weeds to grow, which in turn supports the insects necessary for Partridges and other birds.

Chick survival in Red-legged Partridges *Alectoris rufa* and Pheasants *Phasianus colchicus* was also correlated with the abundance of preferred insects (Green 1984, Hill 1985; **Table 15.6**). In Illinois, Pheasant chick survival fell by 28% between the early 1950s and mid-1970s. As in Partridges, monoculture cropping systems and pesticide-induced insect shortages caused broods to move longer distances than formerly in search of food. Chicks therefore lost more energy in movement, and at the same time were more vulnerable to predation (Warner 1984). Were it not for the fact that Pheasants can be easily reared and released, they would almost certainly have declined over much of their range.

Invertebrate populations on farmland in Britain seem also to have declined in the long term. Many people have noticed massive reductions in some large insect species, such as butterflies and moths, grasshoppers, cockchafers and other large beetles, but the declines extend in lesser degree to smaller insects too. One of the best data-sets is summarised in **Figure 15.10**, based on regular sampling of cereal fields in southeast England (Potts 1991). When this study started, insecticides had already been used for more than 20 years and herbicides for more than 15 years, yet during 1970–89 overall arthropod numbers continued to decline by about 4% per year (equivalent to about 50% per 20 years). Also in southern England, Woiwood (1991) described a 67% loss in total numbers of nocturnal moths as sampled by light trapping between 1933–50 and 1960–89 in arable land, but no change in nearby woodland, while in Germany, Heydemann (1983) documented a loss of 50–80% of species of beetles (Coleoptera) and ants (Formicidae) from arable land during 1951–81. These declines coincided with increases in the numbers of pesticide applications per year and in the area treated, but it is not possible to establish a firm causal link or to exclude other factors. However, some recovery in insect populations occurred within 1–2 years on arable land when pesticide use was experimentally curtailed **(Table 15.6)**.

During the past 30 years, as shown by the monitoring programmes of the British Trust for Ornithology, most of the bird species found on British farmland have declined **(Table 7.6;** Chapter 7). The same seems to have been true over much of western Europe (Tucker & Heath 1994). In the absence of detailed studies, it is hard to say how much these declines were due to pesticide use, and how much to other agricultural changes that occurred at the same time. However, declines in many species steepened from the mid-1970s when pesticide use increased greatly. Over the years, herbicide use has enormously reduced the populations of various farmland

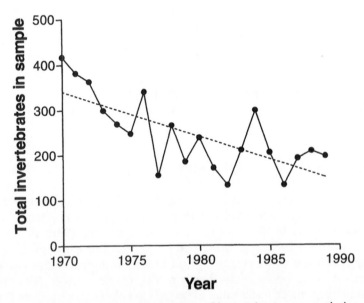

Figure 15.10   Reduction in the mean number of invertebrates, recorded per sample on cereal yields in southeast England, 1970–89. Solid line joins the annual values and dashed line shows the overall trend, calculated by regression analysis. The average annual decline (± S.E.) was 3.8 ± 0.9% (P<0.001). Totals exclude Acari, Collembola, Thysanoptera and Aphidae. Redrawn from Potts 1991.

weeds, on which many seed-eating birds depend. Some once-common arable weeds have now become rare in lowland Britain. Most herbicides are not themselves cumulative, but their effects are, as they lead to progressive depletion of the seed-bank in the soil. Each year seeds turned to the surface germinate to produce plants which are killed before they can seed. Several species of seed-eating birds, notably the Linnet *Carduelis cannabina*, once fed extensively on weed seeds (especially *Polygonum* and *Chenopodium*) turned to the surface each time a field was cultivated (Newton 1972). The old adage 'one year's seeding means seven years weeding' is an understatement, because the seeds of many farmland weed species remain viable for up to several decades (Salisbury 1961). Such seeds could once be counted at the surface of newly-turned farmland soil at thousands per square metre (references in Salisbury 1961), but after more than 30 years of herbicide use, they have now almost disappeared from the soil of cereal-growing regions. It is not surprising, then, that the bird species most dependent on arable weeds have declined too, and that their declines did not become apparent until herbicides had been in use for several years.

In much of western Europe, the effects of herbicide use on seed-eaters must be added to those of another agricultural change, namely the planting of cereal crops in the late summer, soon after harvest, rather than in the following spring. This has meant that spilled grain and weed seeds that once remained available on stubble fields throughout the winter are now buried by ploughing soon after harvest, at once removing a huge potential food-source. Besides the Linnet mentioned above, other bird species likely to have been affected by the decline in seed-supplies are the Skylark *Alauda arvensis*, Tree Sparrow *Passer montanus* and various buntings. But as these other species also require insects for feeding their chicks, it is as yet uncertain to what extent shortage of insects (leading to poor chick production) rather than shortage of seeds (leading to poor adult survival) is responsible for their demise.

Widespread monitoring of bird populations, followed by appropriate research and experiment, has played a crucial role in highlighting some of the long-term consequences of pesticide use and in confirming pesticides as a major factor in reducing, not just pest numbers, but the entire spectrum of biodiversity in farmed landscapes (and beyond in the case of organochlorines). The number of well-documented case-studies is low because declines in bird numbers have mostly been gradual, and in recent years have seldom involved conspicuous large-scale kills. Also, any effects of pesticides have proved hard in retrospect to disentangle from those of other procedural changes.

In other parts of the world, locust spray programmes have greatly reduced the populations of some other bird species, including the Rosy Pastor *Sturnus roseus* on the European steppes (Belik & Mikalevich 1995). Again the main effects are indirect, through removal of food-supply.

Waterbirds and acidification

Combustion of coal and other industrial activities over the last 200 years has greatly increased the amounts of sulphur and other pollutants in the atmosphere. Such pollutants can be blown long distances, affecting areas up to several hundred kilometres from their source. One major identified consequence is the acidifying effect of sulphur dioxide and nitrogen oxides on rain **(Figure 15.11)**. The impact of 'acid rain' on well-buffered soils is probably negligible, but on granitic or other

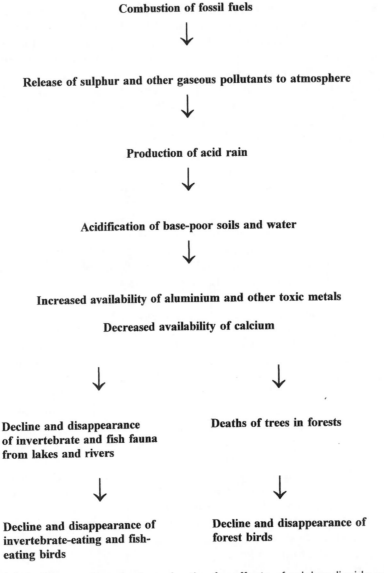

Figure 15.11   Proposed mechanism of action for effects of sulphur dioxide on fauna and flora.

poorly buffered soils, the pH of soils, streams and lakes is lowered from 6–7 to 4–5 or less.[1] This in turn has affected the mobility of toxic metals (notably aluminium, mercury, cadmium and lead), which move from soils to waters, and become

---

[1] *Nowadays acid deposition is a blanket term used to embrace several phenomena, including acid rain (pH less than 5.65), occult precipitation (weak solutions of sulphuric and nitric acid in mist and dew), dry deposition on plants and ground, and even heavy metals, released into the atmosphere from industrial and road traffic emissions.*

generally more available to plants and animals, while calcium and magnesium become less available. The resulting declines in invertebrate and fish populations have led in turn to declines in the bird populations that depend on them. Other identified effects on birds result from metal toxicity and reduced dietary calcium.

Acidification is particularly severe in parts of eastern North America, northwest Europe and southern China, and is gradually spreading. In southern Norway, for example, which has high rainfall and a geologically vulnerable substrate, lakes in more than 1.3 million ha (5000 square miles) are now practically devoid of fish, and in another 1.9 million ha (7500 square miles) have much reduced fish-stocks (Elsworth 1984).

In any one region, the effects on the aquatic fauna are progressive, and involve threshold responses as the pH falls. In general, when the water is around pH 6.0, crustaceans and molluscs with mineralised shells find it hard to survive. Such creatures form an important source of calcium for many birds. At pH 5.5, many insects favoured by birds disappear, including mayflies, caddis and damselflies, though others remain. Among fish, salmonids *Salmo* and roach *Rutilus* start to go as the pH falls below 6.0, whitefish *Coregonus* and grayling *Thymallus* at pH 5.5, followed by perch *Perca* and pike *Esox* at pH 5.0, and eels *Anguilla* at pH 4.5. Different species go at successive stages of acidification, because of differences in their physiology and susceptibility to toxic metals, as well as in their diets. The eradication of fish makes the waters unsuitable for fish-eating birds. As acidification continues, the end result is a sterile ecosystem capable of supporting only a limited range of species, and very few birds.

The main effects on aquatic birds result from the reduced food-supply (Blancher & McAuley 1987, Graveland 1990). Among Dippers *Cinclus cinclus* in Wales, population densities along streams were correlated with the pH of the water, as were clutch sizes and breeding success (Ormerod *et al.* 1988). On some streams, recorded as becoming more acid over the past 40 years, Dipper numbers declined, in association with reduced food-supplies. Among favoured food-items, the crustacean *Gammarus* was absent from acidified streams, while caddis (Trichoptera) and mayfly (Ephemeroptera) nymphs were present in much reduced numbers (Ormerod *et al.* 1985). In eastern North America, Common Loons *Gavia immer* and other fish-eaters have disappeared from many lakes, along with the fish. On 84 lakes in Ontario, only 9% of breeding attempts were successful on lakes of high acidity (with few fish), compared with 57% on lakes of moderate acidity and 59% on lakes with low acidity (with many fish) (Alvo *et al.* 1988). In southwest Sweden, a long-term decline in Osprey *Pandion haliaetus* breeding success was halted in the 1980s following the liming of acidified lakes and the restoration of fish populations (Eriksson & Walter 1994), a process that must be continued if the fish are to stay.

As pH declines, fish disappear before some arthropods which therefore become temporarily very abundant, to the advantage of certain ducks. The main beneficiaries in North America include Goldeneye *Bucephala clangula*, Ring-necked Duck *Aythya collaris* and Hooded Merganser *Mergus cucullatus*, as well as dabbling ducks, such as Mallard *Anas platyrhynchos* and Black Duck *Anas rubripes* (Blancher & McAuley 1987, Parker *et al.* 1992). However, as pH falls further and invertebrates decline, so do the ducks that depend on them. The widespread decline of the Black Duck in eastern North America has been attributed partly to poor duckling survival, resulting from reduced insect supplies (P.W. Hansen 1987, Schreiber & Newman 1988).

Acidification and calcium shortage

Other effects of acidification on birds result from the reduced availability of calcium (through the disappearance of calcium-rich food sources) and increased exposure to toxic metals. Some passerine species that breed in acidified areas have shown non-laying (with birds incubating empty nests) and shell-thinning, which leads to the breakage or desiccation of eggs (Carlsson *et al.* 1991, Graveland *et al.* 1994). In one area of the Netherlands, the proportion of Great Tit *Parus major* females that produced defective eggs rose from 10% in 1983–84 to 40% in 1987–88 (Graveland *et al.* 1994). This was associated with the leaching of calcium from acidified soils, its reduction in vegetation, and consequent loss of snails, whose shells form a major source of calcium for laying songbirds. In one experiment, Graveland *et al.* (1994) provided broken shells from snails and domestic chickens, which resulted in an improvement in the eggshells and breeding success of local titmice. In another experiment, the liming of an acidified area resulted in a recovery of snail populations and the consequent restoration of a calcium source for songbirds.

Similar problems have been identified in Dippers that nest on acidified streams, and in various other insectivorous birds that nest beside acidified lakes, such as the Pied Flycatcher *Ficedula hypoleuca* in Europe, and the Eastern Kingbird *Tyrannus tyrannus* and Tree Swallow *Tachycineta bicolor* in North America (Nyholm 1981, Glooschenko *et al.* 1986, Ormerod *et al.* 1988, 1991, Blancher & McNicol 1991). All these species feed largely on aerial midges and other insects that emerge from aquatic larvae. Their diets are therefore low in calcium and also often high in aluminium (released from acidified soils) which may disrupt calcium metabolism, enhancing the deficiency. Some aquatic bird species also have high mercury contents, a result of the conversion under acid conditions of inorganic mercury to the more readily absorbed methyl mercury.

These various problems, which have so far surfaced only in areas of naturally low pH, are likely to spread more widely as further acid deposition brings more areas below the critical pH threshold. Although limited measures have been taken in some countries to reduce sulphur outputs, these are not enough, and the problem continues. Meanwhile, research has concentrated on finding 'critical loads', the maximum acid deposition that can be sustained by a given region without adverse effects. The critical load naturally varies greatly from region to region depending on the underlying geology.

INDIRECT EFFECTS ON POPULATIONS THROUGH ALTERATION OF HABITATS

Forest dieback

Some bird populations have declined, at least locally, because pollutants have destroyed their habitats. Air pollution is held responsible for forest dieback in several parts of the world, including montane areas in central Europe, eastern North America, West Africa, parts of China and South America. These are again all areas with heavy rainfall on base-poor soils. The problem became evident in the 1970s, and has increased subsequently. The extent to which the trees die from exposure to air pollutants as such, from exposure to acidification or to aluminium and other toxic metals, or from shortages of calcium and other essential nutrients, probably varies with area, but the net result is loss of forest habitat, together with the acidification of streams, as discussed above.

In forests, as in aquatic habitats, the effects are progressive, depending on the initial base-status of the soils and the rates of acid deposition. Some tree species appear more sensitive than others, and die at an earlier stage in the acidification process. In central Europe, Silver Fir *Abies alba*, Norway Spruce *Picea abies* and Beech *Fagus sylvatica* seem most vulnerable, and in North America Red Spruce *Picea rubens*, various pines and Sugar Maple *Acer saccharum*. The geographical scale of the problem is continually changing, but in 1986 in Europe, more than 19 million ha, or 14% of the total forest area, was classed as damaged (Nilsson & Duinker 1987). This damage was at various levels, but included some areas where whole forests had died, mainly near point sources of pollution, as in Silesia (Poland), Bohemia (Czech Republic) and some montane areas of Germany and Switzerland.

Whereas the trees are affected directly, the animal inhabitants are affected indirectly, through destruction of their habitat. Most bird studies have involved comparisons of breeding numbers in plots showing different amounts of tree damage. On the gradient from completely healthy to completely dead, overall bird numbers declined, in studies in both Europe and North America (Desgranges *et al.* 1987, Oelke 1989, Graveland 1990). Species dependent on insects from foliage declined the most, while those dependent on insects from dead wood (such as woodpeckers) may have benefited temporarily, as may have seed-eaters (because trees respond to adversity by seed production), but the net result in all studies was an overall decline in the numbers of species and individuals. In extreme areas, a high density forest community was replaced by a sparse openland one. In some studies, declines of up to 50% in overall bird densities on particular areas have occurred within a five-year period (Flousek 1989).

Heavy metals, derived from industrial and vehicle emissions, are also transported long distances in the atmosphere, eventually to be deposited on vegetation, from which they enter food-chains. Many animals nowadays have elevated metal levels, and in extreme cases show signs of tissue damage, but I know of no documented impacts on bird population levels. The most damaging of these metals is probably lead, the deposition of which declined more than 50% in northern Europe after the introduction of regulations on leaded petrol (Pacyna 1995).

Eutrophication

Not all pollutant effects are entirely negative. Over a certain range, some pollutants may merely result in environmental change that favours some species at the expense of others. In recent decades, the widespread eutrophication of inland and coastal waters, mainly by nitrates and phosphates (derived from fertilisers, sewage, animal wastes and domestic effluent) has led to changes in plant, invertebrate and fish populations, with resulting effects on birds. Impacts vary with the initial nutrient status of the water, because lakes and rivers vary naturally on a gradient from poor (oligotrophic) to rich (eutrophic). As nitrates and phosphates leach into nutrient-poor waters, macrophytic plants (water weeds) may grow, providing new substrates for invertebrates, more food and cover for fish and a greatly increased food-supply for both herbivorous and carnivorous birds. Salmonid and coregonid fish give way to coarse fish, initially percids and then cyprinids (such as Roach *Rutilus rutilus* and Bream *Abramis brama* in European waters). These species mature early, and have high reproductive rates, so can perhaps support larger numbers of fish-eating birds. They also occur in shoals, so are easier than salmonids for most fish-eating birds to catch, despite the murky water in which they often live. Such fish populations can sometimes achieve a biomass exceeding 1000 kg per ha. As nutrients continue to increase, however, algae multiply and cloud the water, reducing light penetration and causing losses of water-weeds, invertebrates and fish. Certain algae and other micro-organisms produce toxins that poison fish and the birds that eat them. Massive bird mortalities have been recorded in both inland and coastal waters, from which affected species can take several years to recover their numbers (Chapter 10). In deep lakes, the dense algal populations may sink and their decomposition may de-oxygenate the water. Under extreme circumstances, the water may also become almost lifeless.

The addition of organic matter to lakes and rivers, and its decomposition to support new life, is a natural process, but the addition of too much human sewage and other organic waste can cause problems because the bacteria that cause decomposition need oxygen to survive. Within limits, a river naturally re-aerates itself, but under heavy organic input the water can become totally devoid of oxygen, resulting in the deaths of fish and other aquatic life. The bacteria that cause decay are sometimes replaced by anaerobic bacteria, which can trigger other processes inimical to most organisms. The problem is increased if river flow is reduced by the extraction of water for irrigation and other purposes. Not surprisingly, eutrophication has also become increasingly apparent in shallow coastal areas.

Global warming

Some other pollutants, in the amounts released, have little or no direct impact on living organisms, but can alter the physical and chemical environment so as to affect population levels. The most far-reaching of these is probably carbon dioxide, whose concentration in the atmosphere is steadily rising, mainly through the burning of fossil fuels. This gas has increased from about 280 ppm in pre-industrial times to about 350 ppm at present, and is predicted to reach about 700 ppm around the year 2100. Its accumulation, together with that of other heat-trapping gases (such as methane and nitrous oxide), is expected to raise global temperatures (Houghton *et al.* 1995), which in turn is expected to cause radical changes in the numbers and distributions of species world-wide. Continuing deforestation is contributing increasingly to the problem because trees contain more fixed carbon per unit area than smaller plants, releasing carbon dioxide when they are burnt; living forest also converts carbon dioxide to oxygen more rapidly than does any other vegetation. Furthermore, as the earth warms, the rates of natural loss of carbon dioxide and methane to the atmosphere are expected to rise. At the same time, moreover, the progressive destruction of stratospheric ozone by chlorofluorocarbons (or CFCs) is steadily increasing the amount of ultra-violet radiation reaching the earth, with effects on plant growth.

The increase of greenhouse gases in the atmosphere is indisputable. It is accepted by scientific consensus that this will lead to a progressive rise in mean global temperature which will be greater at high than at low latitudes. Less certainty surrounds the actual rate of temperature rise, its effects on other aspects of climate, and on regional variation in climate. Most current estimates suggest a mean rate of about 1.0–2.0°C per 100 years, but some suggest rates up to 4.0°C per 100 years. This may not seem much, but as a global average it is close to the 4±1°C change which caused the last transition from glacial to inter-glacial, which took 8000 years. Within 100 years from now, the earth could be warmer than at any time in recent geological history (Webb 1992). The main effects are likely to be on plants, with animals affected secondarily.

In the change from glacial to inter-glacial, the main vegetation belts (together with their fauna) shifted position slowly over the earth's surface, tracking the movements of appropriate climatic zones. The pollen record has given some idea of the maximum speed at which trees have spread in the past. These rates of spread were much slower than needed to keep up with the rapid changes predicted for the future. A mean change of 1.0°C is equivalent to a poleward movement of temperate climatic belts of 200–300 km or an upward altitudinal movement of 200 m. Thus, at the rates of human-induced climate change predicted, many plant species would be left behind, unable to move fast enough. Their passage through modern landscapes would anyway be impeded by human land-use (which will itself have to change). Hence, many plant species are likely to remain stranded in areas of increasingly unsuitable climate, unable to disperse fast enough into more appropriate areas. This is predicted to lead to some regional extinctions. Although many animals (especially birds) are more mobile, they are still tied to areas of suitable vegetation. Island or mountaintop species may be especially vulnerable: if the latitudinal or altitudinal movement required by climate change exceeds the limits of the island or mountain range, extinction may follow, although the sea around oceanic islands is expected to moderate the local change in air temperature.

A second projected consequence of the global warming is a shifting in the distribution of climatic extremes, whether hard winters, drought summers or hurricanes. In some regions, therefore, such extremes will become more or less frequent than at present. Because these extremes influence the distributions and abundance levels of organisms, they can be expected to cause further changes in bird populations. In particular, the more frequent that extreme events become in any one region, the more likely are the most sensitive species to disappear from that region. As a third consequence, rises in sea level can be expected from expansion of sea water, and from the melting of glaciers and polar ice, resulting in loss of land area (Schneider et al. 1992, Wigley & Riper 1992). The rate of mean rise is estimated at about 50 cm per 100 years (range 10–100 cm). Again, this may not sound much, but the associated increase in wave heights and tidal surges can be expected to inundate large areas of coastal habitat.

All these various projections are made on the assumption that insufficient will be done to avoid the problem. Some bird populations in Europe have already shown some apparent response to warming, notably in timing of breeding which (with large annual variations) became progressively earlier between 1970 and 1995 (Winkel 1996, Crick et al. 1997), and in the movements of migrants which since 1960 have arrived progressively earlier in spring and departed progressively later in autumn (Moritz 1993, Vogel & Moritz 1995). Some distributional changes could also be due to climatic change, but as several factors could be involved, the evidence is less clear cut.

Despite the scientific consensus that global warming will occur, it is barely possible to confirm its reality with existing data. The rise in global temperature recorded over the last 100 years is consistent with projected rates, but it could still be construed as within the range of natural variation. However, uncertainty over the reality and rate of the phenomenon should decline greatly in the coming years. Similarly, while many bird and other species are likely to change markedly in abundance and distribution, it is impossible as yet to do more than guess about the precise impacts or about which species will be most affected. Almost certainly, however, bigger changes will result from the human response to climate change, and the alterations in land-use they cause, than from any natural vegetation changes that occur. The enactment of stronger global policy to reduce greenhouse gas emissions could clearly reduce the severity of global warming, but any action now could still be too little and too late to prevent substantial detrimental change.

GENETIC ASPECTS

In nature, the members of any species experience widely differing exposures to the same chemicals, so it is not surprising that some individuals are affected more than others. In addition, however, individuals usually vary in their response to the same dose of a chemical: some are more sensitive than others. For this reason, the toxicities of particular chemicals to a given species are often expressed as LD_{50} values – the lethal dose expected to kill 50% (or any other stated percentage) of a population. If any of this variation in response is under genetic control, an exposed population could in time change genetically, through the selection of resistant genotypes. Resistance is likely to evolve most rapidly when:

(1) the initial genetic variation in response is high,
(2) the proportion of the population exposed and killed is high,
(3) the reproductive rate is high, and
(4) the immigration rate from other unexposed populations is low.

Known examples of the development of resistance (such as many insects to DDT, some plants to toxic metals or some rodents to warfarin) have so far come from only a small proportion of exposed species and from only a small proportion of toxic chemicals. None of the known examples includes birds.

The first known vertebrate to develop resistance to organochlorines was the Mosquito Fish *Gambusia affinis* in the lower Rio Grande valley in Texas, which became resistant to DDE and toxaphene. These fish could then accumulate large amounts in their bodies, and present a greater hazard than before to the predatory fish and birds that ate them. Andreasen (1985) suggested that this change in the Mosquito Fish may have contributed to the shell-thinning and population declines in fish-eating birds observed at the time in southern Texas. This is an instance where resistance in one organism may protect that organism but at the same time harm its predators.

USE OF HISTORICAL MATERIAL

In examining long-term trends in contamination, analysis of dated material from museum collections has helped to distinguish between different possible causes of bird population declines. This procedure proved useful in dating the start of egg-shell thinning in birds **(Figure 15.5)** and in tracing the increase in the contamination of birds with mercury **(Figure 15.12)**. For example, in Sweden, the feathers of Goshawks *Accipiter gentilis* showed a sudden rise in mercury content following the introduction of alkyl-mercury pesticides in the 1950s, whereas those of fish-eating Ospreys *Pandion haliaetus* and Great-crested Grebes *Podiceps cristatus* showed a gradual rise, starting about 1890, reflecting the earlier but progressive contamination from various industries (Jensen *et al.* 1972). Increasing mercury contents were also noted among feathers from Guillemots *Uria aalge* collected in the Baltic Sea between 1830 and 1980 (Appelquist *et al.* 1985).

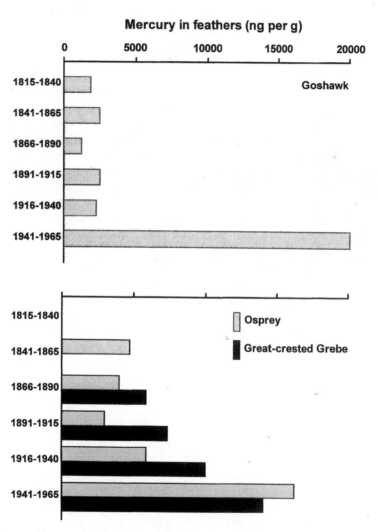

Figure 15.12   Mercury levels in the feathers of Goshawk *Accipiter gentilis*, Osprey *Pandion haliaetus* and Great-crested Grebe *Podiceps cristatus* in Sweden, between 1815 and 1965, showing pattern of contamination of terrestrial and freshwater habitats. Redrawn from Jensen *et al.* 1972.

CONCLUDING REMARKS

The chemical impacts described in this chapter illustrate some key features of pollution, namely that its effects may be indirect, delayed, unexpected and widespread. Pesticides have widescale effects because they are applied to huge areas of land, and some other chemicals because they are dispersed from point sources by wind or water. Other pollutant effects may remain undetected, either because they are not obvious, because they occur in remote areas or because certain familiar phenomena are not recognised as resulting from pollution. Almost all the impacts discussed above are products of the last 150 years, and those of pesticides mainly of the last 50 years. They are therefore new forces in the population ecology of all organisms. One might reasonably ask, then, why pollution has been allowed to become so serious in so short a time.

One reason is that, traditionally, the polluter has not been required to pay the cost of his actions. The owners of factories that pollute the atmosphere have not paid for the deaths of remote forests, for the loss of fish from rivers or for the harm to human health. Nor has the farmer paid for the side-effects of his pesticides: the loss of wildlife, the pollution of water and their effects on human health. This is partly because such costs have been considered acceptable in the development of the modern industrial state, and partly because they are often intangible, difficult to pin down and to value in cash terms. Moreover, many problems, such as acidification, involve threshold effects that do not become apparent until the polluting activity has continued for many years. We may all reasonably wonder how many other ecological time bombs lie in wait.

Another reason for the problems is the application of restricted economics. The value of pesticide use, for example, is judged primarily on whether the extra crop yield that year exceeds the cost of the chemical used. It takes no account of any longer-term effects on soil productivity, future crops, water quality, loss of wildlife and human medical expenses. The cost of all these, where it can be expressed in cash terms, is borne by society at large, and at some later date, often by future generations. Few other modern-day problems offer politicians more excuse for inaction. Remedies are always too uncertain, too difficult, too expensive or too long-term, and obvious disasters to human health are usually needed before urgent action is taken.

SUMMARY

Pesticides and pollutants have reduced bird populations directly through effects on reproduction or survival or indirectly through causing reduction in food-supplies or physical or chemical alteration of habitats.

Some pollutants that have direct effects on birds cause continued breeding failure or mortality, while others cause periodic catastrophic mortality. They usually operate in a density-independent manner. Their impact depends on the frequency and scale of losses, and on the extent to which these losses can be offset by reduced natural mortality or enhanced reproduction. Following the introduction of organochlorine pesticides, the Peregrine Falcon disappeared from large parts of its range, including about one-third of the land area of North America, and several other species were markedly reduced over wide areas. Numbers recovered after the use of these chemicals was reduced. In recent decades, lead poisoning from ingested gunshot has killed millions of waterfowl and oil pollution millions of seabirds, but long-term effects on population levels remain uncertain.

Pesticide use has reduced the food-supplies of many farmland bird species, causing marked and widespread declines in population levels. Seed-eaters have been affected by herbicide use, and insect-eaters by both insecticide use and herbicide use (which removes insect food-plants).

Atmospheric pollution with oxides of sulphur and nitrogen, through leading to acid deposition, has affected many regions that have soils of poor buffering capacity. Its main effects on birds occur through destruction of habitat (as in forest dieback) and reduction in food-supplies, following the loss of fish and invertebrates from acidified waters. Other effects result from reduction in dietary calcium (causing eggshell defects) and perhaps also from accumulation of toxic metals.

Because some pollutants can disperse long distances in air and water currents, they can affect organisms hundreds or thousands of kilometres from areas of production or usage. Moreover, because of threshold effects, some major problems created by atmospheric pollutants did not become widely apparent until many decades after their emission began.

Eutrophication of inland and coastal waters has affected bird populations through changing food-supplies, in some waters upwards and in heavily polluted ones downwards.

Major problems in the future are likely to result from the human-induced rise in atmospheric carbon dioxide and other heat-trapping gases, which is expected to lead to global warming and other climatic change, with pronounced effects on flora and fauna.

New Zealand Great Moa *Dinornis maximus* hunted to extinction.

Chapter 16
Extinction

Extinction is the presumed ultimate fate of all species, as earlier types are superseded by later ones in an evolutionary succession. Species either go extinct or evolve into new ones. During the history of life on earth, these natural processes have probably occurred continuously, but at a variable rate. They have been marked by occasional mass extinctions which on five or more occasions promoted substantial changes in flora or fauna worldwide, as whole groups of organisms disappeared, to be eventually replaced by newer ones. On a geological timescale, such natural extinctions caused only temporary reduction of biodiversity, because the loss of ancestral forms was continually offset by the evolution of new ones. The concern of biologists, however, is that we now seem to be in the midst of a major global pulse of extinctions unlike any previous, in that it is entirely human-driven. Moreover, it is occurring extremely rapidly, on timescales of decades or centuries. In the past 400 years alone, human action has eliminated at least 127 of the approximately 9672 modern species of birds, on average about one every four years. This rate is very much a minimum estimate, yet it is around 100–1000 times faster than the average previous rate, as revealed for birds and other organisms by the fossil record. Moreover, in view of the increasing rapidity with which natural vegetation is being destroyed, especially the species-rich tropical forests, even this high extinction rate is likely to increase in the next 100 years. There is thus a major conservation problem afoot, of how to minimise impending extinctions in the face of the continuing destruction of natural ecosystems. The aim of this chapter is to examine previous extinctions of birds in order to highlight the factors involved.

NATURAL EXTINCTIONS

For any species, extinction is simply distribution and abundance reduced to zero. Hence, those same factors that limit the distribution and abundance of birds may also be capable of causing their demise. For organisms as a whole, speculation on the natural causes of extinction – those that have occurred in the absence of human intervention – has centred on three main processes (Raup 1984). The first is primarily biological: the impact on species of new competitors, new predators and new diseases. This process could result from the evolution of new species which then affected others, or from distributional changes which brought together existing incompatible species that were formerly separated. In theory, any group of organisms might be eliminated by another newly-appearing group that was better adapted to prevailing conditions. Not all species replacements occurred through such interaction, however, for in the fossil record the demise of one group often preceded the radiation of another. In these cases, competition is implied only to the extent that the second group evolved to occupy the same niche space vacated by the first. Extinctions of many organisms were probably linked more closely, however, as when some species disappeared after the loss of others that formed their habitat or food-supply, or after an upsurge in their natural enemies.

The second factor of supposed importance in natural extinctions is climate change, often also associated with sea level change. Such processes could affect some species directly, and others indirectly through altering the extent and distribution of their habitats. Under climate change, some habitats become destroyed or fragmented, causing the loss of species, while other habitats, formerly fragmented, may expand and coalesce, bringing together incompatible species which until then had been kept apart. Climatic changes may thus promote alterations in the distributions and abundance levels of competitors, predators and parasites. In the past, organisms responded to marked climatic change mainly by movement (Chapter 15). Over the generations they spread in one direction and withdrew from another, so that their geographical ranges shifted slowly over the earth's surface, so as to remain within acceptable climatic limits. But species that were trapped in some way, through living on islands or other restricted areas, or through having poor dispersal powers, were left behind. If they were unable to adapt rapidly enough by evolutionary change, extinction was then their likeliest fate. Because most animals were tied to particular plant communities as habitats, the rate at which most animals could move their distribution during a period of climatic change was probably limited by the slower rate at which essential plant species could move.

Thirdly, there are sudden catastrophic changes in the environment caused by rare natural events, such as astrophysical phenomena or volcanic eruptions. In contrast to other mechanisms, which probably promote continuous slow rates of extinction, natural catastrophes could lead to the simultaneous loss of large numbers of species – in theory within a matter of hours, days or years. The most spectacular mass extinction process yet revealed, namely the sudden loss of the large dinosaurs at the end of the Cretaceous period 65 million years ago, has been attributed to a sudden global climatic change (of reduced temperatures and light values), caused by the dust cloud that followed the impact of a large asteroid. It is of course impossible to prove this hypothesis, but evidence is mounting, with extinctions of a large proportion of the plants and animals living at the time. In recent times, on a smaller scale, volcanic eruptions are known to have eliminated at least two types of bird, an unnamed Kermadec megapode in 1876 and the San Benedicto Rock Wren *Salpinctes obsoletus exsul* in 1952, together with many other organisms found on the same islands (Fisher & Orenstein 1985).

These three extinction processes – of species-interaction, climate change and natural disaster – have probably differed in importance between different types of organisms, and at different stages of evolutionary history. To these natural processes, however, the impact of humans has become increasingly added, by 'overhunting', by habitat destruction and by transporting organisms from one part of the world to another.

HUMAN-CAUSED EXTINCTIONS

Prehistoric extinctions

From prehistoric times, modern humans seem to have caused extinctions of birds and mammals, many of which are now known only from their remains or from rock drawings. The evidence for human involvement is entirely circumstantial: such species were known to have been commonly eaten (from midden remains) and in one region after another the synchronous disappearance of many large mammals and birds (the 'megafauna') followed human colonisation, these linked events occurring at markedly different dates in different parts of the world (Martin & Wright 1967, Martin 1984, Diamond 1989, 1991). In some regions, such as North America, climate changed at the same time, 11 000–12 000 years ago, giving an alternative explanation for large-scale extinctions[1], but in other regions, no other

---

[1]*In North America, the wave of extinctions began about 11 000 years ago, roughly coincident with the post-glacial colonisation of the continent by people spreading across a Siberian land bridge. Within a relatively short time, extinction had occurred of at least 27 genera comprising 56 species of large mammals (i.e. larger than 44 kg), two genera comprising 21 species of smaller mammals, and several large birds. Some notable losses included: ten species of horses Equus, the Giant Ground Sloth Gryptotherium listai, four species of camels Camelidae, two species of Bison, a wild cow in the genus Bos, the Saiga Antelope Saiga tatarica (which survived in Asia), four species in the elephant family Proboscidae, including the mastodon Mammut americanum and Mammoth Mammuthus primigenius, the sabretooth tiger Smilodon californicus, the American Lion Pantheron leo and the huge scavenging raptor Terratornis merriami (Martin & Klein 1984). For most of these species an annual kill exceeding 10% is likely to have been unsustainable (Owen-Smith 1987). The large predators may have been killed directly, or died out following the loss of their prey. The climatic explanation for these extinctions is unlikely, because they coincided with the post-glacial period when climate was warming and habitats were expanding as ice retreated. Extinctions progressed from the north into South America, as humans spread southward. Mass extinctions elsewhere also coincided with the first arrivals of aboriginal hunters. In Australia and New Guinea, a wave of extinctions occurred about 50 000 years ago that involved many species of large marsupials, tortoises and large flightless birds (Diamond 1982, Martin & Klein 1984).*

obvious change occurred, leaving human expansion as the only all-pervasive influence. The fact that a part of the megafauna survived in Africa, in the presence of an expanding human population, has been attributed to the fact than on this continent alone the fauna evolved in association with hominoids, providing ample time for the various species to develop effective defences (Owen-Smith 1987). The type of people, and their cultural traditions may also have been more favourable to large animals in Africa than elsewhere. In addition, people were excluded from large parts of Africa by the presence of tsetse flies, the vectors of trypanosomiasis, a disease fatal to humans but endemic in wild ungulates. This disease thus permitted pre-human ecosystems to persist in Africa until the arrival of European people in recent centuries set in train changes that had a much greater destructive impact. Within the last 200 years, most large animals have gone from most of the continent, and a few (such as the Quagga *Equus quagga*) have been totally extinguished.

For the last 1500 years, the link between human colonisation and extinction can be reconstructed in more detail, as people reached remote islands. They colonised Madagascar from Indonesia around 500 AD, and in the ensuing centuries the megafauna gradually disappeared (Martin & Klein 1984). It included 6–12 species of elephant birds (Aepyornithidae), flightless creatures that included the heaviest birds that have ever lived. One species, *Aepyornis maximus*, stood almost three metres tall, and probably weighed around 450 kg when full grown. Its soccer-ball eggs, each equivalent in content to 180 chicken eggs, can still be pieced together from fragments found at archaeological sites. It survived in diminishing numbers well beyond 1600 AD. Other animals erased from Madagascar at the same time included large lemurs and tortoises.

Before the arrival of people around 1000 AD, New Zealand was home to moas (Dinornithidae), large flightless birds unique to these islands. The first Maoris, arriving from the north, found at least 11 species, ranging in size from large turkeys to a 3-m giant (*Dinornis maximus*) of 270 kg (A. Anderson 1990, Cooper *et al.* 1993). Different species of moas lived in forest or open country, and occupied herbivore niches filled elsewhere by medium-sized and large mammals, which were absent from New Zealand. To judge from remains, moas formed a major food of the early Maori colonists, who had effectively eliminated them by 1700 AD. Several Europeans claimed to have seen moas in the 1800s, but their statements were unsubstantiated. The moas and other large birds were preyed upon by a large eagle *Harpagornis moorei*, which would have weighed around 13 kg, nearly twice the weight of any living eagle. Its extinction paralleled the loss of its main prey. Twenty-four other landbirds, including at least nine other flightless forms (notably a goose *Cnemiornis calcitrans* and a giant coot *Fulica chathamensis*), disappeared at the same time. Extinctions progressed in a wave from North Island to South Island as the Maoris spread south. The British settlers from the 1840s caused at least ten further extinctions of smaller bird species, so that at least 40% (45 species) of all New Zealand birds have become extinct since the arrival of humans (Atkinson & Cameron 1993).

In recent years, evidence has emerged from fossil remains of massive destruction of Pacific Island landbirds by the first human colonists, long before the coming of Europeans (Olson & James 1982, 1991, Steadman & Olson 1985, Steadman 1989, 1991, 1995). Relevant deposits were mostly laid down from 8000 years ago to nearly the present, covering the arrival of Polynesian people, 4000–1000 years ago, depending on location. They suggest that in the outer Pacific, in particular from Tonga in the west to Hawaii in the east, the Polynesians extinguished a large proportion of the native bird species found on their arrival (Pimm *et al.* 1994). These people ate many birds, including doves, crakes, rails and starlings, which had never before been exposed to a large predator. Such birds were, therefore, likely to have been tame, and as easily approached as many of the species that still live on the Galapagos Islands, in Antarctica and other places where human presence is relatively recent.

Hawaii may have sustained the greatest losses. When European settlers arrived after 1778, there were about 50 species of native landbirds, of which about 19 have since disappeared (Pimm *et al.* 1994). Yet bone deposits reveal that at least another 35 species identified with certainty, and up to 20 others, had already been eliminated by the Polynesians. These birds included a large eagle, at least seven flying and flightless geese, two flightless ibises, seven flightless rails, three owls with short wings and long legs, and some large flightless birds derived from ducks that had massive legs and beaks (Olson & James 1982). The surviving Hawaiian birds are mostly small inconspicuous species, depleted further after the arrival of European people and now mostly restricted to mountain forests. Twelve are unlikely to survive much longer without special management (Pimm *et al.* 1994).

On a world scale, more than 200 extinct island bird species have been described from fossil remains that post-date human colonisation (Milberg & Tyrberg 1993). In addition, more than 160 other fossils refer to living species (mainly petrels and pigeons) on islands where they no longer occur. Several species, considered today as endemic to single islands or island groups, had much wider distributions in the past. To give one example, from a single archaeological site on Ua Huka in the Marquesas Islands in the Pacific, bones of 36 species were identified. Of 20 seabird species, eight no longer occur on the island; they include Abbot's Booby *Papasula abbotti* which is now found only on Christmas Island in the Indian Ocean, about 11 200 km away. Of 16 landbird species, 13 no longer occur on Huka, including eight that are now totally extinct (Steadman 1991). So not only were many island bird species eliminated completely by Polynesian settlers, but many others were also greatly restricted in their distributions. Moreover, on each group of islands, extinctions again closely followed human settlement as revealed by archaeological evidence. Yet on islands that were not colonised by prehistoric people (such as the Galapagos), subfossil remains have failed to reveal any extinct species from the Holocene period that pre-date the arrival of Europeans (Milberg & Tyrberg 1993). So much for the myth that primitive people lived in a state of 'ecological balance', without appreciable impact on their environment. Over-exploitation of local resources was clearly commonplace and the disappearance of people from several islands could have occurred as a direct result of self-inflicted food-shortages (Milberg & Tyrberg 1993).

Estimates of extinction rates on Pacific Islands are increasing all the time, as new remains are discovered. One recent estimate suggests that, on average, at least 50% of the original bird species may have been eliminated by the early Polynesian settlers, who not only hunted the local birds, but also introduced rats and pigs, and destroyed much of the lowland forest for agriculture (Pimm *et al.* 1994). In general, islands occupied by people for the longest periods have lost the greatest proportion of species. On some islands, all the original species have gone. Single island endemics predominate among the species that were lost, presumably because they were more vulnerable than species that occurred on several islands. In the last 200 years, many other species have been introduced from elsewhere, hastening the demise of some of the remaining endemics (Chapters 9, 10 and 12).

Before the Polynesian occupation, nearly every island group across the Pacific was probably home to one or more species of flightless rails. Today such species survive on only three or four islands, including New Zealand. In fact, through recent excavations, Steadman (1995) estimated that 1–4 endemic rails could have lived on each of approximately 800 larger Pacific islands, plus other unique species, all of which, totalling in the thousands, disappeared after human colonisation. If these estimates prove valid, the rail family (Rallidae), which now holds about 144 species, could once have been one of the most species-rich bird families on earth.

Historical extinctions

Of the birds that have gone within the span of ornithological history, the Dodo *Raphus cucullatus* of Mauritius, last recorded alive in 1662, is perhaps the best known example. The Passenger Pigeon *Ectopistes migratorius* of eastern North America, which declined from hundreds of millions to extinction in less than a century, is perhaps the most spectacular (the last bird, called 'Martha', died in Cincinnati Zoo in 1914). Other well-known vanished birds include the flightless Great Auk *Pinguinus impennis* of the North Atlantic (last seen in 1844), the Labrador Duck *Camptorhynchus labradorius*, which was known only from its winter range and disappeared before its breeding range could be discovered (last seen 1875); the Carolina Parakeet *Conuropis carolinensis*, the only parrot native to eastern North America (last seen 1914); and the Huia *Heteralocha acutirostris* of New Zealand (last seen 1907), famous because the sexes differed remarkably in bill-shape.

In all, the 127 named species known to have died out since 1600 include representatives of 26 non-passerine and 15 passerine families **(Table 16.1)**. Various parrots and rails are especially well represented, at least 11 of the latter being flightless. For most of these species, tangible remains have been preserved, but a few are known only from written descriptions or drawings. Almost certainly the list is incomplete, for other species may have disappeared in the last 400 years before they could be recorded, and others not seen for many years may now be extinct too. Also, estimates of species numbers depend on taxonomy, and some extinct birds, currently classed as single species, might, if they were alive today, be classed as more than one. In addition to full species, at least 60 subspecies and discrete populations have disappeared since 1600.

On the other hand, a few species currently listed as extinct may well prove still to survive. In the past, several species were declared gone, only to be resurrected by new explorations in little known sites. Well-known examples include the New Zealand Takahe *Porphyrio mantelli*, a large flightless rail unrecorded between 1898 and 1948 when it was rediscovered; and the Cahow or Bermuda Petrel *Pterodroma cahow* which went unnoticed from its populous days of the 17th century until its rediscovery as a tiny relict population in 1916 (Fisher & Orenstein 1985). Other species have not been seen for long periods because their haunts have seldom been visited by ornithologists. As an example, the Fiji Petrel *Pterodroma macgillivrayi* was formerly known from only one specimen collected in 1855, and was not seen again until 1984, when another individual was caught on the same island (Watling & Lewanavanua 1985). In addition, some other species now exist only in captivity, or

Table 16.1   Bird species known to have become extinct since 1600. Modified from Greenway (1967), King (1980), Fisher & Orenstein (1985), updated by more recent information from Birdlife International, provided by C.J. Bibby and A. Stattersfield.

Species (with approximate date of extinction)	Range in historical times
Aepyornithidae	
1. Great Elephant Bird	
Aepyornis maximus[1] F (c. 1649)	Madagascar
Dromaiidae	
2. Kangaroo Island Black Emu	
Dromaius minor (diemenianus), S, F (1803–36)	Kangaroo Island, Australia
Dinornithidae	
3. South Island Tokoweka	
Megalapteryx didinus[1] F (1785)	South Island, New Zealand
4. Burly Lesser Moa	
Euryapteryx geranoides (gravis)[1] F (1640)	New Zealand
5. Brawny Great Moa *Dinornis torosus*[1] F (1670)	South Island, New Zealand
Podicipedidae	
6. Atitlán Grebe *Podilymbus gigas* (1987)	Guatamala
7. Colombian Grebe *Podiceps andinus* (1977)	Colombia
Hydrobatidae	
8. Guadalupe Storm Petrel	
Oceanodroma macrodactyla (1912)	Guadalupe Island, Mexico
Phalacrocoracidae	
9. Pallas' Cormorant	
Phalacrocorax perspicillatus (c. 1852)	Bering Islands, Russia
Ardeidae	
10. Reunion Flightless Ibis	
Borbonibis latipes, S, F (1773)	Réunion
11. Mauritius Night Heron	
Nycticorax mauritianus, S (1700)	Mauritius
12. Rodriguez Night Heron	
Nycticorax (?) megacephalus, S, F? (c. 1761)	Rodriguez, Mascarene Islands
13. Dwarf Olive Ibis *Bostrychia bocagei*, S (1928)	São Tome e Principe
Anatidae	
14. Chatham Island Swan	
Cygnus sumnerensis[1] (c. 1590–1690)	Chatham Island, New Zealand
15. Mauritian Shelduck	
Alopochen mauritianus (1698)	Mauritius
16. Mauritian Duck *Anas theodori* (1696)	Mauritius
17. Pink-headed Duck	
Rhodonessa caryophyllacea (1944)	Indian Subregion
18. Labrador Duck	
Camptorhynchus labradorius (1875)	Eastern North America
19. Aukland Merganser *Mergus australis* (1905)	Auckland Island, New Zealand
Falconidae	
20. Guadalupe Caracara *Polyborus lutosus* (1900)	Guadalupe Island, Mexico
Megopodidae	
21. Sula Scrubfowl *Megapodius bernsteinii* (1938)	Indonesia

Table 16.1   continued.

Species (with approximate date of extinction)	Range in historical times
Phasianidae	
22. New Zealand Quail	
Coturnix (n.) novaezelandiae (1875)	New Zealand
23. Himalayan Mountain Quail	
Ophrysia superciliosa (1868)	Himalaya Mountains
24. Double-banded Argus	
Argusianus bipunctatus (1871)	Java
Rallidae	
25. Wake Island Rail *Gallirallus wakensis* F (1945)	Wake Island, Pacific
26. Dieffenbach's Rail	
G. (philippensis) dieffenbachii (c. 1842)	Chatham Island, New Zealand
27. Chatham Rail *G. modestus* F (c. 1900)	Pitt and Mangere Islands, New Zealand
28. New Caledonian Rail	
Tricholimnas lafresnayanus F (1904)	New Caledonia
29. Barred-wing Rail	
Nesoclopeus poecilopterus (1973)	Ovalau, Fiji & Viti Levu
30. Woodford's Rail	
Nesoclopeus woodfordi (1936)	Solomon Islands
31. Ascension Flightless Crake	
Atlantisia elpenor[1] F (1656)	Ascension Island
32. Laysan Crake *Porzana palmeri* F (1944)	Laysan, Hawaiian Islands
33. Hawaiian Crake *P. sandwichensis* F (1884)	Hawaii
34. Kosrae Crake *P. monasa* F (1828)	Kosrae, Caroline Islands
35. Sooty Crake *P. nigra* (date?)	Tahiti
36. Red Rail *Aphanapteryx bonasia*[1] F (1693)	Mauritius, Mascarene Islands
37. Rodriguez Rail *A. leguati*[1] F (1761)	Rodriguez, Mascarene Islands
38. Lord Howe Swamphen	
Porphyrio albus F? (c.1834)	Lord Howe Island, Tasman Sea
39. Gilbert Rail *Gallirallus conditicus* (1928)	Kiribati
40. Red-billed Rail *Gallirallus pacificus* (1773)	Tahiti
41. Sharpe's Rail *Gallirallus sharpei* (1895)	Sunda, Indonesia
42. Tristan Moorhen *Gallinula nesiotis* (1900)	Tristan da Cunha
43. Samoan Moorhen *G. pacifica* F (1908)	Savai'i, Western Samoa
44. Mascarene Coot *Fulica newtoni*[1] (1693)	Mauritius (and Réunion?), Mascarene Islands
Haematopodidae	
45. Canarian Black Oystercatcher	
Haematopus meadewaldoi (1913)	Canary Islands
Charadriidae	
46. Javanese Wattled Lapwing	
Vanellus macropterus (1940)	Java
Scolopacidae	
47. Tahiti Sandpiper *Prosobonia leucoptera* (1773)	Tahiti and Moorea, Society Islands
Alcidae	
48. Great Auk *Pinguinus (Alca) impennis* F (1844)	North Atlantic islands
Columbidae	
49. Red-moustached Fruit-dove	
Ptilinopus mercierii (1922)	Marquesas Islands

Table 16.1   continued.

Species (with approximate date of extinction)	Range in historical times
50. Mauritius Blue Pigeon *Alectroenas nitidissima* (1835)	Mauritius, Mascarene Islands
51. Rodriguez Pigeon *A. rodericana*[1] (1726)	Rodriguez, Mascarene Islands
52. Bonin Wood-pigeon *Columba versicolor* (1829)	Nakondo Shima and Chichi Shima, Bonin Islands
53. Ryukyu Pigeon *C. jouyi* (1936)	Ryukyu Island, Japan
54. Passenger Pigeon *Ectopistes migratorius* (1914)	Eastern North America
55. Tanna Ground Dove *Gallicolumba ferruginea* (1744)	Vanuatu
56. Choiseul Pigeon *Microgoura meeki* (1904)	Choiseul, Solomon Islands

Raphidae

57. Dodo *Raphus cucullatus* F (1662)	Mauritius, Mascarene Islands
58. Réunion Solitaire *Raphus solitarius* (1715)	Réunion

Pezophapidae

59. Rodriguez Solitaire *Pezophaps solitaria*[1] F (1791)	Rodriguez, Mascarene Islands

Psittacidae

60. New Caledonian Lorikeet *Charmosyna diadema* (1900)	New Caledonia
61. Norfolk Kaka *Nestor (meridionalis) productus* (c.1851)	Norfolk and Philip Islands, Tasman Sea
62. Paradise Parrot *Psephotus pulcherrimus* (1927)	Eastern Australia
63. Black-fronted Parakeet *Cyanoramphus zealandicus* (1844)	Tahiti, Society Islands
64. Society Parakeet *C. ulietanus* (c. 1773)	Raiatea, Society Islands
65. Mauritius Parrot *Lophopsittacus mauritianus*[1] F (1638)	Mauritius, Mascarene Islands
66. Rodriguez Parrot *Necropsittacus rodericanus*[1] (1731)	Rodriguez, Mascarene Islands
67. Mascarene Parrot *Mascarinus mascarinus* (c.1834)	Réunion (and Mauritius?), Mascarene Islands
68. Seychelles Parakeet *Psittacula (eupatria) wardi* (1881)	Mahe and Silhouette, Seychelles
69. Newton's Parakeet *P. (krameri) exsul* (1875)	Rodriguez, Mascarene Islands
70. Glaucous Macaw *Anodorhynchus glaucus* (1915)	Brazil, Paraguay, Argentina
71. Carolina Parakeet *Conuropsis carolinensis* (1914; 1938?)	Eastern North America
72. Cuban Macaw *Ara cubensis* (1864)	Cuba
73. Hispaniolan Macaw *Ara tricolor* (1820)	Hispaniola

Cuculidae

74. Snail-eating Coua *Coua delalandei* (c.1834–37; 1930?)	Madagascar

Strigidae

75. Laughing Owl *Sceloglaux albifacies* (1914)	New Zealand
76. Rodriguez Little Owl *Athene murivora*[1] (1726)	Rodriguez, Mascarene Islands
77. Spotted Forest Owlet *Athene blewitti* (1914)	India

Table 16.1   continued.

Species (with approximate date of extinction)	Range in historical times
Caprimulgidae	
78. Jamaican Poorwill	
Siphonorhis americanus (1859)	Jamaica
79. New Caledonian Owlet-nightjar	
Aegotheles savesi (1880)	New Caledonia
Trochilidae	
80. Brace's Emerald *Chlorostilbon bracei* (1877)	Bahamas
Alcedinidae	
81. Ryukyu Kingfisher *Halcyon miyakoensis* (1887)	Miyako Shima, Ryukyu Islands
Picidae	
82. Imperial Woodpecker	
Campephilus imperialis (1958)	Mexico
83. Ivory-billed Woodpecker	
Campephilus principalis	USA and Cuba
Acanthisittidae	
84. Stephen Island Wren *Xenicus lyalli* F? (1874)	Stephen Island, New Zealand
85. Bush Wren *Xenicus longipes* (1972)	New Zealand
Turdidae	
86. Grand Cayman Thrush *Turdus ravidus* (1938)	Grand Cayman, West Indies
87. Bonin Thrush *Zoothera terrestris* (1828)	Chichi Shima, Bonin Islands
Sylviidae	
88. Amaui *Myadestes oahensis* (1825)	Oahu, Hawaiian Islands
89. Aldabra Warbler *Nesillas aldabranus* (1983)	Aldabra
90. Chatham Fernbird *Megalurus rufescens*, S (1920)	Chatham Islands
91. Nihoa Reed Warbler	
Acrocephalus familiaris, S (1923)	Laysan, Hawaiian Islands
Muscicapidae	
92. Lord Howe Gerygone	
Gerygone insularis (date?)	Lord Howe Island
93. Cerulean Paradise-flycatcher	
Eutrichomyias rowleyi (1978)	Sangihe, Indonesia
94. Tahiti Monarch *Pomarea nigra* (1985)	Tahiti
95. Guam Flycatcher *Myiagra freycineti* (1983)	Guam
Dicaeidae	
96. Cebu Flowerpecker *Dicaeum quadricolor* (1906)	Cebu, Philippines
Zosteropidae	
97. Robust White-eye *Zosterops strenua* (1918)	Lord Howe Island, Tasman Sea
98. Chesnut-sided White-eye	
Zosterops mayottensis, S (1940)	Marianne, Seychelles
99. White-chested White-eye	
Zosterops albogularis (1962)	Norfolk Island
Meliphagidae	
100. Hawaii Oo *Moho nobilis* (1898)	Hawaii
101. Oahu Oo *M. apicalis* (1837)	Oahu, Hawaiian Islands
102. Kioea *Chaetoptila angustipluma* (1859)	Hawaii

Table 16.1   continued.

Species (with approximate date of extinction)	Range in historical times
Callaeidae	
103. Huia *Heteralocha acutirostris* (1907)	North Island, New Zealand
Turnagridae	
104. Piopio *Turnagra capensis* (1955)	New Zealand
Sturnidae	
105. Pohnpei Starling *Aplonis pelzelni* (1956)	Pohnpei, Caroline Islands
106. Kosrae Mountain Starling *A. corvina* (1828)	Kosrae, Caroline Islands
107. Norfolk Starling *A. fuscus* (1923)	Lord Howe and Norfolk Islands, Tasman Sea
108. Mysterious Starling *A. mavornata* (c.1774)	Raiatea, Society Islands
109. Rodriguez Starling *Nerospar rodericanus* (1879)	Rodriguez and Ile-au-Mat, Mascarene Islands
110. Réunion Starling *Fregilupus varius* (1860)	Réunion, Mascarene Islands
Ploceidae	
111. São-Tome Canary *Neospiza concolor* (1888)	São Tome, Gulf of Guinea
Parulidae	
112. Bachman's Warbler *Vermivora bachmanii* (1977)	USA – Cuba
113. Semper's Warbler *Leucopeza semperi* (1972)	St Lucia
Fringillidae	
114. Bonin Grosbeak *Chaunoproctus ferreorostris* (1890)	Chichi Shima, Bonin Islands
115. Townsend's Finch *Spiza townsendi* (1833)	USA
Drepanididae	
116. Molokai Creeper *Paroreomyza flammea* (1964)	Molokai, Hawaiian Islands
117. Greater Amakihi *Hemignathus (Loxops) sagittirostris* (1900)	Hawaii
118. Akialoa *H. obscurus* (1965)	Hawaii
119. Kauai Akialoa *H. procerus* (c. 1968–69)	Kauai, Hawaii
120. Lanai Hookbill *Dysmorodrepanis munroi* (1913)	Hawaii
121. Greater Koa-finch *Rhodacanthis palmeri* (1896)	Hawaii
122. Lesser Koa-finch *R. flaviceps* (1891)	Hawaii
123. Kona Grosbeak *Chloridops kona* (1894)	Hawaii
124. Ula-ai-hawane *Ciridops anna* (1892)	Hawaii
125. Hawaii Mamo *Drepanis pacifica* (1898)	Hawaii
126. Black Mamo *D. funerea* (1907)	Molokai, Hawaiian Islands
Icteridae	
127. Slender-billed Grackle *Quiscalus palustris* (1910)	Mexico

[1]Known from subfossil remains, or from subfossil remains and early descriptions.
F flightless.
S may have been better classed as a subspecies, rather than a full species.

In addition to the above list six other species have gone from the wild and currently exist only in captivity or, additionally, with small experimentally released populations: Alagoas Curassow *Mitu mitu*, California Condor *Gymnogyps californianus*, Guam Rail *Rallus owstoni*, Guam Micronesian Kingfisher *Todirhamphus cinnamoninus*, Socorro Dove *Zenaida graysoni*, Spix Macaw *Cyanopsitta spixii*.

additionally with small, experimentally-released populations: California Condor *Gymnogyps californianus*, Alagoa's Curassow *Mitu mitu*, Guam Rail *Gallirallus owstoni*, Guam Micronesian Kingfisher *Todiramphus cinnamominus*, Socorro Dove *Zenaida graysoni* and Spix's Macaw *Cyanopsetta spixii*.

Most of the species lost in the last 400 years were again island endemics, and included a large proportion of species confined to a single small island. However, at least 11 species have gone from larger land areas, including three from Australia, three from Eurasia (India), five from North America and three from Central and South America. They mainly include species confined to small areas (including two grebes found on single large lakes), but also some once widespread species, notably Passenger Pigeon and Carolina Parakeet.[2]

Although the mean rate of extinction in the last 400 years is about one species per four years, this rate increased steadily from about one per decade in the early 17th century to one per two years since 1800. The worst period was the late 19th and early 20th centuries, when many oceanic islands were exploited and altered in various ways **(Table 16.2)**.

If we assume, as a conservative estimate, that half of all island bird species have been eliminated by human activity, there must once have been about 3500 species on islands, instead of the current 1750 (Pimm *et al.* 1994). If we then allow for a number of continental extinctions, known and unknown, at least one-fifth of all bird species worldwide have probably been eliminated by human activity, mostly in the last 1000 years. In the absence of human intervention, then, around 11 500 bird species would live in the world today, instead of the current 9700. Considering that estimates are increasing all the time, as more subfossil material becomes available and that the remains of many small birds will never be found or correctly identified, this remains very much a minimum figure. Moreover, any estimate is of course based on avian taxonomy as it is now, and like any other current estimates, they may be changed as a result of future taxonomic revision, including any resulting from the use of modern molecular taxonomy.

Inequalities in species

In terms of the loss they represent to biodiversity, not all species are equal. Some families of birds have radiated widely around the world, so that many species are closely related and look much the same. Such species richness, coupled with wide distribution, greatly reduces the probability that the whole lineage will become extinct. Other lineages have far fewer members so should perhaps be valued more highly. One of the most extreme examples among living birds is the flightless Kagu

---

[2]*The full list of extinct continental birds and years of last sightings is as follows: North America: Townsend's Finch Spiza townsendi (1833), Cooper's Sandpiper Calidris cooperi (1833), Labrador Duck Camptorhynchus labradorius (1878), Passenger Pigeon Ectopistes migratorius (1914), Carolina Parakeet Conuropsis carolinensis (1914); South & Central America: Glaucous Macaw Anodorhynchus glaucus (1915), Colombian Grebe Podiceps andinus (1977), Atitlán Grebe Podilymbus gigas (1987); Eurasia (Indian region): Himalayan Mountain Quail Ophrysia superciliosa (1868), Forest Spotted Owlet Athene blewitti (1914), Pink-headed Duck Rhodonessa caryophyllacea (1944). The first two species are known only from single specimens collected in New York State in 1833 (King 1980).*

Table 16.2 Chronology of known extinctions of bird species since 1600 AD. From Table 16.1.

	1601–1650	1651–1700	1701–1750	1751–1800	1800–1850	1851–1900	1901–1950	1951–present	Totals
Continental	0	0	0	0	2	2	5	2	11
Island	4	8	5	8	14	31	30	15	115
Totals	4	8	5	8	16	33	35	17	126

Dates were taken from the last dates each species was seen alive; unknown for one species.

Rhynochetos jubatus of New Caledonia, estimated at about 650 individuals (Hunt 1996). This strange bird is the sole member of its family, a degree of evolutionary separation it shares with only 11 other living bird species. For all we know, the Kagu may be the last survivor of a once large family, but in any case its loss would represent the end of a unique evolutionary line. In contrast, 993 species (10% of all birds) belong to the family Fringillidae, which includes finches, buntings and tanagers. Compared to the extinction of the Kagu, the loss of one species of finch from the Fringillidae would clearly be a much less significant deletion from the variety of life on earth (Bibby *et al.* 1992).

On this line of reasoning, the loss of the giant moas and elephant birds, only a few centuries ago, is especially significant because they represented extremely distinct life-forms, in some respects pinnacles of extraordinary evolutionary lines, and quite unlike anything alive today. Three other families that have been lost totally in recent times include the Dodo of Mauritius (Raphidae), the Solitaire of Réunion (Pezophapidae), again peculiar creatures unlike any modern bird, and the much more ordinary New Zealand thrush (Turnagridae). Some other surviving families have lost heavily in representation, notably the waterfowl, parrots, rails, pigeons, starlings and Hawaiian honeycreepers (on the most recent classification included in the Fringillidae, Sibley & Monroe 1990). Moreover, it is often the most extraordinary species that have gone. About 20 other extinct birds were the sole members of their genera, and the remaining 107 species came from genera still represented among living birds.

In another respect not all species extinctions are equal. Among living organisms, certain species have major effects on other organisms in the community. Think of the role of a dominant tree species that forms the structured forests on which many other organisms depend, or of a major herbivore or carnivore which through its feeding habits can affect the whole community. Among birds, the moas and the elephant birds, in their positions as major herbivores, could have played such a role. The loss of such keystone species could lead to many further extinctions and is therefore of much greater biological significance than the loss of (say) some small finch. At the same time, however, the demise of any species – however insignificant in ecosystem function – is likely to trigger the extinction of further species, if only its specific parasites. When the last Passenger Pigeon died, for example, so did the last of its own special lice *Columbicola extincta* and *Campanulotes defectus*.

The particular vulnerability of island species

Because of their isolation, islands provide centres of evolution, which contain far more unique species in proportion to their area than do continents. Speciation is supposedly facilitated by small population sizes and lack of immigration. Although islands make up only about 5% of the earth's land area, about 17% (1750) of all surviving bird species live on islands, and more than 90% of all extinctions recorded since 1600 AD refer to island birds **(Table 16.2)**. To put it another way, island species have proved about 40 times more vulnerable to extinction in the last 400 years than continental ones. The particular vulnerability of island endemics is attributable mainly to their small geographical ranges (making them vulnerable to local destructive events), and associated small population sizes (enabling them to lose fewer individuals before dying out). As species numbers are generally related to habitat area, the destruction of part of the forest or other habitat on a small island would be expected to cause some extinctions, even though some pristine habitat remained (Diamond 1972, 1975, Case 1996). The presence of introduced species in secondary vegetation may prevent the native ones from adapting to it. Interestingly, large and little modified islands, such as New Guinea, are not known to have lost any species in recent centuries, so offer a useful comparison (although they have also been less explored).

On many islands, species have lived in long isolation from predators and pathogens, and so have reduced natural defences. Several island species still present in 1800 were flightless, nested on the ground or in other accessible sites, and showed no fear or other defences against humans or against introduced rats, cats and other predators. Little wonder that, after the arrival of rats and cats, nearly 20 flightless island species, including 11 of the 18 known remaining flightless rails, rapidly disappeared. Some of the most recent extinctions among island birds were on Guam, where the introduction in the 1940s of a predatory tree snake obliterated most of the forest avifauna in less than 30 years (Chapter 9). Of 11 native forest bird species extant at the time of the snake's introduction, seven are now extinct and four are critically endangered (Savidge 1987, Haig & Ballou 1995). Another factor of importance on some islands, notably New Zealand and Madagascar, was the large size of some endemic birds. Gigantism not only made such birds attractive as human food, but was probably also associated with life-history features (long immaturity and low reproductive rates) that made them ill-equipped to withstand high predation.

Comparing different islands from around the world, the proportion of extinct species among the known avifauna is inversely related to island area, with relatively more extinctions on the smaller islands **(Figure 16.1)**. Allowing for island area, however, avifaunas with high levels of endemism have lost the greatest proportions of species, and on the same island, endemics have more often died out than native non-endemic forms (Case 1996). In New Zealand, for example, some 37% of endemic bird species are now extinct, compared with only 6% of native non-endemics. There are at least two likely reasons for this difference. Not only were endemics likely to have been on the island longer than non-endemics and lost their defences against predators and pathogens, but they also lacked recolonisation sources, so that local extirpation meant extinction. Some of them were also highly specialised.

Ultimate causes of extinctions

The three main ways in which people have caused avian extinctions are as follows.

Over-kill. Excessive hunting was probably the main means by which early humans eliminated many species and by which modern humans claimed their first casualties in historical times, namely the Dodo of Mauritius, together with several later ones, such as the Solitaire of Réunion, the Great Auk of the North Atlantic and the Pallas' Cormorant of the North Pacific, all flightless or nearly so. More recently, human predation has eliminated several other edible species, notably waterfowl. The most recent known extinction of an island bird by hunting was that of the Wake Island Rail *Rallus wakensis*, caught and eaten to the last individual by Japanese troops stationed there in 1945 (King 1980). Currently, the birds in most danger from over-harvesting

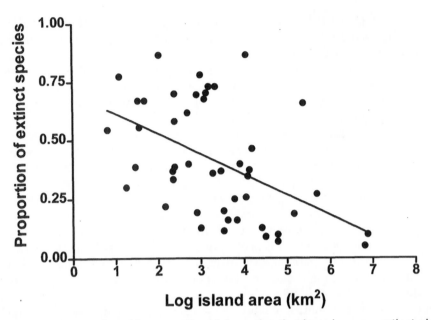

Figure 16.1   Proportion of known native bird species that have become extinct plotted (with a square root transformation) in relation to island area. Only islands with at least one extinct and one extant species are included. Restricting the analysis to islands for which pre-historic (pre 1600 AD) extinctions are also known improves the relationship with island area (not shown). From Case 1996.

include various waterfowl, pheasants, guams, doves and parrots, the latter taken for the pet trade.

Introduced species. Alien species can exterminate native ones through competition, predation, disease transmission, hybridisation or habitat destruction. Of nearly 1200 known introductions of birds, more than 70% have occurred on islands (Long 1981, Lever 1987). Hawaii and New Zealand have been particularly affected, with 47 and 34 introduced species, respectively, that are now well established. On Hawaii, introduced landbird species now outnumber native ones. Not only might these species have outcompeted and eliminated some native species, but being pre-adapted to man-made habitats, they might also have denied native species the opportunity to adjust to such habitats (Chapter 12).

   The spectacular effects of alien predators on island endemics, which evolved in the absence of predators, were discussed in Chapter 9. About half of all recent extinctions of island birds have been attributed to introduced mammals (especially rats and cats) on islands lacking native predators. Endemic birds on islands with either native rats or native land crabs (the invertebrate equivalent of rats) have proved to be virtually immune to introduced mammals (Atkinson 1985). In this category of extinctions is a small wren, found only on Stephen Island, New Zealand, which was notable in three respects. It may have had the smallest range of any bird, restricted to an island only 2.6 km^2 (one square mile) in area; it may have been the only flightless passerine; and thirdly it was both discovered and exterminated single-handedly by the same cat, belonging to a lighthouse keeper (Chapter 9). Because introduced diseases are difficult to establish as a cause of extinction, they may have accounted for the loss of more species than those suspected for Hawaii (Chapter 10). Overall, however, introduced species have probably caused more recent extinctions of island birds than any other single factor.

No bird species is known definitely to have been lost through hybridisation, However, inter-breeding with the Pied-billed Grebe *Podilymbus podiceps* may have contributed to the extinction of the Atitlán Grebe *P. gigas* in Guatemala (Cade & Temple 1995), and other species are currently threatened by hybridisation. For example, the Madagascar Little Grebe *Tachybaptus pelzelnii* and the Alaotra Grebe *T. rufolavatus* are both thought to be disappearing from Madagascar through hybridization with the Little Grebe *T. ruficollis* (Collar & Andrew 1988). Similarly, the Hawaiian Duck *Anas wyvilliana* is now restricted to three islands, but on two it occurs only in hybrid form, as a result of cross-breeding with introduced Mallard *Anas platyrhynchos*. This has not yet occurred to any extent on the third island, Kauai, but it probably will in time (Browne *et al.* 1993). In North America, as a result of human activities, the Mallard has expanded in range, and is now hybridising with the formerly isolated Black Duck *Anas rubripes* in the northeast, the Mexican Duck *Anas diazi* in the southwest, and the Mottled Duck *Anas fulvigula* in the southeast, though large pure populations of all species still remain. In addition, introduced Mallards have hybridised extensively with the native Grey Duck *Anas superciliosa* of New Zealand and with the Pacific Black Duck *Anas superciliosa rogersi* of Australia (Cade 1983). The Ruddy Duck *Oxyura jamaicensis*, introduced from North America to Britain in 1952, has obtained a foothold on the European mainland, where it threatens the continued existence of the native White-headed Duck *Oxyura leucocephala* through hybridisation (Hughes 1993). The western population of *O. leucocephala* is concentrated in Spain and, at the time of writing, numbers only about 800 individuals. Hybrids between the two species are fertile, at least to the third generation, as discovered in captive birds.

In New Zealand, the endemic Black Stilt *Himantopus novaezelandiae* survives as a small population (70 birds in 1993) in one part of South Island, and is inter-breeding with the Pied Stilt *Himantopus leucocephalus* which, after extensive deforestation, colonised New Zealand from Australia in the mid 19th century (Reed *et al.* 1993). Without management, the Black Stilt could soon disappear from the combined impact of predation and hybridisation. Another example involves the Blue-winged Warbler *Vermivora pinus* and Golden-winged Warbler *Vermivora chrysoptera* in North America, which were once allopatric but have come together through mutual range expansion following human-induced habitat changes. About 50 years after contact in any given locality, the Golden-winged and hybrid phenotypes disappear, leaving only the Blue-winged (Gill 1980, Confer & Knapp 1981). In South Africa, the introduced Mozambique Red-eyed Dove *Streptopelia s. semitorquata* has produced a hybrid form with the endemic subspecies *Streptopelia s. australis* (Brooke *et al.* 1986). At least three island subspecies have been completely lost, apparently by genetic swamping by introduced forms of the same species. They include the dove *Streptopelia picturata rostrata* of the Seychelles, whose distinct phenotype has been replaced by that of the introduced nominate race *Streptopelia p. picturata* from Madagascar (Long 1981). In all these examples, genetic material from the original form may survive, but only in combination with that from the invader, which the hybrid stock soon comes to resemble.

Habitat destruction. This has now become the leading cause of extinctions worldwide. Habitat may be completely destroyed, as when forest is felled or wetland is drained, or it may be progressively degraded over a period of years, as when natural swards are overgrazed, leading to gradual loss of vegetation and soil. Either process can lead both to reduction of total habitat area and to fragmentation of remaining habitat, as large tracts are converted piecemeal to secondary habitats. The stepwise process of deforestation, cultivation, overgrazing and soil erosion can be seen over much of the world, with the later stages especially apparent in the Mediterranean Region and the Middle East.

In recent years, attention has focused on tropical rainforest, which covers only 7% of the earth's surface, yet contains more than half of all plant and animal species (estimates range from 50 to 90%) (Wilson 1992). More than half this forest was destroyed in the 20 years from 1970, giving the fastest rate of vegetation change of this magnitude in human history. Current rates of deforestation (derived mainly from satellite images) are staggering, at 20 million hectares per year (equivalent to the area of England or Florida). According to Myers (1979), more than 80% of current tropical deforestation is by 'slash-and-burn' to create cropland, much of which remains productive for only a few years until the soil is degraded or lost by erosion. The felling of forest has already spelt doom for many island species unable to adapt to secondary vegetation. For example, on Cebu Island in the Philippines, which was almost totally deforested in the 19th century, 16 of the 26 forest species had gone by 1947, including eight endemic subspecies and one endemic species (Rabor 1959). As a more extreme example, Easter Island in the central Pacific was completely deforested by the original Polynesian colonists, leading to the loss of an entire avifauna long before any European could set eyes on it.

On many islands, it was not only the devastation of native vegetation by people that led to extinctions, but also its destruction by introduced herbivores, such as goats. For just as some island birds have little protection against introduced predators, some island plants have little defence against introduced herbivores. Rabbits *Oryctolagus cuniculus* released on Laysan in the Hawaiian Islands in 1903 destroyed almost all the island's vegetation in a 20-year period, during which time three of the five resident landbirds disappeared. Another factor on some islands has been competition from alien plants, which are now in the process of replacing the native species on which many local animals depend. Moreover, the deforestation of steep slopes often led to loss of topsoil which once provided sites for burrow-nesting seabirds, many of which are now much less widely distributed than they used to be.

*   *   *

The three main causes of extinctions listed above are seldom species-specific, but vary in the proportion of an avifauna they affect at one time. Extinction from persecution is normally confined to larger species of human interest, while introduced aliens (including predators) can affect anything from single species to whole communities, and habitat destruction almost always affects entire communities, depending on what replaces it. To these various extinctions may be added the linked ones, when species disappear as a result of the loss of others crucial to them. The main known examples are mostly non-avian (e.g. the loss of plants after the demise of their specific pollinators or the loss of parasites after the demise of their specific hosts), but these types of extinctions are probably greatly under-recorded.

Additional causes of extinctions may arise in future. For example, pesticides and pollutants have eliminated species from particular localities or even from whole regions (Chapter 15), so it is probably only a matter of time until they extinguish species entirely. The pesticide DDT came close to eliminating the Mauritius Kestrel *Falco punctatus*, having reduced it by 1974 to only four individuals, before effective conservation measures were taken (Jones *et al.* 1995). The same pesticide, used in mosquito control, is thought to have contributed to the extinction of a distinct sub-species, the Dusky Seaside Sparrow *Ammospiza maritima negrescens*, until 1987 found on the Atlantic coast of Florida, and an unidentified pesticide probably caused the elimination of the Bald Ibis *Geronticus eremita* from Turkey, leaving one small remaining population of about 200 birds in North Africa.

From figures presented by King (1980), some 32% of avian extinctions since 1600 involved habitat modification, 91% involved introduced species and 25% over-kill. As these figures suggest, some species may have succumbed from a combination of factors, acting together or successively. For continental birds, all extinctions so far have been attributed to overhunting, while extinctions from introduced organisms have been confined to island species.

Multiple causes. One of the best documented multi-cause extinctions concerns a subspecies, the Heath Hen *Tympanuchus c. cupido* of the northern United States (Bent 1932, Simberloff 1988). This bird was once common in sandy scrub-oak plains throughout the region, but by 1870 it had been eliminated almost everywhere by a combination of habitat destruction and overhunting, surviving only on Martha's Vineyard, an island off the Massachusetts coast. By 1908, about 50 individuals remained, for which a 670-ha refuge was created. Habitat was improved, and by 1915 the population had grown to about 2000. However, in 1916 a gale-driven fire killed many birds and destroyed their habitat. The next winter was unusually harsh, and marked by an invasion of Goshawks *Accipiter gentilis*. The Heath Hen population fell to 150, mostly males. In addition to the sex ratio imbalance, there was soon evidence of inbreeding depression. In 1920, a disease of poultry killed many birds, and by 1927 only 13 were left, including 11 males. The last died in 1932. This one distinct subspecies was thus affected in turn by almost all the factors known to have caused extinctions of birds, but the predisposing cause was its initial reduction and confinement through human action to a single small area.

For most species, the relative importance of different factors is often hard to assess in retrospect, especially when the factors causing the initial decline are different from those that deal the final blow. For the Passenger Pigeon *Ectopistes migratorius*, which may once have been the most numerous bird in North America, graphic eye-witness accounts describe the huge carnage that was inflicted on the spectacularly dense and extensive flocks, with many 'waggonloads of bodies' sent daily to the growing cities, but at the same time the settlers were felling the forests where the birds nested. These pigeons bred in huge colonies, on which their success may have depended, so that over-kill, forest fragmentation and social dysfunction may all have contributed to their final demise. Even when a species is disappearing before our eyes, unless from something as obvious as habitat destruction or over-kill, it is often hard to discern the cause. All that is recorded is the progressive disappearance of individuals until none is left.

SOME GENERAL POINTS

The long-term security of any organism depends largely on the extent of its geographical range, the number and extent of habitats it can occupy and its average density within those habitats, which together determine its overall numbers. The most numerous species are those that occupy a large geographical area and a wide variety of habitats, and at the same time can live at high density within those habitats. Even widespread and numerous species are not completely immune to rapid extinction, however, as the Passenger Pigeon showed. Conversely, the least numerous species are those that occupy small geographical areas, and a single restricted habitat, within which they can occur only at low density. Different combinations of these three variables give eight different classes of abundance, of which seven contain one or more elements of rarity (Rabinowitz 1981), italicised as follows:

(1) large range, wide habitat, high density,
(2) large range, wide habitat, *low density*,
(3) large range, *narrow habitat, low density*,
(4) large range, *narrow habitat*, high density,
(5) *small range*, wide habitat, high density,
(6) *small range, narrow habitat*, high density,
(7) *small range, narrow habitat, low density*, and
(8) *small range*, wide habitat, *low density*.

Characteristics of species themselves influence in which category they fall. In any well-known avifauna, each species can be placed in one or other of these eight classes, revealing at a glance their potential vulnerabilities. On numbers alone, those in the *small range/narrow habitat/low density* category are most at risk. Most avifaunas would naturally contain species in all eight categories, but the net effect of various human activities is to increase the proportions falling in the least secure categories.

We can make several other generalisations. Firstly, although geographical range and population abundance are two major factors influencing the long-term security of a species, in continental faunas these two variables are not wholly independent of one another (Lawton 1994, Newton 1997). In general, species with large geographical ranges tend also to have greater abundance at sites where they occur than do species with small ranges. This correlation holds only as a crude generalisation, with many exceptions, but it means that more species fall in the *small range/narrow habitat/low density* category than expected if the three attributes were completely independent. As Lawton (1994) put it, nature has 'double jeopardy' built into extinction risk.

Secondly, within bird families it is common for tropical species to occupy smaller geographical ranges and to occur at lower density than their high-latitude counterparts. On these features, then, tropical species should be most vulnerable. The fact that on continents fewer species have so far been lost from tropical than from temperate areas can be attributed to the greater human impact at temperate latitudes. Thirdly, in any species, populations in some parts of the range (especially edges) may not be self-sustaining, but depend on continual immigration from elsewhere. From this, it follows that destruction of populations in the better (usually central) parts of a range will do more to increase extinction risk than the destruction of marginal populations. Fourthly, species that are migratory, with separate breeding and wintering areas, are vulnerable to adverse changes at both ends of their journeys, as well as at staging posts en route. The spectacular decline of the Eskimo Curlew *Numenius borealis* from many millions to fewer than 50 individuals in the late 19th century has been attributed entirely to shooting on migration (Bent 1928). In contrast, the Great Auk *Pinguinus impennis* was eliminated entirely through persecution on its breeding sites and the Labrador Duck *Camptorhynchus labradorius* through persecution on its wintering sites (see above). In some species, current breeding and wintering areas are extremely restricted, as in the Whooping Crane *Grus americana* which now breeds only in a single National Park (Wood Buffalo, on the Alberta–Mackenzie border) and winters only in a single refuge (Arkansas, in Texas) (Fisher *et al.* 1969).

Fifthly, regardless of the extent of the geographical range, the way in which birds distribute themselves can also influence their risk of extinction. Some species of birds are regularly found in huge concentrations. This is true of some migrant waterfowl species, which gather in enormous numbers at particular wintering and staging areas. For example, the 50 000–70 000 Surfbirds *Aphriza virgata* that amass each spring in Prince William Sound are thought to include most of the North American population. The tendency of some species to concentrate has been accentuated in recent decades by the establishment of refuges in regions in which most of the remaining habitat has been destroyed. Such concentrations increase the risks, for when disaster strikes, it can remove at one time a large portion of a population (see Chapter 10 for examples). In conservation terms, large populations are more secure than small ones (see later), but even more secure if they are distributed between several sites rather than concentrated in one.

Finally, other things being equal, large birds are likely to be more extinction-prone than small ones. Not only are large birds more attractive as human food, but in more developed regions they are often perceived as posing greater threats to human interests. Moreover, they tend to live at lower densities than small birds and have more *K*-selected characters, with more extended lifespans and lower natural mortality and reproductive rates (*K*-strategists). This means that they are unable to withstand much extra mortality, and, once reduced in numbers, they take many years to recover. They are thus ill-equipped to cope with any but the lightest forms of human exploitation. Large birds-of-prey are among the most extreme examples, for their carnivorous feeding habits ensure that they live at densities much lower than similar-sized omnivores or herbivores; they are often killed because of depredations on game and livestock, or (by some human societies) for their feathers. Yet they do not breed until several years old, and their low reproductive rates – which in some species amount to less than 0.5 young per pair per year – are among the lowest of any birds. Little wonder that some large raptors are among the rarest and fastest-declining birds in the world. In contrast, the only extinction-promoting features that small birds show more than large ones are the greater year-to-year variability expected in their population levels, and their greater likelihood of disappearing at low numbers because of their naturally higher mortality rates. However, once a species reaches low numbers, whatever the primary cause, other factors intervene to raise the risk of extinction, as explained below.

ROLE OF CHANCE IN EXTINCTIONS

Extinctions can be divided into two kinds, deterministic (or driven) and stochastic (or chance). Deterministic extinctions are those that result from some damaging change external to the population from which there is no escape. They occur when something essential is removed (such as habitat or food), or when something lethal is added (such as cats to an oceanic island). Stochastic extinctions are those that result entirely from chance events. They affect mainly small populations, and result from the risks inherent in restricted distribution or low numbers. In practice, many extinctions probably result from a deterministic process, such as habitat destruction or overhunting, that reduces a population to a size where chance events can finish it.

The chance processes that affect small populations can be divided into demographic, environmental and genetic, which can act alone or together to cause extinction. The first type refers to those random demographic processes, such as fluctuations in births, deaths or sex ratios, that are intrinsic to the population but that can affect future numbers. If a small population consisted almost entirely of elderly post-reproductive individuals, for example, or almost entirely of one sex, this could cause extinction in circumstances when a more demographically balanced population of the same size could survive. It is thus not total population size that influences survival prospects, but 'effective population size', formed of those individuals that in prevailing circumstances are able to reproduce.

Another demographic factor of importance stems from distribution, whether potential breeders are concentrated in one place and have free access to potential mates, or whether they are so widely scattered that they have little chance of finding mates. In this way, a scattered distribution, resulting from widespread decline, may have contributed to the final demise of Bachman's Warbler *Vermivora bachmanii* in the United States (Terborgh 1989). The problem is accentuated if, as in many birds, one sex shows greater dispersal than the other. For then, although males (say) might return each year to their natal or former breeding sites, females may move elsewhere, reducing further the chance of pair-formation. Behaviour evolved in a more-or-less continuous distribution could thus become lethal under an extreme patchiness.

A third demographic factor of importance, which some authors place in a class by itself, centres on social structure. Some organisms in low numbers may fail to thrive because they lack social facilitation; for example, some species may not breed without the stimulus of a 'colony', while others may suffer from lack of group-defence against predators, or from lack of coordinated effort in group foraging. Little information is available on this aspect, but the point is illustrated by an anecdote told by Amotz Zahavi. A pair of pelicans in Tel-Aviv zoo failed to breed for years, but started soon after mirrors placed in their cage created the illusion of a colony. These and other processes, which can cause survival and reproductive rates to decline at low densities, are known collectively as the 'Allee effect' (Allee *et al.* 1949).

Stochastic environmental events, such as extreme weather or volcanic eruptions, can sometimes greatly increase death rates or breeding failures, causing a population to crash. A single disaster, such as a hurricane or volcanic eruption, might totally destroy a small local population, or reduce it so drastically that it is then more vulnerable to any subsequent catastrophes that happen while numbers are still low. In general, the smaller a population, the greater its vulnerability to such events, and the shorter the interval between them, the more likely the population will be pushed to extinction before it can recover to a safe size. Volcanic eruptions greatly decreased the last known colony of the Short-tailed Albatross *Diomedia albatrus* on Torishima Island off Japan, and elsewhere totally eliminated two species, mentioned above. Hurricanes in 1899 are thought to have eliminated altogether the St. Kitts Bullfinch *Loxigella portoricensis grandis* from the Lesser Antilles (Raffaele 1977), and in 1963 greatly reduced the numbers of Laysan Teal *Anas laysanensis* on Laysan and of White-tailed Sabrewings *Campylopterus ensipennis* on Tobago (Greenway 1967, ffrench 1991), and in 1992 of at least four endemic species on Kauai (Hawaii) (Pyle 1993).

Genetic problems arise mainly (1) when excessive inbreeding or outbreeding (hybridisation) leads to offspring of low fertility or viability; or (2) when genetic variation is lost, leaving the species as a whole less able to adapt to future change. Inbreeding is the biggest danger for small populations: the fewer the individuals present, the more likely it is that each will mate with a close relative. Because related individuals are genetically more similar to each other than are others drawn at random from a large population, in genetic terms inbreeding increases the proportion of loci at which offspring are homozygous. The disadvantages of increased homozygosity are that (1) it increases the chances of detrimental recessive genes being expressed, (2) the homozygotes may often be less fit than the heterozygote, and (3) increased homozygosity decreases the variability among offspring, reducing the chance that some will survive through a period of environmental change. These genetic changes can thus reduce fecundity or survival, causing the population to become even smaller or slowing its rate of recovery (Caughley 1994). Eventually, lethal genes may be eliminated from the population through the mortality they cause, but in the meantime numbers are kept low for longer, and hence at risk for longer.

Inbreeding between close relatives does not necessarily lead to increased homozygosity, however, and nor does high homozygosity necessarily result only from inbreeding. None-the-less, the lesson for managing small populations is to try to maintain genetic diversity. This can be done by attempting to ensure that every individual in the population reproduces and, if similar populations exist elsewhere, by importing occasional individuals from these other populations (although there are dangers in this, see below).

The consequences of inbreeding are well known in domestic and captive animals (Darwin 1868, Ralls & Ballou 1983), and have come to light in a few species of wild birds, including the Song Sparrows *Melospiza melodia* on Mandarte Island off western Canada (Keller *et al.* 1994). Over a 30–year period, this population fluctuated between 10 and 140 individuals. The lowest numbers followed two population crashes, each associated with severe winter weather. By the time of the second crash, the mating histories of all birds had been recorded, so that the pedigree of each individual was known. During the crash, birds with the most distantly-related parents survived better, on average, than did those with closely-related parents. However, because of the small number of survivors, inbreeding increased in subsequent years. An occasional immigrant from the mainland (about one per year) helped to maintain genetic diversity, and perhaps reintroduce deleterious alleles lost during the crash. These findings confirmed that inbreeding could affect survival in the wild, and showed that genetic effects were not independent of environmental events. Even though the majority of deaths could be attributed to an environmental cause, the most outbred individuals survived best.

Outbreeding depression refers to the reduction in fitness (usually in fertility or viability) that often follows from hybridisation.[3] Reduced fitness is usually manifest in the first generation hybrids, but sometimes only in subsequent backcrosses or in later generations. It is apparent not only in hybrids between closely-related species, but sometimes also in hybrids between individuals drawn from distinct populations of the same species. The populations may be adapted to different local conditions, so that the hybrids, having intermediate characteristics, are suited to neither parental region. In extreme form, such hybridisation could completely destroy locally-adapted gene-complexes. This was suspected as a problem in some conservation programmes, where small endangered populations were 'supplemented' with individuals from other parts of the range, apparently causing (at least in one mammal) a local extinction (Templeton *et al.* 1986). The phenomenon of local adaptation is therefore an important issue in reintroduction programmes.

Effects of the second major genetic problem faced by small populations, namely reduced genetic variance, are harder to study. The preservation of genetic diversity is assumed to be important because of the long-term evolutionary potential it provides. Rare forms of a gene, which confer no immediate advantage, may turn out to be of great benefit under changed conditions in the future. To my knowledge, no clear link has yet been established between genetic diversity and extinction risk. But one consequence of reduced genetic variance is that a population could become more susceptible to the effects of disease. Any sizeable population will probably contain some individuals that, because of their genetic make-up, could be more resistant to a particular disease than others. They are the ones most likely to survive an epidemic of that disease. But as a population shrinks in size for reasons other than disease, the more likely are rare resistant types to disappear completely, leaving the entire remaining population unprotected. The same could apply to any form of mortality or reproductive failure, the resistance of which has a genetic basis possessed by only a proportion of individuals.

It is hard to judge how important these genetic considerations are in the extinctions of small populations, for in practice once a wild population falls to a size where genetic effects could prejudice its survival, it is also likely to succumb to other chance factors (Lande 1988b). Thus, demographic and environmental factors are usually of more immediate relevance, but genetic aspects could become important in the longer term, as the population faces changing conditions. In theory, small populations could for a long time remain vulnerable from their reduced genetic variance, which can be rectified only slowly, as non-lethal mutations arise and become incorporated in the gene pool of the expanding population. In general, then, genetic problems might reduce the growth potential of a population, lengthen the period when its numbers remain low, and reduce its chance of surviving through a period of environmental change. No species extinctions are known definitely to have occurred through genetic problems resulting from small population size, but

---

[3]*The term hybrid vigour in which the hybrid is more vigorous than either of its parents is not usually used for hybrids between species but is applied instead to crosses between two pure-bred lines of the same species, and is presumed to result from increased heterozygosity.*

until recently the chances of detecting examples would have been almost nil. This is in striking contrast to the 'many known examples of environmentally-driven extinctions. Known cases of species eliminated from certain areas by hybridisation are in a different category, because they each involved a related congener that reached the same area through human action; and not all such cases involved small populations.

In conclusion, the central message from practically all relevant theoretical studies is that extinction risk from chance events increases with shrinking population size. Field studies have given ample support to this prediction (Williamson 1981, Diamond 1984, Bellamy *et al.* 1996). Environmental fluctuation, which leads to big temporal variation in birth and death rates, emerges as the most likely mechanism of extinction in small populations, the risk decreasing with increasing numbers and decreasing variability in numbers. Demographic fluctuation poses a real threat to survival only when the effective population size is small, say fewer than 50 individuals. Inbreeding depression is likely to be important only when a previously large population is suddenly reduced to a very small one.

Minimum viable populations

In wildlife conservation, the term *minimum viable population* has come into vogue (Soulé 1987). The term implies that there exists some minimum threshold for the number of individuals that will ensure (at some acceptable level of risk) that a population will continue in a viable state for a defined period of time, providing that suitable conditions persist. It is frequently defined as the smallest population having at least a 95% chance of persistence for 200 years. This concept has given rise to 'population viability analysis' (PVA), the process that enables estimation of minimum viable populations (MVPs), or of the expected time to extinction of a population of particular size and other characteristics, if nothing external changes (Soulé 1987, Boyce 1992). Because the vulnerability of species to each chance process (demographic, environmental, genetic) will vary, according to habitat, life-history and genetic features, there can be no single generally-applicable MVP.

None-the-less, population viability analyses have become popular with some biologists and MVP values for particular species have been variously estimated from (1) subjective judgements based on expert knowledge, (2) evidence available from long-term studies, or (3) development of population models. Although these approaches have their limitations, the findings could sometimes be useful in planning reserves and breeding programmes for captive specimens. However, adequate data are seldom available for threatened species in the wild, so estimates of MVP are of uncertain value, and I know of no examples of their application to the conservation of any wild population.

In any wild population, factors that reduce the risk of extinction from chance demographic and genetic events do not necessarily reduce the risk from chance environmental events. The concentration of an entire species in one reserve provides at the time the largest breeding unit possible for that species, least prone to demographic and genetic accidents. But a single environmental event, such as a fire, might destroy the entire population. In such cases, the fragmentation of a population into separate reserves may provide better protection. The relative merits of one large population or several small ones thus vary according to the likelihood of chance environmental events, and can again be assessed reliably only for each species separately. The issue has stimulated long debate, christened SLOSS, over whether it is better to have a single large (SL) reserve or several small (SS) ones, totalling the same area. In practice, of course, reserves are seldom established for the protection of a single species, or with much choice in the selection procedure.

PREDICTIONS OF FUTURE EXTINCTIONS

Some biologists have attempted to predict likely extinction rates among birds or other organisms in the coming decades, while others have attempted to predict which species are likely to go. Forecasts of the first type are mostly based on knowledge of species-area relationships and on projected rates of habitat loss.[4] Theory predicts that, if a habitat is reduced in area by 50%, roughly 10% of species will be lost, and if it is reduced by 90%, roughly half its species will be lost, but on unspecified timescales that could vary between taxa and regions. The main concern at present is with the loss of tropical rainforest which, in terms of species numbers per unit area, is the richest of all land habitats (Myers 1989). If current deforestation trends continue, the argument runs, an estimated 5–10% of the world's species will be committed to extinction per decade over the next 30 years (though they will not necessarily all go within that period) (e.g. Wilson 1992). With an estimated 9700 bird species on earth, this would amount to 485–970 bird species per decade. For organisms as a whole, with an estimated 10 million species on earth, the potential commitment would be 0.5–1 million species per decade, producing a wave of extinctions probably unmatched since the last mass extinction event at the end of the Cretaceous era, 65 million years ago (Raven 1988, Reid 1992, Wilson 1992).

Such estimates are, of course, extremely crude and open to criticism. For one thing, the species–area relationship is based primarily on findings from small islands, and may not be directly transferable to large continental forest areas. Secondly, it is not just the proportion of forest lost that is important, but also the sizes and spatial disposition of the patches remaining (Chapter 6). Large patches are likely to preserve more species than small ones, and different patches could hold some different species. The extensive removal of forest in temperate regions has not reduced continental species numbers to anything near the predicted extent. It has, however, revealed the particular vulnerability of species with restricted range or habitat, such as the Ivory-billed Woodpecker *Campephilus principalis* in North America. Such restricted-range species form a much greater proportion of total bird species numbers in tropical than in temperate forests (Bibby *et al.* 1992). So the particular areas that are felled could have a much greater influence on extinction numbers in tropical than in temperate regions, where most forest species are widespread. Further, it is hard to predict accurately what proportion of tropical forest species might adapt to secondary habitats, though current experience in

---

[4] *Species-area relationships are given as $S = CA^z$, where A is the area and S is the number of species. C and z are constants that depend on the type of organism, whether birds, butterflies, grasses and so on, and on region, whether the islands are close to source areas, such as Indonesia, or very remote, such as Hawaii. For different island groups and organisms, z usually falls in the range 0.15–0.35. The rule of thumb, that a 10-fold reduction in area halves the species numbers, is the same as saying that z = 0.30.*

tropical areas suggests not many (e.g. Case 1996). Moreover, other methods of estimating likely extinction rates have given projected rates of roughly similar order (Smith *et al*. 1993, Mace 1995).

Because extinction is most likely to befall small populations, and is usually preceded by a several-year decline, it is not hard to predict likely candidates from among living birds. In one analysis, about 1111 bird species (more than one-tenth of the total), were listed as 'threatened', mainly on the basis of their restricted distributions or habitats, their small populations or downward trends in numbers (Collar *et al*. 1994). Those species on the list for which counts were available mostly numbered fewer than 1000 individuals, and some fewer than 100 (Green & Hirons 1991). At least 22 extant species were known sometime this century to have had total populations of fewer than 50 individuals **(Table 16.3)**. While some such species have continued to decline, most have increased in response to conservation measures.

Any lists of threatened or endangered bird species (so-called 'red-lists') are likely to be incomplete and misleading in some respects, owing to inadequate or wrong information. The lists must be continually up-dated, as fresh data become available and as populations change. Such lists are useful, however, because they give some idea of the overall picture, highlight gaps in knowledge, and pinpoint conservation priorities. They also indicate to what extent future extinctions are likely to differ from previous ones.

As a group, species now classed as threatened show some similar features to those lost in the last 400 years (Collar *et al*. 1994). They include a large proportion of species found entirely or mainly on islands (N=400, 36%), including tube-nosed petrels (Procellariidae), frigate birds (Fregatidae), megapodes (Megapodidae), rails (Rallidae), pigeons (Columbidae), honey-eaters (Meliphagidae), honey-creepers (Drepanididae) and wattlebirds (Callaeidae). Pacific Ocean islands currently have 110 such species, Indonesian islands 84, Indian Ocean islands 66, and other island groups somewhat fewer **(Table 16.4)**. Threatened birds also include many species confined to restricted habitats or localities in mainland areas, such as grebes (Podicipedidae) found on isolated lakes, and scrub-birds (Atrichornithidae) in special forest types. The third main category includes birds taken by people, either for the pot (such as guans (Cracidae), pheasants (Phasianidae) and cranes (Gruidae)) or for the pet trade (parrots (Psittacidae)). Compared with extinct birds, the list of threatened birds includes smaller proportions of island and flightless forms (many having already gone) and larger numbers of continental species (many increasingly threatened by destruction of remaining habitat patches).

Some species are considered vulnerable because, although still numerous, the entire population is naturally confined to a small area. For example, the smallest remaining flightless bird in the world, the Inaccessible Island Rail *Atlantisia rogersi*, is thought to number about 10 000 individuals, but all occur on the 16 km² of Inaccessible Island, in the Tristan da Cuhna group. The Ascension Frigate Bird *Fregata aquila* numbers more than 8000 birds but the whole population now nests only on the 3-ha Boatswainbird Islet, off Ascension Island in the mid-Atlantic (Collar *et al*. 1994). The Seychelles Warbler *Acrocephalus sechellensis* at 250–300 individuals was until recently confined to one 27-ha island (Cousin) in the Seychelles, although historically the species occurred on other islands to which it has since been reintroduced (Komdeur 1994). Such island forms will always be vulnerable to introductions of rats or other predators.

Table 16.3   Extant species whose total populations at some time since 1900 were known to number fewer than 50 individuals, and whose populations have since increased through management measures.
Details mainly from Collar & Andrew (1988), Short & Horne 1990 for Ivory-billed Woodpecker, Komdeur 1994 for Seychelles Scrub Warbler, Jones et al. 1995 for Mauritius Kestrel, Pimm et al. 1994 for Hawaiian Crow, McCulloch 1994 for Seychelles Magpie Robin, Miller & Mullette 1985 for Woodhen.

Species (and lowest known numbers)	Range
Amsterdam Albatross *Diomedia amsterdamensis* (10 in 1980s)	Amsterdam Island, Indian Ocean
Short-tailed Albatross *Diomedea albatrus* (8 in 1953)	Torishima Island, Japan
Cahow *Pterodroma cahow* (70 pairs 1985)	Bermuda
Crested Ibis *Nipponia nippon* (40 in 1987)	South Shaanxi, China
California Condor *Gymnogyps californianus* (22 In 1983)	California
Mauritius Kestrel *Falco punctatus* (4 in 1974)	Mauritius
Laysan Teal *Anas laysanensis* (7 In 1912)	Laysan Island, Hawaii
Whooping Crane *Grus americana* (18 in 1945)	North America
Lord Howe Island Woodhen *Tricholimnas sylvestris* (<20 In 1970s)	Lord Howe Island, Australia
Pink Pigeon *Nesoenas mayeri* (20 in 1980s)	Mauritius
Puerto Rican Parrot *Amazona vittata* (13 In 1975)	Puerto Rico
Mauritius Parakeet *Psittacula eques* (9 In 1986)	Mauritius
Kakapo *Strigops habroptilus* (40 in 1980s)	Stewart Island, New Zealand
Seychelles Magpie Robin *Copsychus sechellarum* (8 In 1965)	Fregate, Seychelles
Seychelles Scrub Warbler *Acrocephalus sechellensis* (30 in 1959)	Seychelles
Aldabra Warbler *Nesillas aldabranus* (5 in 1970s)	Aldabra, Seychelles
Chatham Island Black Robin *Petroica traversi* (7 In 1976)	Chatham Islands, New Zealand
Rarotonga Monarch *Pomarea dimidiata* (21 In 1983)	Cook Islands, New Zealand
Kauai Oo *Moho braccatus* (few in 1990s)	Kauai, Hawaii
Hawaiian Crow *Corvus hawaiiensis* (12 in 1992)	Hawaii'

Other species reduced to low numbers but not yet subject to management include the Po'ouli *Melamprosops phaeosoma* (6 in 1996, Maui, Hawaii) and the Ivory-billed Woodpecker *Campephilus principalis* (few in 1990s, Cuba, already gone from United States).

As with extinct birds, some threatened species are members of distinct taxonomic lineages, whose demise would represent a more significant loss to biodiversity than others. Several such species occur on Madagascar, New Zealand or Australia, areas noted for taxonomic uniqueness in other animals and plants. Such uniqueness is thus one criterion on which conservation priorities could be based, and some formalised procedures have been proposed by which taxonomic distinctness could be assessed (May 1990, Vane-Wright et al. 1991). However, as ever more species join the threatened list, it will become increasingly difficult to conserve them on a species-by-species basis. The only practical means is to preserve large areas of natural habitat, with priority given to regions that are richest in species. Bibby et al. (1992) distinguished what they called 'restricted-range species', namely those that

Table 16.4   Distributions of extinct and threatened birds between major island groups. From Johnson & Stattersfield 1990.

Island groups	Numbers of known extinct species since 1600	Approximate number of extant endemics	Number (%) of endemics considered threatened
Atlantic Ocean	4	50	25(50)
Indian Ocean	31	200	66(33)
Indonesia	2	390	84(22)
Philippines	1	180	34(19)
New Guinea and Melanesia	5	500	50(10)
Pacific Ocean	51	290	110(38)
Caribbean	3	140	31(22)
Totals	97	1750	400(23)

occupied geographical areas less than an arbitrary 50 000 km². Some 2609 (27% of all living species) are in this category. Because certain parts of the world are hotspots of biodiversity and endemism, a large proportion of the world's birds could be saved by safeguarding a relatively small proportion (5%) of the earth's land surface, providing it was in such key tropical areas. So far, only 3% of the earth's land area is in 'protected' areas, and most of these are not in hotspots.

EXAMPLES OF POPULATIONS RECOVERING FROM SMALL NUMBERS

While the fates of many threatened species are probably sealed already, decline to low numbers need not necessarily lead to extinction if action is taken to protect the remaining population and if suitable habitat is still available. One remarkable example is the Mauritius Kestrel *Falco punctatus*, which in 1974 numbered only four individuals. After a successful captive breeding and release programme, accompanied by the provision of nest-boxes, the species had increased to more than 200 individuals by 1993 (Jones *et al.* 1995). Another example is the Chatham Island Black Robin *Petroica traversi* which became restricted to 22 ha of scrub forest on Little Mangere in the Chatham Islands, 600 km from New Zealand. Following deterioration of its scrub habitat, the entire remnant population of seven birds (two breeding pairs) was transferred in 1976–77 to Mangere (a historical stronghold) and later to South East Island, on both of which forest was regrowing after the removal of farm stock. As a result of egg-manipulations and cross-fostering under another species, the population increased, reaching 120 birds in 1992 (Butler & Merton 1992). A third example is the Laysan Teal *Anas laysanensis* which was reduced to only seven individuals in 1912, but had increased to more than 500 by the 1960s (Moulton & Weller 1984, Browne *et al.* 1993). Other examples include the Whooping Crane *Grus americana*, which fell to 18 individuals in 1945, but has since increased under protection to more than 200 (Allen 1954), and the California Condor *Gymnogyps californianus*, reduced to 22 individuals in 1983 and by 1992 increased in captivity to more than 70 (Collar *et al.* 1994).

Not only did these species recover from low numbers, but they survived in low numbers for many years. Several New Zealand species, besides the Black Robin, probably persisted at fewer than 20 pairs for a century or more, including the Forbes Parakeet *Cyanoramphus auriceps* on Little Mangere Island, the Bellbird *Anthornis melanura* on Tiritiri Matange Island, and the Flightless Teal *Anas aucklandica* on Dent Island (Craig 1991). The same was probably true for the Woodhen *Tricholimnas sylvestris* on Lord Howe Island, Australia (Miller & Mullette 1985). Moreover, many species now thriving on oceanic islands must have originally been derived from very small numbers of colonists, as must many introduced species. Such species succeeded despite the problems of small populations discussed above, but are probably few in number compared with those that failed.

Another lesson from species recoveries is that the remnants of once widespread species do not necessarily survive in the most favourable habitat left, and nor do their present lifestyles and feeding habits necessarily represent the norm. The histories of the New Zealand Takahe Rail *Notornis mantelli* (Caughley 1994), the Hawaiian Goose *Branta sandvicensis* (Black 1995), the Lord Howe Woodhen *Tricholimnas sylvestris* (Miller & Mullette 1985) and the Red Kite *Milvus milvus* in Wales (Newton *et al.* 1989), reveal the frequent error in that assumption. Each of these species ended up, not in the habitat most favourable to it, but in the habitat least favourable to the agent of decline.

Other examples of recovery from low numbers exist among mammals, including Northern Elephant Seal *Mirounga angustirostris* once reduced to about 100 individuals, European Bison *Bison bonasus* (17 individuals), Pere David's Deer *Elaphurus davidianus* (16 individuals), Przewalski's Horse *Equus przewalskii* (13 individuals) and others. Some of these species are known to have low genetic variation now, compared with related species, but do not show obvious symptoms of inbreeding depression (Frankel & Soulé 1981). Such examples confirm that small populations are not inevitably doomed. However, the processes relevant to such declining populations were many, and before remedial action could be taken, they had to be examined on a species-by-species basis. It is this laborious procedure that has saved so many species in recent years.

Attempts are now underway to save many of the species currently listed as endangered, so not all are likely to disappear in the immediate future. Such attempts involve not only the protection of populations and habitats, but more expensive procedures, such as manipulations of reproductive rates, captive-breeding and reintroduction programmes. Not all previous conservation measures for species at low numbers have led to population recovery, however. In the 19th century, the Eskimo Curlew *Numenius borealis* was found in millions, breeding on tundra across northern Canada and Alaska and wintering in southern South America, but within a few decades overhunting on migration areas had reduced its numbers to fewer than 50 (Bent 1928). Careful protection prevented extinction, but has not led to spectacular recovery, and this curlew remains one of North America's rarest birds, with only a few sightings each decade. The reasons for lack of recovery are not understood, but may lie in the wintering areas.

Some species of birds have naturally extremely small populations. This is particularly so of raptors that are endemic to small islands. As top predators, individual raptors require large areas over which to hunt, so can live only at low density in the restricted area of an island. An extreme example is the Socorro Hawk *Buteo jamaicensis socorroensis*, an endemic subspecies found only on the 140 km^2 Socorro Island, 460 km from Mexico in the Pacific, where it has remained at about 20 pairs for many generations, and possibly for its whole existence (Walter 1990). A second example is the recently extinct Atitlán Grebe *Podilymbus gigas* which was restricted to Lake Atitlán in Guatemala, and probably never numbered more than around 100 pairs. A third is the Narcondam Hornbill *Aceros narcondami* which is endemic to the island of Narcondam in the Andaman Islands. This island covers only 6.4 km², so could never have supported more than a few hundred individuals. We shall never know how large a population many extinct island birds maintained, but some, such as the Stephen Island Wren *Xenicus lyalli* mentioned earlier, occupied extremely small areas.

CONCLUDING REMARKS

The situation now facing the world's flora and fauna is unprecedented. An estimated 20–50% of the earth's primary productivity is already commandeered each year by the increasing human population, in association with continuing habitat destruction and pollution. Probably never before has a single species occupied such a dominant position among life on earth, with such devastating consequences for other organisms. Extinction rates are increasing all the time, as we head increasingly towards a biologically greatly simplified world.

The fate of much of life on earth ultimately depends on trends in human numbers which are expanding more rapidly than ever. It supposedly took hundreds of thousands of years for our species to reach a population level of 10 million, only 10 000 years ago. With the advent of agriculture, this number grew to perhaps 100 million about 2000 years ago, and to 2.5 billion by 1950. Within less than the span of a single human lifetime, it more than doubled to nearly 6.0 billion in 1995. This accelerated population growth resulted from rapidly lowered death rates (particularly infant rates), combined with sustained high birth rates. If current rates of population growth continue, the world human population is projected to reach 12 billion by 2095. The actual outcome will have enormous implications for the fate of life on earth, including human. At the least, thousands of animal and plant species could be already doomed by the momentum of existing human population growth.

It is not only the increasing numbers of people that are important, but also the increasing resources that individual people in 'developed' countries expect to consume during their lifetimes. Clearly this cannot occur without the earth's resources being stripped at a faster rate than imposed by population growth alone, or without the associated waste production (Ehrlich 1994). So far, advancing technology has done much to alleviate some of the problems, but at the same time has increased the impact of humanity on the earth's resources, by increasing the range, scale and rate of unsustainable activities possible. Besides the reduction of forest cover and associated biodiversity, far-reaching problems stem from greenhouse gas emissions, acid rain, loss of topsoil, and shortages of water, food and fuel-wood in many parts of the world. All these factors could induce further declines in species numbers.

You might reasonably wonder why, in a book about population limitation in birds, I end with the human population. But the fact is that human numbers and activities now ultimately determine the distributions and abundance levels of almost all land species. The natural factors that operate do so within habitats whose extent and productivity are largely influenced by human activity. The reason that birds so starkly illuminate the issues is that, compared with most other organisms, they have been so well and so comprehensively studied.

SUMMARY

For thousands of years, most extinctions have probably been caused directly or indirectly by human action. In various parts of the world, in prehistoric times, mass extinctions of large birds and mammals invariably followed human colonisation. These associated events occurred at markedly different dates in different regions and have been attributed mainly to 'overhunting'. Some of the world's largest and most spectacular birds, the moas of New Zealand and the elephant birds of Madagascar, disappeared as recently as 1000–1600 AD, along with many others. The number of bird species now found on Earth, around 9700, could be up to several thousand less than occurred only 1000–2000 years ago. In addition, many species now found only on a single island or island group were once more widespread, as revealed from their remains. Since 1600 AD, at least 127 recorded bird species are known to have died out, almost all through human action. More than 90% of vanquished species were island forms, vulnerable because of their restricted ranges, small population sizes and lack of defences against introduced competitors, predators and disease organisms. Many were flightless. Habitat destruction and fragmentation have also caused extinctions and are likely to achieve greater importance in future as remaining wild areas, both on islands and on continents, are progressively destroyed. The futures of many organisms are likely to depend increasingly on the numbers and activities of people, and unless action is taken to prevent it, extinction rates could increase greatly in the coming years.

Waxwings *Bombicilla garrulus* with rowan fruits.

Bibliography

Abrams, R.W. (1985). Pelagic seabird community structure in the southern Benguela region: changes in response to man's activities? *Biol. Conserv.* **32:** 33–49.

Adams, L., Hanavan, M.G., Hosley, N.W. & Johnston, D.W. (1949). The effects on fish, birds and mammals of DDT issued in the control of forest insects in Idaho and Wyoming. *J. Wildl. Manage.* **13:** 245–254.

Adrian, W.J. & Spraker, T.R. (1978). Epornitics of aspergillosis in Mallards *Anas platyrhynchos* in north central Colorado. *Wildl. Dis.* **14:** 212–217.

Aebischer, A., Perrins, N., Krieg, M., Struder, J. & Meyer, D.R. (1996). The role of territory choice, mate choice and arrival date on breeding success in the Savi's Warbler *Locustella luscinioides*. *J. Avian Biol.* **27:** 143–152.

Aebischer, N.J. (1986). Retrospective investigation of an ecological disaster in the Shag, *Phalacrocorax aristotelis* – a general method based on long-term marking. *J. Anim. Ecol.* **55:** 613–629.

Aebischer, N.J. (1991). Sustainable yields: gamebirds as a harvestable resource. *Gibier Faune Sauvage* **8**: 353–351.

Aebischer, N.J. (1997). Gamebirds: management of the Grey Partridge in Britain. pp. 131–151 in *'Conservation and the Use of Wildlife Resources'* (ed. M. Bolton). London, Chapman & Hall.

Aebischer, N.J., Coulson, N.C. & Colebrook, J.M. (1990). Parallel long term trends across four marine tropic levels and weather. *Nature* **347**: 753–755.

Åhlund, M. & Götmark, F. (1989). Gull predation on Eider ducklings *Somateria mollissima* – effects of human disturbance. *Biol. Conserv.* **48**: 115–127.

Ainley, D.G. & Lewis, T.J. (1974). The history of the Farallon Island marine bird populations, 1854–1972. *Condor* **76**: 432–464.

Ainley, D.G., Nur, N. & Woehler, E.J. (1995). Factors affecting the distribution and size of pygoscelid penguin colonies in the Antarctic. *Auk* **112**: 171–182.

Al-Joborae, F.F. (1980).The influence of diet on the gut morphology of the Starling (*Sturnus vulgaris* L. 1758). *D.Phil. thesis* (Unpub), University of Oxford.

Alatalo, R., Eriksson, L., Gustafsson, L. & Larsson, K. (1987). Exploitation competition influences the use of foraging sites by tits: experimental evidence. *Ecology* **68**: 284–290.

Alatalo, R.V., Gustafsson, L., Lindén, M. & Lundberg, A. (1985). Interspecific competition and niche shifts in tits and the Goldcrest: an experiment. *J. Anim. Ecol.* **54**: 977–984.

Alatalo, R.V., Gustafsson, L. & Lundberg, A. (1986). Interspecific competition and niche changes in Tits (*Parus* spp) – evaluation of nonexperimental data. *Amer. Nat.* **127**: 819–834.

Alatalo, R.V., Höglund, J., Lundberg, A., Rintamäki, P.T. & Silverin, B. (1996). Testosterone and male mating success on the Black Grouse leks. *Proc. R. Soc. Lond. B.* **263**: 1697–1702.

Alatalo, R.V. & Lundberg, A. (1984). Density dependence in breeding success of the Pied Flycatcher (*Ficedula hypoleuca*). *J. Anim. Ecol.* **53**: 969–977.

Alatalo, R.V., Lundberg, A. & Björklund, M. (1982). Can the song of male birds attract other males? An experiment with the Pied Flycatcher *Ficedula hypoleuca*. *Bird Behav.* **4**: 42–45.

Alatalo, R., Lundberg, A. & Ståhlbrandt, K. (1983). Do Pied Flycatcher males adopt broods of widow-females? *Oikos* **41**: 91–93.

Albert, T.F. (1962). The effect of DDT on the sperm production of the domestic fowl. *Auk* **79**: 104–107.

Alerstam, T. (1985). Fågelsamhället i Borgens lövskogsomåde. *Anser* **24**: 213–234.

Alexander, W.B. (1933). The Swallow mortality in central Europe in September 1931. *J. Anim. Ecol.* **2**: 116–118.

Alexander, W.C. & Hair, J.D. (1979). Winter foraging behaviour and aggression of diving ducks in South Carolina. *Proc. Southeast Assoc. Fish and Wildl. Agencies* **31**: 226–232.

Alisauskas, R.T. (1987). Morphometric correlates of age and breeding status in American Coots. *Auk* **104**: 640–646.

Alisauskas, R.T. & Arnold, T.W. (1994). American Coot. pp. 127–143 in *'Migratory shore and upland game bird management in North America'* (ed. T.C. Tacha & C.E. Braun). Lawrence, Kansas, Allen Press.

Allee, W.C., Emerson, A.E., Park, O., Park, T. & Schmidt, K.P. (1949). Principles of animal ecology. Philadelphia, Saunders.

Allen, D.L. (1954). Our wildlife legacy. New York, Funk & Wagnalls.

Allin, C.C., Chasko, G.G. & Husband, T.P. (1987). Mute Swans in the Atlantic Flyway: a review of the history, population growth and management needs. *Trans. Northeast Sect. Wildl. Soc.* **44**: 32–47.

Allison, A., Newton, I. & Campbell, C.R.G. (1974). Loch Leven National Nature Reserve: a study of waterfowl biology. A WAGBI Conservation Publication. Chester, Wildfowlers Association of Great Britain and Ireland.

Alvo, R., Hussell, D.J.J. & Berrill, M. (1988). The breeding success of Common Loons (*Gavia immer*) in relation to alkalinity and other lake characteristics in Ontario. *Can. J. Zool.* **66**: 746–754.

Ambuel, B. & Temple, S.A. (1983). Area-dependent changes in the bird communities and vegetation of southern Wisconsin forests. *Ecology* **64**: 1057–1068.

Andersen-Harild, P. (1981). Weight changes in *Cygnus olor*. Proc. 2nd Internat. Swan Symp., Sapporo, Japan. IWRB, Slimbridge, UK. Pp. 359–378

Anderson, A. (1990). Prodigious birds: moas and moa-hunting in prehistoric New Zealand. Cambridge: University Press.

Anderson, D.J. (1989). Differential responses of boobies and other seabirds in the Galapagos to the 1986–87 El Niño – Southern Oscillation event. *Mar. Ecol. Prog. Ser.* **52**: 209–216.

Anderson, D.J. (1991). Apparent predator-limited distribution of Galapagos Red-footed Boobies *Sula sula*. *Ibis* **133**: 26–29.

Anderson, D.R. & Burnham, K.P. (1976). Population ecology of the Mallard VI. The effect of exploitation on survival. *U.S. Fish & Wildlife Service Res. Publ.* **128**.

Anderson, D.W. & Gress, F. (1984). Brown Pelicans and the anchovy fishing of southern California. pp. 128–135 in '*Marine birds: their feeding ecology and commercial fisheries relationships*' (ed. D.N. Nettleship, G.A. Sanger & P.F. Springer). Ottawa, Canadian Wildlife Service.

Anderson, R.M., Jackson, H.C., May, R.M. & Smith, A.M. (1981). Population dynamics of fox rabies in Europe. *Nature* **289**: 765–771.

Anderson, R.M. & May, R.M. (1978). Regulation and stability of host–parasite population interactions: 1. Regulatory processes. *J. Anim. Ecol.* **47**: 219–249.

Anderson, R.M. & May, R.M. (1979). Population biology of infectious diseases. Part 1. *Nature* **280**: 361–367.

Anderson, T.R. (1978). Population studies of European sparrows in North America. *Occ. Papers Mus. Nat. Hist. Kansas* **70**: 1–58.

Anderson, T.R. (1990). Excess females in a breeding population of House Sparrow (*Passer domesticus* (L.)). pp. 87–93 in '*Proceedings, General Meetings of the Working Group on Granivorous Birds, INTECOL, 1986*' (ed. J. Pinowski & J.D. Summers-Smith). Warsaw, Polish Scientific Publishers.

Andersson, M. & Wiklund, C. (1978). Clumping versus spacing out: experiments on nest predation in Fieldfares (*Turdus pilaris*). *Anim. Behav.* **26**: 1207–1212.

Andreasen, J.K. (1985). Insecticide resistance in mosquito fish of the Lower Rio Grande Valley of Texas – an ecological hazard. *Arch. Environ. Contam. Toxicol.* **14**: 573–577.

Andrén, H. (1990). Despotic distribution, unequal reproductive success, and population regulation in the Jay *Garrulus glandarius* L. *Ecology* **71**: 1796–1803.

Andrén, H. (1992). Corvid density and nest predation in relation to forest fragmentation: a landscape perspective. *Ecology* **73**: 794–804.

Andrén, H. (1994). Effects of habitat fragmentation on birds and mammals in landscapes with different proportions of suitable habitat – a review. *Oikos* **71**: 355–366.

Andrén, H., Angelstam, P., Lindström, E. & Widén, P. (1985). Differences in predation pressure in relation to habitat fragmentation – an experiment. *Oikos* **45**: 273–277.

Angelstam, P. (1983). Population dynamics of tetraonids, especially the Black Grouse *Tetrao tetrix* L. in boreal forests. Ph.D. thesis (Unpub), Uppsala University, Sweden.

Angelstam, P. (1984). Sexual and seasonal differences in mortality of the Black Grouse *Tetrao tetrix* in boreal Sweden. *Ornis Scand.* **15**: 123–124.

Angelstam, P., Lindström, E.C. & Widén, P. (1984). Role of predation in short-term population fluctuations of some birds and mammals in Fennoscandia. *Oecologia* **62**: 199–208.

Angelstam, P., Lindström, E. & Widén, P. (1985). Synchronous short-term population fluctuations of some birds and mammals in Fennoscandia – occurrence and distribution. *Holarctic Ecology* **8**: 285–298.

Anker-Nilssen, T. (1987). The breeding performance of Puffins *Fratercula arctica* on Røst, northern Norway in 1979–85. *Fauna Norv. Ser C, Cinclus* **10**: 21–38.

Ankney, C.D. & MacInnes, C.D. (1978). Nutrient reserves and reproductive performance of female Lesser Snow Geese. *Auk* **95**: 459–471.

Appelquist, H., Drabaek, I. & Asbirk, S. (1985). Variation in mercury content of Guillemot feathers over 150 years. *Mar. Poll. Bull.* **16:** 244–248.

Arcese, P. (1987). Age, intrusion pressure and defence against floaters by territorial male Song Sparrows. *Anim. Behav.* **35:** 773–784.

Arcese, P. & Smith, J.N.M. (1985). Phenotypic correlates and ecological consequences of dominance in Song Sparrows. *J. Anim. Ecol.* **54:** 817–830.

Arcese, P. & Smith, J.N.M. (1988). Effects of population density and supplemental food on reproduction in Song Sparrows. *J. Anim. Ecol.* **57:** 119–136.

Arcese, P., Smith, J.N.M., Hochachka, W.M., Rogers, C.M. & Ludwig, D. (1992). Stability, regulation and the determination of abundance in an insular Song Sparrow population. *Ecology* **73:** 805–822.

Arendt, W.J. (1985). *Philornis* ectoparasitism of Pearly-eyed Thrashers, II. Effects on adults and reproduction. *Auk* **102:** 281–292.

Armstrong, I.H., Coulson, J.C., Hawkey, P. & Hudson, K.J. (1978). Further mass seabird deaths from paralytic shellfish poisoning. *Brit. Birds* **71:** 58–68.

Arnold, T.W. (1994). Effects of supplemental food on egg production in American Coots. *Auk* **111:** 337–350.

Ashford, R.W., Wyllie, I. & Newton, I. (1990). *Leucocytozoon toddi* in British sparrowhawks *Accipiter nisus*: observations on the dynamics of infection. *J. Nat. Hist.* **24:** 1101–1107.

Ashmole, N.P. (1963). The regulation of numbers of tropical oceanic birds. *Ibis* **103:** 458–473.

Askins, R.A. & Ewert, D.N. (1991). Impact of hurricane Hugo on bird populations on St. John, United States Virgin Islands. *Biotropica* **23:** 481–487.

Askins, R.A., Lynch, J.F. & Greenberg, R. (1990). Population declines in migratory birds in eastern North America. *Curr. Ornithol.* **7:** 1–57.

Atkinson, I.A.E. (1985). The spread of commensal species of *Rattus* to oceanic islands and their effects on island avifaunas. *ICBP Tech. Bull.* **3:** 35–81.

Atkinson, I.A.E. & Cameron, E.K. (1993). Human influence on the terrestrial biota and biotic communities of New Zealand. *Trends Ecol. Evol.* **8:** 447–451.

Babcock, K.M. & Flickinger, E.L. (1977). Dieldrin mortality of Lesser Snow Geese in Missouri. *J. Wildl. Manage.* **41:** 100–103.

Backhouse, F. (1986). Bluebird revival. A heart-warming comeback from the brink of extinction. *Can. Geogr.* **106:** 33–39.

Bacon, P.J. & Andersen-Harild, P. (1989). Mute Swan. pp.363–386 in 'Lifetime Reproduction in Birds' (ed. I. Newton). London, Academic Press.

Baeyens, G. (1981). The role of the sexes in territory defence in the Magpie (*Pica pica*). *Ardea* **69:** 69–82.

Bahat, O. (1989). Aspects in the ecology and biodynamics of the Golden Eagle (*Aquila chrysaetos homeyeri*) in the arid regions of Israel. M.Sc. Thesis (Unpub), Tel-Aviv University.

Bailey, E.P. & Davenport, G.H. (1972). Die-off of Common Murres on the Alaska peninsula and Unimak island. *Condor* **74:** 215–219.

Bailey, E.P. & Kaiser, G.W. (1993). Impacts of introduced predators on nesting seabirds in the northeast Pacific. pp. 218–226 in 'The status, ecology and conservation of marine birds of the North Pacific'. Can. Wildl. Serv. Spec. Publ., Ottawa.

Bailey, R.S. (1966). The seabirds of the southeast coast of Arabia. *Ibis* **108:** 224–264.

Baillie, S.R. (1990). Integrated population monitoring of breeding birds in Britain and Ireland. *Ibis* **132:** 151–166.

Baillie, S.R., Clark, N.A. & Ogilvie, M.A. (1986). Cold weather movements of waterfowl and waders: an analysis of ringing recoveries. *BTO Res. Rep.* **19.**

Baillie, S.R. & Peach, W.J. (1992). Population limitation in Palaearctic–African migrant passerines. *Ibis* **134 (Suppl. 1):** 120–132.

Baines, D. (1990). The roles of predation, food and agricultural practice in determining the feeding success of the Lapwing *Vanellus vanellus* on upland grasslands. *J. Anim. Ecol.* **59:** 915–929.

Baines, D. (1991). Factors contributing to local and regional variation in Black Grouse breeding success in northern Britain. *Ornis Scand.* **22:** 264–269.

Baines, D. (1996). The implications of grazing and predator management on the habitats and breeding success of Black Grouse *Tetrao tetrix. J. Appl. Ecol.* **33:** 54–62.

Baines, D., Baines, M.M. & Sage, R.B. (1995). The importance of large herbivore management to woodland grouse and their habitats. pp. 93–100 in *'Proceedings of the International Symposium on Grouse, Vol. 6.'* Reading, U.K. World Pheasant Association.

Baines, D., Sage, R.B. & Baines, M.M. (1994). The implications of Red Deer grazing to ground vegetation and the invertebrate communities of Scottish native pinewoods. *J. Appl. Ecol.* **31:** 776–783.

Bairlein, F. (1996). Long-term ecological studies on birds. *Verh. Dtsch. Zool. Ges.* **89:** 165–179.

Baker, A.J. (1991). A review of New Zealand ornithology. *Curr. Ornithol.* **8:** 1–67.

Balcomb, R. (1983). Secondary poisoning of Red-shouldered Hawks with carbofuron. *J. Wildl. Manage.* **47:** 1129–1132.

Baldassarre, G.A. & Bolen, E.G. (1994). Waterfowl ecology and management. New York, Wiley.

Balfour, E. & Cadbury, C.J. (1979). Polygyny, spacing and sex ratio among Hen Harriers *Circus cyaneus* in Orkney, Scotland. *Ornis Scand.* **10:** 133–141.

Ball, I.J., Martin, T.E. & Ringelman, J.K. (1994). Conservation of nongame birds and waterfowl: conflict or complement? *Trans. N.A. Wildl. Nat. Res. Conf.* **59:** 337–347.

Ball, K.E. (1950). Breeding behaviour of the Ring-necked Pheasant on Pelee Island, Ontario. *Can. Field Nat.* **64:** 201–207.

Balser, D.J., Dill, H.H. & Nelson, H.K. (1968). Effect of predator reduction on waterfowl nesting success. *J. Wildl. Manage.* **32:** 669–682.

Barker, R.J. (1958). Notes on some ecological effects of DDT sprayed on elms. *J. Wildl. Manage.* **22:** 269–274.

Barnard, C.J. (1979). Interactions between House Sparrows and Sparrowhawks. *Brit. Birds* **72:** 569–573.

Barnard, C.J. (1980). Flock feeding and time budgets in the House Sparrow (*Passer domesticus*). *Anim. Behav.* **28:** 295–309.

Barnes, C.C. & Thomas, V.G. (1987). Digestive organ morphology, diet and guild structure of North American Anatidae. *Can. J. Zool.* **65:** 1812–1817.

Barrett, G.W. (1968). The effects of an acute insecticide stress on a semi-enclosed grassland ecosystem. *Ecology* **49:** 1019–1035.

Barry, T.W. (1962). Effects of late seasons on Atlantic Brant reproduction. *J. Wildl. Manage.* **26:** 19–26.

Barry, T.W. (1968). Observations on natural mortality and native use of Eider Ducks along the Beaufort Sea Coast. *Can. Field Nat.* **82:** 140–144.

Bartholomew, G.A., Howell, T.R. & Cade, T.J. (1957). Torpidity in the White-throated Swift, Anna Hummingbird and Poorwill. *Condor* **59:** 145–155.

Batt, B.D.J., Anderson, M.G., Anderson, C.D. & Caswell, F.D. (1989). The use of prairie potholes by North American ducks. pp. 204–227 in *'Northern prairie wetlands'* (ed. A. van der Valk). Ames, Iowa, Iowa State University Press.

Batten, L.A. (1973). Population dynamics of suburban Blackbirds. *Bird Study* **20:** 251–258.

Bautista, L.M., Alonso, J.C. & Alonso, J.A. (1995). A field test of ideal free distribution in flock-feeding Common Cranes. *J. Anim. Ecol.* **64:** 747–757.

Beasom, S.L. (1974). Intensive short-term predator removals as a game management tool. *Trans. N. A. Wildl. Nat. Res. Conf.* **39:** 230–240.

Beauchamp, W.D., Koford, R.R., Nudds, R.D., Clark, R.G. & Johnson, D.H. (1996a). Long-term declines in nest success of prairie ducks. *J. Wildl. Manage.* **60:** 247–257.

Beauchamp, W.D., Nudds, T.D. & Clark, R.G. (1996b). Duck nest success declines with and without predation management. *J. Wildl. Manage.* **60:** 258–264.

Beebe, F.L. (1960). The marine Peregrines of the northwest Pacific coast. *Condor* **62:** 145–189.

Beecham, J.J. & Kochert, M.N. (1975). Breeding biology of the Golden Eagle in southwestern Idaho. *Wilson Bull.* **87:** 506–513.

Begon, M., Harper, J.L. & Townsend, C.R. (1986). Ecology: individuals, populations and communities. Oxford, Blackwell Scientific Publications.

Beintema, A.J. & Muskens, G.J.D.M. (1987). Nesting success of birds breeding in Dutch agricultural grasslands. *J. Appl. Ecol.* **24:** 743–758.

Beissinger, S.R. & Bucher, E.H. (1992). Sustainable harvesting of parrots for conservation. pp. 73–115 in *'New World parrots in crisis'* (ed. S.R. Beissinger & N.F.R. Snyder). Washington, Smithsonian Institution Press.

Belik, V. & Mikalevich, I. (1995). The pesticides use in the European steppes and its effects on birds. Research Notes on Avian Biology: selected contributions from the 21st International Ornithological Congress. *J. Ornithol.* **135:** 233.

Bell, B.D. (1978). The Big South Cape Islands rat irruption. pp. 33–45 in *'The Ecology and control of rodents in New Zealand Nature Reserves'*, Vol. Information Series No 4, New Zealand. (ed. P.R. Dingwall, I.A.E. Atkinson & C. Hay). New Zealand, Dept. Lands and Survey.

Bellamy, P.E., Brown, N.J., Enoksson, B., Firbank, L.G., Fuller, R.J., Hinsley, S.A. & Schotman, A.G.M. (1998). The influence of habitat, landscape structure and climate on distribution patterns of the Nuthatch *(Sitta europaea* L.). *Oecologia* **67**, in press.

Bellamy, P.E., Hinsley, S.A. & Newton, I. (1996). Factors influencing bird species numbers in woodland fragments in southeast England. *J. Appl. Ecol.* **33:** 249–262.

Bellrose, F.C. (1959). Lead poisoning as a mortality factor in waterfowl populations. *Bull. Illinois Nat. Hist. Surv.* **27:** 235–288.

Bellrose, F.C. (1980). Ducks, geese and swans of North America. Harrisburg, Pennyslvania, Stackpole Books.

Bellrose, F.C. (1990). The history of Wood Duck management. pp. 13–20 in *'Proceedings of the 1988 North American Wood Duck Symposium'* (ed. L.H. Fredrickson, G.V. Burger, S.P. Havera, D.A. Graber, R.E. Kirby & T.S. Taylor). St. Louis, Missouri.

Bellrose, F.C., Hanson, H.C. & Beamer, P.D. (1945). Aspergillosis in Wood Ducks. *J. Wildl. Manage.* **9:** 325–326.

Bellrose, F.C., Johnstone, K.L. & Meyers, T.C. (1964). Relative value of natural cavities and nesting houses for Wood Ducks. *J. Wildl. Manage.* **28:** 661–676.

Bellrose, F.C., Scott, T.G., Hawkins, A.S. & Low, J.B. (1961). Sex ratios and age ratios in North American ducks. *Bull. Illinois Nat. Hist. Surv.* **27:** 391–474.

Belopolskij, L.O. (1971). Migration of Sparrowhawk on Courland Spit. *Notatki ornitologizne* **12:** 111–12.

Bendell, J.F. & Elliott, P.W. (1967). Behaviour and the regulation of numbers in Blue Grouse. *Can. Wildl. Serv. Rep. Ser* **4.**

Bendell, J.F., King, D.G. & Mossop, D.H. (1972). Removal and repopulation of Blue Grouse in a declining population. *J. Wildl. Manage.* **36:** 1153–1165.

Bengston, S.A. (1963). On the influence of snow upon the nesting success in Iceland 1961. *Vår Fågel* **22:** 77–122.

Benkman, C.W. (1997). Feeding behaviour, flock–size dynamics, and variation in sexual selection in Crossbills. *Auk* **114:** 163–178.

Bennett, G.E, Caines, J.R. & Bishop, M.A. (1988). Influence of blood parasites on the body mass of passeriform birds. *J. Wildl. Dis.* **24:** 339–343.

Bennett, G.E, Peirce, M.A. & Ashford, R.W. (1993). Avian Hematozoa: mortality and pathogenicity. *J. Nat. Hist.* **27:** 993–1001.

Bensch, S. & Hasselquist, D. (1991). Territory infidelity in the polygynous Great Reed Warbler *Acrocephalus arundinaceus:* the effect of variation in territory attractiveness. *J. Anim. Ecol.* **60:** 857–871.

Bent, A.C. (1928). Life histories of North American shorebirds. *Smithsonian Inst. U.S. Nat. Mus. Bull.* **146**. Washington D.C., U.S. Nat. Mus.

Bent, A.C. (1932). Life histories of North America gallinaceous birds. *Smithsonian Inst. U.S. Nat. Mus. Bull.* **162**. Washington D.C., U.S. Nat. Mus.

Bent, A.C. (1942). Life histories of North American flycatchers, larks, swallows and their allies. *Smithsonian Institution, U.S. Nat. Mus. Bull.* **179**. Washington D.C., U.S. Nat. Mus.

Berg, H. (1995). Modelling of DDT dynamics in Lake Kariba, a tropical man-made lake, and its implications for the control of tsetse flies. *Ann. Zool. Fenn.* **32**: 331–353.

Bergerud, A.T., Mossop, D.H. & Myrberget, S. (1985). A critique of the mechanics of annual changes in Ptarmigan numbers. *Can. J. Zool.* **63**: 2240–2248.

Bernard, R.F. (1966). DDT residues in avian tissues. *J. Appl. Ecol.* **3 (Suppl.)**: 193–198.

Berndt, R. & Frantzen, M. (1964). Vom Einfluss des strengen Winters 1962/63 auf den Bruthestand der Hohlerbrüter bei Braunschweig. *Ornithol. Mitt.* **16**: 126–130.

Berndt, R. & Henss, M. (1967). Die Kohlmeise, *Parus major*, als Invasionsvogel. *Vogelwarte* **24**: 17–37.

Berner, T.O. & Grubb, T.C. (1985). An experimental analysis of mixed-species flocking in birds of deciduous woodland. *Ecology* **66**: 1229–1236.

Bernstein, C., Krebs, J.R. & Kacelnik, A. (1991). Distribution of birds amongst habitats: theory and relevance to conservation. pp. 317–45 in *'Bird population studies'* (ed. C.M. Perrins, J.D. Lebreton & G.J.M. Hirons). Oxford, Oxford University Press.

Berry, M.P.S. & Crowe, T.M. (1985). Effects of monthly and annual snowfall on game bird populations in the northern Cape Province, South Africa. *S. Afr. J. Wildl. Res.* **15**: 69–76.

Berthold, P. (1993). Bird migration. A general survey. Oxford, University Press.

Bevanger, K. (1995). Estimates and population consequences of tetraonid mortality caused by collisions with high tension power lines in Norway. *J. Appl. Ecol.* **32**: 745–753.

Bibby, C.J. & Green, R.E. (1980). Foraging behaviour of migrant Pied Flycatchers, *Ficedula hypoleuca*, on temporary territories. *J. Anim. Ecol.* **49**: 507–521.

Bibby, C.J., Collar, N.J., Crosby, M.J., Heath, M.F., Imboden, C., Johnson, T.H., Long, A.J., Stattersfield, A.J. & Thirgood, S.J. (1992). Putting biodiversity on the map: priority areas for global conservation. Cambridge. International Council for Bird Preservation.

Bibby, C.J. & Lloyd, C.S. (1977). Experiments to determine the fate of dead birds at sea. *Biol. Conserv.* **12**: 295–309.

Biebach, H. (1996). Energetics of winter and migratory fattening. pp. 280–323 in *'Avian Energetics and Nutritional Ecology'* (ed. C. Carey). London, Chapman & Hall.

Bierregaard, R.O. (1990). Avian communities in the understory of Amazonian forest fragments. pp. 333–343 in *'Biogeography and ecology of forest bird communities'* (ed. A. Keast). The Hague, Netherlands, SPB Academic Publishing.

Bierregaard, R.O., Lovejoy, T.E., Kapos, V., dos Santos, A.A. & Hutchings, R.W. (1992). The biological dynamics of tropical rainforest fragments. *BioScience* **42**: 859–866.

Bijleveld, M. (1974). Birds of prey in Europe. London, Macmillan Press.

Bijlsma, R.G. (1990). Predation by large falcons on wintering waders on the Banc d'Arguin, Mauritania. *Ardea* **78**: 75–82.

Birkhead, M. (1982). Causes of mortality in the Mute Swan on the River Thames. *J. Zool., Lond.* **198**: 15–25.

Birkhead, M. & Perrins, C. (1985). The breeding biology of the Mute Swan *Cygnus olor* on the River Thames with special reference to lead poisoning. *Biol. Conserv.* **32**: 1–11.

Birkhead, T.R. (1977). The effect of habitat and density on breeding success in Common Guillemot (*Uria aalge*). *J. Anim. Ecol.* **46**: 751–764.

Birkhead, T.R., Eden, S.F., Clarkson, K., Goodburn, S.F. & Pellatt, J. (1986). Social organisation of a population of Magpies *Pica pica*. *Ardea* **74**: 59–68.

Birkhead, T.R. & Furness, R.W. (1985). Regulation of seabird populations. pp. 145–167 in *'Behavioural ecology'* (ed. R.M. Sibly & R.H. Smith). Oxford, Blackwell.

Birt, V.L., Birt, T.P., Goulet, D., Cairns, D.K. & Montevecchi, N.A. (1987). Ashmole's halo: direct evidence for prey-depletion by a seabird. *Mar. Ecol. Prog. Ser.* **40**: 205–208.

Black, J.M. (1995). The Nene *Branta sandvicensis* recovery initiative – research against extinction. *Ibis* **137**: 153–160.

Black, J.M. & Owen, M. (1989). Agonistic behaviour in goose flocks: assessment, investment and reproductive success. *Anim. Behav.* **37**: 199–209.

Blackburn, T.M., Brown, V.K., Doube, B.M., Greenwood, J.J.D., Lawton, J.H. & Stork, N.E. (1992). The relationship between abundance and body size in natural animal assemblages. *J. Anim. Ecol.* **62**: 519–528.

Blackburn, T.M., Harvey, P.H. & Pagel, M.D. (1990). Species numbers, population density and body size relationships in natural communities. *J. Anim. Ecol.* **59**: 335–345.

Blackburn, T.M. & Lawton, J.H. (1994). Population abundance and body size in animal assemblages. *Phil. Trans. Roy. Soc. B* **343**: 33–39.

Blais, J.R. (1965). Spruce budworm outbreaks in the past three-centuries in the Laurentide Park, Quebec. *For. Sci.* **11**: 130–138.

Blancher, P.J. & McAuley, D.G. (1987). Influence of wetland acidity on avian breeding success. *Trans. N. A. Wildl. Nat Res. Conf.* **52**: 628–635.

Blancher, P.J. & McNichol, D.K. (1991). Tree Swallow diet in relation to wetland acidity. *Can. J. Zool.* **69**: 2629–2637.

Blancher, P.J. & Robertson, R.J. (1985). Predation in relation to spacing of Kingbird nests. *Auk* **102**: 654–658.

Blancher, P.J. & Robertson, R.J. (1987). Effect of food-supply on the breeding biology of Western Kingbirds. *Ecology* **68**: 723–732.

Blank, T.H., Southwood, T.R.E. & Cross, D.J.H. (1967). The ecology of the Partridge. I. Outline of population processes with particular reference to chick mortality and nest density. *J. Anim. Ecol* **36**: 549–556.

Blankinship, D.R. (1966). The relationship of White-winged Dove production to control of Great-tailed Grackles in the lower Rio Grande Valley of Texas. *Trans. N. A. Wildl. Nat. Res. Conf.* **31**: 45–58.

Blem, C.R. (1990). Avian energy storage. *Curr. Ornithol.* **7**: 59–113.

Blokpoel, H. & Smith, B. (1988). First records of roof nesting by Ring-billed Gulls and Herring Gulls in Ontario. *Ontario Birds* **6**: 15–18.

Blom, R. & Myrberget, S. (1978). Experiments on shooting territorial Willow Grouse *Lagopus lagopus*. *Cinclus*. **1**: 29–33.

Blondel, J., Pradel, R. & Lebreton, J.-D. (1992). Low fecundity insular Blue Tits do not survive better as adults than high fecundity mainland ones. *J. Anim. Ecol.* **61**: 205–213.

Blus, L.J. (1982). Further interpretation of the relation of organochlorine residues in Brown Pelican eggs to reproductive success. *Environ. Pollut.* **28**: 15–33.

Blus, L.J., Henny, C.J. & Grove, R.A. (1989). Rise and fall of endrin usage in Washington State Fruit Orchards: effects on wildlife. *Environ. Poll.* **60**: 331–349.

Blus, L.J., Henny, C.J., Hoffman, D.J. & Grave, R.A. (1991). Lead toxicosis in Tundra Swans near a mining and smelting complex in northern Idaho. *Arch. Environ. Contam. and Toxicol.* **21**: 549–555.

Boag, D.A., McCourt, K.H., Herzog, P.W. & Alway, J.H. (1979). Population regulation in spruce grouse: a working hypothesis. *Can. J. Zool.* **57**: 2275–2284.

Bock, C.E. & Lepthien, L.W. (1972). Winter eruptions of Red-breasted Nuthatches in North America. *Amer. Birds* **26**: 558–561.

Bock, C.E. & Lepthien, L.W. (1976). Synchronous eruptions of boreal seed-eating birds. *Amer. Nat.* **110**: 559–571.

Boersma, P.D. (1978). Breeding patterns of Galapagos Penguins as an indicator of oceanographic conditions. *Science* **200**: 1481–1483.

Bolen, E. (1967). Nesting boxes for Black-bellied Tree Ducks. *J. Wildl. Manage.* **31**: 794–797.

Bollinger, E.K. & Gavin, T.A. (1989). The effects of site quality on breeding site fidelity in Bobolinks. *Auk* **106:** 584–594.

Bolton, M., Houston, D. & Monaghan, P. (1992). Nutritional constraints on egg formation in the Lesser Black-backed Gull: an experimental study. *J. Anim. Ecol.* **61:** 521–532.

Booth, D.T., Clayton, D.H. & Block, B.A. (1993). Experimental demonstration of the energetic cost of parasitism in wild hosts. *Proc. Roy. Soc. B* **253:** 125–129.

Borg, K. Wanntorp, H., Erne, K. & Hanko, E. (1969). Alkyl mercury poisoning in terrestrial Swedish wildlife. *Viltrevy* **6:** 301–379.

Boshell, M.J. (1969). Kysanur forest disease: ecological considerations. *Amer. J. Trop. Med. and Hyg.* **18:** 65–80.

Boström, U. & Nilsson, S.G. (1983). Latitudinal gradients and local variations in species richness and structure of bird communities on raised peat-bogs in Sweden. *Ornis Scand.* **14:** 213–226.

Botzler, R.G. (1991). Epizootiology of avian cholera in wildfowl. J. *Wildl. Dis.* **27:** 367–395.

Boulinier, T. & Danchin, E. (1996). Population trends in Kittiwake *Rissa tridactyla* colonies in relation to tick infestation. *Ibis* **138:** 326–334.

Bourne, W.R.P. (1970). Special review – after the 'Torrey Canyon' disaster. *Ibis* **112:** 120–125.

Bourne, W.R.P., Parrack, J.D. & Potts, G.R. (1967). Birds killed in the Torrey Canyon disaster. *Nature* **215:** 1123–1125.

Boutin, S. (1990). Food supplementation experiments with terrestrial vertebrates: patterns, problems, and the future. *Can. J. Zool.* **68:** 203–220.

Bouvier, G. & Hornung, B. (1965). La pathologie de cygne tubercle (*Cygnus olor* Gmelin) en Suisse. *Mem. Soc. Vaud. Sci. Nat.* **14:** 1.

Bowman, R. & Bird, D.M. (1986). Ecological correlates of mate replacement in the American Kestrel. *Condor* **88:** 440–445.

Bowman, T.D., Schempf, P.F. & Bernatowicz, J.A. (1995). Bald Eagle survival and population dynamics in Alaska after the Exxon Valdez oil spill. *J. Wildl. Manage.* **59:** 317–324.

Boyce, M.S. (1992). Population viability analysis. *Ann. Rev. Ecol. Syst.* **23:** 481–506.

Boyce, M.S. & Miller, R.S. (1985). Ten-year periodicity in Whooping Crane census. *Auk* **102:** 658–660.

Boyd, A.W. (1932). Notes on the Tree Sparrow. *Brit. Birds* **25:** 278–285.

Boyd, H. (1953). On encounters between wild White-fronted Geese in winter flocks. *Behaviour* **5:** 85–129.

Boyd, H. (1964). Wildfowl and other water-birds found dead in England and Wales in January-March 1962. *Ann. Rep. Wildfowl Trust* **15:** 20–22.

Boyd, H. (1990). Duck numbers in the USSR, the western Palaearctic and North America 1967–86: first comparisons. *Wildfowl* **41:** 171–175.

Bradstreet, M.S.W. (1979). Thick-billed Murres and Black Guillemots in the Barrow Strait area, North West Territories, during spring: distribution and habitat use. *J. Zool., Lond.* **57:** 1789–1802.

Braithwaite, L.W. (1976). Breeding seasons of waterfowl in Australia. *Proc. Int. Orn. Congr.* **16:** 235–247.

Brand, C.J. (1984). Avian cholera in the central and Mississippi flyways during 1979–80. *J. Wildl. Manage.* **48:** 399–406.

Braun, C.E. & Rogers, G.E. (1971). The White-tailed Ptarmigan in Colorado. Tech. Publ. No. 27. Div. Game, Fish & Parks, Colorado.

Brawn, J.D. & Balda, R.P. (1988). Population biology of cavity–nesters in northern Arizona: do nest sites limit breeding densities? *Condor* **90:** 61–71.

Brawn, J.D., Boecklen, W.J. & Balda, R.P. (1987). Investigations of density interactions among breeding birds in Ponderosa pine forests – correlative and experimental evidence. *Oecologia* **72:** 348–357.

Bremer, P.E. (1977). Pelican kill. *Loon* **49:** 240–241.

Brinkhof, M.W.G. & Cavé, A.S. (1997). Food supply and seasonal variation in fledging success: an experiment in the European Coot. *Proc. Roy. Soc. B*, **264**: 291–296.

Brittingham, M.C. & Temple, S.A. (1983). Have cowbirds caused forest song birds to decline? *Bioscience* **33**: 31–35.

Brittingham, M.C. & Temple, S.A. (1988). Impacts of supplemental feeding on survival rates of Black-capped Chickadees. *Ecology* **69**: 581–589.

Brockless, M. (1995). Game and shooting at Loddington. *Game Conservancy Rev.* **26**: 67–68.

Brockman, H.J. & Barnard, C.J. (1979). Kleptoparasitism in birds. *Anim. Behav.* **27**: 487–514.

Bromssen, A.V. & Jansson, C. (1980). Effects of food addition to Willow Tit *Parus montanus* and Crested Tit *P. cristatus* at the time of breeding. *Ornis Scand.* **11**: 173–178.

Brooke, M. de L. (1979). Differences in the quality of territories held by Wheatear (*Oenanthe oenanthe*). *J. Anim. Ecol.* **48**: 21–32.

Brooke, R.K., Lloyd, P.H. & de Villiers, A.L. (1986). Alien and translocated terrestrial vertebrates in South Africa. pp. 63–74 in *'The ecology and management of biological invasions in southern Africa'* (ed. I.A.W. Macdonald, F.J. Kruger & A.A. Ferrar). Capetown, Oxford University Press.

Brothers, N. (1991). Albatross mortality and associated bait loss in the Japanese longline fishery in the Southern Ocean. *Biol. Conserv.* **55**: 255–268.

Brown, A.W. & Brown, L.M. (1993). The Scottish Mute Swan census. *Scott. Birds* **17**: 93–102.

Brown, A.W.A. (1978). Ecology of pesticides. New York: Wiley.

Brown, C.R. & Brown, M.B. (1986). Ectoparasitism as a cost of coloniality in Cliff Swallows (*Hirundo pyrrhonata*). *Ecology* **67**: 1206–1218.

Brown, C.R. & Brown, M.B. (1992). Ectoparasitism as a cause of natal dispersal in Cliff Swallows. *Ecology* **73**: 1718–1723.

Brown, C.R. & Brown, M.B. (1996). Coloniality in the Cliff Swallow. Chicago, University Press.

Brown, D. & Rothery, P. (1993). Models in biology: mathematics, statistics and computing. New York: Wiley.

Brown, J.H. & Kodric-Brown, A. (1977). Turnover rates in insular biogeography: effect of immigration on extinction. *Ecology* **58**: 445–449.

Brown, J.H. & Maurer, B.A. (1987). Evolution of species assemblages: effects of energetic constraints and species dynamics on the diversification of the North American avifauna. *Amer. Nat.* **130**: 1–7.

Brown, J.L. (1969). Territorial behaviour and population regulation in birds. *Wilson Bull.* **81**: 293–329.

Brown, J.L., Dow, D.D., Brown, E.R. & Brown, S.D. (1978). Effects of helpers on feeding of nestlings in the Grey-crowned Babbler *Pomatostomus temporalis*. *Behav. Ecol. Sociobiol.* **4**: 43–60.

Brown, R.G.B. (1979). Seabirds of the Senegal upwelling and adjacent waters. *Ibis* **121**: 283–292.

Brown, R.G.B. (1980). Seabirds as marine animals. pp. 1–39 in *'Behaviour of marine animals'*, Vol. 4 (ed. B.O. Burger & H.E. Winn). New York, Plenum Press.

Brown, R.G.B., Cooke, F., Kinnear, P.K. & Mills, E.L. (1975). Summer seabird distributions in Drake Passage, the Chilean fjords and off southern South America. *Ibis* **117**: 339–356.

Brown, W.L. & Wilson, E.O. (1956). Character displacement. *Systematic Zool.* **5**: 49–64.

Browne, R.A., Griffin, C.R., Chang, P.R., Hubley, M. & Martin, A.E. (1993). Genetic divergence among populations of the Hawaiian Duck, Laysan Duck and Mallard. *Auk* **110**: 49–56.

Bruderer, B. & Muff, J. (1979). Bestandesschwankungen schweizerischer Rauch- und Mehlserchwalben, insbesondere im Zusammenhang mit der Schwalbenkatastrophe in Herbst (1974). *Ornithol. Beob.* **76**: 229–234.

Bruggers, R.L. & Elliott, C.C.H. (eds.). (1989). *Quelea quelea* – Africa's bird pest. Oxford, University Press.

Brush, T. (1983). Cavity use by secondary cavity-nesting birds and response to manipulations. *Condor* **85**: 461–466.

Bryant, D.M. (1975). Breeding biology of the House Martin *Delichon urbica*, in relation to aerial insect abundance. *Ibis* **117:** 180–215.

Bryant, D.M. & Jones, G. (1995). Morphological changes in a population of Sand Martins *Riparia riparia* associated with fluctuations in population size. *Bird Study* **42:** 57–65.

Bryant, D.M. & Newton, A.V. (1994). Metabolic costs of dominance in Dippers *Cinclus cinclus. Anim. Behav.* **48:** 449–455.

Bucher, E.H. (1988). Do birds use biological control against nest parasites? *Parasitol. Today* **4:** 1–3.

Buck, N.A., Estesen, B.J. & Ware, G.W. (1983). DDT moratorium in Arizona: residues in soil and alfalfa after 12 years. *Bull. Environ. Contam. Toxicol.* **31:** 66–72.

Buckley, P.A. & McCarthy, M.G. (1994). Insects, vegetation, and the control of Laughing Gulls (*Larus atricilla*) at Kennedy International Airport, New York City. *J. Appl. Ecol.* **31:** 291–302.

Bull, P.C. & Dawson, P.G. (1969). Mortality and survival of birds during an unseasonable snow storm in South Canterbury, November 1967. *Notornis* **14:** 172–179.

Bull, K.R., Every, W.J., Freestone, P., Hall, J.R., Osborn, D., Cooke, A.S. & Stowe, T.J. (1983). Alkyl lead pollution and bird mortalities on the Mersey Estuary, U.K. 1979–81. *Environ. Pollut.* **A31:** 239–259.

Bump, G., Darrow, R.W., Edminster, F.G. & Crissey, W.F. (1947). The Ruffed Grouse. New York, Authority of the New York State Legislature.

Bumpus, H. (1899). The elimination of the unfit as illustrated by the introduced sparrow, *Passer domesticus. Mar. Biol. Lab., Biol. Lect.* (Woods Hole, 1898) 209–228.

Burger, A.E. & Cooper, J. (1982). The effects of fisheries on seabirds in South Africa and Namibia. pp. 150–160 in '*Proceedings of Pacific Seabird Group Symposium*' (ed. D. Nettleship, G.A. Sanger & P.F. Springer). *Can. Wildl. Serv. Spec. Publ.,* Seattle, Washington.

Burger, G.V. (1964). Survival of Ring-necked Pheasants released on a Wisconsin shooting preserve. *J. Wildl. Manage.* **28:** 711–721.

Burger, G.V. (1966). Observations on aggressive behaviour of male Ring-necked Pheasants in Wisconsin. *J. Wildl. Manag.* **30:** 57–64.

Burger, J. (1990). Effects of age on foraging in birds. *Proc. Int. Ornithol. Congr.* **19:** 1127–1140.

Burger, J. & Gochfield, M. (1983). Feeding behaviour in Laughing gulls: compensatory site selection by young. *Condor* **85:** 467–473.

Burgerjon, J.J. (1964). Some census notes on a colony of South African Cliff Swallows *Petrochelidon spilodera* (Sundevall). *Ostrich* **35:** 77–85.

Burgess, E.C., Ossa, J. & Yuill, T.M. (1979). Duck plague: a carrier state in waterfowl. *Avian Dis.* **24:** 940–949.

Buss, I.O. (1942). A managed Cliff Swallow colony in southern Wisconsin. *Wilson Bull.* **54:** 153–161.

Butler, D. & Merton, D. (1992). The Black Robin. Oxford, University Press.

Byrd, G.V., Trapp, J.L. & Zeillemaker, C.F. (1994). Removal of introduced foxes: a case study in restoration of native birds. *Trans. N.A. Wildl. Nat. Res. Conf.* **59:** 317–321.

Cade, T.J. (1974). Plans for managing the survival of the Peregrine Falcon. *Raptor Res. Rep.* **2:** 89–104.

Cade, T.J. (1983). Hybridisation and gene exchange among birds in relation to conservation. pp. 288–309 in '*Genetics and conservation*' (ed. C.M. Schonewald-Cox, S.M. Chambers, B. MacBryde & W.L. Thomas). Menlo Park, California, Benjamin Cummings.

Cade, T.J. & Bird, D.M. (1990). Peregrine Falcons, *Falco peregrinus*, nesting in an urban environment: a review. *Can. Field Nat.* **104:** 209–218.

Cade, T.J., Enderson, J.H., Thelander, C.G. & White, C.M. (1988). Peregrine Falcon populations: their management and recovery. Boise, The Peregrine Fund.

Cade, T.J. & Temple, S.A. (1995). Management of threatened bird species: evaluation of the hands-on approach. *Ibis* **137 (suppl. 1.):** 161–172.

Cairns, D.K. (1989). The regulation of seabird colony size – a hinterland model. *Amer. Nat.* **134:** 141–146.

Cairns, D.K. (1992). Bridging the gap between ornithology and fisheries science – use of seabird data in stock assessment models. *Condor* **94**: 811–824.

Campbell, L., Cayford, J. & Pearson, D. (1996). Bearded Tits in Britain and Ireland. *Brit. Birds* **89**: 335–346.

Campbell, L.H. (1984). The impact of changes in sewage treatment on seaducks wintering in the Firth of Forth, Scotland. *Biol. Conserv.* **28**: 173–180.

Caraco, T. (1979). Time budgeting and group size: a theory. *Ecology* **60**: 618–627.

Caraco, T., Martindale, S. & Pulliam, H.R. (1980). Avian time budgets and distance to cover. *Auk* **97**: 872–875.

Carbone, C. & Owen, M. (1995). Differential migration of the sexes of Pochard *Aythya ferina*: results from a European survey. *Wildfowl* **46**: 99–108.

Carlson, A. & Aulén, G. (eds.). (1990). Conservation and management of woodpecker populations. Report 17, Uppsala, Dept. Wildlife Ecology, Uppsala University.

Carlsson, H., Carlsson, L., Wallin, C. & Wallin, N.-E. (1991). Great Tits incubating empty nest cups. *Ornis Svec.* **1**: 51–52.

Carpenter, E.J. & Smith, K.L. (1972). Plastics on the Sargasso Sea surface. *Science* **175**: 1240–1241.

Carrick, R. (1963). Ecological significance of territory in the Australian Magpie, *Gymnorrhina tibicen*. *Proc. Int. Ornithol. Congr.* **13**: 740–753.

Case, T.J. (1996). Global patterns in the establishment and distribution of exotic birds. *Biol. Conserv.* **78**: 69–96.

Catterall, C.P., Wyatt, W.S. & Henderson, L.J. (1982). Food resources, territory density and reproductive success of an island Silvereye population *Zosterops lateralis*. *Ibis* **124**: 405–421.

Caughley, G. (1970). Eruption of ungulate populations, with emphasis on Himalayan Thar in New Zealand. *Ecology* **51**: 53–72.

Caughley, G. (1977). Analysis of vertebrate populations. New York, Wiley.

Caughley, G. (1994). Directions in conservation biology. *J. Anim. Ecol.* **63**: 215–244.

Cavé, A.J. (1968). The breeding of the Kestrel, *Falco tinnunculus* L., in the reclaimed area Oostelijk Flevoland. *Netherlands J. Zool.* **18**: 313–407.

Cavé, A.J. (1983). Purple Heron survival and drought in tropical West Africa. *Ardea* **71**: 217–224.

Cavé, A.J. & Visser, J. (1985). Winter severity and breeding bird numbers in a Coot population. *Ardea* **73**: 129–138.

Cawthorne, R.A. & Marchant, J.H. (1980). The effects of the 1978/79 winter on British bird populations. *Bird Study* **27**: 163–172.

Cederholm, G. & Ekman, J. (1976). A removal experiment on Crested Tit *Parus cristatus* and Willow Tit *P. montanus* in the breeding season. *Ornis. Scand.* **7**: 207–213.

Chapman, B.R. & George, J.E. (1991). The effects of ectoparasites on Cliff Swallow growth and survival. pp. 69–92 in *'Bird–parasite interactions'* (ed. J.E. Loye & M. Zuk). Oxford, University Press.

Charles, J.K. (1972). Territorial behaviour and the limitation of population size in Crows, *Corvus corone* and *Corvus cornix*. Ph.D. thesis (Unpub), University of Aberdeen, Scotland.

Charnov, E.L., Orians, G.A. & Hyatt, K. (1976). Ecological implications of resource depression. *Amer. Nat.* **110**: 247–259.

Chase, M.K., Nur, N. & Geupel, G.R. (1997). Survival, productivity, and abundance in a Wilson's Warbler population. *Auk* **114**: 354–366.

Cheke, A. (1975). Cyclone Gervoise – an eye witness' comments. *Birds International* **1**: 13–14.

Chesness, R.A., Nelson, M.M. & Longley, W.H. (1968). The effect of predator control on Pheasant reproductive success. *J. Wildl. Manage.* **32**: 683–697.

Cheylan, G. (1973). Notes sur la compétition entré l'Aigle royal *Aquila chrysaëtos* et l'Aigle de Bonelli *Hieraaétus fasciatus*. *Alauda* **41**: 303–312.

Chitty, D. (1967). The natural selection of self-regulatory behaviour in animal populations. *Proc. Ecol. Soc. Aust.* **2**: 51–78.

Choate, J.S. (1967). Factors influencing nesting success of Eiders in Penobscot Bay, Maine. *J. Wildl. Manage.* **31:** 769–777.

Christensen, K.M. & Whitham, T.G. (1993). Impact of insect herbivores on competition between birds and mammals for Pinyon Pine seeds. *Ecology* **74:** 2270–2278.

Cimprich, D.A. & Grubb, T.C. (1994). Consequences for Carolina Chickadees of foraging with Tufted Titmice in winter. *Ecology* **75:** 1615–1625.

Clark, C.W. (1981). Bioeconomics. pp. 387–418 in *'Theoretical ecology'* (ed. R.M. May). Oxford, Blackwell Scientific Publications.

Clark, G.M., O'Meara, D. & van Weelden, J.W. (1958). An epizootic among Eider Ducks involving an acanthocephalid worm. *J. Wildl. Manage.* **22:** 204–205.

Clark, J. A., Baillie, S.R., Clark, N.A. & Langston, R.H.W. (1993). Estuary wader capacity following severe weather mortality. *Thetford, British Trust for Ornithology Res. Rep.* **No 103.**

Clark, L. & Mason, J.R. (1985). Use of nest material as insecticidal and antipathogenic agents by the European Starling. *Oecologia* **67:** 169–176.

Clark, R.G., Meger, D.E. & Ignatiuk, J.B. (1995). Removing American Crows and duck nest success. *Can. J. Zool.* **73:** 518–522.

Clark, W.S. (1985). The migrating Sharp-shinned Hawks at Cape May Point: banding and recovery results. pp. 137–148 in *'Proceedings of the Hawk Migration Conference 4, Rochester 1983'* (ed. M. Harewood). Hawk Migration Association of North America, Rochester, New York.

Clarke, R.B. (1984). Impact of oil pollution on seabirds. *Environ. Poll.* **33:** 1–22.

Clark, R.G. & Weatherhead, P.J. (1987). Influence of population size on habitat use by territorial male Red-winged Blackbirds in agricultural landscapes. *Auk* **104:** 341–315.

Clawson, S.G. & Baker, M.F. (1959). Immediate effects of dieldrin and heptachlor on Bobwhite. *J. Wildl. Manage.* **23:** 215–219.

Clayton, D.H. (1990). Mate choice in experimentally parasitised Rock Doves: lousy males lose. *Amer. Zool.* **30:** 251–262.

Clayton, D.H. (1991). Coevolution of avian grooming and ectoparasite avoidance. pp. 258–289 in *'Bird–parasite interactions'* (ed. J.E. Loye & M. Zuk). Oxford, University Press.

Cline, S.P., Berg, A.B. & Wight, H.M. (1980). Snag characteristics and dynamics in Douglas Fir forests, Western Oregon. *J. Wildl. Manage.* **44:** 773–786.

Cody, M.L. (1983). Bird diversity and density in South African forests. *Oecologia* **59:** 201–215.

Cody, M.L. & Cody, C.B.J. (1972). Territory size, clutch size and food in populations of Wrens. *Condor* **74:** 473–477.

Coleman, J.D. (1974). The use of artificial nest sites erected for Starlings in Canterbury, New Zealand. *N.Z. J. Zool.* **1:** 349–354.

Collar, N.J. & Andrew, P. (1988). Birds to watch. Cambridge, *ICBP. Tech. Publ.* **8.**

Collar, N.J., Crosby, M.J. & Stattersfield, A.J. (1994). Birds to watch 2: the world list of threatened birds. Cambridge, Birdlife International.

Collias, M., Collias, E. & Jennrich, R.I. (1994). Dominant Red Junglefowl (*Gallus gallus*) hens in an unconfined flock rear the most young over their lifetime. *Auk* **111:** 863–872.

Colton, J.B., Knapp, F.D. & Burns, B.R. (1974). Plastic particles in surface waters of the northwestern Atlantic. *Science* **185:** 491–497.

Conan, G. (1982). The long-term effects of the Amoco Cadiz oil spill. *Phil. Trans. Roy. Soc. Lond.* **B 297:** 323–334.

Confer, J.L. & Knapp, K. (1981). Golden-winged Warblers and Blue-winged Warblers – the relative success of a habitat specialist and a generalist. *Auk* **98:** 108–114.

Connell, J.H. (1983). On the prevalence and relative importance of interspecific competition: evidence from field experiments. *Amer. Nat.* **122:** 611–696.

Connors, P.G. & Smith, K.G. (1982). Oceanic plastic particle pollution: suspected effect on fat deposition in Red Phalaropes. *Mar. Poll. Bull.* **13:** 18–20.

Cooch, E.G., Lank, D.B., Rockwell, R.F. & Cooke, F. (1989). Long-term decline in fecundity in a Snow Goose population: evidence for density dependence? *J. Anim. Ecol.* **58:** 711–726.

Cooke, A. (1975). The effects of fishing on waterfowl on Grafham Water. *Cambridge Bird Club* **48**: 40–46.

Cooke, A.S. (1973). Shell-thinning in avian eggs by environmental pollutants. *Environ. Poll.* **4**: 85–152.

Cooke, B.K. & Stringer, A. (1982). Distribution and breakdown of DDT in orchard soil. *Pestic. Sci.* **13**: 545–551.

Cooke, F. (1987). Lesser Snow Goose: a long-term population study. pp. 407–432 in '*Avian Genetics. A population and ecological approach*' (ed. F. Cooke & P.A. Buckley). London, Academic Press.

Coombes, C. (1996). Parasite, biodiversity and ecosystem stability. *Biodiv. Conserv.* **5**: 953–962.

Cooper, A., Atkinson, I.A.E., Lee, W.G. & Worthy, T.H. (1993). Evolution of the Moa and their effect on the New Zealand flora. *Trends Ecol. Evol.* **8**: 433–437.

Cooper, J.E. & Petty, S.J. (1988). Trichomoniasis in free-living Goshawks (*Accipiter gentilis gentilis*) from Great Britain. *J. Wildl. Dis.* **24**: 80–87.

Cooper, R.J., Dodge, K.M., Martinat, P.J., Donahoe, S.B. & Whitmore, R.C. (1990). Effect of diflubenzuron application on eastern deciduous forest birds. *J. Wildl. Manage.* **54**: 486–493.

Coulson, J. (1968). Differences in the quality of birds nesting in the centre and on the edges of a colony. *Nature* **217**: 478–479.

Coulson, J.C. (1983). The changing status of the Kittiwake *Rissa tridactyla* in the British Isles 1969–1979. *Bird Study* **30**: 9–16.

Coulson, J.C., Duncan, N. & Thomas, C. (1982). Changes in the breeding biology of the Herring Gull (*Larus argentatus*) induced by reduction in the size and density of the colony. *J. Anim. Ecol.* **51**: 739–756.

Coulson, J.C., Potts, G.R., Deans, I.R. & Fraser, S.M. (1968). Exceptional mortality of Shags and other seabirds caused by paralytic shellfish poison. *Brit. Birds* **61**: 381–404.

Coulson, J.C. & Wooller, R.D. (1976). Differential survival rates among breeding Kittiwake Gulls *Rissa tridactyla* (L). *J. Anim. Ecol.* **45**: 205–213.

Court, G.S., Bradley, D.M., Gates, C.C. & Boag, D.A. (1988). The population biology of Peregrine Falcons in the Keewater District of the Northwest Territories, Canada. pp. 729–739 in '*Peregrine Falcon populations. Their management and recovery*' (ed. T.J. Cade, J.H. Enderson, G.J. Thelander & C.M. White). Boise, The Peregrine Fund.

Cowardin, L.M., Shaffer, T.L. & Kraft, K.M. (1995). How much habitat management is needed to meet Mallard production objectives. *Wildl. Soc. Bull.* **23**: 48–55.

Cowie, R.J. & Simons, J.R. (1991). Factors affecting the use of feeders by garden birds: 1. The positioning of feeders with respect to cover and housing. *Bird Study* **38**: 145–150.

Cowley, E. (1979). Sand Martin population trends in Britain, 1965–1975. *Bird Study* **26**: 113–116.

Cowley, R.D. (1971). Birds and forest management. *Aust. For.* **35**: 234–250.

Craig, J.L. (1991). Are small populations viable? *Proc. Int. Ornithol. Congr.* **20**: 2546–2551.

Cramp, S. (1973). The effects of pesticides on British wildlife. *Brit. Vet. J.* **129**: 315–323.

Cramp, S. (ed.) (1983). Handbook of the birds of Europe, the Middle East and North Africa. Oxford, University Press.

Cramp, S., Condor, P.J. & Ash, J. (1962). Deaths of birds and mammals from toxic chemicals. Second Report of the Joint Committee of the British Trust for Ornithology, the Royal Society for the Protection of Birds and the Game Research Association.

Cramp, S. & Simmons, K.E.L. (1983). Handbook of the Birds of Europe, the Middle East and North Africa, Vol 3. Oxford, Oxford University Press.

Crawford, H.S. & Jennings, D.T. (1989). Predation by birds on spruce budworm *Choristoneura fumiferana*: functional, numerical and total responses. *Ecology* **70**: 152–163.

Crawford, R.J.M., Allwright, D.M. & Heyl, C.W. (1992). High mortality of Cape Cormorants (*Phalacrocorax capensis*) off western South Africa in 1991 caused by *Pasteurella multocida*. *Colonial Waterbirds* **15**: 236–238.

Crawford, R.J.M. & Dyer, B.M. (1995). Responses by 4 seabird species to a fluctuating availability of Cape Anchovy *Engraulis capensis* off South Africa. *Ibis* **137**: 329–339.

Crawford, R.J.M. & Shelton, P.A. (1978). Pelagic fish and seabird inter-relationships off the coasts of South West and South Africa. *Biol. Conserv.* **14**: 85–109.

Crawford, R.J.M., Shelton, P.A., Cooper, J. & Brooke, R.K. (1983). Distribution, population size and conservation of the Cape Gannet *Morus capensis*. *S. Afr. J. Mar. Sci.* **1**: 153–174.

Crawford, R.J.M., Williams, A.J., Randall, R.M., Randall, B.M., Berruti, A. & Ross, G.J.B. (1990). Recent population trends of Jackass Penguins *Spheniscus demersus* off southern Africa. *Biol. Conserv.* **52**: 229–243.

Cresswell, W. (1994). Age-dependent choice of Redshank (*Tringa totanus*) feeding location – profitability or risk. *J. Anim. Ecol.* **63**: 589–600.

Cresswell, W. & Whitfield, D.P. (1994). The effects of raptor predation on wintering wader populations at the Tyningham Estuary, southeast Scotland. *Ibis* **136**: 223–232.

Creutz, G. (1949). Untersuchungen zur Brutbiologie des Feldsperlings (*Passer m. montanus* L.). *Zool. Jahrb.* **78**: 133–172.

Crick, H.Q.P. & Ratcliffe, D.A. (1995). The Peregrine, *Falco peregrinus* breeding population of the United Kingdom in 1991. *Bird Study* **42**: 1–19.

Crick, H.Q.P., Dudley, C., Glue, D.E. & Thomson, D.L. (1997). UK birds are laying eggs earlier. *Nature* **388**: 526.

Crissey, W.F. & Darrow, R.W. (1949). A study of predator control on Valcour Island. New York State Conservation Dept., Division of Fish and Game. Res. Ser. No.1.

Crook, J.H. (1965). The adaptive significance of avian social organisations. *Symp. Zool. Soc. Lond.* **14**: 181–218.

Crook, J.H. & Ward, P. (1968). The Quelea problem in Africa. pp. 211–229 in *'The problems of birds as pests'* (ed. R.K. Murton & E.N. Wright). London, Academic Press.

Crossner, K.A. (1977). Natural selection and clutch size in the European Starling. *Ecology* **58**: 885–892.

Crouter, R.A. & Vernon, E.H. (1959). Effects of Black-headed Budworm control on salmon and trout in British Columbia. *Can. Fish. Cult.* **24**: 23–40.

Croxall, J.P., Rothery, P., Pickering, S.P.C. & Prince, P.A. (1990). Reproductive performance, recruitment and survival of Wandering Albatrosses *Diomedea exulans* at Bird Island, South Georgia. *J. Anim. Ecol.* **59**: 775–796.

Cruz, A., Manolis, T. & Wiley, J.W. (1985). The Shiny Cowbird: a brood parasite expanding its range in the Caribbean region. *Orn. Monogr.* **36**: 607–620.

Cullen, J.M., Guiton, P.E., Horridge, G.A. & Peirson, J. (1952). Birds on Palma and Gomera (Canary Islands). *Ibis* **94**: 68–84.

Currie, F.A. & Bamford, R. (1982). Songbird nest-box studies in forests in north Wales. *Quart. J. For.* **76**: 250–255.

Curry, R.L. & Grant, P.R. (1989). Demography of the cooperatively breeding Galapagos Mockingbird, *Nesomimus parvulus* in a climatically variable environment. *J. Anim. Ecol.* **58**: 441–463.

Cutforth, B. (1968). Eighth Annual Report of the Brandon Juniors nest box project, 1968. *Blue Jay* **26**: 188.

Dahlsten, D.C. & Copper, W.A. (1979). The use of nesting boxes to study the biology of the Mountain Chickadee *Parus gambeli* and the impact on selected forest insects. pp. 217–260 in *'The role of insectivorous birds in forest ecosystems'* (ed. J.G. Dickson, R.N. Conner, R.R. Fleet, J.C. Kroll & J.A. Jackson). London, Academic Press.

Dallinga, J.H. & Schoenmakers, S. (1989). Population changes of the White Stork *Ciconia ciconia* since the 1850s in relation to food resources. pp. 231–262 in *'White Stork. Status and Conservation'* (ed. G. Rheinwald, J. Ogden & H. Schulz). Bonn, Dachverband Deutscher Avifaunisten, International Council for Bird Preservation.

Dane, D.S. (1948). A disease of Manx Shearwaters (*Puffinus puffinus*). *J. Anim. Ecol.* **17**: 158–164.

Dare, P. (1961). Ecological observations on a breeding population of the Common Buzzard *Buteo buteo*. Ph.D. thesis (Unpub), Exeter University.

Darwin, C. (1859). The origin of species. London, John Murray.

Darwin, C. (1868). The variation of animals and plants under domestication. London, John Murray.

Davies, J.C. & Cooke, F. (1983). Annual nesting production in Snow Geese: Prairie Droughts and Arctic Springs. *J. Wildl. Manage.* **47:** 291–296.

Davies, N.B. (1978). Ecological questions about territorial behaviour. pp. 317–350 in *'Behavioural Ecology. An Evolutionary Approach'* (ed. J.R. Krebs & N.B. Davies). Oxford, Blackwell Scientific Publications.

Davies, N.B. & Houston, A.I. (1983). Time allocation between territories and flocks and owner satellite conflict in foraging Pied Wagtails, *Motacilla alba*. *J. Anim. Ecol.* **52:** 621–634.

Davies, N.B. & Lundberg, A. (1984). Food distribution and a variable mating system in the Dunnock, *Prunella modularis*. *J. Anim. Ecol.* **54:** 895–912.

Davies, N.B. & Lundberg, A. (1985). The influence of food on time budgets and timing of breeding of the Dunnock *Prunella modularis*. *Ibis* **127:** 100–110.

Davis, J. (1973). Habitat preferences and competition of wintering Juncos and Golden-crowned Sparrows. *Ecology* **54:** 174–180.

Davis, P.E. & Newton, I. (1981). Population and breeding of Red Kites in Wales over a 30-year period. *J. Anim. Ecol.* **50:** 759–772.

de Bruijn, O. (1994). Population ecology and conservation of the Barn Owl *Tyto alba* in farmland habitats in Liemers and Achterhoek (The Netherlands). *Ardea* **82:** 1–109.

Deadman, A.J. (1973). A population study of the Coal Tit (*Parus ater*) and the Crested Tit (*Parus cristatus*) in a Scottish pine plantation. Ph.D. thesis (Unpub), Aberdeen University.

Deerenberg, C., Arpanius, V., Daan, S. & Bos, N. (1997). Reproductive effort decreases antibody responsiveness. *Proc. Roy. Soc. B* **264:** 943–1102.

Dein, F.J., Carpenter, J.W., Clark, G.G., Montali, R.J., Crabbs, C.L., Tsai, T.F. & Docherty, D.E. (1986). Mortality of captive Whooping Cranes caused by eastern equine encephalitis virus. *J. Amer. Vet. Med. Assoc.* **189:** 1006–1010.

De Laet, J.F. (1985). Dominance and anti-predator behaviour of Great Tits *Parus major*: a field study. *Ibis* **127:** 372–377.

Delannoy, C.A. & Cruz, A. (1991). *Philornis* parasitism and nestling survival of the Puerto Rican Sharp-shinned Hawk. pp. 93–103 in *'Bird–parasite interactions'* (ed. J.E. Loye & M. Zuk). Oxford, University Press.

Delius, J.D. (1965). A population study of Skylarks *Alauda arvensis*. *Ibis* **167:** 466–492.

Dempster, J.P. (1975). Animal population ecology. London, Academic Press.

Dempster, J.P. (1983). The natural control of populations of butterflies and moths. *Biol. Rev.* **58:** 461–481.

Dennis, R.H. & Dow, H. (1984). The establishment of a population of Goldeneyes (*Bucephala clangula*) breeding in Scotland. *Bird Study* **31:** 217–222.

DeReede, R.H. (1982). A field study on the possible impact of the insecticide diflubenzuron on insectivorous birds. *Agro-Ecosystems* **7:** 327–342.

Desgranges, J.L., Manuffette, Y. & Gagnon, G. (1987). Sugar maple forest decline and implications for forest insects and birds. *Trans. N. A. Wildl. Nat. Res. Conf.* **52:** 677–689.

Desrochers, A. (1989). Sex, dominance and microhabitat use in wintering Black-capped Chickadees: a field experiment. *Ecology* **70:** 636–645.

Desrochers, A., Hannon, S.J. & Nordin, K.E. (1988). Winter survival and territory acquisition in a northern population of Black-capped Chickadees. *Auk* **105:** 727–736.

Desser, S.S., Stuht, J. & Fallis, A.M. (1978). Leucocytozoonosis in Canada Geese in upper Michigan. 1. Strain differences among geese from different localities. *J. Wildl. Dis.* **14:** 124–131.

de Wit, A.A.N. & Spaans, A.L. (1984). Veranderingen in de broedbiologie van de Zilvermeeuw *Larus argentatus* door oegenomen aantallen. *Limosa* **57:** 87–90.

Dhindsa, M.S. & Boag, D.A. (1990). The effect of food supplementation on the reproductive success of Black-Billed Magpies *Pica pica*. *Ibis* **132:** 595–602.

Dhondt, A.A. (1971). Some factors influencing territory in the Great Tit, *Parus major* L. *Gerfaut* **61:** 125–135.

Dhondt, A.A. (1989). Ecological and evolutionary effects of interspecific competition in tits. *Wilson Bull.* **101:** 198–216.

Dhondt, A.A. & Eyckerman, R. (1980). Competition between the Great Tit and the Blue Tit outside the breeding season in field experiments. *Ecology* **61:** 1291–1296.

Dhondt, A.A. & Hublé, J. (1968). Age and territory in the Great Tit (*Parus major* L.). *Angew. Ornithol.* **3:** 20–24.

Dhondt, A.A., Kempenaers, B. & Adriaensen, F. (1992). Density-dependent clutch-size caused by habitat heterogeneity. *J. Anim. Ecol.* **61:** 643–648.

Dhondt, A.A., Kempenaers, B. & Laet, J.D. (1991). Protected winter roosting sites as a limiting resource for Blue Tits. *Proc. Int. Ornithol. Congr.* **20:** 1436–1443.

Dhondt, A.A. & Schillemans, J. (1983). Reproductive success of the Great Tit in relation to its territorial status. *Anim. Behav.* **31:** 902–912.

Dempster, J.P. (1975). Animal population ecology. London, Academic Press.

Diamond, J.M. (1972). Biogeographic kinetics: estimation of relaxation time for avifaunas of southwest Pacific Islands. *Proc. Nat. Acad. Sci.* **69:** 3199–3203.

Diamond, J.M. (1975). Assembly of species communities. pp. 342–444 in '*Ecology and evolution of communities*' (ed. M.L. Cody & J.M. Diamond). Cambridge, Massachusetts, Harvard University Press.

Diamond, J.M. (1982). Man the exterminator. *Nature* **298:** 787–789.

Diamond, J.M. (1984). "Normal" extinctions of isolated populations. pp. 191–246 in '*Extinctions*' (ed. M.H. Nitecki). Chicago, University Press.

Diamond, J.M. (1989). Quaternary megafaunal extinctions: variations on a theme by Paganini. *J. Archaeol. Sci.* **16:** 167–175.

Diamond, J.M. (1991). The rise and fall of the third chimpanzee. London, Vintage.

Diamond, J.M., Bishop, K.D. & Balen, S.V. (1987). Bird survival in an isolated Java woodland: Island or Mirror? *Conserv. Biol.* **1:** 132–142.

Diamond, J.M. & Case, T.J. (1986). Overview: introduction, extinctions, exterminations and invasions. pp. 65–79 in '*Community ecology*' (ed. J.M. Diamond & T.J. Case). New York, Harper Row.

Diamond, J.M. & Pimm, S. (1993). Survival times of bird populations: a reply. *Amer. Nat.* **142:** 1030–1035.

Dietrich, D.R., Schmid, P., Zweifel, U., Schlatter, C., Jenni-Eiermann, S., Bachmann, H., Buhler, U. & Zbinden, N. (1995). Mortality of birds of prey following field application of granular carbofuran: A case study. *Arch. Environ. Contam. and Toxicol.* **29:** 140–145.

Dijkstra, C., Bult, A., Bijlsma, S., Daan, S., Meÿer, T. & Zÿlstra, M. (1990). Brood size manipulations in the Kestrel (*Falco tinnunculus*): effects on offspring and parent survival. *J. Anim. Ecol.* **59:** 269–285.

Dijkstra, C., Vuursteen, L., Daan, S. & Masman, D. (1982). Clutch-size and laying date in the Kestrel *Falco tinnunculus*: Effects of supplementary food. *Ibis* **124:** 210–213.

Dobinson, H.M. & Richards, A.J. (1964). The effects of the severe winter of 1962/63 on birds in Britain. *Brit. Birds* **59:** 373–434.

Dobson, A.P. & Hudson, P.J. (1992). Regulation and stability of a free-living host–parasite system: *Trichostrongylus tenuis* in Red Grouse. II. Population models. *J. Anim. Ecol.* **61:** 487–498.

Dobson, A.P. & May, R.M. (1991). Parasites, cuckoos and avian population dynamics. pp. 391–412 in '*Bird population studies*' (ed. C.M. Perrins, J.D. Lebreton & G.J.M. Hirons). Oxford, University Press.

Dolbeer, R.A. (1990). Ornithology and integrated pest management – Red-winged Blackbirds *Agelaius phoeniceus* and corn. *Ibis* **132:** 309–322.

Donald, P.F. & Forrest, C. (1995). The effects of agricultural change on population size of Corn Buntings *Miliaria calandra* on individual farms. *Bird Study* **42**: 205–215.

Dornbusch, M. (1973). Zur Siedlungsdichte und Ernährung des Feldsperlings in Kiefern-Dickungen. *Der Falke* **20**: 193–195.

Dorney, R. & Kabat, C. (1960). Relation of weather, parasitic disease and hunting to Wisconsin Ruffed Grouse populations. *Wisc. Cons. Dept. Tech. Bull.* **20**: 1064.

Dow, H. & Fredga, S. (1983). Breeding and natal dispersal of the Goldeneye, *Bucephala clangula*. *J. Anim. Ecol.* **52**: 681.

Doyle, F.I. & Smith, J.M.N. (1994). Population responses of Northern Goshawks to the 10-year cycle in numbers of Snowshoe Hares. *Stud. Avian Biol.* **16**: 122–129.

Drent, P.J. (1984). The functional ethology of territoriality in the Great Tit (*Parus major*). Doctoral thesis, Zoological Laboratory, University of Groningen, Netherlands.

Drent, R.H. & Daan, S. (1980). The prudent parent: energetic adjustments in avian breeding. *Ardea* **68**: 225–252.

Duckett, J.E. (1984). Barn Owls (*Tyto alba*) and the second generation rat-baits utilised in oil palm plantations in Peninsular Malaysia. *Planter, Kuala Lumpur* **60**: 3–11.

Duebbert, H.F. (1966). Island nesting of the Gadwall in North Dakota. *Wilson Bull.* **78**: 12–25.

Duebbert, H.F. & Frank, A.M. (1984). Value of prairie wetlands to duck broods. *Wildl. Soc. Bull.* **12**: 27–34.

Duebbert, H.F. & Kantrud, H.A. (1974). Upland duck nesting related to land use and predator reduction. *J. Wildl. Manage.* **38**: 257–265.

Duebbert, H.F. & Lokemoen, J.T. (1980). High duck nesting success in a predator-reduced environment. *J. Wildl. Manage.* **44**: 428–437.

Du Feu, C.R. (1992). How tits avoid flea infestation at nest-sites. *Ring. Migr.* **13**: 120–121.

Duffy, D.C. (1983). Competition for nesting space among Peruvian guano birds. *Auk* **100**: 680–688.

Duffy, D.C., Berruti, A., Randall, R.M. & Cooper, J. (1984). Effects of the 1982–83 warm water event on the breeding of South African seabirds. *S. Afr. J. Sci.* **80**: 65–69.

Dugan, P.J., Evans, P.R., Goodyer, L.R. & Davidson, N.C. (1981). Winter fat reserves in shorebirds: disturbance of regulated levels by severe weather conditions. *Ibis* **123**: 359–363.

Dumke, R.T. & Pils, C.M. (1973). Mortality of radio-tagged Pheasants in the Waterloo wildlife area. Dept. Nat. Res. Madison, Wisconsin, *Tech. Bull.* **72**: 1–52.

Duncan, J.S., Reid, H.W., Moss, R., Philips, J.D.P. & Watson, A. (1979). Ticks, louping ill and Red Grouse on moors in Speyside, Scotland. *J. Wildl. Manage.* **42**: 500–505.

Dunlop, C.L., Blokpoel, H. & Jarvie, S. (1991). Nesting rafts as a management tool for a declining Common Tern (*Sterna hirundo*) colony. *Colonial Waterbirds* **14**: 116–120.

Dunn, E.K. (1977). Predation by Weasels (*Mustela nivalis*) on breeding tits (*Parus* spp.) in relation to the density of tits and rodents. *J. Anim. Ecol.* **46**: 633–652.

Dunnet, G.M. (1982). Oil pollution and seabird populations. *Phil. Trans. R. Soc. Lond. B.* **297**: 413–427.

Dunnet, G.M. & Patterson, I.J. (1968). The Rook problem in north-east Scotland. pp. 119–139 in 'The problems of birds as pests' (ed. R.K. Murton & E.N. Wright). London, Academic Press.

Dunning, J.B. & Brown, J.H. (1982). Summer rainfall and winter sparrow densities – a test of the food limitation hypothesis. *Auk* **99**: 123–129.

East, M.L. & Perrins, C.M. (1988). The effect of nest-boxes on breeding populations of birds in broadleaved temperate woodlands. *Ibis* **130**: 393–401.

Eastin, W.C. & Hoffman, D.J. (1979). Biological effects of petroleum on aquatic birds. pp. 561–582 in 'The proceedings of the conference on assessment of ecological impacts of oil spills' (ed. C.C. Bates). Arlington, Virginia, Amer. Inst. Biol. Sci.

Ebbinge, B.S. (1985). Factors determining the population size of arctic-breeding geese wintering in western Europe. *Ardea* **73**: 121–128.

Ebbinge, B.S. (1989). A multifactorial explanation for variation in breeding performance of Brent Geese *Branta bernicla*. *Ibis* **131**: 196–204.

Ebbinge, B.S. (1992). Regulation of numbers of Dark-bellied Brent Geese *Branta bernicla* on spring staging sites. *Ardea* **80**: 203–228.

Eckert, C.G. & Weatherhead, P.J. (1987). Owners, floaters and competitive asymmetries among territorial Red-winged Blackbirds. *Anim. Behav.* **35**: 1317–1323.

Edington, J.M. & Edington, M.A. (1972). Spatial patterns and habitat partitioning in the breeding birds of an upland wood. *J. Anim. Ecol.* **41**: 331–357.

Edwards, P.J. (1977). "Reinvasion" by some farmland bird species following capture and removal. *Pol. Ecol. Stud.* **3**: 53–70.

Edwards, R. (Chairman) (1996). Sea Empress Environmental Evaluation Committee Initial Report, July 1996. Cardiff, Sea Empress Environmental Evaluation Committee.

Ehrlich, P.R. (1994). Energy use and biodiversity loss. *Phil. Trans. R. Soc. Lond. B* **344**: 1–104.

Einarsen, A.S. (1945). Some factors affecting Ring-necked Pheasant population density. *Murrelet* **26**: 3–9, 39–44.

Einarsen, A.S. (1950). Progress report on studies of the Ring-necked Pheasant on an island under controlled conditions. pp. 164–172 in '*Proceedings of the annual conference of the Western Association of State Game and Fish Commissioners*'. (ed.)

Ekman, J. (1984). Density-dependent seasonal mortality and population fluctuation of the temperate-zone Willow Tit (*Parus montanus*). *J. Anim. Ecol.* **53**: 119–134.

Ekman, J. & Askenmo, C. (1984). Social rank and habitat use in Willow Tit groups. *Anim. Behav.* **32**: 508–514.

Ekman, J., Cederholm, G. & Askenmo, C. (1981). Spacing and survival in winter groups of Willow Tit *Parus montanus* and Crested Tit *Parus cristatus* – a removal study. *J. Anim. Ecol.* **50**: 1–9.

Elgar, M.A. (1989). Predator vigilance and group size in mammals and birds: a critical review of the empirical evidence. *Biol. Rev.* **64**: 13–33.

Elkins, N. (1983). Weather and bird behaviour. Calton, Poyser.

Ellison, L.N. (1991). Shooting and compensatory mortality in tetraonids. *Ornis Scand.* **22**: 229–240.

Ellison, L.N., Léonard, P. & Ménoni, E. (1988). Effect of shooting on a Black Grouse population in France. In 'Atti del I Convegno Nazionale dei Biologi della Selvaggina'. *Supplemento Ricerchi Biol. Selvaggina* **14**: 117–128.

Elsworth, S. (1984). Acid rain. London, Pluto Press.

Elton, C.S. (1942). Voles, mice and lemmings. Oxford, University Press.

Emlen, J.T. (1986). Responses of breeding Cliff Swallows to nidicolous parasite infestations. *Condor* **88**: 110–111.

Emlen, S.T. (1978). The evolution of cooperative breeding in birds. pp. 245–281 in '*Behavioural ecology. An evolutionary approach*' (ed. J.R. Krebs & N.B. Davies). Oxford, Blackwell Scientific Publications.

Emlen, S.T., Rising, J.D. & Thompson, W.L. (1975). A behavioural and morphological study of sympatry in the Indigo and Lazuli Buntings of the Great Plains. *Wilson Bull.* **87**: 145–179.

Enemar, A., Nilsson, L. & Sjöstrand, B. (1984). The composition and dynamics of the passerine bird community in a subalpine birch forest, Swedish Lapland. A 20-year study. *Ann. Zool. Fenn.* **21**: 321–338.

Enemar, A.B. & Sjöstrand, B. (1972). Effects of the introduction of Pied Flycatchers *Ficedula hypoleuca* on the composition of a passerine bird community. *Ornis Scand.* **3**: 79–87.

Enoksson, B. (1990). Autumn territories and population regulation in the Nuthatch *Sitta europaea*: an experimental study. *J. Anim. Ecol.* **59**: 1047–1062.

Enoksson, B. & Nilsson, S.G. (1983). Territory size and population density in relation to food supply in the Nuthatch *Sitta europaea*. *J. Anim. Ecol.* **52**: 927–935.

Ens, B.J. & Goss-Custard, J.D. (1984). Interference among Oystercatchers *Haematopus ostralegus* feeding on mussels on the Exe Estuary. *J. Anim. Ecol.* **53**: 217–231.

Ens, B.J., Kersten, M., Brenninkmeijer, A. & Hulscher, J.B. (1992). Territory quality, parental effort and reproductive success of Oystercatchers (*Haematopus ostralegus*). *J. Anim. Ecol.* **61:** 703–715.

Ens, B.J., Weissing, F.J. & Drent, R.H. (1995). The despotic distribution and deferred maturity: two sides of the same coin. *Amer. Nat.* **146:** 625–650.

Ericksson, M.O.G. (1979). Competition between freshwater fish and Goldeneye *Bucephala clangula* L. for common prey. *Oecologia* **41:** 99–107.

Eriksson, M.O.G. (1982). Differences between old and newly established Goldeneye (*Bucephala clangula*) populations. *Ornis Scand.* **59:** 13–19.

Eriksson, M.O.G. & Götmark, G. (1982). Habitat selection – do passerines nest in association with Lapwings *Vanellus vanellus* as defense against predators. *Ornis Scand.* **13:** 189–192.

Eriksson, M.O.G. & Walter, K. (1994). Survival and breeding success of the Osprey *Pandion haliaetus* in Sweden. *Bird. Conserv. Int.* **4:** 263–277.

Erikstäd, K.E., Blom, R. & Myrberget, S. (1982). Territorial Hooded Crows as predators on Willow Ptarmigan nests. *J. Wildl. Manage.* **46:** 109–114.

Erikstäd, K.E., Moum, T. & Vader, W. (1990). Correlation between pelagic distribution of Common and Brünnich's Guillemots and their prey in the Barents Sea. *Polar Res.* **8:** 77–88.

Erlinge, S., Göransson, G., Hansson, L., Högstedt, G., Liberg, O., Nilsson, I.N., Nilsson, T., von Schantz, T. & Sylvén, M. (1983). Predation as a regulating factor on small rodent populations in southern Sweden. *Oikos* **40:** 36–52.

Erlinge, S., Göransson, G., Högstedt, G., Jansson, G., Liberg, O., Loman, J., Nilsson, I., von Schantz, T. & Sylvén, M. (1984). Can vertebrate predators regulate their prey? *Amer. Nat.* **123:** 125–133.

Errington, P.L. (1934). Vulnerability of Bobwhite populations to predation. *Ecology* **15:** 110–127.

Errington, P.L. (1939). The comparative ability of the Bobwhite and the Ring-necked Pheasant to withstand cold and hunger. *Wilson Bull.* **51:** 22–37.

Errington, P.L. (1946). Predation and vertebrate populations. *Quart. Rev. Biol.* **21:** 145–177, 221–245.

Errington, P.L. & Hammerstrom, F.N. (1936). The Northern Bobwhite's winter territory. *Iowa State Coll. Agric. Res. Bull.* **201:** 305–443.

Erskine, A.J. (1979). Man's influence on potential nesting sites and populations of swallows in Canada. *Can. Field Nat.* **93:** 371–377.

Erskine, A.J. & Temple, S.A. (1970). Nesting activities in a Cliff Swallow colony. *Can. Field Nat.* **84:** 385–387.

Esler, D. & Grand, J.B. (1993). Factors influencing depredation of artificial duck nests. *J. Wildl. Manage.* **57:** 244–248.

Evans, A.D. & Smith, K.W. (1994). Habitat selection of Cirl Buntings *Emberiza cirlus* wintering in Britain. *Bird Study* **41:** 81–87.

Evans, P.G.H. & Nettleship, D.N. (1985). Conservation of the Atlantic Alcidae. pp. 427–488 in 'The Atlantic Alcidae' (ed. D.N. Nettleship & T.R. Birkhead). London, Academic Press.

Evans, P.R. (1969). Winter fat deposition and overnight survival of Yellow Buntings (*Emberiza citrinella*). *J. Anim. Ecol.* **38:** 415–423.

Evans, P.R. (1979). Reclamation of intertidal land: some effects on Shelduck and wader populations in the Tees Estuary. *Verh. orn. Ges. Bayern* **23:** 147–168.

Evans, P.R. (1997). Improving the accuracy of predicting of the local effects of habitat loss on shorebirds: lessons from the Tees and Orwell estuary studies. pp. 35–44 in 'Effect of Habitat Loss and Change on Waterbirds' (ed. J.D. Goss-Custard, R. Rufeno & A. Luis). London, Stationery Office.

Evans, P.R. & Dugan, P.J. (1984). Coastal birds: numbers in relation to food resources – a review. pp. 8–28 in 'Coastal Waders and Wildfowl in Winter' (ed. P.R. Evans, J.D. Goss-Custard & W.G. Hale). Cambridge, University Press.

Ewald, P.W. & Carpenter, F.L. (1978). Territorial responses to energy manipulations in the Anna Hummingbird. *Oecologia* **31**: 277–292.

Ewald, P.W., Hunt, G.L. & Warner, M. (1980). Territory size in Western Gulls: importance of intrusion pressure, defence investments and vegetation structure. *Ecology* **61**: 80–87.

Ewald, P.W. & Rohwer, S. (1982). Effects of supplemental feeding on timing of breeding, clutch size and polygyny in Red-winged Blackbirds *Agelaius phoeniceus*. *J. Anim. Ecol.* **51**: 429–450.

Exo, K.M. (1992). Population ecology of Little Owls *Athene noctua* in Central Europe: a review. pp. 64–75 in *'The ecology and conservation of European Owls'* (ed. C.A. Galbraith, I.R. Taylor & S. Percival). Peterborough, Joint Nature Conservation Committee.

Feare, C.J. (1972). The seasonal pattern of feeding in the Rook (*Corvus frugilegus*) in northwest Scotland. *Proc. Int. Ornithol. Congr.* **15**: 643.

Feare, C.J. (1976a). Desertion and abnormal development in a colony of Sooty Terns *Sterna fuscata* infested by virus-infected ticks. *Ibis* **118**: 112–115.

Feare, C.J. (1976b). The breeding of the Sooty Tern *Sterna fuscata* in the Seychelles and the effects of experimental removal of its eggs. *J. Zool. Lond.* **179**: 317–360.

Feare, C.J. (1982). The Starling. Oxford, University Press.

Feare, C.J. (1991). Control of pest bird populations. pp. 63–78 in *'Bird population studies'* (ed. C.M. Perrins, J.D. Lebreton & G. Hirons). Oxford, University Press.

Feare, C.J. (1994). Changes in numbers of Common Starlings and farming practice in Lincolnshire. *Brit. Birds* **87**: 200–204.

Feare, C.J., Dunnet, M. & Patterson, I.J. (1974). Ecological studies of the Rook (*Corvus frugilegus* L.) in north-east Scotland: food intake and feeding behaviour. *J. Appl. Ecol.* **11**: 867–896.

Feare, C.J., Gill, E.L., McKay, H.V. & Bishop, J.D. (1995). Is the distribution of Starlings *Sturnus vulgaris* within roosts determined by competition? *Ibis* **137**: 379–382.

Ferrer, M. (1993). The Spanish Imperial Eagle. Seville, Quercus.

ffrench, R. (1991). A guide to the birds of Trinidad and Tobago, 2nd edn. London, Christopher Helm.

Ficken, R.W., Ficken, M.S. & Hailman, J.P. (1978). Differential aggression in genetically different morphs of the White-throated Sparrow (*Zonotrichia albicollis*). *Z. Tierpsychol.* **46**: 43–57.

Finlayson, H.H. (1932). Heat in the interior of South Australia and in central Australia. *S. Aust. Ornithol.* **11**: 158–163.

Finley, R.B. (1965). Adverse effects on birds of phosphamidon applied to a Montana forest. *J. Wildl. Manage.* **29**: 580–591.

Fischer, C.A. & Keith, L.B. (1974). Population responses of central Alberta Ruffed Grouse to hunting. *J. Wildl. Manage.* **38**: 585–600.

Fischer, H.I. & Baldwin, P.H. (1946). War and the birds on Midway Atoll. *Condor* **48**: 3–15.

Fisher, C.D., Lindgren, E. & Dawson, W.R. (1972). Drinking patterns and behaviour of Australian desert birds in relation to their ecology and abundance. *Condor* **74**: 111–136.

Fisher, J. (1966). The Fulmar population of Britain and Ireland, 1959. *Bird Study* **13**: 5–76.

Fisher, J. & Orenstein, R.I. (1985). Extinct birds. pp. 196–199 in *'A dictionary of birds'* (ed. B. Campbell & E. Lack). Calton, Poyser.

Fisher, J., Simon, N. & Vincent, J. (1969). The Red Book. Wildlife in Danger. London, Collins.

Fitch, H.S., Swenson, F. & Tillotson, D.F. (1946). Behaviour and food habits of the Red-tailed Hawk. *Condor* **48**: 205–237.

Fleskes, J.P. & Klaas, E.E. (1991). Dabbling duck recruitment in relation to habitat and predators at Union Slough National Wildlife Refuge, Iowa. *U.S. Dept. Interior, Fish & Wildl. Tech. Rep.* **32**: 1–19.

Fleury, B.E. & Sherry, T.W. (1995). Long-term population trends of colonial wading birds in the southern United States: the impact of crayfish aquaculture on Louisiana populations. *Auk* **112**: 613–632.

Flickinger, E.L. (1979). Effects of aldrin exposure on Snow Geese in Texas rice fields. *J. Wildl. Manage.* **43:** 94–101.

Flickinger, E.L. & King, K.A. (1972). Some effects of aldrin-treated rice on Gulf Coast wildlife. *J. Wildl. Manage.* **36:** 706–727.

Flickinger, E.L., Mitchell, C.A., White, D.H. & Kolbe, E.J. (1986). Bird poisoning from misuse of the carbamate Furadan in a Texas rice field. *Wildl. Soc. Bull.* **14:** 59–62.

Flousek, J. (1989). Impact of industrial emissions on bird populations breeding in mountain spruce forests in central Europe. *Ann. Zool. Fenn.* **26:** 255–263.

Fonstad, T. (1984). Reduced territorial overlap between the Willow Warbler *Phylloscopus trochilus* and the Brambling *Fringilla montifringilla* in heath forest: competition or different habitat preferences? *Oikos* **42:** 314–322.

Ford, H.A. (1987). Bird communities on habitat islands in England. *Bird Study* **34:** 205–218.

Fordham, R.A. (1964). Breeding biology of the Southern Black-backed Gull II. Incubation and the chick stage. *Notornis* **11:** 110–126.

Forsell, D.J. (1982). Recolonisation of Baker Island by seabirds. *Bull. Pac. Seabird Group* **9:** 75–76.

Forsman, E.D., Meslow, E.C. & Wight, H.M. (1984). Distribution and biology of the Spotted Owl in Oregon. *Wildl. Monogr.* **87:** 1–64.

Forster, J.A. (1975). Electric fencing to protect Sandwich Terns against Foxes. *Biol. Conserv.* **7:** 85.

Foster, C.C., Forsman, E.D., Meslow, E.C., Miller, G.S., Reid, J.A., Wagner, F.F., Carey, A.B. & Lint, J.B. (1992). Survival and reproduction of radio-marked adult Spotted Owls. *J. Wildl. Manage.* **56:** 91–95.

Foster, W.A. (1968). Total brood mortality in late-nesting Cliff Swallows. *Condor* **70:** 275.

Fournier, M.A. & Hines, J.E. (1994). Effects of starvation on muscle and organ mass of King Eiders *Somateria spectabilis* and the ecological and management implications. *Wildfowl* **45:** 188–197.

Fowler, G.S., Wingfield, J.C. & Boersma, P.D. (1995). Hormonal and reproductive effects of low levels of petroleum fouling in Magellanic Penguins (*Spheniscus magellanicus*). *Auk* **112:** 382–389.

Fox, A.D., Gitay, H., Owen, M., Salmon, D.G. & Ogilvie, M.A. (1989). Population dynamics of Icelandic-nesting geese, 1960–1967. *Ornis Scand.* **20:** 289–297.

Fox, A.D. & Madsen, J. (1997). Behavioural and distributional effects of hunting disturbance on waterbirds in Europe: implications for refuge design. *J. Appl. Ecol.* **34:** 1–13.

Francis, C.M., Richards, M.H., Cooke, F. & Rockwell, R.F. (1992). Long-term changes in survival rates of Lesser Snow Geese. *Ecology* **73:** 1346–1362.

Frank, H. (1970). Die Auswirkung von Raubwild- und Raubzeug minderung auf die Strecken von Hase, Fasan und Rebhuhn in einem Revier mit intensivster landwirtschaftlicher Nutzung. pp. 472–479 in '*Proceedings of the International Congress of Game Biology 1969*'. (ed. A.G. Bannikov).

Frankel, O.H. & Soulé, M.E. (1981). Conservation and evolution. Cambridge, University Press.

Franson, J.C. & Little, S.E. (1996). Diagnostic findings in 132 Great Horned Owls. *J. Raptor Res.* **30:** 1–6.

Franson, J.C., Thomas, N.J., Smith, M.R., Robbins, A.H., Newman, S. & McCartin, P.C. (1996). A retrospective study of postmortem findings in Red-Tailed Hawks. *J. Rapt. Res.* **30:** 7–14.

Franzblau, M.A. & Collins, J.P. (1980). Test of a hypothesis of territory regulation in an insectivorous bird by experimentally increasing prey abundance. *Oecologia* **46:** 164–170.

Fredga, S. & Dow, H. (1984). Factors affecting the size of a local population of Goldeneye *Bucephala clangula* (L.) breeding in Sweden. *Viltrevy* **13:** 225–255.

Freedman, B. (1995). Environmental ecology. London, Academic Press.

Fretwell, S.D. (1969). Dominance behaviour and winter habitat distribution in Juncos (*Junco hyemalis*). *Bird Banding* **40:** 1–25.

Fretwell, S.D. (1972). Populations in a seasonal environment. Princeton, Princeton University Press.

Fretwell, S.D. & Lucas, J.H.J. (1970). On territorial behaviour and other factors influencing habitat distribution in birds. *Acta Biotheoretica* **19**: 16–36.

Friend, M. (1992). Environmental influences on major waterfowl diseases. *Trans. N.A. Wildl. Nat. Res. Conf.* **57**: 517–525.

Friend, M. & Pearson, G.L. (1973). Duck plague: the present situation. *Western Proc. Ann. Conf. Western Assoc. State Game and Fish Commissioners* **53**: 315–325.

Friend, M. & Trainer, D.O. (1970). Polychlorinated byphenyl: interaction with DHV. *Science* **170**: 1314–1316.

Frith, H.J. (1967). Waterfowl in Australia. Sydney, Angus & Robertson.

Fritz, L., Quillian, M.A., Wright, J.L.C., Beale, A.M. & Work, T.M. (1992). An outbreak of domoic acid poisoning attributed to the pennate diatom *Pseudonitzschia australis. J. Phycol.* **28**: 439–442.

Fritz, R.S. (1979). Consequences of insular population structure: distribution and extinction of Spruce Grouse populations. *Oecologia* **42**: 57–65.

Fry, D.M. (1992). Point source and non-point source problems affecting seabird populations. pp. 549–562 in *'Wildlife 2001: populations'* New York, Elsevier.

Fry, D.M., Swenson, J., Addiego, A., Grau, C.R. & Karg, A. (1986). Reduced reproduction of Wedge-tailed Shearwaters exposed to weather and Santa Barbara crude oil. *Arch. Environ. Contam. Toxicol.* **15**: 453–463.

Fuller, R.J. (1982). Bird habitats in Britain. Calton, Poyser.

Fuller, R.J., Gregory, R.D., Gibbons, D.W., Marchant, J.H., Wilson, J.D., Baillie, S.R. & Carter, N. (1995). Population declines and range contractions among lowland farmland birds in Britain. *Conserv. Biol.* **9**: 1425–1441.

Furness, R.W. (1982). Competition between fisheries and seabird communities. *Adv. Mar. Biol.* **20**: 225–307.

Furness, R.W. (1985a). Ingestion of plastic particles by seabirds at Gough Island, South Atlantic Ocean. *Environ. Poll. Ser. A.* **38**: 261–272.

Furness, R.W. (1985b). Plastic particle pollution – accumulation by Procellariiform seabirds at Scottish colonies. *Mar. Poll. Bull.* **16**: 103–106.

Furness, R.W. & Barrett, R.T. (1991). Ecological responses of seabirds to reductions in fish stocks in North Norway and Shetland. *Proc. Int. Ornithol. Congr.* **20**: 2241–2245.

Furness, R.W. & Birkhead, T.R. (1984). Seabird colony distributions suggest competition for food supplies during the breeding season. *Nature* **311**: 655–656.

Furness, R.W., Ensor, K. & Hudson, A.V. (1992). The use of fishing waste by gull populations around the British Isles. *Ardea* **80**: 105–113.

Furness, R.W. & Greenwood, J.J.D. (1993). Birds as monitors of environmental change. London, Chapman & Hall.

Fyfe, R. (1969). The Peregrine Falcon in northern Canada. pp. 101–114 in *'Peregrine Falcon populations: their biology and decline'* (ed. J.J. Hickey). Madison, Wisconsin, University Press.

Galbraith, H. (1988). The effects of territorial behaviour on Lapwing populations. *Ornis Scand.* **19**: 134–138.

Garcia, E.F.J. (1983). An experimental test of competition for space between Blackcaps *Sylvia atricapilla* and Garden Warblers *Sylvia borin* in the breeding season. *J. Anim. Ecol.* **52**: 795–805.

Gardarsson, A. & Einarsson, A. (1994). Responses of breeding duck populations to changes in food supply. *Hydrobiologia* **279–280**: 15–27.

Garden, E.A., Rayski, C. & Thom, V.M. (1964). A parasite disease in Eider Ducks. *Bird Study* **11**: 280–287.

Garnett, M.C. (1981). Body size, its heritability and influence on juvenile survival amongst Great Tits *Parus major. Ibis* **123**: 31–41.

Garrettson, P.R., Rohwer, F.C., Zimmer, J.M., Mense, B.J. & Dion, N. (1996). Effects of mammalian predator removal on waterfowl and non-game birds in North Dakota. *Trans. N. Amer. Wildlife & Nat. Res. Conf.* **65**: 94–101.

Gaskin, J.M. (1989). Psittacine viral diseases: a perspective. *J. Zoo. Wildl. Med.* **20**: 249–264.

Gass, C.L. (1979). Territory regulation, tenure, and migration in Rufous Hummingbirds. *Can. J. Zool.* **57**: 914–923.

Gass, C.L., Angehr, G. & Centa, J. (1976). Regulation of food supply by feeding territoriality in the Rufous Hummingbird. *Can. J. Zool.* **54**: 2046–2054.

Gaston, A.J. (1985). Development of the young in the Atlantic Alcidae. pp. 319–354 in 'The Atlantic Alcidae' (ed. D.N. Nettleship & T.R. Birkhead). London, Academic Press.

Gaston, A.J., de Forest, C.N., Donaldson, G. & Noble, D.G. (1994). Population parameters of Thick-billed Murres at Coats Island, Northwest Territories, Canada. *Condor* **96**: 935–948.

Gaston, A.J. & Elliot, R.D. (1996). Predation by Ravens *Corvus corax* on Brunnich's Guillemot *Uria Lomvia* eggs and chicks and its possible impacts on breeding site selection. *Ibis* **138**: 742–748.

Gates, J.E. & Gysel, L.W. (1978). Avian nest dispersion and fledging success in field-forest ecotones. *Ecology* **59**: 871–883.

Gauhl, F. (1984). Ein Beiträge zur Brutbiologie des Feldsperlings (*Passer montanus*). *Vogelwelt* **105**: 176–187.

Gause, G.F. (1934). The strategy for existence. Baltimore, Williams & Wilkins.

Gauthier, G. & Smith, J.N.M. (1987). Territorial behaviour, nest-site availability, and breeding density in Buffleheads. *J. Anim. Ecol.* **56**: 171–184.

Gauthreaux, S.A. (1978). The ecological significance of behavioural dominance. pp. 17–54 in 'Perspectives in ornithology', Vol. 3 (ed. P.P.G. Bateson & P.H. Klopfer). New York, Plenum Press.

Gauthreaux, S.A. (1982). The ecology and evolution of avian migration systems. pp. 93–167 in 'Avian Biology', Vol. 6 (ed. D.S. Farner & J.R. King). New York, Academic Press.

Geer, T.A. (1978). Effects of nesting Sparrowhawks on nesting tits. *Condor* **80**: 419–422.

George, J.L. & Frear, D.E.H. (1966). Pesticides in the Antarctic. *J. Appl. Ecol.* **Suppl. 3** : 155–167.

Gibb, J. (1954). Feeding ecology of tits, with notes on Treecreeper and Goldcrest. *Ibis* **96**: 513–543.

Gibb, J. (1956). Food, feeding habits and territory of the Rock Pipit, *Anthus spinoletta*. *Ibis* **98**: 506–530.

Gibb, J. (1960). Populations of Tits and Goldcrests and their food supply in pine plantations. *Ibis* **102**: 163–208.

Gibbons, D.W., Reid, J.B. & Chapman, R.A. (1993). The new atlas of breeding birds in Britain and Ireland: 1988–1991. London, Poyser.

Gibbs, H.L. & Grant, P.R. (1987a). Adult survivorship in Darwin's Ground Finch (*Geospiza*) populations in a variable environment. *J. Anim. Ecol.* **56**: 797–813.

Gibbs, H.L. & Grant, P.R. (1987b). Ecological consequences of an exceptionally strong El Niño event on Darwin's Finches. *Ecology* **68**: 1735–1746.

Gibbs, J.P. (1991). Avian nest predation in tropical wet forest: an experimental study. *Oikos* **60**: 155–161.

Gibson, G.G., Broughton, E. & Choquette, L.P.E. (1972). Waterfowl mortality caused by *Cyathocotyle bushiensis* Khan, 1962 (Trematoda: Cyathocotylidae), St. Lawrence River, Quebec. *Can. J. Zool.* **50**: 1351–1356.

Gilbertson, M. (1974). Seasonal changes in organochlorine compounds and mercury in Common Terns of Hamilton Harbour, Ontario. *Bull. Environ. Contam. Toxicol.* **12**: 726.

Gilbertson, M. & Fox, G.A. (1977). Pollutant-associated embryonic mortality of Great Lakes herring gulls. *Environ. Poll.* **12**: 211–216.

Gilbertson, M., Morris, R.D. & Hunter, R.A. (1976). Abnormal chicks and PCB residue levels in eggs of colonial birds on the lower Great Lakes (1971–1973). *Auk* **93**: 434–442.

Giles, N. (1992). Wildlife after gravel. Fordingbridge, The Game Conservancy.

Gill, F.B. (1980). Historical aspects of secondary contact and hybridisation between Blue-winged and Golden-winged Warblers. *Auk* **97**: 1–18.

Gill, F.B., Mack, A.L. & Ray, R.T. (1982). Competition between Hermit Hummingbirds Phaethorninae and insects for nectar in a Costa Rican rain forest. *Ibis* **124**: 44–49.

Gill, F.B. & Wolf, L.L. (1975). Economics of feeding territoriality in the Golden-winged Sunbird. *Ecology* **56**: 333–345.

Gill, J.A., Sutherland, W.J. & Watkinson, A.R. (1996). A method to quantify the effects of human disturbance on animal populations. *J. Appl. Ecol.* **33**: 786–792.

Gjerde, I. & Wegge, P. (1989). Spacing pattern, habitat use and survival of Capercaillie in a fragmented winter habitat. *Ornis Scand.* **20**: 219–225.

Glas, P. (1960). Factors governing density in the Chaffinch (*Fringilla coelebs*) in different types of wood. *Arch. Néere. Zool.* **13**: 466–472.

Glooschenko, V., Blancher, P., Herskowitz, J., Fulthorpe, R. & Rang, S. (1986). Association of wetland acidity with reproductive parameters and insect prey of the Eastern Kingbird (*Tyrannus tyrannus*) near Sudbury, Ontario. *Water, Air Soil Poll.* **30**: 553–567.

Glück, E. (1986). Flock size and habitat dependent food and energy intake of foraging Goldfinches. *Oecologia* **71**: 149–155.

Godfrey, W.E. (1966). The birds of Canada. Ottawa, The Queen's Printers.

Gold, C.S. & Dahlsten, D.L. (1983). Effects of parasitic flies (*Protocalliphora* spp.) on nestlings of Mountain and Chesnut-backed Chickadees. *Wilson Bull.* **95**: 560–572.

Goldstein, M.I., Woodbridge, B., Zaccagnini, M.E. & Canavelli, S.B. (1996). An assessment of mortality of Swainson's Hawks on wintering grounds in Argentina. *J. Rap. Res.* **30**: 106–107.

Goodburn, S.F. (1991). Territory quality or bird quality – factors determining breeding success in the Magpie *Pica-pica*. *Ibis* **133**: 85–90.

Göransson, G., Karlsson, J., Nilsson, S.G. & Ulfstand, S. (1975). Predation on birds' nests in relation to antipredator aggression and nest density: an experimental study. *Oikos* **26**: 117–120.

Gosler, A.G. (1996). Environmental and social determinants of winter fat storage in the Great Tit *Parus major*. *J. Anim. Ecol.* **65**: 1–17.

Gosler, A.G., Greenwood, J.J.D. & Perrins, C. (1995). Predation risk and the cost of being fat. *Nature* **377**: 621–623.

Goss-Custard, J.D. (1970a). Feeding dispersion in some overwintering wading birds. pp. 3–35 in '*Social behaviour in birds and mammals*' (ed. J.H. Crook). London, Academic Press.

Goss-Custard, J.D. (1970b). The responses of Redshank (*Tringa totanus* (L.)) to spatial variations in the density of their prey. *J. Anim. Ecol.* **39**: 91–113.

Goss-Custard, J. (1977a). The ecology of the Wash. III. Density-related behaviour and the effect of loss of feeding grounds on wading birds (*Charadrii*). *J. Appl. Ecol.* **44**: 721–939.

Goss-Custard, J.D. (1977b). Optimal foraging and the size selection of worms by Redshank *Tringa totanus* in the field. *Anim. Behav.* **25**: 10–29.

Goss-Custard, J.D. (1980). Competition for food and interference among waders. *Ardea* **68**: 31–52.

Goss-Custard, J.D. (1985). Foraging behaviour of wading birds and the carrying capacity of estuaries. pp. 169–188 in '*Behavioural ecology: ecological consequences of adaptive behaviour*' (ed. R.M. Sibly & R.H. Smith). Oxford, Blackwell Scientific.

Goss-Custard, J.D., Caldow, R.W.G., Clarke, R.T., Le V. dit. Durell, S.E.A., Urfi, J. & West, A.D. (1995b). Consequences of habitat loss and change to populations of wintering migratory birds: predicting the local and global effects from studies of individuals. *Ibis* **137 (Suppl. 1)**: 56–66.

Goss-Custard, J.D., Clarke, R.T., Briggs, K.B., Ens, B.J., Exo, K.-M., Smit, C., Beintema, A.J., Caldow, R.W.G., Catt, D.C., Clark, N.A., Le V. dit. Durell, S.E.A., Harris, M.P., Hulscher, J.B., Meininger, P.L., Picozzi, N., Prys-Jones, R., Safriel, U.N. & West, A.D. (1995a). Population consequences of winter habitat loss in a migrating shorebird. 1. Estimating model parameters. *J. Appl. Ecol.* **32**: 320–336.

Goss-Custard, J.D., Clarke, R.T. & Le V. dit. Durell, S.E.A. (1984). Rates of food intake and aggression of Oystercatcher *Haematopus ostralegus* on the most and least preferred mussel *Mytilus edulis* beds of the Exe Estuary. *J. Anim. Ecol.* **53**: 233–245.

Goss-Custard, J.D. & Le V. dit. Durrell, S.E.A. (1988). The effect of dominance and feeding method on the intake rates of Oystercatchers *Haematopus ostralegus* feeding on mussels. *J. Anim. Ecol.* **57**: 827–844.

Goss-Custard, J.D., Jenyon, R.A., Jones, R.E., Newbery, P.E. & Williams R. le B. (1977a). The ecology of the Wash. II. Seasonal variation in the feeding conditions of wading birds (Charadrii). *J. Appl. Ecol.* **14**: 701–719.

Goss-Custard, J.D., Jones, R.E. & Newbery, P.E. (1977b). The ecology of the Wash I. Distribution and diet of wading birds (Charadrii). *J. Appl. Ecol.* **14**: 681–700.

Goss-Custard, J.D., Kay, D.G. & Blindell, R.M. (1977c). The density of migratory and overwintering Redshank *Tringa totanus* (L.) and Curlew, *Numenius arquata* (L.) in relation to the density of their prey in south-east England. *Est. Coast. Mar. Sci.* **15**: 497–510.

Goss-Custard, J.D., Le V. dit. Durell, S.E.A., Sitters, H.P. & Swinfen, R. (1982). Age-structure and survival of a wintering population of Oystercatchers. *Bird Study* **29**: 83–98.

Goss-Custard, J.D. & Moser, M.E. (1988). Rates of change in the numbers of Dunlin wintering in British estuaries in relation to the spread of *Spartina anglica*. *J. Appl. Ecol.* **25**: 95–109.

Götmark, F. (1992). The effects of investigator disturbance on nesting birds. *Curr. Ornithol.* **9**: 63–104.

Gould, P.J. (1983). Seabirds between Alaska and Hawaii. *Condor* **85**: 286–291.

Graber, J.W. & Graber, R.R. (1979). Severe winter weather and bird populations in southern Illinois. *Wilson Bull.* **91**: 88–103.

Gradwohl, J. & Greenberg, R. (1981). The formation of antwren flocks on Barro Colorado Island, Panama. *Auk* **97**: 385–395.

Graham, I.M., Redpath, S.M. & Thirgood, S.J. (1995). The diet and breeding density of Common Buzzards *Buteo buteo* in relation to indices of prey abundance. *Bird Study* **42**: 165–173.

Grant, P.R. (1986). Ecology and evolution of Darwin's Finches. Princeton, University Press.

Grant, P.R. & Schluter, D. (1984). Interspecific competition inferred from patterns of guild structure. pp. 201–203 in 'Ecological communities. Conceptual issues and the evidence' (ed. D.R. Strong, D. Simberloff, L.G. Abele & A.B. Thistle). Princeton, University Press.

Graveland, J. (1990). Effects of acid precipitation on reproduction in birds. *Experientia* **46**: 962–970.

Graveland, J., van Derwal, R., van Balen, H. & van Noordwijik, A. (1994). Poor reproduction in forest passerines from decline in snail abundance. *Nature* **368**: 446–448.

Green, G.H., Greenwood, J.J.D. & Lloyd, L.S. (1977). The influence of snow conditions on the date of breeding of wading birds in north-east Greenland. *J. Zool., Lond.* **183**: 311–328.

Green, R.E. (1983). Spring dispersal and agonistic behaviour of the Red–legged Partridge (*Alectoris rufa*). *J. Zool., Lond.* **201**: 541–555.

Green, R.E. (1984). The feeding ecology and survival of Partridge chicks *Alectoris rufa* and *Perdix perdix* on arable farmland in East Anglia. *J. Appl. Ecol.* **21**: 817–830.

Green, R.E. (1988). Stone-curlew conservation. *RSPB Conservation Review* **2**: 30–33.

Green, R.E. (1995). The decline of the Corncrake *Crex crex* in Britain continues. *Bird Study* **42**: 66–75.

Green, R.E. & Hirons, G.J.M. (1991). The relevance of population studies to the conservation of threatened birds. pp. 594–633 in 'Bird population studies' (ed. C.M. Perrins, J.-D. Lebreton & G.J.M. Hirons). Oxford, University Press.

Greenberg, R. (1985). The social behaviour and feeding ecology of neotropical migrants in the non–breeding season. *Proc. Int. Orn. Congr.* **18**: 648–653.

Greenberg, R. (1987). Seasonal foraging specialisation in the Worm-eating Warbler. *Condor* **89**: 158–168.

Greenberg, R. & Gradwohl, J. (1980). Observations of paired Canada Warblers *Wilsonia canadensis* during migration in Panama. *Ibis* **122**: 509–512.

Greenberg, R. & Gradwohl, J. (1986). Constant density and stable territoriality in some tropical insectivorous birds. *Oecologia* **69**: 618–625.

Greenway, J.C. (1967). Extinct and vanishing birds of the world. New York, Dover.

Greenwood, J.J.D. & Baillie, S.R. (1991). Effects of density-dependence and weather on population changes of English passerines using a non-experimental paradigm. *Ibis* **133**: 121–133.

Greenwood, J.J.D., Baillie, S.R. & Crick, H.Q.P. (1994). Long-term studies and monitoring of bird populations. pp. 343–364 in '*Long-term experiments in agricultural and ecological sciences*' (ed. R.A. Leigh & A.E. Johnston). Oxford, CAB International.

Greenwood, P.J. (1980). Mating systems, philopatry and dispersal in birds and mammals. *Anim. Behav.* **28**: 1140–1162.

Greenwood, P.J., Harvey, P.H. & Perrins, C.M. (1979). The role of dispersal in the Great Tit (*Parus major*): the causes, consequences and heritability of natal dispersal. *J. Anim. Ecol.* **48**: 123–142.

Greenwood, R.J. (1986). Influence of Striped Skunk removal on upland duck nest success in North Dakota. *Wildl. Soc. Bull.* **14**: 6–11.

Greenwood, R.J., Arnold, P.M. & McGuire, B.G. (1990). Protecting duck nests from mammalian predators with fences, traps and a toxicant. *Wildl. Soc. Bull.* **18**: 75–82.

Greenwood, R.J., Sargeant, A.B., Johnson, D.H., Cowardin, L.M. & Shaffer, T.L. (1987). Mallard nest success and recruitment in prairie Canada. *Trans. N. A. Wildl. Nat. Res. Conf.* **52**: 298–309.

Greenwood, R.J., Sargeant, A.B., Johnson, D.H., Cowardin, L.M. & Shaffer, T.L. (1995). Factors associated with duck nest success in the prairie pothole region of Canada. *Wildl. Monogr.* **128**: 1–57.

Greenwood, R.J. & Sovada, M.A. (1996). Prairie Duck populations and predation management. *Trans. N. A. Wildl. Nat. Res. Conf.* **65**: 31–42.

Gregory, R.D., Keymer, A.E. & Harvey, P.H. (1991). Life history, ecology and parasite community structure in Soviet birds. *Biol. J. Linn. Soc.* **43**: 249–262.

Grenquist, P. (1951). On the recent fluctuations in numbers of waterfowl in the Finnish archipelago. *Proc. Int. Ornithol. Congr.* **10**: 494–496.

Grenquist, P., Henriksson, K. & Raites, T. (1972). On intestinal occlusion in male Eider ducks. *Suom. Riista* **24**: 91–96.

Griffin, C.R., King, C.M., Savidge, J.A., Cruz, F. & Cruz, J.B. (1988). Effects of introduced predators on island birds: contemporary case histories from the Pacific. *Proc. Int. Ornithol. Congr.* **19**: 688–698.

Griffiths, R.P., Caldwell, B.A. & Morita, R.Y. (1982). Seasonal changes in microbial heterotrophic activity in sub-arctic marine waters as related to phytoplankton primary productivity. *Mar. Biol.* **71**: 121–127.

Groom, M.J. (1992). Sand-coloured Nighthawks parasitise the antipredator behaviour of three nesting bird species. *Ecology* **73**: 785–793.

Grzybowski, J.A., Clapp, R.B. & Marshall, J.T. (1986). History and current population status of the Black-capped Vireo in Oklahoma. *Amer. Birds* **40**: 1151–1161.

Guhl, A.M., Craig, J.V. & Mueller, C.D. (1960). Selective breeding for aggressiveness in chickens. *Poultry Sci.* **39**: 970–980.

Guillemette, M., Reed, A. & Himmelman, J.H. (1996). Availability and consumption of food by Common Eiders wintering in the Gulf of St Lawrence – evidence of prey depletion. *Can. J. Zool.* **74**: 32–38.

Gullion, G.W. & Marshall, W.H. (1968). Survival of Ruffed Grouse in a boreal forest. *Living Bird* **7**: 117–167.

Gustafsson, L. (1987). Interspecific competition lowers fitness in Collared Flycatchers *Ficedula albicollis*: an experimental demonstration. *Ecology* **68**: 291–296.

Gustafsson, L. (1988). Inter- and intraspecific competition for nest-holes in a population of the Collared Flycatcher *Ficedula albicollis*. *Ibis* **130**: 11–15.

Haapanen, A. (1965). Bird fauna of Finnish forests in relation to forest succession. *Ann. Zool. Fenn.* **2**: 153–196.

Haemig, P.D. (1992). Competition between ants and birds in a Swedish forest. *Oikos* **65**: 479–483.

Hagen, Y. (1969). Norwegian studies on the reproduction of birds of prey and owls in relation to micro-rodent population fluctuations. *Fauna* **22**: 73–126.

Hagen, Y. (1952). Rovfuglene og viltpleien. Oslo, Gyldendal Norsk forlag.

Haig, S.M. & Ballou, J.D. (1995). Genetic diversity in 2 avian species formerly endemic to Guam. *Auk* **112**: 445–455.

Haila, Y. (1990). Toward an ecological definition of an island – a Northwest European perspective. *J. Biogeog.* **17**: 561–568.

Haller, H. (1996). Der Steinadler in Graubünden. *Ornithol. Beob.* **9**: 1–167.

Halstead, J.A. (1988). American Dipper nestlings parasitised by blowfly larvae and the northern fowl mite. *Wilson Bull.* **100**: 507–508.

Hamer, K.C., Furness, R.W. & Caldow, R.W.G. (1991). The effects of changes in food availability on the breeding ecology of Great Skuas *Catharacta skua* in Shetland. *J. Zool. Lond.* **223**: 175–188.

Hamer, K.C., Monaghan, P., Uttley, J.D., Walton, P. & Burns, M.D. (1993). The influence of food supply on the breeding ecology of Kittiwakes *Rissa tridactyla* in Shetland. *Ibis* **135**: 255–263.

Hamerström, F. (1969). A harrier population study. pp. 367–383 in '*Peregrine Falcon populations: their biology and decline*' (ed. J.J. Hickey). Madison, Wisconsin, University Press.

Hamerström, F., Hamerström, F.N. & Hart, J. (1973). Nest boxes: an effective management tool for Kestrels. *J. Wildl. Manage.* **37**: 400–403.

Haney, C.J. (1986). Seabird patchiness in tropical oceanic waters: the influence of *Sargassum* reefs. *Auk* **103**: 141–151.

Hannon, S.J. (1983). Spacing and breeding density of Willow Ptarmigan in response to an experimental alteration of sex ratio. *J. Anim. Ecol.* **52**: 807–820.

Hannon, S.J., Mumme, R.L., Koenig, W.D., Spon, S. & Pitelka, F.A. (1987). Poor acorn crop, dominance and decline in numbers of Acorn Woodpeckers. *J. Anim. Ecol.* **56**: 197–207.

Hansen, A.J. (1987). Regulation of Bald Eagle reproductive rates in southeast Alaska. *Ecology* **68**: 1387–1392.

Hansen, H.A. & McKnight, D.E. (1964). Emigration of drought-displaced ducks to the Arctic. *Trans. N. A. Wildl. Nat. Res. Conf.* **29**: 119–126.

Hansen, P.W. (1987). Acid rain and waterfowl: the case for concern in North America. Arlington, Virginia, Izaak Walton League of America.

Hanski, I. (1994). A practical model of metapopulation dynamics. *J. Anim. Ecol.* **63**: 151–162.

Hanski, I., Pakkala, T., Kuussaari, M. & Lei, G. (1995). Metapopulation persistence of an endangered butterfly in a fragmented landscape. *Oikos* **72**: 21–28.

Hanski, I. & Gilpin, M. (1991). Metapopulation dynamics: brief history and conceptual domain. *Biol. J. Linn. Soc.* **42**: 3–16.

Hanski, I., Hansson, L. & Henttonen, H. (1991). Specialist predators, generalist predators, and the microtine rodent cycle. *J. Anim. Ecol.* **60**: 353–367.

Hanson, J. & Howell, J. (1981). Possible Fenthion toxicity in Magpies (*Pica pica*). *Can. Vet. J.* **22**: 18–19.

Hanson, W.R. (1952). Effects of some herbicides and insecticides on biota of North Dakota marshes. *J. Wildl. Manage* **16**: 299–308.

Haramis, G.M. & Thompson, D.Q. (1985). Density-production characteristics of box-nesting Wood Ducks in a northern greentree impoundment. *J. Wildl. Manage.* **49**: 429–436.

Hardin, G. (1968). The tragedy of the commons. *Science* **162**: 1245–1248.

Hardin, G. (1994). The tragedy of the unmanaged commons. *Trends Ecol. Evol.* **9**: 199.

Harris, M.P. (1962). Weights from five hundred birds found dead on Skomer Island in January 1962. *Brit. Birds* **55**: 97–103.

Harris, M.P. (1965). Puffinosis on Skokholm. *Brit. Birds* **58**: 426–433.

Harris, M.P. (1969). Biology of Storm Petrels in the Galapagos Islands. *Proc. Calif. Acad. Sci. Ser 4.* **37**: 95–166.

Harris, M.P. (1970). Territory limiting the size of the breeding population of the Oystercatcher (*Haematopus ostralegus*) – a removal experiment. *J. Anim. Ecol.* **39**: 707–713.

Harris, M.P. (1978). Supplementary feeding of young Puffins *Fratercula arctica*. *J. Anim. Ecol.* **47**: 15–24.

Harris, M.P., Buckland, S.T., Russell, S.M. & Wanless, S. (1994). Year- and age-related variation in the survival of adult European Shags over a 24-year period. *Condor* **96**: 600–605.

Harris, M.P. & Wanless, S. (1990). Breeding status and sex of Common Murres *Uria aalge* at a colony in Autumn. *Auk* **107**: 603–605.

Harris, M.P. & Wanless, S. (1991). Population studies and conservation of Puffins *Fratercula arctica*. pp. 230–248 in '*Bird population studies*' (ed. C.M. Perrins, J.-D. Lebreton & G.J.M. Hirons). Oxford, University Press.

Harris, M.P. & Wanless, S. (1996). Differential responses of Guillemot *Uria aalge* and Shag *Phalocrocorax aristotelis* to a late winter wreck. *Bird Study* **43**: 220–230.

Harris, M.P., Wanless, S., Barton, T.R. & Elston, D.A. (1997). Nest site characteristics, duration of use and breeding success in the Guillemot *Uria aalge*. *Ibis* **139**: 468–476.

Harris, S.W. & Marshall, W.H. (1957). Some effects of a severe windstorm on Coot nests. *J. Wildl. Manage.* **21**: 471–473.

Harrison, J. & Hudson, M. (1964). Some effects of severe weather on wildfowl in Kent in 1962–63. *Wildfowl Trust Ann. Rep.* **15**: 26–32.

Harrison, S. (1991). Local extinction in a metapopulation context: an empirical evaluation. *Biol. J. Linn. Soc.* **42**: 73–88.

Hart, C.M., Glading, B. & Harper, H.T. (1956). The Pheasant in California. pp. 90–158 in '*Pheasants in North America*' (ed. D.L. Allen). Harrisburg, Pennyslvania., Stackpole.

Harvey, P.H., Greenwood, P.J. & Perrins, C.M. (1979). Breeding area fidelity of Great Tits (*Parus major*). *J. Anim. Ecol.* **48**: 305–313.

Hassell, M.P. & Anderson, R.M. (1989). Predator-prey and host-pathogen interactions. pp. 147–196 in '*Ecological concepts*' (ed. J.M. Cherrett). Oxford, Blackwell.

Hatch, S.A. (1987). Did the 1982–1983 El Niño–Southern Oscillation affect seabirds in Alaska? *Wilson Bull.* **99**: 468–471.

Hatchwell, B.J., Chamberlain, D.E. & Perrins, C.M. (1996). The reproductive success of Blackbirds *Turdus merula* in relation to habitat structure and choice of nest site. *Ibis* **138**: 256–262.

Haugen, A.O. (1952). Trichomoniasis in Alabama Mourning Doves. *J. Wildl. Manage.* **16**: 164–169.

Haukioja, E. & Hakala, T. (1975). Herbivore cycles and periodic outbreaks. Formulation of a general hypothesis. *Rep. Kevo Subarctic Res. Stn.* **12**: 1–19.

Hayes, W.J. & Laws, E.R. (1991). Handbook of pesticide toxicology. London, Academic Press.

Heinemann, D., Hunt, G. & Eversson, I. (1989). The distribution of marine avian predators and their prey, *Euphausia superba*, in Bransfield Strait, Southern Drake Passage, Antarctica. *Mar. Ecol. Prog. Ser.* **58**: 3–16.

Held J.D., den (1981). Population changes of the Purple Heron in relation to drought in the wintering area. *Ardea* **69**: 185–191.

Henderson, I.G. & Hart, P.J.B. (1995). Dominance, food acquisition and reproductive success in a monogamous passerine: the Jackdaw *Corvus monedula*. *J. Avian. Biol.* **26**: 217–224.

Hengeveld, R. (1988). Mechanisms of biological invasions. *J. Biogeog.* **15**: 819–828.

Henny, C.J. (1973). Drought displaced movement of North American Pintails into Siberia. *J. Wild. Manage.* **37**: 23–29.

Henny, C.J. (1977). Birds of prey, DDT, and Tussock Moths in Pacific Northwest. *Trans. N. A. . Wildl. Nat. Res. Conf.* **42**: 397–411.

Henny, C.J., Blus, L.J., Kolbe, E.J. & Fitzner, R.E. (1985). Organophosphate insecticide (famphur) topically applied to cattle kills magpies and hawks. *J. Wildl. Manage.* **49**: 648–658.

Henny, C.J., Smith, M.M. & Stotts, V.D. (1974). The 1973 distribution and abundance of breeding Ospreys in the Chesapeake Bay. *Chesapeake Sci.* **15**: 125–133.

Hensley, M.M. & Cope, J.B. (1951). Further data on removal and repopulation of the breeding birds in a spruce-fir forest community. *Auk* **68**: 483–493.

Hepp, G.R. & Hair, J.D. (1984). Dominance in wintering waterfowl (*Anatini*): effects on distribution of sexes. *Condor* **86**: 251–257.

Herman, C.M., Barrow, J.H. & Tarshis, I.B. (1975). Leucocytozoonosis in Canada Geese at the Seney National Wildlife Refuge. *J. Wildl. Dis.* **11**: 404–411.

Hessler, E., Tester, J.R., Siniff, D.B. & Nelson, M.N. (1970). A biotelemetry study of survival of pen-reared Pheasants released in selected habitats. *J. Wildl. Manage.* **34**: 267–274.

Heubeck, M. (1989). Seabirds and sandeels. Proceedings of a seminar held in Lerwick, Shetland, 15–16 October 1988. Lerwick, Shetland Bird Club.

Heusmann, H.W. & Bellville, R. (1978). Effects of nest removal on starling populations. *Wilson Bull.* **90**: 287–290.

Heydemann, B. (1983). Die Beurteilung von Zeikonflikten zwischen Landwirtschaft, Landschaftspflege und Naturschutz aus Sicht der Landschaftspflege und des Naturschutzes. *Schriftenreihe für ländliche Sozialfragen* **88**: 51–78.

Hickey, J.J. (1942). Eastern populations of the Duck Hawk. *Auk* **59**: 176–204.

Hickey, J.J. & Anderson, D.W. (1969). The Peregrine Falcon: life history and population literature. pp. 3–42 in '*Peregrine Falcon populations, their biology and decline*' (ed. J.J. Hickey). Madison, Wisconsin, University Press.

Hickey, J.J., Keith, J.A. & Coon, F.B. (1966). An exploration of pesticides in a Lake Michigan ecosystem. *J. Appl. Ecol.* **3 (Suppl.):** 141–154.

Higgins, K.F. & Johnson, M.A. (1978). Avian mortality caused by a September wind and hail storm. *Prairie Nat.* **10**: 43–48.

Higuchi, H. (1978). Use of nest-boxes by birds according to forest types and the breeding density in forests with and without nest-boxes. *J. Jpn. For. Soc.* **60**: 255–261.

Hildén, O. (1965). Habitat selection in birds. A review. *Ann. Zool. Fenn.* **2**: 53–70.

Hill, D. (1983). Compensatory mortality in the Mallard. *Game Conservancy Ann. Rev.* **14**: 87–92.

Hill, D.A. (1984). Population regulation in the Mallard. *J. Anim. Ecol.* **53**: 191–202.

Hill, D.A. (1985). The feeding ecology and survival of Pheasant chicks on arable farmland. *J. Appl. Ecol.* **22**: 645–654.

Hill, D.A. (1988). Population dynamics of Avocets (*Recurvirostra avosetta* L.) breeding in Britain. *J. Anim. Ecol.* **57**: 669–683.

Hill, E.F., Heath, R.G., Spann, J.W. & Williams, J.D. (1975). Lethal dietary toxicities of environmental pollutants to birds. Washington, D.C., U.S. Dept of Interior, Fish & Wildlife Service.

Hindman, L.J. & Ferrigno, F. (1990). Atlantic flyway goose populations, status and management. *Trans. N.A. Wildl. Nat. Res. Conf.* **55**: 292–311.

Hines, J.E. & Mitchell, G.J. (1983). Gadwall nest-site selection and nesting success. *J. Wildl. Manage.* **47**: 1063–1071.

Hinsley, S.A., Bellamy, P.E. & Newton, I. (1995a). Bird species turnover and stochastic extinction in woodland fragments. *Ecography* **18**: 41–50.

Hinsley, S.A., Bellamy, P.E., Newton, I. & Sparks, T.H. (1995b). Habitat and landscape factors influencing the presence of individual breeding bird species in woodland fragments. *J. Avian Biol.* **26**: 94–104.

Hinsley, S.A., Bellamy, P.E., Newton, I. & Sparks, T.H. (1996). Influence of population size and woodland area on bird species distributions in small woods. *Oecologia* **105**: 100–106.

Hiraldo, F., Negro, J.H., Donazar, J.A. & Gaona, P. (1996). A demographic model for a population of the endangered Lesser Kestrel in southern Spain. *J. Appl. Ecol.* **33**: 1085–1093.

Hirons, G.J.M., Hardy, A.R. & Stanley, P.I. (1979). Starvation in young Tawny Owls. *Bird Study* **26:** 59–63.

Hixon, M.A., Carpenter, F.L. & Paton, D.C. (1983). Territory area, flower density and budgeting in hummingbirds: an experimental and theoretical analysis. *Amer. Nat.* **122:** 366–391.

Hochachka, W.M. & Boag, D.A. (1987). Food shortage for breeding Black-billed Magpies (*Pica pica*): an experiment using supplemental food. *Can. J. Zool.* **65:** 1270–1274.

Hockey, P.A.R. & Cooper, J. (1980). Paralytic shellfish poisoning: a controlling factor in Black Oystercatcher populations? *Ostrich* **51:** 188–190.

Hoffmann, C.H., Townes, H.K., Swift, H.H. & Sailer, R.I. (1949). Field studies on the effects of airplane applications of DDT on forest invertebrates. *Ecol. Monogr.* **19:** 1–46.

Hoffman, D.J. (1990). Embryotoxicity and teratogenicity of environmental contaminants to bird eggs. *Rev. Environ. Contam. Toxicol.* **115:** 40–89.

Hoffman, D.J. & Albers, P.H. (1984). Evaluation of embryotoxicity and teratogenicity of 42 herbicides, insecticides, and petroleum contaminants to Mallard eggs. *Arch. Environ. Contam. Toxicol.* **13:** 15–27.

Hoffman, D.J. & Eastin, W.C. (1981). Effects of malathion, diaxinon and parathion on Mallard embryo development and cholinesterase activity. *Environ. Res.* **26:** 472–485.

Hoffman, D.J., Rattner, B.A., Sileo, L., Docherty, D. & Kubiak, T.J. (1987). Embryotoxicity, Teratogenicity, and Aryl Hydrocarbon Hydroxylase activity in Forster's Terns on Green Bay, Lake Michigan. *Environ. Res.* **42:** 176–184.

Hogstad, O. (1975). Quantitative relations between hole-nesting and open-nesting species within a passerine breeding community. *Norw. J. Zool.* **23:** 261–267.

Hogstad, O. (1987). Social rank in winter flocks of Willow Tits *Parus montanus*. *Ibis* **129:** 1–9.

Hogstad, O. (1989). Subordination in mixed-age bird flocks – a removal study. *Ibis* **131:** 128–134.

Hogstad, O. (1993). Structure and dynamics of a passerine bird community in a spruce-dominated boreal forest – a 12-year study. *Ann. Zool. Fenn.* **30:** 43–54.

Hogstad, O. (1995). Do avian and mammalian nest predators select for different nest dispersion patterns of Fieldfares *Turdus pilaris* – a 15-year study. *Ibis* **137:** 484–489.

Högstedt, G. (1980). Prediction and test of the effects of interspecific competition. *Nature* **283:** 64–66.

Högstedt, G. (1981). Effect of additional food on reproductive success in the Magpie (*Pica pica*). *J. Anim. Ecol.* **50:** 219–229.

Holberton, R.L. (1993). An endogenous basis for differential migration in the Dark-Eyed Junco. *Condor* **95:** 580–587.

Holdgate, M.W. (1971). The Sea Bird Wreck in the Irish Sea Autumn 1969. *London, Nat. Environ. Res. Council.*

Holgersen, H. (1958). Pinkfeet in Europe. The effect of the cold weather of February 1956 on the distribution of Pink-Footed Geese in Northwest Europe. *Wildfowl Trust Ann. Rep.* **9:** 1956–1957.

Hollands, P.K. & Yalden, D.W. (1991). Population dynamics of Common Sandpipers *Actitis hypoleucos* breeding along an upland river system. *Bird Study* **38:** 151–159.

Holling, C.S. (1965). The functional response of predators to prey density and its role in mimicry and population regulation. *Mem. Entomol. Soc. Can.* **45:** 1–60.

Holmes, J.C. & Bethel, W.M. (1972). Modification of intermediate host behaviour by parasites. pp. 123–149 in *'Behavioural aspects of parasite transmission'* (ed. E.U. Canning & C.A. Wright). London, Academic Press.

Holmes, R.T. (1966). Breeding ecology and annual cycle adaptations of the Red-backed Sandpiper *Calidris alpina* in northern Alaska. *Condor* **68:** 3–46.

Holmes, R.T. (1970). Differences in population density, territoriality, and food supply of Dunlin on arctic and subarctic tundra. pp. 303–319 in *'Animal populations in relation to their food resources'* (ed. A. Watson). Oxford, Blackwell.

Holmes, R.T. & Sherry, T.W. (1988). Assessing population trends of New Hampshire forest birds: local vs regional patterns. *Auk* **105:** 756–768.

Holmes, R.T., Sherry, T.W. & Sturges, F.W. (1986). Bird community dynamics in a temperate deciduous forest: long-term trends at Hubbard Brook. *Ecol. Monogr.* **56:** 201–220.

Holmes, R.T., Sherry, T.W. & Sturges, F.W. (1991). Numerical and demographic responses of temperate forest birds to annual fluctuations in their food resources. *Proc. Int. Orn. Congr.* **20:** 1559–1567.

Holmes, R.T., Marra, P.P. & Sherry, T.W. (1996). Habitat-specific demography of breeding Black-throated Blue Warblers (*Dendroica caerulescens*) – implications for population dynamics. *J. Anim. Ecol.* **65:** 183–195.

Holroyd, G.L. (1975). Nest-site availability as a factor limiting population size of swallows. *Can. Field. Nat.* **89:** 60–64.

Holt, R.D. (1977). Predation, apparent competition and the structure of prey communities. *Theoret. Pop. Biol.* **12:** 197–229.

Holt, R.D. (1984). Spatial heterogeneity, indirect interactions, and the coexistence of prey species. *Amer. Nat.* **125:** 377–406.

Holyoak, M. & Baillie, S.R. (1996). Factors influencing detection of density dependence in British birds. *Oecologia* **108:** 47–53.

Hoogland, J.L. & Sherman, P.W. (1976). Advantages and disadvantages of Bank Swallow (*Riparia riparia*) coloniality. *Ecol. Monogr.* **46:** 33–58.

Hoover, J.P., Brittingham, M.C. & Goodrich, L.J. (1995). Effects of forest patch size on nesting success of Wood Thrushes. *Auk* **112:** 146–155.

Hope Jones, P. (1962). Mortality and weights of Fieldfares in Anglesey in January 1962. *Brit. Birds* 55: 178–181.

Hope Jones, P., Monnat, J.-Y., Cadbury, C.J. & Stowe, T.J. 1978. Birds oiled during the Amoco Cadiz incident – an interim report. *Mar. Poll. Bull.* **9:** 307–311.

Hörnfeldt, B. (1978). Synchronous population fluctuations in voles, small game, owls and tularemia in northern Sweden. *Oecologia* **32:** 141–152.

Hörnfeldt, B., Löfgren, O. & Carlson, B.-G. (1986). Cycles in voles and small game in relation to variations in plant production indices in northern Sweden. *Oecologia* **68:** 496–502.

Hotchkiss, N. & Pough, R.H. (1946). Effect on forest birds of DDT used for gypsy moth control in Pennsylvania. *J. Wildl. Manage.* **10:** 202–207.

Houghton, J.T., Meira Filho, L.G.M., Bruce, J., Hoesung Lee, B.A., Callander, B.A., Haites, E., Harris, W. & Maskell, K. (1995). Climate change 1994. Cambridge, University Press.

Houston, C.S. & Francis, C.M. (1995). Survival of Great Horned Owls in relation to the Snowshoe Hare cycle. *Auk* **112:** 44–59.

Houston, D.C. (1988). Competition for food between Neotropical vultures in forest. *Ibis* **130:** 402–417.

Howe, R.W., Davies, G.J. & Mosca, V. (1991). The demographic significance of 'sink' populations. *Biol. Conserv.* **57:** 239–255.

Hudson, P.J. (1986). The effect of a parasitic nematode on the breeding production of Red Grouse. *J. Anim. Ecol.* **55:** 85–92.

Hudson, P.J. (1990). Territorial status and survival in a low density grouse population: preliminary observations and experiments. pp. 20–28 in 'Red Grouse Population Processes' (ed. A.N. Lance & J.H. Lawton). British Ecological Society and Royal Society for the Protection of Birds.

Hudson, P. (1992). Grouse in space and time. The population biology of a managed gamebird. Fordingbridge, Hants, The Game Conservancy.

Hudson, P.J. & Dobson, A.P. (1991). The direct and indirect effects of the caecal nematode *Trichostrongylus tenuis* on Red Grouse. pp. 49–68 in 'Bird–parasite interactions' (ed. J.E. Loye & M. Zuk). Oxford, University Press.

Hudson, P.J., Dobson, A.P. & Newborn, D. (1992a). Do parasites make prey vulnerable to predation? Red Grouse and parasites. *J. Anim. Ecol.* **61:** 681–692.

Hudson, P.J., Newborn, D. & Dobson, A.P. (1992b). Regulation and stability of a free-living host–parasite sytem, *Trichostrongylus tenuis* in Red Grouse. 1: Monitoring and parasite reduction experiments. *J. Anim. Ecol.* **61:** 477–486.

Hudson, R. (1972). Collared Doves in Britain and Ireland during 1965–1970. *Brit. Birds* **65:** 139–155.

Hudson, R.H., Tucker, R.K. & Haegele, M.A. (1984). Handbook of toxicity of pesticides to wildlife, 2nd edn. Washington, U.S. Dept. Interior, Fish & Wildlife Service, Res. Publ. 153.

Hughes, B. (1993). Stiff-tail threat. *BTO News* **185:** 14.

Hughes, R.A. (1985). Notes on the effects of El Niño on the seabirds of the Mollendo district, southwest Peru, in 1983. *Ibis* **127:** 385–388.

Hulscher, J.B. (1982). The Oystercatcher as a predator of the bivalve, *Macoma balthica,* in the Dutch Wadden Sea. *Ardea* **70:** 89–152.

Hunt, E.C. & Bischoff, A.I. (1960). Inimical effects on wildlife of periodic DDD applications to Clear Lake. *Calif. Fish Game.* **46:** 91–106.

Hunt, G.L. (1972). Influence of food distribution and human distribution on the reproductive success of Herring Gulls. *Ecology* **53:** 1057–1061.

Hunt, G.L., Eppley, Z.A. & Schneider, D.C. (1986). Reproductive performance of seabirds: the importance of population and colony size. *Auk* **103:** 306–317.

Hunt, G.L. & Schneider, D.C. (1987). Scale dependent processes in the physical and biological environment of marine birds. pp. 7–41 in *'Seabirds – feeding biology and role in marine ecosystems'* (ed. J.C. Croxall). Cambridge, University Press.

Hunt, G.R. (1996). Environmental variables associated with population patterns of the Kagu *Rhynochetos jubatus* of New Caledonia. *Ibis* **138:** 742–748.

Hunt, K.A., Bird, D.M., Mineau, P. & Schutt, L. (1992). Selective predation of organophosphate-exposed prey by American Kestrels. *Anim. Behav.* **43:** 971–976.

Hunt, W.G. (1988). The natural regulation of Peregrine Falcon populations. pp. 667–676 in *'Peregrine Falcon populations. Their management and recovery'* (ed. T.J. Cade, J.H. Enderson, C.G. Thelander & C.M. White). Boise, Idaho, The Peregrine Fund Inc.

Hunter, B.F., Clark, W., Perkins, P. & Coleman, P. (1970). Applied botulism research, including management recommendations, progress report. Sacramento, California Dept. Fish and Game.

Hunter, M.L., Withan, J.W. & Dow, H. (1984). Effects of a carbaryl induced depression in invertebrate abundance on the growth and behaviour of American Black Duck and Mallard ducklings. *Can. J. Zool.* **62:** 452–456.

Hustler, K. & Howells, W.W. (1989). Habitat preference, breeding success and the effect of primary productivity on Tawny Eagles *Aquila rapax* in the tropics. *Ibis* **131:** 33–40.

Hutchinson, G.E. (1950). Survey of contemporary knowledge of biogeochemistry. *Bull. Amer. Mus. Nat. Hist.* **96:** 1–554.

Hutchinson, G.E. (1957). Concluding remarks. *Cold Spring Harbor Symposium on Quantitative Biology* **22:** 415–427.

Hutchinson, G.E. (1959). Homage to Santa Rosalia, or why are there so many kinds of animals? *Amer. Nat.* **93:** 145–159.

Idyll, C.P. (1973). The anchovy crisis. *Scient. Am.* **1973:** 23–29.

Inglis, I.R., Isaacson, A.J., Thearle, R.J.P. & Westwood, N.J. (1990). The effects of changing agricultural practice upon Woodpigeon *Columba palumbus* numbers. *Ibis* **132:** 262–272.

Inskipp, T.P. & Gammell, A. (1979). The extent of world trade and the mortality involved. pp. 98–103 in *'Thirteenth Bulletin of the International Council for Bird Preservation'* (ed. P.B. Smith & R.D. Chancellor). London, ICBP.

Itamies, J., Valtonen, E.T. & Feagerholm, H.P. (1980). *Polymorphus minutus* infestation in Eiders and its role as a possible cause of death. *Ann. Zool. Fenn.* **17:** 185–289.

Jackson, J.A. & Tate Jr., J. (1974). An analysis of nest-box use by Purple Martins, House Sparrows and Starlings in eastern North America. *Wilson Bull.* **86:** 435–449.

Jaeger, E.C. (1949). Further observations on the hibernation of the Poorwill. *Condor* **51:** 105–109.

Jaeger, M.M., Bruggers, R.L., Johns, B.E. & Erickson, W.A. (1986). Evidence of itinerant breeding of the Red-billed Quelea *Quelea quelea* in the Ethiopian Rift Valley. *Ibis* **128:** 469–482.

James, P. (1956). Destruction of warblers on Padre Island, Texas, in May 1951. *Wilson Bull.* **68:** 224–227.

James, P.C. & Fox, G.A. (1987). Effects of some insecticides on productivity of Burrowing Owls. *Blue Jay* **45:** 65–71.

Jansson, C., Ekman, J. & von Brömssen, A. (1981). Winter mortality and food supply in tits (*Parus* spp.). *Oikos* **37:** 313–322.

Järvinen, A. (1978). Nest-box studies in mountain birch forest at Kilpisjärvi, Finnish Lapland. *Anser* **(Suppl.) 3:** 107–111.

Järvinen, A. (1987). Key-factor analyses of two Finnish hole-nesting passerines: comparisons between species and regions. *Ann. Zool. Fenn.* **24:** 275–280.

Järvinen, O. (1985). Predation causing extended low densities in microtine cycle: implications from predation on hole-nesting passerines. *Oikos* **45:** 157–158.

Järvinen, O. (1990). Changes in the abundance of birds in relation to small rodent density and predation rate in Finnish Lapland. *Bird Study* **37:** 36–39.

Jehl, J.R. (1989). Nest-site tenacity and patterns of adult mortality in nesting California Gulls *Larus californius. Auk* **106:** 102–106.

Jehl, J.R. (1996). Mass mortality events of Eared Grebes in North America. *J. Field Ornithol.* **67:** 471–476.

Jehl J.R. Jnr. (1974). The distribution and ecology of marine birds over the continental shelf of Argentina in winter. *Trans. San Diego Soc. Nat. Hist.* **17:** 217–234.

Jehl, J.R. & Parkes, K.C. (1983). 'Replacements' of landbird species on Socorro Island, Mexico. *Auk* **100:** 551–559.

Jenkins, D. (1961). Population control in protected Partridges (*Perdix perdix*). *J. Anim. Ecol.* **30:** 235–258.

Jenkins, D., Newton, I. & Brown, C. (1976). Structure and dynamics of a Mute Swan population. *Wildfowl* **27:** 77–82.

Jenkins, D., Watson, A. & Miller, G.R. (1963). Population studies on Red Grouse *Lagopus lagopus scoticus* (Lath.) in north-east Scotland. *J. Anim. Ecol.* **32:** 317–376.

Jenkins, D., Watson, A. & Miller, G.R. (1964). Predation and Red Grouse populations. *J. Appl. Ecol.* **1:** 183–195.

Jenni, L. (1987). Mass concentrations of Bramblings *Fringilla montifringilla* in Europe 1900–1983: their dependence upon beech mast and the effect of snow cover. *Ornis Scand.* **18:** 84–94.

Jenni, L. (1993). Structure of a Brambling *Fringilla montifringilla* roost according to sex, age and body mass. *Ibis* **135:** 85–90.

Jennings, A.R. (1961). An analyses of 1,000 deaths in wild birds. *Bird Study* **8:** 25–31.

Jennings, A.R., Soulsby, E.J.L. & Wainwright, C.B. (1961). An outbreak of disease in Mute Swans at an Essex reservoir. *Bird Study* **8:** 19–24.

Jenny, D. (1992). Bruterfolg und Bestandsregulation einer alpinen Population den Steinadlers *Aquila chrysaetos. Orn. Beob.* **89:** 1–43.

Jenny, M. (1990). Nahrungsökologie der Feldlerche *Alauda arvensis* in einer intensiv genutzten Agrarlandschaft des schweizeruschen Mittelandes. *Ornithol. Beob.* **87:** 31–53.

Jensen, B. (1970). Effect of a fox control programme on the bag of some other game species. Page 480 in 'Proceedings of the 9th International Congress on Game Biology 1969' (ed A.G. Bannikov). Moscow.

Jensen, S., Johnels, A.G., Olsson, M. & Westermark, T. (1972). The avifauna of Sweden as indicators of environmental contamination with mercury and chlorinated hydrocarbons. *Proc. Int. Ornithol. Congr.* **15:** 455–465.

Jensen, W.I. (1979). An outbreak of *Streptococcus* in Eared Grebes *Podiceps nigricollis*. *Avian Dis.* **23:** 543–546.

Jensen, W.I. & Cotter, S.E. (1976). An outbreak of erysipelas in Eared Grebes (*Podiceps nigricollis*). *J. Wildl. Dis.* **12:** 583–586.

Jensen, W.I. & Williams, C. (1964). Botulism and waterfowl. pp. 333–341 in *'Waterfowl tomorrow'* (ed. J.P. Linduska). Washington D.C., Government Printing Office.

Joensen, A. (1977). Oil pollution and seabirds in Denmark, 1971–1976. *Dan. Rev. Game Biol.* **10:** 1–31.

Johnson, D.H. (1996). Waterfowl communities in the Northern Plains. pp. 391–418 in *'Long term studies of vertebrate communities'* (ed. M.L. Cody & J.A. Smallwood). London, Academic Press.

Johnson, G., Walker, C.H. & Dawson, A. (1994). Interactive effects of prochloraz and malathion in Pigeon, Starling and hybrid Red-legged Partridge. *Environ. Toxicol. Chem.* **13:** 115–120.

Johnson, L.L. (1967). The Common Goldeneye Duck and the role of nesting boxes in its management in North Central Minnesota. *J. Minn. Acad. Sci.* **34:** 110–113.

Johnson, L.S., Eastman, M.D. & Kermott, L.H. (1991). Effect of ectoparasitism by larvae of the Blow Fly *Protocalliphora parorum* (Diptera, Calliphoridae) on nestling House Wrens, *Troglodytes aedon*. *Can. J. Zool.* **69:** 1441–1446.

Johnson, M.D. & Geupel, G.R. (1996). The importance of productivity to the dynamics of a Swainson's Thrush population. *Condor* **98:** 133–141.

Johnson, N.K. (1975). Controls of number of bird species on montane islands in the Great Basin. *Evolution* **29:** 545–567.

Johnson, T.H. & Stattersfield, A.J. (1990). A global review of island endemic birds. *Ibis* **132:** 167–180.

Johnstone, G.W. (1985). Threats to birds on subantarctic islands. *ICPB Tech. Bull.* **3:** 101–121.

Jones, C.G., Heck, W., Lewis, R.E., Mungroo, Y., Slade, G. & Cade, T. (1995). The restoration of the Mauritius Kestrel *Falco punctatus* population. *Ibis* **137 (suppl.) 1:** 173–180.

Jones, R.E. & Leopold, A.S. (1967). Nesting interference in a dense population of Wood Ducks. *J. Wildl. Manage.* **31:** 221–228.

Jones, L., Gaunt, M., Hails, R.S., Laurenson, K., Hudson, P.J., Reid, H., Henbest, P. & Gould, E.A. (1997). Transmission of louping-ill virus between infected and uninfected ticks cofeeding on Mountain Hares. *Med. & Vet. Ent.* **11:** 172–176.

Jönsson, P.E. (1991). Reproduction and survival in a declining population of the Southern Dunlin *Calidris alpina schinzii*. *Wader Study Group Bull.* **61 (Suppl.):** 56–68.

Jordan, J.J. (1953). Effects of starvation on wild Mallards. *J. Wildl. Manage.* **17:** 304–311.

Jourdain, F.C.R. & Witherby, H.F. (1918). The effect of the winter of 1916–1917 on our resident birds. Part 1. *Brit. Birds* **11:** 266–271; Part 2. *Brit. Birds* **12:** 26–35.

Kalejta, B. & Hockey, P.A.R. (1996). Distribution of shorebirds at the Berg River estuary, South Africa, in relation to foraging mode, food supply and environmental features. *Ibis* **136:** 233–239.

Källander, H. (1974). Advancement of the laying of Great Tits by the provision of food. *Ibis* **116:** 365–367.

Källander, H. (1981). The effects of provision of food in winter on a population of the Great Tit *Parus major* and the Blue Tit *P. caeruleus*. *Ornis Scand.* **12:** 244–248.

Källander, H. & Smith, H.G. (1990). Food storage in birds. An evolutionary perspective. *Curr. Ornithol.* **7:** 147–207.

Kalmbach, E.R. (1939). Nesting success: its significance in waterfowl reproduction. *Trans. N. A. Wildl. Nat. Res. Conf.* **4:** 591–604.

Kalmbach, E.R. & Gunderson, M.F. (1934). Western duck sickness, a form of botulism. *U.S. Dept. Agr. Tech. Bull.* **411:** 1–81.

Kaminski, R.M. & Gluesing, E.A. (1987). Density- and habitat-related recruitment in Mallards. *J. Wildl. Manage.* **51:** 141–148.

Kanyamibwa, S., Bairlein, F. & Schierer, A. (1993). Comparison of survival rates between populations of the White Stork *Ciconia ciconia* in central Europe. *Ornis Scand.* **24**: 297–302.

Kanyamibwa, S., Schierer, A., Pradel, R. & Lebreton, J.D. (1990). Changes in adult survival rates in a western European population of the White Stork *Ciconia ciconia*. *Ibis* **132**: 27–35.

Karlsson, J. & Källander, H. (1977). Fluctuations and density of suburban populations of the Blackbird *Turdus merula*. *Ornis Scand.* **8**: 139–144.

Karr, J.R. (1982). Avian extinction on Barro Colorado Island, Panama: a reassessment. *Amer. Nat.* **119**: 220–239.

Kattan, G.H., Alvarez-López, H. & Giraldõ, M. (1994). Forest fragmentation and bird extinctions: San Antonia eighty years later. *Conserv. Biol.* **8**: 138–146.

Kautz, E.J. & Malecki, R.A. (1990). Effects of harvest on feral rock dove survival, nest success and population size. *U.S. Dept. Interior Fish & Wildl. Tech. Rep.* **31**: 1–16.

Kehoe, F.P. & Ankney, C.D. (1985). Variation in digestive organ size among five species of diving ducks (*Aythya* spp.). *Can. J. Zool.* **63**: 2339–2342.

Keith, L.B. (1961). A study of waterfowl ecology on small impoundments in southeastern Alberta. *Wildl. Monogr.* **6**: 1–88.

Keith, L.B. (1963). Wildlife's ten-year cycle. Madison, Wisconsin, University Press.

Keith, L.B. (1974). Some features of population dynamics in mammals. *Proc. Int. Congr. Game Biol.* **11**: 17–58.

Keith, L.B. & Rusch, D.H. (1988). Predation's role in the cyclic fluctuations of Ruffed Grouse. *Proc. Int. Ornithol. Congr.* **19**: 699–732.

Keith, L.B., Todd, A.W., Brand, C.J., Adamcik, R.S. & Rusch, D.H. (1977). An analysis of predation during a cyclic fluctuation of Snowshoe Hares. *Int. Congr. Game Biol.* **13**: 151–175.

Keller, L.F., Arcese, P., Smith, J.N.M., Hochachka, W.M. & Stearns, S. (1994). Selection against inbred Song Sparrows during a natural bottleneck. *Nature* **372**: 356–357.

Keller, V.E. (1991). Effects of human disturbance on Eider ducklings *Somateria mollissima* in an estuarine habitat in Scotland. *Biol. Conserv.* **58**: 213–228.

Kelly, S.T. & DeCapita, M.E. (1982). Cowbird control and its effect on Kirtland's Warbler reproductive success. *Wilson Bull.* **94**: 363–365.

Kemp, A.C. & Kemp, M.I. (1975). Observations on the breeding biology of the Ovampo Sparrow-hawk, *Accipiter ovampensis* Gurney (*Aves: Accipitridae*). *Ann. Transvaal Mus.* **29**: 185–190.

Kendeigh, S.C. 1941. Territorial and mating behaviour of the House Wren. *Illinois Biol. Monogr.* **18**: 1–120.

Kendeigh, S.C. (1947). Bird population studies in the coniferous biome during a Spruce Budworm outbreak. *Biol. Bull. (Dept. Lands and Forest, Ontario)* **1**: 1–100.

Kennedy, E.D. & White, D.W. (1996). Interference competition from House Wrens as a factor in the decline of Bewick's Wrens. *Conserv. Biol.* **10**: 281–284.

Kennedy, R.J. (1970). Direct effects of rain on birds: a review. *Brit. Birds* **63**: 401–414.

Kenward, R.E. (1974). Mortality and fate of trained birds of prey. *J. Wildl. Manage.* **38**: 751–756.

Kenward, R.E. (1977). Predation on released Pheasants (*Phasianus colchicus*) by Goshawks (*Accipiter gentilis*) in central Sweden. *Viltrevy* **10**: 79–112.

Kenward, R.E. (1978). Hawks and doves: factors affecting success and selection in Goshawk attacks on Woodpigeons. *J. Anim. Ecol.* **47**: 449–460.

Kenward, R.E. (1985). Problems of Goshawk predation on pigeons and other game. *Proc. Int. Ornithol. Congr.* **18**: 666–678.

Kenward, R.E., Marcström, V. & Karlbom, M. (1981). Goshawk winter ecology in Swedish Pheasant habitats. *J. Wildl. Manage.* **45**: 397–408.

Kenward, R.E., Marcström, V. & Karlbom, M. (1991). The Goshawk (*Accipiter gentilis*) as predator and renewable resource. *Gibier Faune Sauvage* **8**: 367–378.

Kenward, R.E., Marcström, V. & Karlbom, M. (1993). Causes of death in radio-tagged northern Goshawks. pp. 57–61 in 'Raptor Biomedicine' (ed. P.T. Redig, J.E. Cooper, J.D. Remple & D.B. Hunter). Minneapolis, University of Minnesota Press.

Kenward, R.E. & Sibly, R.M. (1977). A Woodpigeon *Columba palumbus* L. feeding preference explained by a digestive bottle-neck. *J. Appl. Ecol.* **14:** 815–826.

Kenward, R.E. & Sibly, R.M. (1978). Woodpigeon feeding behaviour at brassica sites. A field and laboratory investigation of Woodpigeon feeding behaviour during adoption and maintenance of a brassica diet. *Anim. Behav.* **26:** 778–790.

Kenyon, K.N. (1947). Breeding populations of the Osprey in lower California. *Condor* **49:** 152–158.

Kepler, C.B. (1967). Polynesian Rat predation on nesting Laysan Albatross and other Pacific seabirds. *Auk* **84:** 426–430.

Kerbes, R.H., Kotanen, P.M. & Jefferies, R.L. (1990). Destruction of wetland habitats by Lesser Snow Geese – a keystone species on the west coast of Hudson Bay. *J. Appl. Ecol.* **27:** 242–258.

Kerlinger, P. & Lein, M.R. (1986). Differences in winter range among age–sex classes of Snowy Owls *Nyctea scandiaca* in North America. *Ornis Scand.* **17:** 1–7.

Ketterson, E.D. & Nolan, V.J. (1976). Geographic variation and its climatic correlates in the sex ratio of eastern-wintering Dark-eyed Juncos (*Junco hyemalis hyemalis*). *Ecology* **57:** 679–693.

Ketterson, E.D. & Nolan, V. (1983). The evolution of differential bird migration. *Curr. Ornithol.* **1:** 357–402.

Keymer, I.F., Fletcher, M.R. & Stanley, P.I. (1981). Causes of mortality in British Kestrels (*Falco tinnunculus*). pp. 143–151 in *'Recent advances in the study of raptor diseases'* (ed. J.E. Cooper & A.G. Greenwood). Keighley, Chiron Publications.

Keymer, I.F., Rose, J.H., Beesley, W.N. & Davies, S.F.M. (1962). A survey and review of parasitic diseases of wild and game birds in Great Britain. *Vet. Rec.* **74:** 887–894.

Keymer, I.F., Smith, G.R., Roberts, T.A., Heaney, S.I. & Hibberd, D.J. (1972). Botulism as a factor in waterfowl morbidity at St. James' Park, London. *Vet. Rec.* **90:** 111–114.

Kikkawa, J. (1980). Winter survival in relation to dominance classes among Silvereyes *Zosterops lateralis chlorocephala* of Heron Island, Great Barrier Reef. *Ibis* **122:** 437–446.

King, J.R. & Farner, D.S. (1966). The adaptive role of winter fattening in the White-crowned Sparrow with comments on its regulation. *Amer. Nat.* **100:** 403–418.

King, K.A. (1976). Bird mortality, Galveston Island, Texas. *Southwest. Nat.* **21:** 414.

King, K.A., Blankenship, D.R., Paul, R.T. & Rice, R.C.A. (1977). Ticks as a factor in the 1975 nesting failure of Texas Brown Pelicans. *Wilson. Bull.* **89:** 157–158.

King, W.B. (1980). Ecological basis for extinction in birds. *Proc. Int. Ornithol. Congr.* **17:** 905–911.

King, W.B. (1984). Incidental mortality of seabirds in gillnets in the north Pacific. *ICBP Tech. Publ.* **2:** 709–715.

King, W.B. (1985). Island birds: will the future repeat the past? *ICBP Tech. Publ.* **3:** 3–15.

Kirby, J.S., Salmon, D.G., Atkinson-Willes, G.L. & Cranswick, P.A. (1995). Index numbers for waterfowl populations. III. Long-term trends in the abundance of wintering wildfowl in Great Britain, 1966/67 to 1991/92. *J. Appl. Ecol.* **32:** 536–551.

Kissam, J.B., Noblet, R.E. & Garris, G.I. (1975). Large scale aerial treatment of an endemic area with abate granular lavicide to control black flies (Diptera: Simulidae) and suppress *Leucocytozoon smithi* of turkeys. *J. Med. Entomol.* **12:** 359–362.

Kjellén, N. (1994). Differences in age and sex ratio among migrating and wintering raptors in southern Sweden. *Auk* **111:** 274–284.

Klett, A.T., Shaffer, T.L. & Johnson, D.H. (1988). Duck nesting success in the Prairie Pothole Region. *J. Wildl. Manage.* **52:** 431–440.

Klomp, A. (1972). Regulation of the size of bird populations by means of territorial behaviour. *Neth. J. Zool.* **22:** 456–488.

Klomp, N.I. & Furness, R.W. (1992). Non-breeders as a buffer against environmental stress: declines in numbers of Great Skuas on Foula, Shetland, and prediction of future recruitment. *J. Appl. Ecol.* **29:** 341–348.

Kluijver, H.N. (1951). The population ecology of the Great Tit, *Parus m. major (L.)*. *Ardea* **39:** 1–135.

Kluijver, H.N. (1966). Regulation of a bird population. *Ostrich* **(Suppl.) 6**: 389.

Kluijver, H.N. (1971). Regulation of numbers in populations of Great Tits (*Parus m. major* L.). pp. 507–524 in '*Dynamics of population*' (ed. P.J.de Boer & G.R. Gradwell). Wageningen, Centre for Agricultural Publishing and Documentation.

Kluijver, H.N. & Tinbergen, L. (1953). Territory and the regulation of density in titmice. *Archives néelandaises de Zoologie* **10**: 265–289.

Knapton, R.W. (1979). Optimal size of territory in the Clay-coloured Sparrow, *Spizella pallida*. *Can. J. Zool.* **57**: 1358–1370.

Knapton, R.W. & Krebs, J.R. (1974). Settlement patterns, territory size and breeding density in the Song Sparrow (*Melospiza melodia*). *Can. J. Zool.* **52**: 1413–1420.

Knight, R.L. (1988). Effects of supplemental food on the breeding biology of the Black-billed Magpie. *Condor* **90**: 956–958.

Kodric-Brown, A. & Brown, J.H. (1978). Influence of economics, interspecific competition and sexual dimorphism on territoriality of migrant Rufous Hummingbirds. *Ecology* **59**: 285–296.

Komdeur, J. (1992). Importance of habitat saturation and territory quality for evolution of cooperative breeding in the Seychelles Warbler. *Nature* **358**: 493–495.

Komdeur, J. (1994). Conserving the Seychelles Warbler *Acrocephalus sechellensis* by translocation from Cousin Island to the islands of Aride and Cousine. *Biol. Conserv.* **67**: 143–152.

Komdeur, J. (1996). Breeding of the Seychelles Magpie Robin *Copsychus sechellarum* and implications for its conservation. *Ibis* **138**: 485–498.

Komdeur, J., Huffstadt, A., Prast, W., Castle, G., Mileto, R. & Wattel, J. (1995). Transfer experiments of Seychelles Warblers to new islands: changes in dispersal and helping behaviour. *Anim. Behav.* **49**: 695–708.

Kornberg, H. (1981). Royal Commission on environmental pollution. London, H.M. Stationery Office. 8th Report.

Korpimäki, E. (1985). Rapid tracking of microtine populations by their avian predators: possible evidence for stabilising selection. *Oikos* **45**: 281–284.

Korpimäki, E. (1988). Effects of territory quality on occupancy, breeding performance and breeding dispersal in Tengmalm's Owl. *J. Anim. Ecol.* **57**: 97–108.

Korpimäki, E. (1989). Breeding performance of Tengmalm's Owl *Aegolius funereus*: Effects of supplementary feeding in a peak vole year. *Ibis* **131**: 51–56.

Korpimäki, E., Hakkarainen, H. & Bennett, G.F. (1993). Blood parasites and reproductive success of Tengmalm's Owls: detrimental effects on females but not males. *Funct. Ecol.* **7**: 420–426.

Korpimäki, E. & Norrdahl, K. (1989). Predation of Tengmalm's Owls: numerical responses, functional responses and dampening impact on population fluctuations of voles. *Oikos* **54**: 154–164.

Korpimäki, E. & Norrdahl, K. (1991). Numerical and functional responses of Kestrels, Short-eared Owls and Long-eared Owls to vole densities. *Ecology* **72**: 814–826.

Korschgen, C.E., Gibbs, H.C. & Mendell, H.L. (1978). Avian cholera in Eider Ducks in Maine. *J. Wildl. Dis.* **14**: 254–258.

Koskimies, J. (1950). The life of the Swift *Micropus apus* (L.) in relation to the weather. *Suom. Tied. Toim. Ann. Acad. Sci. Fenn.* **15**: 1–151.

Kostrzewa, A. & Kostrzewa, R. (1989). The relationship of spring and summer weather with density and breeding performance of the Buzzard *Buteo buteo*, Goshawk *Accipiter gentilis* and Kestrel *Falco tinnunculus*. *Ibis* **132**: 550–559.

Kostrzewa, R. & Kostrzewa, A. (1991). Winter weather, spring and summer density and subsequent breeding success of Eurasian Kestrels, Common Buzzards, and Northern Goshawks. *Auk* **108**: 342–347.

Krapu, G.L., Klett, A.T. & Jorde, D.G. (1983). The effect of variable spring water conditions on Mallard reproduction. *Auk* **100**: 689–698.

Krebs, C.J., Boutin, S., Boonstra, R. & Sinclair, A.R.E. (1995). Impact of food and predation on the Snowshoe Hare cycle. *Science* **269:** 1112–1115.

Krebs, J.R. (1970). Regulation of numbers of the Great Tit (*Aves: Passeriformes*). *J. Zool., Lond.* **162:** 317–333.

Krebs, J.R. (1971). Territory and breeding density in the Great Tit *Parus major* L. *Ecology* **52:** 2–22.

Krebs, J.R. (1977). Song and territory in the Great Tit *Parus major*. pp. 47–62 in *'Evolutionary Ecology'* (ed. B. Stonehouse & C. Perrins). London, Macmillan Press.

Krebs, J.R. (1980). Optimal foraging, predation risk and territory defence. *Ardea* **68:** 83–90.

Krebs, J.R. (1982). Territorial defence in the Great Tit (*Parus major*); do residents always win? *Behav. Ecol. Sociobiol.* **11:** 185–194.

Krebs, J.R., Clayton, N.S., Healy, S., Cristol, D.A., Patel, S.N. & Jolliffe, A.R. (1996). The ecology of the avian brain: food-storing memory and the hippocampus. *Ibis* **138:** 34–46.

Krebs, J.R. & Davies, N.B. (1981). An introduction to behavioural ecology. Oxford, Blackwell Scientific Publications.

Krebs, J.R. & Davies, N.B. (eds.) (1987). An introduction to behavioural ecology, 2nd edn. Oxford, Blackwell Scientific Publications.

Krebs, J.R., MacRoberts, M.H. & Cullen, J.M. (1972). Flocking and feeding in the Great Tit *Parus major* – an experimental study. *Ibis* **114:** 507–530.

Kress, S.W. (1983). The use of decoys, sound recordings, and gull control for re-establishing a tern colony in Maine. *Colonial Waterbirds* **6:** 185–196.

Kruuk, H. (1964). Predators and anti-predator behaviour of the Black-headed Gull (*Larus ridibundus* L.). *Behaviour,* **(Suppl.) 11:** 1–130.

Kubiak, T.J., Harris, H.J., Smith, L.M., Schwartz, T.R., Stalling, D.L., Trick, J.A., Sileo, L., Docherty, D.E. & Erdman, T.C. (1989). Microcontaminants and reproductive impairments of the Forster's Tern on Green Bay, Lake Michigan 1983. *Arch. Environ. Contam. Toxicol.* **18:** 706–727.

Kumari, E. (1974). Past and present of the Peregrine Falcon in Estonia. pp. 230–253 in *'Estonian wetlands and their life'* (ed. E. Kumari). Tallin, Valgus.

Kurth, D. (1970). Der Turmfalke (*Falco tinnunculus*) im Münchener Stadtgebiet. *Anz. orn. Ges. Bayern* **9:** 2–12.

Kvitek, R.G. (1991). Sequestered paralytic shellfish poisoning toxins mediate Glaucous-winged Gull predation on bivalve prey. *Auk* **108:** 381–392.

Labisky, R.F. & Lutz, R.W. (1967). Responses of wild Pheasants to solid-block applications of aldrin. *J. Wildl. Manage.* **31:** 13.

Lack, D. (1947). Darwin's finches. Cambridge, University Press.

Lack, D. (1954). The natural regulation of animal numbers. Oxford, University Press.

Lack, D. (1966). Population studies of birds. Oxford, University Press.

Lack, D. (1968). Ecological adaptions for breeding in birds. London, Methuen.

Lack, D. (1971). Ecological isolation in birds. Oxford, Blackwell.

Lack, P.C. (1987). The structural and seasonal dynamics of the bird community in Tsavo East National Park. *Ostrich* **58:** 9–23.

La Cock, G.D. (1986). The Southern Oscillation environmental anomalies, and mortality of two southern African seabirds. *Clim. Change* **8:** 173–184.

Lahaye, W.J., Gutiérrez, R.J. & Rezit, H. (1994). Spotted Owl metapopulation dynamics in southern California. *J. Anim. Ecol.* **63:** 775–785.

Laird, M. & Bennett, G.F. (1970). The subarctic epizootiology of *Leucocytozoon simondi*. *J. Parasitol.* **56:** 198.

Lambeck, R.H.D., Sandee, A.J.J. & De Wolf, L. (1989). Long-term patterns in the wader usage of an intertidal flat in the Oosterschelde (SW Netherlands) and the impact of the closure of an adjacent estuary. *J. Appl. Ecol.* **26:** 419–431.

Lamberson, R.H., McKelvey, R., Noon, B.R. & Voss, C. (1992). A dynamic analysis of Northern Spotted Owl viability in a fragmented forest landscape. *Conserv. Biol.* **6:** 505–512.

Lambrechts, M. & Dhondt, A.A. (1988). Male quality and territory quality in the Great Tit *Parus major. Anim. Behav.* **36**: 596–601.

Lampio, T. (1946). Game diseases in Finland 1924–43. *Suomen Riista* **1**: 93–142.

Lande, R. (1987). Extinction thresholds in demographic models of territorial populations. *Amer. Nat.* **130**: 624–635.

Lande, R. (1988a). Genetics and demography in biological conservation. *Science* **241**: 1455–1460.

Lande, R. (1988b). Demographic models of the Northern Spotted Owl (*Strix occidentalis caurina*). *Oecologia* **75**: 601–607.

Lande, R. (1991). Population dynamics and extinction in heterogeneous environments: the Northern Spotted Owl. pp. 566–580 in *'Bird population studies'* (ed. C.M. Perrins, J.-D. Lebreton & G.J.M. Hirons). Oxford, University Press.

Lanner, R.M. (1996). Made for each other. A sympiosis of birds and pines. Oxford, University Press.

Lanyon, S.M. & Thompson, C.F. (1986). Site fidelity and habitat quality as determinants of settlement pattern in male Painted Buntings. *Condor* **88**: 206–210.

Latham, R.M. (1947). Differential ability of male and female game birds to withstand starvation and climatic extremes. *J. Wildl. Manage.* **11**: 139–149.

Laursen, K., Gram, I. & Alberto, L.J. (1983). Short-term effect of reclamation on numbers and distribution of waterfowl at Hojer, Danish Wadden Sea. *Proceedings of the Third Nordic Congress of Ornithology.* pp. 97–118.

Laws, R.M. (1977). Seals and whales of the Southern Ocean. *Phil. Trans. R. Soc. Lond.* **279**: 81–96.

Lawton, J.H. (1990). Red Grouse populations and moorland management. Shrewsbury, Field Studies Council.

Lawton, J.H. (1994). Population dynamic principles. *Phil. Trans. R. Soc. Lond., B.* **344**: 61–68.

Leck, C.F. (1979). Avian extinctions in an isolated tropical wet-forest preserve, Ecuador. *Auk* **96**: 343–352.

Lens, L. & Dhondt, A.A. (1992). The effect of a severe storm on a population of Crested Tits *Parus cristatus* in Belgium. *Bird Study* **39**: 31–33.

Lever, C. (1987). Naturalised birds of the world. Harlow, Longman.

Levey, D.J. & Karasov, W.H. (1989). Digestive responses of temperate birds switched to fruit or insect diets. *Auk* **106**: 675–686.

Levins, R. (1969). Some demographic and genetic consequences of environmental heterogeneity for biological control. *Bull. Entomol. Soc. Am.* **15**: 237–240.

Levins, R. (1970). Extinction. pp. 77–107 in *'Some mathematical problems in biology'* (ed. M. Gesternhaber). Providence, Rhode Island, American Mathematical Society.

Lewis, R.A. (1984). Density, movements and breeding success of female Blue Grouse in an area of reduced male density. *Can. J. Zool.* **62**: 1556–1560.

Lewis, R.A. & Zwickel, F.C. (1980). Removal and replacement of male Blue Grouse on persistent and transient territorial sites. *Can. J. Zool.* **58**: 1417–1423.

Ligon, J.D. (1968). Starvation of spring migrants in the Chiricahua Mountains, Arizona. *Condor* **70**: 387–388.

Lima, S.L. (1986). Predation risk and unpredictable feeding conditions: determinants of body mass in birds. *Ecology* **67**: 377–385.

Lima, S.L. & Dill, L.M. (1990). Behavioural decisions made under the risk of predation – a review and prospectus. *Can. J. Zool.* **68**: 619–640.

Lima, S.L. & Valone, T.J. (1991). Predators and avian community organisation: an experiment in a semi-desert grassland. *Oecologia* **86**: 105–112.

Lindén, H. (1988). Latitudinal gradients in predator–prey interactions, cyclicity and synchronism in voles and small game populations in Finland. *Oikos* **52**: 341–349.

Lindén, H. & Wikman, M. (1983). Goshawk predation on tetraonids: availability of prey and diet of the predator in the breeding season. *J. Anim. Ecol.* **52**: 953–968.

Lindström, E.P., Angelstam, P., Widén, P. & Andrén, H. (1987). Do predators synchronize vole and grouse fluctuations? – An experiment. *Oikos* **48**: 121–124.

Lindström, E.R., Andrén, H., Angelstam, P., Cederlund, G., Hornfeldt, B., Jäderberg, L., Lemmell, P.-A., Martinsson, B., Skold, K. & Swenson, J.E. (1994). Disease reveals the predator; sarcoptic mange, Red Fox predation, and prey populations. *Ecology* **75**: 1042–1049.

Littlefield, C.D. (1995). Demographics of a declining flock of Greater Sandhill Cranes in Oregon. *Wilson Bull.* **107**: 667–674.

Lloyd, C., Tasker, M.L. & Partridge, K. (1991). The status of seabirds in Britain and Ireland. London, Poyser.

Lloyd, C.S., Thomas, G.J., MacDonald, J.W., Borland, E.D., Standring, K. & Smart, J.L. (1976). Wild bird mortality caused by botulism in Britain, 1975. *Biol. Conserv.* **10**: 119–129.

Locke, L.N. & Friend, M. (1987). Avian botulism. pp. 83–94 in '*Field guide to wildlife diseases.* Vol. 1. *General field procedures and diseases of migratory birds*' (ed. M. Friend). U.S. Dept. Interior, Fish and Wildlife Service Res. Publ. 167, Washington, D.C.

Locke, L.N., Stotts, V. & Wolfhand, G. (1970). An outbreak of fowl cholera in waterfowl on the Chesapeake Bay. *J. Wildl. Dis.* **6**: 404–407.

Lockie, J.D. (1955). The breeding habits and food of Short-eared Owls after a vole plague. *Bird Study* **2**: 53–69.

Löhrl, H. (1970). Unterschiedliche Bruthöhlenansprüche von Meisenarten und Kleibern als Beitrag zum Nischenproblem. *Verh. Dtsch. Zool. Ges.* **64**: 314–317.

Löhrl, H. (1977). Nistökologische und ethologische Anpassungserscheinungen bei Höhlen-brütern. *Die Vogelwarte* **29**: 92–101.

Lokemoen, J.T., Doty, H.A., Sharp, D.E. & Neaville, J.E. (1982). Electric fences to reduce mammalian predation on waterfowl nests. *Wildl. Soc. Bull.* **10**: 318–323.

Loman, J. (1980). Reproduction in a population of the Hooded Crow *Corvus cornix*. *Holarct. Ecol.* **3**: 26–35.

Lomnicki, A. (1980). Regulation of population density due to individual differences and patchy environment. *Oikos* **35**: 185–193.

Long, J.L. (1981). Introduced birds of the world. New York, Universe Books.

Longcore, J.R. & Stendel, R.C. (1977). Shell thinning and reproductive impairment in Black Ducks after cessation of DDE dosage. *Arch. Environ. Contam. Toxicol.* **6**: 293–304.

Lotka, A.J. (1925). Elements of physical biology. Baltimore, Williams & Wilkins.

Lovat, Lord (ed.) (1911). The Grouse in health and in disease. London, Smith, Elder & Co.

Lovejoy, T.E., Bierregaard, R.O., Rylands, A.B., Malcolm, J.R., Quintela, C.E., Harper Keith, L.H., Brown, S., Powell, A.H., Powell, G.V.N., Schubart, H.O.R. & Hays, M. (1986). Edge and other effects of isolation on Amazon forest fragments. pp. 257–285 in '*Conservation biology*' (ed. M. Soulé, Massachusetts, Sutherland, Seinauer Associates.

Loxton, R.G. & Silcocks, A. (1997). The rise and fall of Bardsey Blackbirds. *Rep. Bardsey Bird Fld. Obs.* **40**: 76–99.

Loyn, R.H. (1987). Effects of patch area and habitat on bird abundances, species numbers and tree health in fragmented Victorian forests. pp. 65–77 in '*Nature Conservation: the role of remnants of native vegetation*' (ed. D.A. Saunders, G.W. Arnold, A.A. Burbidge & A.J.M. Hopkins). Chipping Norton, Australia, Surrey Beatty & Sons.

Lundberg, A. & Alatalo, R. (1992). The Pied Flycatcher. London, Poyser.

Lundberg, A., Alatalo, R.V., Carlson, A. & Ulfstrand, S. (1981). Biometry, habitat distribution and breeding success in the Pied Flycatcher *Ficedula hypoleuca*. *Ornis Scand.* **12**: 68–79.

Lunk, W.A. (1962). The Rough-winged Swallow *Stelgidopteryx ruficollis* (Vieillot). *Publ. Nuttall Ornithol. Club* **4**: 1–155.

Lynch, G.M. (1972). Effect of strychnine control on nest predators of dabbling ducks. *J. Wildl. Manage.* **36**: 436–440.

Lynch, J.F. (1991). Effects of hurricane Gilbert on birds in a dry tropical forest in the Yucatan peninsula. *Biotropica* **23**: 488–496.

Lynch, J.F., Morton, E.S. & van der Voort, M.E. (1985). Habitat segregation between the sexes of wintering Hooded Warblers (*Wilsonia citrina*). *Auk* **112**: 714–721.

Lynch, J.F. & Whigham, D.F. (1984). Effects of forest fragmentation on breeding bird communities in Maryland, USA. *Biol. Conserv.* **28**: 287–324.

MacArthur, R.H. (1959). On the breeding distribution pattern of North American migrant birds. *Auk* **76**: 318–325.

MacArthur, R.H. (1972). Geographical ecology. Patterns in the distribution of species. New York, Harper & Row.

Macdonald, J.W. (1962). Mortality in wild birds with some observations on weights. *Bird Study* **9**: 147–167.

Macdonald, J.W. (1963). Mortality in wild birds. *Bird Study* **10**: 91–108.

Mace, G.M. (1995). Classification of threatened species and its role in conservation planning. pp. 197–213 in '*Extinction rates*' (ed. J.H. Lawton & R.M. May). Oxford, University Press.

Maddock, M. & Geering, D. (1994). Range expansion and migration of the Cattle Egret. *Ostrich* **65**: 191–203.

Madsen, J. (1987). Status and management of goose populations in Europe with special reference to populations nesting and breeding in Denmark. *Danish Rev. Game Biol.* **12**: 1–76.

Madsen, J. (1988). Autumn feeding ecology of herbivorous wildfowl in the Danish Wadden Sea, and impact of food supplies and shooting on movements. *Danish Review of Game Biology* **13**: 1–32.

Madsen, J. (1991). Status and trends of goose populations in the western Palaearctic in the 1980s. *Ardea* **79**: 113–122.

Madsen, J. (1995). Impacts of disturbance on migratory waterfowl. *Ibis* **137 (Suppl.)**: 67–74.

Madsen, J. & Mortensen, C.E. (1987). Habitat exploitation and interspecific competition of moulting geese in East Greenland. *Ibis* **129**: 25–44.

Magnin, G. (1991). Hunting and persecution of migratory birds in the Mediterranean region. pp. 59–71 in '*Conserving migratory birds*' (ed. T. Salathé). Cambridge, International Council for Bird Preservation.

Mannan, R.W., Meslow, E.C. & Wight, H.M. (1980). Use of snags by birds in Douglas fir forests. *J. Wildl. Manage.* **44**: 787–797.

Manuwal, D.A. (1967). Observations on a localised duck sickness in the Delta Marsh; summer 1964. *Wilson Bull.* **79**: 219–222.

Manuwal, D.A. (1974). Effects of territoriality on breeding in a population of Cassin's Auklet. *Ecology* **55**: 1399–1406.

Manville, A.M. (1991). Aleutian islands plastics, pelagic drift and trawl net problems, and their solutions. *Trans. N.A. Wildl. Nat. Res. Conf.* **56**: 205–214.

Marchant, J.H., Hudson, R., Carter, S.P. & Whittington, P. (1990). Population trends in British breeding birds. Tring, British Trust for Ornithology.

Marcström, V., Kenward, R.E. & Engren, E. (1988). The impact of predation on boreal tetraonids during vole cycles: an experimental study. *J. Anim. Ecol.* **57**: 859–872.

Marcström, V. & Mascher, J.W. (1979). Weights and fat in Lapwings *Vanellus vanellus* and Oystercatchers *Haematopus ostralegus* starved to death during a cold spell in spring. *Ornis Scand.* **10**: 235–240.

Marion, L. (1989). Territorial feeding and colonial breeding are not mutually exclusive: the case of the Grey Heron (*Ardea cinerea*). *J. Anim. Ecol.* **58**: 693–710.

Marjakangas, A. (1987). Does the provision of extra food in winter affect the abundance of Finnish Black Grouse? pp. 133–138 in '*Proc. 4th Int. Symp. Grouse*' (ed. T. Lovel & P.J. Hudson). Lam, Germany.

Marquiss, M., Newton, I. & Ratcliffe, D.A. (1978). The decline of the Raven, *Corvus corax*, in relation to afforestation in southern Scotland and northern England. *J. Appl. Ecol.* **15**: 129–144.

Marr, T.G. & Raitt, R.J. (1983). Annual variations in patterns of reproduction of the Cactus Wren (*Campylorhynchus brunneicapillus*). *Southwest. Nat.* **28**: 149–156.

Marra, P.P., Sherry, T.W. & Holmes, R.T. (1993). Territorial exclusion by a long-distance migrant warbler in Jamaica: a removal experiment with American Redstarts (*Setophaga ruticilla*). *Auk* **110**: 565–572.

Martin, P.S. (1984). Prehistoric overkill: the global model. pp. 354–403 in '*Quaternary extinctions*' (ed. P.S. Martin & R.G. Klein). Tucson, Arizona, University of Arizona Press.

Martin, P.S. & Klein, R.G. (eds.) (1984). Quaternary extinctions. Tucson, Arizona, University of Arizona Press.

Martin, P.S. & Wright, E.H. (eds.) (1967). Pleistocene extinctions: the search for a cause. New Haven, Yale University Press.

Martin, T.E. (1987). Food as a limit on breeding birds: a life history perspective. *Ann. Rev. Ecol. Syst.* **18**: 453–487.

Martin, T.E. (1988). Habitat and area effects on forest bird assemblages: is nest predation an influence? *Ecology* **69**: 74–84.

Martin, T.E. (1991). Food limitation in terrestrial breeding bird populations: is that all there is? *Proc. Int. Ornithol. Congr.* **20**: 1595–1602.

Martin, T.E. & Roper, J.J. (1988). Nest predation and nest-site selection of a western population of the Hermit Thrush. *Condor* **90**: 51–57.

Matthysen, E. (1989). Territorial and nonterritorial settling in juvenile Eurasian Nuthatches (*Sitta europaea* L.) in summer. *Auk* **106**: 560–567.

Matthysen, E. (1990). Behavioural and ecological correlates of territory quality in the Eurasian Nuthatch (*Sitta europaea*). *Auk* **107**: 86–95.

Matthysen, E. & Currie, D. (1996). Habitat fragmentation reduces disperser success in juvenile nuthatches *Sitta europaea*: evidence from patterns of territory establishment. *Ecography* **19**: 67–72.

Maurer, B.A. & Heywood, S.G. (1993). Geographic range fragmentation and abundance in neotropical migratory birds. *Conserv. Biol.* **7**: 501–509.

May, R.M. (1988). How many species are there on earth? *Science* **241**: 1441–1449.

May, R.M. (1990). Taxonomy as destiny. *Nature* **347**: 129–130.

May, R.M. (1994). Ecological science and the management of protected areas. *Biodiversity and Conservation* **3**: 437–448.

May, R.M. & Anderson, R.M. (1978). Regulation and stability of host–parasite interactions. II. Destabilising processes. *J. Anim. Ecol.* **47**: 249–267.

May, R.M. & Robinson, S.K. (1985). Population dynamics of avian brood parasitism. *Amer. Nat.* **126**: 475–494.

Mayfield, H. (1965). The Brown-headed Cowbird, with old and new hosts. *Living Bird* **4**: 13–28.

Mayfield, H. (1977). Brown-headed Cowbird: agent of extermination? *Amer. Birds* **31**: 107–113.

Mayfield, H.F. (1983). Kirtland's Warbler, victim of its own rarity? *Auk* **100**: 974–976.

McCleery, R.H., Clobert, J., Julliard, R. & Perrins, C.M. (1996). Nest predation and delayed cost of reproduction in the Great Tit. *J. Anim. Ecol.* **65**: 96–104.

McCleery, R.H. & Perrins, C.M. (1985). Territory size, reproductive success and population dynamics in the Great Tit, *Parus major*. pp. 353–373 in '*Behavioural ecology; ecological consequences of adaptive behaviour*' (ed. R.M. Sibly & R.H. Smith). Oxford, Blackwell Scientific Publications.

McCleery, R.H. & Perrins, C.M. (1991). Effects of predation on the numbers of Great Tits *Parus major*. pp. 129–147 in '*Bird population studies*' (ed. C.M. Perrins, J.-D. Lebreton & G.J.M. Hirons). Oxford, University Press.

McClure, H.E. (1945). Effects of a tornado on bird life. *Auk* **62**: 414–418.

McComb, W.C. & Noble, R.E. (1981). Nest-box and natural cavity use in three mid south forest habitats. *J. Wildl. Manage.* **45**: 93–101.

McCulloch, M.N., Tucker, G.M. & Baillie, S.R. (1992). The hunting of migratory birds in Europe: a ringing recovery analysis. *Ibis* **134 (Suppl. 1):** 55–65.

McCulloch, N. (1994). Rescuing the Seychelles Magpie Robin. *RSPB Conserv. Rev.* **8:** 88–94.

McDonald, M.E. (1969). Catalogue of helminths of waterfowl (Anatidae). *Washington, D.C., U.S. Bureau of Sport Fish. & Wildl. Spec. Sci. Rep.* **126**.

McGilvrey, F.B. & Uhler, F.M. (1971). A starling deterrent wood duck nest-box. *J. Wildl. Manage.* **35:** 793–797.

McGowan, J.D. (1975). Effects of autumn and spring hunting on Ptarmigan population trends. *J. Wildl. Manage.* **39:** 491–495.

McIlhenny, F.A. (1934). Bird City. Boston, Christopher Publishing House.

McIlhenny, E.A. (1940). Effects of excessive cold on birds in southern Louisiana. *Auk* **57:** 408–410.

McKernan, D.L. & Schaffer, V.B. (1942). Unusual numbers of dead birds on the Washington Coast. *Condor* **44:** 264–266.

McLaughlin, C.L. & Grice, D. (1952). The effectiveness of large-scale erection of Wood Duck boxes as a management procedure. *Trans. N. A. Wildl. Nat. Res. Conf.* **17:** 242–259.

McLean, D.D. (1946). Duck disease at Tulare Lake. *Calif. Fish Game* **32:** 71–80.

McMahan, C.A. & Fritz, R.L. (1967). Mortality to ducks from trotlines in Lower Laguna Madre, Texas. *J. Wildl. Manage.* **31:** 783–787.

McNamara, J.M. & Houston, A.I. (1990). The value of fat reserves and the tradeoff between starvation and predation. *Acta Biotheor.* **38:** 37–61.

McNeely, J.A. (1988). Economics and biological diversity. International Union for the Conservation of Nature, Gland, Switzerland.

Mead, C.J. (1991). The missing migrants. *BTO News* **176:** 1.

Meanly, B. & Royall, W.C. (1976). Nationwide estimates of blackbirds and starlings. *Proc. Bird Control. Semin.* **7:** 39–40.

Mearns, R. & Newton, I. (1988). Factors affecting breeding success of Peregrines in south Scotland. *J. Anim. Ecol.* **57:** 903–916.

Meek, E. (1993). Population fluctuations and mortality of Mute Swans on an Orkney loch system in relation to a Canadian pondweed growth cycle. *Scott. Birds* **17:** 85–92.

Meininger, P.L., Blomert, A.-M. & Marteijn, E.C.L. (1991). Watervogelsterfte in het Dellagebied, ZW-Nederland, gedurende de drie koude winters van 1985, 1986 en 1987. *Limosa* **64:** 89–102.

Meltofte, H., Elander, M. & Hjort, C. (1981). Ornithological observations in northeast Greenland between 74° 30' and 76° 00' N. lat, 1976. *Meddr. Grønland, Biosci.* **3:** 1–53.

Meltofte, H. (1996). Are African wintering waders really forced south by competition from northerly wintering conspecifics? Benefits and constraints of northern versus southern wintering and breeding in waders. *Ardea* **84:** 31–44.

Mendelsohn, J.M. (1983). Social behaviour and dispersion of the Black-shouldered Kite. *Ostrich* **54:** 1–18.

Mendelssohn, H. (1972). The impact of pesticides on bird life in Israel. *ICBP Bull.* **11:** 75–104.

Mendelssohn, H. & Paz, U. (1977). Mass mortality of birds of prey caused by azodrin, an organophosphorus insecticide. *Biol. Conserv.* **11:** 163–170.

Metcalfe, N.B. (1984). The effects of habitat on the vigilance of shorebirds: is visibility important. *Anim. Behav.* **32:** 981–985.

Metcalfe, N.B. (1989). Flocking preferences in relation to vigilance benefits and aggression costs in mixed-species shorebird flocks. *Oikos* **56:** 91–98.

Metcalfe, N.B. & Ure, S.E. (1995). Diurnal variation in flight performance and hence potential predation risk in small birds. *Proc. R. Soc. Lond. B.* **261:** 395–400.

Mihelsons, H.A., Mednis, A.A. & Blums, P.B. (1985). Regulatory mechanisms of numbers in breeding populations of migratory ducks. *Proc. Int. Ornithol. Congr. Moscow* **18:** 797–802.

Mikola, J., Miettinen, M., Lehikoinen, E. & Lehtila, K. (1994). The effects of disturbance caused by boating on survival and behaviour of Velvet Scoter *Melanitta fusca* ducklings. *Biol. Conserv.* **67:** 119–124.

Milberg, P. & Tyrberg, T. (1993). Naive birds and noble savages – a review of man-caused prehistoric extinctions of island birds. *Ecography* **16**: 229–250.

Miller, B. & Mullette, K.J. (1985). Rehabilitation of an endangered Australian bird: The Lord Howe Island Woodhen *Tricholimnas sylvestris* (Sclater). *Biol. Conserv.* **34**: 55–95.

Miller, G.R., Jenkins, D. & Watson, A. (1966). Heather performance and Red Grouse populations. 1. Visual estimates of heather performance. *J. Appl. Ecol.* **3**: 313–326.

Miller, G.R., Watson, A. & Jenkins, D. (1970). Responses of Red Grouse populations to experimental improvement in their food. pp. 323–335 in '*Animal populations in relation to their food resources*' (ed. A. Watson). Oxford, Blackwell.

Miller, R.S. (1967). Pattern and process in competition. *Adv. Ecol. Res.* **4**: 1–74.

Miller, R.S. (1968). Conditions of competition between Redwings and Yellowheaded Blackbirds. *J. Anim. Ecol.* **37**: 43–62.

Miller, W. (1970). Factors influencing the status of Eastern and Mountain Bluebirds in southwestern Manitoba. *Blue Jay* **28**: 38–46.

Mills, J. (1973). Some observations on the effects of field applications of fensulfothion and parathion on bird and mammal populations. *Proc. N.Z. Ecol. Soc.* **20**: 65–71.

Mills, J.A. (1989). Red-billed Gull. pp. 387–404 in '*Lifetime reproduction in birds*' (ed. I. Newton). London, Academic Press.

Milne, H. (1976). Body weights and carcass composition of the Common Eider. *Wildfowl* **27**: 115–122.

Mineau, P. (1993). The hazard of carbofuron to birds and other vertebrate wildlife. Environment Canada, Canadian Wildlife Service, Wildlife Toxicology Section, *Ottawa. Tech. Rep.* No. 177.

Minot, E.O. (1981). Effects of interspecific competition for food in breeding Blue and Great Tits. *J. Anim. Ecol.* **50**: 375–385.

Minot, E.O. & Perrins, C.M. (1986). Interspecific interference competition – nest-sites for Blue and Great tits. *J. Anim. Ecol.* **55**: 331–350.

Mock, D.W., Lamey, T.C. & Ploger, B.J. (1987). Proximate and ultimate roles of food amount in regulating egret sibling aggression. *Ecology* **68**: 1760–1772.

Moilanen, P. (1976). Goshawks and Pheasants. *Suomen Luonto* **6**: 315–318.

Møller, A.P. (1982). Characteristics of Magpie *Pica pica* territories of varying duration. *Ornis Scand.* **13**: 94–100.

Møller, A.P. (1989a). Population dynamics of a declining Swallow *Hirundo rustica* population. *J. Anim. Ecol.* **58**: 1051–1063.

Møller, A.P. (1989b). Parasites, predators and nest boxes: facts and artifacts in nest box studies of birds. *Oikos* **56**: 421–423.

Møller, A.P. (1990). Effects of parasitism by the haematophagous mite on reproduction in the Barn Swallow. *Ecology* **71**: 2345–2357.

Møller, A.P. (1991a). Clutch-size, nest predation and distribution of avian unequal competitors in a patchy environment. *Ecology* **72**: 1336–1349.

Møller, A.P. (1991b). Parasites, sexual ornaments, and mate choice in the Barn Swallow. pp. 328–343 in '*Bird-parasite interactions*' (ed. J.E. Loye & M. Zuk). Oxford, University Press.

Møller, A.P. (1993). Ectoparasites increase the cost of reproduction in their hosts. *J. Anim. Ecol.* **62**: 309–322.

Møller, A.P. (1994). Parasites as an environmental component of reproduction in birds as exemplified by the Swallow *Hirundo rustica*. *Ardea* **82**: 161–172.

Møller, A.P., Allander, K. & Dufva, R. (1989). Fitness effects of parasites on passerine birds: a review. pp. 269–280 in '*Population biology of passerine birds*' (ed. J. Blondel, A. Gosler, J.-D. Lebreton & R McCleery). London, Springer-Verlag.

Mönkkönen, M. (1990). Removal of territory holders causes influx of small-sized intruders in passerine bird communities in northern Finland. *Oikos* **57**: 281–288.

Monaghan, P. (1979). Aspects of the breeding biology of Herring Gulls *Larus argentatus* in urban colonies. *Ibis* **121**: 475–481.

Monaghan, P. (1980). Dominance and dispersal between feeding sites in the Herring Gull (*Larus argentatus*). *Anim. Behav.* **28:** 521–527.

Monaghan, P. & Coulson, J.C. (1977). Status of large gulls nesting on buildings. *Bird Study* **24:** 89–104.

Monaghan, P., Uttley, J.D. & Okill, J.D. (1989). Terns and sandeels: seabirds as indicators of marine fish populations. *J. Fish Biol.* **(suppl. A) 35:** 339–340.

Monaghan, P. & Zonfrillo, B. (1986). Population dynamics of seabirds in the Firth of Clyde. *Proc. R. Soc. Edinb.* **90 B:** 363–375.

Montevecchi, W.A. (1993). Birds as indicators of change in marine prey stocks. pp. 217–266 in '*Birds as monitors of environmental change*' (ed. R.W. Furness & J.J.D. Greenwood). London, Chapman & Hall.

Montgomery, R.D., Stein, G., Stotts, V.D. & Settle, F.H. (1979). The 1978 epornitic of avian cholera on the Chesapeake Bay. *Avian Dis.* **23:** 966–978.

Montier, D. (1968). A survey of the breeding distribution of the Kestrel, Barn Owl and Tawny Owl in the London area in 1967. *Lond. Bird Rep.* **32:** 81–92.

Monval, J.-Y. & Pirot, J.-Y. (1989). Results of the IWRB International Waterfowl Census 1967–86. IWRB Spec. Pub. 8. Slimbridge, International Waterfowl & Wetlands Research Bureau.

Moore, J., Freehling, M. & Simberloff, D. (1986). Gastrointestinal helminths of northern Bobwhite in Florida: 1980 and 1983. *J. Wildl. Dis.* **22:** 497–501.

Morant, P.D., Cooper, J. & Randall, R.M. (1981). The rehabilitation of oiled Jackass Penguins *Spheniscus demersus*, 1970–1980. pp. 267–285 in '*Proceeding of a symposium on birds of the sea and shore*' (ed. J. Cooper). Capetown, South Africa, African Seabird Group.

Moreau, R.E. (1927). Quail. *Bull. Zool. Soc. Egypt* **1:** 6–13.

Moreau, R.E. (1966). The bird faunas of Africa and its islands. London, Academic Press.

Moritz, D. (1993). Long-term monitoring of Palaearctic-African migrants at Helgoland/ German Bight, North Sea. *Proc. Pan-African Ornithol. Congr.* **8:** 579–586.

Morozov, N.S. (1993). Short-term fluctuations in a South-Taiga bird assemblage – support for an individualistic view. *Ornis Fenn.* **70:** 177–188.

Morris, R.D., Blokpoel, H. & Tessier, G.D. (1992). Management efforts for the conservation of Common Tern *Sterna hirundo* colonies in the Great Lakes – 2 case histories. *Biol. Conserv.* **60:** 7–14.

Morris, R.F., Cheshire, W.F., Miller, C.A. & Mott, D.G. (1958). The numerical response of avian and mammalian predators during a gradation of the spruce budworm. *Ecology* **39:** 487–494.

Morris, R.J. (1980). Floating plastic debris in the Mediterranean. *Mar. Poll. Bull.* **11:** 125.

Morse, D.H. (1973). Interactions between tit flocks and Sparrowhawks *Accipiter nisus*. *Ibis* **115:** 591–593.

Morse, S.S. (1990). Regulating viral traffic. *Issues in Sci. Technol.* **7:** 81–84.

Morse, S.S. (1991). The origins of new viral diseases. Environmental carcinogenesis and ecotoxicology reviews – Part C of *J. Environ. Sci. Health* **9:** 207–228.

Moser, M.E. (1988). Limits to the numbers of Grey Plover *Pluvialis squatarola* wintering on British estuaries: an analysis of long-term population trends. *J. Appl. Ecol.* **25:** 473–486.

Morton, E.S., Lynch, J.F., Young, K. & Mehlhop, P. (1987). Do male Hooded Warblers exclude females from non-breeding territories in tropical forests? *Auk* **104:** 133–135.

Moss, R. (1972). Social organisation of Willow Ptarmigan on the breeding grounds in interior Alaska. *Condor* **74:** 144–151.

Moss, D. (1978). Song-bird populations in forestry plantations. *Quart. J. For.* **72:** 5–14.

Moss, R. (1986). Rain, breeding success and distribution of Capercaillie *Tetrao urogallus* and Black Grouse *Tetrao tetrix* in Scotland. *Ibis* **128:** 65–72.

Moss, R. (1969). A comparison of Red Grouse (*Lagopus l. scoticus*) stocks with the production and nutritive value of heather (*Calluna vulgaris*). *J. Anim. Ecol.* **38:** 103–122.

Moss, R. & Oswald, J. (1985). Population dynamics of Capercaillie in a northeast Scottish glen. *Ornis Scand.* **16**: 229–238.

Moss, W.W. & Camin, J.H. (1970). Nest parasitism, productivity and clutch size in Purple Martins. *Science* **168**: 1000–1003.

Mossop, D.H. (1985). Winter survival and breeding strategies of Willow Ptarmigan. pp. 330–378 in '*Adaptive strategies and population ecology of northern grouse*' (ed. A.T. Bergerud & M.W. Gratson). Minneapolis, University of Minnesota Press.

Moulding, J.D. (1976). Effects of a low-persistence insecticide on forest bird populations. *Auk* **93** : 692–708.

Moulton, D.W. & Weller, M.W. (1984). Biology and conservation of the Laysan Duck *Anas laysanensis*. *Condor* **86**: 105–117.

Moulton, M.D. & Pimm, S.L. (1983). The introduced Hawaiian avifauna: biogeographic evidence for competition. *Amer. Nat.* **121**: 669–690.

Moulton, M.D. & Pimm, S.L. (1986). The extent of competition in shaping an introduced avifauna. pp. 80–97 in '*Community ecology*' (ed. J. Diamond & T.J. Case). New York, Harper & Row.

Mountainspring, S. & Scott, J.M. (1985). Interspecific competition among Hawaiian forest birds. *Ecol. Monogr.* **55**: 219–239.

Mountford, M.D., Watson, A., Moss, R., Parr, R. & Rothery, P. (1990). Land inheritance and population cycles of Red Grouse. pp. 78–83 in '*Red Grouse population processes*' (ed. A.N. Lance & J.H. Lawton). Sandy, Bedfordshire, British Ecological Society and Royal Society for the Protection of Birds.

Mueller, H.C. & Berger, D.D. (1967). Fall migration of Sharp-shinned Hawks. *Wilson Bull.* **79**: 397–415.

Mueller, H.C., Berger, D.D. & Allez, G. (1977). The periodic invasions of Goshawks. *Auk* **94**: 652–663.

Mullié, W.C. & Keith, J.O. (1993). The effects of aerially applied fenithrothion and chlorpyrifos on birds in the savannah of northern Senegal. *J. Appl. Ecol.* **30**: 536–550.

Mulliken, T.A. (1995). Response to questions posed by The Royal Society for the Protection of Birds regarding the international trade in wild birds. Unpublished report. Cambridge, Traffic International.

Munn, C.A. (1985). Permanent canopy and understory flocks in Amazonia: species composition and population density. *Orn. Mongr.* **36**: 683–712.

Munn, C.A. (1992). Macaw biology and ecotourism, or "when a bird in the bush is worth two in the hand". pp. 47–72 in '*New World parrots in crisis*' (ed. S.R. Beissinger & N.F.R. Snyder). Washington, Smithsonian Institution Press.

Munn, C.A. & Terborgh, J. (1979). Multi-species territoriality in Neotropical foraging flocks. *Condor* **81**: 338–347.

Munro, H.L. & Rounds, R.C. (1985). Selection of artificial nest-sites by five sympatric passerines. *J. Wildl. Manage.* **49**: 264–276.

Murphy, E.C., Springer, A.M. & Roseneau, D.G. (1986). Population status of *Uria aalge* at a colony in western Alaska: results and simulations. *Ibis* **128**: 348–363.

Murphy, R.C. (1936). Oceanic birds of South America. New York, Am. Mus. Nat. Hist.

Murphy, R.C. & Vogt, W. (1933). The Dovekie influx of 1932. *Auk* **50**: 325–349.

Murton, R.K. (1968). Some predator relationships in bird damage and population control. pp. 157–169 in '*The problems of birds as pests*' (ed. R.K. Murton & E.N. Wright). London & New York, Academic Press.

Murton, R.K. (1971). Man and birds. London, Collins.

Murton, R.K., Isaacson, A.J. & Westwood, N.J. (1966). The relationships between Woodpigeons and their clover food supply and the mechanism of population control. *J. Appl. Ecol.* **3**: 55–96.

Murton, R.K. & Westwood, N.J. (1977). Avian breeding cycles. Oxford, Clarendon Press.

Murton, R.K., Westwood, N.J. & Isaacson, A.J. (1964). A preliminary investigation of the factors regulating population size in the Woodpigeon. *Ibis* **106**: 482–507.

Murton, R.K., Westwood, N.J. & Isaacson, A.J. (1974). A study of Woodpigeon shooting: the exploitation of a natural animal population. *J. Appl. Ecol.* **11**: 61–81.

Myers, J.P. (1981). A test of three hypotheses for latitudinal segregation of the sexes in wintering birds. *Can. J. Zool.* **59**: 1527–1534.

Myers, J.P. (1982). Spacing behaviour of non-breeding shorebirds. pp. 271–321 in *'Behaviour of marine animals'*, Vol. 6 (ed. J. Banger). New York, Plenum Press.

Myers, J.P., Conners, P.G. & Pitelka, F.A. (1979). Territory size in wintering sanderlings: the effects of prey abundance and intruder density. *Auk* **96**: 551–561.

Myers, N. (1979). The sinking ark. A new look at the problem of disappearing species. Oxford, Pergamon.

Myers, N. (1989). Deforestation rates in tropical forests and their climatic implications. London, Friends of the Earth.

Myrberget, S. (1972). Fluctuations in a North Norwegian population of Willow Grouse. *Proc. Int. Ornithol. Congr.* **15**: 107–120.

Myrberget, S. (1984). Population dynamics of Willow Grouse *Lagopus lagopus* on an island in north Norway. *Fauna Norv. Ser. C. Cinclus* **7**: 95–105.

Myrberget, S. (1985). Demography of an island population of Willow Ptarmigan in Northern Norway. pp. 379–419 in *'Adaptive strategies and population ecology of Northern Grouse'* (ed. A.T. Bergerud & M.W. Gratson). Minneapolis, University of Minnesota Press.

Nelson, J.B. (1978). The Gannet. Berkhamsted, Poyser.

Nelson, R.W. & Myres, M.T. (1975). Changes in the Peregrine population and its sea bird prey at Langara Island, British Columbia. pp. 13–31 in *'Population Status of Raptors'* (ed. J.R. Murphy, C.M. White & B.E. Harrell). Fort Collins, Colorado, Raptor Research Foundation.

Nelson, T.H. (1907). The birds of Yorkshire. Vol. 1. London, A. Brown & Sons.

Nettleship, D.N. (1975). Effects of *Larus* gulls on breeding performance, nest distribution and winter fattening in Atlantic Puffins. pp. 47–69 in *'Proceedings of Gull seminar, Sackville, New Brunswick'*. Sackville, New Brunswick, Canadian Wildlife Service.

Newmark, W.D. (1991). Tropical forest fragmentation and the local extinction of understorey birds in the eastern Usambara Mountains, Tanzania. *Conserv. Biol.* **5**: 67–78.

Newton, I. (1967a). The adaptive radiation and feeding ecology of some British finches. *Ibis* **109**: 33–98.

Newton, I. (1967b). The feeding ecology of the Bullfinch (*Pyrrhula pyrrhula*) in southern England. *J. Anim. Ecol.* **36**: 721–744.

Newton, I. (1969). Winter fattening in the Bullfinch. *Physiol. Zool.* **42**: 96–107.

Newton, I. (1972). Finches. London, Collins.

Newton, I. (1977). Timing and success of breeding in tundra-nesting geese. pp. 113–126 in *'Evolutionary ecology'* (ed. B. Stonehouse & C.M. Perrins). London, Macmillan.

Newton, I. (1979). Population ecology of raptors. Berkhamsted, Poyser.

Newton, I. (1980). The role of food in limiting bird numbers. *Ardea* **68**: 11–30.

Newton, I. (1985). Lifetime reproductive output of female Sparrowhawks. *J. Anim. Ecol.* **54**: 241–253.

Newton, I. (1986). The Sparrowhawk. Calton, Poyser.

Newton, I. (1988a). A key factor analysis of a Sparrowhawk population. *Oecologia* **76**: 588–596.

Newton, I. (1988b). Population regulation in Peregrines: an overview. pp. 761–770 in *'Peregrine Falcon populations: their management and recovery'* (ed. T.J. Cade, J.H. Enderson, C.G. Thelander & C.M. White). Boise, Idaho, The Peregrine Fund.

Newton, I. (ed.) (1989). Lifetime reproduction in birds. London, Academic Press.

Newton, I. (1991). The role of recruitment in population regulation. *Proc. Int. Ornithol. Congr.* **20**: 1689–1699.

Newton, I. (1992). Experiments on the limitation of bird numbers by territorial behaviour. *Biol. Rev.* **67:** 129–173.

Newton, I. (1993). Predation and limitation of bird numbers. *Curr. Ornithol.* **11:** 143–198.

Newton, I. (1994). Experiments on the limitation of bird breeding densities: a review. *Ibis* **136:** 397–411.

Newton, I. (1995). The contribution of recent research on birds to ecological understanding. *J. Anim. Ecol.* **64:** 675–696.

Newton, I. (1997). Links between the abundance and distribution of birds. *Ecography* **20:** 137–145.

Newton, I., Bell, A.A. & Wyllie, I. (1982). Mortality of Sparrowhawks and Kestrels. *Brit. Birds* **75:** 195–204.

Newton, I. & Campbell, C.R.G. (1975). Breeding of ducks at Loch Leven, Kinross. *Wildfowl* **26:** 83–103.

Newton, I., Campbell, C.R.G. & Allison, A. (1974). Gut analyses of Greylag and Pinkfooted Geese. *Bird Study* **21:** 255–262.

Newton, I., Davis, P.E. & Davis, J.E. (1989). Age of first breeding, dispersal and survival of Red Kites *Milvus milvus* in Wales. *Ibis* **131:** 16–21.

Newton, I., Davis, P.E. & Moss, D. (1994). Philopatry and population growth of Red Kites *Milvus milvus,* in Wales. *Proc. Roy. Soc. Lond. B* **257:** 317–323.

Newton, I. & Kerbes, R.H. (1974). Breeding of Greylag Geese (*Anser anser*) on the Outer Hebrides, Scotland. *J. Anim. Ecol.* **43:** 771–783.

Newton, I. & Marquiss, M. (1981). Effect of additional food on laying dates and clutch sizes of Sparrowhawks. *Ornis Scand.* **12:** 224–229.

Newton, I. & Marquiss, M. (1982). Fidelity to breeding area and mate in Sparrowhawks *Accipiter nisus. J. Anim. Ecol.* **51:** 327–341.

Newton, I. & Marquiss, M. (1986). Population regulation in the Sparrowhawk. *J. Anim. Ecol.* **55:** 463–480.

Newton, I. & Marquiss, M. (1991). Removal experiments and the limitation of breeding density in Sparrowhawks. *J. Anim. Ecol.* **60:** 535–544.

Newton, I. & Perrins, C. (1997). Sparrowhawks and song birds. *Birds* **16:** 65–68.

Newton, I., Rothery, P. & Dale, L.C. (1998). Density dependence in the bird populations of an oakwood over 22 years. *Ibis* **139:** in press.

Newton, I. & Wyllie, I. (1992). Recovery of a Sparrowhawk population in relation to declining pesticide contamination. *J. Appl. Ecol.* **20:** 476–484.

Newton, I., Wyllie, I. & Asher, A. (1991). Mortality causes in British Barn Owls *Tyto alba,* with a discussion of aldrin-dieldrin poisoning. *Ibis* **133:** 162–169.

Newton, I, Wyllie, I. & Asher, A. (1993a). Long term trends in organochlorine and mercury residues in some predatory birds in Britain. *Environ. Poll.* **79:** 143–151.

Newton, I., Wyllie, I. & Rothery, P. (1993b). Annual survival of Sparrowhawks *Accipiter nisus* breeding in three areas of Britain. *Ibis* **135:** 49–60.

Newton, I., Wyllie, I. & Mearns, R.M. (1986). Spacing of Sparrowhawks in relation to food supply. *J. Anim. Ecol.* **55:** 361–370.

Nice, M.M. (1937). Studies in the life history of the Song Sparrow. Part 1. *Trans. Linn. Soc. NY* **4:** 1–247.

Nichols, J.D. (1991). Extensive monitoring programs viewed as long-term population studies – the case of North American waterfowl. *Ibis* **(Suppl.) 133:** 89–98.

Nichols, J.D., Conroy, M.J., Anderson, D.R. & Burham, K.P. (1984). Compensatory mortality in waterfowl populations: a review of the evidence and implications for research and management. *Trans. N. A. Wildl. Nat. Res. Conf.* **49:** 535–554.

Nichols, J.D. & Haramis, G.M. (1980). Sex-specific differences in winter distribution patterns of Canvasbacks. *Condor* **82:** 406–416.

Nichols, J.D. & Johnson, F.A. (1990). Wood Duck population dynamics: a review. pp. 83–105 in 'Proc. 1988 North American Wood Duck Symposium' (ed L.H. Frederickson, G.V. Burger, S.P.

Havera, D.A. Graber, R.E. Kirby & T.S. Taylor). St. Louis, Mo., The 1988 North American Wood Duck Symposium.

Nieman, D.J., Hochbaum, G.S., Caswell, F.D. & Turner, B.C. (1987). Monitoring hunter performance in prairie Canada. *Trans. N.A. Wildl. Nat. Res. Conf.* **52:** 233–245.

Nilsson, L. (1979). Variation in the production of young of swans wintering in Sweden. *Wildfowl.* **30:** 129–134.

Nilsson, S.G. (1986). Density independence and density dependence in the population dynamics of the Wren *Troglodytes troglodytes* and the Goldcrest *Regulus regulus*. *Vår Fågelvärld*, **Suppl. 1 :** 155–160.

Nilsson, S.G. (1975). Clutch-size and breeding success of birds in nest boxes and natural cavities. *Vår Fågelvärld* **34:** 207–211.

Nilsson, S.G. (1979). Seed density, cover, predation and the distribution of birds in a beech wood in southern Sweden. *Ibis* **121:** 177–185.

Nilsson, S.G. (1982). Seasonal variation in the survival rate of adult Nuthatches *Sitta europaea* in Sweden. *Ibis* **124:** 96–100.

Nilsson, S.G. (1984). The evolution of nest-site selection among hole-nesting birds: the importance of nest predation and competition. *Ornis Scand.* **15:** 167–175.

Nilsson, S.G. (1987). Limitation and regulation of population density in the Nuthatch *Sitta europaea* (Aves) breeding in natural cavities. *J. Anim. Ecol.* **56:** 921–937.

Nilsson, S.G., Johnson, K. & Tjernberg, M. (1991). Is avoidance by Black Woodpeckers of old holes due to predators? *Anim. Behav.* **41:** 439–441.

Nilsson, S.G. & Nilsson, I.N. (1978). Breeding bird community densities and species richness in lakes. *Oikos* **31:** 214–221.

Nilsson, S.I. & Duinker, P. (1987). The extent of forest decline in Europe. *Environment* **29:** 4–31.

Nisbet, I.C.T. (1978). Recent changes in gull populations in the western North Atlantic. *Ibis* **120:** 129–130.

Nisbet, I.C.T. (1983). Paralytic shellfish poisoning: effects on breeding terns. *Condor* **85:** 338–345.

Nisbet, I.C.T. & Medway, L. (1972). Dispersion, population ecology and migration of Eastern Great Reed Warblers *Acrocephalus orientalis* wintering in Malaysia. *Ibis* **114:** 451–494.

Nordberg, O. (1936). Biologisch-ökologische Untersuchungen über die Vogelnidicolen. *Acta Zool. Fenn.* **21:** 1–168.

Norberg, R.Å. (1978). Energy content of some spiders and insects on branches of spruce (*Picea abies*) in winter; prey of certain passerine birds. *Oikos* **31:** 222–229.

Nores, A.I. (1995). Botfly ectoparasitism of the Brown Cacholote and the Firewood-gatherer. *Wilson Bull.* **107:** 734–738.

Norris, C.A. (1945). Summary of a report on the distribution and status of the Corncrake (*Crex crex*). *Brit. Birds* **38:** 142–148, 162–168.

Norris, K., Anwar, M. & Read, A.F.J. (1994). Reproductive effort influences the prevalence of haematozoan parasites in Great Tits. *J. Anim. Ecol.* **63:** 601–610.

Norton, D.W., Senner, S.E., Gill, R.E., Martin, R.D., Wright, J.M. & Fiskuyama, A.S. (1990). Shorebirds and herring roe in Prince William Sound, Alaska. *Amer. Birds* **44:** 367–371.

Norton, M.E., Arcese, P. & Ewald, P.W. (1982). Effect of intrusion pressure on territory size in Black-chinned Hummingbirds (*Archilochus alexandri*). *Auk* **99:** 761–764.

Norton-Griffiths, M. (1968). The feeding behaviour of the Oyster-catcher *Haematopus ostralegus*. D.Phil. thesis (Unpub), Oxford University.

Novak, E. (1971). The range expansion of animals and its causes. *Zeszyty Naukowe* **3:** 1–255.

Nuttall, P.A., Perrins, C.M. & Harrap, K.A. (1982). Further studies on puffinosis, a disease of the Manx Shearwater (*Puffinus puffinus*). *Can. J. Zool.* **60:** 3462–3465.

Nyholm, N.E. (1981). Evidence of involvement of aluminium in causation of defective formation of eggshells and of impaired breeding of wild passerine birds. *Environ. Res.* **26:** 363–371.

O'Brien, S.J. & Evermann, J.F. (1988). Interactive influence of infectious disease and genetic diversity in natural populations. *Trends Ecol. Evol.* **3:** 253–259.

O'Connor, R.J. (1980). Population regulation in the Yellowhammer *Emberiza citrinella* in Britain. pp. 190–200 in *'Bird census work and nature conservation'* (ed. H. Oelke). Göttingen, Germany, Dachverbandes Deutscher Avifaunisten.

O'Connor, R.J. (1985). Behavioural regulation of bird populations: a review of habitat use in relation to migration and residence. pp. 105–142 in *'Behavioural Ecology. Ecological consequences and adaptive behaviour'* (ed. R.M. Sibly & R.H. Smith). Oxford, Blackwells.

O'Connor, R.J. (1986). Dynamical aspects of habitat use. pp. 235–240 in *'Wildlife Zoo: Modelling habitat relationships of terrestrial vertebrates'* (ed. J. Verner, M.L. Morrison & C.J. Ralph). Madison, University of Wisconsin Press.

O'Connor, R.J. & Fuller, R.J. (1985). Bird population responses to habitat. pp. 197–211 in *'Bird census and atlas studies'* (ed. K. Taylor, R.J. Fuller & P.C. Lack). Tring, British Trust for Ornithology.

O'Connor, R.J. & Shrubb, M. (1986). Farming and birds. Cambridge, University Press.

Obst, B.S. (1985). Densities of Antarctic seabirds at sea and the presence of the Krill *Euphausia superba. Auk* **102:** 540–549.

Oelke, H. (1989). Effect of acid rain syndrome on bird populations (Hartz Mountains, Lower Saxony, FR Germany). *Beitr. Naturk. Nieders.* **42:** 109–128.

Ogle, C.C. (1987). The incidence of conservation of animals and plant species in remnants of native vegetation within New Zealand. pp. 77–87 in *'Nature conservation'* (ed. D.A. Saunders). Chipping Norton, Australia, Surrey Beatty.

Ojanen, M. (1979). Effect of a cold spell on birds in northern Finland in May 1968. *Ornis Fenn.* **56:** 148–155.

Okoniewski, J.C. & Novesky, E. (1993). Bird poisoning with cyclodienes in suburbs: links to historic use on turf. *J. Wildl. Manage.* **57:** 630–639.

Olendorff, R.R. & Stoddart, J.W. (1974). The potential for management of raptor populations in western grasslands. *Raptor Res. Rep.* **2:** 47–87.

Olsen, P.D. & Olsen, J. (1989). Breeding of the Peregrine Falcon *Falco peregrinus*: III. Weather, nest quality and breeding success. *Emu* **89:** 6–14.

Olson, S.L. & James, H.F. (1982). Fossil birds from the Hawaiian Islands: evidence for wholesale extinction by man before western contact. *Science* **217:** 633–635.

Olson, S.L. & James, H.F. (1991). Descriptions of thirty-two new species of birds from the Hawaiian Islands. Parts 1 & 2. *Ornithol. Monogr.* **45; 46:** 1–88.

Olson, D.P. (1965). Differential vulnerability of male and female Canvasbacks to hunting. *Trans. N. A. Wildl. Nat. Res. Conf.* **30:** 121–134.

Opdam, P. (1978). Feeding ecology of a Sparrowhawk population (*Accipiter nisus*). *Ardea* **66:** 137–155.

Opdam, P., Foppen, R., Reijnen, R. & Schotman, A. (1995). The landscape ecological approach in bird conservation: integrating the metapopulation concept into spatial planning. *Ibis* **137 (Suppl. 1):** 139–146.

Orell, M. (1989). Population fluctuations and survival of Great Tits *Parus major* dependent on food supplied by man in winter. *Ibis* **131:** 113–127.

Orians, G.H. (1961). The ecology of blackbird (*Agelaius*) social systems. *Ecol. Monog.* **31:** 285–312.

Orians, G.H. (1969). Age and hunting success in the Brown Pelican (*Pelecanus occidentalis*). *Anim. Behav.* **17:** 316–319.

Orians, G.H. & Willson, M.F. (1964). Interspecific territories of birds. *Ecology* **45:** 736–745.

Ormerod, S.J., Bull, K.R., Cummins, C.D., Tyler, S.J. & Vickery, J.A. (1988). Egg mass and shell thickness in Dippers (*Cinclus cinclus*) in relation to stream acidity in Wales and Scotland. *Environ. Poll.* **55:** 107–121.

Ormerod, S.J., O'Halloran, J., Griffin, D.D. & Tyler, S.J. (1991). The ecology of Dippers (*Cinclus cinclus* (L)) in relation to stream acidity in upland Wales: breeding performance, calcium physiology and nestling growth. *J. Appl. Ecol.* **28:** 419–433.

Ormerod, S.J., Tyler, S.J. & Lewis, J.M.S. (1985). Is the breeding distribution of Dippers influenced by stream acidity? *Bird Study* **32**: 32–39.

Owen, M. (1984). Dynamics and age structure of an increasing goose population – the Svalbard Barnacle Goose *Branta leucopsis. Norsk Polarinstitutt Skrifter* **181**: 37–47.

Owen, M., Atkinson-Willes, G.L. & Salmon, D.G. (1986). Wildfowl in Great Britain. 2nd Edn. Cambridge, University Press.

Owen, M. & Black, J.M. (1991). The importance of migration mortality in non-passerine birds. pp. 360–372 in *'Bird population studies. Relevance to conservation and management'* (ed. C.M. Perrins, J.-D. Lebreton & G.J.M. Hirons). Oxford, University Press.

Owen-Smith, N. (1987). Pleistocene extinctions: the pivotal role of megaherbivores. *Paleobiology* **13**: 351–362.

Pacyna, J.M. (1995). Emissions inventory for atmospheric lead, cadmium and copper in Europe in 1989. Luxembourg, International Institute for Applied Systems Analyses, Rept No II ASA WP-95-35.

Pain, D.J. (1991). Lead poisoning of waterfowl: a review. pp. 7–13 in *'Lead poisoning in waterfowl'* (ed. D.J. Pain). Slimbridge, Gloucester, Int. Waterfowl and Wetlands Res. Bur. Spec. Publ. 16.

Pain, D.J., Sears, J. & Newton, I. (1995). Lead concentrations in birds of prey in Britain. *Environ. Poll.* **87**: 173–180.

Paine, R.T., Wootton, J.T. & Boersma, P.D. (1990). Direct and indirect effects of Peregrine Falcon predation on seabird abundance. *Auk* **107**: 1–9.

Palmgren, P. 1930. Quantitative Untersuchungen über die Vogelfauna in den Wäldern Sudfinnlands. *Acta Zool. Fenn.* **7**: 1–218.

Panek, M. (1997). Density-dependent brood production in the Grey Partridge *Perdix perdix*, in relation to habitat quality. *Bird Study* **44**: 235–238.

Parker, G.A. & Sutherland, W.J. (1986). Ideal free distributions when individuals differ in competitive ability: phenotype-limited ideal free models. *Anim. Behav.* **34**: 1222–1242.

Parker, G.L., Petrie, M.J. & Sears, D.T. (1992). Waterfowl distribution relative to wetland acidity. *J. Wildl. Manage.* **56**: 268–274.

Parker, H. (1984). Effect of corvid removal on reproduction of Willow Ptarmigan and Black Grouse. *J. Wildl. Manage.* **48**: 1197–1205.

Parker, H. & Holm, H. (1990). Patterns of nutrient and energy-expenditure in female Common Eiders nesting in the high arctic. *Auk* **107**: 660–668.

Parker, J.W. (1974). Populations of the Mississippi Kite in the Great Plains. *Raptor Res. Rep.* **3**: 159–172.

Parr, R. (1979). Sequential breeding by Golden Plovers. *Brit. Birds* **72**: 499–503.

Parr, R. (1992). The decline to extinction of a population of Golden Plover in north-east Scotland. *Ornis Scand.* **23**: 152–158.

Parr, R. (1993). Nest predation and numbers of Golden Plovers *Pluvialis apricaria* and other moorland waders. *Bird Study* **40**: 223–231.

Parrish, J.K. & Paine, R.T. (1996). Ecological interactions and habitat modification in nesting Common Murres *Uria aalge. Bird Conserv. Int.* **6**: 261–269.

Parsons, J. (1976). Nesting density and breeding success in the Herring Gull *Larus argentatus. Ibis* **118**: 537–547.

Partridge, L. (1976). Individual differences in feeding efficiencies and feeding preferences of captive Great Tits. *Anim. Behav.* **24**: 230–240.

Pascal, M. (1980). Population structure and dynamics of feral cats on Kerguelen Islands. *Mammalia* **44**: 161–182.

Paton, P.W.C. (1994). The effect of edge on avian nest success: how strong is the evidence? *Conserv. Biol.* **8**: 17–26.

Paton, P.W.C., Zabel, C.J., Neal, D.L., Steger, G.N., Tilghman, N.G. & Noon, B.R. (1991). Effects of radio tags on Spotted Owls. *J. Wildl. Manage.* **55**: 617–622.

Pattee, O.H. & Hennes, S.K. (1983). Bald Eagles and waterfowl: the leadshot connection. *Trans. N. A. Wildl. Nat. Res. Conf.* **48**: 230–237.

Patterson, I.J. (1965). Timing and spacing of broods in the Black-headed Gull *Larus ridibundus*. *Ibis* **107**: 433–459.

Patterson, I.J. (1970). Food fighting in Rooks. pp. 249–252 in *'Animal populations in relation to their food resources'* (ed. A. Watson). Oxford, Blackwell.

Patterson, I.J. (1977). Aggression and dominance in winter flocks of Shelduck *Tadorna tadorna* (L). *Anim. Behav.* **25**: 447–459.

Patterson, I.J. (1980). Territorial behaviour and the limitation of population density. *Ardea* **68**: 53–62.

Patterson, I.J., Makepeace, M. & Williams, M. (1983). Limitation of local population size in the Shelduck. *Ardea* **71**: 105–116.

Paulus, S.L. (1984). Activity budgets of nonbreeding Gadwalls in Louisiana. *J. Wildl. Manage.* **48**: 371–380.

Payne, R.B. (1997). Avian brood parasitism. pp. 338–369 in *'Host–parasite evolution'* (ed. D.H. Clayton & J. Moore). Oxford, University Press.

Payne, R.B. & Groschupf, K.D. (1984). Sexual selection and interspecific competition: a field experiment on territorial behaviour of nonparental finches (*Vidus* spp.). *Auk* **101**: 140–145.

Peace, T.R. (1962). Pathology of trees and shrubs. Oxford, Clarendon Press.

Peach, W.J., Baillie, S.R. & Underhill, L. (1991). Survival of British Sedge Warblers *Acrocephalus schoenobaenus* in relation to west Africa rainfall. *Ibis* **133**: 300–305.

Peach, W.J., Crick, H.Q.P. & Marchant, J.H. (1995a). The demography of the decline in the British Willow Warbler population. *J. Appl. Statistics* **22**: 905–922.

Peach, W.J., du Feu, C. & McMeeking, J. (1995b). Site tenacity and survival rates of Wrens *Troglodytes troglodytes* and Treecreepers *Certhia familiaris* in a Nottinghamshire wood. *Ibis* **137**: 497–507.

Peach, W.J., Thompson, P.S. & Coulson, J.C. (1994). Annual and long-term variation in the survival rates of British Lapwings *Vanellus vanellus*. *J. Anim. Ecol.* **63**: 60–70.

Peakall, D.B. (1985). Behavioural responses of birds to pesticides and other contaminants. *Residue Rev.* **96**: 45–77.

Peakall, D.B. & Bart, J.R. (1983). Impacts of aerial application of insecticides on forest birds. *CBC Critical Rev. Environ. Control* **13**: 117–165.

Peakall, D.B. & Kiff, L.F. (1988). DDE contamination in Peregrines and American Kestrels and its effects on reproduction. pp. 337–350 in *'Peregrine Falcon populations. Their management and recovery'* (ed. T.J. Cade, J.H. Enderson, C.G. Thelander & C.M. White). Boise, The Peregrine Fund.

Pearce, P.A. & Peakall, D.B. (1977). The impact of fenitrothion on bird populations in New Brunswick. pp. 299–306 in *'Fenitrothion: the long-term effects of its use in forest ecosystems – current status'* (ed. J.R. Roberts, R. Greenhalgh & W.K. Marshall). Ottawa, National Research Council of Canada Publ. No 15389.

Pech, R.P., Sinclair, A.R.E., Newsome, A.E. & Catling, P.C. (1992). Limits to predator regulation of Rabbits in Australia: evidence from predator-removal experiments. *Oecologia* **89**: 102–112.

Pehrson, O. (1976). Duckling production of Long-tailed Duck in relation to spring weather and small rodent fluctuations. Ph.D. thesis (Unpub). University of Gothenburg, Sweden.

Peiponen, V.A. (1962). Über Brütbiologie Nahrung und geographische Verbreitung des Birkenzeisigs (*Carduelis flammea*). *Ornis Fenn.* **39**: 37–60.

Pennycuick, C.J., Einarsson, O., Bradbury, T.A.M. & Owen, M. (1996). Migrating Whooper Swans *Cygnus cygnus*: satellite tracks and flight performance calculations. *J. Avian Biol.* **27**: 118–134.

Percival, S.M. & Houston, D.C. (1992). The effect of winter grazing by Barnacle Geese on grassland yields on Islay. *J. Appl. Ecol* **29**: 35–40.

Perrins, C.M. (1979). British tits. London, Collins.

Perrins, C.M. & Geer, T.A. (1980). The effect of Sparrowhawks on tit populations. *Ardea* **68**: 133–142.

Perrins, C.M. & McCleery, R.H. (1994). Competition and egg-weight in the Great Tit *Parus major. Ibis* **136**: 454–456.

Perrins, C.M., McCleery, R.H. & Ogilvie, M.A. (1994). A study of the breeding Mute Swans *Cygnus olor* at Abbotsbury. *Wildfowl* **45**: 1–14.

Perrins, C.M., Richardson, D. & Stevenson, M. (1985). Deaths of swifts and hirundines. *Bird Study* **32**: 150.

Persson, L., Borg, K. & Falt, H. (1974). On the occurrence of endoparasites in Eider Ducks in Sweden. *Swed. Wildl. Viltrevy* **9**: 1.

Peters, W.D. & Grubb, P. (1983). An experimental analysis of sex-specific foraging in the Downy Woodpecker *Picoides pubescens. Ecology* **64**: 1437–1443.

Peterson, M.J. & Silvy, N.J. (1996). Reproductive stages limiting productivity of the endangered Attwater's Prairie Chicken. *Conserv. Biol.* **10**: 1264–1276.

Petrides, G.A. & Bryant, C.R. (1951). An analysis of the 1949–50 fowl cholera epizootic in Texas Panhandle waterfowl. *Trans. N. A. Wildl. Nat. Res. Conf.* **16**: 193–216.

Pettersson, B. (1985). Extinction of an isolated population of the Middle Spotted Woodpecker *Dendrocopus medius* (L.) in Sweden and its relation to general theories on extinction. *Biol. Conserv.* **32**: 335–353.

Petty, S.J. (1992). Ecology of the Tawny Owl *Strix aluco* in the spruce forests of Northumberland and Argyll. Ph.D thesis (Unpub). Open University, Milton Keynes.

Petty, S.J., Patterson, I.J., Anderson, D.I.K., Little, B. & Davison, M. (1995). Numbers, breeding performance, and diet of the Sparrowhawk *Accipiter nisus* and Merlin *Falco columbarius* in relation to cone crops and seed-eating finches. *For. Ecol. Manage.* **79**: 133–146.

Petty, S.J., Shaw, G. & Anderson, D.I.K. (1994). Value of nest boxes for population studies and conservation of owls in coniferous forests in Britain. *J. Raptor Res.* **28**: 134–142.

Phillips, R.A., Caldow, R.W.G. & Furness, R.W. (1996). The influence of food availability on the breeding effort and reproductive success of Arctic Skuas *Stercorarius parasiticus. Ibis* **138**: 410–419.

Phillips, V.E. (1992). Variation in winter wildfowl numbers on gravel pit lakes at Great Linford, Buckinghamshire, 1974–79 and 1984–91, with particular reference to the effects of fish removal. *Bird Study* **39**: 177–185.

Piatt, J.F. (1990). Aggregative response of Common Murres and Atlantic Puffins to their prey. *Stud. Avian Biol.* **14**: 36–51.

Piatt, J.F., Lensink, C.J., Butler, W., Kendziorek, M. & Nysewander, D.R. (1990). Immediate impact of the 'Exxon Valdez' oil spill on marine birds. *Auk* **107**: 387–397.

Piatt, J.F. & Reddin, D.G. (1984). Recent trends in the west Greenland salmon fishery and implications for Thick-billed Murres. pp. 208–210 in *'Marine birds: their feeding ecology and commercial fisheries relationships'* (ed. D.N. Nettleship, G.A. Sanger & P.F. Springer). Ottawa, Can. Wildl. Serv. Spec. Publ.

Pienkowski, M.W. & Evans, P.R. (1982). Breeding behaviour, productivity and survival of colonial and non-colonial Shelducks, *Tadorna tadorna. Ornis. Scand.* **13**: 101–116.

Pierotti, R. (1980). Spite and altruism in gulls. *American Naturalist* **115**: 290–300.

Piersma, T. (1987). Production by intertidal benthic animals and limits to their predation by shorebirds – a heuristic model. *Mar. Ecol. Prog. Ser.* **38**: 187–196.

Pimental, D., Acquay, H., Biltonen, M., Rice, P., Silva, M., Nelson, J., Lipner, V., Giordans, S., Horowitz, A. & D'Amare, M. (1992). Environmental and economic costs of pesticide use. *BioScience* **41**: 402–409.

Pimm, S.L., Jones, H.L. & Diamond, J. (1988). On the risk of extinction. *Amer. Nat.* **132**: 757–785.

Pimm, S.L., Moulton, M.P. & Justice, L.J. (1994). Bird extinctions in the central Pacific. *Phil. Trans. Roy. Soc. Lond. B* **344**: 27–33.

Pimm, S.L., Rosenzweig, M.L. & Mitchell, W. (1985). Competition and food selection: field tests of a theory. *Ecology* **66:** 798–807.

Pinkowski, B.C. (1976). Use of tree cavities by nesting Eastern Bluebirds. *J. Wildl. Manage.* **40:** 556–563.

Pinkowski, B.C. (1977). Blowfly parasitism of Eastern Bluebirds in natural and artificial tree-sites. *J. Wildl. Manage.* **41:** 272–276.

Pinowski, J. (1967). Die Auswahl des Brutbiotops beim Feldsperling (*Passer m. montanus*). *Ekol. Polska (A)* **15:** 1–30.

Pinowski, J., Mazurkiewicz, M., Malyszko, E., Pawiak, R., Kozlowski, S., Kruszewicz, A. & Indykiewicz, P. (1988). The effect of micro-organisms on embryo and nestling mortality in House Sparrow (*Passer domesticus*) and Tree Sparrow (*Passer montanus*). pp. 273–282 in *'Proc. Int. 100. DO-G Meeting, Current Topics Avian Biol. Bonn 1988'* (ed. J. Pinowski & D. Summers-Smith). Bonn.

Pitelka, F.A., Tomich, P.Q. & Treichel, G.W. (1955). Ecological relations of jaegers and owls near Barrow, Alaska. *Ecol. Monogr.* **25:** 85–117.

Platt, J.B. (1977).The breeding behaviour of wild and captive Gyr Falcons in relation to their environment and human disturbance. Ph.D. thesis (Unpub), Cornell University.

Pocklington, R. (1979). An oceanographic interpretation of seabird distribution in the Indian Ocean. *Mar. Biol.* **51:** 9–21.

Pollard, E. (1981). Resource limited and equilibrium models of populations. *Oecologia* **49:** 377–378.

Pollard, E., Lakhani, K.H. & Rothery, P. (1987). The detection of density dependence from a series of annual censuses. *Ecology* **68:** 2046–2055.

Pons, J.-M. & Migot, P. (1995). Life-history strategy of the herring gull: changes in survival and fecundity in a population subjected to various feeding conditions. *J. Anim. Ecol.* **64:** 592–599.

Poole, A.F. (1989). Ospreys. A natural and unnatural history. Cambridge, University Press.

Portal, M. & Collinge, W.E. (1932). Partridge disease and its causes. London, Country Life Ltd.

Porter, J.M. & Coulson, J.C. (1987). Long-term changes in recruitment to the breeding group, and the quality of recruits at a Kittiwake *Rissa tridactyla* colony. *J. Anim. Ecol.* **56:** 675–689.

Post, W., Cruz, A. & McNair, D.B. (1993). The North American invasion pattern of the Shiny Cowbird. *J. Field Ornithol.* **64:** 32–41.

Post, W. & Wiley, J.W. (1977). Reproductive interactions of the Shiny Cowbird and the Yellow-shouldered Blackbird. *Condor* **79:** 176–184.

Potts, G.R. (1980). The effects of modern agriculture, nest predation and game management on the population ecology of Partridges *Perdix perdix* and *Alectoris rufa*. *Adv. Ecol. Res.* **11:** 2–79.

Potts, G.R. (1986). The Partridge: pesticides, predation and conservation. London, Collins.

Potts, G.R. (1991). The environmental and ecological importance of cereal fields. pp. 3–21 in *'The ecology of temperate cereal fields'* (ed. L.G. Firbank, N. Carter, J.F. Darbyshire & G.R. Potts). Oxford, Blackwell.

Potts, G.R. & Aebischer, N.J. (1991). Modelling the population dynamics of the Grey Partridge: conservation and management. pp. 373–390 in *'Bird population studies'* (ed. C.M. Perrins, J.-D. Lebreton & G.J.M. Hirons). Oxford, University Press.

Potts, G.R. & Aebischer, N.J. (1995). Population dynamics of the Grey Partridge *Perdix perdix* 1793–1993: monitoring, modelling and management. *Ibis* **137 (Suppl):** 29–37.

Potts, G.R., Coulson, J.C. & Deans, I.R. (1980). Population dynamics and breeding success of the Shag, *Phalacrocorax aristotelis* on the Farne Islands, Northumberland. *J. Anim. Ecol.* **49:** 465–484.

Potts, G.R., Tapper, S.C. & Hudson, P.J. (1984). Population fluctuations in Red Grouse: analysis of bag records and a simulation model. *J. Anim. Ecol.* **53:** 21–36.

Powell, G.V.N. (1974). Experimental analysis of the social value of flocking by Starlings (*Sturnus vulgaris*) in relation to predation and foraging. *Anim. Behav.* **22:** 501–505.

Power, H.W. (1975). Mountain Bluebirds: experimental evidence against altruism. *Science* **189**: 142–143.

Powlesland, R.G. (1977). Effects of the haematophagous mite *Ornithonyssus bursa* on nestling Starlings in New Zealand. *N.Z. J. Zool.* **4**: 85–94.

Prater, A.J. (1981). Estuary birds of Britain and Ireland. Calton, Poyser.

Prestrud, P., Black, J.M. & Owen, M. (1989). The relationship between an increasing Barnacle Goose *Branta leucopsis* population and the number and size of colonies in Svalbard. *Wildfowl* **40**: 32–38.

Price, P.W. (1980). Evolutionary biology of parasites. Princeton, University Press.

Prince, P.A., Rothery, P., Croxall, J.P. & Wood, A.G. (1994). Population dynamics of Black-browed and Grey-headed Albatrosses *Diomedia melanophris* and *D. chrysostoma* at Bird Island, South Georgia. *Ibis* **136**: 50–71.

Probst, J.R. (1986). A review of factors limiting Kirtland's Warbler on its breeding grounds. *Amer. Nat.* **117**: 87–100.

Probst, J.R. & Hayes, J.P. (1987). Pairing success of Kirtland's Warbler in marginal vs suitable habitat. *Auk* **104**: 234–241.

Prop, J., van Eerden, M. & Drent, R.H. (1984). Reproductive success of the Barnacle Goose in relation to food exploitation on the breeding grounds, western Spitzbergen. *Norsk. Polarinstitutt Skrifter* **181**: 87–117.

Pulliainen, E. (1978). Influence of heavy snowfall in June 1977 on the life of birds in NE Finnish Forest, Lapland. *Aquila. Ser. Zool.* **18**: 1–14.

Pulliam, H.R. (1985). Foraging efficiency, resource partitioning, and the coexistence of sparrow species. *Ecology* **66**: 1829–1836.

Pulliam, H.R. (1988). Sources, sinks, and population regulation. *Amer. Nat.* **132**: 652–661.

Pulliam, H.R. & Danielson, B.J. (1991). Sources, sinks and habitat selection: a landscape perspective on population dynamics. *Amer. Nat.* **137**: 50–66.

Pulliam, H.R. & Dunning, J.B. (1987). The influence of food supply on local density and diversity of sparrows. *Ecology* **68**: 1009–1014.

Pulliam, H.R., Liu, J.G., Dunning, J.B., Stewart, D.J. & Bishop, T.D. (1995). Modelling animal populations in changing landscapes. *Ibis* **137**: 120–126.

Pulliam, H.R. & Parker, T.A. (1979). Population regulation of sparrows. *Fortschr. Zool.* **25**: 137–147.

Pyle, R.L. (1993). Hawaiian Islands Region. *Am. Birds* **47**: 302–304.

Rabinowitz, D. (1981). Seven forms of rarity. pp. 205–217 in 'The biological aspects of rare plant conservation' (ed. H. Synge). Chichester, Wiley.

Rabor, D.S. (1959). The impact of deforestation on birds of Cebu, Philippines, with new records for that island. *Auk* **76**: 37–43.

Rae, S.R. (1995). Individual dispersion and productivity of Ptarmigan in relation to their selection of food and cover. Ph.D. thesis (Unpub), Aberdeen University, Scotland.

Raffaele, H. (1977). Comments on the extinction of *Loxigilla portoricensis grandis* in St. Kitts. *Condor* **79**: 389–390.

Raitasuo, K. (1964). Social behaviour of the mallard *Anas platyrhynchos*, in the course of the annual cycle. *Papers on Game Research* **24**: 1–72.

Ralls, K. & Ballou, J. (1983). Extinction: lessons from zoos. pp. 164–184 in 'Genetics and conservation' (ed. C.M. Schonewald-Cox, S.M. Chambers, B. MacBryde & W.C. Thomas). Menlo Park, California, Benjamin Cummings.

Ralph, C.J. (1991). Population dynamics of land bird populations on Oahu, Hawaii: fifty years of introductions and competition. *Proc. Int. Ornithol. Congr.* **20**: 1444–1457.

Rands, M.R.W. (1985). Pesticide use on cereals and the survival of partridge chicks: a field experiment. *J. Appl. Ecol.* **22**: 49–54.

Rands, M.R.W. (1986). The survival of gamebird (Galliformes) chicks in relation to pesticide use on cereals. *Ibis* **128**: 57–64.

Ranta, E., Lindstrom, J. & Lindén, H. (1995). Synchrony in tetraonid population dynamics. *J. Anim. Ecol.* **64:** 767–776.

Raphael, M.G., Morrison, M.L. & Yoder-Williams, M.P. (1987). Breeding bird populations during twenty-five years of postfire succession in the Sierra Nevada. *Condor* **89:** 614–626.

Raphael, M.G. & White, M. (1984). Use of snags by cavity-nesting birds in the Sierra Nevada. *Wildl. Monogr.* **86:** 1–66.

Ratcliffe, D.A. (1969). Population trends of the Peregrine Falcon in Great Britain. pp. 239–269 in *'Peregrine Falcon populations'* (ed. J.J. Hickey). Madison, Wisconsin, University Press.

Ratcliffe, D.A. (1970). Changes attributable to pesticides in egg breakage frequency and eggshell thickness in some British birds. *J. Appl. Ecol.* **7:** 67–107.

Ratcliffe, D.A. (1980). The Peregrine Falcon. Calton, Poyser.

Raup, D.M. (1984). Death of species. pp. 1–19 in *'Extinctions'* (ed. M.H. Nitecki). Chicago, University Press.

Raveling, D.G. (1970). Dominance relationships and agonistic behaviour of Canada Geese in winter. *Behaviour* **37:** 291–319.

Raveling, D.G. (1979). The annual cycle of body composition of Canada Geese with special reference to control of reproduction. *Auk* **96:** 234–252.

Raven, P.H. (1988). Our diminishing tropical forests. pp. 119–122 in *'Biodiversity'* (ed. E.O. Wilson). Washington, D.C., National Academic Press.

Ray, C., Gilpin, M. & Smith, A.T. (1991). The effect of interspecific attraction on metapopulation dynamics. *Biol. J. Linn. Soc.* **42:** 123–134.

Redig, P.T. (1978). Raptor rehabilitation: diagnosis, prognosis and moral issues. pp. 29–41 in *'Bird of prey management techniques'* (ed. T.A. Geer). Oxford, British Falconer's Club.

Redondo, T. & Castro, F. (1992). The increase in risk of predation with begging activity in broods of Magpies *Pica pica*. *Ibis* **134:** 180–187.

Redpath, S.M. (1995). Habitat fragmentation and the individual: Tawny Owls *Strix aluco* in woodland patches. *J. Anim. Ecol.* **64:** 632–661.

Redpath, S. & Thirgood, S. (1997). Birds of prey and Red Grouse. London, HMSO Publications.

Reed, A. & Cousineau, J.-G. (1967). Epidemics involving the Common Eider (*Somateria mollissima*) at Ile Blanche, Quebec. *Can. Nat.* **94:** 327–334.

Reed, C.E.M., Nilsson, R.J. & Murray, D.P. (1993). Cross-fostering New Zealand's Black Stilt. *J. Wildl. Manage.* **57:** 608–611.

Reed, J.M. (1990). The dynamics of Red-cockaded Woodpecker rarity and conservation. pp. 37–56 in *'Conservation and management of woodpecker populations'* (ed. A. Carlson & G. Aulén). Uppsala University, Dept. Wildlife Ecology, Report 17.

Reed, T.M. (1982). Interspecific territoriality in the Chaffinch and Great Tit on islands and the mainland of Scotland: playback and removal experiments. *Anim. Behav.* **30:** 171–181.

Reichel, W.L., Cromartie, E., Lamont, T.G., Mulhern, B.M. & Prouty, R.M. (1969). Pesticide residues in Eagles. *Pestic. Mon. J.* **3:** 142–144.

Reichel, W.L., Schmeling, S.K., Cromartie, E., Kaiser, T.E., Krynitsky, A.J., Lamont, T.G., Mulhern, B.M., Prouty, R.M., Stafford, C.J., & Swineford, D.M. (1984). Pesticide, PCB, and lead residues and necropsy data for Bald Eagles from 31 States – 1978–1981. *Environ. Monit. Assess.* **4:** 395–403.

Reid, H.W. (1975). Experimental infection of Red Grouse with louping-ill virus (flavivirus group). I. The viraemia and antibody response. *J. Comp. Pathol.* **85:** 223–229.

Reid, H.W., Duncan, J.S., Phillips, J.D.P., Moss, R. & Watson, A. (1978). Studies of louping ill virus (flavivirus group) in wild Red Grouse (*Lagopus lagopus scoticus*). *J. Hyg.* **81:** 321–329.

Reid, H.W., Moss, R., Pow, I. & Buxton, D. (1980). The response of three grouse species (*Tetrao urogallus, Lagopus mutus, Lagopus lagopus*) to louping-ill virus. *J. Comp. Path.* **90:** 257–263.

Reid, W.V. (1992). How many species will there be? pp. 55–74 in *'Tropical deforestation and species extinction'* (ed. T.C. Whitmore & J.A. Sayer). London, Chapman & Hall.

Reid, W.V. & Miller, K.R. (1989). Keeping options alive: the scientific basis for conserving biodiversity. Washington, D.C., World Resources Institute.

Reinikainen, A. (1937). The irregular migrations of the Crossbill, *Loxia c. curvirostra*, and their relation to the cone-crop of the conifers. *Ornis Fenn.* **14**: 55–64.

Rendell, W.B. & Verbeek, N.A.M. (1996). Old nest material in nest boxes of Tree Swallows: effects on nest-site choice and nest building. *Auk* **113**: 319–328.

Reynolds, R.E. (1987). Breeding duck population, production and habitat surveys, 1979–85. *Trans. N.A. Wildl. Nat. Res. Conf.* **52**: 186–205.

Ribaut, J. (1964). Dynamique d'une population des Merles noirs *Turdus merula* L. *Revue Suisse Zool.* **71**: 815–902.

Rice, J. (1978). Ecological relationships of two interspecifically territorial vireos. *Ecology* **59**: 525–538.

Richner, H. (1989). Habitat specific growth and fitness in Carrion Crows (*Corvus corone corone*). *J. Anim. Ecol.* **58**: 427–440.

Richner, H. (1992). The effect of extra food on fitness in breeding Carrion Crows. *Ecology* **73**: 330–335.

Richner, H., Opplinger, A. & Christe, P. (1993). Effect of an ectoparasite on the reproduction in Great Tits. *J. Anim. Ecol.* **62**: 703–710.

Ricklefs, R.E. (1969). An analysis of nestling mortality in birds. *Smithson. Contr. Zool.* **9**: 1–48.

Ridpath, M.G. & Brooker, M.G. (1986). The breeding of the Wedge-tailed Eagle *Aquila audax* in relation to its food-supply in arid western Australia. *Ibis* **128**: 177–194.

Riggert, T.L. (1977). The biology of the Mountain Duck on Rottnest Island, Western Australia. *Wildl. Monogr.* **52**: 1–67.

Rippin, A.B. & Boag, D.A. (1974). Recruitment to populations of male Sharp-tailed Grouse. *J. Wildl. Manage.* **38**: 616–621.

Risebrough, R.W. (1986). Pesticides and bird populations. *Curr. Ornithol.* **3**: 397–427.

Risebrough, R.W., Sibley, F.C. & Kirven, M.N. (1971). Reproductive failure of the Brown Pelican on Anacapa Island in 1969. *Amer. Birds* **25**: 8–9.

Rising, J.D. (1983). The Great Plains hybrid zones. *Curr. Ornithol.* **1**: 131–157.

Robbins, C.S. (1980). Effect of forest fragmentation on breeding bird populations in the piedmont of the Mid-Atlantic region. *Atlantic Natur.* **33**: 31–36.

Robbins, C.S. & Stewart, R.E. (1949). Effects of DDT on bird population of scrub forest. *J. Wildl. Manage.* **13**: 11–18.

Robbins, C.S., Droege, S. & Sauer, J.R. (1989). Monitoring bird populations with Breeding Bird Survey and atlas data. *Ann. Zool. Fenn.* **26**: 297–304.

Robertson, P.A. & Rosenberg, A.A. (1988). Harvesting gamebirds. pp. 177–201 in 'Ecology and management of gamebirds' (ed. P.J. Hudson & M.R.W. Rands). London, BSP Professional Books.

Robinson, S. & Terborgh, J. (1995). Interspecific aggression and habitat selection by Amazonian birds. *J. Anim. Ecol.* **64**: 1–11.

Rockenbauch, D. (1968). Zur Brutbiologie des Turmfalken (*Falco tinnunculus* L.). *Anz. orn. Ges. Bayern* **8**: 267–276.

Rodenhouse, N.L. & Holmes, R.T. (1992). Results of experiments and natural food reductions for breeding Black-throated Blue Warblers. *Ecology* **73**: 357–372.

Rogers, C.A., Robertson, R.J. & Stutchbury, B.J. (1990). Patterns and effects of parasitism by *Protocalliphora sialia* on Tree Swallow (*Tachycineta bicolor*) nestlings. pp. 123–139 in 'Bird-parasite interactions' (ed. J.E. Loye & M. Zuk). Oxford, University Press.

Rogers, J.P. (1964). Effect of drought on reproduction of the Lesser Scaup. *J. Wildl. Manage.* **28**: 213–222.

Rohner, C. (1994). The numerical response of Great Horned Owls to the Snowshoe Hare cycle in boreal forest. PhD thesis (unpublished), University of British Columbia.

Rohner, C. (1995). Great Horned Owls and Snowshoe Hares – what causes the time-lag in the numerical response of predators to cyclic prey. *Oikos* **74**: 61–68.

Rohner, C. (1996). The numerical response of Great Horned Owls to the Snowshoe Hare cycle: consequences of non-territorial 'floaters' on demography. *J. Anim. Ecol.* **65:** 359–370.

Rolstad, J. & Wegge, P. (1987). Distribution and size of Capercaillie leks in relation to old forest fragmentation. *Oecologia* **72:** 389–394.

Rolstad, J. & Wegge, P. (1989). Capercaillie *Tetrao urogallus* populations and modern forestry – a case for landscape ecological studies. *Finnish Game Res.* **46:** 43–52.

Roseberry, J.L. (1979). Bobwhite population responses to exploitation: real and simulated. *J. Wildl. Manage.* **43:** 285–305.

Roselaar, C.S. (1979). Fluctuaties in aantallen Krombek strandlopers *Calidris ferruginea. Watervogels* **4:** 202–210.

Rosen, M.N. & Bischoff, A. (1949). The 1948–49 outbreak of fowl cholera in birds in the San Francisco Bay area and surrounding counties. *Cal. Fish & Game.* **35:** 185–192.

Rosen, M.N. (1971). Avian Cholera. pp. 59–74 in '*Infectious and parasitic diseases of wild birds*' (ed. J.W. Davis, R.C. Anderson, L. Karstad & D.O. Trainer). Ames, Iowa, University Press.

Rosen, M.N. & Morse, E.E. (1972). An interspecies chain in a fowl cholera epizootic. *Calif. Fish Game.* **45:** 51–56.

Rosenzweig, M.L. & Clark, C.W. (1994). Island extinction rates from regular censuses. *Conserv. Biol.* **8:** 491–494.

Ross, G. (1971). Our Jackass Penguins: are they in danger? *African Wildl.* **25:** 130–134.

Rotenberry, J.T. & Wiens, J.A. (1991). Weather and reproductive variation in shrubsteppe sparrows: a hierarchical analysis. *Ecology* **72:** 1325–1335.

Roth, R.R. & Johnson, R.K. (1993). Long-term dynamics of a Wood Thrush population breeding in a forest fragment. *Auk* **110:** 37–48.

Rothery, P., Newton, I., Dale, L. & Wesolowski, T. (1997). Testing for density dependence allowing for weather effects. *Oecologia:* in press.

Rothschild, M. & Clay, T. (1952). Fleas, flukes and cuckoos. London, Collins.

Rothstein, S.I. (1990). A model system for coevolution: avian brood parasitism. *Ann. Rev. Ecol. Syst.* **21:** 481–508.

Rothstein, S.I. & Robinson, S.K. (1994). Conservation and coevolutionary implications of brood parisitism by cowbirds. *Trends Ecol. Evol.* **9:** 162–164.

Rowan, M.K. (1962). Mass mortality among European Common Terns in South Africa in April-May 1961. *Brit. Birds* **55:** 103–114.

Rowan, M.K. (1965). Regulation of sea-bird numbers. *Ibis* **107:** 54–59.

Rowan, M.K. (1966). Territory as a density-regulating mechanism in some south African birds. *Ostrich Supplement* **6:** 397–408.

Rowan, W. (1921–22). Observations on the breeding habits of the Merlin. *Brit. Birds* **15:** 122–129, 194–202, 222–231, 246–253.

Rowley, I. & Russell, E.R. (1991). Demography of passerines in the temperate southern hemisphere. pp. 22–44 in '*Bird population studies*' (ed. C.M. Perrins, J.D. Lebreton & G.J.M. Hirons). Oxford, University Press.

Royal, W.C. (1966). Breeding of the Starling in central Arizona. *Condor* **68:** 196–205.

Royama, T. (1992). Analytical population dynamics. London, Chapman & Hall.

Rudd, R.L. (1964). Pesticides and the living landscape. Madison, Wisconsin, University Press.

Ruiz, G.M., Connors, P.G., Griffin, S.E. & Pitelka, F.A. (1989). Structure of a wintering Dunlin population. *Condor* **91:** 562–570.

Ryder, J.P. & Alisauskas, R.T. (1995). 162. Ross Goose. In '*The birds of North America: life histories for the 21st century. Numbers 161–200*' (ed. A.F. Poole & F.B. Gill). Washington, D.C., The American Ornithologist's Union and the Academy of Natural Sciences of Philadelphia.

Ryel, L.A. (1981). Population change in the Kirtland's Warbler. *Jack-pine Warbler* **59:** 76–91.

Salisbury, E. 1961. Weeds and Aliens. London, Collins.

Samson, F.B. & Lewis, J.J. (1979). Experiments on population regulation in two North American parids. *Wilson Bull.* **91:** 222–233.

Sanderson, G.C. & Bellrose, F.C. (1986). A review of the problem of lead poisoning in waterfowl. *Illinois Nat. Hist. Surv. Spec. Publ. 4.*

Sandström, U. (1991). Enhanced predation rates on cavity bird nests at deciduous forest edges – an experimental study. *Ornis Fenn.* **68:** 93–98.

Sargeant, A.B., Allen, S.H. & Eberhardt, R.T. (1984). Red Fox predation on breeding ducks in midcontinent North America. *Wildl. Monogr.* **89:** 1–41.

Sargeant, A.B., Sovada, M.A. & Schaffer, T.L. (1995). Seasonal predator removal relative to hatch rate in duck nests in waterfowl production areas. *Wildl. Soc. Bull.* **23:** 507–513.

Saunders, D.A. (1989). Changes in the avifauna of a region, district and remnant as a result of fragmentation of native vegetation: the wheatbelt of Western Australia. *Biol. Conserv.* **50:** 99–135.

Saunders, D.A. & Curry, P.J. (1990). The impact of agricultural and pastoral industries on birds in the southern half of western Australia: past, present and future. *Proc. Ecol. Soc. Aust.* **16:** 303–321.

Saunders, D.A. & Hobbs, R.J. (eds.) (1991). Native conservation: the role of corridors. Chipping Norton, Australia, Surrey Beatty.

Saunders, D.A., Smith, G.T. & Rowley, I. (1982). The availability and dimensions of tree hollows that provide nest sites for cockatoos (*Psittaciformes*) in Western Australia. *Aust. Wildl. Res.* **9:** 541–556.

Saurola, P. (1976). Mortality of Finnish Goshawks. *Suomen Luonto* **6:** 310–314.

Saurola, P. (1989). Ural Owl. pp. 327–345 in *'Lifetime reproduction in birds'* (ed. I. Newton). London, Academic Press.

Savard, J.-P.L. (1988). Winter, spring and summer territoriality in Barrow's Goldeneye: characteristics and benefits. *Ornis Scand.* **19:** 119–128.

Savard, J.-P.L. & Smith, G.E.J. (1991). Duckling mortality in Barrow's Goldeneye and Bufflehead broods. *Auk* **108:** 568–577.

Savidge, J.A. (1987). Extinction of an island forest avifauna by an introduced snake. *Ecology* **68:** 660–668.

Savory, C.J. (1978). Food consumption of Red Grouse in relation to the age and productivity of heather. *J. Anim. Ecol.* **47:** 269–282.

Savory, C.J. & Gentle, M.J. (1976). Effects of dietary dilution with fibre on the food intake and gut dimensions of Japanese Quail. *Brit. Poult. Sci.* **17:** 561–570.

Schaefer, M.B. (1970). Men, birds and anchovies in the Peru current – dynamic interactions. *Trans. Amer. Fish. Soc.* **99:** 461–467.

Schifferli, L. (1978). Die Rolle des Manchers während der Bebrütung der Eier beim Hausperling *Passer domesticus. Ornithol. Beob.* **75:** 44–47.

Schifferli, L. (1979). Waterfowl counts and duck wing analysis in Switzerland. pp. 129–136 in *'Proceedings of the 2nd Technical Meeting on western Palaearctic migratory bird management'.* Slimbridge, International Waterfowl Research Bureau.

Schipper, W.J.A. (1979). A comparison of breeding ecology in European harriers (*Circus*). *Ardea* **66:** 77–102.

Schläpfer, A. (1988). Populationsökologie der Feldlerche *Alauda arvensis* in der intensiv genutzten Agrarlandschaft. *Ornithol. Beob.* **85:** 309–371.

Schluter, D. (1986). Character displacement between distantly related taxa? Finches and bees in the Galapagos. *Amer. Nat.* **127:** 95–102.

Schluter, D. (1988). The evolution of finch communities on islands and continents – Kenya vs Galapagos. *Ecol. Monogr.* **58:** 229–249.

Schluter, D. & Repasky, R.R. (1991). World-wide limitations of finch densities by food and other factors. *Ecology* **72:** 1763–1774.

Schneider, F. (1966). Some pesticide-wildlife problems in Switzerland. *J. Appl. Ecol.* **3 (Suppl.):** 15–20.

Schneider, S.H., Mearns, L. & Gleick, P.H. (1992). Climate change scenarios for impact assessment. pp. 18–55 in *Global warming and biological diversity* (ed. R.L. Peters & T.E. Lovejoy). New Haven, Yale University Press.

Schoener, T.W. (1983). Field experiments on interspecific competition. *Amer. Nat.* **122:** 240–285.

Schönfeld, M. & Brauer, P. (1972). Ergebnisse der 8 jährigen Untersuchungen an der Höhlen – brüterpopulation eines Eichen-Haenbuchen-Linden-Waldes in der 'Alten Göhle' bei Freyburg/Unstrut. *Hercynia N.F.* **9:** 40–68.

Schrank, B.W. (1972). Waterfowl nest cover and some predation relationships. *J. Wildl. Manage.* **36:** 182–186.

Schreiber, E.A. (1994). El Niño–Southern Oscillation events may be affecting your field data as they alter weather patterns worldwide. Research Notes on Avian Biology: selected contributions from the 21st International Ornithological Congress. *J. Ornithol.* **135:** 210.

Schreiber, R.K. & Newman, J.R. (1988). Acid precipitation effects on forest habitat: implications for wildlife. *Conserv. Biol.* **2:** 249–259.

Schreiber, R.W. & Schreiber, E.A. (1984). Central Pacific seabirds and the El Niño Southern Oscillation: 1982 to 1983 retrospectives. *Science* **225:** 731–736.

Schultz, A.M. (1964). The nutrient recovery hypotheses for arctic microtine cycles. II. Ecosystem variables in relation to arctic microtine cycles. *Symp. Br. Ecol. Soc.* **4:** 57–68.

Sciple, G.W. (1953). Avian botulism: information on earlier research. *U.S. Dept. Interior Spec. Sci. Rep. Wildlife No. 23.*

Scott, V.E. (1979). Bird responses to snag removal in Ponderosa Pine. *J. For.* **77:** 26–28.

Seabloom, R.W., Pearson, G.L., Oring, L.W. & Reilly, J.R. (1973). An incident of fenthion mosquito control and subsequent avian mortality. *J. Wildl. Dis.* **9:** 18–20.

Sears, J. (1988). Regional and seasonal variations in lead poisoning in the Mute Swan *Cygnus olor* in relation to the distribution of lead and lead weights, in the Thames area, England. *Biol. Conserv.* **46:** 115–134.

Sedgwick, J.A. & Knopf, F.L. (1992). Cavity turnover and equilibrium cavity densities in a Cottonwood bottomland. *J. Wildl. Manage.* **56:** 477–484.

Seiskari, P. (1962). On the winter ecology of the Capercaillie, *Tetrao urogallus*, and the Black Grouse, *Lyrurus tetrix*, in Finland. *Pap. Game Res.* **22:** 1–119.

Selander, R.K. & Giller, D.R. (1961). Analysis of sympatry of Great-tailed and Boat-tailed Grackles. *Condor* **63:** 29–86.

Serventy, D.L. & Whittell, H.M. (1976). Birds of Western Australia, 5th edn. Perth, University of Western Australia Press.

Seymour, N.R. (1974). Territorial behaviour of wild Shovellers at Delta, Manitoba. *Wildfowl* **25:** 49–55.

Shaffer, M.L. (1985). The metapopulation and species conservation: the special case of the Northern Spotted Owl. *U.S. Dept. Agriculture, Forest Service, Gen. Tech. Rep. PNW-185:* 86–99.

Shannon, L.V., Crawford, R.J.M. & Duffy, D.C. (1984). Pelagic fisheries and warm events: a comparative study. *S. Afr. J. Sci.* **80:** 51–60.

Sharp, B.E. (1996). Post-release survival of oiled, cleaned seabirds in North America. *Ibis* **138:** 222–228.

Shaw, J. (1988). Arrested development of *Trichostrongylus tenuis* as third stage larvae in Red Grouse. *Res. Vet. Sci.* **45:** 256–258.

Shepherd, J.G. & Cushing, D.H. (1990). Regulation in fish populations: Myth or Mirage? *Phil. Trans. R. Soc. Lond. B.* **330:** 151–164.

Sherman, K., Jones, C., Sullivan, L., Smith, W., Berrien, P. & Ejsymont, L. (1981). Congruent shifts in sand eel abundance in western and eastern North Atlantic ecosystems. *Nature* **291:** 486–489.

Sherrod, S.K., White, C.M. & Williamson, F.S.L. (1977). Biology of the Bald Eagle (*Haliaeetus leucocephalus alascanus*) on Amchitka Island, Alaska. *Living Bird* **15:** 143–182.

Sherry, T.W. & Holmes, R.T. (1988). Habitat selection by breeding American Redstarts in response to a dominant competitor, the Least Flycatcher. *Auk* **105**: 350–364.

Sherry, T.W. & Holmes, R.T. (1991). Population fluctuations in a long distance neotropical migrant: demographic evidence for the importance of breeding season events in the American Redstart. pp. 431–442 in '*Ecology and conservation of Neotropical migrant landbirds*' (ed. D.W. Johnston & J.M. Hagan). Washington D.C., Smithsonian Institution Press.

Shields, W.M. & Crook, J.R. (1987). Barn Swallow coloniality: a net cost for group breeding in the Adirondacks. *Ecology* **68**: 1373–1386.

Shields, W.M., Crook, J.R., Hebblewaite, M.L. & Wiles-Ehmann, S.S. (1988). Ideal free coloniality in the swallows. pp. 189–228 in '*The ecology of social behaviour*' (ed. C.N. Slobodchikoff). New York, Academic Press.

Short, L.L. & Horne, J.F.M. (1990). The Ivory-billed Woodpecker – the costs of specialisation. pp. 93–98 in '*Conservation and management of woodpecker populations*' (ed. A. Carlson & G. Aulén). Uppsala University, Dept Wildlife Ecology. Report 17.

Sibley, C. & Monroe, B. (1990). Distribution and taxonomy of birds of the world. New Haven, Yale University Press.

Siegfried, W.R. & Frost, P.G.H. (1975). Continuous breeding and associated behaviour in the Moorhen *Gallinula chloropus*. *Ibis* **117**: 102–109.

Sih, A., Crowley, P., McPeek, M., Petranka, J. & Stohmeier, K. (1985). Predation, competition, and prey communities: a review of field experiments. *Ann. Rev. Ecol. Syst.* 16: 269–311.

Siivonen, L. (1948). Structure of short-cycle fluctuations in numbers of mammals and birds in the northern parts of the northern hemisphere. *Pap. Game Res.* **1**: 1–166.

Simberloff, D. (1988). The contribution of population and community biology to conservation science. *Ann. Rev. Ecol. Syst.* **19**: 475–511.

Simberloff, D. (1994). Habitat fragmentation and population extinction of birds. *Ibis* **137 (suppl.)**: 105–111.

Simmons, K.E.C. (1966). Anting and the problem of self-stimulation. *Nature* **219**: 690–694.

Simmons, R.E. (1996). Population declines, viable breeding areas and management options for Flamingos in southern Africa. *Conserv. Biol.* **10**: 504–514.

Simons, L.S. & Martin, T.E. (1990). Food limitation of avian reproduction – an experiment with the cactus wren. *Ecology* **71**: 869–876.

Sirén, M. (1951). Increasing the Goldeneye population with nest-boxes. *Suomen Riista.* **6**: 189–190.

Skeel, M.A. (1983). Nesting success, density, philopatry and nest-site selection of the Whimbrel (*Numenius phaeopus*) in different habitats. *Can. J. Zool.* **61**: 218–225.

Slagsvold, T. (1979). Competition between the Great Tit (*Parus major*) and the Pied Flycatcher (*Ficedula hypoleuca*): an experiment. *Ornis Scand.* **9**: 46–50.

Small, M.F. & Hunter, M.L. (1988). Forest fragmentation and avian nest-predation in forested landscapes. *Oecologia* **76**: 62–64.

Small, R.J., Holzwart, J.C. & Rusch, D.H. (1991). Predation and hunting mortality of Ruffed Grouse in central Wisconsin. *J. Wildl. Manage.* **55**: 512–520.

Small, R.J., Marcström, V. & Willebrandt, T. (1993). Synchronous and nonsynchronous population fluctuations of some predators and their prey in central Sweden. *Ecography* **16**: 360–364.

Smallwood, J.A. (1988). A mechanism of sexual segregation by habitat in American Kestrels (*Falco sparverius*). *Auk* **105**: 36–46.

Smith, A.G. & Webster, H.R. (1955). Effects of hail storms on waterfowl populations in Alberta, Canada. *J. Wildl. Manage.* **19**: 368–374.

Smith, F.D.M., May, R.M., Pellew, R., Johnson, T.H. & Walter, K.R. (1993). How much do we know about the current extinction rate? *Trends Ecol. Evol.* **8**: 375–378.

Smith, G.R. (1975). Recent European outbreaks of botulism in waterfowl. *I.W.R.B. Bull.* **39/40**: 72–74.

Smith, G.R., Oliphant, J.C. & Evans, E.D. (1983). Diagnosis of bolutism in water birds. *Vet. Rec.* **24**: 457–458.

Smith, H. 1913. Partridges: yesterday and today. London, The Field Press Ltd.

Smith, J.N.M. (1981). Cowbird parasitism, host fitness and age of the host female in an island Song Sparrow population. *Condor* **83**: 153–161.

Smith, J.N.M. & Arcese, P. (1989). How fit are floaters? Consequences of alternative territorial behaviours in a non-migratory sparrow. *Amer. Nat.* **133**: 830–845.

Smith, J.N.M., Arcese, P. & Hochachka, W.M. (1991). Social behaviour and population regulation in insular bird populations: implications for conservation. pp. 148–167 in '*Bird Population Studies: Relevance to Conservation and Management*' (ed. C.M. Perrins, J.-D. Lebreton & G.J.M. Hirons). Oxford, University Press.

Smith, J.N.M., Montgomerie, R.D., Taitt, M.J. & Yom-Tov, Y. (1980). A winter feeding experiment on an island Song Sparrow population. *Oecologia* **47**: 164–170.

Smith, J.N.M., Taitt, M.J., Rogers, C.M., Arcese, P., Keller, L.F., Cassidy, A.L.E.V. & Hochachka, W.M. (1996). A metapopulation approach to the population biology of the Song Sparrow *Melospiza melodia*. *Ibis* **138**: 120–128.

Smith, K.G. (1982). Drought-induced changes in avian community structure along a mountain sere. *Ecology* **63**: 952–961.

Smith, R.D., Marquiss, M., Rae, R. & Metcalfe, N.B. (1993). Age and sex variation in choice of wintering site by Snow Buntings: the effect of altitude. *Ardea* **81**: 47–52.

Smith, R.H. (1973). The analysis of intra-generation change in animal populations. *J. Anim. Ecol.* **42**: 611–622.

Smith, R.I. (1970). Response of Pintail breeding populations to drought. *J. Wildl. Manage.* **34**: 943–946.

Smith, S.M. (1976). Ecological aspects of dominance hierarchies in Black-capped Chickadees. *Auk* **93**: 95–107.

Smith, S.M. (1978). The 'underworld' in a territorial sparrow: adaptive strategy for floaters. *Amer. Nat.* **112**: 571–582.

Smith, S.M. (1984). Flock switching in chickadees: why be a winter floater? *Amer. Nat.* **123**: 81–98.

Smith, T.B. (1987). Bill size, polymorphism and intraspecific niche utilisation in an African finch. *Nature* **329**: 717–719.

Smith, T.M. & Shugart, H.H. (1987). Territory size variation in the Ovenbird: the role of habitat structure. *Ecology* **68**: 695–704.

Snow, B. & Snow, D. (1988). Birds and berries. Calton, Poyser.

Snow, D.W. (1954). The habitats of Eurasian tits (*Parus* spp.). *Ibis* **96**: 565–585.

Snow, D.W. (1958). A Study of Blackbirds. London, Allen & Unwin.

Snow, D.W. (1968). Movements and mortality of British Kestrels *Falco tinnunculus*. *Bird Study* **15**: 65–83.

Snyder, B., Thilsted, J., Burgess, B. & Richard, M. (1985). Pigeon herpesvirus mortalities in foster reared Mauritius Pink Pigeons. pp. 69–70 in '*Proceedings of the American Association of Zoo Veterinarians*', Scottsdale, Arizona (ed. M.S Silbermann & S.D. Silbermann). Philadelphia, PA: AAZV.

Snyder, N.F.R. & Snyder, H.A. (1989). Biology and conservation of the California Condor. *Curr. Ornithol.* **5**: 175–267.

Snyder, N.F.R., Wiley, J.W. & Kepler, C.B. (1987). The parrots of Luquillo: natural history and conservation of the Puerto Rican Parrot. Los Angeles, Western Foundation of Vertebrate Zoology.

Snyder, W.D. (1985). Survival of radio-marked hen Ring-necked Pheasants in Colorado. *J. Wildl. Manage.* **49**: 1044–1050.

Sodhi, N.S., Didiuk, A. & Oliphant, L.W. (1990). Differences in bird abundance in relation to proximity of Merlin nests. *Can. J. Zool.* **68**: 852–854.

Soikkeli, M. *(1970)*. Dispersal of Dunlin *Calidris alpina* in relation to sites of birth and breeding. *Ornis Fenn.* **47**: 1–9.

Soler, M. & Soler, J.J. (1996). Effects of experimental food provisioning on reproduction in the Jackdaw *Corvus monedula*, a semi–colonial species. *Ibis* **138**: 379–383.

Solomon, M.E. (1949). The natural control of animal populations. *J. Anim. Ecol.* **18**: 1–35.

Soloman, M.E., Glen, D.M., Kendall, D.A. & Milson, N.F. (1976). Predation of overwintering larvae of Codling Moth (*Cyclia pomonella* (L)) by birds. *J. Appl. Ecol.* **13**: 341–352.

Solonen, T. (1997). Effect of Sparrowhawk predation on forest birds in southern Finland. *Ornis Fenn.* **74**: 1–14.

Sonerud, G. (1993). Reduced predation by nest box relocation: differential effect in Tengmalm's Owl nests and artificial nests. *Ornis Scand.* **24**: 249–253.

Sotherton, N.W. (1991). Conservation headlands: a practical combination of intensive cereal farming and conservation. pp. 373–397 in *'The ecology of temperate cereal fields'* (ed. L.G. Firbank, N. Carter, J.F. Darbyshire & G.R. Potts). Oxford, Blackwell.

Soulé, M.E. (ed.) (1987). Viable populations for conservation. Cambridge, University Press.

Souliere, G.J. (1988). Density of suitable Wood Duck nest cavities in a northern hardwood forest. *J. Wildl. Manage.* **52**: 86–89.

Southern, H.N. (1970). The natural control of a population of Tawny Owls *(Strix aluco)*. *J. Zool., Lond.* **162**: 197–285.

Spaans, A.L. & Swennen, C. (1968). De Vogels van Vlieland Wetenschappelijke. *Mededelinge* **75**: 1–104.

Spalding, M.G., Smith, J.P. & Forrester, D.J. (1994). Natural and experimental infections of *Eustrongyloides ignotus:* effect on growth and survival of nestling wading birds. *Auk* **111**: 328–336.

Spitzer, G. (1972). Jahreszeitliche Aspeckte der Biologie der Bartmeise *(Panurus biarmicus)*. *J. Ornithol.* **113**: 241–275.

Spray, C.J., Crick, H.Q.P. & Hart, A.D.M. (1987). Effects of aerial applications of fenitrothion on bird populations of a Scottish pine plantation. *J. Appl. Ecol.* **24**: 29–47.

Spray, C.J. & Milne, H. (1988). The incidence of lead poisoning among Whooper and Mute Swans *Cygnus cygnus* and C. *olor* in Scotland. *Biol. Conserv.* **44**: 265–281.

Stabler, R.M & Herman, C.M. (1951). Upper digestive tract *Trichomoniasis* in Mourning Doves and other birds. *Trans. N. A. Wildl. Nat. Res. Conf.* **16**: 145–162.

Stacey, P.B. & Koenig, W.D. (1990). Cooperative breeding in birds: long term studies of ecology and behaviour. Cambridge, University Press.

Stanley, P.I. & Bunyan, P.J. (1979). Hazards to wintering geese and other wildlife from the use of dieldrin, chlorfenvinphos and carbophenothion as wheat seed treatments. *Proc. R. Soc. Lond. B* **205**: 31–45.

Steadman, D.W. (1989). Extinction of birds of Eastern Polynesia: a review of the record and comparisons with other Pacific Island groups. *J. Archaeol. Sci.* **16**: 177–205.

Steadman, D.W. (1991). Extinction of species: past, present and future. pp. 156–169 in *'Global climate change and life on earth'* (ed. R.C. Wyman). New York, Routledge, Chapman & Hall.

Steadman, D.W. (1995). Prehistoric extinctions of Pacific island birds: biodiversity meets zooarchaeology. *Science* **267**: 1123–1131.

Steadman, D.W. & Olson, S.R. (1985). Bird remains from an archaeological site on Henderson Island, South Pacific; man-caused extinctions on an 'uninhabited' island. *Proc. Nat. Acad. Sci.* **82**: 6191–6195.

Steenhof, K., Kochert, M.N. & MacDonald, T.L. (1997). Interactive effects of prey and weather on Golden Eagle reproduction. *J. Anim. Ecol.* **66**: 350–362.

Steenhof, K., Kochert, M.N. & Poppe, J.A. (1993). Nesting by raptors and Common Ravens on electrical transmission line towers. *J. Wildl. Manage.* **57**: 271–281.

Stempniewicz, L. (1994). Marine birds drowning in fishing nets in the Gulf of Gdansk (southern Baltic): numbers, species composition, age and sex structure. *Ornis Svecica* **4**: 123–132.

Stenger, J. (1958). Food habits and available food of Ovenbirds in relation to territory size. *Auk* **75**: 335–346.

Stenning, M.J., Harvey, P.H. & Campbell, B. (1988). Searching for density-dependent regulation in a population of Pied Flycatchers *Ficedula hypoleuca* Pallas. *J. Anim. Ecol.* **57**: 307–317.

Stewart, R.E. & Aldrich, J.W. (1951). Removal and repopulation of breeding birds in a spruce-fir forest community. *Auk* **68**: 471–482.

Steyn, P. & Brooke, R.K. (1970). Cold induced mortality of birds in Rhodesia during November 1968. *Ostrich (Suppl.)* **8**: 271.

Stickel, L.F. (1975). The costs and effects of chronic exposure to low-level pollutants in the environment. pp. 716–728 in *'Hearings before the sub-committee on the environment and the atmosphere'* (ed. Committee on Science and Technology). Washington, D.C., U.S. House of Representatives.

Stiles, F.G. (1992). Effects of a severe drought on the population biology of a tropical hummingbird. *Ecology* **73**: 1375–1390.

Still, E., Monaghan, P. & Bignal, E. (1987). Social structuring at a communal roost of Choughs *Pyrrhocorax pyrrhocorax*. *Ibis* **129**: 398–403.

Stoate, C. & Szczur, J. (1994). Game management and songbirds. *Game Conservancy Rev.* **25**: 55–56.

Stock, T.M. & Holmes, J.C. (1987). Host specificity and exchange of intestinal helminths among four species of grebes (Podicipedidae). *Can. J. Zool.* **65**: 669–676.

Stonehouse, B. (1962). Ascension Island and the British Ornithologists' Union Centenary Expedition 1957–1959. *Ibis* **103B**: 107–123.

Storaas, T., Wegge, P. & Sonerud, G. (1982). Destruction des nids de grand tetras et cycle des petits rongeurs dans l'est de la Norvege. pp. 166 in *'Actes du Colloque International sur le Grand Tetras* (Tetrao urogallus major)' (ed. C. Kempf). Colmar, France. Union Nationale des Associations Ornithologiques.

Stott, R.J. & Olson, D.P. (1972). Differential vulnerability patterns among three species of sea ducks. *J. Wildl. Manage.* **36**: 775–783.

Strain, J.G. & Mumme, R.L. (1988). Effects of food supplementation, song playback and temperature on vocal territorial behavior of Carolina Wrens. *Auk* **105**: 11–16.

Strann, K.B., Vader, W. & Barrett, R.T. (1991). Auk mortality in fishing nets in North Norway. *Seabird* **13**: 22–29.

Stribling, H.L. & Smith, H.R. (1987). Effects of dimilin on diversity and abundance of forest birds. *North J. Appl. For.* **4**: 37–38.

Stutchbury, B.J. (1994). Competition for winter territories in a neotropical migrant: the role of age, sex and colour. *Auk* **111**: 63–69.

Stutchbury, B.J. & Robertson, R.J. (1985). Floating populations of female Tree Swallows. *Auk* **102**: 651–654.

Stutchbury, B.J. & Robertson, R.J. (1987). Behavioural tactics of sub-adult female floaters in the Tree Swallow. *Behav. Ecol. Sociobiol.* **20**: 413–419.

Sugden, L.G. & Beyersbergen, G.W. (1986). Effect of density and concealment on American Crow predation on simulated duck nests. *J. Wild. Manage.* **50**: 9–14.

Suhonen, J., Alatalo, R.V., Carlson, A. & Höglund, J. (1992). Food resource distribution and the organisation of the *Parus* guild in a spruce forest. *Ornis Scand.* **23**: 467–474.

Suhonen, J., Norrdahl, K. & Korpimäki, E. (1994). Avian predation risk modifies breeding bird community on a farmland area. *Ecology* **75**: 1626–1634.

Summers, R.W. & Cooper, J. (1977). The population ecology and conservation of the Black Oystercatcher *Haematopus moquini*. *Ostrich* **48**: 28–40.

Summers, R.W. & Underhill, L. (1987). Factors related to breeding production of Brent Geese *Branta b. bernicla* and waders (Charadrii). *Bird Study* **37**: 161–171.

Summers, R.W. & Underhill, L.G. (1991). The growth of the population of Dark-bellied Brent Geese *Branta b. bernicla* between 1955–1988. *J. Appl. Ecol.* **28**: 574–585.

Summers-Smith, D. (1988). The sparrows. Calton, Poyser.

Summers-Smith, J.D. (1996). The Tree Sparrow. Guisborough, J.D. Summers-Smith.

Suter, W. (1995). Are Cormorants *Phalacrocorax carbo* wintering in Switzerland approaching carrying capacity? An analysis of increase patterns and habitat choice. *Ardea* **83**: 255–266.

Suter, W. & van Eerden, M.R. (1992). Simultaneous mass starvation of wintering diving ducks in Switzerland and the Netherlands: a wrong decision in the right strategy. *Ardea* **80**: 229–242.

Sutherland, W.J. (1996a). From individual behaviour to population ecology. Oxford, University Press.

Sutherland, W.J. (1996b). Predicting the consequences of habitat loss for migratory populations. *Proc. Roy. Soc. B* **263**: 1325–1327.

Sutherland, W.J. & Allport, G.A. (1994). Spatial depletion model of the interaction between Bean Geese and Wigeon with the consequences for habitat management. *J. Anim. Ecol.* **63**: 51–59.

Sutherland, W.J. & Anderson, C.W. (1993). Predicting the distribution of individuals and the consequences of habitat loss: the role of prey depletion. *J. Theoretical Biol.* **160**: 223–230.

Sutherland, W.J. & Koene, P. (1982). Field estimates of the strength of interference between Oystercatchers. *Oecologia* **55**: 108–109.

Sutherland, W.J. & Parker, S.A. (1985). Distribution of unequal competitors. pp. 255–274 in 'Behavioural ecology. Ecological consequences of adaptive behaviour' (ed. R.M. Sibly & R.H. Smith). Oxford, Blackwells.

Svärdson, G. (1957). The 'invasion' type of bird migration. *Brit. Birds* **48**: 425–428.

Swanson, G.A. (1980). Techniques to improve nesting success in birds. *Proc. Intl. Ornithol. Congress* **17**: 918–922.

Swanson, G.A. & Ryder, R.A. (1979). Conservation and management of nongame birds. Unpublished Class syllabus. Colorado State University, Fort Collins, CO. 286 pp.

Swartz, L.G., Walker, W., Roseneau, D.G. & Springer, A.M. (1974). Populations of Gyrfalcons on the Seward Peninsula, Alaska, 1968–1972. *Raptor Res. Rep.* **3**: 71–75.

Swatzman, G.L. & Haar, R.T. (1983). Interactions between Fur Seal populations and fisheries in the Bering Sea. *Fish. Bull.* **81**: 121–132.

Swennen, C. (1972). Chlorinated hydrocarbons attacked the Eider population in The Netherlands. *TNO nieuws* **27**: 556–560.

Swennen, C. (1984). Differences in quality of roosting flocks of Oystercatchers. pp. 177–189 in 'Coastal waders and wildfowl in winter' (ed. P.R. Evans, J.D. Goss-Custard & W.G. Hale). Cambridge, University Press.

Swennen, C. (1989). Gull predation upon Eider *Somateria mollissima* ducklings: destruction or elimination of the unfit? *Ardea* **77**: 21–45.

Swennen, C. & Duiven, P. (1983). Characteristics of Oystercatchers killed by cold-stress in the Dutch Wadden Sea area. *Ardea* **71**: 155–159.

Swennen, C. & Smit, Th. (1991). Pasteurellosis among breeding Eiders *Somateria mollissima* in the Netherlands. *Wildfowl* **42**: 94–97.

Swenson, J.E. (1995). Habitat requirement of Hazel Grouse. pp. 115–159 in 'Proceedings of the 6th International Grouse Symposium' (ed. D. Jenkins). Udine, Norway.

Swingland, I.R. (1977). The social and spatial organisation of winter communal roosting in Rooks (*Corvus frugilegus*). *J. Zool., Lond.* **182**: 509–528.

Sylven, M. (1979). Interspecific relations between sympatrically wintering Common Buzzards *Buteo buteo* and Rough-legged Buzzards *Buteo lagopus*. *Ornis. Scand.* **9**: 197–206.

Symens, P. & Salamah M.I. Al. (1993). The impact of the Gulf War oil spills on wetlands and waterfowl in the Arabian Gulf. pp. 25–28 in 'Wetland and waterfowl conservation in south and west Asia', IWRB Spec. Pub. No 25 (ed. M.L. Moser & J. van Vessem). International Waterfowl & Wetlands Research Bureau, Slimbridge, Glos. & Asian Wetland Bureau, Kuala Lumpur.

Székely, T., Szép, T. & Juhász, T. (1989). Mixed species flocking of tits (*Parus* spp.): a field experiment. *Oecologia* **78:** 490–495.

Szuba, K.J. & Bendell, N.F. (1988). Nonterritorial males in populations of Spruce Grouse. *Condor* **90:** 492–496.

Takekawa, J.E., Carter, H.R. & Harvey, T.E. (1990). Decline of the Common Murre in central California, 1980–1986. *Stud. Avian Biol.* **14:** 149–163.

Tanner, E.V.J., Kapos, V. & Healey, J.R. (1991). Hurricane effects on forest ecosystems in the Caribbean. *Biotropica* **23:** 513–521.

Tapper, S.C., Brockless, M. & Potts, G.R. (1991). The Salisbury Plain experiment: the conclusion. *Game Conservancy Rev.* **22:** 87–91.

Tapper, S.C., Potts, G.R. & Brockless, M.H. (1996). The effect of an experimental reduction in predation pressure on the breeding success and population density of Grey Partridges *Perdix perdix*. *J. Appl. Ecol.* **33:** 965–978.

Tate, J.L. (1972). The changing seasons. *Amer. Birds* **26:** 828–831.

Taylor, I. (1994). Barn Owls. Predator–prey relationships and conservation. Cambridge, University Press.

Taylor, R.H. (1979). How the Macquarie Island parakeet became extinct. *N.Z. J. Ecol.* **2:** 42–45.

Taylor, R.H. & Wilson, P.R. (1990). Recent increase and southern expansion of Adelie Penguin populations in the Ross Sea, Antarctica, related to climatic warming. *N. Z. J. Ecol.* **14:** 25–29.

Techlow, A.F. (1983). Forster's Tern nesting platform study. Madison, Wisconsin, Dept. Natural Resources, Prog. Rep., U.S.A.

Tella, J.L., Forest, M.G., Gajón, A., Hiraldo, F. & Donázar, J.A. (1996). Absence of blood parasitisation effects on Lesser Kestrel fitness. *Auk* **113:** 253–256.

Temple, S.A. (1987). Do predators always capture substandard individuals disproportionately from prey populations? *Ecology* **68:** 669–674.

Temple, S.A. & Cary, J.R. (1988). Modelling dynamics of habitat-interior bird populations in fragmented landscapes. *Conserv. Biol.* **2:** 340–347.

Templeton, A.R., Hemner, H., Mace, G., Seal, U.S., Shields, W.M. & Woodruff, D.S. (1986). Local adaptation, coadaptation, and population boundaries. *Zoo Biol.* **5:** 115–125.

Tenaza, R. (1971). Behaviour and nesting success relative to nest location in Adélie Penguins (*Pygoscelis adeliae*). *Condor* **73:** 81–92.

Tennyson, A. (1990). Seabirds in strife. *Forest & Birds, November 1990*: 23–30.

Terborgh, J. (1975). Faunal equilibria and the design of nature reserves. pp. 369–380 in '*Tropical ecological systems: trends in terrestrial and aquatic research*' (ed. F.B. Golley & E. Medina). New York, Springer Verlag.

Terborgh, J. (1989). Where have all the birds gone? Princeton, University Press.

Terborgh, J. & Faaborg, J. (1973). Turnover and ecological release in the avifauna of Mona Island, Puerto Rico. *Auk* **90:** 759–779.

Terborgh, J. & Weske, J.S. (1975). The role of competition in the distribution of Andean birds. *Ecology* **56:** 562–576.

Terborgh, J.W. & Winter, B. (1980). Some causes of extinction. pp. 119–133 in '*Conservation biology: an evolutionary-ecological perspective*' (ed. M.E. Soulé & B.A. Wilcox). Sunderland, Massachusetts, Sinauer.

Thiollay, J.-M. (1988). Comparative predation pressure on solitary and colonial breeding passerines. *Proc. Int. Ornithol. Congr.* **19:** 660–673.

Thomas, J.W., Forsman, E.D., Lint, J.B., Meslow, E.C., Noon, B.R. & Verner, J. (1990). A conservation strategy for the Northern Spotted Owl. Report of the Interagency Scientific Committee to address the conservation of the Northern Spotted Owl. U.S. Dept. Agriculture, Forest Service and Fish & Wildlife Service, U.S. Dept. Interior, Bureau of Land Management and National Park Service, Portland, Oregon.

Thompson, D.B. & Whitfield, D.P. (1993). Research on mountain birds and their habitats. *Scott. Birds* **17:** 1–8.

Thomson, D.L., Baillie, S.R. & Peach, W.J. (1997). The demography and age-specific survival of Song Thrushes during periods of population stability and decline. *J. Anim. Ecol.* **66:** 414–424.

Thorne, E.T. & Williams, E.S. (1988). Disease and endangered species: the Black–footed Ferret as a recent example. *Conserv. Biol.* **2:** 66–74.

Threlfall, W., Eveleigh, E. & Maunder, J.E. (1974). Seabird mortality in a storm. *Auk* **91:** 846–849.

Tiainen, J., Hanski, I.K, Pakkala, 1, Piiruoinen, J. & Yrjölä, R. (1989). Clutch–size, nestling growth and nestling mortality of the Starling *Sturnus vulgaris* in south Finnish agroenvironments. *Ornis Fenn.* **66:** 41–48.

Ticehurst, C.B. 1908–1911. The Wood–pigeon diphtheria. *Brit. Birds* **1:** 243–245; **2:** 69–77; **3:** 213–214; **4:** 304–305.

Ticehurst, N.E & Hartley, P.H.T. 1948. Report on the effect of the severe winter of 1946–1947 on bird-life. *Brit. Birds* **41:** 322–334.

Tinbergen, J.M., van Balen, J.H. & van Eck, H.M. (1985). Density-dependent survival in an isolated Great Tit population: Kluijver's data reanalysed. *Ardea* **73:** 38–48.

Tinbergen, L. (1946). Sperver als Roofvijand van Zangvogels. *Ardea* **34:** 1–123.

Tinbergen, N. (1957). The functions of territory. *Bird Study* **4:**14–27.

Tinbergen, N., Impekoven, M. & Frank, D. (1967). An experiment on spacing out as a defence against predation. *Behaviour* **28:** 307–321.

Tjernberg, M., Johnsson, K. & Nilsson, S.G. (1993). Density variation and breeding success of the Black Woodpecker *Dryocopus martius* in relation to forest fragmentation. *Ornis Fenn.* **70:** 155–162.

Toft, C.A. (1991). Current theory of host-parasite interactions. pp. 3–15 in '*Bird-parasite interactions*' (ed. J.E. Loye & M. Zuk). Oxford, University Press.

Tomiałojc, L. (1980). The impact of predation on urban and rural Woodpigeon *(Columba palumbus* (L.)) populations. *Polish Ecol. Stud.* **5:** 141–220.

Tomiałojc, L. & Wesolowksi, T. (1990). Bird communities of the primaeval temperate forest of Bialowieza, Poland. pp. 141–165 in '*Biogeography and ecology of forest bird communities*' (ed. A. Keast). The Hague, Netherlands, SPB Academic Publishing.

Tomiałojc, L., Wesolowski, T. & Walankiewicz, W. (1984). Breeding bird community of a primaeval temperate forest (Bialowieza National Park, Poland). *Acta Ornithol.* **20:** 241–310.

Tomkins, R.J. (1985). Breeding success and mortality of Dark-rumped Petrels in the Galapagos and control of their predators. *ICBP Tech. Bull* **3:** 159–175.

Tompa, F.S. (1964). Factors determining the number of Song Sparrows *Melospiza melodia* (Wilson) on Mandarte Island, B.C., Canada. *Acta Zool. Fenn.* **109:** 1–73.

Tompa, F.S. (1967). Reproductive success in relation to breeding density in Pied Flycatchers, *Ficedula hypoleuca* (Pallas). *Acta Zool. Fenn.* **118:** 1–28.

Török, J. (1987). Competition for food between Great Tit *Parus major* and Blue Tit *P. caeruleus* during the breeding season. *Acta. Reg. Soc. Sci. Litt. Gothoburgensis. Zoologica* **14:** 149–152.

Török, J. & Tóth, L. (1988). Density dependence in reproduction of the Collared Flycatcher *(Ficedula albicollis)* at high population levels. *J. Anim. Ecol.* **57:** 251–258.

Trail, P.W. & Baptista, L.F. (1993). The impact of Brown-headed Cowbird parasitism on populations of the Nuttall's White-crowned Sparrow. *Conserv. Biol.* **7:** 309–315.

Trautman, C.G., Fredrickson, L.F. & Carter, A.V. (1974). Relationship of Red Foxes and other predators to populations of Ring-necked Pheasants and other prey, South Dakota. *Trans. N. A. Wildl. Nat. Res. Conf.* **39:** 241–255.

Trautman, M.B., Bills, W.E. & Wickliff, E.L. (1939). Winter losses from starvation and exposure of waterfowl and upland game birds in Ohio and other northern states. *Wilson Bull.* **51:** 86–104.

Trivelpierce, W., Butler, R.G., Miller, D.S. & Peakall, D.B. (1984). Reduced survival of chicks of oil-dosed adult Leach's Storm Petrels. *Condor* **86:** 81–82.

Tuck, L. (1968). Laughing Gulls *Larus atricilla* and Black Skimmers *Rynchops nigra* brought to Newfoundland by hurricane. *Bird-Banding* **39**: 200–208.

Tucker, G. & Heath, M. (1994). Birds in Europe: their conservation status. Cambridge, Birdlife International.

Tucker, R.K. & Crabtree, D.G. (1970). Handbook of toxicity of pesticides to wildlife. Resource Pub. 84. Washington, D.C., U.S. Dept. of Interior, Fish & Wildlife Service.

Tuomenpuro, J. (1991). Effect of nest site on nest survival in the Dunnock *Prunella modularis*. *Ornis Fenn.* **68**: 49–56.

Turnbull, R.E., Brakhage, D.H. & Johnson, F.A. (1986). Lesser Scaup mortality from commercial trotlines on Lake Okeechobee, Florida. *Proc. Southeast. Assoc. Fish and Wildl. Agencies* **40**: 465–469.

Tye, A. (1986). Economics of experimentally-induced territorial defence in a gregarious bird, the Fieldfare *Turdus pilaris*. *Ornis Scand.* **17**: 151–164.

Underhill, L.G. & Summers, R.W. (1990). Multivariate analyses of breeding performance in Dark-bellied Brent Geese *Branta b. bernicla*. *Ibis* **132**: 477–480.

Underwood, L.A. & Stowe, T.J. (1984). Massive wreck of seabirds in eastern Britain, 1983. *Bird Study* **31**: 79–88.

Vader, W., Barrett, R.T., Erikstad, K.E. & Strann, K.-B. (1990). Differential response of Common and Thick-billed Murres to a crash in the capelin stock in the southern Barents Sea. *Stud. Avian Biol.* **14**: 175–180.

van Balen, J.H. (1980). Population fluctuations of the Great Tit and feeding conditions in winter. *Ardea* **68**: 143–164.

van Balen, J.H., Booy, C.J.H., van Franeker, J.A. & Osieck, E.R. (1982). Studies on hole nesting birds in natural nest sites, 1. Availability and occupation of natural nest sites. *Ardea* **70**: 1–24.

van Dobben, W.H. (1952). The food of the Cormorants in the Netherlands. *Ardea* **40**: 1–63.

van Horne, B. (1983). Density as a misleading indicator of habitat quality. *J. Wildl. Manage.* **47**: 893–901.

van Impe, J. (1996). Long term reproductive performance in White-fronted Geese *Anser a. albifrons* and Tundra Bean Geese *A. fabalis rossicus* wintering in Zeeland (The Netherlands). *Bird Study* **40**: 280–289.

van Noordwijk, A.J. (1987). Quantitative ecological genetics of Great Tits. pp. 363–380 in 'Avian Genetics. A population and ecological approach' (ed. F. Cooke & P.A. Buckley). London, Academic Press.

van Riper, C.I. (1984). The influence of nectar resources on nesting success and movement patterns of the Common Amakihi *Hemignathus virens*. *Auk* **101**: 38–46.

van Riper, C.I., van Riper, S.G., Goff, M.G. & Laird, M. (1986). The epizootiology and ecological significance of malaria in Hawaiian land birds. *Ecol. Monogr.* **56**: 327–344.

Vander Wall, S.B. (1988). Foraging of Clark's Nutcrackers on rapidly changing pine seed resources. *Condor* **90**: 621–631.

Vane-Wright, R.I., Humphries, C.J. & Williams, P.H. (1991). What to protect? Systematics and the agony of choice. *Biol. Conserv.* **55**: 235–254.

Varley, G.C. (1947). The natural control of population balance in the Knapweed Gallfly (*Urophora jaceana*). *J. Anim. Ecol.* **16**: 139–187.

Varley, G.C. & Gradwell, G.R. (1960). Key factors in population studies. *J. Anim. Ecol.* **29**: 399–401.

Vaughan, G.E. & Coble, P.W. (1975). Sublethal effects of three ectoparasites on fish. *J. Fisheries Biol.* **7**: 283–294.

Vaught, R.W., McDougle, H.C. & Burgess, H.H. (1967). Fowl cholera in waterfowl at Squaw Creek National Wildlife Refuge, Missouri. *J. Wildl. Manage.* **31**: 248–253.

Vaurie, C. (1951). Adaptive differences between two sympatric species of nuthatches (*Sitta*). *Proc. Int. Ornithol. Congr.* **10**: 163–166.

Veit, R.R., Silverman, E.D. & Everson, I. (1993). Aggregation patterns of pelagic predators and their principal prey, Antarctic krill, near South Georgia. *J. Anim. Ecol.* **62:** 551–564.

Veitch, C.R. (1985). Methods of eradicating feral cats from offshore islands in New Zealand. *ICBP Tech. Bull.* **3:** 125–141.

Verbeek, N.A.M. (1982). Egg predation by Northwestern Crows: its association with human and Bald Eagle activity. *Auk* **99:** 347–352.

Verboom, J., Schotman, A., Opdam, P. & Metz, J.A.J. (1991). European Nuthatch metapopulations in a fragmented agricultural landscape. *Oikos* **61:** 149–156.

Verhulst, S. (1992). Effects of density, beech crop and winter feeding on survival of juvenile Great Tits; an analysis of Kluijver's removal experiment. *Ardea* **80:** 285–292.

Verhulst, S. (1994). Supplementary food affects reproductive success in Pied Flycatchers (*Ficedula hypoleuca*). *Auk* **111:** 714–716.

Vermeer, K. (1968). Ecological aspects of ducks nesting in high densities among larids. *Wilson Bull.* **80:** 78–83.

Vermeer, K., Power, D. & Smith, G.E.J. (1988). Habitat selection and nesting biology of roof-nesting Glaucous-winged Gulls. *Colonial Waterbirds* **11:** 189–201.

Vermeer, K., Risebrough, R.W., Spaans, A.L. & Reynolds, L.M. (1974). Pesticide effects on fishes and birds in rice fields of Surinam, South America. *Environ. Poll.* **7:** 217–236.

Verner, G. & Ritter, L.V. (1983). Current status of the Brown-headed Cowbird in the Sierra National Forest. *Auk* **100:** 355–368.

Vespäläinen, K. (1968). The effect of the cold spring 1966 upon the Lapwing *Vanellus vanellus* in Finland. *Ornis Fenn.* **45:** 33–47.

Vickery, J. (1991). Breeding density of Dippers *Cinclus cinclus*, Grey Wagtails *Motacilla cinerea* and Common Sandpipers *Actitis hypoleucos* in relation to the acidity of streams in south-west Scotland. *Ibis* **133:** 178–187.

Vickery, J.A. & Ormerod, S.J. (1991). Dippers as indicators of stream acidity. *Proc. Int. Ornithol. Congr.* **20:** 2494–2502.

Vickery, J.A., Sutherland, W.J., Watkinson, A.R., Lane, S.J. & Rowcliffe, J.M. (1995). Habitat switching by Dark-bellied Brent Geese *Branta b. bernicla* (L.) in relation to food depletion. *Oecologia* **103:** 499–508.

Vickery, W.L. & Nudds, T.D. (1984). Detection of density-dependent effects in annual duck censuses. *Ecology* **65:** 96–104.

Viestola, S., Lehikoinen, E., Eeva, T. & Iso-Iivari, L. (1996). The breeding biology of the Redstart *Phoenicurus phoenicurus* in a marginal area of Finland. *Bird Study* **43:** 351–355.

Village, A. (1983). The role of nest-site availability and territorial behaviour in limiting the breeding density of Kestrels. *J. Anim. Ecol.* **52:** 635–645.

Village, A. (1987). Numbers, territory size and turnover of Short-eared Owls *Asio flammeus* in relation to vole abundance. *Ornis Scand.* **18:** 198–204.

Village, A. (1990). The Kestrel. Calton, Poyser.

Virkhala, R. (1991). Annual variation of northern Finnish forest and fen bird assemblages in relation to spatial scale. *Ornis Fenn.* **68:** 193–203.

Virolainen, M. (1984). Breeding biology of the Pied Flycatcher *Ficedula hypoleuca* in relation to population density. *Ann. Zool. Fenn.* **21:** 187–197.

Vogel, C. & Moritz, D. (1995). Langjährige Änderungen von Zugzeiten auf Helgoland. *Jber Inst. Vogelforshung* **2:** 8–9.

Vogt, W. (1942). Ares guaneras. *Bol. Compāria Administradora Guano* **18:** 3–132.

Volterra, V. (1926). Variations and fluctuations of the number of individuals in animal species living together. *J. Cons. perm. int. Ent. Mer.* **3:** 3–51.

von Haartman, L. (1957). Adaptation in hole-nesting birds. *Evolution* **11:** 339–347.

von Haartman, L. (1971). Population dynamics. pp. 391–459 in 'Avian biology', Vol. 1 (ed. D.S. Farner & J.R. King). London, Academic Press.

von Haartman, L. (1973). Talgmespopulationen Lemsjöholm. *Lintumies* **8:** 7–9.

Voous, K.H. (1960). Atlas of European birds. London, Nelson.

Wagner, F.H., Beradny, C.D. & Kabat, C. (1965). Population ecology and management of Wisconsin Pheasants. *Wis. Cons. Dept., Madison, Wis. Tech. Bull.* **No. 34**.

Walker, C.H. (1983). Pesticides and birds – mechanisms of selective toxicity. *Agric. Ecosys. Environ.* **9**: 211–226.

Walkinshaw, L.H. (1983). The Kirtland's Warblers. Broom Field Hills, Michigan, Cranbrook Institute of Science.

Wallace, G.J., Nickell, W.P. & Bernard, R.F. (1961). Bird mortality in the Dutch Elm Disease program in Michigan. *Cranbrook Inst. Sci. Bull.* **41**: 1–44.

Walter, H.S. (1990). Small viable population: the Red-tailed Hawk of Socorro Island. *Conserv. Biol.* **4**: 441–443.

Walters, J.R. (1991). Application of ecological principles to the management of endangered species: the case of the Red-cockaded Woodpecker. *Ann. Rev. Ecol. Syst.* **22**: 505–523.

Walters, J.R., Hansen, S.K., Carter, J.H. & Manor, P.M. (1988). Long distance dispersal of an adult Red-cockaded Woodpecker. *Wilson Bull.* **100**: 494–496.

Ward, F.P. (1973). A clinical evaluation of parasites in birds of prey. *Zoo. Anim. Med.* **6**: 3–8.

Ward, P. (1965). Feeding ecology of the Black-faced Dioch *Quelea quelea* in Nigeria. *Ibis* **107**: 173–214.

Ward, P. (1968). Origin of the avifauna of urban and suburban Singapore. *Ibis* **110**: 239–255.

Ward, P. (1979). Rational strategies for the control of Queleas and other migrant bird pests of Africa. *Phil. Trans. Roy. Soc. Lond. B* **287**: 289–300.

Warham, J. (1990). The petrels. London, Academic Press.

Warner, R.E. (1968). The role of introduced diseases in the extinction of the endemic Hawaiian Avifauna. *Condor* **70**: 101–120.

Warner, R.E. (1984). Declining survival of Ring-necked Pheasant chicks in Illinois agricultural ecosystems. *J. Wildl. Manage.* **48**: 82–88.

Wasserman, F. (1983). Territories of Rufous-sided Towhees contain more than minimal food resources. *Wilson Bull.* **95**: 664–667.

Waters, J.R., Noon, B.R. & Verner, J. (1990). Lack of nest site limitation in a cavity-nesting bird community. *J. Wildl. Manage.* **54**: 239–245.

Watkinson, A.R. & Sutherland, W.J. (1995). Sources, sinks and pseudo-sinks. *J. Anim. Ecol.* **64**: 126–130.

Watling, D. & Lewanavanua, R.F. (1985). A note to record the continuing survival of the Fiji (Macgillivray) Petrel *Pseudobulweria macgillivrayi*. *Ibis* **127**: 230–233.

Watson, A. (1965). A population study of Ptarmigan *Lagopus mutus* in Scotland. *J. Anim. Ecol.* **34**: 135–172.

Watson, A. (1985). Social class, socially-induced loss, recruitment and breeding of Red Grouse. *Oecologia* **67**: 493–498.

Watson, A. & Jenkins, D. (1968). Experiments on population control by territorial behaviour in Red Grouse. *J. Anim. Ecol.* **37**: 595–614.

Watson, A. & Moss, R. (1970). Dominance, spacing behaviour and aggression in relation to population limitation in vertebrates. pp. 167–200 in '*Animal populations in relation to their food resources*' (ed. A. Watson). Oxford, Blackwell Scientific Publications.

Watson, A. & Moss, R. (1979). Population cycles in the Tetraonidae. *Ornis Fenn.* **56**: 87–109.

Watson, A. & Moss, R. (1990). Spacing behaviour and winter loss in Red Grouse. pp. 35–52 in '*Red Grouse population processes*' (ed. A.N. Lance & J.H. Lawton). British Ecological Society & Royal Society for the Protection of Birds.

Watson, A., Moss, R. & Parr, R. (1984). Effects of food enrichment on numbers and spacing of Red Grouse. *J. Anim. Ecol.* **53**: 663–678.

Watson, A., Moss, R., Parr, R., Mountford, M.D. & Rothery, P. (1994). Kin landownership, differential aggression between kin and non-kin, and population fluctuations in red grouse. *J. Anim. Ecol.* **63**: 39–50.

Watson, A., Moss, R., Parr, R., Treadholm, I.B. & Robertson, A. (1988). Preventing population decline of Red Grouse *Lagopus lagopus scoticus* by manipulating density. *Experientia* **44:** 274–275.

Watson, A. & O'Hare, P.J. (1979). Red grouse populations on experimentally treated and untreated Irish bog. *J. Appl. Ecol.* **16:** 433–452.

Watson, J. (1996). The Golden Eagle. London, Academic Press.

Watson, J., Rae, S.R. & Stillman, R. (1992). Nesting density and breeding success of Golden Eagles in relation to food supply in Scotland. *J. Anim. Ecol* **61:** 543–550.

Wauer, R.H. & Wunderle, J.M. (1992). The effect of hurricane Hugo on bird populations on St Croix, United States Virgin Islands. *Wilson Bull.* **104:** 656–673.

Weatherhead, P.J. & Bennett, G.T. (1991). Ecology of Red-winged Blackbird parasitism by haematozoa. *Can. J. Zool.* **69:** 2352–2359.

Weatherhead, P.J. & Bennett, G.F. (1992). Ecology of parasitism of Brown-headed Cowbirds by haematozoa. *Can. J. Zool.* **70:** 1–7.

Weatherhead, P.J. & Hoysak, D.J. (1984). Dominance structuring of a Red-winged Blackbird roost. *Auk* **101:** 551–555.

Webb, T. (1992). Past changes in vegetation and climate: lessons for the future. pp. 59–75 in 'Climate change scenarios for impact assessment' (ed. R.L. Peters & T.E. Lovejoy). New Haven, Yale University Press.

Webster, H.M. (1944). A survey of the Prairie Falcon in Colorado. *Auk* **61:** 609–616.

Weeden, R.B. & Theberge, J.B. (1972). The dynamics of a fluctuating population of Rock Ptarmigan in Alaska. *Proc. Int. Ornithol. Congr.* **15:** 90–106.

Weimerskirch, H. & Jouventin, P. (1987). Population dynamics of the Wandering Albatross *Diomedea exulans*, of the Crozet Islands: causes and some consequences of the population decline. *Oikos* **49:** 315–322.

Weise, J.H., Davidson, W.R. & Nettles, V.F. (1977). Large-scale mortality of nestling ardeids caused by nematode infection. *J. Wildl. Dis.* **13:** 376–382.

Weller, M.W. (1979). Density and habitat relationships of Blue-winged Teal nesting in northwestern Iowa. *J. Wildl. Manage.* **43:** 367–374.

Wellicome, T.I. (1993). Reproductive performance of Burrowing Owls (*Athene cunicularia*): effects of supplementary food. *J. Raptor Res.* **27:** 94–95.

Wesolowski, T. (1980). Population restoration after removal of Wrens *Troglodytes troglodytes* breeding in primaeval forest. *J. Anim. Ecol.* **50:** 809–814.

Wesolowski, T., Tomialojc, L. & Stawarczyk, T. (1987). Why low numbers of *Parus major* in Bialowieza Forest-removal experiments. *Acta Ornithologica* **23:** 303–316.

White, D.H., King, K.A., Mitchell, C.A., Hill, E.T. & Lamont, T.G. (1979). Parathion causes secondary poisoning in a Laughing Gull breeding colony. *Bull. Environ. Contam. Toxicol.* **23:** 281–284.

Whiteley, P.L. (1989). Effects of environmental contaminants, particularly selenium, on waterfowl disease and immunity. Unpublished MSc (Veterinary Science) Thesis, University of Wisconsin, Madison.

Whitfield, D.P. (1985). Raptor predation on wintering waders in southeast Scotland. *Ibis* **127:** 544–558.

Whitmore, R.C., Mosher, J.A. & Frost, H.H. (1977). Spring migrant mortality during unseasonable weather. *Auk* **94:** 778–781.

Whittle, C.L. (1926). Notes on the nesting habits of Tree Swallow. *Auk* **43:** 247.

Widén, P. (1987). Goshawk predation during winter, spring and summer in the boreal forest area of central Sweden. *Holarctic Ecology* **10:** 104–109.

Widén, P. (1994). Habitat quality for raptors: a field experiment. *J. Avian Biol.* **25:** 219–223.

Wiedenfeld, D.A. & Wiedenfeld, M.G. (1995). Large kill of neotropical migrants by tornado and storm in Louisiana, April 1993. *J. Field Ornithol.* **66:** 70–80.

Wiemeyer, S.N., Scott, J.M., Anderson, M.P., Bloom, P.H. & Stafford, C.J. (1988). Environmental contaminants in California Condors. *J. Wildl. Manage.* **52:** 238–247.

Wiens, J.A. (1989). The ecology of bird communities. Vol. 2. Processes and variations. Cambridge, University Press.

Wiens, J.A. & Rotenberry, J.T. (1986). A lesson in the limitations of field experiments: shrubsteppe birds and habitat alteration. *Ecology* **67**: 365–376.

Wigley, T.M.L. & Riper, S.C.B. (1992). Implications for climate and sea level of revised IPCC emissions scenarios. *Nature* **357**: 293–300.

Wikman, M. & Lindén, H. (1981). The influence of food-supply on Goshawk population size. pp. 105–113 in *'Understanding the Goshawk'* (ed. R.E. Kenward & I.M. Lindsay). Oxford, International Association for Falconry and Conservation of birds of prey.

Wilcove, D.S. (1985). Nest predation in forest tracts and the decline of migratory songbirds. *Ecology* **66**: 1211–1214.

Willebrand, T. (1988). Demography and ecology of a Black Grouse (*Tetrao tetrix* L.) population. Ph.D. thesis (Unpub), Uppsala University, Sweden.

Williams, J.B. & Batzli, G.O. (1979). Competition among bark-foraging birds in central Illinois: experimental evidence. *Condor* **81**: 122–132.

Williams, M. (1979). The social structure, breeding and population dynamics of Paradise Shelduck in the Gisborne-east coast District. *Notornis* **26**: 213–272.

Williams, T.D., Cooch, E.G., Jefferies, R.L. & Cooke, F. (1993). Environmental degradation, food limitation and reproductive output. *J. Anim. Ecol.* **62**: 766–777.

Williamson, K. (1969). Habitat preferences of the Wren on English farmland. *Bird Study* **16**: 53–59.

Williamson, M. (1981). Island populations. Oxford, University Press.

Willis, E.O. (1979). The composition of avian communities in remanescent woodlots in southern Brazil. *Papeis Avulsos Zoologia* **33**: 1–25.

Willis, E.O. (1980). Species reduction in remanescent woodlots in southern Brazil. *Proc. Int. Ornithol. Congr.* **19**: 783–786.

Willis, E.O. & Oniki, Y. (1978). Birds and army ants. *Ann. Rev. Ecol. Syst.* **9**: 243–263.

Willson, M.F., de Santo, T.L., Sabag, C. & Armesto, J.J. (1994). Avian communities of fragmented south-temperate rainforests in Chile. *Conserv. Biol* **8**: 508–520.

Wilson, E.O. (1992). The diversity of life. Cambridge, Massachusetts, Belknap Press.

Wilson, M.H., Kepler, C.B., Snyder, N.F.R., Derrickson, S.R., Dein, F.J., Wiley, J.W., Wunderle Jr, J.M., Lugo, A.E., Graham, D.L. & Toone, W.D. (1994). Puerto Rican Parrots and potential limitations of the metapopulation approach to species conservation. *Conserv. Biol.* **8**: 114–123.

Wimberger, P.H. (1988). Food supplemental effects on breeding time and harem size in the Red-Winged Blackbird (*Agelaius phoeniceus*). *Auk* **105**: 799–802.

Winkel, W. (1996). Langzeit-Erfassung brutbiologischer Parameter bei Höhlenbrüter (*Parus* spp. *Sitta europaea*) – gibt es signifikante varänderungen? *Verh. Dtsch. Zool. Ges.* **89**: 1–130.

Winker, K., Rappole, J.H. & Ramos, M.A. (1990). Population dynamics of the Wood Thrush in southern Veracruz, Mexico. *Condor* **92**: 444–460.

Winstanley, D.R., Spencer, R. & Williamson, K. (1974). Where have all the Whitethroats gone? *Bird Study* **21**: 1–14.

Witherby, H.F., Jourdain, F.C.R., Ticehurst, N.F. & Tucker, B.W. (1939). The handbook of British birds. London, Witherby.

Witter, M.S., Cuthill, I.C. & Bonser, R. (1994). Experimental investigations of mass-dependent predation risk in the European Starling *Sturnus vulgaris*. *Anim. Behav.* **48**: 201–222.

Wobeser, G.A. (1981). Diseases of wild waterfowl. New York, Plenum Press.

Woffinden, N.D. & Murphy, J.R. (1977). Population dynamics of the Ferruginous Hawk during a prey decline. *Great Basin Nat.* **37**: 411–425.

Woiwood, I.P. (1991). The ecological importance of long-term synoptic monitoring. pp. 275–304 in *'The ecology of temperate cereal fields'* (ed. L.G. Firbank, N. Carter, J.F. Darbyshire & G.R. Potts). Oxford, Blackwell.

Wong, M. (1983). Effect of unlimited food availability on the breeding biology of wild Eurasian Tree Sparrows in West Malaysia. *Wilson Bull.* **95**: 287–294.

Woodburn, M. (1995). Do parasites alter Pheasant breeding success? *Game Conservancy Rev.* **26**: 96–97.

Woolfenden, G.E. & Fitzpatrick, J.W. (1984). The Florida Scrub Jay: Demography of a cooperative-breeding bird. New Jersey, USA, Princeton University Press.

Woolfenden, G.E. & Fitzpatrick, J. (1991). Florida Scrub Jay ecology and conservation. pp. 542–565 in 'Bird population studies' (ed. C.M. Perrins, J.-D. Lebreton & G.J.M. Hirons). Oxford, University Press.

Wormington, A. & Leach, J.H. (1992). Concentrations of migrant Diving Ducks at Point Pelee National Park, Ontario, in response to invasion of Zebra Mussels, *Dreissena polymorpha*. *Can. Field Nat.* **106**: 376–380.

Wrånes, E. (1988). Massedød av aerfugl på Sørlandet vinteren 1981/82. *Vår Fuglefauna* **11**: 71–74.

Wright, R.G., Paisley, R.N. & Kubisiak, J.F. (1996). Survival of wild Turkey hens in southwestern Wisconsin. *J. Wildl. Manage.* **60**: 313–320.

Wright, R.M. & Phillips, V.E. (1990). Mallard response to increased food-supply in the wildfowl reserve. *Game Conservancy Ann. Rev.* **1**: 105–108.

Wunderle, J.M. (1991). Age specific foraging proficiency in birds. *Curr. Ornithol.* **8**: 273–324.

Wunderle, J.M., Lodge, J.M. & Waide, R.B. (1992). Short-term effects of hurricane Gilbert on terrestrial bird populations on Jamaica. *Auk* **109**: 148–166.

Wyllie, I., Dale, L. & Newton, I. (1996). Unequal sex-ratio, mortality causes and pollutant residues in Long-eared Owls in Britain. *Brit. Birds* **89**: 429–436.

Wyllie, I. & Newton I. (1991). Demography of an increasing population of Sparrowhawks. *J. Anim. Ecol.* **60**: 749–766.

Yahner, R.H. & Wright, A.L. (1985). Depredation on artificial ground nests – effects of edge and plot age. *J. Wildl. Manage.* **49**: 508–513.

Yalden, D.W. & Pearce-Higgins, J.W. (1997). Density dependence and winter weather as factors affecting the size of a population of Golden Plovers *Pluvialis apricaria*. *Bird Study* **44**: 227–234.

Yamamoto, J.T., Fry, D.M., Wilson, B.W., Stein, R.W., Ottum, N.D., Seiber, J.N. & McChessney, M.M. (1993). Kestrel habitat use and pesticide exposure during winter in agricultural areas of the central valley of California. *J. Raptor Res.* **27**: 95.

Yates, M.G., Goss-Custard, J.D., McGrosty, S., Lakhani, K., Le V. Dit, Durell, S.E.A., Clarke, R.T., Rispin, W.E., Moy, I., Yates, T., Plant, R.A. & Frost, A.J. (1993). Sediment characteristics, invertebrate densities and shorebird densities on the inner banks of the Wash. *J. Appl. Ecol.* **30**: 599–614.

Ydenberg, R.C. (1984). The conflict between feeding and territorial defence in the Great Tit. *Behav. Ecol. Sociobiol.* **15**: 103–108.

Yom-Tov, Y. (1974). The effect of food and predation on breeding density and success, clutch size and laying date of the Crow (*Corvus corone* L.). *J. Anim. Ecol.* **43**: 479–498.

Yorio, P. & Quintana, F. (1997). Predation by Kelp Gulls *Larus dominicanus* at a mixed-species colony of Royal Terns *Sterna maxima* and Cayenne Terns *Sterna eurygnatha* in Patagonia. *Ibis* **139**: 536–541.

Yosef, R. & Grubb, T.C. (1994). Resource dependence and territory size in Loggerhead Shrikes (*Lanius ludovicianus*). *Auk* **111**: 465–469.

Young, C.M. (1970). Territoriality in the Common Shelduck *Tadorna tadorna*. *Ibis* **112**: 330–335.

Young, E.C. (1972). Territory establishment and stability in McCormick's Skua. *Ibis* **114**: 234–244.

Yue-Hua, S. & Yu, F. (1994). Winter habitat selection and analysis of Hazel Grouse in the Changbai Mountains. Research notes on Avian Biology: selected contributions from the 21st International Ornithological Congress. *J. Ornithol.* **135**: 191.

Zahavi, A. (1971). The social behaviour of the White Wagtail *Motacilla alba alba* wintering in Israel. *Ibis* **113:** 203–211.

Zahavi, A. (1989). Arabian Babbler. pp. 253–275 in *'Lifetime reproduction in birds'* (ed. I. Newton). London, Academic Press.

Zarnowitz, J. & Manuwal, D. (1985). The effect of forest management on cavity-nesting in northwestern Washington. *J. Wildl. Manage.* **49:** 255–263.

Zimmerman, J.L. (1971). The territory and its density dependent effect in *Spiza americana. Auk* **88:** 591–612.

Zimmerman, J.L. (1982). Nesting success of Dickcissels (*Spiza americana*) in preferred and less preferred habitats. *Auk* **99:** 292–298.

Zumeta, D.C. & Holmes, R.T. (1978). Habitat shift and roadside mortality of Scarlet Tanagers during a cold wet New England spring. *Wilson Bull.* **90:** 575–586.

Zwank, P.J., Wright, V.L., Shealy, P.M. & Newsom, J.D. (1985). Lead toxicosis in waterfowl on two major wintering areas in Louisiana. *Wildl. Soc. Bull.* **13:** 17–26.

Zwarts, L. (1976). Density related processes in feeding dispersion and feeding activity of Teal (*Anas crecca*). *Ardea* **66:** 192–209.

Zwarts, L. & Drent, R.H. (1981). Prey depletion and regulation of predator density: Oystercatchers (*Haematopus ostralegus*) feeding on mussels (*Mytilus edulis*). pp. 193–216 in *'Feeding and survival strategies of estuarine organisms'* (ed. N.V. Jones & W.J. Wolf). London, Plenum.

Zwickel, F.C. (1972). Removal and repopulation of Blue Grouse in an increasing population. *J. Wildl. Manage.* **36:** 1141–1152.

Zwickel, F.C. (1980). Surplus yearlings and the regulation of breeding density in Blue Grouse. *Can. J. Zool.* **58:** 896–905.

Golden Eagle *Aquila chrysaetos*.

Index